Chemical Engineering Practice — GATE and PSUs

Chemical Engineering Practice — GATE and PSUs

Edited by
Dr. Harendra Kumar
B.Sc., Engg. Chamical, M.Tech, PhD
Retired Professor and Head
Department of Chemical Engg.
National Institute of Technology, Raipur- 492010
(Chhattisgarh)

KALPAZ

ISBN : 978-93-5324-222-0

First Published, 2020

Published by

Kalpaz Publications
C-30, Satyawati Nagar,
Delhi – 110052
Ph.: 011-47034999, 9811692060
www.kalpazpublications.com
E-mail: kalpaz@hotmail.com

Printed at: G. Print Process, Delhi

Contents

Preface

During my long service in technical education spanning over four decades, I noticed several changes in generations of students. The emphasis on GATE to be eligible for admission to higher studies and employment in Public Sector Undertakings (PSUs) has overridden the importance of university system of education. Students securing very good marks in B.E./B.Tech. become nervous in appearing in GATE. The present book is the outcome to instill confidence in the minds of students for competitive examinations. Every chapter consists of brief theory followed by large number of objective questions. Previous years GATE questions with solutions have also been provided. Objective questions having multiple choices are designed to test the theoretical understanding of the topic in depth. Since these questions carry negative marks for wrong attempts, the students need to have extensive practice. GATE questions also include short questions, which need numerical calculations. The present book fulfils all these criteria. These questions have been tested in classrooms with good result. I hope aspiring students will find the book useful. Suggestions for improvement are welcome.

Dr. Harendra Kumar

Synopsis: "Chemical Engineering Practice – GATE and PSUs"

Graduate Aptitude Test Examination (GATE) has come to be recognized as the real gateway not only to higher studies like M.Tech, Ph.D. etc but also for a direct entry to various PSUs. In addition, most of state and central level Public Service Examinations, Electricity boards, Pollution control boards and other competitive examinations follow the GATE Syllabus and GATE pattern examinations.

Since last few years the awareness about GATE has increased and now even average students are aspiring for a good GATE score. GATE score supersedes all the University scores and provides an excellent opportunity to the students to wash off their feelings that they have not graduated from very reputed or national level institutes.

Graduating students study various core subjects in a textbook style in the University pattern. This is true for IIT/NIT and other autonomous Institutes also. The University pattern study follows a well defined systematic approach towards understanding the subject as in standard textbooks. Naturally, the examination pattern is also descriptive and long answer type. However, GATE pattern of examination is of short answer or multiple choice types. Even within a subject the relative emphasis in GATE is quite different. For example, in Fluid Mechanics valves and pumps find more weightage in GATE in comparison to semester examinations and similarly in Heat Transfer, radiation finds more weightage than in semester examination. This is true for all the subjects. The present book has been designed and organized with this viewpoint. A more difficult situation faced by the students is how to pinpoint the correct choice in a multiple choice question. In most of such cases two or more options may appear alike. This is where a clear concept and a sense of scientific logic makes all the difference. In several questions there may be only a hairline difference.

The questions in the book have been designed so as to develop the concept of the subject and scientific logic systematically. This helps the students in improving their scores. Tricks and blind answers may be disastrous and should not be practiced.

The book has been divided into ten chapters. Brief conceptual outlines of the concerned subject has been given followed by about 350 multiple short answer type questions including ten years of previous GATE questions of the subject. Correct answer key has been provided at the end of each chapter. Solutions of numerical type of questions have been incorporated. For multiple type questions of conceptual reasoning, a brief note has been provided in many cases along with the correct answer.

1. General Aptitude

- ➢ VERBAL ABILITY
- ➢ NUMERICAL ABILITY

VERBALABILITY

- English Grammar
- Sentence Completion
- Synonyms
- Antonyms
- Reasoning Ability

1. English Grammar:

Articles

Types of Articles:

1. Definite (the)

2. Indefinite (a/an)

Where to use 'a' and 'an'

'a' is used before singular countable noun started with a consonant alphabet or which are sound consonant.

Ex: chair, BA etc

'an' is used before the singular countable noun which are started countable noun which are started with vowels or give sound of vowels.

Ex : Heir, Apple etc.

Use of indefinite article:

1. with Proper Noun:

Example: This book is written by a Chetan.

This book is written by Chetan.

{ Difference between above two sentence is that in the 1st sentence we are talking about 'Chetan' who is not a particular and in the next sentence 'Chetan' is described as famous or a particular special person.}

Where to use 'The'

GENERAL RULES

1. Use *the* to refer to something which has already been mentioned.
2. Use *the* when you assume there is just one of something in that place, even if it has not been mentioned before.
3. Use *the* in sentences or clauses where you define or identify a particular person or object.
4. Use *the* to refer to people or objects that are unique.
5. Use *the* before superlatives and ordinal numbers.
6. Use *the* with adjectives, to refer to a whole group of people.
7. Use *the* with decades.
8. Use *the* with clauses introduced by *only*
9. *PROPER NOUNS:*Use *the* with names of geographical areas, rivers, mountain ranges, groups of islands, canals, and oceans.
10. Use *the* with countries that have plural names
11. Use *the* with countries that include the words "republic", "kingdom", or "states" in their names.
12. Use *the* with newspaper names.
13. Use *the* with the names of famous buildings, works of art, museums, or monuments.
14. Use *the* with the names of hotels & restaurants, unless these are named after a person.
15. Use *the* with the names of families, but not with the names of individuals.

Examples

1. My father enjoyed **the book** you gave him.
2. He is **the doctor** I came to see.
3. This is a painting from **the 1820's**.
4. She is visiting **the United States**.
5. I would like to visit **the Eiffel Tower**.

The Noun

A noun is a part of speech that is used to name a person, place, thing, quality, or action. A noun can function as a subject, object, complement, appositive, or object of a preposition.

Plural vs singular

Nouns can be singular or plural. The plural form of a noun is usually formed by adding s at the end of the noun. But this is not always the case. There are exceptions to the rule. Some plurals are irregular:

Singular	*Plural*
fish	fish
tooth	teeth
man	Men
woman	women

Different types of nouns

There are different types of nouns:

1. An **abstract noun** names an idea, event, quality, or concept (*freedom, love, courage...*) **Concrete nouns** name something recognizable through the sense (*table, dog, house...*)
2. **Animate nouns** refer to a person, animal, or other creature (*man, elephant, chicken...*) An **inanimate noun** refers to a material object (*stone, wood, table...*)
3. **A collective noun** describes a group of things or people as a unit (*family, flock, audience...*)
4. **Common noun** is the name of a group of similar things (*table, book, window...*) **Proper nouns**, however, refer to the name of a single person, place or thing (*John, Joseph, London...*)
5. **Compound nouns** refer to two or more nouns combined to form a single noun (*sister-in-law, schoolboy, fruit juice*)
6. **Countable (or count) nouns** have a singular and a plural form. In plural, these nouns can be used with a number- they can be counted. (*friends, chairs, houses, boys...*)**Uncountable (or non count) nouns**, however, can only be used in singular. They can't be counted. (*money, bread, water, coffee...*)

PRONOUNS

Pronouns are words that substitute a noun or another pronoun. Examples of pronouns are ***he, she, who, themselves..***

In the example:

Mike *likes* **his daughter**.

Mike and **his daughter** can be replaced by **he** and **her**:

He *likes* **her**

Types of pronouns

Subject pronouns (I, you, he, she, it, we, they)

Object pronouns (me, you, him, her, it, us, them)

Reflexive pronouns (myself, yourself, himself, herself, itself, ourselves, themselves)

Possessive pronouns (mine, yours, his, hers, ours, theirs)

Demonstrative pronouns (this, that, those, these)

Relative pronouns (who, which, that, whose)

Comparison between Noun and Pronoun:

	Noun	*Pronoun*
Meaning	Nouns are used to name people, places, objects, events, occurrences, etc.	A pronoun is a word that suitably replaces the noun in a sentence.
Types	Common, proper, countable, abstract, etc.	Demonstrative, indefinite, interrogative, etc.
Example of usage	Meet my neighbor ***Sheila***, who is a dancer.	Meet my neighbor Sheila, ***who*** is a dancer.

Adjectives

1. An **adjective** is a word which modifies a noun or a pronoun. In the example below:

   He's got a ***beautiful*** *car*

   beautiful is an adjective modifying *car* (it gives more information about the noun *car*.)
2. Adjectives may come before a noun

   He found a ***nice*** job
3. Adjectives may also come after certain verbs like ***be, feel, seem, look:***

   He is ***intelligent***

   I feel ***happy***

   She seems ***unhappy***

   They look ***fantastic***

The English Verb

Verbs in English

A verb in syntax is a part of speech which conveys

1. action (*bring*, *read*, *walk*, *run*, *learn*)
2. or state of being (*exist*, *stand*)

Sentences in English have a main verb which is stated in a tense (simple present, simple past, simple future...)

Inflections

Verbs are inflected, modified in form, when conjugated. For example, verbs take *s, ed or ing* in some of its forms depending on the tense and the subject-verb agreement.

Agreement

In English a verb may agree with the person and number of its subject. For example, verbs take *s*in the third person singular of the simple present:

Bare Infinitive	*Third Person Singular*
play	he she it plays
work	he, she, it works

When the verb *to have* conjugates in the third person singular of the simple present, the right inflection is *has* NOT *haves*

The verb *to be* has different inflections:

to	*be*
I	*am*
he, she, it	*Is*
we, you, they	*are*

2. SENTENCE COMPLETION

WHAT ARE SENTENCE COMPLETION PROBLEMS?

Sentence completion problems test you vocabulary skills as well as, your reading ability. Unlike the reading comprehension questions, which require you to read long passages, these problems contain a single sentence expressing a complete idea that can be understood without any additional information. Each sentence contains one or two blanks, indicating that one or two world are missing. You are then presented with five words or phrases, or five pairs of words or phrases if there are two blanks in the sentence. From these choices, you need to select the words or phrases that fit into the blanks to best, complete the sentence. To make the correct choice, you will need to be able to understand the main idea of the sentence of the logical structure of the sentence. Next, you will need to know the definitions of the words in the answer choices.

BASIC STRATEGY FOR SENTENCE COMPLETION PROBLEMS MORE STRATEGIES

Strategy 1. The first strategy is to break the sentence dawn into the main clause and introductory clause or phrase in order to identify the relationship it expresses. For example, consider the following:

In a hospital setting, hand washing is primarily a____ measure.

(a) rophylactic (b) pseudocyetic

(c) protensive (d) propaedeutic

The main clause declares that hand washing is primarily a measure. However, preceding the declaration is a condition or constraint expressed in an introductory phrase "a hospital setting". This phrase seta the contextual field within which the main clause is to be understood. If you generalize from your knowledge, it would seem that any measure promoted in a hospital should result in good health. Only one of the answers would result in this. That is answer (a) in which *prophylactic* means "prevents disease."

Strategy 2. A second strategy is to attend to sets, whether a set of synonyms or antonyms or special word groups. For example, consider the following:

Even when a______ mother was made available to the infant chimpanzee, the infant detected the_____ and reacted with high anxiety.

(a) hedonistic.. contrivance

(b) vagrant apostasy

(c) veteran ruse

(d) pretender.. interference

Nothing in a common sense association supports the pairing of *hedonistic* and *contrivanor*, *vagrant* and *apostasy*, *veteran* and *ruse,* or pretender and *interference*. The only set of related words is surrogate (meaning substitute) and substitution.

Strategy 3 Pay attention to the words that control the figurative field of the sentence. Consider the following passage: In seventeenth-century France, private and public opinion portrayed the physician as a vulture of greed, circling the sick and dying, battening off the suffering of the afflicted, and ______ the remains of his prey.

(a) spending (b) devouring
(c) retreating from (d) decrying

Clearly the correct answer is (b), devouring. The key words that control the figurative field are vulture of greed, circling, and *battening*. Read the items carefully and make sure that the completed sentence is meaningful, that it makes sense.

3. SYNONYMS

Some important word with their synonyms

Amazing — incredible, unbelievable, improbable, fabulous, wonderful, fantastic, astonishing, astounding, extraordinary

Anger — enrage, infuriate, arouse, nettle, exasperate, inflame, madden

Angry — mad, furious, enraged, excited, wrathful, indignant, exasperated, aroused, inflamed

Answer — reply, respond, retort, acknowledge

Ask– — question, inquire of, seek information from, put a question to, demand, request, expect, inquire, query, interrogate, examine, quiz

Awful — dreadful, terrible, abominable, bad, poor, unpleasant

Bad — evil, immoral, wicked, corrupt, sinful, depraved, rotten, contaminated, spoiled, tainted, harmful, injurious, unfavorable, defective, inferior, imperfect, substandard, faulty, improper, inappropriate, unsuitable, disagreeable, unpleasant, cross, nasty, unfriendly, irascible, horrible, atrocious, outrageous, scandalous, infamous, wrong, noxious, sinister, putrid, snide, deplorable, dismal, gross, heinous, nefarious, base, obnoxious, detestable, despicable, contemptible, foul, rank, ghastly, execrable

Beautiful — pretty, lovely, handsome, attractive, gorgeous, dazzling, splendid, magnificent, comely, fair, ravishing, graceful, elegant, fine, exquisite, aesthetic, pleasing, shapely, delicate, stunning, glorious, heavenly, resplendent, radiant, glowing, blooming, sparkling

Begin — start, open, launch, initiate, commence, inaugurate, originate

Big — enormous, huge, immense, gigantic, vast, colossal, gargantuan, large, sizable, grand, great, tall, substantial, mammoth, astronomical, ample, broad, expansive, spacious, stout, tremendous, titanic, mountainous

Brave — courageous, fearless, dauntless, intrepid, plucky, daring, heroic, valorous, audacious, bold, gallant, valiant, doughty, mettlesome

Break — fracture, rupture, shatter, smash, wreck, crash, demolish, atomize

Bright — shining, shiny, gleaming, brilliant, sparkling, shimmering, radiant, vivid, colorful, lustrous, luminous, incandescent, intelligent, knowing, quick-witted, smart, intellectual

Calm — quiet, peaceful, still, tranquil, mild, serene, smooth, composed, collected, unruffled, level-headed, unexcited, detached, aloof

Come — approach, advance, near, arrive, reach

Cool — chilly, cold, frosty, wintry, icy, frigid

Crooked — bent, twisted, curved, hooked, zigzag

Cry — shout, yell, yowl, scream, roar, bellow, weep, wail, sob, bawl

Cut — gash, slash, prick, nick, sever, slice, carve, cleave, slit, chop, crop, lop, reduce

Dangerous — perilous, hazardous, risky, uncertain, unsafe

Dark — shadowy, unlit, murky, gloomy, dim, dusky, shaded, sunless, black, dismal, sad

Decide — determine, settle, choose, resolve

Definite — certain, sure, positive, determined, clear, distinct, obvious

Delicious — savory, delectable, appetizing, luscious, scrumptious, palatable, delightful, enjoyable, toothsome, exquisite

Describe — portray, characterize, picture, narrate, relate, recount, represent, report, record

Destroy — ruin, demolish, raze, waste, kill, slay, end, extinguish

Difference — disagreement, inequity, contrast, dissimilarity, incompatibility

Do — execute, enact, carry out, finish, conclude, effect, accomplish, achieve, attain

Dull — boring, tiring,, tiresome, uninteresting, slow, dumb, stupid, unimaginative, lifeless, dead, insensible, tedious, wearisome, listless, expressionless, plain, monotonous, humdrum, dreary

Eager — keen, fervent, enthusiastic, involved, interested, alive to

End — stop, finish, terminate, conclude, close, halt, cessation, discontinuance

Enjoy — appreciate, delight in, be pleased, indulge in, luxuriate in, bask in, relish, devour, savor, like

Explain — elaborate, clarify, define, interpret, justify, account for

Fair — just, impartial, unbiased, objective, unprejudiced, honest

Fall — drop, descend, plunge, topple, tumble

False — fake, fraudulent, counterfeit, spurious, untrue, unfounded, erroneous, deceptive, groundless, fallacious

Famous — well-known, renowned, celebrated, famed, eminent, illustrious, distinguished, noted, notorious

Fast — quick, rapid, speedy, fleet, hasty, snappy, mercurial, swiftly, rapidly, quickly, snappily, speedily, lickety-split, posthaste, hastily, expeditiously, like a flash

Fat — stout, corpulent, fleshy, beefy, paunchy, plump, full, rotund, tubby, pudgy, chubby, chunky, burly, bulky, elephantine

Fear — fright, dread, terror, alarm, dismay, anxiety, scare, awe, horror, panic, apprehension

Fly — soar, hover, flit, wing, flee, waft, glide, coast, skim, sail, cruise

Funny — humorous, amusing, droll, comic, comical, laughable, silly

Get — acquire, obtain, secure, procure, gain, fetch, find, score, accumulate, win, earn, rep, catch, net, bag, derive, collect, gather, glean, pick up, accept, come by, regain, salvage

Go — recede, depart, fade, disappear, move, travel, proceed

Good — excellent, fine, superior, wonderful, marvelous, qualified, suited, suitable, apt, proper, capable, generous, kindly, friendly, gracious, obliging, pleasant, agreeable, pleasurable, satisfactory, well-behaved, obedient, honorable, reliable, trustworthy, safe, favorable, profitable, advantageous, righteous, expedient, helpful, valid, genuine, ample, salubrious, estimable, beneficial, splendid, great, noble, worthy, first-rate, top-notch, grand, sterling, superb, respectable, edifying

Great — noteworthy, worthy, distinguished, remarkable, grand, considerable, powerful, much, mighty

Gross — improper, rude, coarse, indecent, crude, vulgar, outrageous, extreme, grievous, shameful, uncouth, obscene, low

Happy — pleased, contented, satisfied, delighted, elated, joyful, cheerful, ecstatic, jubilant, gay, tickled, gratified, glad, blissful, overjoyed

Hate — despise, loathe, detest, abhor, disfavor, dislike, disapprove, abominate

Have — hold, possess, own, contain, acquire, gain, maintain, believe, bear, beget, occupy, absorb, fill, enjoy

Help — aid, assist, support, encourage, back, wait on, attend, serve, relieve, succor, benefit, befriend, abet

Hide — conceal, cover, mask, cloak, camouflage, screen, shroud, veil

Hurry — rush, run, speed, race, hasten, urge, accelerate, bustle

Hurt — damage, harm, injure, wound, distress, afflict, pain

Idea — thought, concept, conception, notion, understanding, opinion, plan, view, belief

Important — necessary, vital, critical, indispensable, valuable, essential, significant, primary, principal, considerable, famous, distinguished, notable, well-known

Interesting — fascinating, engaging, sharp, keen, bright, intelligent, animated, spirited, attractive, inviting, intriguing, provocative, though-provoking, challenging, inspiring, involving, moving, titillating, tantalizing, exciting, entertaining, piquant, lively, racy, spicy, engrossing, absorbing, consuming, gripping, arresting, enthralling, spellbinding, curious, captivating, enchanting, bewitching, appealing

Keep — hold, retain, withhold, preserve, maintain, sustain, support

Kill — slay, execute, assassinate, murder, destroy, cancel, abolish

Lazy — indolent, slothful, idle, inactive, sluggish

Little — tiny, small, diminutive, shrimp, runt, miniature, puny, exiguous, dinky, cramped, limited, itsy-bitsy, microscopic, slight, petite, minute

Look — gaze, see, glance, watch, survey, study, seek, search for, peek, peep, glimpse, stare, contemplate, examine, gape, ogle, scrutinize, inspect, leer, behold, observe, view, witness, perceive, spy, sight, discover, notice, recognize, peer, eye, gawk, peruse, explore

Love — like, admire, esteem, fancy, care for, cherish, adore, treasure, worship, appreciate, savor

Make — create, originate, invent, beget, form, construct, design, fabricate, manufacture, produce, build, develop, do, effect, execute, compose, perform, accomplish, earn, gain, obtain, acquire, get

Mark — label, tag, price, ticket, impress, effect, trace, imprint, stamp, brand, sign, note, heed, notice, designate

Mischievous — prankish, playful, naughty, roguish, waggish, impish, sportive

Move — plod, go, creep, crawl, inch, poke, drag, toddle, shuffle, trot, dawdle, walk, traipse, mosey, jog, plug, trudge, slump, lumber, trail, lag, run, sprint, trip, bound, hotfoot, high-tail, streak, stride, tear, breeze, whisk, rush, dash, dart, bolt, fling, scamper, scurry, skedaddle, scoot, scuttle, scramble, race, chase, hasten, hurry, hump, gallop, lope, accelerate, stir, budge, travel, wander, roam, journey, trek, ride, spin, slip, glide, slide, slither, coast, flow, sail, saunter, hobble, amble, stagger, paddle, slouch, prance, straggle, meander, perambulate, waddle, wobble, pace, swagger, promenade, lunge

Moody — temperamental, changeable, short-tempered, glum, morose, sullen, mopish, irritable, testy, peevish, fretful, spiteful, sulky, touchy

Neat — clean, orderly, tidy, trim, dapper, natty, smart, elegant, well-organized, super, desirable, spruce, shipshape, well-kept, shapely

New — fresh, unique, original, unusual, novel, modern, current, recent

Old — feeble, frail, ancient, weak, aged, used, worn, dilapidated, ragged, faded, broken-down, former, old-fashioned, outmoded, passe, veteran, mature, venerable, primitive, traditional, archaic, conventional, customary, stale, musty, obsolete, extinct

Part — portion, share, piece, allotment, section, fraction, fragment

Place — space, area, spot, plot, region, location, situation, position, residence, dwelling, set, site, station, status, state

Plan — plot, scheme, design, draw, map, diagram, procedure, arrangement, intention, device, contrivance, method, way, blueprint

Popular — well-liked, approved, accepted, favorite, celebrated, common, current

Predicament — quandary, dilemma, pickle, problem, plight, spot, scrape, jam

Put — place, set, attach, establish, assign, keep, save, set aside, effect, achieve, do, build

Quiet — silent, still, soundless, mute, tranquil, peaceful, calm, restful

Right — correct, accurate, factual, true, good, just, honest, upright, lawful, moral, proper, suitable, apt, legal, fair

Run — race, speed, hurry, hasten, sprint, dash, rush, escape, elope, flee

Say/Tell — inform, notify, advise, relate, recount, narrate, explain, reveal, disclose, divulge, declare, command, order, bid, enlighten, instruct, insist, teach, train, direct, issue, remark, converse, speak, affirm, suppose, utter, negate, express, verbalize, voice, articulate, pronounce, deliver, convey, impart, assert, state, allege, mutter, mumble, whisper, sigh, exclaim, yell, sing, yelp, snarl, hiss, grunt, snort, roar, bellow, thunder, boom, scream, shriek, screech, squawk, whine, philosophize, stammer, stutter, lisp, drawl, jabber, protest, announce, swear, vow, content, assure, deny, dispute

Scared — afraid, frightened, alarmed, terrified, panicked, fearful, unnerved, insecure, timid, shy, skittish, jumpy, disquieted, worried, vexed, troubled, disturbed, horrified, terrorized, shocked, petrified, haunted, timorous, shrinking, tremulous, stupefied, paralyzed, stunned, apprehensive

Show — display, exhibit, present, note, point to, indicate, explain, reveal, prove, demonstrate, expose

Slow — unhurried, gradual, leisurely, late, behind, tedious, slack

Stop — cease, halt, stay, pause, discontinue, conclude, end, finish, quit

Story — tale, myth, legend, fable, yarn, account, narrative, chronicle, epic, sage, anecdote, record, memoir

Strange — odd, peculiar, unusual, unfamiliar, uncommon, queer, weird, outlandish, curious, unique, exclusive, irregular

Take — hold, catch, seize, grasp, win, capture, acquire, pick, choose, select, prefer, remove, steal, lift, rob, engage, bewitch, purchase, buy, retract, recall, assume, occupy, consume

Tell — disclose, reveal, show, expose, uncover, relate, narrate, inform, advise, explain, divulge, declare, command, order, bid, recount, repeat

Think — judge, deem, assume, believe, consider, contemplate, reflect, mediate

Trouble — distress, anguish, anxiety, worry, wretchedness, pain, danger, peril, disaster, grief, misfortune, difficulty, concern, pains, inconvenience, exertion, effort

True — accurate, right, proper, precise, exact, valid, genuine, real, actual, trusty, steady, loyal, dependable, sincere, staunch

Ugly — hideous, frightful, frightening, shocking, horrible, unpleasant, monstrous, terrifying, gross, grisly, ghastly, horrid, unsightly, plain, homely, evil, repulsive, repugnant, gruesome

Unhappy — miserable, uncomfortable, wretched, heart-broken, unfortunate, poor, downhearted, sorrowful, depressed, dejected, melancholy, glum, gloomy, dismal, discouraged, sad

Use — employ, utilize, exhaust, spend, expend, consume, exercise

Wrong — incorrect, inaccurate, mistaken, erroneous, improper, unsuitable.

REASONING

There are two types

1. Verbal reasoning
2. Non-verbal reasoning

Verbal reasoning is understanding and reasoning using concepts framed in words. It aims at evaluating ability to think constructively, rather than at simple fluency or vocabulary recognition.

Nonverbal reasoning Is ability to understand and analyse visual information and solve problems using visual reasoning. For example: identifying relationships, similarities and differences between shapes and patterns, recognizing visual sequences and relationships between objects, and remembering these.

NUMERICAL ABILITY

1. Problems on number
2. Permutation and combination
3. Ratio and proportion
4. Logarithm
5. Percentage
6. Probability
7. Time and work
8. Simple Interest
9. Compound Interest
10. Profit and Loss
11. Allegation and mixture

1. Problems on number

1. What is the two-digit number?

   I The difference between the two-digit number and the number formed by interchanging the digits is 27.

   II The difference between the two digits is 3.

   III The digit at unit's place is less than that at ten's place by 3.

   A. I and II only

   B. I and III only

   C. I, and either II or III

   D. even with all I, II and III, answer cannot be give

Sol. **A.** let the tens and units digit be x and y respectively.

I. $(10x + y) - (10y + x) \Leftrightarrow x - y = 3.$

II. $x - y = 3.$

III. $x - y = 3$

thus, even all the given three statements together do not give the answer

2. What is the two-digit number?

   I. Sum of the digits is 7.

   II. Difference between the number and the number obtained by interchanging the digits is 9.

   III. Digit in the ten's place is bigger than the digit in the unit's place by 1.

   A. I and II only

   B. II and III only

   C. I, and either II or III

   D. none of these

Sol. **A.** let the tens and units digit be x and y respectively.

I. $x + y = 7$

II. $(10x + y) - (10y + x) = 9 \Leftrightarrow x - y = 1$

III. $x - y = 1$

thus, I and II as well as I and III give the answer.

3. What is the number?

   I The sum of the two digits is 8. The ratio of the two digits is 1 : 3

   II The product of the two digit of a number is 12. The quotient of two digits is 3.

   A. I alone sufficient while II alone not sufficient to answer

   B. II alone sufficient while I alone not sufficient to answer

   C. Either I or II alone sufficient to answer

   D. Both I and II are not sufficient to answer

Sol. **C.** let the tens and units digit be x and y respectively.

Then,

I. $X+Y = 8$ and $\frac{X}{Y} = \frac{1}{3}$

$\therefore$ I gives, $4y = 24 \Leftrightarrow y = 6$

So, $x + 6 = 8 \Leftrightarrow x = 2.$

II. $XY = 12$ and $\frac{X}{Y} = \frac{3}{1}$

$\therefore$ II gives, $x^2 = 36 \Leftrightarrow x = 6$

So, $3y = 6 \Leftrightarrow y = 2$

therefore, Either I or II alone sufficient to answer.

4. What is the two-digit number?

   I The difference between the two digits is 9.

   II The sum of the digits is equal to the difference between the two digits.

A. I alone sufficient while II alone not sufficient to answer

B. II alone sufficient while I alone not sufficient to answer

C. Either I or II alone sufficient to answer

D. Both I and II are necessary to answer

***Sol.* D.** let the tens and unit digits be x and y respectively. then

I. $x - y = 9$

II. $x + y = x - y$

from I and II, we get $x - y = 9$ and $x + y = 9$

on solving, we get $x = 9$ and $y = 0$

$\therefore$ required number is 90

Thus, both I and II are needed to get the answer

5. What is the two-digit number whose first digit is a and the second digit is b?. The number is greater than 9.

I The number is multiple of 51.

II The sum of the digits a and b is 6.

A. I alone sufficient while II alone not sufficient to answer

B. II alone sufficient while I alone not sufficient to answer

C. either I or II alone sufficient to answer

D. both I and II are not sufficient to answer

***Sol.* A.** from statement I:A two digit number, greater than 9 and multiple of 51 should be 51 itself. because, 2 x 51 = 102 (3 digit number). Therefore, I alone sufficient to answer. from statement II:A two digit number, greater than 9 and sum of the digit is 6.

It can be 15, 24, 33, 42, 51. So we cannot determine the required answer from the statement II alone.

thus, I alone give the answer while II alone not sufficient to answer.

6. If one-third of one-fourth of a number is 15, then three tenth of that number is

A. 35

B. 36

C. 45

D. 54

***Sol.* D.** let the number be x.

then, $\frac{1}{3}$ of $\frac{1}{4}$ of $x = 15 \Leftrightarrow x = 15 \times 12 = 180$

so, required number $= \left(\frac{3}{10} \times 180\right) = 54$.

7. Three times the first of three consecutive odd integers is 3 more than twice the third. The third integer is

A. 9

B. 11

C. 13

D. 15

***Sol.* D.** let the three integers be x, $x + 2$ and $x + 4$

then, $3x = 2(x + 4) + 3 \Leftrightarrow x = 11$

$\therefore$ Third integer $= x + 4 = 15$.

8. The difference between a two-digit number and the number obtained by interchanging the positions of its digits is 36. What is the difference between the two digits of that number?

A. 3

B. 4

C. 9

D. cannot be determined

***Sol.* B.** let the ten's digit be x and unit's digit be y

then, $(10x + y) - (10y + x) = 36$

$9(x - y) = 36$

$x - y = 4$

9. The ratio between the digit of the number is 1 : 2?

A. 4

B. 8

C. 16

D. none of these

***Sol.* B.** since the number is greater than the number obtained on reversing the digits, so the ten's digit is greater than the unit's digit

let ten's and unit's digits be $2x$ and x respectively

then, $(10 \times 2x + x) - (10x + 2x) = 36$

$9x = 36$

$x = 4$

$\therefore$ required difference $= (2x + x) - (2x - x) = 2x = 8$.

10. A two-digit number is such that the product of the digits is 8. When 18 is added to the number, then the digits are reversed. The number is:

A. 18

B. 2

C. 24

D. 42

Sol. **B.** let the ten's unit digit $\frac{8}{x}$ respectively.

$$\text{then, } \left(10x + \frac{8}{x}\right) + 18 = 10 \times \frac{8}{x} + x$$
$$10x^2 + 8 + 18x = 80 + x^2$$
$$9x^2 + 18x - 72 = 0$$
$$x^2 + 2x - 8 = 0$$
$$(x + 4)(x - 2) = 0$$
$$x = 2$$

2. Permutation and Combination

1. In how many ways can a group of 5 men and 2 women be made out of a total of 7 men and 3 women?

   A. 63
   B. 90
   C. 126
   D. 45

Sol. **A.** Required number of ways = $({}^7C_5 \text{ x } {}^3C_2)$
$= ({}^7C_2 \text{ x } {}^3C_1) = \left(\frac{7\times6}{2\times1} \times 3\right) = 63$

2. How many 4-letter words with or without meaning, can be formed out of the letters of the word, 'LOGARITHMS', if repetition of letters is not allowed?

   A. 40
   B. 400
   C. 5040
   D. 2520

Sol. **C.** 'LOGARITHMS' contains 10 different letters.
required number of words = Number of arrangements of 10 letters, taking 4 at a time.
$= {}^{10}P_4$
$= (10 \times 9 \times 8 \times 7)$
$= 5040$

3. In how many different ways can the letters of the word 'mathematics' be arranged so that the vowels always come together?

   A. 10080
   B. 4989600
   C. 120960
   D. none of these

Sol. **C.** in the word 'MATHEMATICS' we treat vowels AEAI as one letter.
thus, we have MTHMTCS (AEAI)
now, we have to arrange 8 letters, out of which M occurs twice, T occurs twice and the rest are different.

$\therefore$ number of ways of arranging these letters

$$= \frac{8!}{(2!)(2!)} = 10080.$$

now, AEAI has 4 letters in which A occurs 2 times and the rest are different
number of ways of arranging these letters
$= \frac{4!}{2!} = 12$

$\therefore$ Required number of words = (10080×12)
= 120960.

4. In how many different ways can the letters of the word 'optical' be arranged so that the vowels always come together?

   A. 120
   B. 720
   C. 4320
   D. none of these

Sol. **B.** the word 'OPTICAL' contains 7 different letters.
when the vowels OIA are always together, they can be supposed to form one letter.
then, we have to arrange the letters PTCL (OIA).
now, 5 letters can be arranged in 5! = 120 ways.
the vowels (OIA) can be arranged among themselves in 3! = 6 ways.
$\therefore$ Required number of ways = $(120 \times 6) = 720$.

5. In a group of 6 boys and 4 girls, four children are to be selected. In how many different ways can they be selected such that at least one boy should be there?

   A. 159
   B. 194
   C. 205
   D. 209

Sol. **D.** we may have (1 boy and 3 girls) or (2 boys and 2 girls) or (3 boys and 1 girl) or (4 boys).
Required number of ways
$= ({}^6C_1 \text{ x } {}^4C_3) + ({}^6C_2 \text{ x } {}^4C_2) + ({}^6C_3 \text{ x } {}^4C_1) + ({}^6C_4)$
$= ({}^6C_1 \text{ x } {}^4C_1) + ({}^6C_2 \text{ x } {}^4C_2) + ({}^6C_3 \text{ x } {}^4C_1) + ({}^6C_2)$
$= (6\times 4) + \left(\frac{6\times5}{2\times1} \times \frac{4\times3}{2\times1}\right) + \left(\frac{6\times5\times4}{3\times2\times1} \times 4\right) + \left(\frac{6\times5}{2\times1}\right)$
$= (24 + 90 + 80 + 15)$
$= 209$

6. How many 3-digit numbers can be formed from the digits 2, 3, 5, 6, 7 and 9, which are divisible by 5 and none of the digits is repeated?

A. 5
B. 10
C. 15
D. 20

Sol. D. since each desired number is divisible by 5, so we must have 5 at the unit place. So, there is 1 way of doing it.

The tens place can now be filled by any of the remaining 5 digits (2, 3, 6, 7, 9). So, there are 5 ways of filling the tens place.

The hundreds place can now be filled by any of the remaining 4 digits. So, there are 4 ways of filling it.

$\therefore$ Required number of numbers = $(1 \times 5 \times 4) = 20$.

7. In how many ways a committee, consisting of 5 men and 6 women can be formed from 8 men and 10 women?

A. 266
B. 5040
C. 11760
D. none of these

Sol. A. required number of ways = $({}^8C_5 \times {}^{10}C_6)$

$= ({}^8C_3 \times {}^{10}C_4)$

$= \left(\frac{8\times7\times6}{3\times2\times1} \times \frac{10\times9\times8\times7}{4\times3\times2\times1}\right)$

$= 11760$

8. A box contains 2 white balls, 3 black balls and 4 red balls. In how many ways can 3 balls be drawn from the box, if at least one black ball is to be included in the draw?

A. 32
B. 48
C. 64
D. 96

Sol. C. we may have (1 black and 2 non-black) (2 black and 1 non-black) or (3 black)

Required number = $({}^3C_1 \times {}^6C_2) + ({}^3C_2 \times {}^6C_1) + ({}^3C_3)$

$= \left(3 \times \frac{6\times5}{2\times1}\right) + \left(\frac{3\times2}{2\times1} \times 6\right) + 1$

$= (45 + 18 + 1)$

$= 64$

9. In how many different ways can the letters of the word 'detail' be arranged in such a way that the vowels occupy only the odd positions?

A. 32
B. 48
C. 36
D. 60

Sol. C. there are 6 letters in the given word, out of which there are 3 vowels and 3 consonants.

let us mark these positions as under:

(1) (2) (3) (4) (5) (6)

now, 3 vowels can be placed at any of the three places out 4, marked 1, 3, 5

number of ways of arranging the vowels

$= {}^3P_3 = 3! = 6$

also, the 3 consonants can be arranged at the remaining 3 positions

number of ways of these arrangements

$= {}^3P_3 = 3! = 6$

total number of ways = $(6 \times 6) = 36$

3. Ratio and Proportion

1. A and B together have ₹ 1210. If $\frac{4}{15}$ of A's amount is equal to $\frac{2}{5}$ of B's amount, how much amount does B have?

A. ₹ 460
B. ₹ 484
C. ₹ 550
D. ₹ 500

Sol. B. $\frac{4}{5}A = \frac{2}{5}B$

$A = \left(\frac{2}{5} \times \frac{15}{4}\right)B$

$A = \frac{3}{2}B$

$\frac{A}{B} = \frac{3}{2}$

$A : B = 3 : 2$

$\therefore$ B's share = ₹ $\left(1210 \times \frac{2}{5}\right) = 484$

2. Two numbers are respectively 20% and 50% more than a third number. The ratio of the two numbers is

A. 2 : 5
B. 3 : 5
C. 4 : 5
D. 6 : 7

Sol. C. let the third number be x

then, first number = 120% of $x = \frac{120x}{100} = \frac{6x}{5}$

second number = 150% of $x = \frac{150x}{100} = \frac{3x}{2}$

$\therefore$ ratio of first two numbers $= \frac{6x}{5} : \frac{3x}{5} = 12x : 15x$

$= 4:5$

3. A sum of money is to be distributed among A, B, C, D in the proportion of 5 : 2 : 4 : 3. If C gets ₹ 1000 more than D, what is B's share?

A. ₹ 500

B. ₹ 1500

C. ₹ 2000

D. none of these

***Sol.* C.** let the shares of A, B, C and D be ₹ $5x$, ₹ $2x$, ₹ $4x$ and ₹ $3x$ respectively

then, $4x - 3x = 1000$

$x = 1000$

$\therefore$ B's share = ₹ $2x$ = ₹ (2×1000) = ₹ 2000

4. Seats for Mathematics, Physics and Biology in a school are in the ratio 5 : 7 : 8. There is a proposal to increase these seats by 40%, 50% and 75% respectively. What will be the ratio of increased seats?

A. 2 : 3 : 4

B. 6 : 7 : 8

C. 6 : 8 : 9

D. none of these

***Sol.* A.** originally, let the number of seats for mathematics, Physics and Biology be $5x$, $7x$ and $8x$ respectively

number of increased seats are (140% of $5x$), (150% of $7x$) and (175% of $8x$)

$\left(\frac{140}{100} \times 5x\right), \left(\frac{150}{100} \times 7x\right)$ and $\left(\frac{175}{100} \times 8x\right)$

$7x, \frac{21x}{2}$ and $14x$

$\therefore$ the required ratio $= 7x : \frac{21x}{2} : 14x$

$14x : 21x : 28x$

$2 : 3 : 4$

5. In a mixture 60 litres, the ratio of milk and water 2 : 1. If this ratio is to be 1 : 2, then the quantity of water to be further added is

A. 20 litres

B. 30 litres

C. 40 litres

D. 60 litres

***Sol.* D.** quantity of milk $= \left(60 \times \frac{2}{3}\right)_{litre = 40\ litres}$.

quantity of water in it = (60- 40) litres = 20 litres

new ratio = 1 : 2

let quantity of water to be added further be x litres

then, milk : water $= \left(\frac{40}{20+x}\right)$

now $\left(\frac{40}{20+x}\right) = \frac{1}{2}$

$20 + x = 80$

$x = 60$

$\therefore$ Quantity of water to be added = 60 litres

6. The salaries A, B, C are in the ratio 2 : 3 : 5. If the increments of 15%, 10% and 20% are allowed respectively in their salaries, then what will be new ratio of their salaries?

A. 3 : 3 : 10

B. 10 : 11 : 20

C. 23 : 33 : 60

D. none of these

***Sol.* C.** Let A = $2k$, B = $3k$ and C = $5k$

A's new salary $= \frac{115}{100}$ of $2k = \left(\frac{115}{100} \times 2k\right) = \frac{23k}{10}$

B's new salary $= \frac{110}{100}$ of $3k = \left(\frac{110}{100} \times 3k\right) = \frac{33k}{10}$

C's new salary $= \frac{120}{100}$ of $5k = \left(\frac{120}{100} \times 5k\right)$

$\therefore$ New ratio $\left(\frac{23k}{10} : \frac{33k}{10} : 6k\right)$

7. If 40% of a number is equal to two-third of another number, what is the ratio of first number to the second number?

A. 2 : 5

B. 3 : 7

C. 5 : 3

D. 7 : 3

***Sol.* C.** let 40% of A = B

then, $\frac{40A}{100} = \frac{2B}{3}$

$\frac{2A}{5} = \frac{2B}{3} = \frac{A}{B} = \left(\frac{2}{3} \times \frac{5}{2}\right) = \frac{5}{3}$

$\therefore$ A : B = 5 : 3.

8. The fourth proportional to 5, 8, 15 is

A. 18

B. 24

C. 19

D. 20

Sol. **B.** let the fourth proportional to 5, 8, 15 be x

Then, $5 : 8 : 15 : x$

$5x = (8 \times 15)$

$x = \frac{(8\times15)}{5} = 24$

9. Two number are in the ratio 3 : 5. If 9 is subtracted from each, the new numbers are in the ratio 12 : 23. The smaller number is

A. 27

B. 33

C. 49

D. 55

Sol. **B.** let the numbers be $3x$ and $5x$

Then, $\frac{3x-9}{5x-9} = \frac{12}{23}$

$23(3x - 9) = 12(5x - 9)$

$9x = 99$

$x = 11$

the smaller number = $(3 \times 11) = 33$

10. In a bag, there are coins of 25 p, 10 p and 5 p in the ratio of 1 : 2 : 3. If there is ₹30 in all, how many 5 p coins are there?

A. 50

B. 100

C. 150

D. 200

Sol. **C.** let the number of 25 p, 10 p and 5 p coins be x, $2x$, $3x$ respectively

sum of their values

$= ₹\left(\frac{25x}{100} + 10 \times \frac{2x}{100} + 5 \times \frac{3x}{100}\right) = \frac{60x}{100}$

$\frac{60x}{100} = 30 \Leftrightarrow x = \frac{30\times100}{60} = 50$

hence, the number of 5 p coins

$= (3 \times 50) = 150$

4. Logarithm

1. Which of the following statements is not correct?

A. $\log_{10} 10 = 1$

B. $\log (2 + 3) = \log (2 \text{ x } 3)$

C. $\log_{10} 1 = 0$

D. $\log (1 + 2 + 3) = \log 1 + \log 2 + \log 3$

Sol. **B.** Here

A. Since $\log_a a = 1$, so $\log_{10} 10 = 1$

B. $\log (2 + 3) = \log 5$ and $\log (2 \times 3) = \log 6$ $= \log 2 + \log 3$

$\therefore \log (2 + 3) \neq \log (2 \times 3)$

C. Since $\log_a 1 = 0$, so $\log_{10} 1 = 0$

D. $\log(1 + 2 + 3) = \log 6 = \log (1 \times 2 \times 3)$ $= \log 1 + \log 2 + \log 3.$

So, **B.** is incorrect

2. If $\log 2 = 0.3010$ and $\log 3 = 0.4771$, the value of $\log_5 512$ is

A. 2.870

B. 2.967

C. 3.876

D. 3.912

Sol. **C.** $\log^5 512 = \frac{\log 512}{\log 5}$

$= \frac{\log 2^9}{\log 10/2}$

$= \frac{9 \log 2}{\log 10 - \log 2}$

$= \frac{9\times0.3010}{1-0.3010}$

$= \frac{2.709}{0.699}$

$= \frac{2709}{699}$

$= 3.876$

3. If $\log 27 = 1.431$, then the value of $\log 9$ is

A. 0.934

B. 0.945

C. 0.954

D. 0.958

Sol. **C.** $\log 27 = 1.431$

$3 \log 3 = 1.431$

$\log 3 = 0.477$

$\log 9 = \log(3^2) = 2 \log 3 = (2 \times 0.477) = 0.954.$

4. If log + log =log (a+b), then

A. $a + b = 1$

B. $a - b = 1$

C. $a = b$

D. $a^2 - b^2 = 1$

Sol. **A.** $\log\frac{a}{b} + \log\frac{b}{a} = \log (a+b)$

$\log (a+b) = \log \left(\frac{a}{b} \times \frac{b}{a}\right) = \log 1$

So, $a + b = 1$.

5. If $\log 2 = 0.30103$, the number of digits in 2^{64} is

A. 18

B. 19

C. 20

D. 21

Sol. **C.** $\log(2^{64}) = 64 \times \log 2$

$= (64 \times 0.30103)$

$= 19.26592$

Its characteristic is 19.

Hence, then number of digits in 2^{64} is 20

6. If $\log 2 = .30103$, the number of digits in 2^{64} is

A. -3/4

B. 3/4

C. 18/256

D. 256/81

Sol. **D.** $\log_x = (\frac{9}{16}) = \frac{1}{2}$

$x^{-1/2} = \frac{9}{16}$

$\frac{1}{x} = \frac{9}{16}$

$x = \frac{16}{9}$

$x = \left(\frac{16}{9}\right)^2$

$x = \frac{256}{81}$

7. If $a^x = b^y$, then

A. $\log \frac{a}{b} = \frac{x}{y}$

B. $\log \frac{a}{b} = \frac{x}{y}$

C. $\log \frac{a}{b} = \frac{y}{x}$

D. none of these

Sol. **C.** $a^x = b^y$

$\log a^x = \log b^y$

$x \log a = y \log b$

$\frac{\log a}{\log b} = \frac{y}{x}$.

8. If $\log_x y = 100$ and $\log_2 x = 10$, then the value of y is

A. 2^{10}

B. 2^{100}

C. 2^{1000}

D. 2^{10000}

Sol. **C.** $\log_2 x = 10 \Rightarrow x = 2^{10}$

$\therefore \log_x y = 100$

$y = x^{100}$

$y = (2^{10})^{100}$ [put value of x]

$y = 2^{1000}$.

9. The value of $\log_2 16$ is

A. 1/8

B. 4

C. 8

D. 16

Sol. **B.** let $\log_2 16 = n$

Then, $2^n = 16 = 2^4 \Rightarrow n = 4$.

$\therefore \log_2 16 = 4$.

10. If $\log_{10} 5 + \log_{10}(5x + 1) = \log_{10}(x + 5) + 1$, then x is equal to

A. 1

B. 3

C. 5

D. 10

Sol. **B.** $\log_{10} 5 + \log_{10}(5x + 1) = \log_{10}(x + 5) + 1$

$\log_{10} 5 + \log_{10}(5x + 1) = \log_{10}(x + 5) + \log_{10} 10$

$\log_{10}[5(5x + 1)] = \log_{10}[10(x + 5)]$

$5(5x + 1) = 10(x + 5)$

$5x + 1 = 2x + 10$

$3x = 9$

$x = 3$

5. Percentage

1. If 20% of $a = b$, then b% of 20 is the same as

A. 4% of a

B. 5% of a

C. 20% of a

D. none of these

Sol. **A.** 20% of $a = b \Rightarrow \frac{20}{100} a = b$

$\therefore$ b% of 20 $\left(\frac{b}{100} \times 20\right) = \left(\frac{20}{100} a \times \frac{1}{100} \times 20\right)$

$= \frac{4}{100} a = 4\%$ of a.

2. In a certain school, 20% of students are below 8 years of age. The number of students above 8 years of age is $\frac{2}{3}$ of the number of students of 8 years of age which is 48. What is the total number of students in the school?

A. 80

B. 120

C. 150

D. 100

Sol. **E.** let the number of students be x. Then number of students above 8 years of age

$\therefore$ 80% of $x = 48 + \frac{2}{3}$ of 48 of x.

$\Rightarrow \frac{80}{100} x = 80$

$\Rightarrow x = 100$

3. Two numbers A and B are such that the sum of 5% of A and 4% of B is two-third of the sum of 6% of A and 8% of B. Find the ratio of A : B

A. 2 : 3
B. 1 : 1
C. 3 : 4
D. 4 : 3

***Sol.* D.** 5% of A + 4% of B = (6% of A + 8% of B)

$$\Rightarrow \frac{5}{100}A + \frac{4}{100}B = \frac{2}{3}\left(\frac{6}{100}A + \frac{8}{100}B\right)$$
$$\Rightarrow \frac{1}{20}A + \frac{1}{25}B = \frac{1}{25}A + \frac{4}{75}B$$
$$\Rightarrow \left(\frac{1}{20} - \frac{1}{25}\right)_A = \left(\frac{4}{75} - \frac{1}{25}\right)_B$$
$$\Rightarrow \frac{1}{100}A = \frac{1}{75}B$$
$$\Rightarrow \frac{100}{75} = \frac{4}{3}$$

∴ Required ratio = 4 : 3

4. A student multiplied a number by $\frac{3}{5}$ instead of $\frac{5}{3}$ what is the percentage error in the calculation?

A. 34 %
B. 44 %
C. 54 %
D. 64 %

***Sol.* D.** let the number be x

then, error $= \frac{5}{3}x - \frac{3}{5}x = \frac{16}{15}x$

error% $= \left(\frac{16}{15} \times \frac{3}{5x} \times 100\right)_{\%} = 64\%$

5. In an election between two candidates, one got 55% of the total valid votes, 20% of the votes were invalid. If the total number of votes was 7500, the number of valid votes that the other candidate got, was

A. 2700
B. 2900
C. 3000
D. 3100

***Sol.* A.** number of valid votes = 80% of 7500 = 6000.

∴ valid votes polled by other candidate = 45% of 6000

$= \left(\frac{45}{100} \times 6000\right) = 2700$

6. Three candidates contested an election and received 1136, 7636 and 11628 votes respectively. What percentage of the total votes did the winning candidate get?

A. 57 %
B. 60 %
C. 65 %
D. 90 %

***Sol.* A.** total number of votes polled

= (1136 + 7636 + 11628) = 20400

∴ Required percentage $= \left(\frac{11628}{20400} \times 100\right)_{\%} = 57\%$

7. Two tailors X and Y are paid a total of ₹ 550 per week by their employer. If X is paid 120 percent of the sum paid to Y, how much is Y paid per week?

A. ₹ 200
B. ₹ 250
C. ₹ 300
D. none of these

***Sol.* B.** let the sum paid to Y per week be ₹ z.

then, z + 120% of z = 550.

$$\Rightarrow z + \frac{120}{100}z = 550$$
$$\Rightarrow \frac{120}{100}z = 550$$
$$\Rightarrow z = \left(\frac{550 \times 5}{1}\right) = 250$$

8. Gauri went to the stationers and bought things worth ₹ 25, out of which 30 paise went on sales tax on taxable purchases. If the tax rate was 6%, then what was the cost of the tax free items?

A. ₹15
B. ₹15.70
C. ₹19.70
D. ₹20

***Sol.* C.** let the amount taxable purchases be ₹ x.

then, 6% of $x = \frac{30}{100}$

$$\Rightarrow x = \left(\frac{30}{100} \times \frac{100}{6}\right) = 5$$

Cost of tax free items = ₹ [25 – (5 + 0.30)]
= ₹ 19.70

9. Rajeev buys goods worth ₹6650. He gets a rebate of 6% on it. After getting the rebate, he pays sales tax @ 10%. Find the amount he will have to pay for the goods.

A. ₹6876.10
B. ₹6999.20
C. ₹6654
D. ₹7000

***Sol.* A.** rebate = 6% of ₹ 6650

$= ₹\left(\frac{6}{100} \times 6650\right) = ₹\ 399$

sales tax = 10% of ₹ (6650 – 399)

$= ₹\left(\frac{10}{100} \times 6251\right) = ₹\ 625.10$

final amount = ₹ (6251 + 625.10)
= ₹ 6876.10

10. The population of a town increased from 1,75,000 to 2,62,500 in a decade. The average percent increase of population per year is

 A. 4.37%

 B. 5%

 C. 6 %

 D. 8.75 %

Sol. **B.** increase in 10 years

$= (262500 - 175000) = 87500$

increase% $= \left(\frac{87500}{175000} \times 100\right) = 50\%$

$\therefore$ required average $= \left(\frac{50}{10}\right) = 5\%$.

6. Probability

1. A bag contains 2 red, 3 green and 2 blue balls. Two balls are drawn at random. What is the probability that none of the balls drawn is blue?

 A. 1/2

 B. 2/5

 C. 8/15

 D. 9/20

Sol. **D.** here, S = {1, 2, 3, 4, ..., 19, 20}

let E = event of getting a multiple of 3 or 5

= {3, 6 , 9, 12, 15, 18, 5, 10, 20}

$\therefore$ P(E) $= \frac{n(E)}{n(S)} = \frac{9}{20}$

2. A bag contains 2 red, 3 green and 2 blue balls. Two balls are drawn at random. What is the probability that none of the balls drawn is blue?

 A. 10/21

 B. 11/21

 C. 2/7

 D. 5/7

Sol. **A.** total number of balls = (2 + 3 + 2) = 7

let S be the sample space

then, n(S) = Number of ways of drawing 2 balls out of 7= $^7C_2$

$= \frac{7\times6}{2\times1} = 21$

let E = Event of drawing 2 balls, none of which is blue

$\therefore$ n(E) = Number of ways of drawing 2 balls out of (2 + 3) balls.

$= {}^5C_2$

$= \frac{5\times4}{2\times1}$

$= 10$

$\therefore$ P(E) $= \frac{n(E)}{n(S)} = \frac{10}{21}$

3. In a box, there are 8 red, 7 blue and 6 green balls. One ball is picked up randomly. What is the probability that it is neither red nor green?

 A. 1/3

 B. 3/4

 C. 7/19

 D. 8/21

Sol. **A.** total number of balls = (8 + 7 + 6) = 21

let E = event that the ball drawn is neither red nor green = event that the ball drawn is blue.

$\therefore$ n(E) = 7.$\therefore$

$\therefore$ P(E) $= \frac{n(E)}{n(S)} = \frac{7}{21} = \frac{1}{3}$

4. What is the probability of getting a sum 9 from two throws of a dice?

 A. 1/6

 B. 1/8

 C. 1/9

 D. 1/12

Sol. **C.** in two throws of a dice, n(S) = (6 x 6) = 36

let E = event of getting a sum

={(3, 6), (4, 5), (5, 4), (6, 3)}

$\therefore$ P(E) $= \frac{n(E)}{n(S)} = \frac{4}{36} = \frac{1}{9}$

5. Three unbiased coins are tossed. What is the probability of getting at most two heads?

 A. 3/4

 B. 1/4

 C. 3/8

 D. 7/8

Sol. **D.** here S = {TTT, TTH, THT, HTT, THH, HTH, HHT, HHH}

let E = event of getting at most two heads.

then E = {TTT, TTH, THT, HTT, THH, HTH, HHT}

$\therefore$ P(E) $= \frac{n(E)}{n(S)} = \frac{7}{8}$

6. A card is drawn from a pack of 52 cards. The probability of getting a queen of club or a king of heart is:

 A. 1/13

 B. 2/13

 C. 1/26

 D. 1/52

Sol. **C.** here, n(S) = 52

let E = event of getting a queen of club or a king of heart

then, $n(E) = 2$

$\therefore P(E) = \frac{n(E)}{n(S)} = \frac{2}{52} = \frac{1}{26}$

7. A bag contains 4 white, 5 red and 6 blue balls. Three balls are drawn at random from the bag. The probability that all of them are red, is

A. 1/22

B. 3/22

C. 2/91

D. 2/77

Sol. **C.** let S be the sample space

then, $n(S)$ = number of ways of drawing 3 balls out of 15= $^{15}C_3$

$= \frac{(15\times14\times13)}{(3\times2\times1)} = 455$

let E = event of getting all the 3 red balls.

$\therefore n(E) = {}^5C_3 = {}^5C_2 = \frac{(5\times4)}{(2\times1)} = 10.$

$P(E) = \frac{n(E)}{n(S)} = \frac{10}{455} = \frac{2}{91}$

8. Two cards are drawn together from a pack of 52 cards. The probability that one is a spade and one is a heart, is:

A. 3/20

B. 29/34

C. 47/100

D. 13/102

Sol. **D.** let S be the sample space

then, $n(S) = {}^{52}C_2 = \frac{(52\times51)}{(2\times1)} = 1326$

let E = event of getting 1 spade and 1 heart.

$\therefore n(E)$ = number of ways of choosing 1 spade out of 13 and 1 heart out of 13

$= ({}^{13}C_1 \times {}^{13}C_1)$

$= (13 \times 13)$

$= 169.$

$\therefore P(E) = \frac{n(E)}{n(S)} = \frac{169}{1326} = \frac{13}{102}$

9. One card is drawn at random from a pack of 52 cards. What is the probability that the card drawn is a face card (Jack, Queen and King only)?

A. 1/13

B. 3/13

C. 1/4

D. 9/52

Sol. **B.** clearly, there are 52 cards, out of which there are 12 face cards

$\therefore$ P (getting a face card) $= \frac{12}{52} = \frac{3}{13}$

10. A bag contains 6 black and 8 white balls. One ball is drawn at random. What is the probability that the ball drawn is white?

A. 3/4

B. 4/7

C. 1/8

D. 3/7

Sol. **B.** let number of balls = (6 + 8) = 14

number of white balls = 8

P (drawing a white ball) $= \frac{8}{14} = \frac{4}{7}$

7. Time and Work

1. 8 men and 14 women are working together in a field. After working for 3 days, 5 men and 8 women leave the work. How many more days will be required to complete the work?

I 19 men and 12 women together can complete the work in 18 days

II 16 men can complete two-third of the work in 16 days

III In 1 day, the work done by three men in equal to the work done by four women

A. I only

B. II only

C. III only

D. I or II or III

Sol. **D.** clearly, I only give the answer

similarly, II only gives the answer

and, III only gives the answer

$\therefore$ Correct answer is **D.**

2. In how many days can 10 women finish a work?

I. 10 men can complete the work in 6 days

II. 10 men and 10 women together can complete the work in $3\frac{3}{7}$ days

III. If 10 men work for 3 days and thereafter 10 women replace them, the remaining work in completed in 4 days

A. Any two of the three

B. I and II only

C. II and III only

D. I and III only

Sol. **A.** I. (10 × 6) men can complete the work in 1 day

$\Rightarrow$ 1 man's 1 day's work $= \frac{1}{60}$

II. $10\times\frac{24}{7}$ men + $10\times\frac{24}{7}$ women can complete the work in 1 day.

$\Rightarrow \frac{240}{7}$ men's 1 day work + $\frac{240}{7}$ women's 1 day work =1.

$\Rightarrow \frac{240}{7} \times \frac{1}{60} + \frac{240}{7}$ women's 1 day's work = 1.

$\Rightarrow \frac{240}{7}$ women's 1 day's work = 1 $- \frac{4}{7} = \frac{3}{7}$

$\Rightarrow$ 10 women's 1 day's work = $\frac{3}{7} \times \frac{7}{240} \times 10 = \frac{1}{8}$

So, 10 women's can finish the work in 8 day's

III. (10 men's work for 3days) + (10 women's work for 4days) = 1

$\Rightarrow$ (10 x 3) men's 1 day's work + (10 x 4) women's 1 day's work = 1

$\Rightarrow$ 30 men's 1 day's work + 40 women's 1 day's work = 1

Thus, I and III will give us the answer

And, II and III will give us the answer

$\therefore$ Correct answer is **A.**

3. How many workers are required for completing the construction work in 10 days?

   I. 20% of the work can be completed by 8 workers in 8 days

   II. 20 workers can complete the work in 16 days

   III. One-eighth of the work can be completed by 8 workers in 5 days

   A. I only

   B. II and III only

   C. III only

   D. Any one of the three

Sol. **D.** I. work can be completed by (8 × 8) workers in 1 day.

$\Rightarrow$ Whole work can be completed by (8 × 8 × 5) workers in 1 day.

$= \frac{8\times8\times5}{10}$ workers in 10 days

= 32 workers in 10 days.

II. (20 × 16) workers can finish it in 1 day.

$= \frac{20\times16}{10}$ workers can finish it in 10 days.

$\Rightarrow$ 32 workers can finish it in 10 days.

$\frac{1}{8}$ work can be completed by (8×5) workers in 1 day.

$\Rightarrow$ whole work can be completed by (8×5× 8) workers in 1 day.

$= \frac{8\times5\times8}{10}$ = workers in 10 days

= 32 workers in 10 day.

$\therefore$ Any one of the three gives the answer.

$\therefore$ Correct answer is **D.**

4. A and B together can complete a task in 7 days. B alone can do it in 20 days. What part of the work was carried out by A?

   I. A completed the job alone after A and B worked together for 5 days

   II. Part of the work done by A could have been done by B and C together in 6 days

   A. I alone sufficient while II alone not sufficient to answer

   B. II alone sufficient while I alone not sufficient to answer

   C. Either I or II alone sufficient to answer

   D. Both I and II are not sufficient to answer

Sol. **A.** B's 1 day's work = $\frac{1}{20}$

(A+ B)'s 1 day's work = $\frac{1}{7}$

I. (A + B)'s 5 day's work = $\frac{5}{7}$

Remaining work = $(1 - \frac{5}{7}) = \frac{2}{7}$

$\frac{2}{7}$ Work was carried by A.

II. is irrelevant.

$\therefore$ Correct answer is **A.**

5. How long will Machine Y, working alone, take to produce x candles?

   I. Machine X produces x candles in 5 minutes

   II. Machine X and Machine Y working at the same time produce x candles in 2 minutes

   A. I alone sufficient while II alone not sufficient to answer

   B. II alone sufficient while I alone not sufficient to answer

   C. Either I or II alone sufficient to answer

   D. Both I and II are necessary to answer

Sol. **D.** **I.** gives, Machine X produces $\frac{x}{5}$ candles in 1 min

II. gives, Machine X and Y produce $\frac{x}{2}$ candles in 1 min

From I and II, Y produces $\left(\frac{x}{2} - \frac{x}{5}\right) = \frac{3x}{10}$ candles in 1 min

$\frac{3x}{10}$ Candles are produced by Y in 1 min

x candles will be produced by Y in $\frac{10}{3x} \times x$

Thus, I and II both are necessary to get the answer.

$\therefore$ Correct answer is **D.**

6. A can do a work in 15 days and B in 20 days. If they work on it together for 4 days, then the fraction of the work that is left is

A. $\frac{1}{4}$

B. $\frac{1}{10}$

C. $\frac{7}{15}$

D. $\frac{8}{15}$

***Sol.* D.** A's 1 day's work = $\frac{1}{15}$;

B's 1 day's work = $\frac{1}{20}$

(A + B)'s 1 day's work = $\left(\frac{1}{15}+\frac{1}{20}\right)=\frac{7}{60}$

(A + B)'s 4 day's work = $\left(\frac{7}{60}\times 4\right)=\frac{7}{15}$

Therefore, Remaining work = $\left(1-\frac{7}{15}\right)=\frac{8}{15}$

7. A can lay railway track between two given stations in 16 days and B can do the same job in 12 days. With help of C, they did the job in 4 days only. Then, C alone can do the job in

A. $9\frac{1}{5}$ days

B. $9\frac{2}{5}$ days

C. $9\frac{3}{5}$ days

D. 10 days

***Sol.* C.** (A + B + C)'s 1 day's work = $\frac{1}{4}$

A's 1 day's work = $\frac{1}{16}$

B's 1 day's work = $\frac{1}{12}$

C's 1 day's work = $\frac{1}{4}-\left(\frac{1}{16}+\frac{1}{12}\right)$

$=\left(\frac{1}{4}-\frac{1}{48}\right)=\frac{5}{48}$

So, C alone can do the work in 48 = 93 days.

8. A, B and C can do a piece of work in 20, 30 and 60 days respectively. In how many days can A do the work if he is assisted by B and C on every third day?

A. 12 days

B. 15 days

C. 16 days

D. 18 days

***Sol.* B.** A's 2 day's work = $\left(\frac{1}{20}\times 2\right)=\frac{1}{10}$

(A + B + C)'s 1 day's work = $\left(\frac{1}{20}+\frac{1}{30}+\frac{1}{60}\right)$

$=\frac{6}{60}=\frac{1}{10}$

Work done in 3 days = $\left(\frac{1}{10}+\frac{1}{10}\right)=\frac{1}{5}$

Now $\frac{1}{5}$ work is done in 3 days.

Whole work will be done in (3 × 5) = 15 days

9. A is thrice as good as workman as B and therefore is able to finish a job in 60 days less than B. Working together, they can do it in

A. 20 days

B. $22\frac{1}{2}$ days

C. 25 days

D. 30 days

***Sol.* B.** Ratio of times taken by A and B = 1 : 3.

The time difference is (3 – 1) 2 days while B take 3 days and A takes 1 day.

If difference of time is 2 days, B takes 3 days.

If difference of time is 60 days,

B takes $\left(\frac{3}{2}\times 60\right)=$ 90 days.

So, A takes 30 days to do the work.

A's 1 day's work = $\frac{1}{30}$

B's 1 day's work = $\frac{1}{90}$

(A + B)'s 1 day's work = $\left(\frac{1}{30}+\frac{1}{90}\right)=\frac{4}{90}=\frac{2}{45}$

A and B together can do the work in 45

= 221 days.

10. A alone can do a piece of work in 6 days and B alone in 8 days. A and B undertook to do it for ₹3200. With the help of C, they completed the work in 3 days. How much is to be paid to C?

A. ₹375

B. ₹400

C. ₹600

D. ₹800

***Sol.* D.** C's 1 day's work = $\frac{1}{30}-\left(\frac{1}{6}+\frac{1}{8}\right)=\frac{1}{3}-\frac{1}{24}=\frac{1}{24}$

A's wages : B's wages : C's wages = $\frac{1}{6}:\frac{1}{8}:\frac{1}{24}$

= 4:3:1.

C's share (for 3 days) = ₹ $\left(3\times\frac{1}{24}\times 3200\right)$

= ₹400

8. Simple Interest

1. A sum of money at simple interest amounts to ₹ 815 in 3 years and to ₹854 in 4 years. The sum is

A. ₹650

B. ₹690

C. ₹698

D. ₹700

***Sol.* C.** S.I. for 1 year = ₹ (854 – 815) = ₹39.

S.I. for 3 years = ₹(39 × 3) = ₹117.

∴ Principal = ₹(815 – 117) = ₹698.

2. Mr. Thomas invested an amount of ₹13,900 divided in two different schemes A and B at the simple interest rate of 14% p.a. and 11% p.a. respectively. If the total amount of simple interest earned in 2 years be ₹3508, what was the amount invested in Scheme B?

A. ₹6400
B. ₹6500
C. ₹7200
D. ₹7500

Sol. **A.** Let the sum invested in Scheme A be ₹x and that in Scheme B be ₹(13900 – x).

Then, $\left(\frac{x \times 14 \times 2}{100}\right) + \left(\frac{(13900 - X) \times 11 \times 2}{100}\right) = 3508$

$\Rightarrow 28x - 22x = 350800 - (13900 \times 22)$
$\Rightarrow 6x = 45000$
$\Rightarrow x = 7500.$

So, sum invested in Scheme B
= ₹ (13900 – 7500) = ₹ 6400.

3. A sum fetched a total simple interest of ₹4016.25 at the rate of 9 p.c.p.a. in 5 years. What is the sum?

A. ₹4462.50
B. ₹8032.50
C. ₹8900
D. ₹ 8925

Sol. **D.** Principal= ₹ $\left(\frac{100 \times 4016.25 \times}{9 \times 5}\right)$
= ₹ $\left(\frac{401625 \times}{45}\right)$
= ₹ 8925

2. How much time will it take for an amount of ₹450 to yield ₹81 as interest at 4.5% per annum of simple interest?

A. 3.5 years
B. 4 years
C. 4.5 years
D. 5 years

Sol. **B.** Time = $\left(\frac{100 \times 81}{450 \times 4.5}\right)$ = 4 years.

3. Reena took a loan of ₹1200 with simple interest for as many years as the rate of interest. If she paid ₹432 as interest at the end of the loan period, what was the rate of interest?

A. 3.6
B. 6
C. 18
D. none of these

Sol. **B.** Let rate = R% and time = R years.

Then, $\left(\frac{1200 \times R \times R}{100}\right) = 432$

$\Rightarrow 12R^2 = 432$
$\Rightarrow R^2 = 36$

$\Rightarrow R = 6.$

4. A sum of ₹12,500 amounts to ₹15,500 in 4 years at the rate of simple interest. What is the rate of interest?

A. 3%
B. 4%
C. 5%
D. 6%

Sol. **D.** S.I. = ₹(15500 – 12500) = ₹3000

Rate = $\left(\frac{100 \times 3000}{12500 \times 4}\right) = 6\%$

5. An automobile financier claims to be lending money at simple interest, but he includes the interest every six months for calculating the principal. If he is charging an interest of 10%, the effective rate of interest becomes

A. 10%
B. 10.25%
C. 10.5%
D. None of these

Sol. **B.** Let the sum be ₹100. Then,

S.I. for first 6 months

= ₹ $\left(\frac{100 \times 10 \times 1}{100 \times 2}\right)$ = ₹ 5.

S.I. for last 6 months = ₹ $\left(\frac{105 \times 10 \times 1}{100 \times 2}\right)$ =₹ 5.25

So, amount at the end of 1 year

= ₹ (100 + 5 + 5.25) = ₹ 110.25

∴ Effective rate = (110.25 – 100) = 10.25%

8. A lent ₹5000 to B for 2 years and ₹3000 to C for 4 years on simple interest at the same rate of interest and received ₹2200 in all from both of them as interest. The rate of interest per annum is

A. 5%
B. 7%
C. $7\frac{1}{8}\%$
D. 10%

Sol. **D.** Let the rate be R% p.a.

$\left(\frac{500 \times R \times 2}{100}\right) + \left(\frac{3000 \times R \times 4}{100}\right) = 2200$

$\Rightarrow 100R + 120R = 2200$

$\Rightarrow R = \left(\frac{2200}{220}\right) = 10.$

$\therefore$ Rate = 10%.

9. A sum of ₹ 725 is lent in the beginning of a year at a certain rate of interest. After 8 months, a sum of Rs. 362.50 more is lent but at the rate twice the former. At the end of the year, ₹33.50 is earned as interest from both the loans. What was the original rate of interest?

A. 3.6%
B. 4.5%
C. 5%
D. none of these

Sol. D. Let the original rate be R%. Then,

new rate = (2R)%.

Note:

Here, original rate is for 1 year(s); the new rate is for only 4 months i.e.$\frac{1}{3}$ year(s).

$\therefore \left(\frac{725 \times R \times 1}{100}\right) + \left(\frac{362.50 \times R \times 1}{100 \times 3}\right) = 33.50$

$\Rightarrow (2175 + 725)\ R = 33.50 \times 100 \times 3$

$\Rightarrow (2175 + 725)\ R = 10050$

$\Rightarrow (2900)R = 10050$

$\Rightarrow R\ \frac{10050}{2900} = 3.46$

$\therefore$ Original rate = 3.46%

10. A man took loan from a bank at the rate of 12% p.a. simple interest. After 3 years he had to pay ₹5400 interest only for the period. The principal amount borrowed by him was

A. ₹2000
B. ₹10,000
C. ₹15,000
D. ₹20,000

Sol. C. Principal = $\left(\frac{100 \times 5400}{12 \times 3}\right)$ = ₹15000.

9. Compound Interest

1. Mr. Gupta borrowed a sum of money on compound interest. What will be the amount to be repaid if he is repaying the entire amount at the end of 2 years?

I The rate of interest is 5 p.c.p.a.

II Simple interest fetched on the same amount in one year is ₹600

III The amount borrowed is 10 times the simple interest in 2 years

A. I only
B. III only
C. I or II only
D. II and Either I or III only

Sol. D. **I.** gives, Rate = 5% p.a.

II. gives, S.I. for 1 year = ₹600.

III. gives, sum = 10 × (S.I. for 2 years).

Now I, and II give the sum.

For this sum, C.I. and hence amount can be obtained.

Thus, III is redundant.

Again, II gives S.I. for 2 years = ₹(600 × 2)
= ₹1200.

Now, from III, Sum = ₹(10 × 1200) = ₹12000.

Thus, Rate = $\frac{100 \times 1200}{2 \times 12000}$ = 5% p.a.

Thus, C.I. for 2 years and therefore, amount can be obtained.

Thus, I is redundant.

Hence, I or III redundant.

2. What is the compound interest earned at the end of 3 years?

I Simple interest earned on that amount at the same rate and for the same period is ₹4500

II The rate of interest is 10 p.c.p.a.

III Compound interest for 3 years is more than the simple interest for that period by ₹465.

A. I and II only
B. II and III only
C. I and III only
D. Either II or III only

Sol. D. **I.** gives, S.I for 3 years = ₹4500.

II. gives, Rate = 10% p.a.

III. gives, (C.I.) × (S.I.) = ₹465.

Clearly, using I and III we get C.I. = ₹(465 + 4500).

Thus, II is redundant.

Also, from I and II, we get sum

$= \left(\frac{100 \times 4500}{10 \times 3}\right) = 15000.$

Now C.I. on ₹15000 at 10% p.a. for 3 years may be obtained.

Thus, III is redundant.

Either II or III is redundant.

3. What is the rate of compound interest?

I The principal was invested for 4 years

II The earned interest was ₹1491

A. I alone sufficient while II alone not sufficient to answer

B. II alone sufficient while I alone not sufficient to answer

C. Either I or II alone sufficient to answer

D. Both I and II are not sufficient to answer

Sol. **D.** Let Principal = ₹P and Rate = R% p.a. Then,

$$\text{Amount} = ₹\left[P\left(1+\frac{R}{100}\right)^4\right]$$

$$\therefore \quad \text{C. I.} = P\left[\left(1+\frac{R}{100}\right)^4 - 1\right]$$

$$\Rightarrow \quad P\left[\left(1+\frac{R}{100}\right)^4 - 1\right] = 1491$$

Clearly, it does not give the answer.

∴ Correct answer is **D.**

4. What will be compounded amount?

I ₹200 was borrowed for 192 months at 6% compounded annually.

II ₹200 was borrowed for 16 years at 6%.

A. I alone sufficient while II alone not sufficient to answer

B. II alone sufficient while I alone not sufficient to answer

C. Either I or II alone sufficient to answer

D. Both I and II are not sufficient to answer

Sol. **C.** $\text{Amount} = ₹\left[200\times\left(1+\frac{6}{100}\right)^{16}\right]$

$$\text{Amount} = ₹\left[200\left(1+\frac{R}{100}\right)^{16}\right]$$

Thus, I as well as II gives the answer.

∴ Correct answer is **C.**

5. An amount of money was lent for 3 years. What will be the difference between the simple and the compound interest earned on it at the same rate?

I The rate of interest was 8 p.c.p.a.

II The total amount of simple interest was Rs. 1200.

A. I alone sufficient while II alone not sufficient to answer

B. II alone sufficient while I alone not sufficient to answer

C. Either I or II alone sufficient to answer

D. Both I and II are not sufficient to answer

Sol. **C.** $\text{Amount} = ₹\left[200\times\left(1+\frac{6}{100}\right)^{16}\right]$

$$\text{Amount} = ₹\left[200\left(1+\frac{R}{100}\right)^{16}\right]$$

Thus, I as well as II gives the answer.

∴ Correct answer is **C.**

6. The difference between simple and compound interests compounded annually on a certain sum of money for 2 years at 4% per annum is ₹1. The sum (in ₹) is:

A. 625

B. 630

C. 640

D. 650

Sol. **C.** Let the sum be ₹ x. Then,

$$\text{C.I.} = \left[x\left(1+\frac{4}{100}\right)^2 - x\right] = \left(\frac{676}{625}x - x\right) = \frac{51}{625}x$$

$$\text{S.I.} = \left(\frac{x\times4\times2}{100} - x\right) = \frac{2x}{25}$$

$$\therefore \frac{51x}{625} - \frac{2x}{25} = 1$$

$x = 625.$

7. There is 60% increase in an amount in 6 years at simple interest. What will be the compound interest of ₹12,000 after 3 years at the same rate?

A. ₹ 2160

B. ₹ 3120

C. ₹ 3927

D. ₹ 6240

Sol. **C.** Let P = ₹ 100. Then, S.I. ₹ 60 and T = 6 years.

$$R = \left(\frac{100\times60}{100\times6}\right) = 10\%\text{ p.a.}$$

Now, P = ₹ 12000. T = 3 years and R = 10% p.a.

$$\therefore \text{C.I.} = ₹\left[12000\times\left\{\left(1+\frac{10}{100}\right)^3 - 1\right\}\right]$$

$$= ₹\left(12000\times\frac{331}{1000}\right)$$

$= 3972.$

8. What is the difference between the compound interests on ₹5000 for $1\frac{1}{2}$ years at 4% per annum compounded yearly and half-yearly?

A. ₹2.04

B. ₹3.06

C. ₹4.80

D. ₹8.30

Sol. **A.** C.I. when interest compounded yearly

$$= ₹\left[500\times\left(1+\frac{4}{100}\right)\times\left(1+\frac{\frac{1}{2}\times4}{100}\right)\right]$$

$$= \left(5000\times\frac{26}{25}\times\frac{51}{50}\right)$$

$= 5304.$

C.I. when interest is compounded half-yearly

$$= ₹\left[5000\times\left(1+\frac{2}{100}\right)^3\right]$$

$= ₹\left(5000 \times \frac{51}{50} \times \frac{51}{50} \times \frac{51}{50}\right)$
$= ₹\ 5306.04$
$\therefore$ Difference = ₹ (5306.04 - 5304) = ₹ 2.04

9. The compound interest on ₹30,000 at 7% per annum is ₹4347. The period (in years) is:
   A. 2
   B. $2\frac{1}{2}$
   C. 3
   D. 4

***Sol.* A.** Amount = ₹(30000 + 4347) = ₹34347.
Let the time be *n* years.
Then, $3000\left(1+\frac{7}{100}\right)^n = 34347$
$\Rightarrow \left(\frac{107}{100}\right)^n = \frac{34347}{30000} = \frac{11449}{10000} = \left(\frac{107}{100}\right)^n$
$\therefore$ $n = 2$ years.

10. A bank offers 5% compound interest calculated on half-yearly basis. A customer deposits ₹1600 each on 1st January and 1st July of a year. At the end of the year, the amount he would have gained by way of interest is:
   A. ₹120
   B. ₹121
   C. ₹122
   D. ₹123

***Sol.* B.** Amount $= ₹\left[\begin{array}{l}1600 \times \left(1+\frac{5}{2\times 100}\right)^2 \\ +1600 \times \left(1+\frac{5}{2\times 1000}\right)\end{array}\right]$
$= ₹\left[1600 \times \frac{41}{40} \times \frac{41}{40} 1600 \times \frac{41}{40}\right]$
$= ₹\left[\frac{1600\times 41\times 81}{40\times 40}\right]$
$= ₹\ 3321.$
$\therefore$ C.I. = ₹ (3321 - 3200) = ₹ 121

10. Profit and Loss

1. On selling 17 balls at ₹720, there is a loss equal to the cost price of 5 balls. The cost price of a ball is
   A. ₹45
   B. ₹50
   C. ₹55
   D. ₹60

***Sol.* D.** (C.P. of 17 balls) – (S.P. of 17 balls) = (C.P. of 5 balls)
C.P. of 12 balls = S.P. of 17 balls = ₹720
C.P. of 1 ball $= ₹\left(\frac{720}{12}\right) = ₹60$

2. When a plot is sold for ₹18,700, the owner loses 15%. At what price must that plot be sold in order to gain 15%?
   A. ₹21,000
   B. ₹22,500
   C. ₹25,300
   D. ₹25,800

***Sol.* C.** 85 : 18700 = 115 : *x*
$x = \left(\frac{18700\times 115}{85}\right) = 25300.$
hence, S.P. = ₹25,300

3. 100 orange are bought at the rate of ₹350 and sold at the rate of ₹48 per dozen. The percentage of profit or loss is
   A. $14\frac{15}{17}$% gain
   B. 15% gain
   C. $14\frac{15}{17}$% loss
   D. 15% loss

***Sol.* A.** C.P. of 1 orange $= ₹\left(\frac{350}{100}\right) = ₹\ 3.50$
S.P. of 1 orange $= ₹\left(\frac{48}{12}\right) = ₹\ 4$
$\therefore$ gain% $= \left(\frac{0.50}{3.50}\times 100\right)_{\%} = \frac{100}{7}\% = 14\frac{2}{7}\%$

4. A shopkeeper sells one transistor for ₹840 at a gain of 20% and another for ₹960 at a loss of 4%. His total gain or loss percent is
   A. $5\frac{15}{17}$% loss
   B. $5\frac{15}{17}$% gain
   C. $6\frac{2}{3}$% gain
   D. none of these

***Sol.* B.** C.P. of 1st transistor $= ₹\left(\frac{100}{120}\times 840\right) = ₹\ 700$
C.P. of 2nd transistor $= ₹\left(\frac{100}{96}\times 960\right) = ₹\ 1000$
So, total C.P. = ₹ (700 + 1000) = ₹ 1700
Total S.P. = ₹ (840 + 960) = ₹ 1800.
$\therefore$ gain % $= \left(\frac{100}{1700}\times 100\right)_{\%} = 5\frac{15}{17}\%$

5. A trader mixes 26 kg of rice at ₹20 per kg with 30 kg of rice of other variety at ₹36 per kg and sells the mixture at ₹30 per kg. His profit percent is
   A. no profit, no loss
   B. 5%

C. 8%

D. none of these

***Sol.* B.** C.P. of 56 kg rice = ₹ (26 × 20 + 30 × 36)

= ₹ (520 + 1080) = ₹1600

S.P. of 56 kg rice = ₹ (56 × 30) = ₹1680

gain = $\left(\frac{80}{1600} \times 100\right)_{\%}$ = 5%

6. The percentage profit earned by selling an article for ₹1920 is equal to the percentage loss incurred by selling the same article for ₹1280. At what price should the article be sold to make 25% profit?

A. ₹ 2000

B. ₹2200

C. ₹2400

D. data inadequate

***Sol.* A.** let C.P. be ₹ x

then, $\frac{1920-x}{x} \times 100 = \frac{x-1280}{x} \times 100$

$1920 - x = x - 1280$

$2x = 3200$

$x = 1600$

∴ required S.P. = 125% of ₹ 1600

= ₹ $\left(\frac{125}{100} \times 1600\right)$ = ₹ 2000

7. A shopkeeper expects a gain of 22.5% on his cost price. If in a week, his sale was of ₹392, what was his profit?

A. ₹18.20

B. ₹70

C. `72

D. `88.25

***Sol.* C.** C.P.=₹$\left(\frac{100}{122.5} \times 392\right)$ =₹ $\left(\frac{1000}{1225} \times 392\right)$

= ₹ 320

∴ profit = ₹ (392 - 320) = ₹. 72.

8. A man buys a cycle for ₹1400 and sells it at a loss of 15%. What is the selling price of the cycle?

A. ₹1090

B. ₹1160

C. ₹1190

D. ₹1202

***Sol.* C.** S.P. = 85% of ₹ 1400

= ₹ $\left(\frac{85}{100} \times 1400\right)$

= ₹ 1190

9. Sam purchased 20 dozens of toys at the rate of ₹375 per dozen. He sold each one of them at the rate of ₹33. What was his percentage profit?

A. 35

B. 4.5

C. 5.6

D. 6.5

***Sol.* C.** cost price of 1 toy = ₹ $\left(\frac{375}{12}\right)$ = ₹ 31.25

selling price of 1 toy = ₹ 33

so, Gain = ₹ (33 - 31.25) = ₹ 1.75

∴ profit % = $\left(\frac{1.75}{31.25} \times 100\right)_{\%}$ = 5.6%

10. Some article were bought at 6 articles for ₹5 and sold at 5 articles for ₹6. Gain percent is

A. 30 %

B. $33\frac{1}{3}$%

C. 35 %

D. 44 %

***Sol.* D.** suppose, number of articles bought

= L.C.M. of 6 and 5 =30

C.P. of 30 articles = ₹ $\left(\frac{5}{6} \times 30\right)$ = ₹ 25

S.P. of 30 articles = ₹ $\left(\frac{6}{5} \times 30\right)$ = ₹ 36

Gain % = $\left(\frac{11}{25} \times 100\right)_{\%}$ = 44%

GATE 2015-2010

(1 Mark Questions)

GATE 2015

1. If $\log_x \frac{5}{7} = -\frac{1}{3}$, then the value of x is

 A. 343/125

 B. 125/343

 C. -25/49

 D. -49/25

Sol. **A.** $\frac{5}{7} = x^{\frac{-1}{3}}$

or $\frac{7}{5} = x^{\frac{1}{3}}$

So, $x = \left(\frac{7}{5}\right)^3 = \frac{343}{125} = (2.744)$

2. Choose the most appropriate word from the option given below to complete the following sentence.

 Principal presented chief guest with ______ as a token of gratitude.

 A. momentum

 B. moment

 C. momento

 D. memento

Sol. **D.** memento

3. What is meaning of word "Educe"?

 A. extract

 B. educate

 C. exert

 D. training

Sol. **A.** extract

4. Choose the appropriate word/phase, out of the four option given below, to complete the following sentence;

 Frog .

 A. croak

 B. roar

 C. hiss

 D. patter

Sol. **A.** croak

5. If □ ◇ and ⟶ defined as

 a □ b = $\frac{a-b}{a+b}$, a ◇ b = $\frac{a+b}{a-b}$ and $a \longrightarrow b = ab$

 Then calculate the value of (66 □ 6) ⟶ (66 ◇ 6)

Sol. 66 □ 6 $= \frac{66-6}{66+6} = \frac{60}{72} = \frac{5}{6}$

66 ◇ 6 $= \frac{66+6}{66-6} = \frac{72}{60} = \frac{6}{5}$

(66 □ 6) ⟶ (66 ◇ 6) $= \frac{5}{6} \times \frac{6}{5} = 1$

GATE 2014

6. A student is required to demonstrate a high level of comprehension of the subject, especially in the social science.

 The word closest in meaning to comprehension

 A. understanding

 B. meaning

 C. concentration

 D. stability

Sol. **A.** understanding

7. Choose the most appropriate word from the options given below to complete the following sentence. One of his biggest _________was his ability to forgive.

 A. vice

 B. virtue

 C. choice

 D. stability

Sol. **B.** virtue

8. Rajan was not happy that Sajan decided to do the project on his own. On observing his unhappiness, Sajan explained to Rajan that he preferred to work independently.

 Which one of the statements below is logically valid and can be inferred from the above sentence?

 A. Rajan has decided to work only in a group

 B. Rajan and Sajan were formed into a group against their wishes

C. Sajan had decided to given into Rajan's request to work with him

D. Rajan had believed that Sajan and he would be working together

***Sol.*A.** Rajan was not happy because believed that he would work with Sajan.

9. If y = $5x^2 + 3$, then the tangent at x = 0, y = 3

A. passes through x = 0, y = 0

B. has a slope of +1

C. is parallel to the x-axis

D. has a slope of – 1

***Sol.*C.** y = $5x^2 + 3$

Tangent at (0,3)

$$\frac{1}{2}(y+3) = 5 \times x \times 0 + 3$$

$$\frac{1}{2}(y+3) = 3$$

$$y = 3$$

So, this is parallel to x-axis.

10. A foundry has a fixed daily cost of ₹50000 whenever it operates and a variable cost of ₹800 Q, where Q is the daily production in tone. What is the cost of production in ₹ per tonne. for a daily production of 100 tonne?

Sol. Cost of production per tonne,

$$C = \frac{\text{Total cost}}{\text{Production (Q)}}$$

or $$C = \frac{\text{Fixed cost + variable cost}}{Q}$$

or $$C = \frac{50000 + 800\,Q}{Q} = \frac{50000}{Q} + 800$$

or $$C = \frac{50000}{100} + 800 = 1300$$

GATE 2013

11. If $3 \le X \le 5$ and $8 \le Y \le 11$ then which of the following options is true?

A. $\frac{3}{5} \le \frac{X}{Y} \le \frac{8}{5}$

B. $\frac{3}{11} \le \frac{X}{Y} \le \frac{5}{8}$

C. $\frac{3}{11} \le \frac{X}{Y} \le \frac{8}{5}$

D. $\frac{3}{8} \le \frac{X}{Y} \le \frac{5}{11}$

***Sol.*D.** $3 \le x \le 5$ and $8 \le y \le 11$

Now for minimum value of $\frac{x}{y}$

$$= \frac{\text{Minimum Value of x}}{\text{Maximum Value of y}} = \frac{3}{11}$$

Similarly for maximum value of $\frac{x}{y}$

$$= \frac{\text{Maximum Value of y}}{\text{Minimum Value of x}} = \frac{5}{8}$$

on arranging the above two results

$$= \frac{3}{8} \le \frac{x}{y} \le \frac{5}{8}$$

12. The Headmaster to speak to you. Which of the following options is incorrect to complete the above sentence?

A. is wanting

B. wants

C. want

D. was wanting

***Sol.*C.** Want

13. Mahatma Gandhi was known for his humility as

A. he played an important role in humiliating exit of British from India

B. he worked for humanitarian causes

C. he displayed modesty in his interactions

D. he was a fine human being

***Sol.*C.** He displayed modesty in his interactions

14. Which of the above underlined parts of the sentence in not appropriate?

A. I

B. II

C. III

D. IV

***Sol.*D.**

15. Select the pair that best expresses a relationship similar to that expressed in the pair:

water : pipe ::

A. cart : road

B. electricity : wire

C. sea : beach

D. music : instrument

***Sol.*B.** Electricity : wire

GATE 2012

16. Which one of the following options is the closest in meaning to the word given below?

A. diminish

B. divulge

C. dedicate

D. denote

***Sol.*A.** diminish

17. Choose the most appropriate alternative from the options given below to complete the following sentence:

 Despite several ________ the mission succeeded in its attempt to resolve the conflict.

 A. attempts
 B. setbacks
 C. meetings
 D. delegations

Sol. B. Setbacks

18. The cost function for a product in a firm is given by $5q^2$, where q is the amount of production. The firm can sell the product at a market price of Rs 50 per unit. The number of units to be produced by the firm such that the profit is maximized is

 A. 5
 B. 10
 C. 15
 D. 25

Sol. A. $p = 50q - 5q^2$

$$\frac{dp}{dq} = 50 - 10q; \ \frac{d^2p}{dq^2} < 0$$

∴ p is maximum at 50 – 70 q = or, q = 5

19. Choose the most appropriate alternative from the options given below to complete the following sentence:

 Suresh's dog is the one ……… was hurt in the stampede.

 A. that
 B. which
 C. who
 D. whom

Sol. A. that

20. Choose the grammatically incorrect sentence among the following sentences.

 A. they gave us the money back less the service charges of three hundred rupess
 B. this country's expenditure is not less than that of Bangladesh
 C. the committee initially asked for a funding of fifty lakh rupees but later settled for a lesser sum
 D. this country's expenditure on educational reforms is very less

Sol. D.

GATE 2011

21. Choose the word from the option given below that is most nearly opposite in meaning to the given word. Amalgamate

 A. merge
 B. split
 C. collect
 D. separate

Sol. B. amalgamate means combine or unite to form an organization or structure when two companies or organizations merge and become finally one organization, process to becoming a one is amalgamation. So, the best option here is split. Separate on the other hand, although a close synonym, it is too general to be the best antonyms in the given question while merge is the synonyms, collect is not related.

22. Which of the following option is the closest in the meaning of the word given below?

 Inexplicable

 A. incomprehensible
 B. indelible
 C. inextricable
 D. infallible

Sol. A. inexplicable means not explicable, that cannot be explained, understood, or accounted for. So, the best synonym here is incomprehensible.

23. If $\log P = \left(\frac{1}{2}\right)\log Q = \left(\frac{1}{3}\right)\log R$, then which of the following option is true?

 A. $P^2 = Q^3R^2$
 B. $Q^2 = PR$
 C. $Q^2 = R^3P$
 D. $R = P^2Q^2$

Sol. B. Log P= $\frac{1}{2}$ log Q= $\frac{1}{3}$ log R = k

∴ P= b^k, Q= b^{2k}, R= b^{3k}

Now, $Q^2 = b^{4k} = b^k$ = PR

24. Choose the most appropriate words(s) from the option given below to complete the following sentence. I contemplated _Singapore for my vacation but decided against it.

 A. to visit
 B. having to visit
 C. visiting
 D. for a visit

Sol. C. contemplate is a transitive verb and hence is followed by a gerund. Hence, the correct usage of contemplate is verb + ing form.

25. Choose the most appropriate word from the option given below to complete the following sentence. If you are trying to make a strong impression on your audience, you cannot do so by being understated , tentative or......

 A. hyperbolic
 B. restrained
 C. argumentative
 D. indifferent

***Sol.*B.** the tone of the sentence clearly indicates a word that is similar to understated s needed for the blank. Alternatively, the word should be antonym of strong (fail to make strong impression). Therefore, the best choice is restrained which means controlled/ reserved/timid.

GATE 2010

26. Which of the following option is the closest in meaning of the word given below?

 Circuitous

 A. cyclic
 B. indirect
 C. confusing
 D. crooked

***Sol.*B.** circuitous means round about or indirect

27. The question below consists of a pair of related words followed by four pairs of words. Select the pair that best expresses the relation in the original pair.

 Unemployed : Worker

 A. Fallow : Land
 B. Unaware : sleeper
 C. Wit : Jester
 D. Renovated : House

***Sol.*A.** Relationship between the given words is that first word tells the negative quality of record. So, option A. is correct because fallow means uncultivated land which is a negative quality.

28. Choose the most appropriate word from the option given below to complete the following sentence.

 If we manage to ..... Our nature resources would leave a batter planet for our children.

 A. uphold
 B. restrain
 C. cherish
 D. conserve

***Sol.*D.** conserve is the most appropriate word because it means protect or safe.

29. Choose the most appropriate word from the option given below to complete the following sentence. His rather casual remarks on politics ..... his lack of seriousness about the subject.

 A. masked
 B. belied
 C. betrayed
 D. conserve

***Sol.*C.** betrayed is the most appropriate word because it means to tell identity unintentionally.

30. 25 persons are in a room, 15 of them play hockey 17 of them played football and 10 of them played both hockey and football. Then the number of persons playing neither hockey not football is

 A. 2
 B. 17
 C. 13
 D. 3

***Sol.*A.**

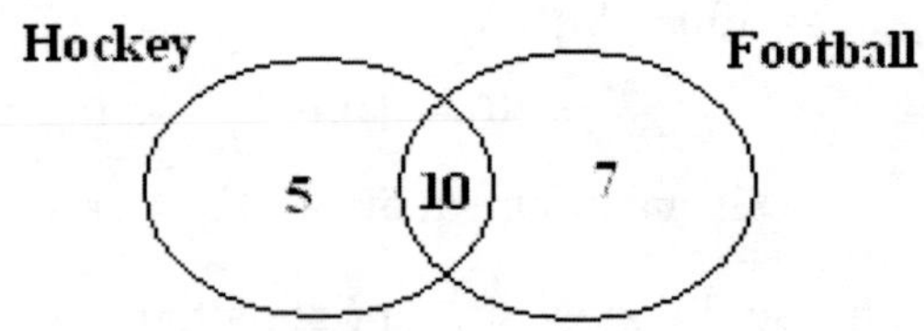

From the Venn-diagram, it is clear that 5 persons play only hockey. 7 persons play only football.

Total number of persons playing neither game = 5+10+7 = 22

So, the number of persons playing neither hockey not football

= 25 – 22 = 3

(2 Mark Questions)

GATE 2015

31. Cube of 3 side unit is constructed using cube of side 1 unit then ratio of number of visible faces to non-visible face is

 A. 1 : 2
 B. 2 : 3
 C. 1 : 3
 D. 3 : 4

***Sol.*A.**

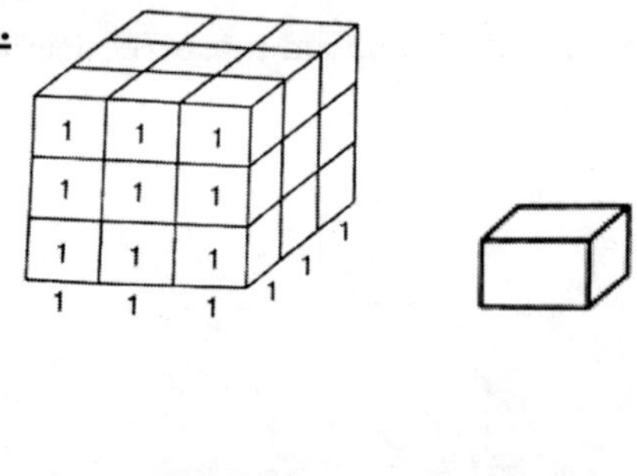

Number of faces per cube = 6

Total number of cubes = 9 × 3 = 27

∴ Total number of faces = 27 × 6 = 162

So, Total number of non – visible faces = 162 – 54 = 108

$$\frac{\text{Number of visible faces}}{\text{Number of non} - \text{visible faces}} = \frac{54}{108} = \frac{1}{2}$$

32. The missing term in the following arrangement is ________.

		6	5	4		
	7	4	7	2	1	
1	9	2	8	1	2	1
	4	1	5	2	3	
		3	⊗	3		

***Sol.*A.** $\frac{6+4}{2} = 5$ and $\frac{7+4+2+1}{2} = 7$

$\frac{1+9+2+1+2+1}{2} = 8$

and $\frac{4+1+2+3}{2} = 5$

$\frac{3+3}{2} = 3$

33. Ram and Ramesh attends interview for two vacancies in same department, Ram has probability of selection of $\frac{1}{6}$ and Ramesh has $\frac{1}{8}$, The probability of selection of either of them interview is ________.

A. $\frac{4}{3}$

B. $\frac{3}{4}$

C. $\frac{1}{4}$

D. $\frac{1}{3}$

***Sol.*C.** Probability of selection of Ram = $\frac{1}{6}$

∴ Probability of not selection of Ram = $1 - \frac{1}{6} = \frac{5}{6}$

Probability of selection of Ramesh = $\frac{1}{8}$

∴ Probability of not selection of Ramesh

$= 1 - \frac{1}{8} = \frac{7}{8}$

So, The probability of selection either Ram or Ramesh

$= \frac{1}{6} \times \frac{7}{8} + \frac{1}{8} \times \frac{5}{6} = \frac{7+5}{48} = \frac{12}{48} = \frac{1}{4}$

34. A tiger is 50 leaps of its own behind the deer as it starts chasing it. Tiger takes 5 leaps per min as compared to 4 leaps per min of the deer. If the tiger covers a distance of 8 m in one leap while deer covers 5 m in one leap, than how much distance tiger needs to run in order to catch the deer?

A. 50 min

B. 40 min

C. 1 h 20 min

D. 1 h 30 min

***Sol.*A.** in 1 min, tiger covers 5 leaps

And 1 min, deer covers 4 leaps

Deer is 50 m ahead of tiger.

Distance covered by tiger in one leap = 8 m

Distance covered by deer in one leap = 5 m

So, Time taken by tiger to catch the deer

$= \frac{50 \times 5}{5} = 50$ min

35. If $a^2 + b^2 + c^2 = 1$, then the value of $ab + bc + ca$ will lie in the interval?

A. $\left[-\frac{1}{2}, 1\right]$

B. $\left[\frac{1}{2}, -1\right]$

C. $\left[-\frac{1}{2}, -1\right]$

D. None of these

***Sol.*A.** $(a+b+c)^2 = a^2 + b^2 + c^2 + 2(ab + bc + ac)$

or $(a+b+c)^2 = 1 + 2(ab + bc + ac)$

or $(a+b+c)^2 > 0$

So, $(ab + bc + ca) > -\frac{1}{2}$ using AM-GM inequality

$$\frac{a^2 + b^2}{2} \geq \sqrt{a^2b^2} \Rightarrow a^2 + b^2 \geq 2ab$$

Similarly, $b^2 + c^2 \geq 2bc$

$\Rightarrow c^2 + a^2 \geq 2ac$

or $2(a^2 + b^2 + c^2) \geq 2(ab + bc + ca)$

or $(ab + bc + ca) \leq 1$

Here, it will lie in range $\left[-\frac{1}{2}, 1\right]$

GATE 2014

36. Find the odd one the following group: **ALRVX, EPVZB, ITZDF, OYEIK**

A. ALRVX

B. EPVZB
C. ITZDF
D. OYEIK

***Sol.*D.** OYEIK

37. Anuj, Bhola, Chandan, Dilip, Eswar and Faisal live on different floors in a six-storied building (the ground floor is numbered 1, the floor above it 2, and so on). Anju lives on an even numbered floor. Bhola does not live on any of the floors below Faisal's floor. Dilip does not live on a floor immediately above or immediately below Bhola. Faisal live three floors above Dilip. Which of the following floor-person combinations is correct?

	Anju	*Bhola*	*Chandan*	*Dilip*	*Eswar*	*Faisal*
(a)	6	2	5	1	3	4
(b)	2	6	5	1	3	4
(c)	4	2	6	3	1	5
(d)	2	4	6	1	3	5

***Sol.*B.** Name Possible cases

Anuj $\longrightarrow$ 2, 4, 6
Bhola $\longrightarrow$ 2, 4, 6
Chandan > Faisal
Dilip $\neq 2$
Eswar $\neq$ Bhola ± 1
Faisal = Dilip + 3

It matches with only option **B.**

38. The smallest angle of a triangle is equal to two thirds of the smallest angle of a quadrilateral. The ratio between the angle of a quadrilateral is 3 : 4 : 5 : 6. The largest angle of the triangle is twice its smallest angle. What is the sum, in degree, of the second largest angle of the quadrilateral?

Sol. For quadrilateral,

$$3x + 4x + 5x + 6x = 360$$
$$x = 20$$

Smallest angle = 3 × 20 = 60
Largest angle (A) = 6 × 20 = 120°
For triangle PQR,

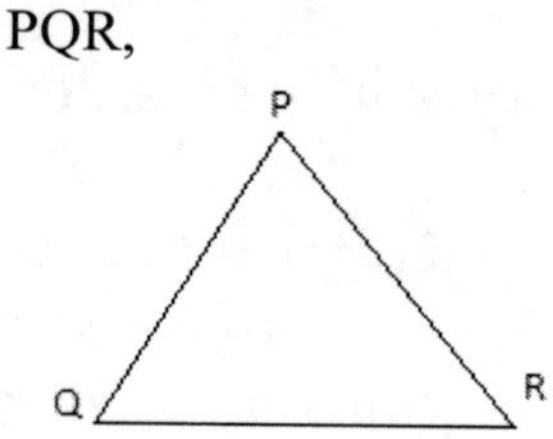

Smallest angle (p) = $\frac{2}{3} \times 60^0$
$= 40^0$

39. 1% of the people of country X are taller than 6ft. 2% of the people of country Y are taller than 6 ft. There are thrice as many people in country X as in country Y. Taking both countries together, what is the percentage of people taller than 6 ft?

A. 3.0
B. 2.5
C. 1.5
D. 1.25

***Sol.*D.** suppose number of people in X = x
And number of people in Y = y
Given, x = 3y
% of people taller than 6 ft.

$$= \frac{\text{Number of people taller than 6 ft in X and Y}}{\text{Total number of people in X and Y}}$$

$$= \frac{0.01x+0.02\,y}{x+y} \times 100$$

$$= \frac{0.01 \times 3y+0.02\,y}{3\,y+y} \times 100$$

$$= \frac{5}{4} = 1.25\%$$

40. The monthly rainfall chart based on 50 years of rainfall in Agra is shown in the following figure. Which of the following are true? (k percentile is the value such that k% of the data fall below that value)

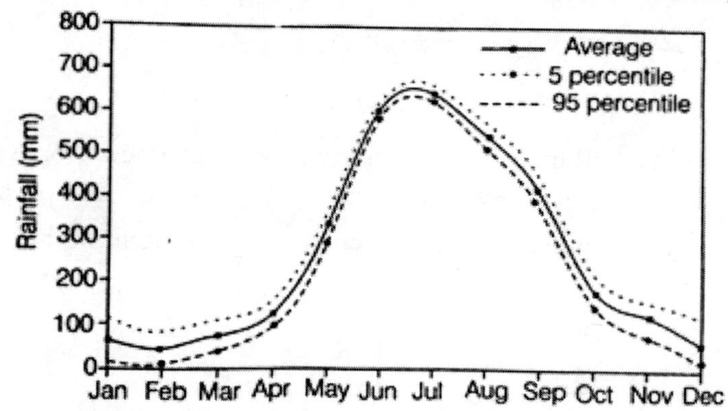

(i) On average, if rains more in July than in December
(ii) Every year, the amount of rainfall in August is more than that in January
(iii) July rainfall can be estimated with better confidence then February rainfall
(iv) In August, there is at least 500 mm of rainfall

A. (i) and (ii)
B. (i) and (iii)
C. (ii) and (iii)
D. (iii) and (iv)

***Sol.*C.** statement (i) is true as it rained more in July on an average compared to December,

Statement (ii) is not true for each year as the data is given for 50 years on an average.

Statement (iii) is also true.

Statement (iv) is not true as 5% of data is less than 500 mm of rainfall in August.

GATE 2013

41. Velocity of an object fired directly in upward direction is given by u = 80 – 32 t, where t (time) is in seconds. When will the velocity be between 32 m/s and 64 m/s?

A. (1, 3/2)
B. (1/2, 1)
C. (1/2, 3/2)
D. (1,3)

Sol. **C.** Given,

velocity (V) = 80 – 32 t

where, t is time

when v = 32 m/s, then

$$32 = 80 - 32t$$

or $32t = 48$

or $t = \frac{48}{32} = \frac{3}{2}$

When v = 64 m/s, then

$$64 = 80 - 32t$$

or $32t = 16$

or $t = \frac{16}{32} = \frac{1}{2}$

Hence, interval be $\left(\frac{1}{2}, \frac{3}{2}\right)$

42. In a factory, two machines M1 and M2 manufacture 60% and 40% of the auto components respectively. Out of the total production, 2% of M1 and 3% of M2 are found to be defective. If a randomly drawn auto components from the combined lot is found defective, what is the probability that it was manufactured by M2?

A. 0.35
B. 0.45
C. 0.5
D. 0.4

Sol. **C.** probability that the defective auto component is manufactured by $M_1 = \frac{2}{100} \times \frac{60}{100}$

$$= \frac{120}{10000}$$

$$= \frac{12}{1000}$$

Probability that the defective auto component is manufacture by $M_2 = \frac{3}{100} \times \frac{40}{100}$

$$= \frac{120}{1000}$$

$$= \frac{12}{1000}$$

Now, probability when randomly a auto component is picked and it belongs to M_1.

$$= \frac{\frac{12}{1000}}{\frac{12}{100} + \frac{12}{1000}}$$

$$\frac{\frac{12}{1000}}{\frac{24}{1000}} = 0.5$$

43. Following table gives date on tourists from different countries visiting India in the year 2011.

Country	*Number of tourists*
USA	2000
England	3500
Germany	1200
Italy	1100
Japan	2400
Australia	2300
France	1000

Which two countries contributed to the one third of the total number of tourists who visited India in 2011?

A. USA and Japan
B. USA and Australia
C. England and France
D. Japan and Australia

Sol. **C.** Total number of visitors

= 2000 + 3500 + 1200 + 1100 + 2400 + 230

= 13500

Number of visitors of England and France

= 3500 + 1000 = 4500

But 4500 is one-third of 13500.

44. If $|-2X + 9| = 3$ then the possible value of $|-X| - X^2$ would be

A. 30
B. – 30
C. – 42
D. 42

Sol. **B.** Given,

$$|-2X + 9| = 3$$

Case I When $-2X + 9 = 3$

or $X = 3$

Then, $|-X| - X^2 = |-3| - 3^2$

$= 3 - 9$

$= -6$

Case II When $-2X + 9 = -3$

or $X = 6$

Then, $|-X| - X^2 = |-6| - 6^2$

$= 6 - 36$

$= -30$

Hence, possible value of $|-X| - X^2$ is -30

45. All professors are researchers.
Some scientists are professors.
Which of the given conclusions is logically valid and is inferred from the above arguments?
A. all scientists are researchers.
B. all professors are scientists.
C. some researchers are scientists
D. no conclusion follows.

***Sol.*C.** some researchers are scientists

GATE 2012

46. An automobile plant contracted to buy shock absorbers from two suppliers X and Y. X supplies 60% and Y supplies 40% of the shock absorbers. All shock absorbers are subjected to a quality test. The ones that pass the quality test are considered reliable. Of X's shock absorbers, 72% are reliable. The probability that a randomly chosen shock absorber which is found to be reliable, is made by Y is
A. 0.288
B. 0.334
C. 0.667
D. 0.720

***Sol.*B.**

	x	y
Supply	60%	40%
Reliable	96%	72%
Overall	0.576	0.288

$$\therefore p(x) = \frac{0.288}{0.576 + 0.288} = 0.334$$

47. A political party orders an arch for the entrance to the ground in which the annual convention is being held. The profit of the arch follows the equation y = 2x – $0.1x^2$, where y is the height of the arch in metre. The maximum possible height of the arch is
A. 8 m
B. 10 m
C. 12 m
D. 14 m

***Sol.*B.** $y = 2x - 0.1x^2$

$$\frac{dy}{dx} = 2 - 0.2x$$

$\frac{d^2y}{dx^2} < 0$ ∴ y max imises at 2 – 0.2x = 0

$x = 10$

$\therefore y = 20 - 10 = 10\text{m}$

48. Wanted temporary, part-time persons for the post of field interviewer to conduct personal interviews to collect collate economic data. Requirements: High school pass, must be available for day, evening and Saturday work. Transportation paid, expenses reimbursed.
Which one of the following is the best inference from the above advertisement?
A. gender discriminatory
B. xenophobic
C. not designed to make the post attractive
D. not gender discriminatory

***Sol.*C.**

49. Given the sequence of terms, AD CG FK JP, the next term is
A. OV
B. OW
C. PV
D. PW

***Sol.*A.**

50. Which of the following assertions are correct?
P. adding 7 to each entry in a list adds 7 to the mean of the list.
Q. adding 7 to each entry in a list adds 7 to the standard deviation of the list.
R. doubling each entry in a list doubles the mean of the list
S. doubling each entry in a list leaves the standard deviation of the list unchanged.
A. P,Q
B. Q, R
C. P, R
D. R, S

***Sol.*C.**

GATE 2011

51. A container originally contains 10 L of pure spirit. From this container 1 L of spirit is replaced with 1 L of water. Subsequently, 1 L of the mixture is again replaced with 1 L of water and this process is repeated one more time. Howe much spirit is now left in the container?
A. 7.58 L
B. 7.84 L
C. 7 L
D. 7.29 L

Sol.**D.** $10\left(\frac{10-1}{10}\right)^3 = 10\left(\frac{9}{10}\right)^3 = 10\times\frac{729}{1000} = \frac{729}{100}$

$\therefore \frac{729}{100} \times 1 = 7.29$ L

52. Few school curriculums include a unit on how to deal with bereavement and grief, and yell all students at some point in their lives suffer from losses through death and parting.

Based on the above passage which topic would not be included in a unit of bereavement?

A. how to write a letter of condolence

B. what emotional stages are passed through in the healing process

C. what the leading causes of death are

D. how to give support to a grieving friend

Sol.**C.** the given passage clearly deals with how to deal with bereavement and grief. So, after the tragedy occurs and not about precautions. Therefore, irrespective of the causes of death, a school student rarely gets into details of causes – which is beyond the scope of the context. Rests all are important in dealing with grief.

53. *P, Q, R* and *S* are four type of dangerous microbes recently found in a human habitat. The area of each circle with its diameter printed in brackets represents the growth of single microbe surviving human immunity system within 24 h of entering the body. The danger to human beings varies proportionately with the toxicity, potency and growth attributed to a microbe shown in the figure below.

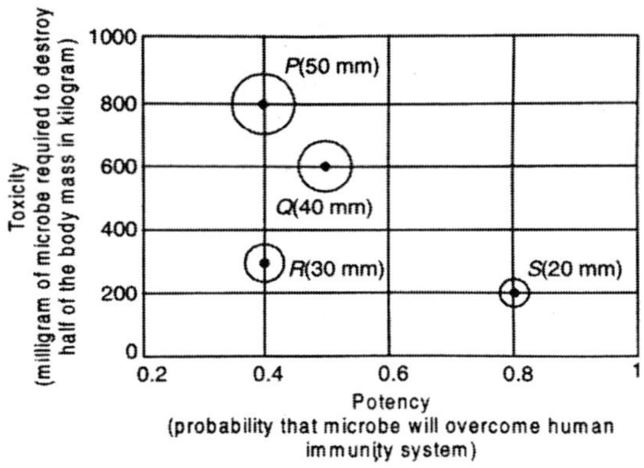

A pharmaceutical company is contemplating the development of a vaccine against the most dangerous microbe. Which microbe should the company target in its first attempt?

A. *P*

B. *Q*

C. *R*

D. *S*

Sol.**D.** By observation of the table, we can say S.

	P	*Q*	*R*	*S*
Requirement	800	600	300	200
Potency	0.4	0.5	0.4	0.8

54. The variable cost (V) of manufacturing a product varies according to the equation $= 4a$, where q is the quantity produced. The fixed cost (F) of production of same produce reduces with q according to the equation $F = 100/q$. how many units should be product to minimize the total cast $(V + F)$?

A. 5

B. 4

C. 7

D. 6

Sol.**A.** Checking with all options in formula

(4q + 100/q) i.e. (V + F)

Option **A.** gives the minimum cost.

55. A transporter receives the same number of order each day. Currently, he has some pending orders (backlog) to be shipped. If he uses 7 trucks, then at the end of the 4th day hi can clear all the orders are cleared at the end of the 4th day he can clear all the orders. Orders are cleared at the end of the 10th day. What is the minimum number of trucks, so that there will be no pending order at the end of the 5th day?

A. 4

B. 5

C. 6

D. 7

Sol.**C.** let each truck carries 100 units.

$2800 = 4n + e \quad (n = \text{normal})$

$3000 = 10n + e \quad (e = \text{express/pending})$

$\therefore \quad n = \frac{100}{3}, e = \frac{8000}{3}$

5 days $\quad 500x = \frac{5\times100}{3} + \frac{8000}{3}$

$X = \frac{8500}{3\times500} = 5.6$ days

So, $\quad x > 5$ days

So, minimum possible days = 6

Hint 7 truck × 4days × 100 units = 2800

GATE 2010

56. Modern warfare has changed from larger scale clashes of armies to suppression, of civilian population. Chemical agents that do their work silently appear to be suited to such warfare; and regretfully, there exits people in military establishment

who think that chemical agents are useful tools for their cause.

Which of the following statement best sums up the meaning of the above passage?

A. modern warfare has resulted in civil strife

B. chemical agents are useful in modern warfare

C. use of chemical agents in warfare would be undesirable

D. people in military establishments like to use chemical agents in war

***Sol.*D.** from the statement of the passage, it is clear that people in military establishments think that chemical agents are useful tool their cause. It is not given in the passage that chemical agents are useful or undesirable in modern warfare.

57. If 137 + 276 = 435, how much is 731 + 672?

A. 534

B. 1403

C. 1623

D. 1531

***Sol.*A.** 137 + 276 = 435

If 137 and 276 are written in reverse order, then 731 and 672 are obtained.

So, 731 + 672 reverse order of 435 = 534

58. 5 skilled workers can build a wall in 20 days; 8 semi-skilled workers can be build a wall in 25 days; 10 unskilled workers can build a wall in 30 days; if a term has 2 skilled, 6 semi-skilled and 5 unskilled workers, how long will it take to build the wall?

A. 20 days

B. 18 days

C. 16 days

D. 15 day

***Sol.*D.** 5 skilled workers can build a wall in 20 days.

∴ 2 skilled workers can build the wall in $\frac{20\times5}{2}$ days = 50 days

8 semi-skilled workers can build a wall in 25 days.

∴ 6 semi-skilled workers can build the wall in $\frac{20\times8}{6}$ days = $\frac{100}{3}$ days

10 unskilled workers can build a wall in 30 days.

∴ 6 unskilled workers can build the wall in 60 days.

One day work of a team of 2 skilled, 6 semi-skilled and 5 unskilled workers

$$=\frac{1}{50}+\frac{3}{100}+\frac{1}{60}$$

$$=\frac{6+9+5}{300}=\frac{20}{300}=\frac{1}{15}$$

So, this team can complete the work in 15 days.

59. Given digits 2, 2, 3, 3, 3, 4, 4, 4, 4 how many distinct 4 digits number greater than 3000 can be formed?

A. 50

B. 51

C. 52

D. 54

***Sol.*B.** as the number should be greater than 3000.

So, thousand's place can be filled by 3 and 4, i.e., 2 ways.

Hundred's place can be filled by 2, 3 or 4, i.e., 3 ways.

Tenth place can be filled by 2, 3 or 4, i.e., 3 ways.

First place can be filled by 2, 3 and 4, i.e., 3 ways

But when all 2's and 3's are used, i.e., 333, 322 and 422 then there is a limitation for first place.

So, total three such types of numbers have limitation.

Total number of numbers greater than 3000

$= 2 \times 3 \times 3 \times 3 - 3$

$= 54 - 3 = 51$

60. Hari (H), Gita (G), Irfan (I) and Sarita (S) are siblings (*i.e,,* brothers and sister). All were born on 1st January. The age different between any two successive siblings (that is born one after another) is less than 3 yr. Given the following facts:

(i) hari's age + Gita's age > Irfan's age + Sarita's age.

(ii) the age different between Gita and Sarita is 1 yr. however, Gita is not the oldest and Sarita is not the youngest.

(iii) there are not twins.

in what order were they born (oldest first)?

A. HSIG

B. SGHI

C. IGSH

D. ISHG

***Sol.*B.** the age difference between any two successive siblings should be at least 1 year and at most 2 year.

So, Gita and Saira are successive siblings.

If Gita is older than Saira,

Then, Hari's age + Gita's age > Irfan's age + (Gita's age – 1)

Hari's age > Irfan's age – 1

As there is at least 1 year age difference.

So, Hari is older than Saira,

Then, Hari's age + Gita's age > Irfan's age + (Gita's age + 1)

Hari's age > Irfan's age + 1

So, Hari is older than Irfan in this case also.

As, Gita is not the oldest.

So, correct order is SGHI.

2. Engineering Mathematics

(1 Mark Questions)

GATE 2015

1. The following set of three vectors

$$\begin{pmatrix}1\\2\\1\end{pmatrix}\begin{pmatrix}x\\6\\x\end{pmatrix} \text{ and } \begin{pmatrix}3\\4\\2\end{pmatrix}$$

is linearly dependent when x is equal to

A. 0
B. 1
C. 2
D. 3

Sol. **D.**

2. For the matrix $\begin{bmatrix}4 & 3\\3 & 4\end{bmatrix}$, if $\begin{bmatrix}1\\1\end{bmatrix}$ is an eigenvector, the corresponding eigenvalue is ____________.

Sol. 7

3. A scalar function in the x-y plane is given by $\phi\,(x,y) = x^2 + y^2$. If $\hat{\imath}$ and $\hat{\jmath}$ are unit vectors in the x and y directions, the direction of maximum increase in the value of ϕ at (1,1) is along

A. -2î+2ĵ
B. 2î+2ĵ
C. -2î-2ĵ
D. 2î-2ĵ

Sol. **B.**

4. Consider a linear ordinary differential function equation; $\frac{dy}{dx} + p(x)y = r(x)$. Function $p(x)$ and $r(x)$ are defined and have a continuous first derivative. The integrating factor of this equation is non-zero. Multiplying this equation by its factor covert this into a:

A. homogeneous differential equation
B. non-linear differential equation
C. second order differential equation
D. exact differential equation

Sol. **D.**

5. A complex-valued function $f(z)$,given below is analytic domain D, $f(z) = u(x,y) + iv(x,y)$; $z = x + fy$.
Which of the following is not correct?

A. $\frac{df}{dz} = \frac{\partial v}{\partial y} + i\frac{\partial u}{\partial y}$
B. $\frac{df}{dz} = \frac{\partial u}{\partial x} + i\frac{\partial v}{\partial x}$
C. $\frac{df}{dz} = \frac{\partial v}{\partial y} - i\frac{\partial u}{\partial y}$
D. $\frac{df}{dz} = \frac{\partial v}{\partial y} + i\frac{\partial v}{\partial x}$

Sol. **A.**

GATE 2014

6. Gradient of a scalar variable is always

A. a vector
B. a scalar
C. a dot product
D. zero

Sol. **A.**

7. If $f*(x)$ is the complex conjugate of $f(x) = cosx + isinx$, then for real a and b, $\int_a^b f * f(x)dx$is always

A. positive
B. negative
C. real
D. imaginary

Sol. **C.**

8. If $f(x)$ is a real and continuous function of x, the Taylor series expansion of $f(x)$ about its minima will never have a term containing

A. first derivative
B. second derivative
C. third derivative
D. any higher derivative

Sol. **A.**

9. For the time domain function $f(t) = t^2$, which one of the following is the Laplace transform of $\int_0^t f(t)dt$?
   A. $\frac{3}{s^4}$
   B. $\frac{1}{4s^2}$
   C. $\frac{2}{s^3}$
   D. $\frac{2}{s^4}$

Sol. **D.**

GATE 2013

10. The number of emails received on six consecutive days is 11, 9, 18, 18, 4, and 15, respectively. What are the median and the mode for these data?
    A. 18 and 11, respectively
    B. 13 and 18, respectively
    C. 13 and 12.5, respectively
    D. 12.5 and 18, respectively

Sol. **B.**

11. For two rolls of a fair die, the probability of getting a 4 in the first roll and a number less than 4 in the second roll, up to 3 digits after the decimal point, is
    A. 0.083
    B. 0.089
    C. 0.075
    D. 0.085

Sol. **A.**

12. Which of the following statements are true?
    P. The eigenvalues of a symmetric matrix are real.
    Q. The value of the determinant of an orthogonal matrix canonly be +1.
    R. The transpose of a square matrix A has the same eigenvalues as those of A.
    S. The inverse of an n n matrix exists if and only if the rank is less than n.
    A. P and Q
    B. P and R
    C. Q and R
    D. P and S

Sol. **B.**

13. Evaluate $\int \frac{dx}{e^x - 1}$(Note C is a constant of integration.)
    A. $\frac{e^x}{e^x-1} + C$
    B. $\frac{\ln\,(e^x-1)}{e^x} + C$
    C. (c) $\ln\left(\frac{e^x}{e^x-1}\right) + C$
    D. $\ln(1 - e^{-x}) + C$

Sol. **D.**

GATE 2012

14. Consider the following set of linear algebraic equations:
$$x_1 + 2x_2 + 3x_3 = 2$$
$$x_2 + x_3 = -1$$
$$2x_2 + 2x_3 = 0$$
    The system has
    A. a unique solution
    B. no solution
    C. an infinite number of solutions
    D. only the trivial solution

Sol. **B.**

15. If a and b are arbitrary constants, then the solution to the ordinary differential equation $\frac{d^2y}{dx^2} - 4y = 0$ is
    A. $y = ax + b$
    B. $y = ae^{-x}$
    C. $y = \text{a}\sin 2x + b\cos 2x$
    D. $y = a\cos h2x + b\sin h2x$

Sol. **D.**

16. For the function $f(t) = e^{-t/\tau}$; the Taylor series approximation for t <<τ is
    A. $1 + \frac{t}{\tau}$
    B. $1 - \frac{t}{\tau}$
    C. $1 - \frac{t}{2\tau^2}$
    D. $1 + t$

Sol. **B.**

17. A box containing 10 identical compartments has 6 red balls and 2 blue balls. If each compartment can hold only one ball, then the numbers of different possible arrangements are
    A. 1026
    B. 1062
    C. 1260
    D. 1620

Sol. **C.**

18. Consider the following (2) matrix:
$$\begin{bmatrix} 4 & 0 \\ 0 & 4 \end{bmatrix}$$

Which one of the following vectors is not a valid eigen vector of the above matrix?

A. $\begin{bmatrix}1\\0\end{bmatrix}$

B. $\begin{bmatrix}-2\\1\end{bmatrix}$

C. $\begin{bmatrix}4\\-3\end{bmatrix}$

D. $\begin{bmatrix}0\\0\end{bmatrix}$

Sol. **D.**

GATE 2011

19. R is a closed planar region as shown by the shaded area in the figure below. Its boundary C consists of the circles C_1 and C_2.

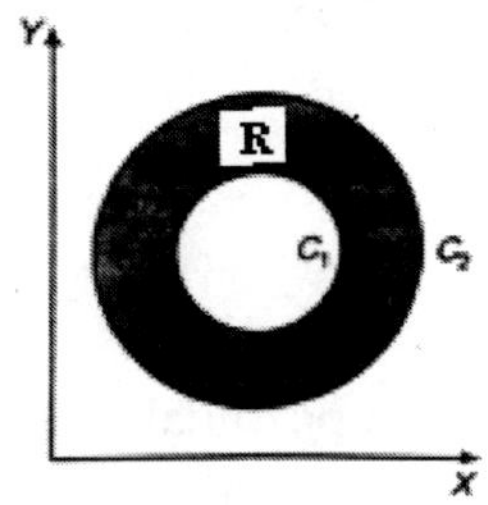

If F_1(x,y),F_2(x,y), $\frac{\partial F_1}{\partial y}$ and $\frac{\partial F_2}{\partial x}$ are all continuous everywhere in R, Green's theorem states that

$$\iint \left(\frac{\partial F_2}{\partial x} - \frac{\partial F_1}{\partial y}\right) dxdy = \oint (F_1 dx + F_2 dy)$$

Which one of the following alternatives correct depicts the direction of integration along C?

A. 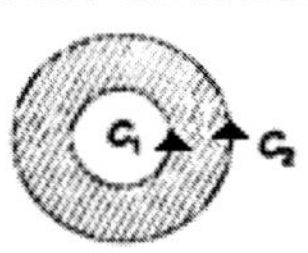

B. 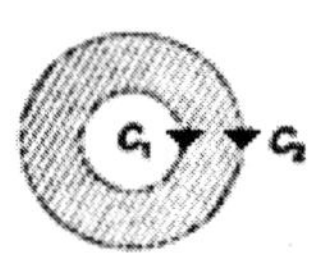

C. 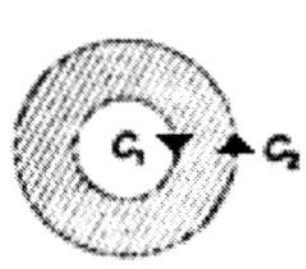

D. 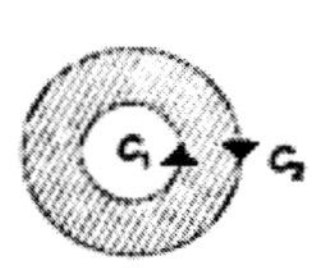

Sol. **C.**

20. Which one of the following function y(x) has slope of its tangent equal to $\frac{ax}{y}$?

Note and are real constants.

A. $y = \frac{x+b}{a}$

B. $y = ax + b$

C. $y = \sqrt{\frac{x^2+b}{a}}$

D. $y = \sqrt{ax^2 + b}$

Sol. **D.**

21. Let $\lambda_1 = -1$ and $\lambda_2 = 3$ be the eigen values and $\overline{V_1} = \begin{pmatrix}1\\0\end{pmatrix}$ and $\overline{V_2} = \begin{pmatrix}1\\1\end{pmatrix}$ be the corresponding eigen vectors of a real 2×2 matrix $\bar{\bar{R}}$. Given that $\bar{\bar{p}} = (\overline{V_1}\overline{V_2})$, which one of the following matrices represents $\bar{\bar{p}}^{-1}\bar{\bar{R}}\bar{\bar{P}}$?

A. $\begin{pmatrix}0 & -1\\3 & 0\end{pmatrix}$

B. $\begin{pmatrix}0 & 3\\-1 & 0\end{pmatrix}$

C. $\begin{pmatrix}3 & 0\\0 & -1\end{pmatrix}$

D. $\begin{pmatrix}-1 & 0\\0 & 3\end{pmatrix}$

Sol. **D.**

GATE 2010

22. The inverse of the matrix $\begin{bmatrix}1 & 2\\3 & 4\end{bmatrix}$ is

A. $\begin{bmatrix}-2 & -1\\-3/2 & -1/2\end{bmatrix}$

B. $\begin{bmatrix}-2 & 3/2\\1 & -1\end{bmatrix}$

C. $\begin{bmatrix}-2 & 1\\3/2 & -1/2\end{bmatrix}$

D. $\begin{bmatrix}2 & -3/2\\-1 & 1/2\end{bmatrix}$

Sol. **C.**

23. The Laplace transform of the function shown in the figure below is

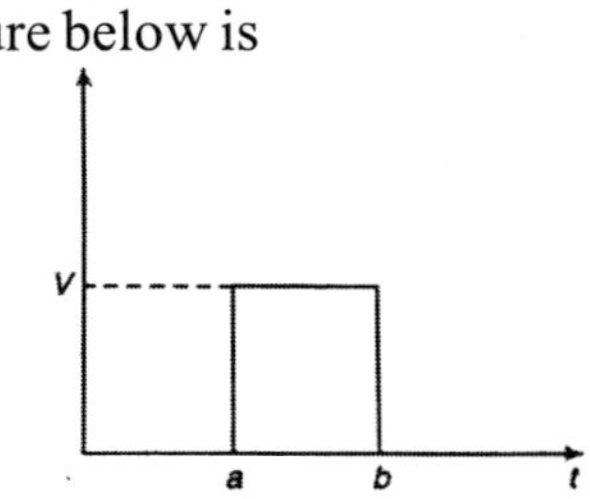

A. $\frac{V_e^{(a-b)s}}{s}$

B. $\frac{V}{s}\left(e^{-bs} - e^{-bs}\right)$

C. $\frac{V}{s}\left(e^{-as} - e^{-bs}\right)$

D. $\frac{V}{s^2}\left(e^{as} - e^{bs}\right)$

Sol. **C.**

24. Given that $i = \sqrt{-1}$, i^i is equal to

A. $\frac{\pi}{2}$

B. -1

C. $i \ln i$

D. $e^{-\pi/2}$

Sol. D.

25. A root of the equation $x^4 - 3x + 1 = 0$ needs to be found using the Newton-Raphson method. If the initial guess x_0 is taken as 0, then the new estimate x_1 after the fist iteration is

A. $\frac{1}{3}$

B. $-\frac{1}{3}$

C. 3

D. -3

Sol. A.

26. The stream function in -plane is given below.

$$\Psi = \frac{1}{2}x^2y^3$$

The velocity vector for this stream function is

A. $Xy^3\hat{\imath} - \frac{3}{2}x^2y^2\hat{\jmath}$

B. $\frac{3}{2}x^2y^2\hat{\imath} - xy^3\hat{\jmath}$

C. $\frac{3}{2}x^2y^2\hat{\imath} + xy^3\hat{\jmath}$

D. $xy^3\hat{\imath} + \frac{3}{2}x^2y^2\hat{\jmath}$

Sol. C.

GATE 2009

27. The value of $\lim_{x\to\pi/2} \frac{\cos x}{(x-\pi/2)^3}$ is

A. $-\infty$

B. 0

C. 1

D. ∞

Sol. A.

28. The general solution of the differential equation $\frac{d^2y}{dx^2} - \frac{dy}{dx} - 6y = 0$, where C_1 and C_2 are constants of integration, is

A. $C_1e^{-3x} + C_2e^{-2x}$

B. $C_1e^{3x} + C_2e^{-2x}$

C. $C_1e^{3x} + C_2e^{2x}$

D. $C_1e^{-3x} + C_2e^{2x}$

Sol. B.

29. Using the residue theorem, the value of the integral (counter-clockwise) $\oint \frac{8-7z}{z-4} dz$ around a circle with centre at z = 0 and radius = 8 (where z is a complex number and $i = \sqrt{-1}$), is

A. $-20\,\pi i$

B. $-40\,\pi$

C. $-40\,\pi i$

D. $40\,\pi i$

Sol. C.

30. Consider the integral $\iint (2x\hat{\imath} - 2y\hat{\jmath} + 5z\hat{k}) \cdot \hat{n}dS$ over the surface of a sphere of radius = 3 with centre at the origin and surface unit normal $\hat{n}$ pointing away from the origin. Using the Gauss's divergence theorem, the value of this integral is

A. -180π

B. zero

C. $90\,\pi$

D. $180\,\pi$

31. Using the trapezoidal rule and 4 equal intervals (n = 4), the calculated value of the integral (rounded to the place of decimal) $\int_0^\pi \sin\theta\, d\theta$ is

A. 1.7

B. 1.9

C. 2.0

D. 2.1

Sol. D.

32. The eigen values of matrix A = $\begin{bmatrix} 1 & 2 \\ 4 & 3 \end{bmatrix}$ are 5 and -1.

Then the eigen values of $-2A + 3I (I\ is\ a\ 2 \times 2\ identify\ matrix)$ are

A. -7 and 5

B. 7 and -5

C. $-\frac{7}{5}$ and $\frac{1}{5}$

D. $\frac{1}{7}$ and $-\frac{1}{2}$

33. A fair die is rolled. Let R denotes the event of obtaining a number less than or equal to 5 and 5 denotes the event of obtaining an odd number. Then which one of the following about the probability (p) is true?

A. *P(R/S)=1*

B. *P(R/S)=0*

C. *P(S/R)=1*

D. *P(S/R)=0*

GATE 2008

34. Which one of the following is not an integrating factor for the differential equation *xdy–ydx=0*

A. $\frac{1}{x^2}$
B. $\frac{1}{y^2}$
C. $\frac{1}{xy}$
D. $\frac{1}{(x+y)}$

Sol. **D.**

35. Which one of the following is not a solution of the differential equation $\frac{d^2y}{dx^2} + y = 1$?

A. $y = 1$
B. $y = 1 + cosx$
C. $y = 1 + sinx$
D. $y = 2 + sinx + cosx$

Sol. **D.**

36. The limit of $\frac{sinx}{x}$ as x → ∞ is

A. −1
B. 0
C. 1
D. ∞

Sol. **B.**

37. The unit normal vector to the surface of the sphere $x^2 + y^2 + z^2 = 1$ at the point $\left(\frac{1}{\sqrt{2}}, 0, \frac{1}{\sqrt{2}}\right)$ is ($\hat{\imath}, \hat{\jmath}, \hat{k}$ are unit normal vectors in the cartesian coordinate system)

A. $\frac{1}{\sqrt{2}} i + \frac{1}{\sqrt{2}} j$
B. $\frac{1}{\sqrt{2}} i + \frac{1}{\sqrt{2}} \hat{k}$
C. $\frac{1}{\sqrt{2}} \hat{\jmath} + \frac{1}{\sqrt{2}} \hat{k}$
D. $\frac{1}{\sqrt{3}} i + \frac{1}{\sqrt{3}} j + \frac{1}{\sqrt{3}} \hat{k}$

Sol. **B.**

38. A non-linear function $f(x)$ is defined in the interval $-1.2 < x < 4$ as illustrated in the figure below. The equation $f(x) = 0$ is solved for x within this interval by using the Newton-Raphson iterative scheme. Among the initial guesses (I_1, I_2, I_3 *and* I_4), the guess that is likely to lead to the root most rapidly is

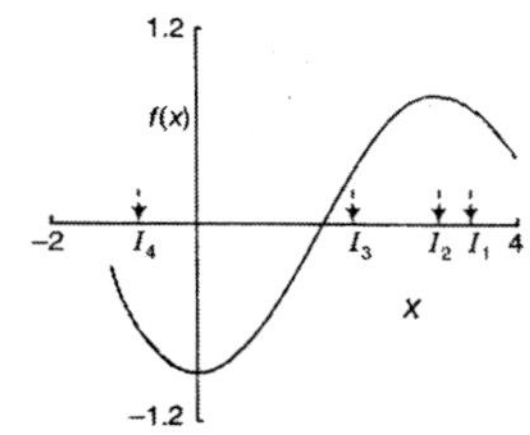

A. I_1
B. I_2
C. I_3
D. I_4

Sol. **C.**

GATE 2007

39. Given i = $\sqrt{-1}$, the ratio $\frac{(i+3)}{(i+1)}$ is given by

A. I
B. −2
C. $-i + 2$
D. $i + 1$

Sol. **C.**

40. The value of "a" for which the following set of equations

$$y + 2z = 0$$
$$2x + y + z = 0$$
$$ax + 2y = 0$$

have non-trivial solution, is

A. 0
B. 8
C. – 2
D. 3

Sol. **B.**

41. The initial condition for which the following equation

$(x^2 + 2x)\frac{dy}{dx} = 2(x + 1)y$; $y(x_0) = y_0$

has infinitely many solutions, is

A. $y(x = 0) = 5$
B. $y(x = 0 = 1$
C. $y(x = 2) = 1$
D. $y(x = -2) = 0$

Sol. **D.**

42. Given that the Laplace transform of the function below over a single period 0 < t < 2 is $\frac{1}{S^2}(1 - e^{-S})^2$, the Laplace transform of the periodic function over $0 < t < \infty$ is

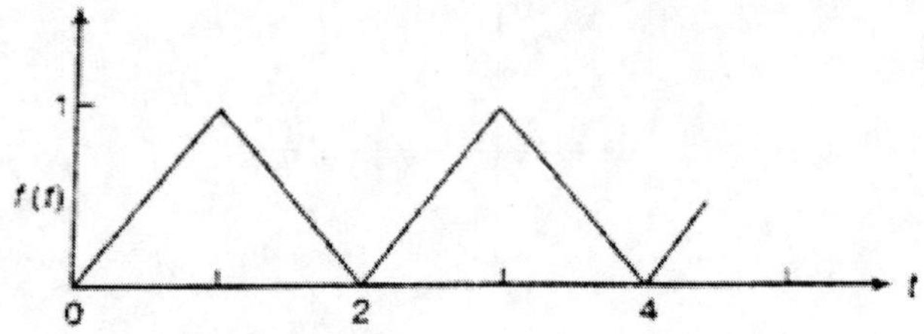

A. $\frac{1}{S}(1-e^{-S})^2$

B. $\frac{1}{S}(1+e^{-S})^2$

C. $\frac{1(1-e^{-S})}{S^2(1-e^{-S})}$

D. $\frac{1}{S}\tan h\frac{S}{2}$

Sol. **C.**

GATE 2006

43. The ordinary differential equation $dY/dt = f(Y)$ is solved using the approximation $Y(t+\Delta t) = Y(t) + f(t)\Delta t$. the numerical error introduced by the approximation at each step is

A. Proportional to Δt

B. Proportional to $(\Delta t)^2$

C. Independent of Δt

D. Proportional to (1/Δt)

Sol. **B.**

44. The trapezoidal rule of integration, when applied to $\int_a^b f(x)dx$, will give the exact value of the integral,

A. if f(x) is a linear function of x

B. if f(x) is a quadratic function of x

C. for any f(x)

D. for no f(x)

Sol. **A.**

45. The value of ∝ for which the following three vectors are coplanar is

$$a = \hat{\imath} + 2\hat{\jmath} + \hat{k}$$
$$b = 3\hat{\jmath} + \hat{k}$$
$$c = 2\hat{\imath} + \propto \hat{\jmath}$$

A. 4

B. zero

C. – 2

D. – 10

Sol. **C.**

46. The derivative of |x| w.r.t. x, when x ≠ 0, is

A. $\frac{|x|}{x}$

B. – 1

C. 1

D. Undefined

Sol. **D.**

GATE 2005

47. Match the following, where x is the spatial coordinate and t is time.

Group I	*Group II*
P. Wave equation	1. $\frac{\partial c}{\partial t} = \propto \frac{\partial c}{\partial x}$
Q. Heat equation	2. $\frac{\partial c}{\partial t} = \propto^2 \frac{\partial^2 c}{\partial x^2}$
	3. $\frac{\partial^2 c}{\partial t^2} = \propto^2 \frac{\partial c}{\partial x}$
	4. $\frac{\partial^2 c}{\partial t^2} = \propto^2 \frac{\partial^2 c}{\partial x^2}$

A. P-4, Q-2

B. P-2, Q-4

C. P-3, Q-1

D. P-1, Q-3

Sol. **A.**

48. The bags contain ten coins each, and the coins in each bag are numbered from 1 to 10. One coin is drawn at random from each bag. The probability that one of the coins has value 1, 2, 3, or 4, while the other has value 7, 8, 9, or 10, is

A. 2/5

B. 4/25

C. 8/25

D. 16/25

Sol. **C.**

49. Given $i = \sqrt{-1}$, the ratio $\frac{(1+2j)}{(i-2)}$ is given by

A. 1

B. – 1

C. I

D. – i

Sol. **D.**

GATE 2004

50. The inverse Laplace transform of the function $F(s) = \frac{1}{S(1+S)}$ is

A. $1 + e^t$

B. $1 - e^t$

C. $1 + e^{-t}$

D. $1 - e^{-t}$

Sol. **D.**

51. The function $f(x) = 3x(x - 2)$ has a

A. minimum at x = 1
B. maximum at x = 1
C. minimum at x = 2
D. maximum at x = 2

Sol. A.

52. The complex number 2 (1 + i) can be represented in polar form as

A. $2\sqrt{2}e^{i\pi/4}$
B. $\frac{\pi}{4} e^{i2\sqrt{2}}$
C. $\sqrt{2}e^{i\pi/4}$
D. $\frac{\pi}{4} e^{i\sqrt{2}}$

Sol. A.

53. The differential equation

$\frac{d^2Y}{dx^2} + \sin x\frac{dy}{dx} + ye^x = \sinh x$ is

A. first order and linear
B. first order and non-linear
C. second order and linear
D. second order and non-linear

Sol. C.

54. For the time domain function $f(t) = t$, the Laplace transform of $\int_0^t f(t)dt$ is given by

A. $\frac{1}{(2S^3)}$
B. $\frac{2}{S^3}$
C. $\frac{1}{S^3}$
D. $\frac{2}{S^2}$

Sol. C.

GATE 2003

55. A box contains 6 red balls and 4 green balls, one ball is randomly picked and then a second ball is picked without replacement of the first ball. The probability that both the balls are green is

A. 1/15
B. 2/25
C. 2/15
D. 4/25

Sol. C.

56. The directional derivative of f (x, y, z) = $x^2 + y^2 + z^2$ at the point (1, 1, 1) in the direction $\hat{\imath} - \hat{k}$ is

A. 0
B. 1
C. $\sqrt{2}$
D. $2\sqrt{2}$

Sol. A.

57. The Taylor series expansion of the function F(x) = x/(1 + x) around x = 0 is

A. $x + x^2 + x^3 + x^4$ ....
B. $1 = x + x^2 + x^3 + x^4$ ......
C. $2x + 4x^2 + 8x^3 + 16x^4$ ......
D. $x - x^2 + x^3 - x^4$ ....

Sol. D.

GATE 2002

58. In the complex plane, the angle between line $1 + i$ and $-1 + i$ (where, i = $\sqrt{-1}$) is

A. $\frac{\pi}{4}$
B. $\frac{\pi}{2}$
C. $\frac{3\pi}{4}$
D. Π

Sol. B.

59. The inverse of the matrix $\begin{bmatrix} 0.2 & 0 & 0 \\ 0 & 1 & 0 \\ 0 & 0 & 0.5 \end{bmatrix}$ is

A. $\begin{bmatrix} 2 & 0 & 0 \\ 0 & 1 & 0 \\ 0 & 0 & 5 \end{bmatrix}$
B. $\begin{bmatrix} -0.2 & 0 & 0 \\ 0 & -1 & 0 \\ 0 & 0 & -0.5 \end{bmatrix}$
C. $\begin{bmatrix} 5 & 0 & 0 \\ 0 & 1 & 0 \\ 0 & 0 & 2 \end{bmatrix}$
D. $\begin{bmatrix} 0.5 & 0 & 0 \\ 0 & 1 & 0 \\ 0 & 0 & 0.2 \end{bmatrix}$

Sol. D.

60. Which of the following holds for any non-zero vector $\vec{a}$?

A. $\nabla . \vec{a} = 0$
B. $\nabla \times \vec{a} = 0$
C. $\nabla . (\nabla \times \vec{a}) = 0$
D. $\nabla (\nabla \times \vec{a}) = 0$

Sol. C.

GATE 2001

61. The value of the following determinant

$$\begin{vmatrix} 1 & 0 & 0 & 0 & 0 \\ -2 & 2 & 0 & 0 & 0 \\ 3 & 5 & 3 & 0 & 0 \\ -1 & 4 & 7 & 4 & 0 \\ -5 & -6 & 2 & 1 & 1 \end{vmatrix} \text{ is}$$

A. 24

B. 32

C. −112

D. 0

Sol. **A.**

62. The value of $(1+i)^8$, where i = $\sqrt{-1}$, is

A. 8 + 4i

B. 8 – 4i

C. 16

D. 8

Sol. **C.**

GATE 2000

63. A pair of fair dice is rolled simultaneously. The probability that the sum of the numbers from the dice equals six is

A. $\frac{1}{6}$

B. $\frac{7}{36}$

C. $\frac{5}{36}$

D. $\frac{1}{12}$

Sol. **C.**

64. From an even function $f(x)$.

A. $\int_{-n}^{a} f(x)dx = 0$

B. $\int_{-a}^{a} f(-x)dx = 0$

C. $f(x) = -f(-x)$

D. $f(x) = f(-x)$

Sol. **D.**

65. The integrating factor for the differential equation,

$(cos^2 x)\frac{dy}{dx} + y = tanx$, is

A. e^{tanx}

B. $cos2x$

C. e^{-tanx}

D. $sin2x$

Sol. **A.**

(2 Mark Questions)

GATE 2015

66. A vector $u = -2y\hat{\imath} + 2x\hat{\jmath}$, where $\hat{\imath}$ and $\hat{\jmath}$ are unit vector in x and y directions respectively. Evaluate the line integral.

$$I = \oint_C u.dr$$

Where, C is a closed loop formed by connecting point (1, 1), (3, 1), (3, 2) and (1, 2) in that order. The value of I is .

Sol. 8

67. The solution of the non-linear equation

$x^3 - x = 0$

Is to be obtained using Newton-Raphson method. If the initial guess is x = 0.5, the Method converges to which one of the following values.

A. −1

B. 0

C. 1

D. 2

Sol. **A.**

68. For complex variable z, the value of the clockwise contour integral $\frac{1}{2\pi i}\int_C \frac{e^{-2z}}{z(z-3)}dz$ along the contour C; $|z| = 2$ (up to two decimal places) is .

Sol. 0.31 to 0.35

GATE 2014

69. Consider the following differential equation

$\frac{dy}{dx} = x + lny$; $y = 0$ at $x = 0$

The solution of this equation at 0.4 x= using Euler method with a step size of h = 0.2 is___________

Sol. 2.331

70. The integrating factor for the differential equation $\frac{dy}{dx} - \frac{y}{1+x} = (1+x)$ is

A. $\frac{1}{(1+x)}$

B. $(1+x)$

C. $x(1+x)$

D. $\frac{x}{(1+x)}$

Sol. **A.**

71. The differential equation $\frac{d^2y}{dx^2} + x^2\frac{dy}{dx} + x^3y = e^x$ is a

A. non-linear differential equation of first degree
B. linear differential equation of first degree
C. linear differential equation of second degree
D. non-linear differential equation of second degree

Sol. **B.**

72. Consider the following two normal distribution.

$$f_1(x) = \exp(-\pi x^2)$$
$$f_2(x) = \frac{1}{2\pi} exp\left[-\frac{1}{4\pi}(x^2 + 2x + 1)\right]$$

If μ and σ denoted the mean and standard deviation, respectively, then

A. $\mu_1 < \mu_2$ and $\sigma_1^2 < \sigma_2^2$
B. $\mu_1 < \mu_2$ and $\sigma_1^2 > \sigma_2^2$
C. $\mu_1 > \mu_2$ and $\sigma_1^2 < \sigma_2^2$
D. $\mu_1 > \mu_2$ and $\sigma_1^2 > \sigma_2^2$

Sol. **A.**

73. In rolling of two match dice, the outcome of an experimental is considered to be the sum of numbers appearing on the dice. The probability is highest for outcome of .

Sol. $\frac{6}{36}$ *or* $\frac{1}{6}$

GATE 2013

74. For function

$$f(z) = \frac{1}{(2-z)(z+2)}$$

The residue at $z = 2$ is .

Sol. -0.25

75. The solution of the differential equation $\frac{dy}{dx} - y^2 = 0$, given $y = 1$ at $x = 0$ is

A. $\frac{1}{1+x}$
B. $\frac{1}{1-x}$
C. $\frac{1}{(1-x)^2}$
D. $\frac{x^3}{3} + 1$

Sol. **A.**

76. The solution of the differential equal $\frac{d^2y}{dx^2} - \frac{dy}{dx} + 0.25y = 0$, given $y = 0$ at $x = 0$ and $\frac{dy}{dx} = 1$ at $x = 0$ is

....

A. $xe^{0.5x} - xe^{-0.5x}$
B. $0.5xe^x - 0.5xe^{-x}$
C. $xe^{0.5x}$
D. $-xe^{0.5x}$

Sol. **C.**

77. The value of integral

$$\int_{0.1}^{0.5} e^{-x^3}\, dx$$

Evaluated by the Simpson's rule using subintervals (up to 3 digits after the decimal point) is .

A. 0.40
B. 0.42
C. 0.38
D. 0.97

Sol. **C.**

GATE 2012

78. If "a" is a constant, then the value of the integral $a^2\int_0^\infty xe^{-ax}\,dx$ is

A. $\frac{1}{a}$
B. a
C. 1
D. 0

Sol. **C.**

79. The Newton-Raphson method is used to find the roots of the equation

$$f(x) = x - cos\pi x \qquad 0 \le x \le 1$$

If the initial guess for the root is 0.5, then the value of x after the first iteration is

A. 1.02
B. 0.62
C. 0.55
D. 0.38

Sol. **D.**

80. If $i = \sqrt{-1}$, the value of the integral

$$\oint_C \frac{7z+i}{z(z^2+1)}dz \qquad |z| < 2,$$

Using the Cauchy residue theorem is

A. $2\pi i$
B. 0
C. -6π
D. 6π

Sol. **B.**

GATE 2011

81. Unit vectors in X and Z directions are $\hat{\imath}$ and $\hat{k}$ respectively. Which one of the following is the directional derivative of the function $F(x, z) = In(x^2 + z^2)$ at the point P (4, 0), in the direction of $(\hat{\imath} - \hat{k})$?

A. $\frac{\hat{\imath}}{2\sqrt{2}}$

B. $\hat{\imath}$

C. 1

D. $\frac{1}{2\sqrt{2}}$

Sol. **D.**

82. Which one of the following choices is a solution of the differential equation given below?

$$\frac{dy}{dx} = \frac{y^2}{x} + \frac{y}{x} - \frac{2}{x}$$

Note c is a real constant.

A. $y = \frac{c - x^2}{c + 2x^2}$

B. $y = \frac{c + 2x^2}{c - x^2}$

C. $y = \frac{c - x^3}{c + 2x^3}$

D. $y = \frac{c + 2x^3}{c - x^3}$

Sol. **D.**

83. The value of the improper integral $\int_{-\infty}^{\infty} \frac{dx}{(1+x^2)}$ is

A. -2π

B. 0

C. π

D. 2π

Sol. **C.**

84. In the fixed point iteration method for solving equations of the form x = g(x), the (n + 1)th iteration value is $x_{n+1} = g(x)$, where,x_n represents the nth iteration value, g(x) and corresponding initial guess value x_0 in the domain of interest are shown in the following choices. Which one of these choices leads to a converged solution for x?

A.

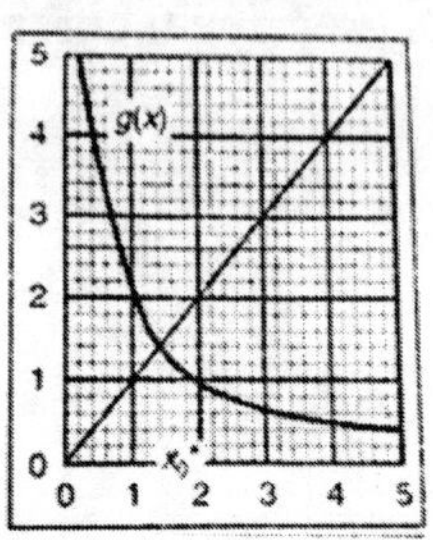

B.

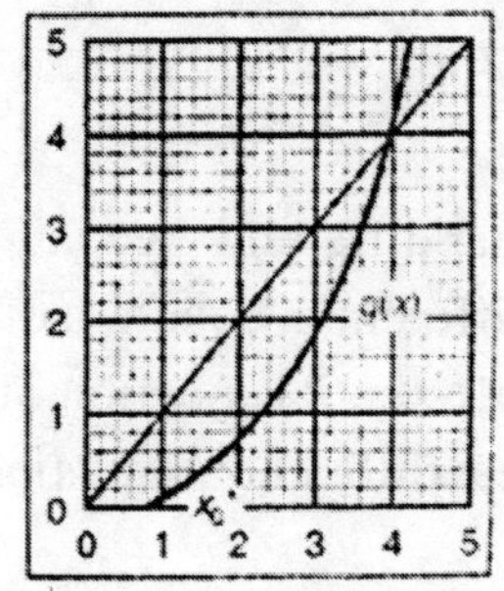

C.

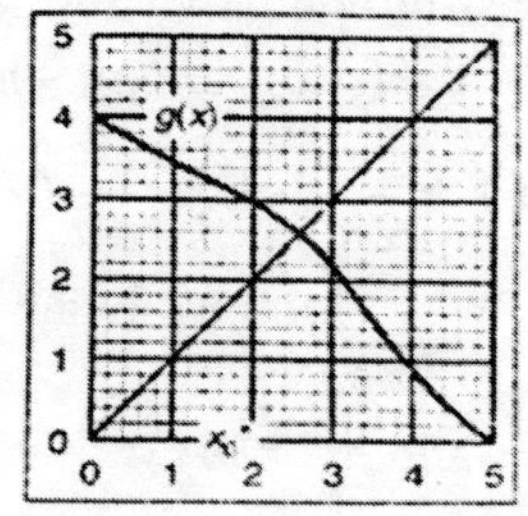

D.

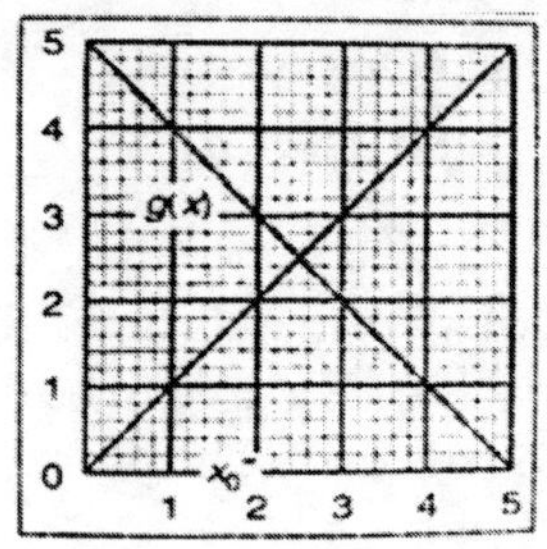

Sol. **C.**

GATE 2010

85. The solution of the differential equation

$$\frac{d^2y}{dt^2} = 2\frac{dy}{dt} + 2y = 0$$

With the initial conditions

$$y(0) = 0, \left.\frac{dy}{dt}\right|_{t=0} = -1, \text{ is}$$

A. $-t \sin t$

B. $-e^{-t}(1 - \cos t)$

C. $-(t + \sin t)/2$

D. $-e^{-t} \sin t$

Sol. **D.**

86. If y $\vec{u}$ = y $\hat{\imath}$ + xy $\hat{\jmath}$and $\vec{v} = x^2\hat{\imath} + xy^2\hat{\jmath}$, then curl $(\vec{u} \times \vec{v})$ is

A. $(2xy^2)\hat{\imath} - (x + y^2)\hat{\jmath}$

B. $(xy - x^2)\hat{\imath} - (y - 3xy)\hat{\jmath}$

C. $(2x^2y^2 - 3x^3)\hat{\imath} - (y^3 - 3xy^2)\hat{k}$

D. $(3xy^2 - x^3)\hat{\imath} - (y^3 - 3x^2y)\hat{\jmath}$

Sol. **D.**

87. X and Y are independent random variables. X follows a binomial distribution with N = 5 and p = 1/2. Y takes integer values 1 and 2 with equal probability. Then the probability that X = Y, is

A. $\frac{15}{64}$

B. $\frac{15}{32}$

C. $\frac{1}{2}$

D. $\frac{15}{16}$

Sol. **A.**

88. A box contains three red and two black balls. Four balls are removed from the box one by one without replacement. The probability of the ball remaining in the box being red, is

A. $\frac{609}{625}$

B. $\frac{3}{5}$

C. $\frac{2}{5}$

D. $\frac{81}{625}$

Sol. **B.**

89. For a function g(x), if g(0) = 0 and g'(0) = 2, then $\lim_{x \to 0} \int_0^{g(x)} \frac{2t}{x}\, dt$ is equal to

A. ∞

B. 2

C. 0

D. $-\infty$

Sol. **C.**

GATE 2009

90. For an incompressible flow, the x and y components of the velocity vector are $V_x = 2(x + y)$, $V_y = 3(y + z)$, where x, y, z are in metre, all velocities are in m/s. Then the z component of the velocity vector (V_Z) of the flow for the boundary condition $V_Z = 0$ at z = 0, is

A. 5z

B. $-5z$

C. $2x + 3z$

D. $-2x - 3z$

91. The inverse Laplace transform of $\frac{1}{2s^2+3s+1}$ is

A. $e^{-t/2} - e^{-t}$

B. $2e^{-t/2} - e^{-t}$

C. $e^{-t} - 2e^{-t/2}$

D. $e^{-t} - e^{-t/2}$

GATE 2008

92. Which one of the following transformations {u = f(y)} reduces $\frac{dy}{dx} + Ay^3 + By = 0$ to a linear differential equation? (A and B are positive constants)

A. $u = y^{-3}$

B. $u = y^{-2}$

C. $u = y^{-1}$

D. $u = y^2$

Sol. **B.**

93. The Laplace transform of the function f(t) = t sin t is

A. $\frac{2s}{(s^2+1)^2}$

B. $\frac{1}{s^2(s^2+1)}$

C. $\frac{1}{s^2} + \frac{1}{(s^2+1)}$

D. $\frac{1}{(s-1)^2+1}$

Sol. **A.**

94. The value of surface integral $\oint\int_S (x\hat{i} + y\hat{i}).\hat{n}dA$ evaluated over the surface of a cube having sides of length a is ($\hat{n}$ is unit normal vector)

A. 0

B. a^3

C. $2a^3$

D. $3a^3$

Sol. **C.**

95. The first terms of the Taylor series expansion of cos(x) about the point x = 0 are

A. $1 + x + \frac{x^2}{2!} + \frac{x^3}{3!}$

B. $1 - x + \frac{x^2}{2!} - \frac{x^3}{3!}$

C. $1 - \frac{x^2}{2!} + \frac{x^4}{4!} - \frac{x^6}{6!}$

D. $x - \frac{x^3}{3!} + \frac{x^5}{5!} - \frac{x^7}{7!}$

Sol. **C.**

96. If $A = \begin{bmatrix} 1 & 2 \\ 2 & 1 \end{bmatrix}$, then the eigen values of A^3 are

A. 5, 4

B. 3, −1

C. 9, −1

D. 27, −1

Sol. **D.**

97. An analytic function w(z) is defined as $w = u + iv$, Where, i = $\sqrt{-1}$ and z = x + iy. If the real part is given by $u = \frac{Y}{x^2+y^2}$, $w(z)$ is

A. $\frac{1}{z}$

B. $\frac{1}{z^2}$

C. $\frac{i}{z}$

D. $\frac{1}{iz}$

Sol. **C.**

98. The normal distribution is given by

$$f(x) = \frac{1}{\sqrt{2\pi\sigma}} exp\left(-\frac{(x-\mu)^2}{2\sigma^2}\right), \quad -\infty < x < \infty$$

The points of inflexion to the normal curve are

A. $x = -\sigma, +\sigma$

B. $x = \mu + \sigma, \mu - \sigma$

C. $x = \mu + 2\sigma,\ \mu - 2\sigma$

D. $x = \mu + 3\sigma,\ \mu - 3\sigma$

Sol. **B.**

99. Using Simpson's 1/3 rule and four equally spaced intervals (n = 4), estimate the value of the integral $\int_0^{\pi/4} \frac{\sin x}{cos^3 x} dx$.

A. 0.3887

B. 0.4484

C. 0.5016

D. 0.5527

Sol. **C.**

100. The following differential equation is to be solved numerically by the Euler's explicit method.

$$\frac{dy}{dx} = x^2y - 1.2y \text{ with } y(0) = 1$$

A step size of 0.1 is used. The solution for y at x = 0.1 is

A. 0.880

B. 0.905

C. 1.000

D. 1.100

Sol. **A.**

101. The Poisson distribution is given by $P(r) = \frac{m^r}{r!}$ exp(− m). The first moment about the origin for the distribution is

A. 0

B. m

C. 1/m

D. m^2

Sol. **B.**

GATE 2007

102. If $z = x + iy$ is a complex number, where $i = \sqrt{-1}$ then the derivative of z $\bar{z}$ at $2 + i$ is

A. 0

B. 2

C. 4

D. Does not exist

Sol. **C.**

103. $\bar{\bar{A}}$ and $\bar{\bar{B}}$ are two 3 × 3 matrix such that

$$\bar{\bar{A}} = \begin{bmatrix} -2 & 4 & 6 \\ 1 & 2 & 1 \\ 0 & 4 & 4 \end{bmatrix}, \bar{\bar{B}} = \bar{\bar{0}}$$

And $\bar{\bar{A}}\bar{\bar{B}} = \bar{\bar{0}}$. Then the rank of matrix is $\bar{\bar{B}}$ is

A. r = 2

B. r < 3

C. r ≤ 3

D. r = 3

Sol. **B.**

104. The solution of the following differential equation

$$x\frac{dy}{dx} + y(x^2 - 1) = 2x^3 \text{ is}$$

A. 0

B. $2 + ce^{\frac{x^2}{2}}$

C. $c_1x + c_2x^2$

D. $2x + cxe^{\frac{x^2}{2}}$

Sol. **D.**

105. The directional derivative of

$$f = \frac{1}{2}\sqrt{x^2 + y^2}$$

at (1, 1) in the direction of $\vec{b} = \vec{i} - \vec{j}$ is

A. 0

B. $\frac{1}{\sqrt{2}}$

C. $\sqrt{2}$

D. 2

Sol. **D.**

106. Evaluate the following integral, where (n ≠ 0),

$$\int_C(-xy^n dx + x^n ydy),$$

Within the area of a triangle with vertices (0, 0), (1, 0) and (1, 1) counter – clockwise.

A. 0
B. $\frac{1}{(n+1)}$
C. $\frac{1}{2}$
D. $\frac{n}{2}$

Sol. **C.**

107. The family of curves that is orthogonal to $xy = c$ is
A. $y = c_1 x$
B. $y = c_1/x$
C. $y^2 + x^2 = c_1$
D. $y^2 - x^2 = c_1$

Sol. **D.**

108. The Laplace transform of $f(t) = \frac{1}{\sqrt{t}}$ is
A. $\sqrt{\frac{\pi}{s}}$
B. $\frac{1}{\sqrt{s}}$
C. $\frac{1}{s^{\frac{3}{2}}}$
D. Does not exist

Sol. **A.**

GATE 2006

109. If the following represents the equation of a line
$$\begin{vmatrix} x & 2 & 4 \\ y & 8 & 0 \\ 1 & 1 & 1 \end{vmatrix} = 0,$$
then the line passes through the point
A. (0, 0)
B. (3, 4)
C. (4, 3)
D. (4, 4)

Sol. **B.**

110. If $A = \begin{vmatrix} 2 & 1 \\ 2 & 3 \end{vmatrix}$, then the eigen values of A^3 are
A. 27 and 8
B. 64 and 1
C. 12 and 3
D. 4 and 1

Sol. **B.**

111. With $y = e^{ax}$, if the sum
$$S = \frac{dy}{dx} + \frac{d^2y}{dx^2} + \cdots + \frac{d^ny}{dx^n}$$
approaches 2y as $n \to \infty$, then the value of a is
A. $\frac{1}{3}$
B. $\frac{1}{2}$
C. $\frac{2}{3}$
D. 2

Sol. **C.**

112. Determine the following integral $I = \int_S \vec{r}.d\vec{s}$

Where, $\vec{r}$ is the position vector field $(\vec{r} = \hat{\imath}x + \hat{\jmath}y + \hat{k}z)$ and S is the surface of a sphere of radius R.
A. $4\pi R^2$
B. $\frac{3}{4}\pi R^3$
C. πR^3
D. $4\pi R^3$

Sol. **D.**

113. The liquid surface in a cylindrical bucket of radius R rotating about its axis acquires a parabolic profile given by the equation $y = a + br^2$, where y is the height of the liquid surface form the bottom of the bucket at a radial distance r from the bucket axis. If the liquid has density ρ, then the mass of the liquid in the bucket is
A. $\pi\rho R^2\left(\frac{a+bR^2}{2}\right)$
B. $\pi\rho R^2\left(a + \frac{bR^2}{2}\right)$
C. $\pi\rho R^2 a$
D. $\pi\rho R^2(a + bR^2)$

Sol. **B.**

114. The solution to the following equation
$x^2\frac{d^3y}{dx^3} + 2x\frac{d^2y}{dx^2} - 2\frac{dy}{dx} = 0$ is given by
A. $y = C_1x + C_2x^{-2} + C_3$
B. $y = C_1x^2 + C_2x^{-2} + C_3$
C. $y = C_1x^2 + C_2x^{-1} + C_3$
D. $y = C_1x + C_2x^{-1} + C_3$

Sol. **C.**

115. The value of the contour integral
$$I = \frac{1}{2\pi i}\oint_C \frac{e^z}{(z+1)(z+3)}dz$$
Where, C is the circle $|z| = 2$ is

A. $\frac{1}{2e}$

B. $\frac{1}{2}\left(\frac{1}{e}-\frac{1}{e^3}\right)$

C. Zero

D. $\frac{1}{(2\pi i)e^3}$

Sol. A.

116. The Newton-Raphson method is used to solve the equation $(x-1)^2 + x - 3 = 0$. The method will failin the very first iteration, if the initial guess is

A. Zero

B. 0.5

C. 1

D. 3

Sol. B.

117. A pair of fair dice is rolled three times. What is the probability that 10 (sum of the numbers on the two faces) will show up exactly once?

A. $\frac{121}{1728}$

B. $\frac{363}{1728}$

C. $\frac{121}{576}$

D. $\frac{363}{576}$

Sol. C.

118. A company purchased components from three firms P, Q and R as shown in the table below.

Firm	*Total number of components purchased*	*Number of components likely to be defective*
P	1000	5
Q	2500	5
R	500	2

The components are stored together. One of the components is selected at random and found to be defective. What is the probability that it was supplied by firm R?

A. $\frac{1}{250}$

B. $\frac{1}{12}$

C. $\frac{1}{8}$

D. $\frac{1}{6}$

Sol. C.

GATE 2005

119. How many solutions does the following system of equations have?

$$4x + 2y + z = 7$$
$$x + 3y + z = 3$$
$$3x + 4y + 2z = 2$$

A. 0

B. 1

C. 2

D. ∞

Sol. B.

120. The matrix A is given by

$$A = \begin{bmatrix} 1 & 4 \\ a & 2 \end{bmatrix}$$

The eigen values of the matrix A are real and non-negative for the condition

A. $-\frac{1}{16} \le a \le \frac{1}{16}$

B. $-\frac{1}{2} \le a \le \frac{1}{2}$

C. $-\frac{1}{2} \le a \le \frac{1}{16}$

D. $-\frac{1}{16} \le a \le \frac{1}{2}$

Sol. D.

121. The divergence of a vector field A is always equal to zero, if the vector field A can be expressed as

A. the gradient of any scalar field $\emptyset$

B. the divergence of any scalar field $\emptyset$

C. the divergence of any vector field B

D. the curl of any vector field B

Sol. D.

122. In the limit x → 0, what is the limiting value of the function F(x) given below.

$$F(x) = \frac{1 - \cos(2x)}{x^2}$$

A. 0

B. 1

C. 2

D. ∞

Sol. C.

123. If z = x + iy is a complex number, where i = $\sqrt{-1}$, then which of the following is an analytic function of z?

A. $x^2 + y^2$

B. $2ixy$

C. $x^2 + y^2 - 2ixy$

D. $x^2 - y^2 + 2ixy$

Sol. D.

124. What condition is to be satisfied so that the solution of the differential equation

$$\frac{d^2y}{dx^2} + a\frac{dy}{dx} + by = 0$$

is of the from y = $(C_1 + C_2\,x)e^{mx}$, where C_1 and $C_2$are constants of integration?

A. $a^2 = b$

B. $b^2 = a$

C. $a^2 = 4b$

D. $b^2 = 4a$

Sol. **C.**

125. In the domain $-\infty < x < \infty$, the function y(x) = x^3e^{-x} has

A. no maximum and no minimum

B. one maximum and no minimum

C. no maximum and one minimum

D. one maximum and one minimum

Sol. **B.**

126. If f(x) is the solution of the equation

$$\frac{dy}{dx} + 2xy + 2x = 0$$

nd g(x) is the solution of the equation

$$\frac{dy}{dx} + 2xy - 2x = 0$$

and the constant of integration in f(x) is equal to that in g(x), then which of the following is true?

A. g(x) = f(x) + 2

B. g(x) = f(x) + 1

C. g(x) = f(x) – 1

D. g(x) = f(x) – 2

Sol. **A.**

127. If the constant A is suitably chosen so as to satisfy the initial condition and n is an integer, which of the following is a valid solution for the unsteady one dimensional diffusion equation $\frac{\partial c}{\partial t} = D\frac{\partial^2 c}{\partial x^2}$ in the domain $0 \leq x \leq L$ with boundary conditions $c = 0$ at $x = 0$ and $c = 0$ at $x = L$?

A. $\sum_n A_n \sin\left(\frac{n\pi x}{L}\right)e^{\left(\frac{Dn^2\pi^2 t}{L^2}\right)}$

B. $\sum_n A_n \cos\left(\frac{n\pi x}{L}\right)e^{\left(\frac{Dn^2\pi^2 t}{L^2}\right)}$

C. $\sum_n A_n \sin\left(\frac{n\pi x}{L}\right)e^{-\left(\frac{Dn^2\pi^2 t}{L^2}\right)}$

D. $\sum_n A_n \cos\left(\frac{n\pi x}{L}\right)e^{-\left(\frac{Dn^2\pi^2 t}{L^2}\right)}$

Sol. **C.**

128. The function f(x) satisfies the equation f(x) = 0 at x $=x_e$. The Newton-Raphson iterative method converges to the solution in one step, regardless of the initial guess, if

A. f(x) is a linear function of x

B. f(x) is a quadratic function of x

C. f(x) is a cubic function of x

D. f(x) is an exponential function of x

Sol. **A.**

GATE 2004

129. The sum of the eigen values of the matrix $\begin{bmatrix} 3 & 4 \\ x & 1 \end{bmatrix}$ for real and negative values of x is

A. greater than zero

B. less than zero

C. zero

D. dependent on the value of x

Sol. **A.**

130. The system of equations

$$4x + 6y = 8$$
$$7x + 8y = 9$$
$$3x + 2y = 1$$

has

A. no solution

B. only one solution

C. two solutions

D. infinite number of solutions

Sol. **B.**

131. A box contains three blue balls and four red balls. Another identical box contains two blue balls and five red balls. One ball is picked at random from one of the two boxes and it is red. The probability that it came from the first box is

A. $\frac{2}{3}$

B. $\frac{4}{9}$

C. $\frac{4}{7}$

D. $\frac{2}{7}$

***Sol.* B.**

132. The series $\sum_{n=1}^{\infty}\frac{(z-2)^n}{n!}$ converges for

A. all z

B. z > 2

C. $|z| > 2$

D. $|z| < 2$

***Sol.* A.**

133. The differential equation for the variation of the amount of salt x in a tank with time t is given by $\frac{dx}{dt}+\frac{x}{20}=10$. x is in kg and t is in minute. Assuming that there is no salt in the tank initially, the time (in minute) at which the amount of salt increases to 100 kg is

A. 10 In 2

B. 20 In 2

C. 50 In 2

D. 100 In 2

***Sol.* B.**

134. Value of the integral $\int_{-2}^{2}\frac{dx}{x^2}$ is

A. 0

B. 0.25

C. 1

D. ∞

***Sol.* D.**

135. The differential equation $\left(\frac{dy}{dx}\right)^2+y\frac{d^2y}{dx^2}=0$ can be reduced to (where, ∝ is a constant)

A. $\left(\frac{dy}{dx}\right)^3=\propto-\frac{3y^2}{2}$

B. $\left(\frac{dy}{dx}\right)^2=\propto-2y$

C. $\frac{dy}{dx}=\frac{\propto}{y^2}$

D. $\frac{dy}{dx}=\frac{\propto}{y}$

***Sol.* D.**

136. The value of $\lim_{x\to 9}\frac{\sqrt{x}-3}{x^2-81}$ is

A. 0

B. $\frac{1}{27}$

C. $\frac{1}{108}$

D. ∞

***Sol.* C.**

GATE 2003

137. The range of values for a constant K to yield a stable system in the following set of time dependent differential equations is $\frac{dy_1}{dt}=-5y_1+(4K)y_2$

$\frac{dy_2}{dt}=y_1-2y_2$

A. 0 < K < 7

B. 6.25 < K < 10

C. – 6 < K ≤ 6.25

D. $0 \le K \le 7$

***Sol.* C.**

138. The value of y as t → ∞ for the following differential equation for an initial value of y(1) = 0 is

$$(4t^2+1)\frac{dy}{dt}+8yt-t=0$$

A. 1

B. 1/2

C. 1/4

D. 1/8

***Sol.* D.**

139. The equilibrium data of component A in two phases B and C are given below.

X (mol of A/mol of B)	*Y mol of A/mole of B*
1	0.5
2	4.125

The estimate of Y for X = 4 by fitting a quadratic expression of a form Y = mX^2 for the above data is

A. 15.5

B. 16

C. 16.5

D. 17

***Sol.* C.**

140. A fluid element has a velocity v = $-y^2x\hat{i}+2yx^2\hat{j}$. The motion at (x, y) = $(1/\sqrt{2},1)$ is

A. rotational and incompressible

B. rotational and compressible

C. irrotational and compressible

D. irrotational and incompressible

***Sol.* A.**

141. The most general complex analytical function f(z) = u(x, y) + iv (x, y) for u = x^2-y^2 is

A. Z
B. z^2
C. z^3
D. $1/z^2$

Sol. B.

142. The differential equation $\frac{d^2x}{dt^2} + 10\frac{dx}{dt} + 25x = 0$ will have a solution of the form

A. $(C_1 + C_2t)e^{-5t}$
B. C_1e^{-2t}
C. $C_1e^{-5t} + C_2e^{5t}$
D. $C_1e^{-5t} + C_2e^{2t}$

where, C_1 and C_2 are constants.

Sol. A.

143. Find the Laplace transform of the following input function shown in figure.

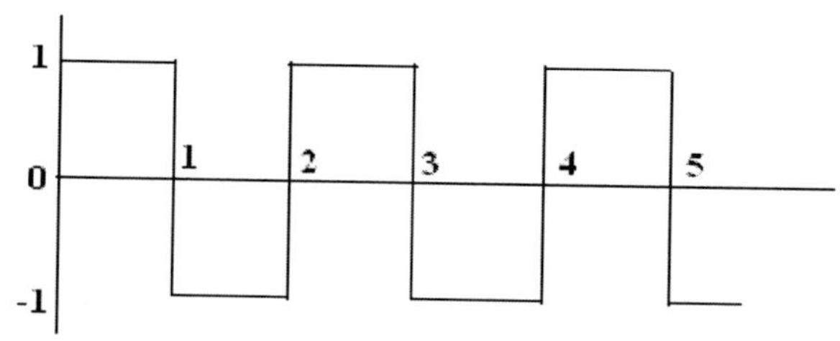

A. $\frac{(1-e^{-s})}{s(1-e^{-2s})}$
B. $\frac{2(1-e^{-s})}{s}$
C. $\frac{(1-e^{-s})}{s(1+e^{-s})}$
D. $\frac{(1-e^{-s})e^{-s}}{s(1-e^{-2s})}$

Sol. C.

GATE 2002

144. The coefficient x^2 in the Taylor series of cos^2 x about 0 is

A. 2
B. 0
C. 1
D. – 1

Sol. D.

145. Three grades of paint (A, B and C, production rates 12, 24 and 18 batches per day respectively) are produced in independent batch production lines and stored in separate areas. The number of off-specification batches in a day are 1, 3 and 2 for grades A, B and C respectively. The probability of picking an off-specification batch from a randomly chosen storage area is

A. $\frac{23}{216}$
B. $\frac{24}{216}$
C. $\frac{19}{216}$
D. $\frac{18}{216}$

Sol. A.

146. If an amount R is paid at the end of every year for n years, then the net present value of the annuity at an interest rate of i is

A. $R\left[\frac{(1+i)^n-1}{i}\right]$
B. $R\left[\frac{(1+i)^n-1}{i(1+i)^n}\right]$
C. $R(1+i)^n$
D. $\frac{R}{(1+i)^n}$

Sol. B.

GATE 2001

147. An investment of ₹1000 is carrying an interest of 10% compounded quarterly. The value of the investment at the end of five years will be

A. $1000\left(1+\frac{0.1}{4}\right)^{20}$
B. $1000\ (1+0.10)^{20}$
C. $1000\left(1+\frac{0.1}{4}\right)^{5}$
D. $1000\left(1+\frac{0.1}{2}\right)^{5}$

Sol. B.

148. The function f(x, y) = $x^2 + y^2 - xy - x - y + 5$ has

A. Maximum at (1, 1)
B. saddle point at (1, 1)
C. minimum at (1, 1)
D. None of the above at (1, 1)

Sol. C.

149. A fair die is rolled four times. Find the probability that six shows up twice.

A. $\frac{1}{2}$
B. $\frac{16}{325}$
C. $\frac{1}{36}$
D. $\frac{25}{216}$

Sol. D.

GATE 2000

150.The line integral of $\int_C \left\{ \frac{Y}{x^2+y^2} dx - \frac{x}{x^2 y^2} dy \right\}$

where, C is the unit circle around the origin traversed once in the counter-clockwise direction, is

A. -2π
B. zero
C. 2π
D. π

Sol. **B.**

151.The inverse of the matrix $\begin{bmatrix} 1 & -1 \\ -1 & -1 \end{bmatrix}$

A. Does not exist
B. $\begin{bmatrix} 1 & 1 \\ 1 & -1 \end{bmatrix}$
C. $\begin{bmatrix} 0.5 & -0.5 \\ -0.5 & -0.5 \end{bmatrix}$
D. $\begin{bmatrix} -0.5 & 0.5 \\ 0.5 & 0.5 \end{bmatrix}$

Sol. **C.**

152.The complex conjugate of $\frac{1}{1+i}$ is

A. $\frac{1}{1-i}$
B. (1 – i)
C. 0.5 (1 – i)
D. in the first quadrant of the complex plane

Sol. **A.**

153. The general solution of $\frac{d^4y}{dx^4} + 2\frac{d^2y}{dx^2} + y = 0$ is

A. $(C_1x + C_2)e^x + (C_3 + C_4x)e^{-x}$
B. $C_1 \cos x + C_2 \sin x + C_3 e^x + C_4 e^{-x}$
C. $C_1 e^{ix} + C_2 e^{-ix}$
D. $(C_1 + C_2x)\cos x + (C_3 + C_4x)\sin x$

Sol. **D.** 7

3. Chemical Reaction Engineering

1. Basic Concept of Chemical Reaction Engineering: A chemical reaction may be defined as a process where one set of chemical substance may undergo rearrangement of molecules to produce another set of substance. In a chemical reaction there is only change in positions of electrons in the forming and breaking of chemical bonds without any change in the nucleus.

1.1. Rate of Homogeneous Reactions: The rate of a homogeneous reaction is defined as the change in number of moles of a reactant or product per unit time per unit volume of the reaction mixture. For the reaction.

$$A + B \rightarrow C + D$$

$$r_A = -\frac{1}{V}\frac{dN_A}{dt} = -\frac{1}{V}\frac{dN_B}{dt} = \frac{1}{V}\frac{dN_C}{dt} = \frac{1}{V}\frac{dN_D}{dt}$$

Where,

$$r_A = rate\ of\ reaction$$

$N_A, N_B, N_C\ and\ N_D$ are number of moles of A,B,C and D respectively at any time t since beginning of the reaction.

Similarity, for the reaction

$$aA + bB \rightarrow cC + dD$$

$$r_A = -\frac{1}{V}\left(\frac{1}{a}\right)\frac{dN_A}{dt} = -\frac{1}{V}\left(\frac{1}{b}\right)\frac{dN_B}{dt} = \frac{1}{V}\left(\frac{1}{c}\right)\frac{dN_C}{dt} = \frac{1}{V}\left(\frac{1}{d}\right)\frac{dN_D}{dt}$$

1.2. Order of Reaction: The order of a reaction with respect to a given reactant is defined as the index to which its concentration term in the rate equation is raised. For a typical rate equation $r = KC_A^a C_B^b$, the reaction order is 'a' with respect to component A and 'b' with respect to component B. The reaction may also be termed to have (a + b) as the overall order of reaction. For many reactions the reaction order is not equal to the stoichiometric coefficient. Order of a reaction is an experimental concept and order of the same reaction may be different under different experimental conditions. Numerical order may be equal to zero, positive, negative or fractional. Such reactions whose actual order is different from that otherwise expected are known as pseudo order reactions. An example of this is hydrolysis reaction where one of the reactants i.e. water is present in large excess is known as pseudo first order reaction.

1.3. Molecularity of reaction: Molecularity of reaction is the minimum total number of molecules which must combine in order that the reaction may proceed. It is mostly equal to the sum of stoichiomertric coefficients of reactions. Reactions may be unimolecular, bimolecular or trimolecular. Such reactions in which molecularity and order are same are known as elementary reactions.

1.4. Rate constant: A reaction rate constant quantifies the rate of the chemical reaction.

For a nth order reaction $A \xrightarrow{k} B$

Rate of reaction $r_A \propto C_A^n$

or $r_A = kC_A^n$

Where K is defined as the rate constant or specific reaction rate of the reaction. Rate constant depends on the nature of reactants, the temperature and pressure and also on the presence of a catalyst.

The relationship between reaction rate constant and temperature is given by Arrhenius equation

$$k = Ae^{-E/RT}$$

Where, E is activation energy of the reaction which is defined as the minimum amount of energy required by reactant molecules in order that the reaction may proceed. A is frequency factor. Activation energy can be determined by plotting *ln k* versus *1/T* in which slope of the line will be E/R and the y – axis intercept is *ln A*. It is to be noted that the effect of presence of a catalyst is only to lower the activation energy so that more number of molecules are in a positive to participate in the reaction. A catalyst does not change the equilibrium point.

1.5. Reaction Rate Theories: Several theories of reaction rates have been postulated to explain how and why reaction occurs at the molecular level. The two more popular and accepted theories are (i) Collision theory and

(ii) Activated complex theory. The collision theory postulates that reactions occur as a result of random collisions of reactant molecules. An explanation to this is provided by kinetic theory of gases which states that molecules are in continuous random motions and as a result of the collisions there is exchange of energy. Subsequent to these collisions there may be molecules who will acquire energy as a result of which they cross the barrier of activation energy level and hence react. This postulation has led to a rate expression based on frequency of molecular collisions and the fraction of the collision that evolve the required minimum energy.

For many reactions the predicted rate by collision theory are much larger. For complex reactions this deviation increases. Furthers by collision theory unimolecular reactions are difficult to be explained. The activated complex theory postulater that an activated complex is formed from the reactant which subsequently decomposes to form the products. The activated complex is assumed to be in Thermodynamic equilibrium with the reactants. It is the answer to the question why all collision may not be effective in producing a reaction.

1.6. Mechanism of reaction: Considerable numbers of reactions do not occurs in the way they are represented stoichiometrically. An example is

$$H_2 + Br_2 \rightarrow 2HBr$$

Apparently, the above reaction indicates that one molecule of H_2 reacts with one molecule of Br_2 react with one molecule of to directly produce two molecules of HBr. However, in actual practice it is much more complex.

$$Br_2 \xrightarrow{k_1} 2\,Br \qquad \ldots(i)$$

$$Br + H_2 \xrightarrow{k_2} HBr + H \qquad \ldots(ii)$$

$$H + Br_2 \xrightarrow{k_3} HBr + Br \qquad \ldots(iii)$$

$$H + HBr \xrightarrow{k_4} H_2 + Br \qquad \ldots(iv)$$

$$2Br \xrightarrow{k_5} Br_2 \qquad \ldots(v)$$

Overall, $H_2 + Br_2 \rightarrow 2HBr$

The rate equation for the above set of reactions is given as

$$\frac{d[C_{HBr}]}{dt} = \frac{K_{11}[C_{H_2}][C_{Br_2}]^{1/2}}{1 + K_{12}\frac{[C_{HBr}]}{[C_{Br_2}]}}$$

Where K_{11} and K_{12} are new constants made up of the specific reaction rates of the five elementary reactions (i) to (v). This equation perfectly fits the experimental results. Experimentally it is known that the equilibrium represented by reactions (i) and (v) is achieved rapidly and that only a small amount of atomic hydrogen is present in the reaction mixture. Reactions involve energy consideration and therefore without any external considerations they follow the route where there is minimum energy expenditure. Outwardly, this may appear to be a complicated route but it must be recognized that reactions easiest to write is not necessarily the one taking place most easily. Deriving the mathematical expression, smce the equilibrium of Br_2 represented by steps (i) and (v) is attained rapidly

$$\frac{d[C_{HBr}]}{dt} = k_1[C_{Br_2}] + k_5[C_{Br_2}]^2 = 0 \qquad \ldots(i)$$

Further, only small amount of C_H is present

$$\therefore \frac{dC_H}{dt} = k_2(C_{Br})(C_{H_2}) - k_3(C_H)(C_{Br_2})$$
$$- k_4(C_H)(C_{HBr}) = 0 \qquad \ldots(ii)$$

Now, the rate of formation of (C_{HBr}) is

$$\frac{d(C_{HBr})}{dt} = k_2(C_{Br})(C_{H_2}) + k_3(C_H)(C_{Br_2})$$
$$- k_4(C_H)(C_{HBr}) = 0 \qquad \ldots(iii)$$

Solving equations (i) and (ii) for (C_{Br}) and (C_H)

$$(C_{Br}) = \sqrt{\frac{k(C_{Br_2})}{k_5}} \qquad \ldots(iv)$$

And $$(C_H) = \frac{k_2(C_{Br})(C_{H_2})}{k_3(C_{Br_2}) + k_4(C_{HBr})} \qquad \ldots(v)$$

Now, substitute (iv) and (v) in equation (iii)

$$\frac{d(C_{HBr})}{dt} = k_2(C_{H_2})\sqrt{\frac{k_1(C_{Br_2})}{k_5}}$$

$$+[k_3(C_{Br_2}) - k_4(C_{HBr})]\frac{k_2(C_{H_2})\sqrt{\frac{k_1(C_{Br_2})}{k_5}}}{k_3(C_{Br_2}) + k_4(C_{HBr})}$$

$$= \frac{2k_2k_3\sqrt{\frac{k_1}{k_5}}(C_{H_2})(C_{Br_2})^{0.5}}{k_3 + k_4\frac{(C_{HBr})}{(C_{Br_2})}}$$

Or $$\frac{d(C_{HBr})}{dt} = \frac{2k_2k_3\sqrt{\frac{k_1}{k_5}}(C_{H_2})(C_{Br_2})^{0.5}}{k_3\left[1 + \frac{k_4(C_{HBr})}{k_3(C_{Br_2})}\right]}$$

Hence, $$\frac{d(C_{HBr})}{dt} = \frac{k_6(C_{H_2})(C_{Br_2})^{0.5}}{1 + k_7\frac{(C_{HBr})}{(C_{Br_2})}}$$

Where $k_6 = 2k_2\sqrt{\frac{k_1}{k_5}}$

And $k_7 = \frac{k_4}{k_3}$

The final rate expressions consists of only the reactent concentration terms along with reaction rate constant combinations and is the ultimate rate equation for the reaction

$$H_2 + Br_2 \rightarrow 2HBr$$

From the above analysis it is clear that a general rule to be followed in such derivations is that the net rate of formation of intermediate substances are zero. In chain reaction, generally it may be observed that the specific reaction rate of the chain terminating step is negligible in comparison with those of the chain propagation steps.

1.7. Series and Parallel reaction: A series reaction may be defined as the one in which the product formed by an elementary reaction acts as the reactant in another elementary reaction in a sequential manner. A typical series reaction may be represented as

$$A \xrightarrow{K_1} R \xrightarrow{K_2} S$$

Series reactions may also be known as successive reactions. There are numerous examples of series reactions of industrial importance e.g. chlorination of benzene.

$$C_6H_6 + Cl_2 \rightarrow C_6H_5Cl + HCl$$
$$C_6H_5Cl + Cl_2 \rightarrow C_6H_4Cl_2 + HCl$$
$$C_6H_4Cl_2 + Cl_2 \rightarrow C_6H_3Cl_3 + HCl$$
$$C_6H_3Cl_3 + Cl_2 \rightarrow C_6H_2Cl_4 + HCl$$
$$C_6H_2Cl_4 + Cl_2 \rightarrow C_6HCl_5 + HCl$$
$$C_6HCl_5 + Cl_2 \rightarrow C_6Cl_6 + HCl$$

In such series reactions it may be required to identify the reaction condition and time when a particular intermediate product e.g. dichloro or trichloro benzene is having maximum production.

Or $\frac{C_R}{C_{A0}} = \left[\frac{K_1}{K_2 - K_1}\right]\left[e^{-K_1 t} - e^{-K_2 t}\right]$ ...(i)

Considering a typical series reaction

$$A \xrightarrow{k_1} R \xrightarrow{k_2} S$$

This reaction may be considered as

$$A \xrightarrow{k_1} R$$
$$R \xrightarrow{k_2} S$$

For the first reaction the rate equation is

$$-\frac{dC_A}{dt} = k_1 C_A$$

or $C_A = C_{A0}e^{-k_1 t}$ ...(ii)

For the second reaction

$$-\frac{dC_A}{dt} = k_2 C_R - k_1 C_A$$
$$= k_2 C_R - k_1 C_{A0} e^{-k_1 t}$$

or $\frac{dC_R}{dt} + k_2 C_R = k_1 C_{A0} e^{-k_1 t}$

Solving this linear differential equation by finding the integrating factor and solving

$$(C_R)(e^{k_2 t}) = \left[\frac{k_1 C_{A0}}{k_2 - k_1}\right]\left[e^{(k_2 - k_1)t} - 1\right]$$

Or $\frac{C_R}{C_{A0}} = \left[\frac{k_1}{k_2 - k_1}\right]$

Writing material balance for C_s

$$C_s = C_{A0} - C_A - C_R$$

Substituting from (i) and (ii)

$$\therefore \quad C_s = C_{A0} - C_{A0}e^{-k_1 t}$$
$$-C_{A0}\left[\frac{k_1}{k_2 - k_1}\right]\left[e^{-k_1 t} - e^{-k_2 t}\right] \quad \text{...(iii)}$$

Now for determination of $C_{R_{max}}$ and t_{max} differentiating equation (ii) with respect to t and putting it equal to zero for maximum

$$\frac{dC_R}{dt} = \left[\frac{k_1 C_{A0}}{k_2 - k_1}\right]\left[-k_1 e^{-k_1 t} + k_2 e^{-k_2 t}\right]$$

or $k_1 e^{-k_1 t_{max}} = k_2 t_{max} e^{-k_2}$

or $\frac{k_2}{k_1} = \frac{e^{-k_1 t_{max}}}{e^{-k_2 t_{max}}}$

or $t_{max} = \left[\frac{1}{k_2 - k_1}\right] ln^{k_2/k_1}$

$$t_{max} = \frac{1}{K_{\log mean}} \quad \text{...(iv)}$$

Continuing from equation (ii) at maximum condition

$$\frac{C_{R_{max}}}{C_{A0}} = \left[\frac{k_1}{k_2 - k_1}\right]\left[e^{-k_1 t_{max}} - e^{-k_2 t_{max}}\right]$$

$$= \left[\frac{k_1}{k_2 - k_1}\right]\left[e^{-k_1\left(\left[\frac{1}{k_2 - k_1}\right] ln^{\frac{k_2}{k_1}}\right)} - e^{-k_2\left(\left[\frac{1}{k_2 - k_1}\right] ln^{k_2/k_1}\right)}\right]$$

$$= \left[\frac{k_1}{k_2 - k_1}\right]\left[\left(\frac{k_2}{k_1}\right)^{\frac{-k_1}{k_2 - k_1}} - \left(\frac{k_2}{k_1}\right)^{\frac{-k_2}{k_2 - k_1}}\right]$$

$$= \frac{k_1}{k_2 - k_1}\left(\frac{k_2}{k_1}\right)^{\frac{-k_2}{k_2 - k_1}}\left[\left(\frac{k_2}{k_1}\right)^{\frac{-k_1 + k_2}{k_2 - k_1}} - 1\right]$$

$$= \frac{k_1}{k_2 - k_1}\left[\frac{k_1}{k_2}\right]^{\frac{k_2}{k_2 - k_1}}\left[\frac{k_2 - k_1}{k_1}\right]$$

$$\frac{C_{R_{max}}}{C_{A0}} = \left[\frac{k_1}{k_2}\right]^{\frac{k_2}{k_2 - k_1}}$$

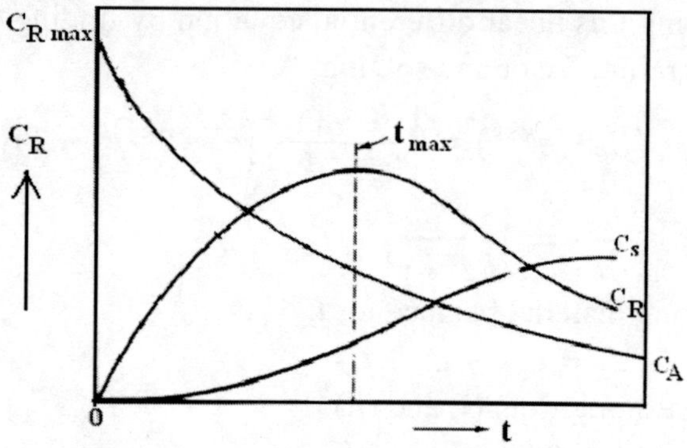

Initially, there is only A, which gradually falls off continuously. The concentration of R increases, as A keeps on decompressing and rises to a maximum. During their ascending portion as the concentration of R increases the second reaction becomes more and more prominent. Thereby C_R starts decreasing whereas C_s keeps on increasing.

Analyzing a set of parallel reaction on similar lines

$$A \overset{k_1}{\rightarrow} R$$

$$R \overset{k_2}{\rightarrow} S$$

The rate of equation $-\frac{dC_A}{dt} = k_1 C_A + k_2 C_R$

On integration

$$\ln {C_{Ao}}/{C_A} = (k_1 + k_2)\, t$$

Or $\quad C_A = C_{Ao} e^{-(k_1+k_2)\,t}$ ...(v)

Writing; the rate equation for R

$$\frac{dC_R}{dt} = k_1 C_A$$

Substituting for C_A from equation (v)

$$\frac{dC_R}{dt} = k_1 C_{Ao} e^{-(k_1+k_2)\,t}$$

Integrating

$$C_R = \left[\frac{k_1 C_{Ao}}{(k_1+k_2)}\right]\left[1 - e^{-(k_1+k_2)\,t}\right] \quad \text{...(vi)}$$

Similarly for C_s

$$\frac{dC_s}{dt} = k_2 C_A$$

$$\frac{dC_s}{dt} = k_2 C_{Ao} e^{-(k_1+k_2)\,t}$$

Integrating

$$C_S = \left[\frac{k_2 C_{Ao}}{(k_1+k_2)}\right]\left[1 - e^{-(k_1+k_2)\,t}\right] \quad \text{...(vii)}$$

The expression for C_S could also have been derived by writing overall material balance

$$C_{Ao} = C_A + C_R + C_S$$

Or $C_S = C_{Ao} - C_A - C_R$

$$C_S = C_{Ao} - C_{Ao} e^{-(k_1+k_2)\,t} - \left[\frac{k_1 C_{Ao}}{(k_1 + k_2)}\right]\left[1 - e^{-(k_1+k_2)\,t}\right]$$

$$= C_{Ao}\left[1 - e^{-(k_1+k_2)\,t} - \frac{k_1}{(k_1 + k_2)} + \frac{K_1}{(k_1 + k_2)} e^{-(k_1+k_2)\,t}\right]$$

$$= C_{Ao}\left[\begin{matrix} 1 - \frac{k_1}{(k_1+k_2)} - e^{-(k_1+k_2)\,t} \\ \left(1 - \frac{k_1}{(k_1+k_2)}\right) \end{matrix}\right]$$

$$= C_{Ao}\left[\frac{k_1+k_2-k_1}{k_1+k_2} - \left(\frac{k_1+k_2-k_1}{k_1+k_2}\right) e^{-(k_1+k_2)\,t}\right]$$

$$C_S = C_{Ao}\frac{k_2}{k_1 + k_2}\left[1 - e^{-(k_1+k_2)\,t}\right]$$

1.8. Reactor design: A reactor is heart of any process and therefore all the foregoing discussions revolve around designing an economic reactor to carry out a given reaction achieving the desired conversion. Since all reaction will involve heat changes, reactor must be equipped to maintain the desired temperature by incorporating suitable heat transfer system. Reactors are classified under three categories i.e. Batch, Flow and Semibatch.

1.9 Ideal Batch Reactor: A batch reactor in its simplest form may be a tank type with a mixture and suitable provision for heating or cooling as the case may be depending on whether the reaction is exothermic or endothermic. The operation of a batch reactor is characterized by feeding the raw material batchwise and after the desired conversion, emplying the reactor. In between the batches there may be clearing, charging and discharging etc. which together us with conversion time is known as time required for one cycle of operation. The daily or monthly capacity may be calculated on this basis. A batch reactor presumes that the operation is under unsteady state and that there is perfect mixing. The condition of perfect mixing is necessary to ensure that at any given time of operation the composition of the reactor mixture remains the same in all parts of the reactor. This is the condition of ideality otherwise the reactor may become a non – ideal batch reactor.

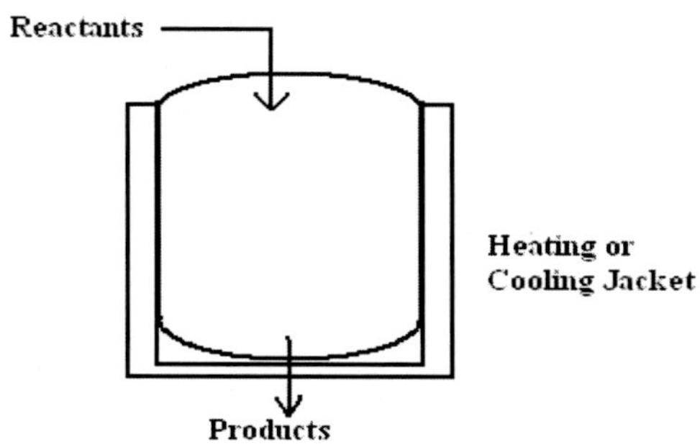

Writing material balance around the limiting component (A)

Input – [Output + Disappearance due to reaction]

= Accumulation

Writing the rate equation assuming a first order irreversible reaction.

Since for a batch system

Input and Output are both zero and the rate of disappears of the reactant is equal to accumulation. For constant volume reactions.

$$-\frac{dC_A}{dt} = k_1 C_A$$

or $$t = -\frac{1}{k}\int \frac{dC_A}{C_A}$$

For n[th] order reaction

$$t = -\frac{1}{k}\int \frac{dC_A}{C_A^n}$$

A batch reactor is most suitable for small operations and they are easy to operate and control.

1.10. Flow Reactor or Plug Flow Reactor:A p-lug flow reactor differs from a batch reactor in that it operator continuously under steadystate and the flow of reactants as well as that of products are continuous. A plug flow reactor also known as a turbular or piston flow reactor is further characterised by a plug like flow condition prevailing all along the reactor with zero backmixing. To ensure there conditions it has a tubular configuration having high L/D ratio. It has mostly useful in large scale production with gaseous reactants.

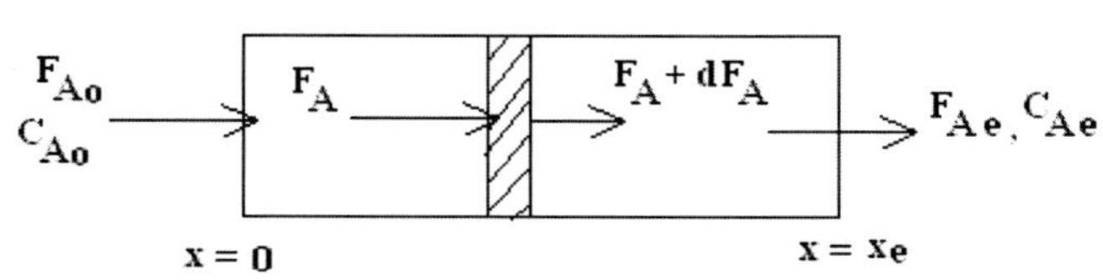

Referring to the figure a molar federate F of reactant stream undergoes reaction starting from zero conversion at the inlet to final conversion x_e at the outlet.

The conversion progressively increases along the passage of the reactor without any backmixing. For the sake of analyser let us consider which an elemental reaction volume dVR in which an elemental conversion dx is taking place. Since the operation is under steady state the accumulation will be zero.

Input = [Output + Disappearance due to chemical reaction]

or $FA = (FA + dFA) + (-r_A)dV_R$

or $-dFA = (-r_A)dV_R$

But $FA = F_{Ao}\,dx_A$

∴ $dFA = -F_{Ao}\,dx_A$

Hence, $-F_{Ao}\,dx_A = (-r_A)dV_R$

or $$\int_o^{V_R} \frac{dV_R}{F_{Ao}} = \int_o^{x_A} \frac{dx_A}{(-r_A)}$$

or $$\frac{dV_R}{F_{Ao}} = \int_o^{x_A} \frac{dx_A}{(-r_A)}$$

The term $\frac{dV_R}{F_{Ao}}$for a plug flow reactor is an important characteristic. The larger its numerical value the greats must be the size of the reactor necessary to process a given feed F_{Ao}. The space velocity may be defined as the permissible feed rate per unit volume of reactor. The reciprocal of space velocity is called residence time and has unit of time.

1.11. Continuous Stirred Tank Reactor or Tank Flow Reactor:A continuous stirred tank reactor (CSTR) deriver its name due to the reactor being of tank in and outflow being continuous in nature. A CSTR incorporates the features of both, a batch reactor in the way that the geometrical a PFR in as much as the operation being continuous and under steady state. A CSTR is usually operated in a battery form having several stages. Since the reaction rate in a CSTR is governed by the level of concentration of reactants in exit stream, its efficiency is at the lowest. In comparison, a batch reactor utiliser the higher concentration level of reactants as it starts from zero conversion moving to final conversion and a PFR, similarity, user the higher concentration levels progressively decreasing towards the exit.

1.12.Non-Ideal Flow: In all type of reactor system it is presumed that there is either complete mixing in the case of a batch reactor or CSTR or else there is no mixing in PFR. However, in reality these conditions are difficult to be achieved. The deviations from ideality may be due to channeling, short-circuiting or creation of dead space in the reactor.

1.13. Residence Time Distribution: The time spent by anelement in the reactor is called its residence time.

Since different fluid elements may spend different time in the reactor because of deviations in ideal flow pattern, the distribution of these times is called the exit age distribution or residence time distribution.

Mathematically $\int_0^{\infty} E\,dt = 1$

The curve between E and time is known as $E - curve$

Further, $\int_0^{t_1} E\,dt + \int_{t_1}^{\infty} E\,dt = 1$

The fraction of fluid elements younper

Than $t_1 = \int_0^{t} E\,dt$

The fraction of fluid elements order than time

$$t_1 = \int_{t_1}^{\infty} E\,dt$$

The fraction of material which has spent a time between t and $t + dt$ is Edt. If v and C represent volumetric flowrate and concentration of an input tracer, then amount of tracer that has a resistance time between t and $t + dt$ is $vcdt$.

Hence total amount of tracer is $\int_0^{\infty} v\,Cdt$

Fraction of tracer material which has spent a time between t and $t + dt = \frac{vCdt}{\int_0^{\infty} v\,Cdt}$

Hence, $E\,dt = \frac{vCdt}{\int_0^{\infty} v\,Cdt}$

or $\quad E = \frac{C}{\int_0^{\infty} Cdt}$

Cumulative Age Distribution function is defined as the fraction of elements in the exit stream which has been in the system for a time less than t and $F(t)$

$$F(t) = \int_0^{t} E(t)\,dt$$

and $(0) = 0;\quad F(\infty) = 1$

or $\quad E(t) = \frac{dF(t)}{dt}$

The fraction of material inside a reactor with age between t and $t + dt$ is defined as internal age distribution function I(t).

Writing material balance

$$1 - F(t) = \bar{t}I(t)$$

or

$$1 - \int_0^{t} E(t)\,dt = \bar{t}I(t)$$

where $\bar{t}$ is mean residence time

$$\bar{t} = \frac{\int_0^{\infty} tCdt}{\int_0^{\infty} Cdt}$$

1.14. Heterogeneous Reactions: Basically a heterogeneous reaction may be defined as a class of reaction in which two or more phases are present. Since more than two phase i.e. gas- solid or liquid –solid are present the mass transfer aspects should also be incorporated along with kinetic considerations.

The definition of rate of reaction for such reactions may be represented based on weight of catalyst pellets.

$$-r_A = -\frac{1}{W}\frac{dN_A}{dt}$$

Similarly based on catalyst surface

$$-r_A = -\frac{1}{S}\frac{dN_A}{dt}$$

Based on total reactor volume

$$-r_A = -\frac{1}{V_r}\frac{dN_A}{dt}$$

Based on volume of voids in the reactor

$$-r_A = -\frac{1}{V}\frac{dN_A}{dt}\left[\frac{moles\ of\ reacted}{volume\ of\ voids - sec.}\right]$$

There are heterogeneous reactions other than those involving solid catalysts for example, reactions between gases and solid such as in the reduction of metallic over. The sequence of steps for converting reactants to products in

(i) Transport of reactants from the bulk fluid to the fluid solid interface

(ii) Intraparticle transport of reactants into the process catalyst particle.

(iii) Adsorption of reactant at interior sites of the catalyst particle.

(iv) Chemical reaction of adsorbed reactant of adsorbed products i.e. surface reaction.

(v) Desorption of adsorbed products.

(vi) Transport of products from the interior sites to the outer surface of the catalyst particle.

(vii) Transport of product from the fluid – solid interface into the bulk fluid stream. At steady state the rates of all the individual step will be the same.

Multiple Choice Questions

1. A plug flow reactor is used to conduct a reversible liquid phase endothermic reaction. For minimum reactor volume its operation should be such that the temperature along the length __________.

   A. increases.

   B. decreases.

   C. is at the highest allowable temperature throughout.

   D. first increases and then decreases.

2. Which of the following statements is correct for a heterogeneous catalytic reaction?

   A. free energy of activation is lowered in the presence of catalyst which remains unchanged at the end of reaction

   B. the surface of the catalyst does not play an important role during reaction

   C. a relatively small amount of catalyst can cause the conversion of large amount of reactants which does not mean that catalyst concentration is important

   D. the catalyst does not form an intermediate complex with the reactant

3. The diffusivity under the condition when catalyst pore size is small in comparison with the mean free path and when collision with the pore wall controls the process is called 'Knudsen diffusivity'. This process is affected by the

   A. pressure

   B. temperature

   C. neither (a) nor (b)

   D. both (a) & (b)

4. Which of the following statements is wrong?

   A. space time in a flow reactor is a measure of its capacity and is equal to the residence time when the density of reaction mixture is constant

   B. the vessel dispersion number (*D*/*UL*) for plug flow and mixed flow approaches zero and infinity respectively

   C. mixed reactor is always smaller than the plug flow reactor for all positive reaction orders for a particular duty

   D. in an ideal tubular flow reactor, mixing takes place in radial direction and there is no mixing in longitudinal direction

5. For very high pressure gas phase reactions the most suitable reactor is?

   A. stirred tank reactor

   B. tubular flow reactor

   C. fluidised bed reactor

   D. batch reactor

6. For the irreversible elementary reactions in parallel *e.g.* $A \rightarrow R$ *and* $A \rightarrow S$, the rate of disappearance of A is given by

   A. $C_A(K_1+K_2)$

   B. $C_A.K_1/2$

   C. $C_A(K_1 + K_2)/2$

   D. $C_A.K_2/2$

7. The reactants will have __________ in comparison to products for an exothermic reaction

   A. less energy

   B. more energy

   C. higher temperature

   D. same energy

8. In which of the following processes sulphuric acid is used as a catalyst?

   A. gas phase oxidation of SO_2 in chamber process

   B. hydrogenation of oils

   C. alkylation of hydrocarbons

   D. none of these

9. In which order of chemical reaction the time required to complete a definite fraction of reaction varies inversely as the concentration of the reactants?

A. 1
B. 0
C. 2
D. 3

10. When the temperature is increased from 400 °K to 500 °K the rate constant of chemical reaction increases by 100 times. What is the value of E/R if transition state theory is valid?
A. 8621°K
B. 9210°K
C. 8987°K
D. 8764°K

11. As thetemperature increases the rate of reaction also may increase due to
A. increase in the average kinetic energy of the reacting molecules
B. decrease in activation energy
C. increase in the number of effective collisions
D. none of these

12. Which order kinetics is applicable for radioactive decay?
A. first
B. zero
C. second
D. third

13. For which order of chemical reaction the dimensions of rate constant is lit/gm mole/min.
A. 3
B. 0
C. 2
D. 1

14. The dissociation of reactants into products is caused by the reactants possessing excess energy in chemical reaction which is called __________ energy.
A. activation
B. binding
C. threshold
D. potential

15. For a first order irreversible reaction conducted in an isothermal variable volume batch reactor which of the following relation is applied?
A. $-\ln(1 - X_A) = kt$
B. $\frac{C_{A0}}{1+\varepsilon_A X_A}\left(\frac{dX_A}{dt}\right) = k$
C. $\varepsilon_A \cdot \ln(1 - X_A) = k \cdot t$
D. $X_A = kt$

16. Which of the following statements is correct?
A. a catalyst initiates a chemical reaction
B. in catalytic reactions the catalyst reacts with the reactants
C. a catalyst lowers the activation energy of the reacting molecules
D. a catalyst cannot be recovered chemically unchanged at the end of the chemical reaction

17. The activity and effectiveness of solid catalyst depends upon __________ adsorption.
A. physical
B. chemical
C. neither (a) nor (b)
D. both (a) and (b)

18. From operational point of view a batch reactor is best suited for
A. large scale gaseous phase reactions
B. achieving cent percent conversion of reactants into products
C. liquid phase reactions
D. obtaining uniform polymerisation products in highly exothermic reactions

19. If a first order irreversible chemical reaction $A \rightarrow B$ is carried out separately in a constant volume as well as in a variable volume reactor for a particular period then
A. both the conversion as well as concentrations will be different
B. conversion in both will be the same but concentrations will be different
C. both conversion as well as concentration are same
D. none of these.

20. Which of the following processes is exothermic in nature?
A. ammonia synthesis
B. catalytic cracking
C. oxidation of sulphur
D. methanol synthesis

21. A catalyst may be defined as a substance which
A. cannot be recovered chemically unchanged at the end of a chemical reaction
B. lowers the activation energy of reacting molecules
C. initiates a reaction

D. is capable of reacting with any one of the reactants

22. Which of the following statements is correct?

A. reactions with high activation energies are very temperature sensitive

B. a photochemical reaction is catalysed by light

C. a chemical reaction occurs when the energy of the reacting molecule is less than the activation energy of the reaction

D. chemical equilibrium is a static state

23. For ammonia synthesis reaction *i.e.*

$N_2 + 3H_2 \rightleftharpoons 2NH_3$. The fractional conversion __________ with increase in pressure.

A. increases

B. remains unchanged

C. decreases

D. unpredictable from the data

24. Which of the following reactions should be used for conducting a highly exothermic solid catalysed reaction so that maximum conversion is achieved?

A. fluidised

B. fluidised bed reactor followed by a fixed

C. fixed

D. fixed bed reactor followed by a fluidized

25. The expression for rate of reaction for a series reaction $V \xrightarrow{K_1} X \xrightarrow{K_2} Y \xrightarrow{K_3} Z$, $(-r_x)$ is represented by

A. $(K_1\text{-}K_2)C_X$

B. $(K_1+K_1)C_X$

C. K_1C_V - K_2C_X

D. $(K_1+K_2+K_3)C_X$

26. A catalyst increases the rate of reaction because it

A. decreases the energy barrier for reaction

B. decreases the molecular collision diameter

C. increases the activation energy

D. none of these

27. What is the unit of rate constant for a gas phase reaction with a rate equation $-\frac{dpA}{dt} = K_p p_A^2$

A. atm $(\text{hr})^{-1}$

B. $(\text{atm})^{-1}$

C. $(\text{atm})^{-1}$. $(\text{hr})^{-1}$

D. $(\text{hr})^{-1}$

28. A first order reaction requires two equal sized CSTR. The conversion is

A. more when they are connected in parallel

B. more when they are connected in series

C. less when they are connected in series

D. same whether they are connected in series or in parallel

29. By decreasing the __________, the rate constant of a chemical reaction decreases

A. concentration of reactants

B. pressure

C. temperature

D. duration of reaction

30. Which of the following theories explain the mechanism of catalytic reaction?

A. activated complex theory

B. thermodynamics

C. collision theory

D. none of these

31. The heat of adsorption is likely to be of the order of _________ Kcal " Kg.mole for physical adsorption

A. 100000

B. 1000

C. 10000

D. 100

32. An equipment for conducting BET method

A. is made entirely of stainless steel

B. measures the catalyst surface area directly

C. operates at very high pressure

D. none of these

33. For a chemical reaction with zero order kinetics the

A. product concentration increases linearly with time

B. half life period is directly proportional to the initial concentration of the reactants

C. plot of products concentration with time is a straight line through the origin.

D. all (a), (b) and (c)

34. For a second order reaction thespecific reaction rate constant

A. varies with temperature

B. depends on the nature of the reactants

C. independent of temperature

D. both (a) and (b)

35. Investigations on kinetics of a chemical reaction $A + 2B \rightarrow C$ indicate that the rate of formation

of C increases 2.82 times on doubling the concentration of A whereas it increases by 9 times on trebling the concentration of B. What order these data suggest?

A. 7/2

B. 5/4

C. 7/4

D. 5/2

36. Air containing 18 gm oxygen is used to burn 6 gm of combustible charcoal with the generation of 16.5 gm Co_2 and 2.8 gm CO besides other constituents. What is the degree of conversion on the basis of disappearance of limiting reactant?

A. 75%

B. 95%

C. 100%

D. 20%

37. A series combination of *N* plug flow reactors with a total volume *V* gives the same conversion as a single plug flow reactor of volume '*V*' for __________ order reactions.

A. third

B. second

C. first

D. any

38. The decrease in activity of a catalyst is largely due to

A. covering of the catalytic active sites by a foregin substance

B. loss in surface area of the active component

C. agglomeration of metal particles caused by thermal sintering of the solid surface

D. all (a), (b) and (c)

39. For areaction in which high reactant concentration favors high yields is the most unsuitable reactor?

A. backmix reactor

B. series of CSTR

C. plug flow reactor

D. PFR in series

40. An autoclave is nothing but a __________,

A. high pressure batch reactor

B. high pressure tubular reactor

C. atmospheric pressure tank reactor

D. high pressure tubular reactor

41. For the reaction $A + 2B \rightarrow R + S$, the reaction rate is shown to be $\frac{dCA}{dt} = KC_AC_B^2$. If *B* at is present in large excess what would be the order of this reaction?

A. two

B. one

C. zero

D. three

42. For the case of space velocity of flow reactors which of the following statements is false?

A. the space velocity of 3 hr^{-1} means that three reactor volumes of feed at specified conditions are being fed into the reactor every hour.

B. the unit of space velocity is $(time)^{-1}$.

C. the space velocity of 3 hr^{-1} means that one third reactor volume of feed at specified conditions are being fed into the reactor

D. none of these

43. Non-ideal flow model is described by the single parameter model __________ ,

A. dispersion

B. tank in series

C. both (a) & (b)

D. neither (a) nor (b).

44. When oxygen reacts with organic material the reaction falls under the category of _______reaction.

A. exothermic

B. biochemical

C. endothermic

D. photochemical

45. For a particular reaction the rate of reaction doubles for every 10°C rise in temperature. If the temperature is increased from 30 to 70°C for how many times the rate of reaction will increase?

A. 32

B. 12

C. 16

D. 8

46. For an isothermal gas phase reaction $A \rightarrow 3B$, the fractional volume change between no conversion and complete conversion is

A. 1

B. 0.5

C. 2

D. 3

47. The equilibrium conversion of a reversible endothermic reaction __________ with decrease in temperature

A. decreases

B. increases

C. increases linearly with temperature
D. remains unaffected

48. For a gas phase reaction the maximum gas conversion will take place in a
A. fluidised bed reactor
B. plug-flow catalytic reactor
C. semi-fluidised bed reactor
D. fixed bed reactor

49. In a consecutive reaction system $A \xrightarrow{E_1} B \xrightarrow{E_2} C$ when E_1 is much greater than E_2 (E_1 and E_2 being heat of reaction), the yield of B increases with the
A. increase in temperature.
B. increase in initial concentration of A.
C. decrease in initial concentration of A.
D. decrease in temperature.

50. The response curve for a step input signal from a reactor is represented by a
A. *I*-curve
B. *C*-curve
C. *S*-curve
D. none of these

51. The rate of reaction doubles as the concentration of reactant is doubled for the chemical reaction $A \rightarrow B$. What is the value of 'n' for this chemical reaction If the reaction rate is proportional to concentration of reactant
A. 1
B. 0
C. 2
D. 3

52. Transition state theory relates the below quantities as
A. $k \propto T.e^{E/RT}$
B. $k \propto e^{-E/RT}$
C. $k \propto T$
D. $k \propto T^{1.5}$

53. The equilibrium constant For an ideal gas mixture undergoing a reversible gaseous phase chemical reaction is
A. decreases with pressure.
B. increases with pressure.
C. increases /decreases with pressure depending on the stoichiometric co-efficients of the reaction.
D. is independent of pressure.

54. A batch reactor falls under which of the following categories
A. suitable for gas-phase reactions on commercial scale.
B. suitable for liquid phase reactions involving small production rate.
C. most suitable for very large production rate.
D. least expensive to operate for a given rate.

55. In a chemical reactionThreshold energy is equal to the
A. normal energy of reactants
B. activation energy
C. sum of (a) & (b)
D. neither (a) nor (b)

56. ___________ are conservedin a chemical reaction,
A. ions
B. atoms
C. masses
D. both(b) & (c)

57. Which of the following statement is true if a spherical porous catalyst particle of radius R is subjected to reactant A which reacts to form B by a zero order surface reaction $A \rightarrow B$. Film mass transfer resistance is negligible and pore diffusion of A is rate controlling. The effectiveness factor of the catalyst is reported as 7/8.
A. Inner catalyst core of radius R/8 does not participate in reaction.
B. Inner catalyst core of radius 7R/8 does not participate in reaction.
C. Inner catalyst core of radius R/2 does not participate in reaction.
D. Effectiveness factor for a zero order reaction cannot be 7/8 as it must always be 1.

58. A constant volume isothermal batch reactor is being used to carry out a second order couseention irreversible reaction $A \xrightarrow{k_1} B \xrightarrow{k_2} C$ with different initial feed composition. In one of the test runs the ratio of concentration B to that of A in the initial feed was less than 0.5. The concentration of B increased first, reached a maximum and then declined with time. However for all experiments when this concentration ratio was 0.5 or above, concentration of B decreased uniformly with the time right from the beginning. Determine the ratio of the two rate constant (k_1/k_2)?

A. 1/4
B. 1/2
C. 2
D. 4

59. Which of the following statements is incorrect?
A. Catalytic activity of enzyme catalysed reactions which is affected by temperature, pH value & chemical agents, is maximum at a temperature of about 45°C.
B. Equilibrium concentrations in enzyme catalysed reactions can be calculated by using the thermodynamic properties of substrates & products.
C. Enzymes help in increasing the activation energy of the reaction.
D. Most of the enzyme catalysed reactions involve at least two substrates.

60. What will be the space time for a first order reaction carried out in a plug flow reactor?
A. $\frac{1}{k}\ln\frac{C_0}{C}$
B. $\frac{1}{k}\ln\frac{C}{C_0}$
C. $k\ln\frac{C_0}{C}$
D. $k\ln\frac{C}{C_0}$

61. For calcination of limestone and dolomite in industrial practice which of the following is used?
A. Fixed bed reactor
B. Moving bed reactor
C. Fluidised bed reactor
D. None of these

62. Calculate the ratio of the time required for 90% conversion to the time required for 45% conversionFor an isothermal second order aqueous phase reaction, $A \rightarrow B$, is
A. 22
B. 4
C. 11
D. 2

63. An imbalancedequation of a chemical reaction is against the law of
A. constant proportion
B. conservation of mass
C. multiple proportion
D. none of these

64. What iulli condition that a chemical reaction of zero order has a reaction rate if C is concentration of reactant?
A. $\propto 1/C_A$.
B. independent of temperature.
C. $\propto C_A$.
D. none of these.

65. For s reversible chemical reaction of the type, $A + B \rightleftharpoons AB$ a general statement is that
A. both forward and backward reactions will be exothermic.
B. neither of the reactions will be endothermic.
C. the combination reaction will be exothermic, while the dissociation reaction will be endothermic.
D. the combination reaction will be endothermic, while the dissociation reaction will be exothermic.

66. An exothermic liquid phase first order reaction, $A \rightarrow R$, is taking place a CSTR, which will be provided in a cooling jacket. What will be the amount of heat is to be removed. If its ρ and C_P are same for the reactant and product streams. Data given in rate constant at 40°C = 1 min^{-1}, X_A = 0.5; C_{A0} = 5 kmol/m^3, Feed and reactor temperature = 40°C, (ΔH) = -40 KJ / mole; ρ = 1000kg/m^3 C_P = 4 J/gm.°C ; q = 10^{-3} m^3/min.
A. 1kW
B. 2/3 kW
C. 5/3 kW
D. 4kW

67. In case of a/an __________ reaction There is no correspondence between stoichiometry and the rate equation
A. elementary
B. autocatalytic
C. multiple
D. non-elementary

68. An ideal plug flow reactor show characteristic of?
A. Uniform mixing
B. Flat velocity profile
C. Axial dispersion
D. None of these

69. Which of the following statements is correct about a semi-batch reactor
A. residence time is constant
B. velocity of reaction can be controlled

C. mixing takes place in axial direction only
D. condition similar to plug flow reactor exists

70. The dependences of a photochemical reaction on temperature is not significant because
A. it is an exothermic reaction.
B. it is a reversible reaction.
C. the energy of reacting molecules exceeds the activation energy by absorption of light.
D. none of these.

71. The half-life period of rate of a chemical reaction may be defended as
A. the time required to reduce the concentration of the reacting substance to half its initial value.
B. half of the residence time of a reaction.
C. half of the space time of a reaction.
D. none of these

72. In which of the following conditions enzymes are destroyed?
A. temperature is very high
B. reactant's concentration is very low
C. reaction rate is independent of the reactant's concentration
D. reactant's concentration is very high

73. The value of the equilibrium constant in the gaseous phase ammonia formation reaction
$(N_2 + 3H_2 \rightleftharpoons 2NH_3)$, depends on the
A. volume of the reactor.
B. initial concentration of N_2 and H_2.
C. temperature.
D. total pressure of the system.

74. A first order reaction has a half life period of (where, K = rate constant).
A. 6.93 K
B. 1.5 K
C. 0.693/K
D. 2.5 K

75. The preparation of batch process over continuous process may be required when the product
A. sales demand is fluctuating
B. quality & yield cannot be achieved in continuous processes because of long residence time
C. both (a) & (b)
D. neither (a) nor (b)

76. Velocity of any reaction depend upon
A. concentration of the reactants.
B. nature of the reactants.
C. temperature at which the reaction is carried.
D. all (a), (b) and (c).

77. Ionic reactions take place in
A. liquid state only
B. any state
C. solutions
D. solid state only

78. The controlling factor in very fast heterogeneous reaction is?
A. Heat and mass transfer effects
B. Temperature
C. Composition of reactant
D. Pressure

79. The most suitable reactor set up for an autocatalytic reactoris
A. CSTR in series
B. PFRs followed by CSTR
C. CSTR followed by PFR
D. PFRs in series

80. In which order reaction the time required to change the concentration of reactant to half its original value is independent of the initial concentration
A. three
B. one
C. two
D. zero

81. The primary characteristics of gas-liquid reactor are
A. liquid hold-up
B. specific inter-facial area
C. both (a) and (b)
D. none of these

82. Pure A at a concentration of 2 Kg mole/m^3 is fed to a adiabatic plug flow reactor (PFR). Where the reaction $A \rightarrow B$ takes place at a temperature of 500 K and flowrate of 0.01 m^3/s. For an exit conversion of 20%, what will be the exit temperature in °K assuming heat of reaction at 298 K = - 50000 kJ/ k mole of A reacted .Heat capacities $C_{PA} = C_{PB}$ = 100kJ/k mole.

A. 400
B. 1000

C. 600

D. 500

83. The rate of reaction isdirectly proportional to the _______, which appear in the 'law of mass action'

A. volume of the reaction vessel.

B. nature of the reactants.

C. equilibrium constant.

D. molar concentration of the reactants.

84. In a chemical reaction $X + Y + Z \xrightarrow{k} P$, it is observed that the

(i) rate of formation of 'P' is doubled on doubling the concentration of 'X'.

(ii) rate of formation of 'P' is quadrupled on doubling the concentration of 'Y'.

(iii) doubling the concentration of 'Z' does not affect the rate of formation of 'P'.

What is the order of the above chemical reaction?

A. Zeroth order

B. First order

C. Second order

D. Third order

85. What will be the temperature rise for a conversion of 50% when thereaction, $A \rightarrow B$ is carried out in abatch adiabatic reactor at an initial temperature of 373°K.The heat of reaction is - 1kJ/mole at 373°K and the heat capacity of both A and B to be constant and equal to 50J/mole°K.

A. 100°C

B. 10°C

C. 5°C

D. 20°C

86. Which of the following statements is not correct?

A. a flat velocity profile exists in a plug flow reactor

B. chemical reactions with high activation energy are ery temperature sensitive

C. the residence time for all the elements of fluid in case of a PFR need not be same

D. half life of a reaction increases with increased initial concentration for reaction orders more than one.

87. For a chemical reaction what will be the quantity of reactants left after n half periods if C_A is the quantity of reactant initially present

A. $(C_A)^{1/2n}$

B. $\left(\frac{1}{2}\right)^n . C_A$

C. $\left(\frac{C_A}{2}\right)^n$

D. $\left(\frac{C_A}{2}\right)^{1/n}$

88. 100 Kg.mole / hr of pure ethanol vapors are fed to a packed bed reactor where the following reactions take place $C_2H_5OH \rightarrow C_2H_4 + H_2O$

$2C_2H_5OH \rightarrow C_2H_5 - O - C_2H_5 + H_2O$

The product composition is ethylene 95 Kg.mole / hr, water vapour: 97.5 k mole / hr and diethyl ether 2.5 Kg.mole/hr. What is the percent conversion of ethanol in the reactor?

A. 100

B. 95

C. 97.5

D. 2.5

89. Which of the following statements is wrong?

A. the differential method of analysing kinetic data requires more accurate or larger amounts of data

B. the integral method of analysing kinetic data is used when the data is scattered

C. when the reaction rate is independent of temperature, the reaction is said to be of zero order

D. The ratio of volumes of plug flow reactor to that of mixed reactor is always less than one for identical feed composition, flow rate, conversion and for all positive reaction orders.

90. Which of the following theories is not applicableto a homogeneous reaction?

A. chain reaction theory

B. radiation hypothesis

C. collision theory and activated complex theory

D. none of these

91. A catalyst may be termed a negative catalyst ifit

A. retards the rate of reaction.

B. does not initiate the reaction.

C. reduces the value of equilibrium constant.

D. all (a), (b) and (c).

92. For particles of different unchanging sizes,for the same fractional conversion, for the non-catalytic reaction of particles which is the controlling resistance

A. diffusion through ash layer
B. film diffusion
C. chemical reaction
D. either (a), (b) or (c)

93. Which of the following does not influence the rate of a chemical reaction
A. reactants concentration
B. catalyst
C. temperature
D. number of molecules of reactants taking part in a reaction

94. What will be the best feed arrangement for a system of reactors for optimum performance if the reactors are operated in parallel
A. space time for each parallel line is same.
B. larger reactors have more space time compared to smaller ones.
C. space time for parallel lines is different.
D. none of these.

95. What will be the conversion for the second order irreversible reaction, $A \xrightarrow{k_2} B$, conducted in a constant volume batch reactor
A. $\frac{1}{1+k_2 \cdot C_{A0} \cdot t}$
B. $\frac{k_2 \cdot C_{A0} \cdot t}{1+k_2 \cdot C_{A0} \cdot t}$
C. $\frac{(k_2 \cdot C_{A0} \cdot t)^2}{1+k_2 \cdot C_{A0} \cdot t}$
D. $\frac{k_2 \cdot C_{A0} \cdot t}{(1+k_2 \cdot C_{A0} \cdot t)^2}$

96. A mixed flow reactor
A. is most suitable for gas phase reactions
B. is same as ideal stirred tank reactor.
C. employs mixing in axial direction only.
D. is same as plug-flow reactor.

97. In a gas phase catalytic (gas-solid) reaction which of the following resistances is not involved?
A. ash resistance
B. surface phenomenon resistance
C. gas film and pore surface diffusion resistances for reactants
D. gas film and pore surface diffusion resistances for products

98. In a CSTR a first order homogeneous reaction of the type $X \rightarrow Y \rightarrow Z$ (consecutive reaction) is carried out. Which of the following curves respectively show the variation of the concentration of X, Y and Z with time?

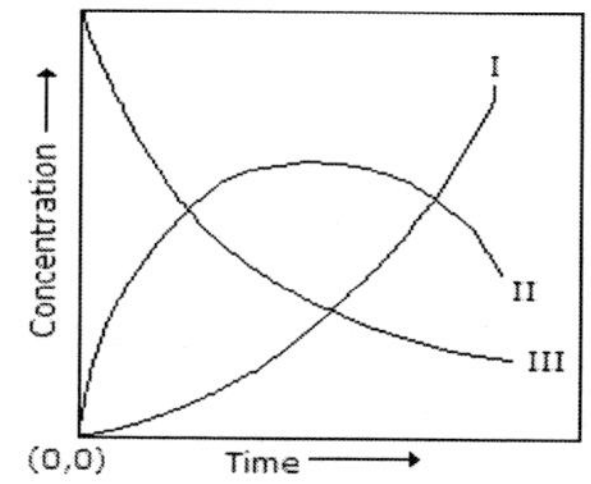

A. II, III, I
B. III, II, I
C. I, II, III
D. III, I, II

99. For a heterogeneous irreversible gas phase catalytic reaction $A(g) + B(g) \rightarrow C(g)$the rate constant of the rate controlling step is the surface reaction between absorbed A with absorbed B. What will be the rate expression for this reaction if K_A, K_B and K_C are the equilibrium constant
A. $\frac{k\, K_A K_B \rho_A \rho_B}{[1+K_A \rho_A+K_B \rho_B+K_C \rho_C]}$
B. $\frac{k\, K_A \rho_A \rho_B}{[1+K_A \rho_A+K_B \rho_B]}$
C. $\frac{k\, K_A K_B \rho_A \rho_B}{[1+K_A \rho_A+K_B \rho_B]}$
D. $\frac{k\, K_A K_B \rho_A \rho_B}{[1+K_A \rho_A+K_B \rho_B+K_C \rho_C]^2}$

100. The characteristics of photochemical reaction is that it is
A. catalysed by light
B. accompanied with emission of light
C. initiated by light
D. all (a), (b) & (c)

101. A chemical reaction having a higher free energy of activation is
A. slower rate of reaction.
B. higher equilibrium conversion.
C. higher rate of reaction.
D. both (b) and (c)

102. Identify the independent variable for a tubular flow reactor with uniform concentration and temperature,.
A. diameter.
B. length.

C. time
D. none of these

103. The operation of a rotary kiln is such that The calcination reaction of limestone ($CaCO_3 \rightarrow CaO + CO_2$) goes to completion because
A. CaO is more stable than $CaCO_3$
B. CO_2 escapes continuously
C. of high calcination temperature
D. CaO is not dissociated

104. For an elementary reaction $A + B \rightarrow C + D$ the molecularity is
A. 3
B. 4
C. 1
D. 2

105. For homogeneous reaction the rate of reaction is a function of
A. pressure and composition only
B. temperature and composition only
C. temperature and pressure only
D. all temperature, pressure and composition.

106. For a chemical reaction, the energy of activation.
A. varies as fifth power of the temperature
B. is the minimum energy which the molecules must have before the reaction can take place
C. is same as heat of reaction at constant pressure.
D. both (b) and (c)

107. For the unimolecular type first order reaction $A \xrightarrow{K} 4R$ the volume change increases ______ with time.
A. logarithmically
B. exponentially
C. parabolically
D. linearly

108. For an irreversible first order reaction the half-life period for decomposition of a liquid 'A' is 12 minutes what will the time required for 75% conversion of 'A'?
A. 18
B. 24
C. 6
D. 12

109. Which of the following reaction is not endothermic in nature?
A. Combustion of sulphur
B. Steam reforming of naphtha
C. Thermal cracking of fuel oil
D. Gasification of carbon

110. What will be the rate of disappearance of A in gm mole/litre.hr, for the reaction $2A + B \xrightarrow{K} C$ if the rate of formation of Cis 0.2 am mole /litre.hr?
A. 0.4
B. 0.2
C. 0.1
D. none of this

111. The __________ of a chemical reaction can be predicted by chemical kinetics.
A. rate
B. feasibility
C. neither(a) nor (b)
D. both (a) and (b)

112. In a PFR if space time for an irreversible isothermal reaction is increased the conversion will
A. increase
B. remain same
C. decrease
D. data in insufficient; can't be predicted

113. Which of the following statements is wrong
A. Arrehenius equation describing the effect of temperature on rate constant is given by, $K = A.e^{-Ea/RT}$
B. Two mixed reactors of unequal size are available for producing a specified product, formed by a homogenous second order reaction. To achieve maximum production rate, the smaller reactor should be placed in series before the larger reactor.
C. For the same conversion, the holding time required in a batch reactor, is always equal to the space time required in a PFR.
D. The mechanism for the decomposition of CH_3CHO into CH_4 and CO in presence of I_2 is:
$CH_3CHO + I_2 \rightarrow CH_3I + HI + CO$; slow
$CH_3I + HI \rightarrow CH_4 + I_2$; fast
Then the rate of disappearance of CH_3CHO is equal to $K.C_{CH3I}.C_{HI}$ and acts as a catalyst.

114. The dimension-less group which is not used in catalysis is (where, D = dispersion co-efficient, cm^2 /sec. D_1 = diffusion co-efficient; cm^2/sec L = length of the reactor, cm t = time, sec, v = volumetric flow rate, cm^3/sec. V = volume, cm^3.)

A. Thiele modulus $\left(L\sqrt{\frac{K}{D_1}}\right)$

B. Reactor dispersion number (D/vL)

C. Reduced time $\left(\frac{V_t}{V}\right)$

D. None of these

115. The function of a catalyst prompter that it

A. improves the activity of a catalyst

B. itself has very high activity

C. acts as a catalyst support

D. all (a), (b) and (c)

116. Which of the following will be the controlling resistance if for the non-catalytic reaction of particles with surrounding fluid the time needed to achieve the same fractional conversion for particles of different but unchanging sizes is proportional to the square of particle diameter.

A. chemical reaction

B. diffusion through ash layer

C. film diffusion

D. either (a), (b) or (c)

117. An irreversible first order reaction is carried out in a CSTR and PFR of the same volume. If the liquid felonries are same is both the case the relative conversion will be

A. depend on the temperature

B. be more in PFR than in CSTR

C. be more in CSTR tan in PFR

D. be more in CSTR than in PFR

118. Which of the following statements is wrong?

A. the slowest step has the greatest influence on the overall reaction rate in case of an irreversible series reaction.

B. autocatalytic reactions are exemplified by microbial fermentation reactions.

C. the fractional conversion at any time is same for both the constant as well as the variable volume system in case of anirreversible unimolecular type first order reaction.

D. hydrolysis of ester in presence of alkali or acid is a zero order reaction

119. The kinetics of zero order reaction concentration dose with change with time and the

A. time for half change is half the time taken for completion of the reaction.

B. time for completion of the reaction is independent of the initial concentration.

C. time for half change is independent of the initial concentration.

D. reaction rate is trebled when the initial concentration is trebled.

120. Promoters are added to the catalyst to improve its ________

A. Surface area

B. Sensitivity

C. Porosity

D. none of these

121. Which of the following statement in wrong?

A. For a first order consecutive reaction, a tubular flow reactor as compared to a stirred tank reactor provides higher overall selectivity.

B. For an ideal mixed reactor at steady state, the exit stream has the same composition as fluid within the reactor and the space time is equivalent to holding time for constant density system.

C. Plug flow reactor (PFR) is always smaller than mixed reactor for all positive reaction orders for a particular duty.

D. Reaction rate does not decrease appreciably as the reaction proceeds in case of an autocatalytic reaction.

122. A chemical reaction is __________ if the free energy change ΔG s very large and negative

A. unpredictable as ΔG is no measure of feasibility of a reaction.

B. not feasible

C. very much feasible

D. just feasible

123. What is rate controlling step for the heterogeneous catalytic reaction $A + B \rightarrow C$, with equimole feed of A and B

A. surface reaction between A in the gas phase and absorbed B.

B. surface reaction between absorbed A and absorbed B

C. surface $K_c/(1+T_s)$ reaction between absorbed A and B in the gas phase

D. desorption of C

124. The energy of the reacting substances is __________ in compared to that of that of product for an exothermic reaction

A. more

B. same

C. Less

D. either (a) or (b), depends on order of reaction

125. What is the reaction rate completed for a first dent chemical reaction which in half completed in 23 minutes

A. 0.03 sec^{-1}

B. 0.03 min^{-1}

C. 0.05 min^{-1}

D. 0.03 hr^{-1}

126. A cording to the design and based on its functioning in a PRF there

A. may be lateral mixing of fluid

B. should not be any mixing along the flow path

C. both (a) and (b).

D. neither (a) nor (b)

127. For constant density systems the performance equations are identical for

A. P.F.R, batch reactor and backmix reactor

B. P.F.R. and batch reactor

C. batch reactor and backmix reactor

D. P.F.R. and backmix reactor

128. Which of the following statement is correct?

A. A catalyst does not change ΔH of the reaction.

B. Law of mass action was suggested by Le-Chatelier

C. A catalyst changes the equilibrium point

D. The difference between the energy of reactants and that of the products is the activation energy

129. If $-r_A = k.C_A.C_B$ is the rate of an autocatalytic reaction A+B $\rightarrow$ B+B which of the following statements is correct?

A. rate of disappearance of reactant A is maximum, where $C_A = C_B$

B. plot of $-r_A$ *Vs* C_A is a hyperbola.

C. plot of $-r_A$ *Vs* C_A is a straight line with slope k.

D. both 'a' & 'b'

130. For the reaction system $X \xrightarrow{K_1} A, X \xrightarrow{K_2} B, X \xrightarrow{K_3} C$ the rate of decomposes of liquid 'X' is given by

A. $(K_1 + K_2 + K_3)C_X$

B. $(K_2 + K_3)C_X$

C. $(K_1 + K_2)C_X$

D. $K_1 .C_X$

131. For particular conversion and production rate, which of the following fixes the volume of a botch reactor

A. rate constant

B. operating conditions (e.g. pressure and temperature)

C. density of mixture

D. none of these

132. The addition of more of CaO, in case of calcinations of limestone, CaCO3 CaO + CO_2, will result in __________ in the conc. Of CO_2

A. no change

B. unpredictable from the data

C. decrease

D. increase

133. Which of the following reactors in used for hydrogenation using a nickel catalyst

A. Fluidised bed

B. Slurry

C. batch

D. fixed bed

134. For homogeneous reaction the rate of reaction is a function of

A. pressure and composition only

B. temperature and composition only

C. temperature and pressure only

D. all temperature, pressure and composition.

135. Of the following which is the half period of a 1st order irreversible reaction $A \rightarrow B$

A. ln $k/2$

B. $k/2$

C. ln $2/k$

D. ln $0.5/k$

136. In the fluid catalytic cracker (FCC) the cracking and reaction are (A) __________ and (B)__________ respectfully

A. (A) exothermic (B) exothermic

B. (A) exothermic (B) endothermic

C. (A) endothermic (B) endothermic

D. (A) endothermic (B) exothermic

137. A __________ coloum is used when a high liquid hold up is required in reactor for gas liquid reaction

A. tray
B. packed
C. spray
D. bubble

138. What is the extent of a comical reaction?
A. dimensionless
B. different for reactants and products
C. depends on the stoichiometric co-efficient
D. all of the above

139. In a first order chemical reaction the catalyst changes the
A. equilibrium constant
B. activation energy
C. heat of reaction
D. heat of formation of the product

140. What the point selectivity of the product Y in the reaction

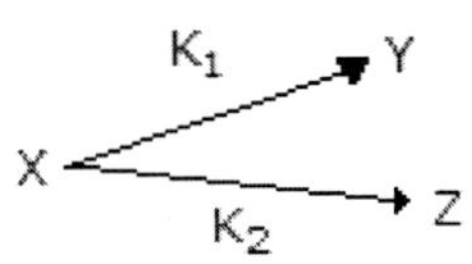

A. K_1/K_2
B. K_1-K_2
C. K_2/K_1
D. K_2-K_1

141. Which of the following expressions represents the rate of change of conversion with time for me first order gas phase reaction $A \overset{k_1}{\rightarrow} 2B$, conducted isothermally batch mode.
A. $\frac{dX_A}{dt} = k_1(1 - X_A)^2(1 + 2X_A$
B. $\frac{dX_A}{dt} = k_1(1 - X_A)(1 + 0.5X_A)$
C. $\frac{dX_A}{dt} = k_1(1 - X_A)$
D. $\frac{dX_A}{dt} = \frac{k_1(1-X_A)}{(1+X_A)}$

142. If the total pressure is increase, in which of the following reactions, the-equilibrium will shift to the right
A. $H_2 + I_2 \rightleftharpoons 2HI$
B. $N_2 + O_2 \rightleftharpoons 2NO$
C. $H_2 + Cl_2 \rightleftharpoons 2HCl$
D. $N_2O_4 \rightleftharpoons 2\,NO_2$

143. The function of a catalyst is to increases the rate of a chemical reaction what is its effect on rate constant
A. remains constant
B. increases
C. decreases
D. becomes infinite

144. Such reactions having low activation energy are ________
A. slow
B. non-spontaneous
C. fast
D. always spontaneous

145. for a zero order reaction, $A \overset{K}{\rightarrow}$ Products the half-life period 't' is equal to
A. $C_{A0}/2K$
B. $1/K$
C. C_{A0}/K
D. $0.693/K$

146. For on exothermic reversible reaction taking place in plug-flow reactor which of the following is the optimum operating condition.
A. temperature should be high in the beginning and decreased towards the end of the reaction
B. very high temperature should be used throughout the reaction.
C. very low temperature should be used throughout the reaction.
D. none of these.

147. In a steady state mixed reactor, the space time is equivalent to the holding time, for
A. variable fluid density systems.
B. gas reactions with changing no. of moles.
C. constant fluid density systems.
D. non-isothermal gas reaction

148. The variation of ________ with temperature is mathematically expressed by Arrhenius
A. energy of activation
B. rate constant
C. reaction rate
D. frequency factor

149. Which of the following quantities represent the dispersion number for prefect mixed flow is?
A. > 150
B. 0

C. ∞
D. <2100.

150. What is the order of the reaction

$H_2 + Cl_2 \xrightarrow{sunlight} 2HCL$?

A. 0
B. 2
C. 1
D. 3

151. A reaction which is known as shift conversion reaction.

A. is non-catalytic.
B. converts CO to CO_2 with steam.
C. is highly exothermic.
D. converts N_2 and H_2 into NH_3.

152. Which of the following statements is wrong?

A. Alumina is added as a promoter to iron catalyst in ammonia synthesis reaction.
B. Use of different catalysts in a reversible catalytic chemical reaction does not change the equilibrium composition.
C. Activation energy for a reaction is obtained from the intercept of the Arrhenius plot.
D. Presence of inerts affects the equilibrium conversion of reactants in a chemical reaction.

153. Which of the following statements represents a backmix reactor?

A. ideal at very low conversion.
B. same as plug flow reactor (PFR).
C. suitable for gas phase reactions.
D. same as ideal stirred tank reactor.

154. The rate of reaction will__________when an exothermic reversible reaction is conducted adiabatically,

A. continuously decreases.
B. passes through a minimum.
C. passes through a maximum.
D. continuously increases

155. Which of the following is not possible for the gas phase reaction $CO + 2H_2 \rightleftharpoons CH_3OH$, if its rate is given by $r = k_1 \, P_{CO}^{\alpha} \, . \, P_{H_2}^{\beta} - k_2 \, . \, P_{CH_3OH}^{\gamma}$

A. $\alpha = 1, \beta = 1, \gamma = 1$
B. $\alpha = 1/3, \beta = 2/3, \gamma = 1/3$
C. $\alpha = 1, \beta = 2, \gamma = 1$
D. $\alpha = 1/2, \beta = 1, \gamma = ½$

156. For a solid-gas non-catalytic reactionoccurring at very high temperature, the rare controlling step is.

A. Pore diffusion
B. Film diffusion
C. Chemical reaction
D. Ash layer diffusion

157. What will be the approximate value of the effectiveness factor for a first order chemical reaction in a porous catalyst?

A. 0.5
B. 1
C. 0.1
D. 0

158. For a given chemical reaction doubling the initial concentration of the reactants results in half time being double also what is the order the reaction?

A. 0
B. 2
C. 1
D. 3

159. In a catalytic reaction like importance of diffusion is increased when the catalyst has .

A. an active surface of the catalyst.
B. large catalyst particle size.
C. small pore diameter.
D. all (a), (b) and (c).

160. Which of the following statement is correct for a semi batch reactor?

A. velocity of reaction can be controlled.
B. residence time is constant.
C. both the reactants flow counter-currently.
D. maximum conversion can be controlled.

161. For chemical reaction $A \rightarrow B$, the reaction rate is proportional to (C_A^n). If on doubling the concentrate of A. The rate of reaction increases four times then what is the value of n?

A. 4
B. 16
C. 1/4
D. 2

162. Backmixing is most predominant in

A. a single CSTR
B. a well stirred batch reaction
C. plug flow reactor
D. CSTR connected in series

163. For an endothermic reaction maximum equilibrium conversion is obtained at the .

A. highest possible
B. intermediate
C. room
D. lowest possible

164. In which of the following reaction conditions for solid catalysed reactions. The diffusion effects are more likely to affect the overall rate of reaction?

A. slow reactions in catalyst of large pore diameter.
B. fast reactions in catalyst of small pore diameter.
C. slow reactions in catalyst of small pore diameter.
D. fast reaction in catalyst of large pore diameter.

165. The composition of the exit stream in continuous stirred tank reactor.

A. is same as that in the reactor.
B. depends upon the flow rate of inlet stream.
C. is different than that in the reactor.
D. none of these.

166. The heterogeneous catalytic reaction $A(g) + B(g) \rightarrow C(g)$ has a rate of reaction $-r_A = k.K_A.p_A.P_B/(1 + K_A.P_A + K_c.p_c)$, where K_A and K_c are the adsorption equilibrium constants. Which of the following is rate controlling step?

A. surface reaction between absorbed *A* and absorbed *B*.
B. absorption of *A*.
C. surface reaction between absorbed *A* and *B* in the gas phase.
D. surface reaction between *A* in the gas phase and absorbed *B*.

167. Which of the following statements is the experimentally determined overall under of the reaction $A + B \rightarrow C + D$, is true.

A. reaction is elementary with a molecularity of 2.
B. reaction is elementary but the molecularity may not be 2.
C. reaction may be elementary with molecularity of 2.
D. molecularity of the reaction is 2, but the reaction may not be elementary.

168. Which of the following statements correctly interprets the meaning of a space time of 3 hours for a flow reactor.

A. the time required to process one reactor volume of feed (measured at specified conditions) is 3 hours.
B. it takes three hours to dump the entire volume of the reactorwith feed.
C. conversion is cent per cent after three hours.
D. three reactor volumes of feed can be processed every hour.

169. Which of the following represents the ratio of moles of a reactant converted into the desired product to that converted into unwanted product?

A. relative yield
B. operational yield
C. selectivity
D. none of these

170. The flow pattern with the vessel in which multiple reaction are taking place affects the .

A. distribution of reaction products
B. size requirement
C. both (a) and (b)
D. neither (a) nor (b)

171. Which of the following statements is correct for a trickle bed reactor?

A. processes three reactants at different flow rates.
B. has altogether three streams either entering or leaving.
C. processes three reactants with same flow rate.
D. employs all the three phases (i.e... solid, liquid and gas).

172. If the half-life period of the reaction $A \xrightarrow{K}$ product is independent of the initial concentration of the reactant A what is the under of the reaction?

A. second
B. first
C. third
D. zero

173. The rate constant may be expressed as according to the transition state theory.

A. $K \alpha e^{-E/RT}$
B. $K \alpha T . e^{-E/RT}$
C. $K \alpha T . e^{-E/RT}$
D. $K \alpha e^{E/RT}$

174. The reaction rate for any reactant depends on
A. temperature of the system.
B. pressure of the system.
C. composition of the component only.
D. all (a), (b) and (c).

175. For a turbulent reactor operation under transient condition, the energy balance equation will be in the form of
A. an algebric differential equation.
B. a non-linear partial differential equation.
C. a linear partial differential equation.
D. an ordinary non-linear differential equation.

176. In which following catalyst properties the Heluim-mercury method used?
A. solid density
B. porosity
C. pore volume
D. all (a), (b), & (c).

177. A chemical reaction of the type $A \rightarrow 3R$ is being conducted in a mixed reactor where the conversions is 50% of a gaseous reactants A is introduced at a rate of 1 liter/second. If the output steam flow is 2 liter/second. What will be the holding for this operation is second?
A. 0.5
B. 3
C. 1
D. 2

178. An ideal continuous stirred tank reactor is used to carry out a first order liquid phase reactor. What will be the result of recycling back of outlet steam to the reactor on conversion?
A. increase
B. decrease
C. no change
D. either (a) or (b), depends on the type of reaction

179. Predicate the order of a reaction which fulfils the following condition (1) The time required for half change is inversely proportional to the square of initial concentration and (2) the velocity depends on the unit in which the concentration term is expressed.
A. 0
B. 1
C. 3
D. 2

180. Under the kinetics domain equilibrium of a chemical reaction is viewed as a __________ state.
A. dynamic steady
B. dynamic unsteady
C. static steady
D. none of these

181. Comparing the performance of a plug flow reactor with that of a mixed reactor, for all positive reaction orders the volume of the former will always be _______, the later.
A. equal to
B. greater than
C. smaller than
D. unpredictable from the data

182. The order of a reaction may be
A. zero
B. whole number
C. fraction
D. any one of the above depending upon the type of reaction

183. An adiabatic plug flow reactor is used to conduct an endothermic second order reaction. The rate of heat generation is .
A. maximum at the inlet of the reactor.
B. maximum at the centre of the reactor.
C. maximum at the exit of the reactor.
D. constant throughout the reactor.

184. Which of the following statements is wrong?
A. Exit age description function (E) and internal age distribution function (I) are related as, $\left(E = -\frac{dI}{dO}\right)$.
B. A higher temperature favours the reaction of higher activation energy.
C. Chmisorption studies are useful in the determination of catalyst surface area and pore size distribution.
D. A catalyst increases the potential energy barrier over which the reactants must pass to form products.

185. What is the value of Peclet number for an ideal plug flow reactor,
A. ∞
B. 0
C. 10
D. 1

186. The deactivation of a porous catalyst pellet is control by
  A. Form of surface attack by poison.
  B. Decay reactions.
  C. Pore diffusion.
  D. All (a), (b) and (c).

187. An adiabatic plug flow reactor is used to carry out an endothermic aqueous phase first order irreversible reaction. The rate of reaction
  A. is maximum at the inlet of the reactor.
  B. is maximum at the exit of the reactor.
  C. goes through a maximum along the length of the reactor.
  D. goes through a minimum along the length of the reactor.

188. It is known that the response curve for a step input signal from a reactor is called C-curve. For a series of stirred tank reactors of m tank variance of C-curve is equal to .
  A. $\sqrt{m}$
  B. $1/m$
  C. m
  D. m^2

189. A liquid in plug flow will have eddy diffusivity.
  A. ∞
  B. 0
  C. between 0 and 1
  D. 1

190. Threshold energy is the minimum energy required to allow a chemical reaction to proceed. Such chemical reaction with low activation energy are
  A. highly temperature sensitive.
  B. insensitive to temperature changes.
  C. always irreversible.
  D. mostly irreversible.

191. The following plot shows the variation of reaction of rate as a function of conversion for a liquid phase reaction carried out under isothermal conditions. For a given conversion of 0.9 what choice of reactor/ reactor combustion will requires the minimum?

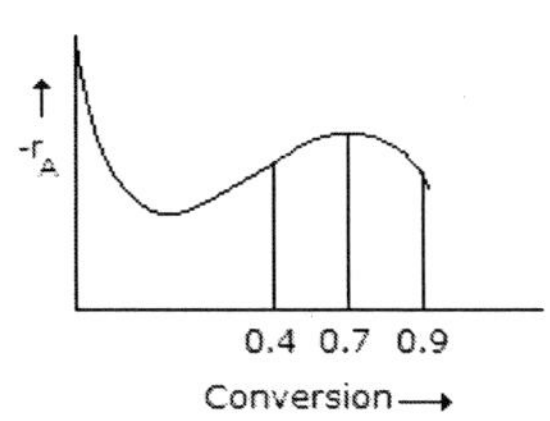

  A. CSTR followed by a PFR
  B. CSTR followed by a PFR followed by a CSTR.
  C. PFR followed by a CSTR
  D. PFR followed by a CSTR followed by a PFR.

192. A reversible reaction is being conducted in the pressure of a catalyst increases the rate of forward reaction .
  A. only.
  B. and decreases that of the backward reaction.
  C. to a greater extent than that of the backward reaction.
  D. and the backward reaction equally.

193. The phenomena of adsorption as postulated by Langmuir implies that
  A. all the adsorption does not take place by the same mechanism.
  B. whole surface of the catalyst is essentially uniform and the adsorbed molecule has no effect on the rate of adsorption per site.
  C. extent of adsorption is more than one complete monomolecular layer on the surface.
  D. whole surface of the catalyst does not have the same activity for adsorption and there is attraction between the adsorbed molecules.

194. The difference between organic catalyst and inorganic catalyst is that the former is .
  A. prohibitively costly.
  B. active at cryogenic temperatures only.
  C. proteinous in nature.
  D. active at very high temperature only.

195. Which of the following expression is valid for the reaction $A + B \rightarrow 2B + C$?
  A. $r_A = 2r_B$
  B. $r_A = -r_B$
  C. $r_A = r_B/2$
  D. $r_A = r_B$

196. Which of the following expressions represents the correct expression for the rate of reaction for a mixed flow reactor operating at steady state?
  A. $\frac{FA_0}{V} - \frac{dC_A}{dt}$
  B. $\frac{FA_0}{V} + \frac{dC_A}{dt}$
  C. $\frac{FA_0}{V} . X_A$
  D. $-\frac{dC_A}{dt}$

197. _________ energy is the excess energy of the reactants required to dissociate in to products.

A. threshold
B. activation
C. thermal
D. binding

198. The ratio of moles of a reactant converted in to the desired product to that converted in to unwanted product is called

A. operational yield
B. selectively
C. relative yield
D. none of these

199. If the pure diffusion resistance in a catalyst is negligible then the Thiele modulus is .

A. ∞
B. 0
C. < 0.5
D. > 0.5

200. In a catalytic chemical reaction thecatalytic action follows from the ability of catalyst to change the.

A. activation energy
B. heat of reaction
C. equilibrium constant
D. none of these

201. Which of the following statement is incorrect?

A. A very high value of equilibrium constant, K ($K >> 1$) indicates that the reaction is practically irreversible in nature.
B. Non-ideal flow takes place in reactors due to recycling channeling or by creation of stagnant regions.
C. The intercept of the Arrhenious plot is called the 'activation energy'.
D. A particular chemical reaction is more temperature sensitive art low temperatures.

202. The fractional conversion of a first order reaction in a given time _______ With increase in initial concentration

A. unpredictable
B. increases
C. remains constant
D. decreases

203. The dispersion number of a plug flow reactor is .

A. 0
B. –1
C. ∞
D. 1

204. When the elementary consecutive reaction $A \xrightarrow{K_1} R \xrightarrow{K_2} S$ is carried out in mixed flow reactor then fractional yield of 'R' _________ with increase in K_2/K_1.

A. decreases
B. remains same
C. increases
D. increases linearly

205. For a first order reaction half-life period is _________ the initial concentration of the reaction.

A. inversely proportional to
B. directly proportional to
C. independent of
D. none of these

206. To achieve maximum decomposition it is desirable to carry out the exothermic reaction A $2B$,which represents decomposition of A with B.

A. at high P and high T.
B. at low P and high T.
C. at low P and low T.
D. at high P and low T.

207. According to Arhenius law, what will be the expression for temperature dependence of reaction rate constant (K)?

A. $K \alpha e^{-E/RT}$
B. $K \alpha T.e^{-E/RT}$
C. $K \alpha e^{E/RT}$
D. $K \alpha T.e^{-E/RT}$

208. For a bimolecular reaction the rate constant is represented by according to Collision theory?

A. $K \alpha e^{-E/RT}$
B. $K \alpha e^{E/RT}$
C. $K \alpha T.e^{-E/RT}$
D. none of these

209. If the fractional conversion of reactant A for the chemical reaction $A + B \xrightarrow{K} C$ is proportional to time what will be the order of the reaction?

A. zero
B. third
C. first
D. second

210. Conversion increases with rise in temperature for a/an__________ Chemical reaction.

A. irreversible exothermic

B. reversible exothermic

C. irreversible endothermic

D. reversible endothermic

211. If the rate equation for the chemical reaction $2A + B \xrightarrow{K} C$ is given by $r_A = K.\ C_A^2.\ C_B$ the order of reaction is

A. 1

B. 0

C. 2

D. 3

212. In which of the following rector systems Backmixing is most predominant?

A. CSTR's connected in series.

B. a well stirred batch reactor.

C. a single CSTR.

D. a plug-flow reactor.

213. Which of the following is the correct statement for apparent activation energy E_a if the pore diffusion is the controlling step in a catalytic reaction?

A. the intrinsic activation energy E.

B. $(E + E_D)$ where E_D is activation due to diffusion.

C. $(E + E_D)/2$

D. $E_D/2$

214. According to the collision theory for bimolecular reactions the relation between rate constant K and absolute temperature T is

A. $k \propto T$

B. $k \propto T^{1.5}$

C. $k \propto T$

D. $k \propto \exp(-E/RT)$

215. In contact process the catalytic converter for conversion of SO_2 to SO_3 the feed should preferably have SO_2 content between

A. 12-15%

B. 7-10%

C. 20-25%

D. 2-5%

216. What is the name of the Biocatalyst which occur in the synthesis of proteins and metabolism in biological objects?

A. haemoglobin

B. a ferment or an enzyme

C. fungi

D. none of these

217. For the irreversible liquid phase reaction $A \rightarrow$ products, what will be the our all order of the reaction if the following hall life data in given

Initial concentration (K.mol/m^3)	Half-life
2	2
8	1

The overall order of reaction is

A. 1

B. 2

C. 1.5

D. 0.5

218. Which of the following mechanism is consistent with the rate equation and reaction $2A + B \rightarrow A_2B$, $(-r_A) = k.C_A.C_B$

A. $A + B \rightarrow AB; AB + A \rightarrow A_2B$

B. $A + A \rightarrow AA; AA + B \rightarrow A_2B$

C. $A + B \rightleftharpoons AB;\ AB + A \rightarrow A_2B$

D. $A + A \rightleftharpoons AA; AA + B \rightarrow A_2B$

219. Which of the following statements regarding velocity of a chemical reaction is correct?

A. decreases with increase in temperature.

B. increases with increase of pressure of reactants for all reactions.

C. decreases with increase of reactant concentration.

D. none of these.

220. For a first order irreversible liquid phase decomposition the half-life period is 8 minutes. What will be the time required for 75 % conversion of the liquid in minutes?

A. 12

B. 8

C. 4

D. 16

221. What will be the space time in seconds when gaseous reactant is introduced in a mixed reactor of 3 litres volume at the rate of 1 litre/second?

A. 1/3

B. 3

C. 32

D. 1

222. In a reactor containing SO_3, SO_2 and O_2 at equilibrium if helium is introduced the total pressure increase while volume and temperature remains constant. According to Le Chatlier principle the dissociation of SO_2

A. decreases

B. remains unaltered

C. increases

D. changes unpredictably.

223. The maximum and minimum permissible temperature exothermic reactions being conducted in plug flow reactor are T_{max} and T_{min} respectively. Which of the following temperature (T) profiles will require the shortest residence time to achieve the desired conversion?

A.

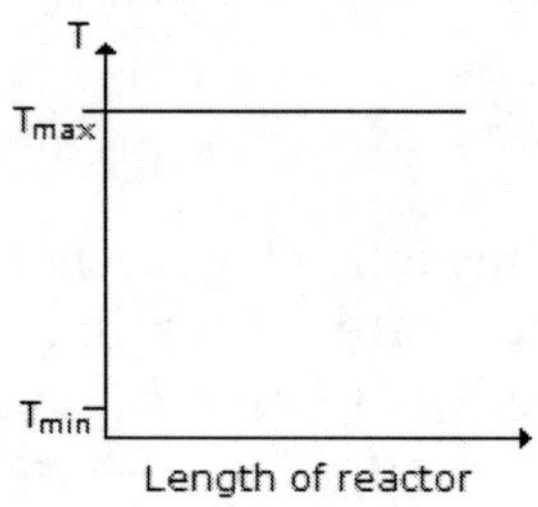

B.

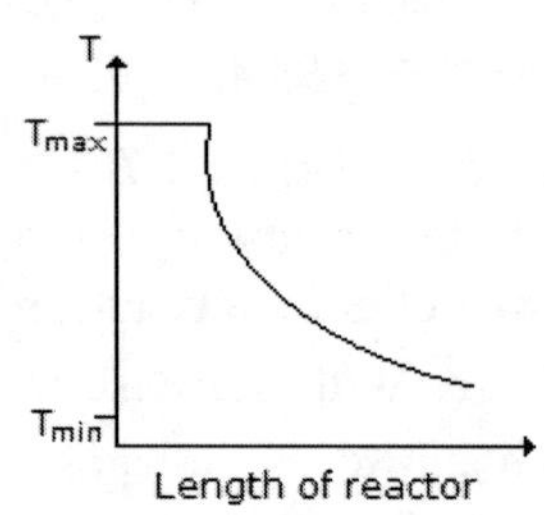

C.

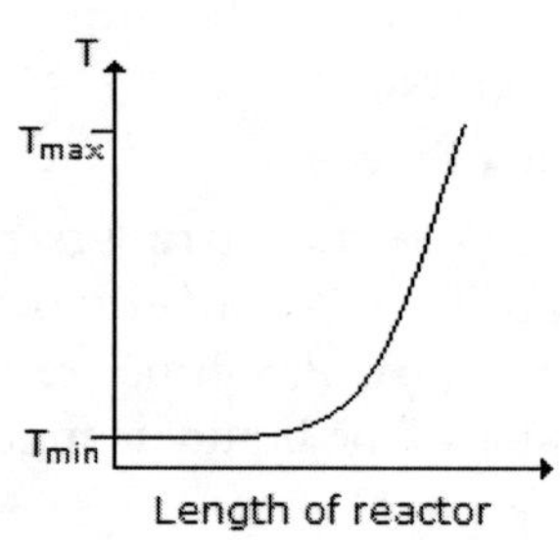

D.

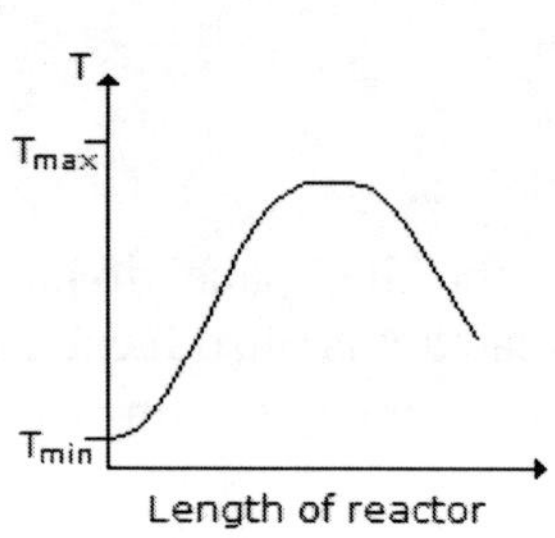

224. For an autocatalytic reactor, the suitable reactor setup is

A. P.F. reactor is series

B. CSTR is series

C. CSTR is series

D. P.F. reactor followed by CSTR

225. Which of the following statements typically characterised a fluidised bed reactor?

A. uniformity of temperature.

B. absence of continuous catalyst regeneration facility.

C. comparatively smaller equipment.

D. very small pressure drop.

226. If the reaction on $A \xrightarrow{+Z,K_1} B \xrightarrow{+Z,K_2} C \xrightarrow{+Z,K_3} D$ is a non-catalytic chemical reaction it may be called .

A. series-parallel

B. series

C. parallel

D. none of these

227. The rate of reaction of which of the type of the following reaction in not affected by temperature rise.

A. zero order.

B. photochemical

C. consecutive

D. autocatalytic

228. In which of the following gaseous phase reversible reaction the yield of product cannot be increased by the application of high pressure?

A. $N_2 + O_2 \rightleftharpoons 2NO$

B. $N + 3H_2 \rightleftharpoons 2NH_3$

C. $PCl_3 + Cl_2 \rightleftharpoons PCl_5$

D. $2SO_2 + O_2 \rightleftharpoons 2SO_3$

229. When benzene reach whit chlorine gas to produce tri-chlorobenzene it a/an example of ________ reaction.

A. parallel

B. elementary

C. consecutive

D. none of these

230. Which of the following conditions will apply when space time become equal to the mean residence time?

A. when the density of the reaction mixture is constant.
B. for CSTR.
C. for large diameter tubular reactor.
D. for narrow diameter tubular reactor.

231. Which of the following criteria is applicable to an ideal mixed reactor
A. exit stream has the same composition as the fluid within the reactor.
B. space time is equivalent to holding time for constant density systems.
C. composition throughout the reactor remains same.
D. all (a), (b) and (c).

232. In the gas oil cracking of carbon particles are accumulated on the catalyst, it falls under the category of poison.
A. deposited
B. stability
C. chemisorbed
D. selectivity

233. When the feed consists of 50 mole % A and 50 mole % inerts for a gaseous reaction $2A \rightarrow B$, the expansion factor will be
A. 0
B. 1
C. –0.25
D. -0.5

234. Identify the correct relationship for time required of complete a definite fraction of a reaction (t) assuming a is initial concentration of reactant and n is overall order of the reaction.
A. $t \propto \frac{1}{a^n}$
B. $t \propto \frac{1}{a^{n-1}}$
C. $t \propto \frac{1}{a^{n+1}}$
D. $t \propto a^n$

235. The controlling factor in the design of a fluid-solid reactor is.
A. Flow patterns of solids and fluid in the reactor.
B. Reaction kinetics for single particle.
C. Size distribution of solids being treated.
D. All (a), (b) and (c).

236. In gaseous phase chemical reactions with an increase in pressure, the fractional conversion __________ when the total number of moles decreases.
A. increases
B. unpredictable from the data
C. decreases
D. remains unaffected

237. A reversible reaction way behave line an irreversible reaction when
A. conversion of the reactant at equilibrium condition is zero.
B. equilibrium constant is 1.
C. reactant concentration at equilibrium condition is zero.
D. all (a), (b) and (c).

238. Which of the following reaction can be used for shady of kinetics of a catalytic reaction
A. integral (plug flow)
B. differential (flow)
C. mixed
D. either (a), (b) and (c).

239. Predict the order of the reaction $A + B + C \xrightarrow{K} R$, if itis represented that on doubling the concentration of *B* the reaction rate doubles also the reaction rate doubles when the concentration of both A and B are simultaneously doubled. Further when the concentrations of both *B* & *C* are doubled the reaction rate is quadrupled.
A. 3
B. 2
C. 4
D. 1

240. Which of the following statements represent a space velocity of 5 hr^3
A. five reactor volumes of feed (at specified conditions) are being fed into the reactor per hour.
B. a fixed conversion of a given batch of feed takes 5 hours.
C. cent per cent conversion can be achieved in at least 5 hours.
D. after every 5 hours, reactor is being filled with the feed.

241. __________ reactions have high activation energy.
A. very temperature sensitive
B. temperature insensitive
C. always irreversible
D. always reversible

242. In which of the following types of reactor of the composition in the rector and exit of the reactor is the same.

A. tubular

B. batch

C. semi-batch

D. back-mix

243. A__________ order reaction requires the knowledge of initial concentration and rate constant for determine the of its half life

A. first

B. zero

C. second

D. none of these

244. An aqueous stream having a pollutant degrading with first order rate with a volumetric flow rate 1 m^3/h at 2 kg mole/ m^3requires a mixed flow 0.5 kg mole/m^3. If the inlet concentration of the pollutant is now doubled and the volumetric flow rate is tripled and the pollutant level is to be brought down to the same level of 0.5 Kg.mole/ $m^3$by what factor the volume of the mixed flow reactor should be increased?

A. 7

B. 6

C. 6

D. 7/3

245. The relation between reaction rate constants at two different temperatureT_1 and T_2 are expressed as

A.$\ln\left(\frac{K_2}{K_1}\right) = \frac{E}{R}\left(\frac{1}{T_2} - \frac{1}{T_1}\right)$

B.$\ln\left(\frac{K_2}{K_1}\right) = \frac{E}{R}\left(\frac{1}{T_1} - \frac{1}{T_2}\right)$

C.$\exp\left(\frac{K_2}{K_1}\right) = \frac{E}{R}\left(\frac{1}{T_1} - \frac{1}{T_2}\right)$

D.$\exp\left(\frac{K_2}{K_1}\right) = \frac{E}{R}\left(\frac{1}{T_2} - \frac{1}{T_1}\right)$

246. The hydrolysis of methyl acetate in presence of acids has a molecularity the and order of reaction ________ & _________ respectively

A. 2 & 1

B. 1 & 1

C. 1& 2

D. 2:& 2

247. The ratio of the volume of mixed reactor to the volume of plug flow reactor for identical feed composition, flow rate and conversion, The increase in order f reaction for all positive reaction orders __________.

A. increases

B. remains same

C. increases linearly

D. decreases

248. The frequency factor in Arhenious equation will have the unit

A. same as that of rate constant.

B. dimensionless.

C. same as that of activation energy.

D. none of these.

249. The most suitable reactor/reactor system for carrying out exothermic reactions is

A. CSTR in series.

B. a plug flow reactor followed by CSTR.

C. a CSTR.

D. CSTR followed by a plug flow reactor.

250. Which of the following statement is valid for a catalyst

A. changes the equilibrium constant of the reaction.

B. increases the equilibrium concentration of the product.

C. shortens the time to reach the equilibrium.

D. none of these.

Answer

1. C
2. A
3. B
4. C
5. B
6. A
7. B
8. C
9. C
10. B
11. B
12. A
13. C
14. A
15. B
16. C
17. B
18. C
19. B
20. B
21. B
22. A
23. A
24. B
25. C
26. A
27. C
28. B
29. C
30. A
31. B
32. D
33. D
34. D
35. A
36. B
37. D
38. D
39. A
40. A
41. B
42. C
43. C
44. A
45. C
46. C
47. A
48. C
49. A
50. B
51. A
52. B
53. B
54. B
55. C
56. D
57. A
58. A
59. C
60. A
61. B
62. C
63. B
64. D
65. C
66. C
67. D
68. B
69. B
70. C
71. A
72. A
73. C
74. C
75. C
76. D
77. C
78. A
79. C
80. B
81. C
82. C
83. D
84. D
85. B
86. C
87. B
88. A
89. C
90. D
91. A
92. C
93. D
94. A
95. B
96. B
97. A
98. B
99. D
100. C
101. A
102. B
103. B
104. B
105. B
106. B
107. B
108. B
109. A
110. A
111. A
112. A
113. D
114. D
115. A
116. B
117. B
118. D
119. A
120. B
121. A
122. C
123. D
124. A
125. B
126. C
127. D
128. A
129. D
130. C
131. C
132. A
133. B
134. D
135. C
136. D
137. D
138. D
139. B
140. A
141. C
142. B
143. B

144. C
145. A
146. A
147. C
148. B
149. C
150. A
151. B
152. C
153. D
154. C
155. A
156. B
157. C
158. A
159. D
160. A
161. A
162. A
163. A
164. C
165. A
166. C
167. A
168. A
169. C
170. C
171. D
172. B
173. C
174. D
175. C
176. D
177. A
178. C
179. C
180. A
181. C
182. D
183. A
184. D
185. A
186. D
187. A
188. B
189. B
190. B
191. D
192. D
193. B
194. C
195. B
196. C
197. B
198. B
199. C
200. A
201. C
202. C
203. A
204. A
205. C
206. C
207. A
208. C
209. A
210. D
211. D
212. C
213. D
214. C
215. B
216. C
217. C
218. D
219. D
220. D
221. B
222. A
223. B
224. D
225. A
226. C
227. B
228. A
229. C
230. A
231. D
232. A
233. C
234. B
235. D
236. A
237. C
238. D
239. B
240. A
241. A
242. D
243. C
244. A
245. B
246. A
247. A
248. A
249. D
250. C

GATE 2015-2000

(1Mark Questions)

GATE 2015

1. For which reaction order, the half-life of the reactant is half of the full-life time (time for 100% conversion) of the reactant?

   A. zero order
   B. half order
   C. first order
   D. second order

***Sol.* A.**

For zero order reaction,

$$-\frac{dC_A}{dt} = KC_{A0} = K$$

Where, K = rate constant

$$-C_{A0} - C_A = Kt$$

For full-life time, $C_A = 0$

$$t = \frac{C_{A0}}{K}$$

and for half-life time,

$$C_A = \frac{C_{A0}}{2} \Rightarrow t_{1/2} = \frac{C_{A0}}{2K}$$

So, t = $2t_{1/2}$

2. An irreversible homogeneous reaction A → products, has the rate expression, rate $= \frac{2C_A^2 + 0.1C_A}{1+50C_A}$, where C_A is the concentration of A. C_A varies in the range 0.5-50 mol/m^3. For very high concentration of A, the reaction order tends to

   A. 0
   B. 1
   C. 1.5
   D. 2

***Sol.* B.**

According to the question,

$$\text{Rate} = \frac{2C_A^2 + 0.1C_A}{1+50C_A}$$

$0.5 < C_A < 50$ (mol/m^3)

For very high value of C_A (say 50 mol/m^3),

$$0.1C_A << 2C_A^2$$

and $50C_A >> 1$

So, Rate $= \frac{2C_A^2}{50C_A} = \frac{1}{25}C_A$

So, reaction order is 1.

3. Which of the following ca change if only the catalyst is changed for a reaction system?

   A. enthalpy of reaction
   B. activation energy
   C. free energy of the reaction
   D. equilibrium constant

***Sol.* B.**

GATE 2014

4. In order to achieve the same conversion under identical reaction conditions and feed flow rate for a non-autocatalytic reaction of positive order, the volume of an ideal CSTR is

   A. always greater than that of an ideal PFR
   B. always smaller than that of an ideal PFR
   C. same as that of an ideal PFR
   D. smaller than that of an ideal PFR only for first order reaction

***Sol.* A.**

For CSTR (Continuous Stirred Tank Reactor)

$$\frac{V}{F_{A0}} = \frac{\Delta X_A}{-r_A}$$

For PFR (Plus Flow Reactor), $\frac{V}{F_{A0}} = \int_0^{X_A} \frac{\Delta X_A}{-r_A}$

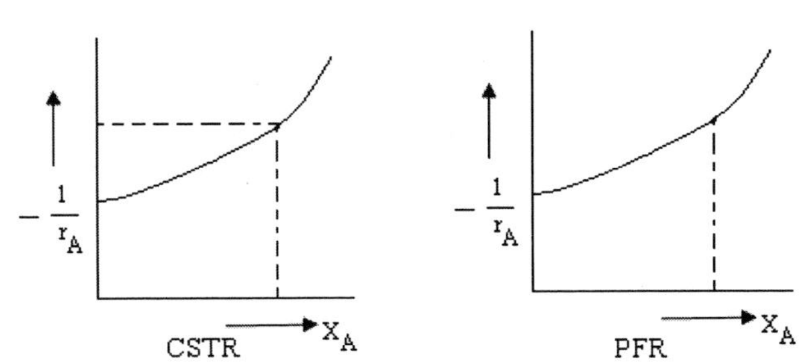

If we plot a curve between $\frac{1}{-r_A}$ and the fractional conversion (X_A) for two cases, the shaded area as shown in figures gives the value of $\frac{V}{F_{A0}}$

It is clear that the shaded area is greater for CSTR.

$\therefore V_{CSTR} > V_{PFR}$, for same F_{A_0}.

5. The vessel dispersion number for an ideal CSTR is

A. -1

B. 0

C. 1

D. ∞

***Sol.* D.**

For ideal CSTR, vessel dispersion number ∞

GATE 2013

6. An isothermal liquid phase zero order reaction A $\rightarrow$ B (k = 0.5 mol/m^3-s) is carried out in a batch reactor. The initial concentration of A is 2 mol/m^3. At 3 seconds from the start of the reaction, the concentration of A in mol/m^3 is

A. 0.8 mol/m^3

B. 0. Mol/m^3

C. 0.5 mol/m^3

D. 0.1 mol/m^3

***Sol.* C.**

7. The overall rates of an isothermal catalytic reaction using spherical catalyst particles of diameters 1 mm and 2 mm arer_{A1} and r_{A2} (in mol$(kg - catalyst)^{-1}h^{-1}$), respectively. The other physical properties of the catalyst particles are identical. If pore diffusion resistance is very high, the ratio r_{A2}/r_{A1} is

A. 0.5

B. 0.4

C. 0.2

D. 0.6

***Sol.* A.**

For batch reactor, zero order reaction,

$$-\frac{dC_A}{dt} = k$$

or $\int_{C_{A0}}^{C_A} dC_A = \int_0^t dt$

$$C_A = C_{A0} = -\text{kt}$$

$$C_A = C_{A0} - kt$$

$$= 2 - 0.5 \times 3 = 0.5 \text{ mol/m}^3$$

8. In the manufacture of sulphuric acid by the contact process, the catalyst oxidation of SO_2 is carried out in multiple stages mainly to

A. increase the reaction rate by providing inter-stage heating

B. increase the overall conversion by providing inter-stage heating

C. increase the overall conversion by providing inter-stag cooling

D. decrease the overall conversion by removing sulphur trioxide between stages

***Sol.* C.**

In contact process, the catalytic oxidation of is carried out to increase the overall conversion by providing interstate cooling.

GATE 2012

9. For an exothermic reversible reaction, which one of the following correctly describes the dependence of the equilibrium constant (*K*) with temperature (*T*) and pressure (*P*)?

A. *K* is independent of *T* and *P*

B. *K* increases with an increase in *T* and *P*

C. *K* increase with *T* and decreases with *P*

D. *K* decreases with an increase *T*and is independent of *P*

***Sol.* C.**

For change in temperature and pressure, the equilibrium shift to minimize the effect as per LeChatlier principle, but equilibrium constant is independent of *T* and ρ.

10. The half-life of an *n* th order reaction in a batch reactor depends

A. only on the rate constant

B. on the rate constant and the order of the reaction

C. on the rate constant and initial reaction concentration

D. on the rate constant, initial reactant concentration and the order of the reaction

***Sol.* D.**

For *n* th order reaction,

$$-\frac{dC}{dT} = kC^n$$

or $\int_0^C \frac{dC}{C^n} = \int_0^t -k\,dt$

$$\text{or} \quad \frac{1}{1-n}\left[\frac{1}{C^{n-1}} - \frac{1}{C_0^{n-1}}\right] = -kt$$

For half-life, $t = t_{1/2}$, $C = \frac{C_0}{2}$

$$\frac{1}{C_0^{n-1}} \cdot \frac{1}{(n-1)}(2^{n-1} - 1) = kt_{1/2}$$

It can be seen that the half-life depends on rate constant (K), initial concentration (C_0) and order of the reaction (n).

11. Consider the reaction scheme shown below:

$$A \xrightarrow{k_1} B \xrightarrow{k_2} C$$

Both the reactions are of first-order. The activation energies for k_1 and k_2 are 80 and 20 kJ/mol respectively. To maximize the yield of *B*, it is preferable to use.

A. CSTR and high temperature
B. PFR and high temperature
C. CSTR and low temperature
D. PFR and low temperature

Sol. **B.**

$$A \xrightarrow{k_1} B \xrightarrow{k_2} C$$

Since, $k = Ae^{-E/RT}$

The activation energy for k_1 (80 kJ/mol) is higher than that for k_2 (20 kJ/mol). Thus, higher temperature is preferable to maximize the yield of *B*. Also, Plug Flow Reactor (PFR) should be used for better conversion rate.

GATE 2011

12. Reactant *R* forms three products *X, Y* and *Z* irreversibly, as shown below.

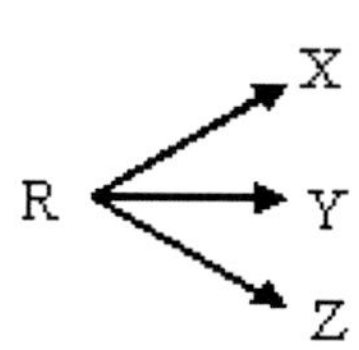

The reaction rates are given are by $r_X = k_X C_R$, $r_Y = k_Y C_R^{1.5}$ and $r_Z = k_Z C_R$. The activation energies for formation of *X,Y* and *Z* are 40,40 and 5 kJ/mol respectively. The pre-exponential factors for all reaction are nearly same. The desired conditions for maximizing the yield of X are

A. high temperature, high concentration of *R*
B. high temperature, low concentration of *R*
C. low temperature, high concentration of *R*
D. low temperature, low concentration of *R*

Sol. **B.**

By Arrhenius equation

$K = e^{-\frac{E}{RT}}$

For x and y E is 40 kJ/mole

Whereas for z E = 5 kJ/mole

$K_x = e^{-\frac{40}{RT}} = K_y$

And $K_z = e^{-\frac{5}{RT}}$

Hence at high temperatures $K_x > K_z$

Further $r_x = K_x C_R, r_y = K_y C_R^{1.5}$

And $r_z = K_z C_R$

Thus, at high concentration more of y will be formed than *x* Hence, for maximizing the yield of *x* reaction should be carried out at high temperature and low concentration.

13. Consider an irreversible, solid catalysed, liquid phase first order reaction. The diffusion and the reaction resistances are comparable. The overall rate constant (k_c) is related to the overall mass transfer coefficient (k_m) and the reaction rate constant (k) as

A. $k_c = \frac{kk_m}{k+k_m}$
B. $k_c = \frac{k+k_m}{kk_m}$
C. $k_c = \frac{k+k_m}{2}$
D. $k_c = k + k_m$

Sol. **A.**

For the given reaction the overall rate constant can be expressed as

$$\frac{1}{K_C} = \frac{1}{K_m} + \frac{1}{k}$$

$$\text{or} \quad \frac{1}{K_C} = \frac{k\,k_m}{k\,K_m}$$

GATE 2010

14. For a first order isothermal catalytic reaction $A \rightarrow P$, occurring in an infinitely long cylindrical pore, the relationship between effectiveness factor ε and Thiele modulus ϕ is

A. $\varepsilon = \frac{1}{\phi^2}$
B. $\varepsilon = \phi$
C. $\varepsilon = 1$
D. $\varepsilon = \frac{1}{\phi}$

Sol. C.

For the first order catalytic reaction A → P occurring is an
infinitely long cylindrical pore
Thiele modulus, $\emptyset \propto 1$

And effectiveness factor is nearly one$\epsilon = 1$

GATE 2009

15. For a homogenous reaction system, where

C_i is the concentration of j at a time t

N_i is the number of moles of j at time t

V is the reaction volume at time t

t is the reaction time.

The rate of reaction for species j is defined as

A. $\frac{dC_i}{dt}0$

B. $-\left(\frac{dC_i}{dt}\right)$

C. $\frac{1}{V}\frac{dN_i}{dt}$

D. $-\left(\frac{1}{V}\frac{dN_i}{dt}\right)$

Sol. C.

By definition rate of reaction $=\frac{1}{V}\cdot\frac{dN_j}{dt}$

16. The half-life of a first order liquid phase reaction is 30 s. Then the rate constant in min^{-1}, is

A. 0.0231

B. 0.602

C. 1.386

D. 2.0

Sol. C.

Half-life for a first order reaction

$$t_{1/2} = \frac{0.693}{k}$$

$$K = \frac{0.693}{t_{1/2}}$$

Given

$t_{1/2} = 0.5$ min

$\therefore$ k $= 1.386 min^{-1}$

17. For a solid-catalyzed reaction, the Thiele modulus is proportional to

A. $\sqrt{\frac{intrinsic\ reaction\ rate}{Diffusion\ rate}}$

B. $\sqrt{\frac{Diffusion\ rate}{intrinsic\ reaction\ rate}}$

C. $\frac{intrinsic\ reaction\ rate}{Diffusion\ rate}$

D. $\frac{Diffusion\ rate}{intrinsic\ reaction\ rate}$

Sol. A.

For solid catalyzed reaction

Thiele modulus $\propto \sqrt{\frac{Intrinsic\ reaction\ rate}{Diffussion\ rate}}$

GATE 2008

18. A species (A) reacts on a solid catalyst to produce *R* and *S* as follows

1. A → R $r_R = k_1C_A^2$
2. A → S $r_S = k_2C_A^2$

Assume film resistance to mass transfer is negligible. The ratio of instantaneous fractional yield of *R* in the presence of pore diffusion to that in the absence of pore diffusion is

A. 1

B. > 1

C. < 1

D. Zero

Sol. C.

Since for the reaction A → R

$r_R = k_1C_A^2$

In the case of pore diffusion

$$\frac{(C_A)_{pore\ diffusion}}{(C_A)_{without\ pore\ diffusion}} < 1$$

GATE 2007

19. The reaction $2A + B \rightarrow 2C$ occurs on a catalyst surface. The reactants *A* and *B* diffuse to the catalyst surface and get converted completely to the product*C*, which diffuses back. The steady state molar fluxes of *A,B* and *C* are related by

A. $N_A = 2N_B = N_C$

B. $N_A = -\left(\frac{1}{2}\right)N_B = -N_C$

C. $N_A = 2N_B = -N_C$

D. $N_A = (1/2)N_B = N_C$

Sol. C.

For the reaction 2A + B→ C

Under steady state the relationship between molar fluxes will be

$N_A = 2N_B = -N_C$

20. An irreversible gas phase reaction $A \rightarrow 5$B is conducted in an isothermal batch reactor at constant pressure in the presence of an inert. The feed contains no B. If the volume of the gas at complete conversion must not exceed three

times the initial volume, the minimum mol% of the inert in the feed must be

A. Zero

B. 20

C. 33

D. 50

***Sol.* C.**

The reaction is A → 5B

Let the initial number of moles of inert in the feed = N_i

After complete conversion $N_B = 5N_A$

Number of moles of A initially = N_A

∴ total number of moles = $(5N_A + N_i)$

However $\frac{(V)_{x=1}}{(V)_{x=0}} = 3$

or $\frac{5N_A + N_i}{N_A + N_i} = 3$

or $N_A = N_i$

Hence on equimolar basis minimum mole percent of inert will be 50%

21. A first order reversible reaction $A \underset{k_2}{\overset{k_1}{\rightleftharpoons}} B$ occurs in a batch reactor. The exponential decay of the concentration of A has the time constant

A. $\frac{1}{k_1}$

B. $\frac{1}{k_2}$

C. $\frac{1}{k_1-k_2}$

D. $\frac{1}{k_1+k_2}$

***Sol.* D.**

For the reaction $A \underset{k_2}{\overset{k_1}{\rightleftharpoons}} B$

Expression for the rate of reaction is

$$-\frac{dC_A}{dt} = \frac{dC_B}{dt} = k_1 C_A - k_2 C_B$$

Since

$$C_A = C_{A_0}(1 - x_A)$$

$$-\frac{dC_A}{dt} = \frac{dC_B}{dt} = -C_{A_0}\frac{dx_A}{dt}$$

Hence C_{A_0} rate can be expressed as

$$C_{A_0}\frac{dx_A}{dt} = k_1 C_{A_0}(1 - x_A) - k_2\left(C_{B_0} + C_{A_0}x_A\right)$$

$$= k_1 C_{A_0} - k_2 C_{B_0} - (k_1 + k_2)C_{A_0}x_A$$

At equilibrium $\frac{dX_{Ae}}{dt} = 0$ .... (i)

∴ $k_1 C_{A_0} - k_2 C_{B_0} = (k_1 + k_2)C_{A_0}x_{Ae}$ ....(ii)

Combining equation (i) and (ii)

$$C_{A_0}\frac{dx_A}{dt} = (k_1 + k_2)C_{A_0}(x_{Ae} - x_A)$$

Integrating $\int_0^{x_A}\frac{dx_A}{x_{Ae} - x_A} = \int_0^t (k_1 + k_2)dt$

or $[-In(x_{Ae} - x_A)]_0^{x_A} = (k_1 + k_2)t$

or $In\left(\frac{x_{Ae} - x_A}{x_{Ae}}\right) = -(k_1 + k_2)t$

or $\frac{x_{Ae} - x_A}{x_{Ae}} = -e^{(k_1+k_2)t} = e^{-t/\tau}$

Where τ is time constant

∴ Time constant for exponential decay

$$\tau = \frac{1}{k_1 + k_2}$$

GATE 2005

22. For the reaction $2R + S \rightarrow T$, the rates of formation r_R, r_S and r_T of the substance *R,S* and *T* respectively are related by

A. $2r_R = r_s = r_T$

B. $2r_R = r_s = -r_T$

C. $r_R = 2r_s = 2r_T$

D. $r_R = 2r_s = -2r_T$

***Sol.* D.**

The reaction is 2R + S → T

From the material balance approach the rate of decomposition of r is twice of that of S whereas the rate of formation of T is on equimolal basis to that of S

Hence $r_R = 2r_S = -2r_T$

23. For the liquid phase reaction $A \rightarrow P$, in a series of experiments in a batch reactor, the half-life ($t_{1/2}$) was found to be inversely proportional to the square root of initial concentration of A. The order of this reaction is

A. $\frac{3}{2}$

B. 1

C. $+\frac{1}{2}$

D. $-\frac{1}{2}$

***Sol.* A.**

For the reaction A → P

Let the order of reaction be n

Hence, $-\frac{dC_A}{dt} = kC_A^n$

or $-\int_{C_{A_0}}^{C_A}\frac{dC_A}{dt} = \int_0^t k\,dt$

or integration

$C_A^{1-n} - C_{A_0}^{1-n} = k(n-1)t$

At half life, $C_A = C_{A_o}/2$

$\therefore \quad t_{1/2} = \frac{2^{n-1}-1}{k(n-1)} C_{A_o}^{1-n}$ ....(i)

Given that $t_{1/2} \propto \frac{1}{\sqrt{C_{A_o}}}$ ....(ii)

Comparing (i) and (ii)

$C_{A_o}^{1-n} = C_{A_o}^{-1/2}$

or $1 - n = -\frac{1}{2}$

$\therefore \quad n = 1 + \frac{1}{2}$

$\therefore \quad n = \frac{3}{2}$

24. Which is the correct statement from the following statements on the Arrhenius model of the rate constant $k = Ae^{-E/RT}$?

A. A is always dimensional

B. For two reactions 1 and 2 , if $A_1 = A_2$ and $E_1 > E_2$ then $k_1(T) > k_2(T)$

C. Fora given reaction, the percentage change of **k** with respect to temperature is higher at lower temperature

D. the percentage change of **k** with respect to temperature is higher for higher A

Sol. **C.**

For $k = Ae^{-E/RT}$

It can be seen that since $Ae^{-E/RT}$ is dimensionless, hence A cannot be dimensionless. Since k is not dimensionless. For second option if for two separate reactions if $A_1 = A_2$

Then $\frac{k_1}{k_2} = e^{-\frac{1}{RT}(E_1 - E_2)}$

Since $E_1 > E_2$

Hence $k_1 < k_2$

Does not satisfy the condition for option (b) for option (c)

since $\ln k = \ln A - \frac{E}{RT}$

and $\frac{d \ln k}{dT} = \frac{E}{RT^2}$

For lower temperature

$\frac{d \ln k}{dT}$ will be higher, hence this is the correct option

Examining option (d) we can see that A does not depend on temperature. As such the percentage change of k with respect to temperature is not related to A.

GATE 2004

25. The rate expression for the gaseous phase reaction

$CO + 2H_2 \rightarrow CH_3OH$ is given by

$r = k_1 p_{CO}^{\alpha} p_{H_2}^{\beta} - k_2 p_{CH_3OH}^{\gamma}$

Which of the following is not possible?

A. $\alpha = 1, \beta = 1, \gamma = 1$

B. $\alpha = 1, \beta = 2, \gamma = 1$

C. $\alpha = \frac{1}{3}, \beta = \frac{2}{3}, \gamma = \frac{1}{3}$

D. $\alpha = 1/2, \beta = 1, \gamma = 1/2$

Sol. **A.**

For gas phase reaction

$$CO + 2H_2 \rightleftharpoons CH_3OH$$

The rate of reaction

$r = k_1 p_{CO}^{\alpha} p_{H_2}^{\beta} - k_2 p_{CH_3OH}^{\gamma}$

Usually the indices α, β and γ should be in proportion to their stoichiometric ratios for simple reactions. All the options except (a) i.e. 1:1:1 are in the ratio of 1:2:1. Hence (a) is not possible.

26. The rate of ammonia synthesis for the reaction

$N_2 + 3H_2 \rightleftharpoons 2NH_3$ is given by

$r = 0.8\, p_{N_2} p_{H_2}^3 - 0.6\, p_{NH_3}^2$. If the reaction is represented as, $\frac{1}{2}N_2 + \frac{3}{2}H_2 \rightleftharpoons \frac{1}{2}NH_3$ the rate of ammonia synthesis is

A. $r = 0.8\, p_{N_2}^{0.5} p_{H_2}^{1.5} - 0.6\, p_{NH_3}$

B. $r = 0.8\, p_{N_2} p_{H_2}^3 - 0.6\, p_{NH_3}^2$

C. $r = 0.5(0.8\, p_{N_2} p_{H_2}^3 - 0.6\, p_{NH_3}^2)$

D. $r = 0.5\ (0.8\, p_{N_2}{}_{N_2}^{0.5} p_{H_2}^{1.5} - 0.6 NH_3)$

Sol. **B.**

For ammonia synthesis reaction

$N_2 + 3H_2 \rightleftharpoons 2NH_3$ the rate is given as

$r = 0.8\, p_{N_2} p_{H_2}^3 - 0.6 p_{NH_3}^2$

It is know that the rate of reaction does not depend on the way a reaction is represented. Even, for a different soichiometric representation the rate of reaction will be the same i.e. (b).

27. An endothermic aqueous first order irreversible reaction is carried out in an adiabatic plug flow reactor. The rate of reaction
   a. is maximum at the inlet of the reactor
   b. goes through a maximum along the length of the reactor
   c. goes through a minimum along the length of the reactor
   d. is maximum at the exit of the reactor

Sol. **A.**

For adiabatic PFR carrying out an endothermic aqueous phase reaction the rate of reaction will be maximum at the inlet because the reaction is being under adiabatic and endothermic conditions the temperature will go on decreasing, thus decreasing the reaction rate. Further, in a PFR the reactant concentration is maximum at the inlet and goes on decreasing progressively towards exit thus giving maximum reaction rate at the inlet.

28. A first order gaseous phase reaction is catalyzed by a non-porous solid. The kinetic rate constant and the external mass transfer coefficient are k and k_g, respectively. The effective rate constant (k_{eff}) is given by

A. $k_{eff} = k + k_g$

B. $k_{eff} = \frac{(k+k_g)}{2}$

C. $k_{eff} = (kk_g)^{\frac{1}{2}}$

D. $\frac{1}{k_{eff}} = \frac{1}{k} + \frac{1}{k_g}$

Sol. **D.**

For solid catalyzed gas phase reactions the effective rate constant can be expressed as

$$k_{eff} = \frac{1}{\left(\frac{1}{k}+\frac{1}{k_g}\right)}$$

$$\therefore \ k_{eff} = \frac{1}{k} + \frac{1}{k_g}$$

29. For a packed bed reactor, the presence of a long tail in the residence time distribution curve is an indication of

A. ideal plug flow

B. bypass

C. dead zone

D. channeling

Sol. **C.**

For a packed bed reaction the RTD curve in as follows

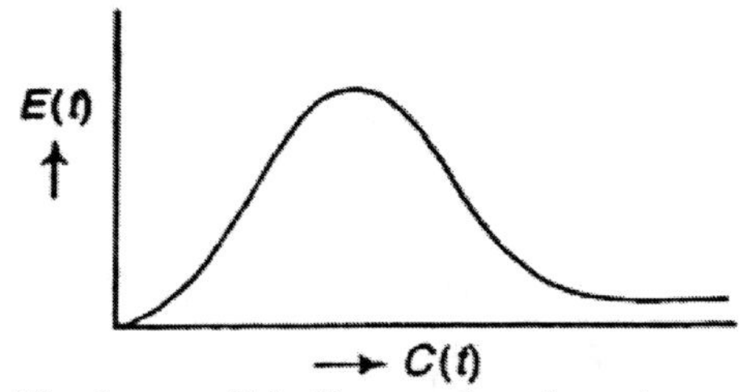

The long tail indicates that there is no reaction in that zone and it is a dead zone.

GATE 2003

30. For a series reaction

$$A \xrightarrow{k_1} B \xrightarrow{k_2} C \text{ having } k_1 << k_2,$$

The reaction system can be approximated as

A. $A \xrightarrow{k_1} B$

B. $A \xrightarrow{k_2} B$

C. $A \xrightarrow{k_2} C$

D. $A \xrightarrow{k_1} C$

Sol. **D.**

For the reaction system

$$A \xrightarrow{k_1} B \xrightarrow{k_2} C$$

Since $k_1 << k_2$ thus, A →B becomes the slowest and the rate controlling step. If also means that as soon as B is formed it will be transformed into C.

Hence $A \xrightarrow{k_1} C$ is correct.

31. An elementary liquid phase decomposition reaction $A \xrightarrow{k} 2B$ is to be carried out in a CSTR. The design equation is

A. $k\tau = \frac{X_A}{(1-X_A)}$

B. $k\tau = \frac{X_A(1+X_A)}{(1-X_A)}$

C. $k\tau = \frac{X_A}{(1-X_A)^2}$

D. $k\tau\, C_{AB} = \frac{X_A/(1+X_A)^2}{(1-X_A)^2}$

Sol. **A.**

A CSTR always operates under steady state conditions and

Input – Output – Disappearance due to reaction = 0

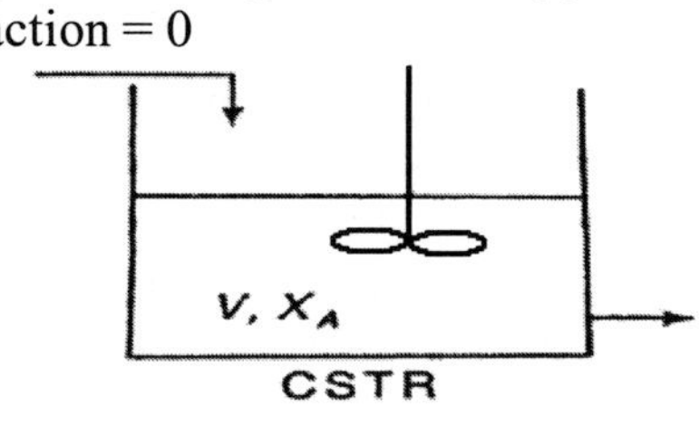

$FC_{A0} - FC_A = rV = (KC_A)\,V$

or $F\left(FC_{A0} - FC_A\right) = (K\,V)C_A$

or $\left(C_{A0} - C_A\right) = KC_A\frac{V}{F} = KC_A\tau$

or $K\tau = \frac{C_{A0} - C_A}{C_A}$

$= \frac{C_{A0} - C_A(1-x)}{C_A(1-x)}$

or $K\tau = \frac{x}{(1-x)}$

32. Find a mechanism that is consistent with the rate equation and reaction given below
$2A = B \rightarrow A_2B\,,\ (-r_A) = kC_AC_B$

A. $A + B \rightleftharpoons AB;\ AB + A \rightarrow A_2B$

B. $A + B \rightarrow AB;\ AB + A \rightarrow A_2B$

C. $A + A \rightarrow AA;\ AA + B \rightarrow A_2B$

D. $A + A \rightleftharpoons AA;\ AA + B \rightarrow A_2B$

Sol. **C.**

Since the reaction is first order with respect to each reactant A and B option (C) indicating

$A + A \longrightarrow AA;\ AA + B \longrightarrow A_2B$

Is the correct option

GATE 2002

33. For an ideal plug flow reactor the value of the Peclet number is

A. Zero

B. infinity

C. 1

D. 10

Sol. **A.**

For ideal PFR Peclet number = 0

34. The extent of a reaction is

A. different for reactants and products

B. dimensionless

C. dependent on the stoichiometric coefficients

D. all of the above

Sol. **D.**

Analyzing the different options, for (a) the extent of reaction for reactants will be just inverse of the products and they will be different. For (b), the extent of reaction will be a fractional number and will be dimensionless. For (c) the extent of reaction will obviously depend on the stoichiometric relationship as it will decide that in which proportion the reactants will be consumed or the products will be formed. Hence all the options are correct. Answer is (d).

35. An exothermic reaction takes place in an adiabatic reactor. The product temperature.......the reactor feed temperature.

A. is always equal to

B. is always greater then

C. is always less then

D. may be greater or less then

Sol. **B.**

For exothermic reaction being conducted in an adiabatic reactor with increase in conversion more and more heat will be evolved as heat of reaction. Since the reactor is adiabatic this heat will be absorbed by the reaction mixture and thus the product temperature will always be greater than feed temperature.

GATE 2001

36. The conversion for a second order, irreversible reaction (constant volume) $A \xrightarrow{k_2} B$, in batch mode is given by

A. $\frac{1}{1+k_2C_{A0}\,t}$

B. $\frac{k_2C_{A0}t}{1+k_2C_{A0}t}$

C. $\frac{(k_2C_{A0}\,t)^2}{1+k_2C_{A0}t}$

D. $\frac{k_2C_{A0}t}{(1+k_2C_{A0}t)^2}$

Sol. **B.**

For the reaction $A \xrightarrow{k_2} B$ the rate of reaction can be written as

$-\frac{dC_A}{dt} = K_2C_A^2$

or $-\int_{C_{Ao}}^{C_A}\frac{dC_A}{C_A^2} = \int_0^t k_2dt$

or $\left[\frac{1}{C_A} - \frac{1}{C_{Ao}}\right] = k_2^t$

Expressing in terms of conversion x

Since $C_A = C_{A0}\,(1 - x)$

$\frac{1}{C_{Ao}\,(1-x)} - \frac{1}{C_{Ao}} = k_2^t$

on simplification

$x_A = (1 - x_A)(k_2C_{A0}\,t)$

or $x_A = \frac{k_2C_{A0}\,t}{1+k_2C_{A0}\,t}$

37. The reaction rate constant at two different temperatures T_1 and T_2 are related by

A. $\ln\left(\frac{k_2}{k_1}\right) = \frac{E}{R}\left(\frac{1}{T_2} - \frac{1}{T_1}\right)$

B. $\ln\left(\frac{k_2}{k_1}\right) = \frac{E}{R}\left(\frac{1}{T_1} - \frac{1}{T_2}\right)$

C. $\exp\left(\frac{k_2}{k_1}\right) = \frac{E}{R}\left(\frac{1}{T_1} - \frac{1}{T_2}\right)$

D. $\exp\left(\frac{k_2}{k_1}\right) = \frac{E}{R}\left(\frac{1}{T_2} - \frac{1}{T_1}\right)$

***Sol.* B.**

Since by Arrhenius equation A = $e^{-E/RT}$

At two different temperatures

$$\frac{k_1}{k_2} = e^{-\frac{E}{T}\left(\frac{1}{T_1} - \frac{1}{T_2}\right)}$$

or $\ln\frac{k_2}{k_1} = \frac{R}{T}\left(\frac{1}{T_1} - \frac{1}{T_2}\right)$

38. The E-curve for a non-ideal reactor defines the fraction of a fluid having age between t and t + dt

A. at the inlet

B. at the outlet

C. in the reactor

D. averaged over the inlet and outlet

***Sol.* B.**

For a non ideal reactor the E - curve denotes the exit age distribution of the time element for the stream of fluid leaving the reactor. Hence E – curve represents the fraction of fluid having age between t and t + dt at the outlet of the reactor.

GATE 2000

39. The experimentally determined overall order for the reaction A + B → C + D is two. Then the

A. reaction is elementary with a molecularity of 2

B. molecularity of the reaction is 2 but the reaction may not be elementary

C. reaction may be elementary with a molecularity of 2

D. reaction is elementary but the molecularity may not be 2

***Sol.* A.**

For the reaction $A + B \rightarrow C + D$ it is given that the order of the reaction is 2. The molecularity of the reaction is also 2 as the reaction will proceed only when one molecule each of A and B will react. Now, elementary reactions are defined as the reactions in which molecularity and order is same. Hence, the reaction is elementally with a molecularity of 2.

40. The reaction A→B is conducted in a isothermal batch reactor. If the conversion of A increases linearly with holding time, then the order of the reaction is

A. Zero

B. 1

C. 1.5

D. 2

***Sol.* A.**

The design equation for an isothermal batch reactor is

$$t = C_{Ao}\int_0^{X_A}\frac{dX_A}{-r_A}$$

Assuming the reaction to be zero order the rate of reaction

$$-r_A = k\,C_{Ao}^0 = k$$

hence $t = C_{Ao}\int_0^{X_A}\frac{dX_A}{k}$

or $t = \frac{C_{Ao}X_A}{k}$

hence $X_A = \left(\frac{k}{C_{Ao}}\right)t$

It can be seen that X_A vs t graph will give a straight line passing through the origine and X_A increases linearly with T

∴ The reaction is of zero order

41. For the liquid phase parallel reactions

$A \rightarrow R,\quad r_R = k_1C_A^2\ ;\ E_1$ 80kJ/mol

$A \rightarrow S,\quad r_S = k_2C_A; E_2 = 120$ kJ/mol

The desired product is R. A higher selectivity of R will be achieved, if the reaction is conducted at

A. low temperature in a CSTR

B. high temperature in a CSTR

C. low temperature in a PFR

D. high temperature in a PFR

***Sol.* C.**

For parallel reactions

$A \rightarrow R;\quad r_R = k_1C_A^2; E_1 = 80$ kJ/mol

$A \rightarrow S;\quad r_S = k_2C_A; E_2 = 120$ kJ/mol

A PFR will give higher conversion of R since initially higher concentration of A will be utilised and the reaction being second order with respect to A, rate of formation of R wil be higher.

Further, since activation energy for first reaction is lower higher formation of R will be achieved at lower temperature.

42. In solid catalysed reactions, the diffusional effects are more likely to affect the overall rate of reaction for

A. fast reactions in catalysts of small pore diameter

B. fast reactions in catalysts of large pore diameter

C. slow reactions in catalysts of small pore diameter

D. fast reactions in catalysts of large pore diameter

Sol. C.

For solid catalysed reactions there are diffusion, adsorption and surface reaction steps which control the overall rate of reaction. It is given that diffusion effects are more likely to affect the overall rate of reaction. This will mean that surface reaction is slow and the adsorption step is also slow. Slow adsorption will occur in catalysts of small pore diameter, Hence option (c) starting slow reactions in catalysts of small pore diameter is correct.

(2 Mark Questions)

GATE 2015

43. An isothermal steady state mixed flow reactor (CSTR) of 1 m^3 volume is used to carry out the first order liquid-phase reactor A → products. Fresh feed at a volumetric flow rate of Q containing reactant A at a concentration C_{A0} mixes with the recycle steam at a volumetric flow rate RQ as shown in the figure below. It is observer that when the recycle ratio R = 0.5, the exit conversion X_{Af} = 50%.When the recycle ratio is increased to R = 2, the new exit conversion (in per cent) will be

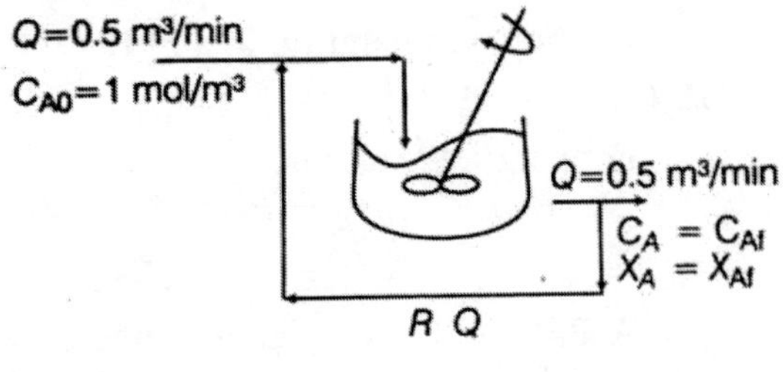

A. 50

B. 54.3

C. 58.7

D. 63.2

Sol. A.

Mole balance

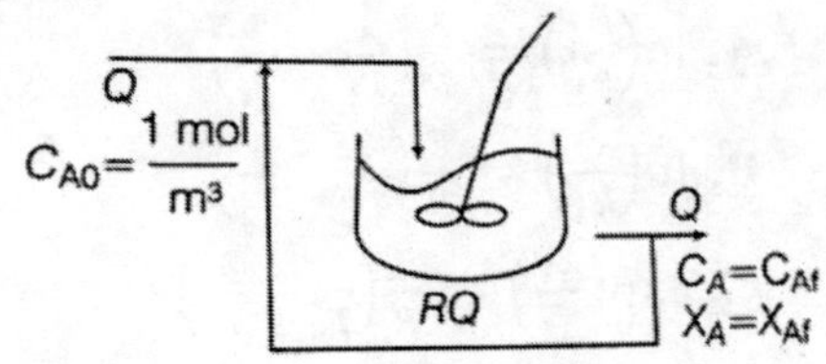

Input – Output

$$or\ \ \tau=\frac{X_A}{k(1-X_A)}\Rightarrow \tau k=\frac{X_A}{1-X_A}$$

$$or\ \ \ X_A=\frac{\tau k}{1+\tau k}$$

So, conversion X_A is independent of recycle ratio R.

So, final conversion = 50%

44. A catalyst slab of half-thickness L (the width and length of the slab >> L) is used to conduct the first order reaction A → B. At 450 K, the Thiele modulus for this system is 0.5. The activation energy for the first order rate constant is 100 kJ/mol.The effective diffusivity of the reactant in the slab can be assumed to be independent of temperature, and external mass transfer resistance can be neglected. If the temperature of the reaction is increased to 470 K, then the effectiveness factor at 470 K (upto two decimal place) will be ________.
[Value of universal gas constant = 8.314 J/mol-K]

Sol. Thiele modulus, $\emptyset \propto \sqrt{k}$

$$\frac{\emptyset_2}{\emptyset_1}=\sqrt{\frac{k_2}{k_1}}=\sqrt{\frac{A_e-(E/RT_2)}{A_e-(E/RT_1)}}=\sqrt{e^{-\frac{E}{k}\left(\frac{1}{T_2}-\frac{1}{T_1}\right)}}$$

$$=\sqrt{e^{-\frac{100}{8.314}\left(\frac{1}{470}-\frac{1}{450}\right)}}=1.766$$

$$\emptyset_2=1.766\times 0.5=0.883$$

Effectiveness factor, ϵ

$$=\frac{\tan h\ \emptyset_2}{\emptyset_2}=\frac{\tan h\ (0.883)}{0.883}=0.38$$

45. Consider two steady isothermal flow configuration shown schematically as Case I and Case II below. In Case I, a CSTR of volume V_1 is followed by a PFR of volume V_2, while in Case II a PFR of volume V_2 is followed by a CSTR of volume V_1. In each case, a volumetric flow rate Q of liquid reactant is flowing through the two units in series. An irreversible reaction v A → products (order n) takes place in both cases, with a reactant concentration C_{A0} being fed into the first unit.

Case 1.

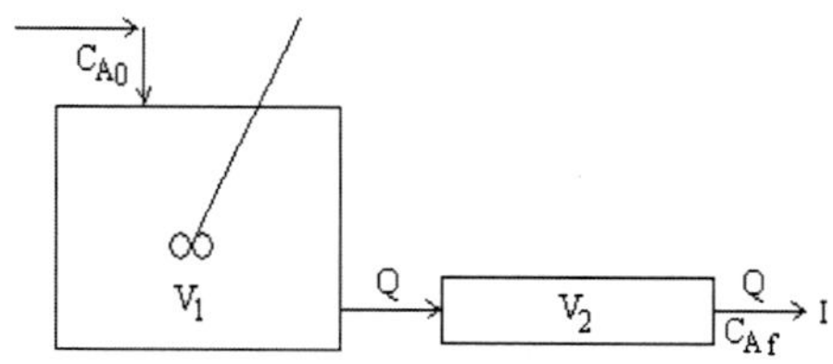

Case 2.

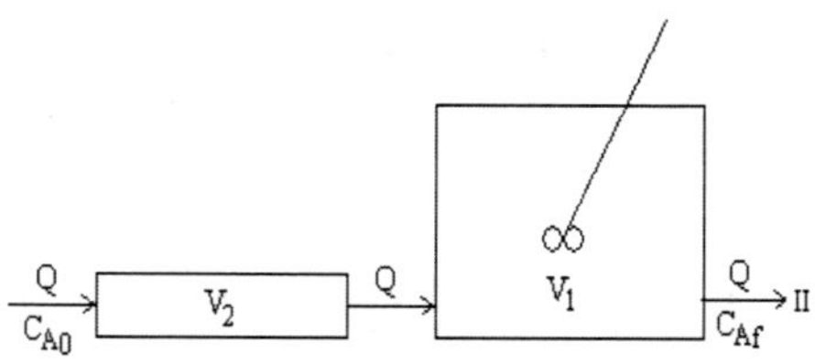

Choose the correct option

A. $\frac{C_{Af}^{I}}{C_{Af}^{II}} > 1, \text{for } n = 1$

B. $\frac{C_{Af}^{I}}{C_{Af}^{II}} = 1, \text{for } n = 1$

C. $\frac{C_{Af}^{I}}{C_{Af}^{II}} < 1, \text{for } n = 1$

D. $\frac{C_{Af}^{I}}{C_{Af}^{II}} = 1, \text{for } n > 1$

***Sol.* B.**

By taking 1st reaction (n = 1)

Case I

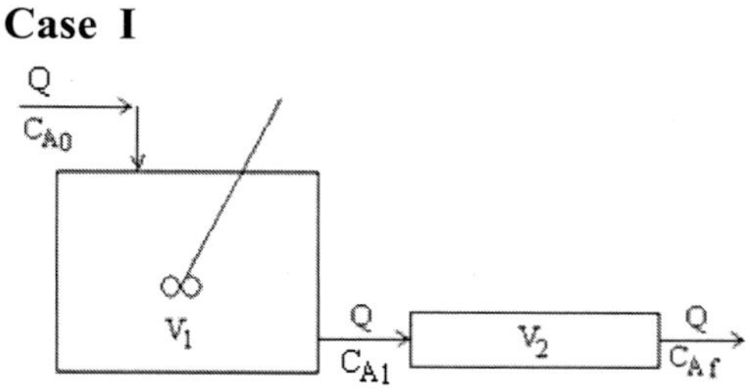

Consider, $V_1 = V_2 = 1\text{m}^3$, $k = 1\text{S}^{-1}$

$Q = 1\text{m}^3/\text{S}$ and $C_{A0} = 1\text{mol/m}^3$

For CSTR, $\tau K = \frac{C_{A0}-C_{A1}}{C_{A1}}$

$$1 \times 1 = \frac{1-C_{A1}}{C_{A1}}$$

$$\Rightarrow C_{A1} = 0.5 \text{ mol/m}^3$$

Case II

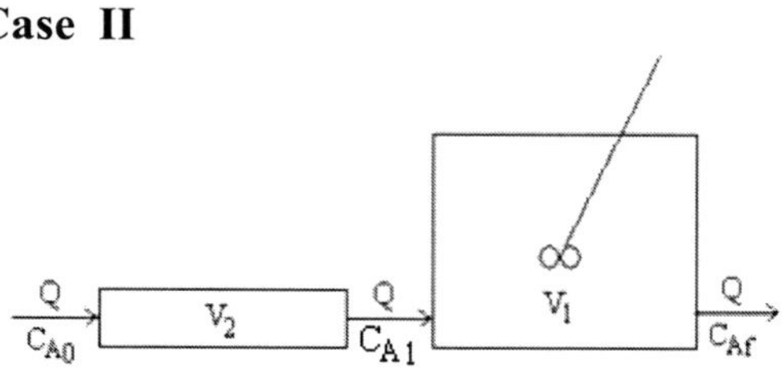

With same data,

For PFR,

$$\tau = -\text{In}\left(\frac{C_{A1}}{C_{A0}}\right) = 1$$

$$\Rightarrow C_{A1} = 0.3680 \text{ mol/m}^3$$

and for CSTR,

$$\tau K = 1 = \frac{C_{A1}-C_{Af}}{C_{Af}} = \frac{0.3680-C_{Af}}{C_{Af}}$$

$$C_{Af} = 0.184 \text{ mol/m}^3$$

Clearly for both the cases, C_{Af} is equal

i.e $\frac{C_{Af}^{I}}{C_{Af}^{II}} = 1$, for $n = 1$

GATE 2014

46. A homogeneous reaction (R → P) occurs in a batch reactor. The conversion of the reactant R is 67% after 10 min and 80% after 20 min. The rate equation for this reaction is

A. $-r_R = k$

B. $-r_R = kC_R^2$

C. $-r_R = kC_R^3$

D. $-r_R = C_R^{0.5}$

***Sol.* B.**

For second order reaction,

$$-\frac{dC_A}{dt} = kC_4^2$$

Or $-\frac{d}{dt}C_{A_0}(1-X_A) = kC_{A_0}^2(1-X_A)^2$

$$\frac{dX_4}{dt} = kC_{A_0}(1-X_A)^2$$

$$\int_0^{X_A} \frac{d\,X_A}{(1-X_A)^2} = \int_0^t k\,C_{A_0}dt$$

$$\frac{1}{1-X_A} - 1 = kC_{A_0}t$$

Or $\frac{X_A}{(1-X_A)t} = kC_{A_0}$

For case I, $kC_{A_0} = \frac{\frac{1}{2}}{\left(1-\frac{2}{3}\right)\times 10} = \frac{1}{5}$

For case II, $kC_{A_0} = \frac{0.8}{(1-0.8)\times 20} = \frac{1}{5}$

The value of kC_{A_0} is equal in both cases which justifies it to be a second order reaction.

Thus, the rate equation is, $-r_R = kC_R^2$

47. A vapour phase catalytic reaction (Q + R → S) follows Rideal mechanism (R and S are not adsorbed). Initially, the mixture contains only the reactants in equimolar ratio. The surface reaction step is rate controlling. With constants a and b, the initial rate of reaction $(-r_o)$ in terms of total pressure (P_T) is given by

A. $-r_o = \frac{ap_T}{1+bp_T}$

B. $-r_o = \frac{ap_T}{1+bp_T^2}$

C. $-r_o = \frac{ap_T^2}{a+bp_T}$

D. $-r_o = \frac{ap_T^2}{(1+bp_T)^2}$

Sol. **C.**

For Q + R → S

For first order radial mechanism,

Initial rate,

$$-r_o = \frac{ap_T^2}{a + bp_T}$$

48. Match the following.

Group I	*Group II*
P. Tank in series model	I. Non-isothermal reaction
Q.Liquid-liquid extraction	II.Mixer-settler
R.Optimum temperature progression	III. PFR with axial mixing
S. Thiele modulus	IV.Solid catalysed reaction

A. P – II, Q-IV, R-I, S-III

B. P-I, Q-II, R-III, S-IV

C. P-III, Q-I, R-II, S-IV

D. P-III, Q-II, R-I, S-IV

Sol. **D.**

Tank in series model – PFR with axial mixingLiquid – liquid extraction – mixer settler Optimum temperature progression – Non-isothermal reaction

Thiele modulus – Solid catalysed reaction

GATE 2013

Common data for Question 50-51

Liquid reactant A decomposes as follows

$A \rightarrow R \quad r_R = k_1C_A^2 \qquad k_1 = 0.5\, m^3/mol\text{-}s$

$A \rightarrow S \quad r_S = k_2C_A \qquad k_2 = 1\, s^{-1}$

An aqueous feed of composition $C_{A0} = 30\, mol/m^3$, $C_{R0} = 2\, mol/m^3$, and $C_{S0} = 1\, mol/m^3$ enters a CSTR in which the above reactions occur. Assume isothermal and steady state conditions.

49. If the conversion of A is 80%, the concentration of R in the exit stream in mol/m^3, to the nearest integer, is

A. 40 mol/m^3

B. 50 mol/m^3

C. 30 mol/m^3

D. 20 mol/m^3

Sol. **D.**

Given, C_{A_0} = 30 mol/m^3

C_{R_0} = 2 mol/m^3

$X_A = 0.8$

∴ $C_A = C_{A_0}(1 - X_A)$

= 30 (1 – 0.8) = 6 mol/m^3

Applying material balance for A,

$VC_{A_0} = VC_A + (-r_A)V$

$VC_{A_0} = VC_A + (k_1C_A^2 + k_2C_A)V$

or $VC_{A_0} - C_A = (k_1C_A^2 + k_2C_A)V$

$V(30 - 6) = (0.5 \times 6 \times 6 + 1 \times 6)V$

$\frac{V}{V} = \tau =$ 1 s

Applying material balance for R,

$VC_{R_0} = VC_R - r_R.V$

$C_{R_0} = C_R - k_1C_A^2\,\tau$

Or $C_R = C_{R_0} + k_1C_A^2\,\tau$

$C_R = 2 + 0.5 \times 6 \times 6 + 1$

C_R = 20 mol/m^3

50. What is the % conversion of A, to the nearest integer, so that the concentration of S in the exit stream is 11.8 mol/m^3?

A. 90%

B. 40%

C. 60%

D. 50%

Sol. **A.**

Applying material balance for S,

$$VC_{S_0} = VC_S - (-r_S)V$$
$$VC_{S_0} = C_S - k_2C_A\tau$$
$$1 = 11.8 - 1 \times C_A\tau$$
$$C_A\tau = 10.8$$

Now, applying material balance for A,

$$VC_{A_0} = VC_A + (-r_A)V$$
$$C_{A_0} = C_A + (k_1C_A^2 + k_2C_A)\tau$$
or $$C_{A_0} = C_A + (0.5C_A + 1)C_A\,\tau$$
or $$30 = C_A + (0.5C_A + 1) \times 10.8$$
$$C_A = 3$$

% conversion of A, $X_A = 1 - \frac{C_A}{C_{A0}} = 1 - \frac{3}{30} = 0.9$

= 90%

51. A first order liquid phase reaction is carried out isothermally at steady state in a CSTR and 90%

conversion is attained. With a same inlet conditions and for the same overall conversion, if the CSTR replaced by two smaller and identical isothermal CSTRs in series, the % reduction in total volume, to the nearest integer, is

A. 30%

B. 45%

C. 52%

D. 40%

Sol. **C.**

For (Mixed Flow Reactor) MFR,

$$\frac{V}{V_0} = \frac{C_{A0}X_A}{-r_A}$$

For first order reaction,

$$\frac{V}{V_0} = \frac{C_{A0}X_A}{kC_A} = \frac{C_{A0}-C_A}{kC_A}$$

$$\text{or } k\tau = \frac{C_{A0}-C_A}{kC_A} = \frac{X_A}{1-X_A}$$

$$\text{In first case, } k\tau = \frac{0.9}{1-0.9} = 9 \qquad \text{.... (i)}$$

In second case, for first MFR,

$$k\tau_1 = \frac{C_{A0}-C_{A1}}{C_{A1}}$$

$$= \frac{C_{A0}}{C_{A1}} - 1$$

$$\frac{C_{A0}}{C_{A1}} = 1 + k\tau_1 \qquad \text{....(ii)}$$

Similarly, for 2nd MFR,

$$\frac{C_{A1}}{C_{A2}} = 1 + k\,\tau_2 \qquad \text{....(iii)}$$

By Eq. (ii) and (iii)

$$\frac{C_{A0}}{C_{A2}} = (1 + k\tau_1)(1 + k\tau_2)$$

$$\text{But } \frac{C_{A2}}{C_{A0}} = 1 - X_A = 1 - 0.9 = 0.1$$

and $\tau_1 = \tau_2$ for identical reactors,

$\therefore$ $10 = (1 + k\tau_1)^2$

$1 + k\tau_1 = 3.16$

$k\tau_1 = 2.16$

$$\% \text{ Reduction in reactor volume} = \frac{V-(V_1-V_2)}{V} \times 100$$

$$= \frac{V-2V_1}{V} \times 100$$

$$= \frac{k\tau - 2k\tau_1}{k\tau} \times 100 \quad (\because V \propto k\tau)$$

$$= \frac{9-2\times 2.16}{9} \times 100$$

$$= 52\%$$

GATE 2012

52. The rate controlling step for the solid catalyzed irreversible reaction

$$A+B \rightarrow C$$

is known to be the reaction of adsorbed A with adsorbed B to give adsorbed C. If p_i is the partial pressure of component i with k_i is the adsorption equilibrium constant of component i, then the form of the Langmuir- Hinshelwood rate expression will be

A. rate $\propto \dfrac{\rho_A\rho_B}{1+K_A\rho_B+K_B\rho_B+K_C\rho_C}$

B. rate $\propto \dfrac{\rho_A\rho_B}{(1+K_A\rho_B+K_B\rho_B+K_C\rho_C)^2}$

C. rate $\propto \dfrac{\rho_A\rho_B}{(1+K_A\rho_B+K_B\rho_B+K_C\rho_C)^{0.5}}$

D. rate $\propto \dfrac{\rho_A\rho_B}{\rho_C}$

Sol. **A.**

For solid catalyzed 2nd order reaction

$$A+B \longrightarrow C$$

The rate of reaction,

$$r \propto \frac{\rho_A\rho_B}{(1+K_A\rho_B+K_B\rho_B+K_C\rho_C)^2}$$

53. The elementary reversible exothermic gas-phase reaction

$$A + 3B \rightleftharpoons 2C$$

is to be conducted in a non-isothermal, non-adiabatic plug flow reactor. The maximum allowable reactor temperature is T_{max}. to minimize the total reactor volume, the variation of reactor temperature (T) with axial distance from inlet (Z) will be

A. 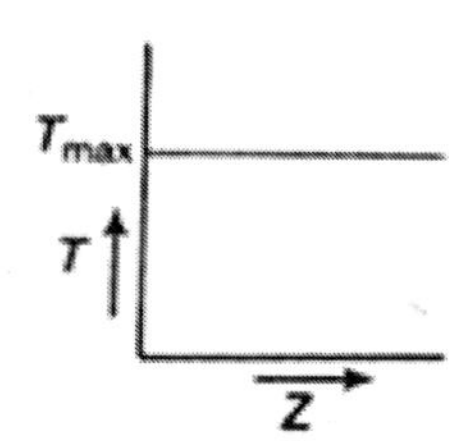

B. 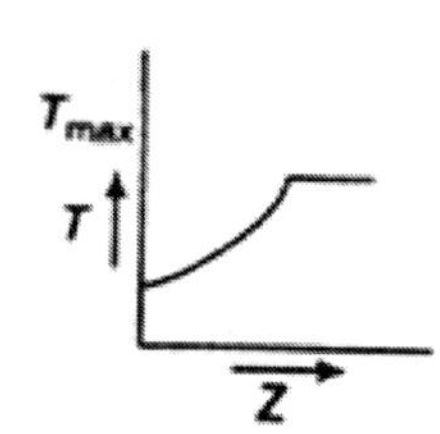

C. 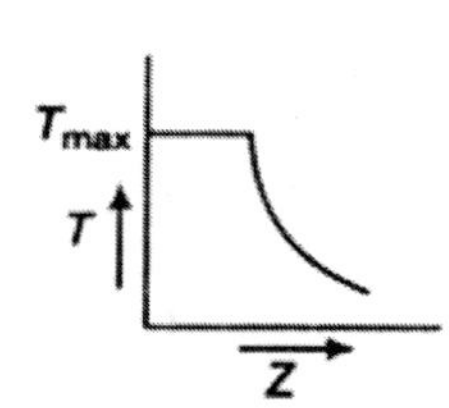

D. 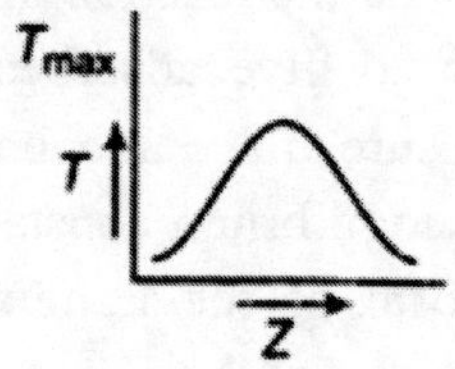

Sol. **B.**

For exothermic gas-phase reaction in NINA PFR. The reaction progress along the axial length and the temperature increase due to heat released. The temperature becomes constant on completion of reaction which is called the maximum allowable temperature. Thus, the correct plot of *Tversus Z* is given below.

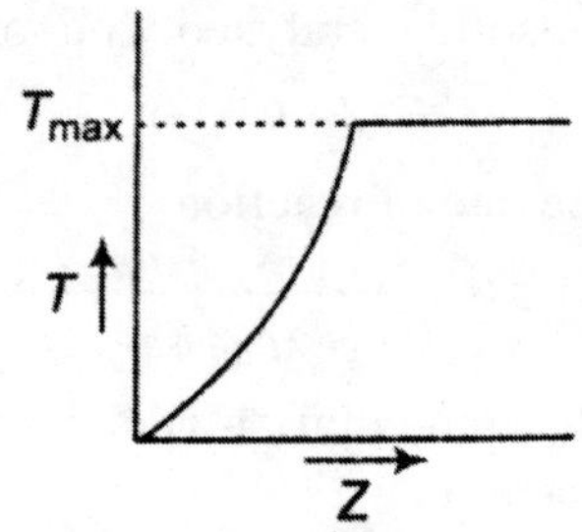

Common data for Question 55-56

The first order liquid phase reaction A→P is conducted isothermally in a plug flow reactor of 5 liter volume. The inlet volumetric flow rate is 1 liter/min and the inlet concentration of A is 2 mole/liter.

54. If the exit concentration of *A* is 0.5 mol/L, then the rate constant (in min^{-1}) is

A. 0.06
B. 0.28
C. 0.42
D. 0.64

Sol. **B.**

$A \longrightarrow P$

For lst order reaction in PFR,

$$\frac{V}{F_{A0}} = \int \frac{dX_A}{-r_A}$$

$$\frac{V}{F_{A0}} = \int_0^{x_A} \frac{dx_A}{kC_{A0}(1-X_A)}$$

Or $$\frac{V}{F_{A0}} = -\frac{1}{kC_{A0}}.\ln(1-X_A) \quad ....(i)$$

Here, $V = 5\ L, C_{A0} = 2\text{mol/L}, v_0 = 1\ \text{L/min}$

$$F_{A0} = C_{A0}.v_0 = 2\ \text{L/min}$$

$$C_A = 0.5\ \text{mol/L}$$

Since, $C_A = C_{A0}(1-X_A)$

$$0.5 = 2(1-X_A)$$

Putting values in Eq. (i),

$$\frac{5}{2} = -\frac{1}{k\times 2}\log_e(0.25)$$

$$k \approx 0.28\ \text{min}^{-1}$$

55. The plug flow reactor is replaced by 3 mixed flow reactors in series, of 2.0 L volume. The exact conversion of *A* (in %) is

A. 35.9
B. 52.5
C. 73.7
D. 94.8

Sol. **C.**

For N-MFR in series,

$$\frac{C_{AN}}{C_{AO}} = \frac{1}{(1+k\tau_i)^N}$$

Or $$1 - X_A = \frac{1}{(1+k\tau_i)^N}$$

Here, $\tau_i = \frac{2}{1} = 2\ \text{min},\ N = 3$

Putting values,

$$1 - X_A = \frac{1}{(1+0.28\times 2)^3}$$

Or $$X_A = 0.737 = 73.7\%$$

GATE 2011

56. For a first order catalytic reaction, the Thiele modulus (ϕ) of a spherical pellet is defined as

$$\phi = \frac{R_p}{3}\sqrt{\frac{k\rho_p}{D_e}}$$

Where ρ_p = pellet density
R_p = pellet radius
D_e = effective diffusivity
k = first order reaction rate constant

If $\phi > 5$, then the apparent activation energy (E_a) is related to the intrinsic (or true) activation energy (E) as

A. $E_a = E^{0.5}$
B. $E_a = 0.5\ E$
C. $E_a = 2E$
D. $E_a = E^2$

Sol. **B.**

Thiele modulus $\emptyset$ for a first order catalytic reaction is defined as

$$\emptyset = \frac{R_p}{3}\sqrt{\frac{k\rho_p}{D_e}}$$

Where ρ_p is pellet density, R_p is pellet redius, D_e is effective diffusivity and K is first order reaction rate constant

The above relationship can be written as

$$\emptyset = \propto \sqrt{\frac{k}{D_e}}$$

When $\emptyset > 5$ then $E_a = 0.5$ E

57. The following figures show the outlet tracer concentration profiles (cvst) for a pulse input.

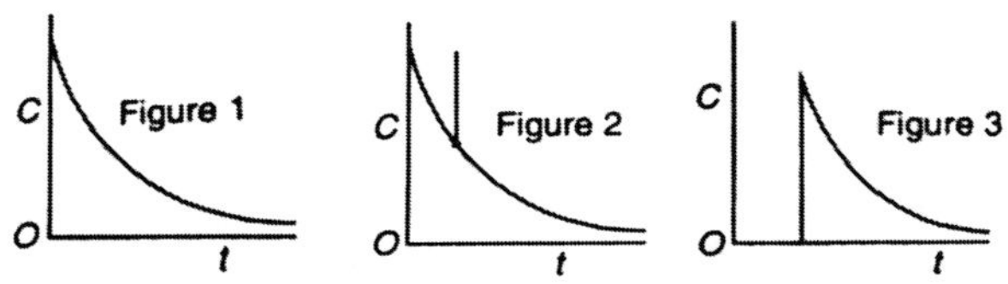

Match the figures in a Group I with the reactor configuration in group II

Group I	*Group II*
P. Figure 1 Q. Figure 2 R. Figure 3	I. PFR II. CSTR III. PFR and CSTR in series IV. PFR and CSTR in parallel

A. P-II, Q-IV, R-III
B. P-IV, Q-III, R-I
C. P-III, Q-IV, R-II
D. P-I, Q-III, R-II

***Sol.* A.**

Figure 1 – CSTR
Figure 2 – PFR and CSTR in parallel
Figure 3 – PFR and CSTR in series

Linked Questions 59- 60

58. In an aqueous solution, reaction $P \rightarrow Q$ occurs under isothermal conditions following first order kinetics. The feed rate is 500 cm^3/min and concentration of P in the feed is 1.5×10^{-4} mol/cm^3. The reaction is carried out in a 5 L CSTR. At steady state, 60% conversion in observed. The rate constant (in min^{-1}) is

A. 0.06
B. 0.15
C. 0.21
D. 0.28

***Sol.* B.**

Reaction $P \longrightarrow Q$ being conducted in a CSTR

$V_0 = 500$ cc /min

$C_{A0} = 1.5 \times 10^{-4}$ mol/sec

Vol. V = 5 lit = 500cc

$X_A = 0.6$

For a CSTR

$$\frac{V}{F_{Ao}} = \frac{X_A}{(-r_A)}$$

For first order kinetics

$$\frac{V}{V_o(C_{Ao})} = \frac{X_A}{kC_{Ao}(1-X_A)}$$

$$\text{or } k = \frac{V_o X_A}{V(1-X_A)}$$

$$= \frac{(500)(0.6)}{(5000)(1-0.6)} = 0.15 \text{ min}^{-1}$$

59. The 5 L CSTR is replaced by five CSTR in series. If the capacity of each new CSTR is 1 L, then the overall conversion (in percentage) is

A. 65
B. 67
C. 73
D. 81

***Sol.* C.**

For a CSTR battery of 5 reactors each having I lit volume

For 1st stage $C_{A1} = \frac{C_{Ao}}{(1+K\tau)}$ for first aids erasable reaction

Similarly for second stage

$$C_{A2} = \frac{C_{A1}}{(1+K\tau)} = \frac{C_{Ao}}{(1+K\tau)^2}$$

Similarly for fifth stage

$$C_{A5} = \frac{C_{Ao}}{(1+K\tau)^5}$$

Putting values and taking k = 0.15 min^{-1} from previous linked problem

$$T = \frac{V}{V_0} = \frac{1000}{500} = 2 \text{ min}$$

Now

$$C_{A5} = \frac{0.15 \times 10^{-4}}{[1+(0.15)(2)]}$$

$$= 0.403 \times 10^{-4}$$

$$\therefore x = \frac{C_{Ao} - C_{A5}}{C_{Ao}}$$

$$= \frac{(1.5\times10^{-4})-(0.403\times10^{-4})}{1.5\times10^{-4}}$$

$$= \frac{1.5-0.403}{1.5}$$

$$= 0.73$$

$\therefore$ overall conversion = 73%

GATE 2010

60. Two reactors (reactor 1 and reactor 2) with average resistance times τ_1 and τ_2 respectively, are placed in series. Reactor 1 has zero dispersion and reactor 2 has infinite dispersion. The residence-time distribution $E(t)$ of this system, is given by

A. $\begin{cases} 0 & \text{for } t \le \tau_1 \\ \frac{1}{\tau_2} exp\left(-\frac{t-\tau_1}{\tau_2}\right) & \text{for } t > \tau_1 \end{cases}$

B. $\begin{cases} 0 & \text{for } t \le \tau_2 \\ \frac{1}{\tau_1} exp\left(-\frac{t-\tau_2}{\tau_1}\right) & \text{for } t > \tau_2 \end{cases}$

C. $\begin{cases} 0 & \text{for } t \le \tau_2 \\ \frac{1}{\tau_1} exp\left(-\frac{t-\tau_1}{\tau_2}\right) & \text{for } t > \tau_2 \end{cases}$

D. $\begin{cases} 0 & \text{for } t \le \tau_1 \\ \frac{1}{\tau_2} exp\left(-\frac{t}{\tau_1}\right) & \text{for } t > \tau_2 \end{cases}$

Sol. **A.**

Reactor 1 has zero dispersion which occurs in a PFR because there is no back mixing whereas reactor 2 has infinite dispersion which occurs in a CSTR in which there is complete mixing

E(t) $= 0$ for $t \le \tau_1$

And E(t) $\frac{1}{\tau_2} \exp\left[-\left\{\frac{t-\tau_1}{\tau_2}\right\}\right]$, for $t > \tau_2$

An autocatalytic liquid phase reaction $A+R \longrightarrow 2R$, is conducted in an isothermal batch reactor with a small initial concentration of R. Assume that the order of reaction with respect to both reactants is positive. The rate of reaction $(-r_A)$ *versus* concentration C_A, as the reaction proceeds is depicted by

A.

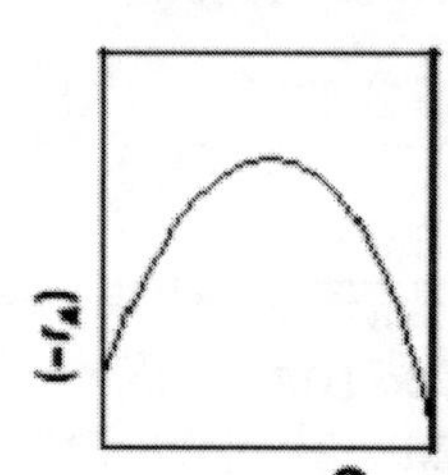

B.

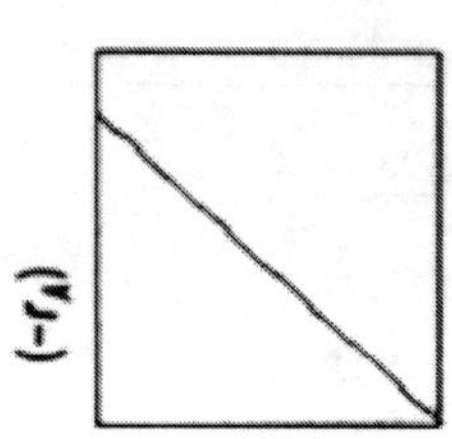

C.

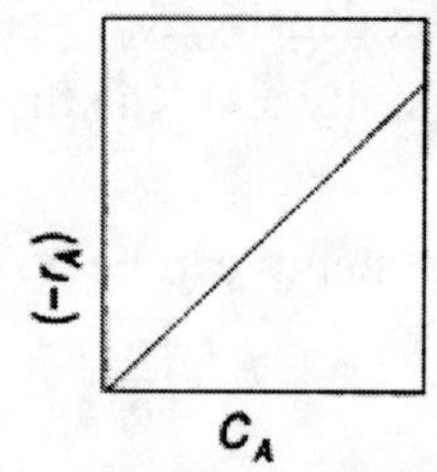

D.

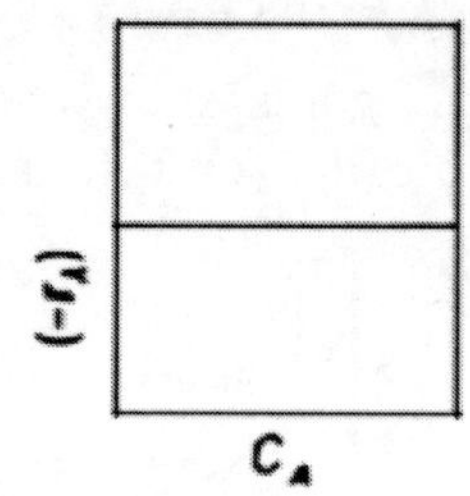

Sol. **A.**

For autocatalytic reaction $A+R \longrightarrow R+R$

Initially concentration of R is small giving small value of rate of reaction. However, as the concentration of R increases the rate of reaction will increase rapidly reaching a maximum thereafter, as concentration of A starts decreasing the rate will fall continuously and the curve will look as below.

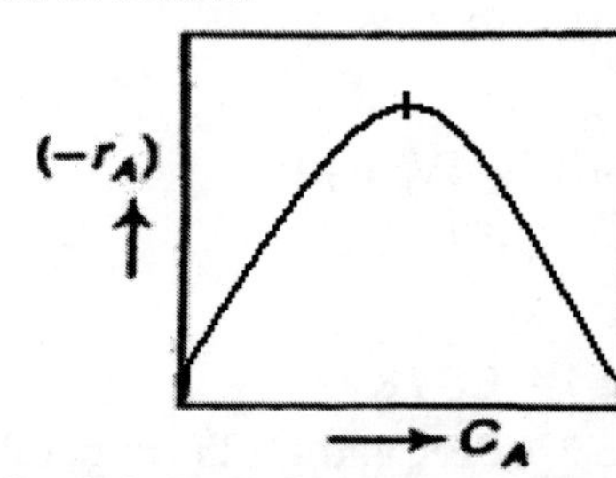

Statement for Linked Questions 68 and 69

A liquid phase reaction $A \rightarrow B$ is conducted isothermally in a CSTR having a residence time of 2 s. The inlet concentration of species A is 2 mol/L, and the outlet concentration is 1 mol/L. The rate law for the reaction is $-r_A = \frac{kC_A}{k+C_A}$ where, k = 5 mol/L-s.

62. The value of *K* in mol/L, is

A. 11

B. 9

C. 5

D. 2

Sol. **B.**

Given $\tau = 2$sec. $C_{A_0} = 2$ mol/lit

$C_{A_1} = 1$ mol/lit K $= 5$ mol/(lit)(sec)

For a CSTR the design equation is

$FC_{A_0} - (FC_{A_1} + rV) = 0$ under steady state

Given the rate of reaction is

$$-r_A = \frac{kC_A}{K+C_A}$$

$$FC_{A_0} - \left[FC_{A_1} + \left(\frac{kC_{A_1}}{K+C_{A_1}}\right)V\right] = 0$$

Since $\tau = \frac{V}{F} = 2$ sec.

$$\therefore C_{A_0} - \left[C_{A_1} + \left(\frac{kC_{A_1}}{K+C_{A_1}}\right)\tau\right] = 0$$

or $C_{A_0} - C_{A_1}\left[1 + \left(\frac{kC_{A_1}}{K+C_{A_1}}\right)\tau\right] = 0$

Substituting numerical values

$$2 - 1\left[1 + \left(\frac{(5)(2)}{K+1}\right)\right] = 0$$

Solving K = 9

63. If the same reaction is conducted in a series of two CSTRs with residence times 1s and 0.2 s, then the inlet concentration of A in mol/L, required to attain an outlet concentration of A of 1 mol/L, is

<u>A.</u> 2.64

<u>B.</u> 2.00

<u>C.</u> 1.64

<u>D.</u> 0.54

Sol. <u>C.</u>

Same reaction is now conducted in a series of two CSTRs

$\tau_1 = 1$ sec. $\tau_2 = 0.2$ sec. $C_{A_2} = 1$ mol/lit

From previous linked question R = 5, K = 9

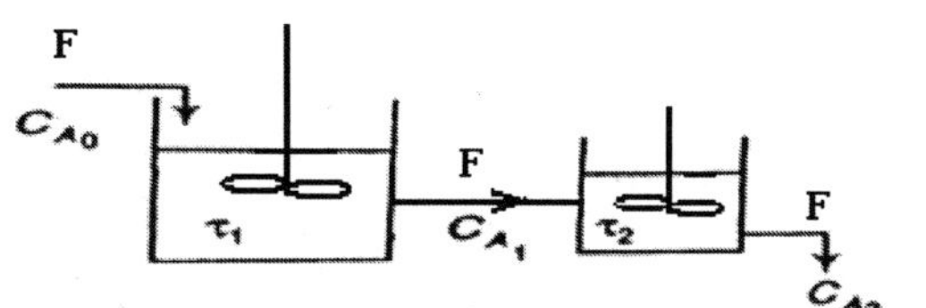

Material balance for 2nd stage

$$C_{A_1} - C_{A_2}\left[1 + \left(\frac{RC_{A_2}}{K+C_{A_2}}\right)\tau_2\right] = 0$$

$$\therefore C_{A_1} = 1\left[1 + \left\{\frac{(5)(1)}{(9+1)}\right\}0.2\right]$$

$$= 1\left[1 + \frac{1}{10}\right]$$

$$\therefore C_{A_1} = 1.1$$

Similarly for 1st stage

$$C_{A_0} = C_{A_1}\left[1 + \left(\frac{kC_{A_1}}{K+C_{A_1}}\right)\tau_1\right]$$

$$= 1.1\left[1 + \left\{\frac{(5)(1.1)}{(9+1.1)}\right\}\Delta\right]$$

= 1.64 mole/lit

GATE 2009

64. The liquid-phase reaction $A \longrightarrow B$, is conducted in an adiabatic plug flow reactor.

Data

Inlet concentration of A = 4.0 kmol/m³

Density of reaction mixture (independent of temperature) = 1200 kg/m³

Average heat capacity of feed stream (independent of temperature) = 2000 j/kg K

Heat of reaction (independent of temperature) = –120 kJ/mol of A reacting

If the maximum allowable temperature in the reactor is 800 K, then the feed temperature (in ⁰K) should not exceed.

<u>A.</u> 400

<u>B.</u> 500

<u>C.</u> 600

<u>D.</u> 700

Sol. <u>C.</u>

For the reaction $A \rightarrow B$ being conducted is an adiabatic PFR

$$C_{A_0} = 4.0 \text{ kmol/m}^3$$

$$\rho_{(mixture)} = 1200 \text{ kg/m}^3$$

$$C_p = 2000 \text{ J/(kg)(K)}$$

$$\Delta HR = -120 \text{ kJ/mol of A reacting}$$

Maximum allowable temperature

$$= 800 = K = T_{2_{max}}$$

Writing the heat balance

Input – output + heat generated due to reaction = 0

$$(V_0)(\rho)(C_p)T_1 - (V_0)\rho C_p T_2 + (V_0)(C_{A_0})(\Delta H_x) = 0$$

$$(V_0)(1200)(1200)(T_1 - T_2) + (V_0)(4)(120)(10^6) = 0$$

or $24 \times 10^5 (T_1 - T_2) + (480)(10^6) = 0$

Since $T_{2_{max}} = 800$ K

$\therefore$ $T_{1_{max}} = 600$ K

65. An isothermal pulse test is conducted on a reactor and the variation of the outlet tracer concentration with time is shown below.

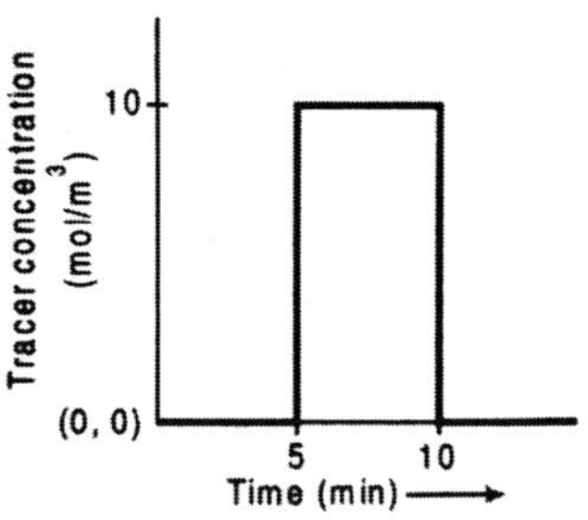

The mean residence time of the fluid in the reactor
(in min) is

A. 5.0

B. 7.5

C. 10.0

D. 15.0

***Sol.* C.**

For the isothermal pulse test

Mean residence time of the fluid in the reactor

$$\bar{t} = \frac{\int_0^{\infty} tCdt}{\int_0^{\infty} Cdt} = \frac{\Sigma tC\Delta T}{\Sigma C\Delta T}$$

$$\therefore \bar{t} = \frac{10\times 5}{5} = 10\,\text{min}$$

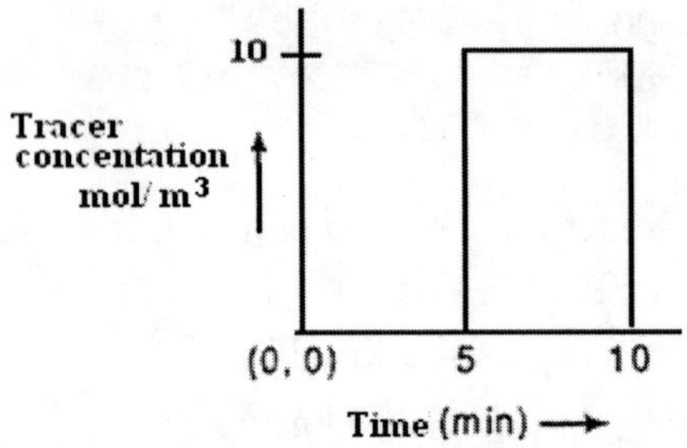

Statement for Linked Questions 72 and 73

The liquid-phase reaction $A \rightarrow B + C$, is conducted isothermally at 50^0 C in a continuous Stirred Tank Rector (CSTR). The inlet concentration of A is 8.0 g-mol/L. At a space time of 5 min, the concentration of A at the exit of CSTR is 4.0 g-mol/L. the kinetics of the reaction is

$$-r_A = kC_A^{0.5} \frac{g - mol}{L - min}$$

A plug flow reactor of the same volume is added in series after the existing CSTR.

66. The rate constant (k) for this reaction at 50°C is

A. $0.2\left(\frac{g-mol}{L}\right)^{0.5}.min^{-1}$

B. $0.2\left(\frac{L}{g-mol}\right)^{0.5}.min^{-1}$

C. $0.4\left(\frac{g-mol}{L}\right)^{0.5}.min^{-1}$

D. $0.4\left(\frac{L}{g-mol}\right)^{0.5}.min^{-1}$

***Sol.* C.**

Reaction $A \longrightarrow B + C$

Given $C_{A_0} = 8$ gmol/lit $\quad \tau = 5$ min

$C_{A_1} = 4$ g-mol/lit

And $\quad -r_A = -kC_A^{0.5}$

Writing material balance equation for a CSTR

$$FC_{A_0} - [FC_{A_1} + rV] = 0$$

$$FC_{A_0} - [FC_{A_1} + kVC_{A_1}^{0.5}] = 0$$

Dividing both sidesby F

$$C_{A_0} - \left[C_{A_1} + \left(K\tfrac{V}{F}\right)C_{A_1}^{0.5}\right] = 0$$

or $(C_{A_0} - C_{A_1}) - (K\tau)C_{A_1}^{0.5} = 0$

$$\frac{C_{A0}-C_{A1}}{\tau} = kC_{A_1}^{0.5}$$

or $\frac{8-4}{5} = \text{K}\,(4)^{0.5} = 2\text{ K}$

$$\frac{4}{5} = 2\text{ K}$$

or $\text{K} = \frac{4}{10} = 0.4\left(\frac{gmole}{lit}\right)^{0.5}\text{min}^{-1}$

67. The concentration of A (in g-mol/L) at the exit of the plug flow reactor is

A. 0.5

B. 1.0

C. 2.0

D. 2.5

***Sol.* B.**

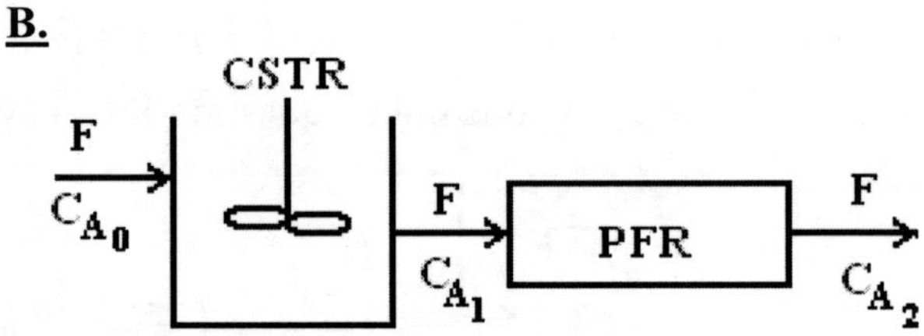

Given PFR is of same volume

For a PFR; $\tau = -\int_{C_{A1}}^{C_{A2}} \frac{dC_A}{(-r_A)}$

or $\tau = -\int_{C_{A1}}^{C_{A2}} \frac{dC_A}{kC_A^{0.5}} = \frac{1}{0.5K}\left[\sqrt{C_{A_1}} - \sqrt{C_{A_2}}\right]$

Since volume of PFR and CSTR are same

Hence τ is also same $\tau_{CSTR} = \tau_{PFR}$

$$\therefore \quad 5 = \left[\frac{\sqrt{4} - \sqrt{C_A}}{0.5\times 4}\right]$$

Solving $C_A = 1$ gmol/lit.

GATE 2008

68. The gas phase reaction $A + 3B \longrightarrow 2C$, is conducted in a PFR at constant temperature and pressure. The PFR achieves a conversion of 20% of A. The feed is a mixture of A, B and an inert

I. It is found that the concentration of A remains the same throughout the reactor.

Which one of the following ratios of inlet molar rate $\left(F_{A,in} : F_{B,in} : F_{I,in}\right)$ is consistence with this observation? Assume the reaction mixture is an ideal gas mixture.

A. 2 : 3 : 0
B. 2 : 2 : 1
C. 3 : 2 : 1
D. 1 : 2 : 1

***Sol.* C.**

Gas phase reaction in a PFR

$A+3B \longrightarrow 2C$ where feed consists of inerts

Material balance at x = 0 and x = 0.2

When x = 0

No. of moles of A, B and C are F_A, F_B, $F_C = 0$ and inerts are F_I

Where x = 0.2

A becomes $F_A(1-0.2) = 0.8\ F_A$

B becomes $F_B - (3\times0.2\times F_A)$

C becomes $(2)(0.2)\ F_A$

Inerts F_I

Hence $C_{A_o} = \frac{F_A}{F_A+F_B+F_C+F_I}$ since $F_C = 0$ initially

$= \frac{F_A}{F_A+F_B+F_I}$

At x = 0.2; $C_A = \frac{0.8\ F_A}{0.8\ F_A+F_B-0.6F_A+F_I+0.4F_A}$

$\therefore \quad C_A = \frac{0.8\ F_A}{0.6F_A+F_B+F_I}$

Since concentration of A remains the same throughout the reactor

$$\frac{F_A}{F_A+F_B+F_I} = \frac{0.8\ F_A}{0.6F_A+F_B+F_I}$$

or $0.8\ (F_A+F_B+F_I) = 0.6\ F_A + F_B + F_I$

or $0.2F_A = 0.2\ F_B +\ 0.2\ F_I$

This condition is satisfied only when

$F_A : F_B : F_C = 3:2:1$

Option C is valid.

69. The elementary liquid phases series parallel reaction scheme

$$A \longrightarrow B \longrightarrow C$$

$$A \longrightarrow R$$

is to be carried out in an isotheral CSTR. The rate laws are given by

$r_R = k'C_A$

$r_B = kC_A - kC_B$

Feed is pure A. The space time of the CSTR which results in the maximum exit concentration of B is given by

A. $\frac{1}{\sqrt{kk'}}$

B. $\frac{1}{\sqrt{k'(k+k')}}$

C. $\frac{1}{\sqrt{(k+k')}}$

D. $\frac{1}{\sqrt{k(k+k')}}$

***Sol.* D.**

Series parallel reaction

$A \rightarrow B \rightarrow C$

$A \rightarrow R$

Given $r_R = K'C_A$ and $r_B = KC_A - KC_B$

Hence $(-r_A) = KC_A + k'C_A(K+K')C_A$

Writing material balance for A

$FC_{A_0} = FC_A + (-r_A)V$

$FC_{A_0} = FC_A + (K+K')C_A V$

$C_{A_0} = C_A + (K+K')C_A\tau \quad \left(since, \frac{V}{F} = \tau\right)$

$\therefore \quad C_A = \frac{C_{A0}}{1+(K+K')\tau}$ .... (1)

Now applying material balance for B,

$FC_{B_0} +\ (r_B)V = FC_B$

or substituting $\left(\tau = \frac{V}{F}\right)$

or $\tau(KC_A - KC_B) = C_B$

$\therefore \quad C_B = \frac{(KC_A\ \tau)}{(1+K\tau)}$

Substituting for C_A from equation (i)

$C_B = \frac{K\tau C_{A0}}{(1+K\tau)[1+(K+K')\tau]}$

For maximum concentration of B differeating C_B with respect to τ

$\frac{dC_B}{d\tau} = 0$

or $KC_{A_o}(1+K\tau)[K+K'\tau] - K\tau C_{A_0}\left[(1+K\tau)(K+K') + K[1+(K+K')\tau]\right] = 0$

or $1 + K(K+K')\tau^2 + (2K+K')\tau$

$= (K+K')\tau + K(K+K')\tau^2$

$+ K\tau + K(K+K')\tau^2(K+K')\tau^2 = 1$

$\tau = \frac{1}{\sqrt{K(K+K')}}$

70. The liquid phases reaction A → Products is governed by the kinetics $(-r_A) = k\ C_A^{1/2}$

If the reaction undergoes 75% conversion of A in 10 min in an isothermal batch reactor, the time (in min) for complete conversion of A is

A. 40/3
B. 20
C. 30
D. Infinite

***Sol.* B.**

$A \rightarrow$ Products

$-r_A = kC_A^{1/2}$

For isothermal batch reactor

$(-r_A) = -\frac{dC_A}{dt}$

Or $-\frac{dC_A}{dt} = k\sqrt{C_A}$ ....(i)

Since $x_A = \frac{C_{A0} - C_A}{C_{A0}}$

$-\frac{dC_A}{dt} = C_{A_0}\frac{dx_A}{dt}$

or $C_{A_0}\frac{dx_A}{dt} = k\sqrt{C_{A_0}(1-x_A)}$

or $(\sqrt{C_{A_0}})\int_0^{0.75}\frac{dx_A}{\sqrt{1-x_A}} = \int_0^{10} k\,dt$ ....(ii)

or $10\text{ k} = \sqrt{C_{A_0}}\left[-2\sqrt{1-x_A}\right]_0^{0.75}$

or $10\text{ k} = \sqrt{C_{A_0}}[-1+2]$

or $10\text{ k} = \frac{\sqrt{C_{A0}}}{k}$ ....(iii)

Now equation (ii) becomes

$\sqrt{C_{A0}}\int_0^1\frac{dx_A}{\sqrt{1-x_A}} = \int_0^t k\,dt$

or $t = \frac{\sqrt{C_{A0}}}{K}\left[-2\sqrt{1-x_A}\right]_0^1$

$\therefore$ $t = 10 \times 2$

$= 20$ min.

71. The homogenous reaction $A+B \rightarrow C$ is conducted in an adiabatic CSTR at 800 K so as to achieve a 30% conversion of A. The relevant specific heats and enthalpy change of reaction are given by

$C_{p^A} = 100$ J/(mol-k) , $C_{p^c} = 50$ J/(mol-k)

$C_{p^B} = 50$ J/mol-k, $\Delta h^{rxn} = -100$ kJ/mol

If the feed, a mixture of A and B, is available at 550 K, the mole fraction of A in the feed that is consistent with the above data is

A. 5/7
B. 1/4
C. 1/2
D. 2/7

***Sol.* A.**

For the reaction $A + B \longrightarrow C$

Let the initial number of moles be

N_{A_0} , N_{B_0} and $(N_{C_0} = 0)$

After 30% conversion

$A = N_{A_0}(1-0.3) = 0.7\,N_{A_0}$

Similarly $B = N_{B_0} - 0.7\,N_{A_0}$

And $C = 0.3\,N_{A_0}s$

Further energy balance equation.

$\sum(Ni_{in})(C_p)T_i = \sum(Ni_{out})(C_p)(T_o) + \varepsilon\Delta H$

Writing $\varepsilon = \frac{N_{A0} - 0.7N_A}{1} = 0.3\,N_{A_0}$

Thus putting the values

(N_{A_0}) (100) (550) + (N_{B_0}) (50) (550)

$= (0.7N_{A_0})$ (100) (800) + $(N_{B_0} - 0.3N_{A_0})$ (50) (800) + $(0.3N_{A_0})$ (150) (800) – 0.3 (N_{A_0}) (100) (10^3)

Since $\Delta H = (-100)\,10^3$ J/mol

Simplifying

$5000\,(N_{A_0}) = 12500\,(N_{B_0})$

or $N_{A_0} = \frac{5}{2}N_{B_0}$

Hence, the mole fraction of A in the feed

$= \frac{N_{A0}}{N_{A0}+N_{B0}}$

$= \frac{5}{7}$

72. The irreversible zero order reaction $A \rightarrow B$ takes place in a porous cylindrical catalyst that is sealed at both ends as shown in the figure. Assume dilute concentration and neglect any variations in the axial direction.

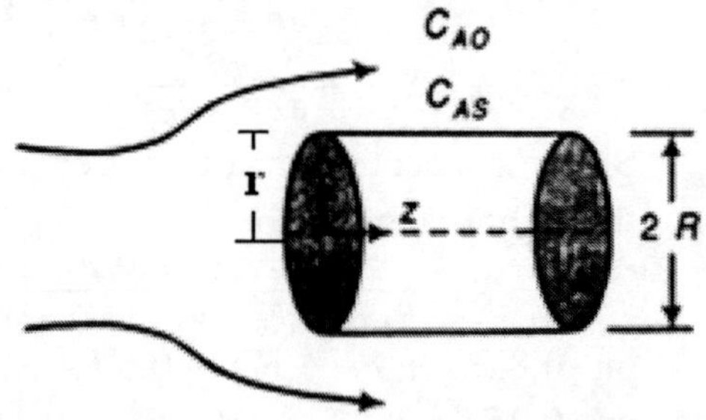

The steady state concentration profiles is

$\frac{C_A}{C_{As}} = 1 + \frac{\varphi_0^2}{4}\left[\left(\frac{r}{R}\right)^2 - 1\right]$, where ϕ_0 is the Thiele modules. For $\phi_0 = 4$, the range of r, where $C_A = 0$, is

A. $0 < r < \frac{R}{4}$
B. $0 < r < \frac{R}{2}$
C. $0 \le r \le R\sqrt{\frac{3}{4}}$
D. $0 \le r \le R$

Sol. **C.**

Given $\frac{C_A}{C_{As}} = 1+\frac{\phi_0^2}{4}\left[\left(\frac{r}{R}\right)^2 - 1\right]$

Data given is

$\phi_0 = 4, C_A = 0$

Hence $0 = 1+\frac{16}{4}\left[\left(\frac{r}{R}\right)^2 - 1\right]$

or $\left(\frac{r}{R}\right)^2 - 1 = -\frac{1}{4}$

or $\left(\frac{r}{R}\right) = \sqrt{3/4}$

Therefore range of r would be

$0 \leq r \leq R\sqrt{3/4}$

Common Data for Questions 79 and 80

A liquid is flowing through a reactor at a constant flow rate. A step input of tracer at a molar flow rate of 1 mol / min is given to the reactor at time t = 0. The time variation of the concentration (C) of the tracer at the exit of the reactor is as shown in the figure.

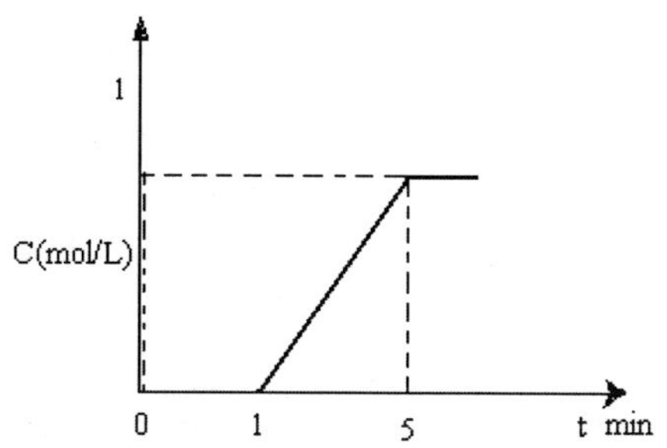

73. The volumetric flow rate of the liquid through the reactor (in L/min) is
A. 1
B. 2
C. 1.5
D. 4

Sol. **A.**

We know $C_{max} = \frac{m}{V}$

Since $C_{max} = 1$ mol/lit and $m = 1$ mol / min

$V = \frac{m}{C_{max}} = \frac{1}{1} = 1$ lit / min

74. The mean residence time of the fluid in the reactor (in min) is
A. 1
B. 2
C. 3
D. 4

Sol. **C.**

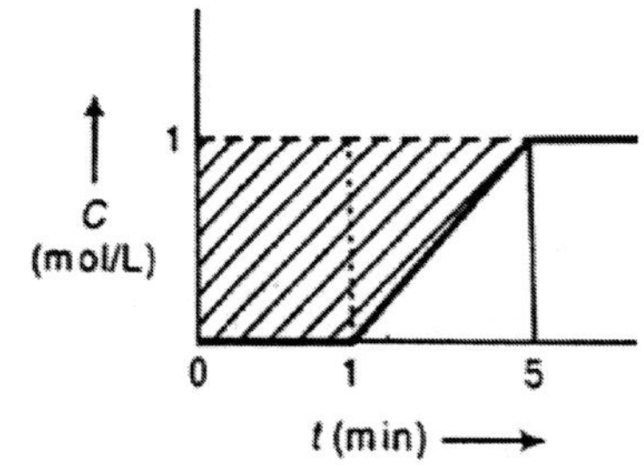

Shaded portion area = $(\bar{t})(C_{max})$

Where $\bar{t}$ is mean residence time

Putting the numerical values

$(1)(1) + \left(\frac{1}{2}\right)(4)(1) = (\bar{t})\ (1)$

$t = 3$ min

Statement for Linked Questions 75 and 76

The liquid phase reaction A → products is to be carried out at constant temperature in a CSTR followed by a PFR in series. The overall conversion of A achieved by the reactor system (CSTR + PFR) is 95%. The CSTR has a volume of 75 L. pure A is fed to the CSTR at a concentration C_{A0} = 2 mol/L and a volumetric flow rate of 4 L/min. the kinetics of the reaction is given by

$$-r_A = 0.1C_A^2 \frac{mol}{L - min}$$

75. The conversion achieved by the CSTR is
A. 40%
B. 50%
C. 60%
D. 80%

Sol. **C.**

Data given

$C_{A_o} = 2$ mol/lit. F = 4 lit/min, V = 75 lit

$-r_A = 0.1\ C_A^2$

Design equation for a CSTR

$\frac{V}{F} = \frac{C_{A_o} - C_{A_1}}{r}$

$= \frac{C_{A_o} - C_{A_1}}{KC_{A_1}^2}$

Since $C_{A_1} = C_{A_0}(1 - x_A)$

$\therefore \frac{V}{F} = \tau = \frac{C_{A_o} - C_{A_0}(1-x_A)}{KC_{A_0}^2(1-x_A)^2}$

or $\tau = \frac{1-1+x_A}{KC_{A_o}(1-x_A)^2}$

Since $\tau = \frac{V}{F} = \frac{75}{4}$

$\frac{75}{4} = \frac{x_A}{0.1C_{A_o}(1-x_A)^2}$

Putting C_{A_o} = 2 mol/lit and
Solving x_A = 1.67 or 0.6
Since x_A will always be less than 1
∴ x_A = 60 = 60%

76. The volume of the PFR required (in litre) is

A. 380
B. 350
C. 75
D. 35

***Sol.* B.**

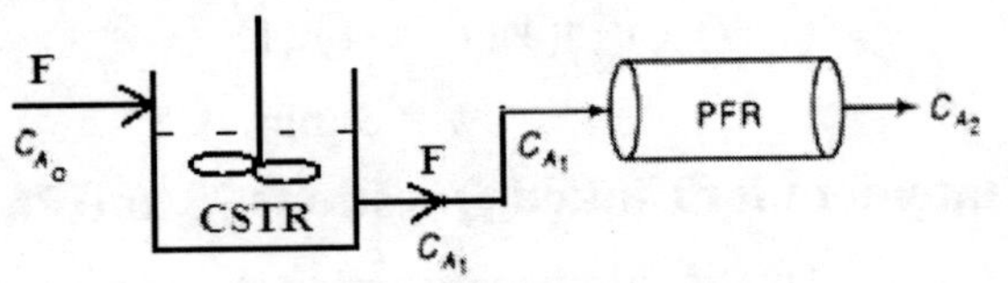

CSTR followed by a PFR in series
Conversion achieved in CSTR is 0.6

$C_{A_1} = C_{A_0}(1 - x_A) = 2(1 - 0.6) = 0.8$ mol/lit

Data given, overall conversion is
CSTR + PFR = 0.95

∴ $C_{A_2} = 2(1 - 0.95)$
= 0.1 mol/lit

For PFR; C_{A_1} = 0.8 mol/lit; C_{A_2} = 0.1 mol/lit
Conversion achieved in PFR only in

$$x = \frac{C_{A_1} - C_{A_2}}{C_{A_1}} = 1 - \frac{0.1}{0.8}$$

= 0.875

$$\text{For PFR } \frac{V}{F} = \int_0^{0.875} \frac{dx_A}{(-r_A)}$$

$$= \int_0^{0.875} \frac{dX_A}{0.1C_{A_1}^2(1-x_A)^2}$$

$$\text{or } \frac{V}{F}(0.1C_{A_1}^2) = \int_0^{0.875} \frac{dX_A}{(1-x_A)^2}$$

$$\frac{V}{4}(0.1) = \left[\frac{1}{1-x_A}\right]_0^{0.875}$$

$$= 8 - 1$$

Hence, V = 350 lit

GATE 2007

77. A well-stirred reaction vessel is operated as a semi batch reactor in which it is proposed to conduct a liquid phase first order reaction of the type A → B. the reactor is fed with the reactant A at a constant rate 1 L/min having feed concentration equal to 1 mol/L. the reactor is initially empty. Given k = 1 min^{-1}, the conversion of reactant A based on moles of A fed at t = 2 min is

A. 0.136
B. 0.43
C. 0.57
D. 0.864

***Sol.* C.**

Semi-batch reactor

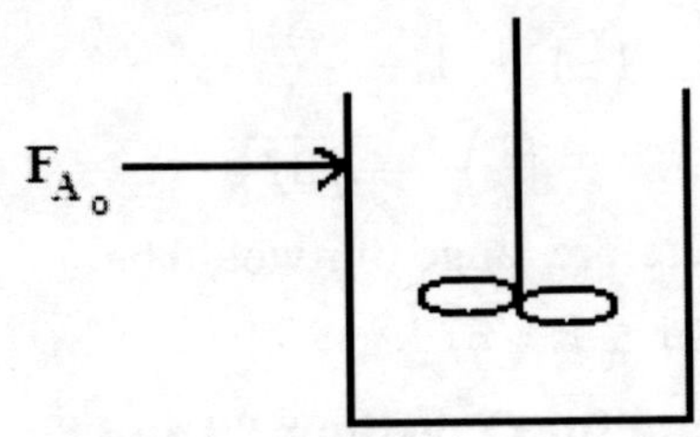

Initial volume $V_0 = 0$

$$-r_A = KC_A$$

Since the reactor is being filled continuously, the volume changes with time and at anytime t, V(t) can be found from overall material balance.

$$V_0\rho_0 = \frac{d}{dt}(\rho_0 V)$$

Since output and generation are zero.

$$\text{or } \frac{dV}{dt} = V_0$$

$$\text{or } V = V_0 + \text{constant}$$

Boundary condition is t = 0; V = 0

∴ constant = 0

Hence V = (V_0)t ....(i)

Now during material balance for A only

$$F_{A_0} = (-r_A)V + \frac{dN_A}{dt}$$

$$\text{or } V_0C_{A_o} = KC_AV + \frac{d(C_AV)}{dt}$$

$$V_0C_{A_o} = KC_AV + \frac{d(C_AV)}{dt}$$

$$= (KC_AV_0)\,t + V\frac{dC_A}{dt} + C_A\frac{dV}{dt}$$

$$\text{Or } V_0C_{A_o} = (KC_AV_0)\,t + V\frac{dC_A}{dt} + C_A\frac{dV}{dt}$$

$$\text{Or } V_0C_{A_o} = (KC_AV_0)\,t + V_0t + \frac{dC_A}{dt} + C_AV_0$$

$$\text{or } \frac{dC_A}{dt} + \left(\frac{1+kt}{t}\right)C_A = \frac{C_{A_o}}{t} \quad \text{....(ii)}$$

We need to solve the equation

$$\text{Integrating factor} = \exp\left[\int\left(\frac{1+kt}{t}\right)dt\right]$$

$$= \exp(kt + \ln t) = t\,e^{kt}$$

Solution of Eq. (ii) can be written as

$$C_Ate^{kt} = \int \frac{C_{A0}}{k} e^{kt}\,dt + C_2 \quad \text{....(iii)}$$

At, t = 0, $C_A = 0$

$$\therefore \quad C_2 = -\frac{C_{A_0}}{k}$$

From Eq. (iii),

$$C_A t\, e^{kt} = \frac{C_{A_o}}{k}\left(e^{kt}-1\right)$$

$$\frac{C_A}{C_{A_0}} = \frac{1}{kt}\left(1-e^{-kt}\right)$$

or $1 - X_A = \frac{1}{kt}\left(1-e^{-kt}\right)$

Putting values, $k = 1\, min^{-1}$, t = 2 min

$$1 - X_A = 0.43$$

$$X_A = 0.57$$

78. A liquid phase exothermic first order reaction is being conducted in a batch reactor under isothermal conditions by removing heat generated in the reactor with the help of cooling water. The cooling water flow at a very high rate through a coil immersed in the reactor such that is negligible rise in heats temperature from inlet to outlet of the coil. If the rate constant is given as k. heat of reaction $(-\Delta H)$, volume of the reactor V, initial concentration as C_{A0}, overall heat transfer coefficient U, heat transfer area of the coil is equal to A, the required cooling water inlet temperature T_{ci} is given by the following equation

A. $T_{ci} = T - \frac{(-\Delta H)VkC_{A0}}{UA}$

B. $T_{ci} = T - \frac{(-\Delta H)VkC_{A0}e^{-kt}}{UA}$

C. $T_{ci} = T - \frac{(-\Delta H)VC_{A0}e^{-kt}}{UAt}$

D. $T_{ci} = T - \frac{(-\Delta H)VC_{A0}}{UAt}$

***Sol.* B.**

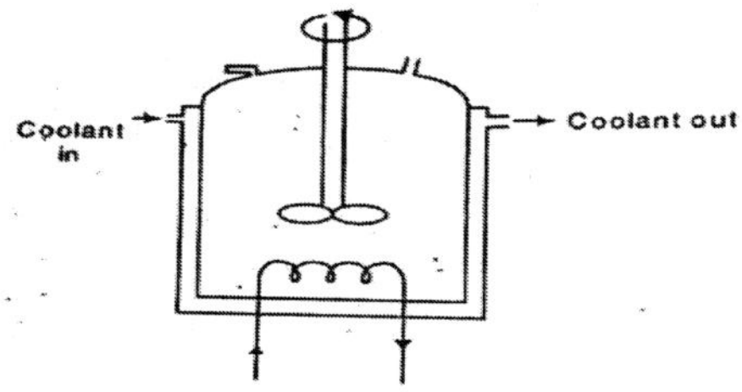

For a first order reaction in a batch reactor

$$-\frac{dC_A}{dt} = kC_A$$

or $C_A = C_{A_o}e^{-kt}$

Heat balance equation is

Heat generated during reaction = heat removed by coolant

or $(r_A)(V)(-\Delta H) = UA\,(T - T_{ci})$

or $(kC_A)(V)(-\Delta H) = UA\,(T - T_{ci})$

or $(T - T_{ci}) = \frac{KV}{UA}\left(C_{A_o}e^{-kt}\right)(-\Delta H)$

or $T_{ci} = T - \left[\frac{(-\Delta H)kV C_{A_o}e^{-kt}}{UA}\right]$

79. The following liquid phase reaction is taking place in an isothermal CSTR

$$A \xrightarrow{k_1} B \xrightarrow{k_2} C$$

$$2A \xrightarrow{k_3} D$$

Reaction mechanism is same as the stoichiometry given above. Given,

$k_1 = 1\, min^{-1}; k_2 = 1\, min^{-1}$;

$k_3 = 0.5 \frac{L}{mol-min}$; $C_{A_0} = 10$ mol/L, $C_{B_0} = 0$ mol/L and $C_B = 1$ mol/L, the solution for *F/V* (flow rate / reactor volume in min^{-1}) yields

A. 6.7

B. 6 and 0.5

C. 2 and 4/3

D. 8

***Sol.* B.**

The reaction is

$$A \xrightarrow{K_1} B \xrightarrow{K_2} C$$

$$2A \xrightarrow{K_3} D$$

$$(-r_A) = k_1 C_A + k_3 C_A^2$$

$$(-r_B) = -k_1 C_A + k_2 C_B$$

$$r_C = k_2 C_B$$

$$r_D = k_3 C_A$$

Data given is

$k_1 = 1\ \text{min}^{-1}$, $k_2 = 1\text{min}^{-1}$,

$k_3 = 0.5 \frac{lit}{(mol)(min)}$

$C_{A_o} = 10$ mol/lit, $C_{B_o} = 0$, $C_B = 1$ mol/lit

For a CSTR the material balance equation is

$$FC_{A_o} = F\,C_A + (r_A)V$$

or $(F)(10) = F\,C_A + (k_1 C_A + k_3 C_A^2)V$

or $\frac{F}{V} = \frac{k_1 C_A + k_3 C_A^2}{(10 - C_A)}$

$$= \frac{C_A + 0.5C_A^2}{(10 - C_A)} \quad \text{.... (i)}$$

Applying material balance for B

$$FC_{B_o} = F\,C_B + (-r_B)V$$

$$0 = (F)(1) + (1 \times 1 - 1 \times C_A)V \quad \text{....(ii)}$$

Or $\frac{F}{V} = C_A - 1$

Combining equations (i) and (ii)

$$(C_A + 0.5C_A^2) = (10 - C_A)(C_A - 1)$$

Or $1.5C_A^2 - 10\,C_A + 10 = 0$

Solving, C_A will have two roots

$C_A = 5.44$ and 1.23

Hence by equation (ii)

$\frac{F}{V} = 4.44$ and 0.23

The option closest to this value is 6 and 0.5.

80. A pulse of concentrated KCI solution is introduced as tracer into the fluid entering a reaction vessel having volume equal to 1 m^3 and flow rate equal to 1 m^3/min. the concentration of tracer measured in the fluid leaving the vessel is shown in the figure given below. The flow model parameters that fit the measured RTD in terms of one or all of the following mixing elements, namely, volume of plug flow reactor V_p, mixed flow volume V_m, and dead space V_d, are

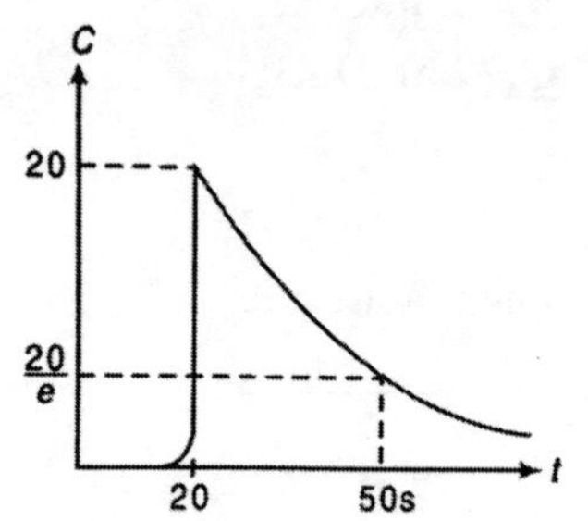

A. $V_p = \frac{1}{6}m^3, V_m = \frac{1}{2}m^3, V_d = \frac{1}{3}m^3$

B. $V_p = V_m = V_d = \frac{1}{3}m^3$

C. $V_p = 1/3m^3, V_m = 1/2\,m^3, V_d = 1/6m^3$

D. $V_m = \frac{5}{6}m^3, V_d = \frac{1}{6}m^3$

***Sol.* C.**

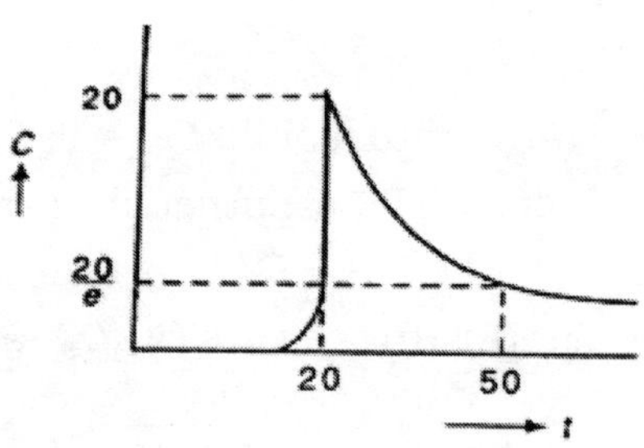

Data given is

$V = 1\text{ m}^3$

$F_{A_0} = 1\text{ m}^3/\text{min} = \frac{1}{60}\text{ m}^3/\text{sec}.$

For PFR $\tau_p = 20$ sec

$$\frac{V_p}{V_0} = \tau_p$$

$$V_p = \frac{1}{60} \times 20 = \frac{1}{3}\text{ m}^3$$

Further, for PFR + MFR system,

$$\frac{V_p + V_m}{F_{A_0}} = 50\text{ sec.}$$

$$\frac{1}{3} + V_m = \left(\frac{1}{60}\right)(50)$$

$$\therefore \quad V_m = \frac{1}{2}\text{ m}^3$$

For dead space,

$$V_d = V_{Total} - (V_p + V_m)$$

$$= 1 - \left(\frac{1}{3} + \frac{1}{2}\right)$$

$$V_d = \frac{1}{6}\text{ m}^3$$

81. The first order reaction of A to R is run in an experimental mixed flow reactor. Find the role played by pore diffusion in the run given below. C_{A0} is 100 and W is fixed. Agitation rate was found to have no effect on conversion

d_p	$F_{A_0}X_A$	
4	2	0.8
6	4	0.4

A. Strong pore diffusion control

B. Diffusion free

C. Intermediate role by pore diffusion

D. External mass transfer

***Sol.* A.**

The trend of the data indicates that with the increases in particle diameter (d_p) the conversion decreases. Hence, the reaction is strongly controlled by pore diffusion.

82. A packed bed reactor converts A to R by first order reaction with 9 mm pellets in strong pore diffusion regime to 63.2% level. If 18 mm pellets are used what is the conversion.

A. 0.39

B. 0.61

C. 0.632

D. 0.865

***Sol.* D.**

For a first order reaction taking place is a packed bed reactor

$$\omega = \frac{1}{(k\propto)d_p}\ln(1 - x_A) \qquad \text{....(i)}$$

Where $\propto$ is an arbitrary constant and d_p is particle diameter

Data given is

$d_{p_1} = 9$ mm; $d_{p_2} = 18$ mm

$x_{A_1} = 0.632$; $x_{A_2} = ?$

From equation (i)

$$\frac{ln(1-x_{A1})}{d_{p1}} = \frac{ln(1-x_{A2})}{d_{p2}}$$

Substituting the values

$$ln(1-x_{A_2}) = -1.99$$

$\therefore \quad x_{A_2} = 0.865$

83. The following rate concentration data are calculated from experiment. Find the activation energy temperature (E/R) of the first order reaction.

d_p	C_A	$-r_A$	T
1	20	1	480
2	40	2	480
2	40	3	500

A. 2432.8
B. 4865.6
C. 9731.2
D. 13183.3

Sol. **B.**

For a first order reaction

$$-r_A = k\,C_A$$

From Arrhenius equation; $k = Ae^{-E/RT}$

$\therefore \quad -r_A = (Ae^{-E/RT})\,C_A$

or $\quad \ln\frac{(-r_A)}{C_A} = \ln A - \frac{E}{RT}$

Plotting a graph between $\ln\frac{(-r_A)}{C_A}$ versus $\frac{1}{T}$ using the given data, slope of the graph is

$$E/R = 4865.6$$

84. Determine the level of C_{A0} (high, low, Intermediate), temperature profiles (high, low, increasing, decreasing), which will favour the formation of the desired product indicated in the reaction scheme given below

$$A \xrightarrow{1} R \xrightarrow{3} S \text{ (desired)}$$
$$A \xrightarrow{2} U$$

$n_1 E_1 n_2 E_2 n_3 E_3$

2 25 1 35 3 45

A. High C_{A_0}, Increasing T, plug flow reactor
B. Low C_{A_0}, increasing T, plug flow reactor
C. High C_{A_0}, decreasing T, mixed flow reactor
D. High C_{A_0}, decreasing T, plug flow reactor

Sol. **A.**

$$A \xrightarrow{1} R \xrightarrow{3} S \text{ (desired)}$$
$$A \xrightarrow{2} U$$

As is evident from the given data, the activation energy is highest for the desired product reaction. This means, increasing temperature will favor this desired reaction. For the same volume of PFR and MFR, PFR is chosen for higher conversion rate. Finally, higher initial concentration of A will give higher rate of reaction resulting in formation of higher amount of product.

In summary, high C_{A_o} increasing T plug flow reactor is the correct answer.

Common Data for Questions 91 and 92

The following liquid phase reaction is taking place in an isothermal batch reactor

$$\xrightarrow{k_1\,(first\ order)} B \xrightarrow{k_2\,(zero\ order)} C$$

Feed concentration = 1 mol/L

85. The time at which the concentration of B will reach its maximum value is given by

A. $t = \frac{1}{k_1}\,in\left(\frac{k_1}{k_2}\right)$
B. $t = \frac{1}{k_2-k_1}\,in\left(\frac{k_2}{k_1}\right)$
C. $t = \frac{1}{k_2}\,in\left(\frac{k_2}{k_1}\right)$
D. $t = \frac{1}{k_2}\,in\left(\frac{k_1}{k_2}\right)$

Sol. **A.**

$$A \xrightarrow{1\ Order} B \xrightarrow{zero\ Order} C$$

Feed concentration = 1 mol/lit

For first order reaction

$$-\frac{dC_A}{dt} = k_1 C_A$$

$$\int k_1 dt = -\int \frac{dC_A}{C_A}$$

or $\quad C_A = C_{A_o} e^{-k_1 t}$

$\quad C_A = e^{-k_1 t}$ (since $C_{A_o} = 1$)

For B;

$$-\frac{dC_B}{dt} = k_1 C_A - k_2$$

$$= k_1 e^{-k_1 t} - k_2 \qquad \text{.... (i)}$$

Applying condition

$\frac{dC_B}{dt} = 0$ when C_B is maximum

$$k_1 e^{-k_1 t} - k_2 = 0$$

or $\quad -k_1 t = \text{In}\,(k_2/k_1)$

or $\quad t = \frac{1}{k_1}\text{In}\,(k_1/k_2)$

86. The time at which the concentration of B will become zero is given by the following equation

A. $1 - e^{-k_1 t} = k_2 t$

B. t = infinite

C. $t = 1/k_2$

D. $t = 1/k_1$

***Sol.* A.**

$$\frac{dC_B}{dt} = k_1 e^{-k_1 t} - k_2$$

$$\int_0^{C_B} dC_B = \int_0^t (k_1 e^{-k_1 t} - k_2)\, dt$$

or $C_B = 1 - k_2 t - e^{-k_1 t}$

Determining t for $C_B = 0$

$$0 = 1 - k_2 t - e^{-k_1 t}$$

or $1 - e^{-k_1 t} = k_2 t$

GATE 2006

87. The rate at which an antiviral drug acts , increases with is concentration in the blood C, according to the equation

$$r = \frac{kC}{C_{50} + C}$$

Where, C_{50} is the concentration at which the rate is 50% of the maximum rate k. Often, the concentration C_{90}, when the rate is 90% of the maximum, is measured instead of C_{50}. The rate equation then becomes

A. $r = \frac{18kC}{C_{90} + C}$

B. $r = \frac{kC}{\left(\frac{C_{90}}{9} + C\right)}$

C. $r = \frac{kC}{C_{90}}$

D. $r = \frac{0.9\, kC}{C_{90}}$

***Sol.* B.**

Given $r = \frac{kC}{C_{50} + C}$ ....(i)

Further r_{max} = k and r_{90} = 0.9 k

or $\frac{kC_{90}}{C_{50} + C_{90}} = 0.9$ k

or $kC_{90} = 0.9\text{ k }(C_{50} + C_{90})$ ....(ii)

$C_{90} = 9C_{50}$

Combining equations (i) and (ii)

$$r = \frac{kC}{\left(\frac{C_{90}}{9} + C\right)}$$

88. Consider the following reactions between gas A and two solid spherical particles B and C of the same size.

$A + B \xrightarrow{k_1}$ gaseous product,

$A + C \xrightarrow{k_2}$ ash

The ash does not leave the particle C. Let t_1 and t_2 be the times required for A to completely consume particles B and C, respectively. If k_1 and k_2 are equal at all temperatures and the gas phases mass transfer resistance is negligible, then

A. $t_1 = t_2$ at all temperatures

B. $t_1 = t_2$ at high temperatures

C. $t_1 > t_2$ at high temperatures

D. $t_1 < t_2$ at high temperatures

***Sol.* D.**

$A + B \xrightarrow{k_1}$ gaseous product .... (i)

$A + C \xrightarrow{k_2}$ ash .... (ii)

Gaseous product is formed in reaction (i) which will favor easier penetration of reactant A and B as no resistance will be offered in penetration. On the other hand in reaction (ii) ash is formed which will offer resistance to further penetration of reactant A and C due to solid layer in between which is being formed continuously. The resistance will increases with time and this will specially be more significant at high temperature where the rate of reaction is higher.

Hence it can be concluded that $t_1 > t_2$ at higher temperature.

89. A reaction A→ B is to be conducted in two CSTR in series. The steady state conversion desired is X_f. The reaction rate as the function of conversion is given by $r = -1/(1 + X)$. If the feed contains no B, then the conversion in the first reactor that minimizes the total volume of the two reactors is

A. $1 - X_f$

B. $0.2\, X_f$

C. $0.5\, X_f$

D. $0.5(1 - X_f)$

***Sol.* C.**

Rate equation is given as

$r = \frac{-1}{(1+X)}$ when X is conversion achieved

or $-r = \frac{1}{(1+X)}$

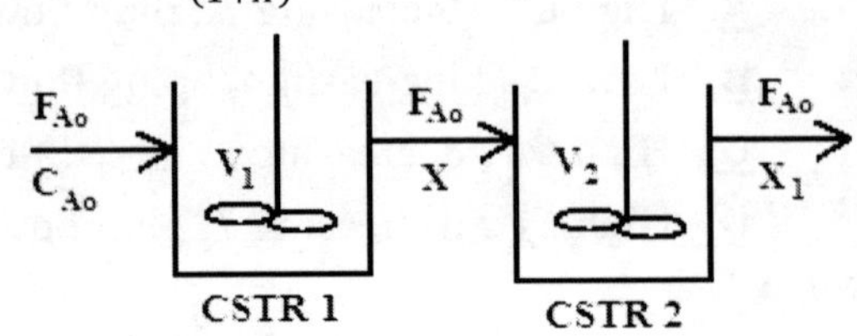

For CSTR 1

$$\frac{V_1}{F_{A0}} = \frac{X}{\frac{1}{(1+X)}}$$

or $V_1 = F_{A_0}(X)(1+X)$ ....(i)

For CSTR 2

$$\frac{V_2}{F_{A0}} = \frac{X_1 - X}{\frac{1}{(1+X_1)}}$$

or $V_2 = F_{A_0}(X_1 - X)(1 + X_1)$

$= F_{A_0}[X_1^2 + X_1(1 - X) - X]$

Total volume of the two CSTR

$V = V_1 + V_2$

$= (F_{A_0})(X)(1+X) + F_{A_0}[X_1^2 - X_1(1 - X) - X]$

For total volume to be minimum

$$\frac{dV}{dX} = 0$$

or $F_{A_0}(1+2X) + F_{A_0}(0 - X_1 - 1) = 0$

or $1 + 2X - X_1 - 1 = 0$

or $X = 0.5X_1$

90. Consider the following elementary reaction network

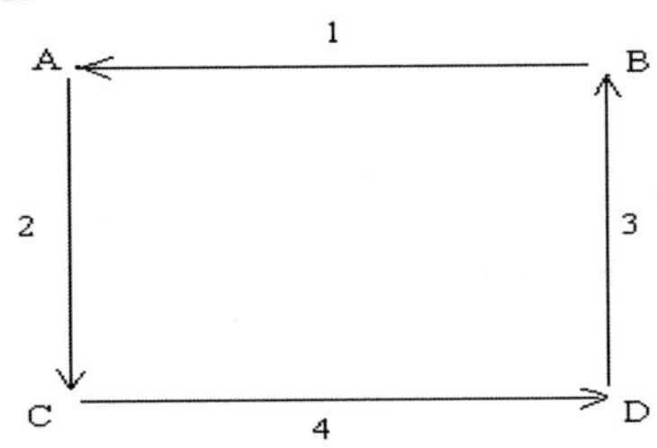

The activation energies for the individual reactions are $E_1 = 100$ kJ/mol, $E_2 = 150$ kJ/mol, $E_3 = 100$ kJ/mol and $E_4 = 200$ kJ/mol. If the feed is pure A and the desired product is C, then the desired temperature profile in a plug flow reactor in the direction of flow should be

A. constant at low temperature
B. constant at high temperature
C. increasing
D. decreasing

***Sol.* D.**

Given $E_1 = 100$ kJ/mol; $E_2 = 150$ kJ/mol,
$E_3 = 100$ kJ/mol; $E_4 = 200$ kJ/mol,

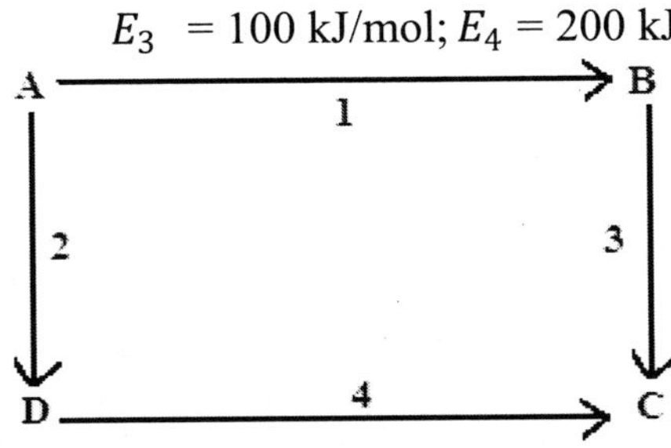

Starting with pure A, higher temperature of the reactor will favour reaction 2 since $E_2 > E_1$. After substantial amount of C is formed reaction 4 should be suppressed which means temperature should be lower. Since E_4 is largest a lower temperature will lower the formation of D from reaction 4.

91. The exit gage distribution in a stirred reactor is given by $E(t) = \frac{1}{\tau}e^{-t/\tau}$. Fluid elements e_1 and e_2 enter the reactor at a time $t = 0$ and $t = \theta > 0$, respectively. The probability that e_2 exits the reactor before e_1 is

A. ½
B. $\frac{1}{2}e^{-\theta/\tau}$
C. (c) $e^{-\theta/\tau}$
D. Zero

***Sol.* B.**

Given $E(t) = \frac{1}{\tau}e^{-t/\tau}$

e_1 enters at $t = 0$

e_2 enters at $t = \theta > 0$

P(exit of e_2) $= \left(\frac{1}{2}\right)\left(\frac{1}{\tau}\right)e^{-\theta/\tau}$

P(exit of e_1) $= \frac{1}{\tau}$

$$\text{P(exit of } e_2 \text{ before } e_1) = \frac{\left(\frac{1}{2}\right)\left(\frac{1}{\tau}\right)e^{-\theta/\tau}}{\frac{1}{\tau}} = \frac{1}{2}e^{-\theta/\tau}$$

GATE 2005

92. The rate expression for the reaction of A is given by $-r_A = \frac{k_1 C_A^2}{1+k_2 C_A^{1/2}}$. The units of k_1 and k_2 are respectively,

A. ($mol^{-1}\ m^3\ s^{-1}$), ($mol^{-1/2}\ m^{3/2}$)
B. ($mol^{-1}\ m^3\ s^{-1}$), ($mol^{1/2}\ m^{3/2}$)
C. ($mol\ m^{-3}\ s^{-1}$), ($mol^{1/2}\ m^{-3/2}\ s^{-1}$)
D. ($mol^{-1}\ m^3\ s^{-1}$), ($mol^{-1/2}\ m^{3/2}\ s^{-1/2}$)

***Sol.* A.**

Given $-r_A = \frac{k_1 C_A^2}{1+k_2 C_A^{1/2}}$

or $-r_A\left(1 + k_2 C_A^{1/2}\right) = k_1 C_A^2$

or $(-r_A) + (-r_A)\left(k_2 C_A^{1/2}\right) = k_1 C_A^2$

Mathematically, each term of this equation will have the same units or dimension as $(-r_A)$.

Since the unit of $(-r_A)$ is $\frac{mol}{(m^3)(sec)}$

Unit of $k_1 C_A^2$ will also be mol/$(m^3)(sec)$

or $k_1 = \left[\frac{mol}{(m^3)(sec)}\right]\left[\frac{m^3}{mol}\right]^2$

or unit of $k_1 = mol^{-1}m^3sec^{-1}$

Further, unit of $(-r_A)k_2C_A^{1/2}$ = unit of $(-r_A)$

$\therefore \quad k_2 = \left(\frac{mol}{m^3}\right)^{1/2} = 1$

$k_2 = mol^{-1/2}m^{3/2}$

93. The first order liquid phase reaction A→ P is to be carried out isothermally in the following ideal reactor configurations

(P) A 1m³ CSTR followed by a 1m³ PFR,

(Q) A 2m³ CSTR

(R) A 1m³ PFR followed by a 1m³ CSTR

(S) A 1m³ CSTR followed by a 1 m³ CSTR

The overall exit conversions X, for the above configurations *P, Q, R* and *S*, assuming indentical inlet conditions and temperature, are related as

A. $X_p > X_R > X_S > X_Q$

B. $X_p = X_R > X_S > X_Q$

C. $X_p = X_S = X_Q = X_R$

D. $X_Q = X_P > X_R > X_S$

***Sol.* B.**

For a CSTR we can write

$$\frac{V}{F_{A0}} = \frac{X_A}{(-r_A)}$$

or $\quad \frac{V}{V_0C_{A0}} = \frac{X_A}{kC_{A0}(1-X_A)}$

or $\quad \tau = \frac{X_A}{k(1-X_A)} \qquad \left(since\ \tau = \frac{V}{V_0}\right)$

$\therefore \quad X_A = \frac{k\tau}{1+k\tau}$ .... (i)

Similarly for a PFR

$X_A = 1 - e^{-k\tau}$ ....(ii)

Since inlet conditions are identical as given in the problem. Assuming $V_0 = 1\ m^3/s$

For option p

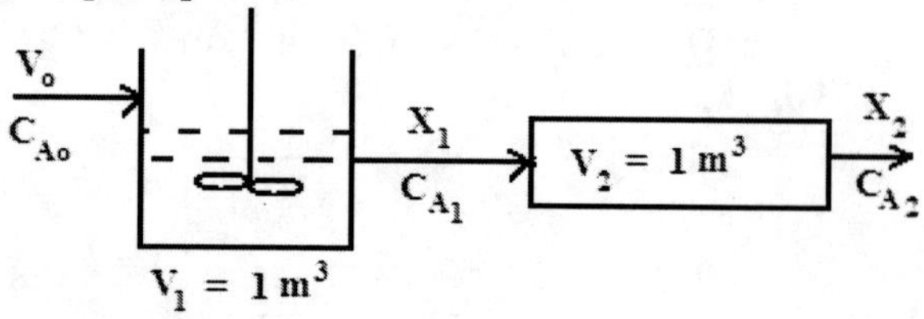

For the CSTR

$\tau = \frac{V}{V_0} = \frac{1}{1} = 1$ sec

$\therefore X_1 = \frac{k}{k+1}$

Now since

$X_1 = \frac{C_{A0}-C_{A1}}{C_{A0}} = 1 - \frac{C_{A1}}{C_{A0}}$

or $\frac{C_{A1}}{C_{A0}} = 1 - \frac{k}{k+1} = \frac{k+1-k}{k+1} = \frac{1}{k+1}$

For PFR $\tau = \frac{1}{1} = 1$sec

$X_2 = 1 - e^{-k}$

or $\quad X_2 = 1 - \frac{C_{A1}}{C_{A0}} = 1 - e^{-k}$

or $\quad \frac{C_{A1}}{C_{A0}} = e^{-k}$

Hence overall exit concentration

$X_p = 1 - \frac{C_{A2}}{C_{A0}}$

$= 1 - \left\{\left(\frac{C_{A2}}{C_{A1}}\right)\left(\frac{C_{A1}}{C_{A0}}\right)\right\}$

$\therefore \quad X_p = 1 - \left(\frac{1}{k+1}\right)e^{-k}$ ....(iii)

Option Q

For 2 m³ CSTR

$\tau = \frac{2}{1} = 2$sec.

From equation (i)

$X_Q = \frac{2k}{1+2k}$ ....(iv)

Option R

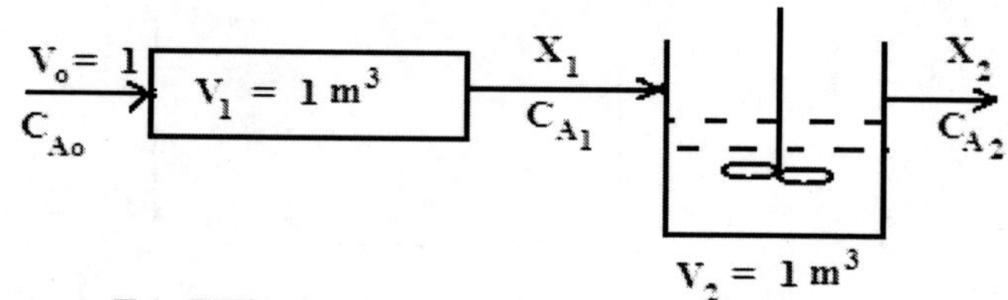

For PFR;

$\tau = 1$ sec.

$X_1 = 1 - e^{-k}$

or $\frac{C_{A1}}{C_{A0}} = e^{-k}$

For CSTR

$\tau = 1$ sec.

$X_2 = 1 - \frac{C_{A2}}{C_{A1}} = \frac{k}{1+k}$

or $\frac{C_{A2}}{C_{A1}} = \frac{1}{1+k}$

The overall conversion

$X_R = 1 - \frac{C_{A2}}{C_{A0}}$

$= 1 - \left(\frac{1}{k+1}\right)e^{-k}$ .... (v)

Option S

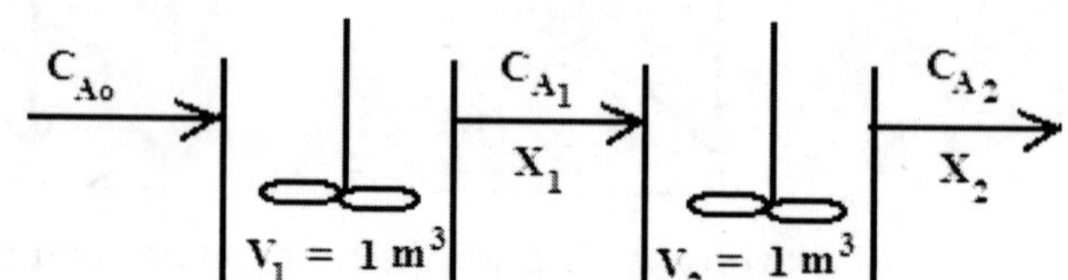

Here $\tau_1 = \tau_2 = 1$ sec.

$$X_1 = \frac{k}{1+k}$$

or $1 - \frac{C_{A1}}{C_{A0}} = \frac{k}{1+k}$

or $\frac{C_{A1}}{C_{A0}} = \frac{k}{1+k}$

And $X_2 = \frac{1}{1+k}$

$\therefore \quad \frac{C_{A2}}{C_{A1}} = \frac{1}{1+k}$

The overall conversion

$$X_S = 1 - \frac{C_{A2}}{C_{A_0}} = 1 - \left\{\frac{k}{(1+k)}\right\}^2$$

$$= \frac{k(k+2)}{(1+k)^2} \quad \ldots.(vi)$$

Now comparing equation (iii), (iv), (v) and (vi)

$$X_p = X_R$$

Further, two CSTRs in series always give higher conversion than one single CSTR with same total value.

$$\therefore \; X_S > X_Q$$

It can be verified by putting any value of k in these equations

For example let us put

k = 1

$X_p = X_R = 0.82$

$X_S = 0.75$

$X_Q = 0.67$

Hence, $X_p = X_R > X_S > X_Q$ is the correct answer.

94. The gas phase reaction A →B + C is carried out in an ideal PFR achieving 40% conversion of A. The feed has 70 mol% A and 30 mol% inserts. The inlet temperature is 300 K and the outlet temperature is 400 K. The ratio of the outlet to inlet molar concentration of A (assuming ideal gas mixture and uniform pressure) is

A. 0.60

B. 0.30

C. 0.47

D. 0.35

***Sol.* C.**

Gas phase reaction

$$A + (Inerts) \longrightarrow B + C + (Inerts)$$

Initially 70 30 0 0 30

After 40% 70 (1 – 0.4) 70×0.4 70×0.4 30

Conversion = 42 = 28 = 28

Is assumed to be 100 moles

Total moles in outlet = 42 + 28 + 28 + 30

= 128

∴ Mole fraction of A in the outlet is

$\frac{42}{128} = 0.33$

Hence, ratio of outlet to the inlet molar concentration of A is

$$\frac{0.33}{0.7} = 0.47$$

95. Match the item in Group I with those in Group II.

Group I	Group II
P. Porous catalyst	1. Selectivity
Q. Parallel reactions	2.Shrinking core modal
R. Non-ideal tubular reactor	3. Thiele modulus
S. Gas-solid non-catalytic reaction	4. Dispersion number

A. P-3, Q-1, R-4, S-2

B. P-1, Q-3, R-2, S-4

C. P-1, Q-4, R-2, S-3

D. P-3, Q-4, R-1, S-2

***Sol.* A.**

Porous catalyst – Thiele modulus

Parallel reactions – selectivity

Non – ideal tubular reactor – Dispersion number

Gas – solid non – catalyst reaction – Shrinking core model

96. The rate of the liquid phase reversible reaction

$A \rightleftarrows 2B$ in (kmol/m^3-min) at 298 K, is

$-r_A = 0.02\, C_A - 0.01\, C_B$ where the concentrations C_A and C_B are expressed in (kmol/m^3). What is the maximum limiting conversion of A achievable in an isothermal CSTR at 298 K, Assuming pure A is fed at the inlet?

A. 1

B. 2/3

C. 1/2

D. 1/3

***Sol.* C.**

Given reaction A $\rightleftarrows$ 2 B

Initial mole 1 0

At equilibrium $(1 - X_e)$ $2X_e$

Rate equation is $(-r_A) = 0.02\, C_A - 0.01 C_B$

Since the reaction is reversible

The maximum limiting conversion of A will be at equilibrium where rate of forwarded reaction = rate of backward reaction

or $0.02\, C_{A_e} - 0.01 C_{A_B} = 0$

or $X_e = {}^1/_2$

Statement for Linked Questions 103 and 104

The residence time distribution E(t) (as shown below) of a reactor is zero until 3 min and then increases linearly to a maximum value E_{max} at 8 min after which it decreases linearly back to zero at 15 min.

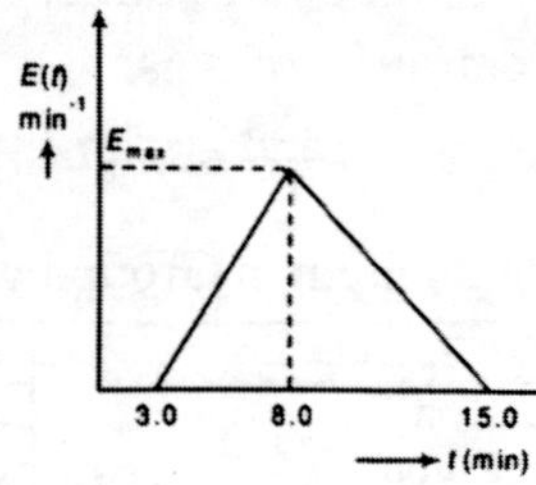

97. What is the value of ?

A. 1/6
B. 1/8
C. 1/4
D. 1/3

Sol. **A.**

It is given that

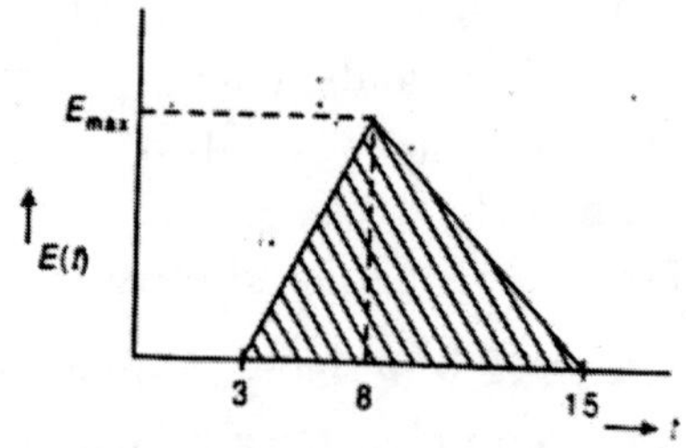

$\int_0^{\infty} E(t)dt = 1$

Area of shaded portion = 1

Calculating the area of shaded portion

$\frac{1}{2}(8-3)\,E_{max} + \frac{1}{2}(15-8)\,E_{max} = 1$

or $E_{max} = \frac{1}{6}$

98. What is the value of the mean residence time in min?

A. 5.7
B. 8
C. 8.7
D. 12

Sol. **B.**

Mean residence time is

$$t = \frac{\int_0^{\infty} t\,E(t)dt}{\int_0^{\infty} E(t)dt}$$

substituting $\int_0^{\infty} E(t)dt = 1$

$\bar{t} = \int_0^{\infty} t\,E(t)dt$

$= 8$ min.

GATE 2004

99. The following gas phases reaction is taking place in a plug flow reactor ,

$$A + \frac{1}{2}B \rightarrow C$$

A stoichiometric mixture of A and B at 300 K is fed to the reactor. At 1 m along the length of the reactor, the temperature is 360 K. The pressure drop is negligible and an ideal gas behavior can be assumed. Identify the correct expression relating the concentration of A at 1m (C_A) and the corresponding conversion of A (X).

A. $C_A = 1.2\,C_{A0}\frac{(1-X)}{(1-0.33\,X)}$

B. $C_A = 1.2\,C_{A0}\frac{(1-X)}{(1-0.5\,X)}$

C. $C_A = 0.83\,C_{A0}\frac{(1-X)}{(1-0.33\,X)}$

D. $C_A = 0.83\,C_{A0}\frac{(1-X)}{(1-0.5\,X)}$

Sol. **B.**

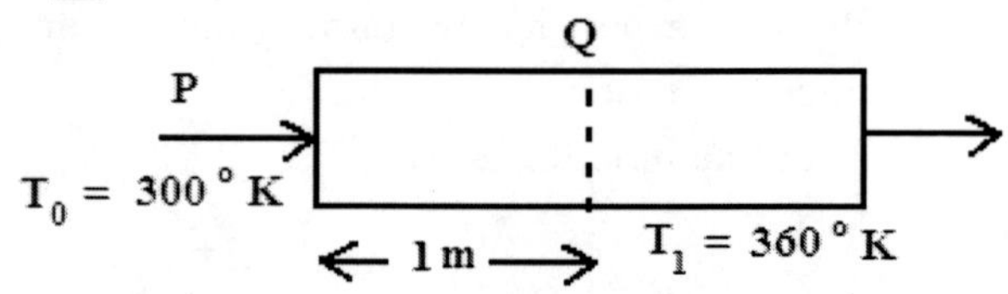

The reaction is $A + \frac{1}{2}B \longrightarrow C$

Initial moles 1 0.5 0

$\therefore \quad \frac{1-\left(1+\frac{1}{2}\right)}{\left(1+\frac{1}{2}\right)} = -\frac{1}{3}$

Calculating volume at point Q

We know that

$V_1 = V_0(1 + \varepsilon_A x_A)$

or $V_1 = V_0\left(1 - \frac{1}{3}x_A\right)$

Since temperature is also changing, thus we need to calculate the volume at Q with temperature $T_1 = 360$ K, at constant pressure

$\frac{V_1}{T_1} = \frac{V_2}{T_2}$

$\frac{V_0\left(1-\frac{1}{3}x_A\right)}{300} = \frac{V_2}{360}$

Or $V_2 = 1.2\,V_0\left(1 - \frac{1}{3}x_A\right)$

Hence the concentration at point Q

$$C_A = \frac{N_A}{V_2}$$

$$= \frac{N_{A_0}(1-x_A)}{1.2V_0(1-0.33\ x_A)}$$

or $C_A = 0.83C_A\left[\frac{(1-x_A)}{1-0.333\ x_A}\right]$

100. A second order liquid phase reaction A → B is carried out in a mixed flow reactor operated in semi-batch mode (no exit stream). The reactant A at concentration C_{AF} is fed to the reactor at a volumetric flow rate of F. The volume of the reacting mixture is V and the density of the liquid mixture is constant. The mass balance for A is

A. $\frac{d(VC_A)}{dt} = -F\ (C_{AF} - C_A) - kC_A^2\ V$

B. $\frac{d(VC_A)}{dt} = F\ (C_{AF} - C_A) - kC_A^2\ V$

C. $\frac{d(VC_A)}{dt} = -FC_{FA} - kC_A^2\ V$

D. $\frac{d(VC_A)}{dt} = FC_{FA} - kC_A^2\ V$

***Sol.* A.**

$$A \longrightarrow B$$

For second order reaction $(-r_A) = kC_A^2$

Writing material balance for A

Input – (Output + Reaction) = Accumulation

$$F_{A_0} = 0 + (-r_A)V + \frac{dN_A}{dt}$$

or $\quad V_0C_{A_0} = kC_A^2 + \frac{d(C_AV)}{dt}$

Since $\quad V_0 = F$ and $C_{A_0} = C_{A_F}$

$\therefore \quad \frac{d(V\ C_A)}{dt} = FC_{A_F} - (kC_A^2)\ V$

101. For an isothermal second order aqueous phase reaction A → B, the ratio of the time required for 90% conversion to the time required for 45% conversion is

A. 2

B. 4

C. 11

D. 22

***Sol.* C.**

For second order reaction

$$A \longrightarrow B$$

$$(-r_A) = \frac{-dC_A}{dt} = KC_A^2$$

and $\quad C_A = C_{A_0}(1 - x_A)$

or $\quad C_{A_0}\frac{dx_A}{dt} = KC_{A_0}^2(1 - x_A)^2$

or $\quad KC_{A_0}\int_0^t dt = \int_0^{x_A}\frac{dx}{(1-x_A)^2}$

or $\quad t = \frac{1}{KC_{A_0}}\left(\frac{x_A}{1-x_A}\right)$

Time required for 90% conversion

$$t = \frac{1}{KC_{A_0}}\left(\frac{0.9}{1-0.9}\right) = \frac{9}{KC_{A_0}}$$

Time required for 45% conversion

$$t_{45} = \frac{45}{55}\left(\frac{1}{KC_{A_0}}\right)$$

Hence the ratio $\frac{t_{90}}{t_{45}} = 9 \times \frac{45}{55}$

$= 11$

102. An isothermal aqueous phase reversible reaction $P \rightleftarrows R$ is to be carried out in a mixed flow rector. The reaction rate in (kmol/m^3-h) is given by r = 0.5 C_p – 0.125 C_R. A stream containing only P enters the reactor. The residence time required (in hour) for 40% conversion of P is

A. 0.80

B. 1.33

C. 1.60

D. 2.67

***Sol.* C.**

The reaction is P	$\rightleftarrows$	R
Initial mole	C_{P_0}	0
After 40% conversion	C_{P_0} (1 – 0.4)	0.4 C_{P_0}

Rate of reaction is given as

$$r = 0.5\ C_P - 0.125\ C_P$$

For a mixed flow reactor

$$\frac{F}{F_{A_0}} = \frac{X_A}{(-r_A)}$$

or $\quad \frac{V}{(V_0)(C_{P_0})} = \frac{X_P}{0.5C_P - 0.125\ C_R}$

or $\quad \frac{\tau}{C_{P_0}} = \frac{0.4}{0.5C_P - 0.125\ (4\ C_{P_0})}$

$\therefore \quad \tau = 1.6$ hr.

103. A pollutant P degrades according to first order kinetics. An aqueous stream containing P at 2 kmol/m^3 and volumetric flow rate 1 m^3/h requires a mixed flow reactor of volume V to bring down the pollutant level to 0.5 kmol/m^3. The inlet concentration of the pollutant is now doubled and the volumetric flow rate is tripled. If the pollutant level is to be brought down to the same level of 0.5 mol/m^3, the volume of the mixed flow reactor should be increased by a factor of

A. 7
B. 6
C. 3
D. 7/3

***Sol.* A.**

For 1st case $C_{A_0} = 2$ kmol/m^3

$V_0 = 1$ m^3/hr; $C_A = 0.5$ kmol/m^3

For a mixed flow reactor

$$\frac{V}{F_{A0}} = \frac{X_A}{(-r_A)}$$

$$\text{or} \quad \frac{V}{(V_0)(C_{A0})} = \frac{X_A}{kC_A} \quad \text{....(i)}$$

$$\text{Since} \quad X_A = 1 - \frac{C_A}{C_{A0}} = 0.75$$

$$\therefore \quad \frac{V}{(1)(2)} = \frac{0.75}{0.5\,k}$$

$$\text{or} \quad kV = 3 \quad \text{....(ii)}$$

For 2nd case

$C_{A0_2} = 4$ kmol/m^3; $V_{02} = 3$ m^3/hr

$C_{A_2} = 0.5$ kmol/m^3

$$\therefore \quad X_A' = 1 - \frac{0.5}{4} = 0.875$$

Substituting in equation (i)

$$\frac{V_2}{(3)(4)} = \frac{0.875}{0.5\,k}$$

$$\text{or} \quad kV_2 = 21 \quad \text{....(iii)}$$

From equation (ii) and (iii)

$$\frac{V_2}{V} = 7$$

104. Consider a reversible exothermic reaction in a plug flow reactor. The maximum and minimum permissible temperatures are and T_{max} and T_{min} respectively. Which of the following temperature (T) profiles will require the shortest residence tie to achieve the desired conversion?

A. 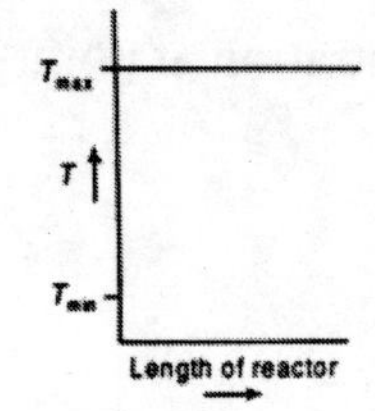

B. 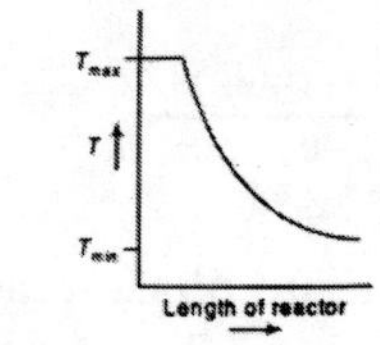

C. 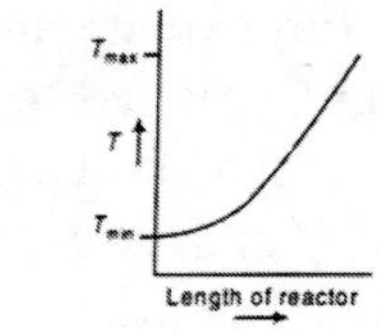

D. 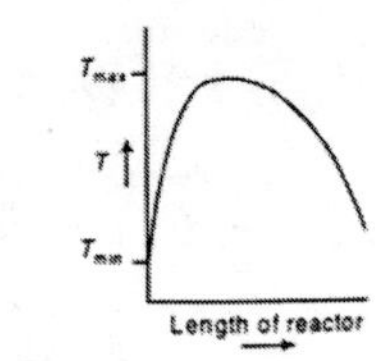

***Sol.* B.**

For the shortest residence time it should be noted that maximum rate for forward reaction which will happen in the beginning at maximum temperature should be utilized and thereafter when substantial concentration of B is formed the reverse reaction will be more prominent at that time temperature should be minimum. From an inspection of the four given temperature profiles it is evident that this criteria is satisfied in option (b).

Option (a) is ruled out because during regime of high concentration of B i.e. when the reverse reaction is prominent the temperature remains in the maximum zone.

Option (c) and (d) indicate initial lower temperature with lower rate of forward reaction and with lower rate of forward reaction and with increase in temperature and product concentration the backward rate of reaction increases resulting in decrease in overall reaction rate

Hence **B.** is the correct answer.

105. An irreversible aqueous phase reaction $A + B \rightarrow P$ is carried out in an adiabatic mixed flow reactor. A feed containing 4 kmol/m^{3}of each A and B enters the reactor at 8 m^3/h. If the temperature of the exit stream is never to exceed 390 k, what is the maximum feed inlet temperature allowed?

Data: heat of reaction = –50 KJ/mol density of the reacting mixture = 1000 kg/m^3, specific heat of reacting mixture = 2 J/Kg-K.

The above data can be assumed to be independent of composition and temperature

A. 190 K

B. 290 K

C. 390 K

D. 490 K

Sol. **B.**

Given C_{A_0} = 4 kmol/m^3 = 4000 mol/m3

V_0 = 8 m^3/hr; ρ = 1000 kg/m^3

$(-\Delta H)$ = 50kJ/mol; C_p = 2 kJ/kg-K

Writing energy balance equation when all A is reacted i.e. the criteria when maximum heat is generated

8 (m^3/hr)×1000 (kg/m^3) ×2 (kJ/kg-K) × ΔT_{max}

= 8 (m^3/hr)×4000 (mol//m^3) ×50 (kJ/mol)

or ΔT_{max} = 100

ΔT_{max} = 390 – T_{inlet} = 100

∴ T_{inlet} = 290K

106. Pick the wrong design guideline for a reactor in which the reactions $A \rightarrow R$(desired)and $A \rightarrow S$(undesired) are to take place. The ratio of the reaction rate is

$$\frac{r_R}{r_S} = \left(\frac{k_1}{k_2}\right) C_A^{a-b}$$

A. use high pressure and eliminate inerts when a > b

B. avoid recycle when a > b

C. use batch reactor or plug flow reactor when a> b

D. use CSTR with a high conversion when a > b

Sol. **B.**

The reactions are

A → R (desirable)

A → S (undesirable)

Further $\frac{r_R}{r_S} = \left(\frac{k_1}{k_2}\right) C_A^{a-b}$

When a > b selectively of R is higher when C_A is higher. All the options given can be used to keep C_A higher except option (b). It is required to use recycle of unreacted A to convert it into R.

GATE 2003

107. A CSTR is to be designed in which an exothermic liquid phase first order reaction of the type A→R is taking place. The reactor is to be provided with a jacket in which coolant is flowing. Following data is given

C_{A0} = 5 kmol/m^3; X_A = 0.5; feed temperature = reactor temperature = 40^{0}C; rate constant at 40^{0}C = 1 min^{-1}; (ΔH) = – 40 KJ/mol;

ρ = 1000 kg/m^3, C_ρ = 4 J/g –^{0}C; q = 10^{-3} m^3/min (ρ and C_ρ are same for the reactant and product streams. The amount of heat to be removed is

A. 2/3 kW

B. 1 kW

C. 5/3 kW

D. 4 kW

Sol. **C.**

Data given

C_{A_0} = 5 kmol/m^3 $\quad X_A$ = 0.5

T = 40^{0}C; $\quad$ k = 1 min^{-1}

ρ = 1000 kg/m^3; $\quad (-\Delta H)$ = – 40 kJ/mol;

C_p = 4 J/kg-K ; $\quad$ q = 10^{-3} m^3/min

Amount of heat removed

$= [V_0 C_{A_0}(1 - X_A)]\Delta H$

Substituting values

$= \frac{10^{-3}}{60}\left(\frac{m^3}{sec}\right) \times 5 \times 10^3 \left(\frac{mol}{m^3}\right) \times 0.5 \times 40\ kJ/mol$

$= \frac{5}{3}(kJ/sec)$

$= \frac{5}{3}$ kW

108. A liquid phase reaction is to be carried out under isothermal conditions. The reaction rate as a function of conversion has been determined experimentally and is shown in figure given below. What choice of reactor or combination of reactors will require the minimum overall reactor volume, if a conversion of 0.9 is desired?

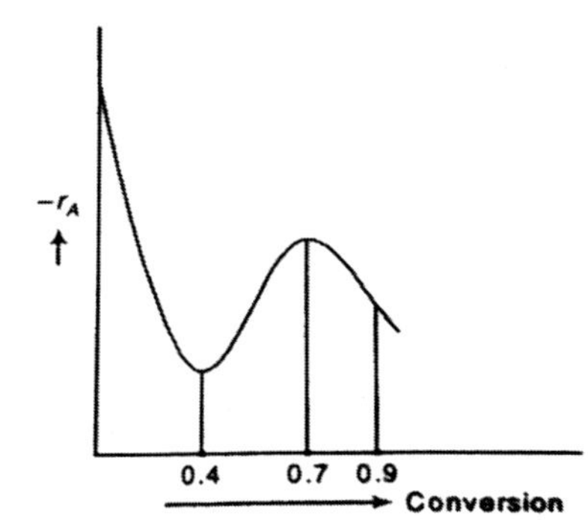

A. CSTR followed by a PFR
B. PFR followed by a CSTR
C. CSTR followed by a PFR followed by a CSTR
D. PFR followed by a CSTR followed by a PFR

***Sol.* D.**

An inspection of the reaction rate versus conversion curve indicates three distinct zones i.e. falling rate followed by rising trend reaching a peak and again falling thereafter. This indicates three reactor systems in series. Obviously the two falling regime curves are typical PFR performances intercepted by a CSTR.

Hence, for maximum overall reactor volume the reactor configuration should be PFR followed by a CSTR followed again by a PFR.

Common Data for Questions 115 and 116

The following gas phase reactions are carried out isothermally in a CSTR

$$A \rightarrow 2R; \quad r_1 = k_1 p_A;$$
$$k_1 = 20 \text{ mol/(s} - \text{m}^3 - \text{bar)}$$
$$A \rightarrow 3S; \quad r_2 = k_2 p_A;$$
$$k_2 = 40 \frac{\text{mol}}{(\text{s} - \text{m}^3 - \text{bar})}$$

Total pressure = 1 bar;
$F_{A0} = 1$ mol/s; feed is pure A

109. What is the maximum possible value of F_R (mol/s)?

A. 1/3
B. 1/2
C. 2/3
D. 2

***Sol.* C.**

Given reactions are

A → 2R, $r_1 = k_1 p_A$,
k_1 = 20 mol/s-m^3-bar
A → 3S, $r_2 = k_2 p_A$,
k_2 = 40 mol/s-m^3-bar

Total rate of reaction for A,

$$r_A = r_1 + r_2 = (k_1 + k_2) p_A$$

And $$\frac{r_1}{r_A} = \frac{k_1}{k_1 + k_2} = \frac{20}{20 + 40} = \frac{1}{3}$$

For the reaction A → 2R each mole of A being reacted 2 moles of R is formed.

Hence maximum possible value of

$$F_R = \left(\frac{1}{3}\right) \times 2$$
$$= \frac{2}{3}$$

110. The volume of a CSTR required for a fractional conversion of A equal to 0.3 due to the first reaction is

A. 0.011
B. 0.021
C. 0.275
D. 0.375

***Sol.* B.**

For CSTR

$$\frac{V}{F_{A0}} = \frac{X_A}{-r_A}$$

Given F_{A_0} = 1 mol/sec; X_A = 0.3

$$\therefore \quad \frac{V}{F_{A0}} = \frac{X_A}{k_1 P_A} = \frac{X_A}{k_1 P_{A0}(1 - X_A)}$$

$$\therefore \quad \frac{V}{1} = \frac{0.3}{(20)(1)(1 - 0.3)}$$

$$\therefore \quad V = 0.021 \, m^3$$

111. A step input tracer test is used to explore the flow pattern of fluid through a vessel of total volume equal to 1 m³ having a feed rate of 1 m³/min.

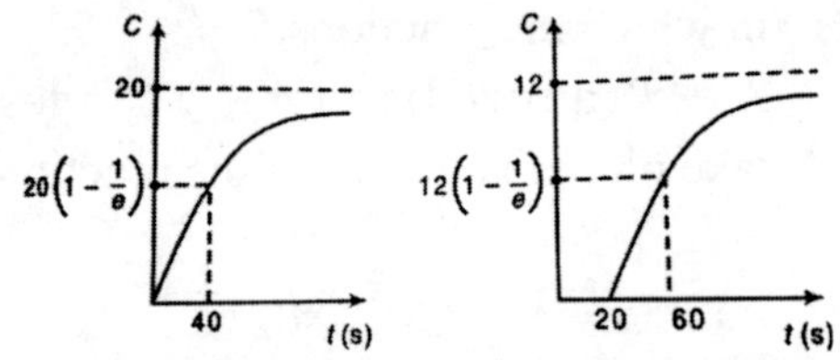

Curve 1 Curve 2

Identify for each curve in Group I a suitable flow model from the list given under Group II.

Group I	*Group II*
Curve 1	1. PFR and CSTR in series
Curve 2	2. CSTR with dead space
	3. PFR in series with a CSTR and dead space
	4. CSTR

A. P-4, Q-3
B. P-4, Q-1
C. P-2, Q-3
D. P-1, Q-1

***Sol.* A.**

Curve 1 — CSTR
Curve 2 — PFR in series with CSTR and dead space

112. Following isothermal kinetic data are obtained in a basket type of mixed flow reactor for a porous catalyst. Determine the role of pore diffusion and external mass transfer process.

Pellet Diameter	*Leaving concentration of the reactant*	*Spinning rate of basket*	*($-r_A'$)*
1	1	high	2
2	1	low	1
2	1	high	1

A. Strong pore diffusion control and mass transfer not controlling

B. Both pore diffusion and mass transfer not controlling

C. Both pore diffusion and mass transfer controlling

D. Mass transfer controlling

***Sol.* A.**

The given data clearly suggest it to be a case of strong pore diffusion control where mass transfers is not controlling.

GATE 2002

113. A pulse tracer is introduced in an ideal CSTR (with a mean residence time τ) at time = 0. The time taken for the exit concentration of the tracer to reach half its initial value will be

A. 2τ

B. 0.5τ

C. τ/0.693

D. 0.693τ

***Sol.* A.**

Given at $t = 0 \quad C_A = \frac{C_{A0}}{2}$

$$\bar{t} = \tau = \frac{\Sigma t_i C_i}{\Sigma C_i}$$

$$\tau = \frac{(t)(C_{A0}/2)}{C_{A0}}$$

$$\therefore \quad t = 2\tau$$

114. A batch adiabatic reactor at an initial temperature of 373 K is being used for the reaction $A \longrightarrow B$. Assume the heat of reaction is – 1 kJ/mol at 373 K and the heat capacity of both A and B to be constant and equal to 50 J/mol-K. The temperature rise after a conversion of 0.5 will be

A. 5°C

B. 10°C

C. 20°C

D. 100°C

***Sol.* A.**

For the reaction $A \longrightarrow B$

Given $C_{p_A} = C_{p_B} = 50$ J/mol-K

$(-\Delta H) = -1$ kJ/mol $= -1000$ J/mol

For $X_A = 0.5$

Starting with 1 mole of A

	A ⟶	B
at X = 0	1	0
at X = 0.5	0.5	0.5

Taking heat balance after 50% conversion

$(0.5)(C_{p_A})(\Delta T) + (0.5)(C_{p_B})\Delta T = (0.5)$ (1(

Since $C_{p_A} = C_{p_B}$

$\therefore$ Simplifying $\Delta T = \frac{(0.5)(1000)}{(50+50)}$

$\therefore \quad \Delta T = 5^0$C

115. In the hydro dealkylation of toluene to benzene, the following reactions occur

$$C_7H_8 + H_2 \longrightarrow C_6H_6 + CH_4$$

$$2C_6H_6 \rightleftharpoons C_{12}H_{10} + H_2$$

Toluene and hydrogen are fed to a reactor in a molar ratio 1.5. 80% of the toluene gets converted and the selectivity of benzene (defined as moles of benzene formed/per moles of toluene converted) is 90%. The fractional conversion of hydrogen is

A. 0.16

B. 0.144

C. 0.152

D. 0.136

***Sol.* C.**

	C_7H_8 +	$H_2 \longrightarrow$	C_6H_6 +	CH_4
$At\ x = 0$	a	5a	0	0
$No. of\ moles\ when\ x = 0.8$	a(1 – 0.8)	(5a – 0.8a)	0.8a	0.8a

and $2C_6H_6 \rightleftharpoons C_{12}H_{10} + H_2$

0.8 a – 2b   b   b

Given $\frac{Moles\ of\ benzene\ formed}{Moles\ of\ toluene\ converted} = 0.9$

$\frac{0.8a - 2b}{0.8a} = 0.9$

Or $b = 0.04a$

Fractional conversion of hydrogen $= \frac{0.8a - b}{5a}$

$= \frac{0.8a - 0.04a}{5a} = 0.152$

GATE 2001

116. The mean conversion in the exit stream, for a second order, liquid phase reaction in a non-ideal flow reactor is given by

A. $\int_0^\infty \frac{k_2 C_{A0}\, t}{1+k_2 C_{A0}\, t} E(t)dt$

B. $\int_0^\infty \frac{1}{1+k_2 C_{A0}\, t} E(t)dt$

C. $\int_0^\infty \frac{1}{1+k_2 C_{A0}\, t} [1 - E(t)]dt$

D. $\int_0^\infty \frac{exp\,(-k_2 C_{A0}\, t)}{1+k_2 C_{A0}\, t} E(t)dt$

***Sol.* A.**

$$A \xrightarrow{k_2} B$$

Since the reaction is second order

$$-\frac{dC_A}{dt} = k_2 C_A^2 \quad \text{(for a batch reactor)}$$

$$\int_{C_{A0}}^{C_A} -\frac{dC_A}{C_A^2} = \int_0^t k_2 dt$$

or $$\frac{1}{C_A} - \frac{1}{C_{A0}} = k_2 t$$

or $$C_A = \frac{C_{A0}}{1+k_2\, C_{A0} t}$$

Now for non – ideal reactor

$$\frac{\overline{C_A}}{C_{A0}} = \int_0^\infty \left(\frac{C_A}{C_{A0}}\right)_{batch} \text{E(t) dt}$$

$$1 - \overline{X_A} = \int_0^\infty \left(\frac{1}{1+k_2\, C_{A0}\, t}\right) \text{E(t) dt}$$

$$\overline{X_A} = 1 - \int_0^\infty \left(\frac{1}{1+k_2\, C_{A0}\, t}\right) \text{E(t) dt}$$

$$\overline{X_A} = \int_0^\infty E(t)dt - \left(\frac{1}{1+k_2\, C_{A0}\, t}\right) \text{E(t) dt}$$

$$\overline{X_A} = \int_0^\infty \left[1 - \left(\frac{1}{1+k_2\, C_{A0}\, t}\right)\right] \text{E(t) dt}$$

$$\overline{X_A} = \int_0^\infty \left(\frac{k_2\, C_{A0}\, t}{1+k_2\, C_{A0}\, t}\right) \text{E(t) dt}$$

117. For a vapour phase catalytic reaction A + B → P which follows the Rideal mechanism and the reaction step is rate controlling, the rate of reaction is given by (reaction step is irreversible, product also adsorbs)

A. $-r_A = \frac{kp_A p_B}{1+K_A\, p_A + K_P\, p_p}$

B. $-r_A = \frac{kp_A^2 - k_1 p_p}{1+K_A\, p_A + K_P\, p_p}$

C. $-r_A = \frac{kp_A p_B}{1+K_A\, p_A + K_B\, p_B + K_P\, p_p}$

D. $-r_A = \frac{kp_A p_B}{1+K_A\, p_A}$

***Sol.* A.**

Catalytic reaction A + B → P

Since the reaction step is the rate controlling step hence using the equilibrium method

$$-r_A = \frac{k\square P_A P_B}{1+k_A\, P_A + k_p P_p}$$

118. The first order, gas phase reaction $A \xrightarrow{k_1} 2B$ is conducted isothermally in batch mode. The rate of change of conversion with time is given by

A. $\frac{dX_A}{dt} = k_1(1 - X_A)^2\ (1 + 2X_A)$

B. $\frac{dX_A}{dt} = k_1(1 - X_A)(1 + 0.5X_A)$

C. $\frac{dX_A}{dt} = k_1(1 - X_A)$

D. $\frac{dX_A}{dt} = \frac{k_1\,(1 - X_A)}{(1 + X_A)}$

***Sol.* C.**

The reaction $A \xrightarrow{k_1} 2B$ is first order being conducted in batch mode

$$-r_A = -\frac{dC_A}{dt} = k_2 C_A$$

Since $C_A = C_{A_0}(1 - x_A)$

$$\frac{dC_A}{dt} = -C_{A_0}\frac{dx_A}{dt}$$

Hence, $C_{A_0}\frac{dx_A}{dt} = \text{k}\ C_{A_0}(1 - x_A)$

or $$\frac{dx_A}{dt} = k_1(1 - x_A)$$

GATE 2000

119. The conversion for a first-order liquid-phase reaction A → B in CSTR is 50%. If another CSTR of the same volume is connected in series, then the percentage conversion at the exit of the second reactor will be

A. 60

B. 75

C. 90

D. 100

***Sol.* B.**

For a first order irreversible reaction in a CSTR

Material balance equation

$$FC_{A_0} - (FC_{A_1} + rV) = 0$$

$$FC_{A_0} - FC_{A_1} + (kC_{A_1})V = 0$$

Since $\tau = V/F$

$$C_{A_1} = \frac{C_{A0}}{1+k_1\tau}$$

Writing $C_{A_1} = C_{A_0}(1 - X_A)$

And $X_A = 0.5$

∴ $k_1\tau = 1$

For two tanks of same volume in series

$$C_{A_2} = \frac{C_{A1}}{1+k_1\tau} = \frac{C_{A0}}{(1+k_1\tau)^2}$$

$$\therefore \quad C_{A_2} = \frac{C_{A1}}{(1+1)^2} = \frac{C_{A0}}{4}$$

Hence $C_{A_2} = C_{A_0}(1 - x_{A_2})$

$$C_{A_0}(1 - x_{A_2}) = \frac{C_{A0}}{4}$$

or $x_{A_2} = \frac{3}{4} = 0.75$

120. The following half-life data are available for the irreversible liquid phase reaction, A→products :

Initial concentration (kmol/m³)	***Half-life (min)***
2	2
8	1

The overall order of the reaction is

A. 0.5

B. 1

C. 1.5

D. 2

***Sol.* A.**

For n[th] order reaction

$$-\frac{dC_A}{dt} = k_2 C_A^n$$

$$-\int_{C_{A0}}^{C_A} \frac{dC_A}{C_A^n} = \int_0^t k\, dt$$

or $C_A^{1-n} - C_{A_0}^{1-n} = k(n-1)t$

For half life $C_A = C_{A_0}/2$; $t = t_{1/2}$

$$\therefore \quad t_{1/2} = \frac{2^{n-1}-1}{k(n-1)} C_{A_0}^{1-n}$$

Putting the given values

When $C_{A_0} = 2$ Kmol/m³; $t_{1/2} = 2$ min

and when $C_{A_0} = 8$ Kmol/m³; $t_{1/2} = 1$ min

$$2 = \left[\frac{(2^{n-1}-1)}{k(n-1)}\right] 2^{n-1} \quad ....(i)$$

$$\text{and} \quad 1 = \left[\frac{(2^{n-1}-1)}{k(n-1)}\right] 8^{n-1} \quad ....(ii)$$

Dividing (ii) by (i)

$$\frac{1}{2} = \left(\frac{8}{2}\right)^{1-n} = (4)^{1-n}$$

$\therefore$ n = 0.5

121. The first order series reaction $A \xrightarrow{k_1} B \xrightarrow{k_2} C$ is conducted in a batch reactor. The initial concentrations of A, B and C (C_{A0}, C_{B0}, C_{c0} respectively) are all non zero. The variation of C_B with reaction time will not show a maximum, if

A. $K_2 C_{B_0} > K_1 C_{A_0}$

B. $K_1 C_{A_0} > K_2 C_{B_0}$

C. $C_{B_0} > C_{A_0}$

D. $C_{A_0} > C_{B_0}$

***Sol.* A.**

First order series reaction

$$A \xrightarrow{k_1} B \xrightarrow{k_2} C$$

$$r_B = k_1 C_A - k_2 C_B$$

A normal series reaction is always characterized by a maximum for B, the reason being that initially B will be increasing at first but after maximum point second reaction will become more prominent as the concentration of B becomes appreciable. However, if $k_2 C_{B_0} > k_1 C_{A_0}$ then the resulting effect on B will be that since the rate of disappearance of B is much larger than its formation B will disappear as soon as it is formed thereby no appreciable C_B will be present at any given time. Under this condition B will not show any maximum.

122. The reaction A → B is conducted in an adiabatic Plug Flow Reactor (PFR). Pure A at a concentration of 2kmol/m³ is fed to the reactor at the rate of 0.01 m³/s and at a temperature of 500 K. If the exit conversion is 20 % then the exit temperature (in Kelvin) is

A. 400

B. 500

C. 600

D. 1000

Given, heat of reaction at 298 K = –50000 kJ/kmol of A reacted

Heat capacities, $C_{PA} = C_{PB} = 100$ kJ/Kmol-k (may be assumed to be independent of temperature).

***Sol.* C.**

Data given is

$C_{A_0} = 2$ kmol/m³; $V_0 = 0.01$ m³/sec

$T_1 = 500$ K, $x_A = 0.2$

$C_{P_A} = C_{P_B} = 100$ kmol-K

$\Delta H = 50{,}000$ kJ/kmol of A reacted.

writing energy balance equation

$$(V_0)(C_{A_0})(C_{PB})T_2 - (V_0)(C_{A_0})(C_{PA})T_1 = (\Delta H)(V_0)(C_{A_0}x_A)$$

or $C_{PB}T_2 - C_{PA}T_1 = (\Delta H)x_A$

Substituting the values

$100\ T_2 - (100)\ (500) = (50{,}000)\ (0.2)$

$\therefore\ T_2 = 600$ K

123. The rate controlling step for the heterogeneous irreversible catalytic reaction

$$A\ (g) + B\ (g) \rightarrow C\ (g)$$

is the surface reaction of adsorbed A with adsorbed B to give adsorbed C. The rate expression for this reaction can be written as

A. $\dfrac{KK_A P_A P_B}{[1+K_A P_A+K_B P_B]}$

B. $\dfrac{KK_A K_B P_A P_B}{[1+K_A P_A+K_B P_B]}$

C. $\dfrac{KK_A K_B P_A P_B}{[1+K_A P_A+K_B P_B+K_C P_C]}$

D. $\dfrac{KK_A K_B P_A P_B}{[1+K_A P_A+K_B P_B+K_C P_C]}$

Sol. **D.**

For irreversible catalytic reaction

$$A(g) + B(g) \longrightarrow C(g)$$

For surface reaction controlling

$$r_C = \frac{K\,K_A K_B P_A\,P_B}{(1+K_A P_A + K_B P_B + K_C P_C)^2}$$

4. Chemical Technology

Multiple Choice Questions

1. Highly porous refractory bricks are processed without which of the following properties?
   A. having very high thermal conductivity
   B. less susceptible to chemical attack by molten fluxes and gases etc.
   C. very strong
   D. none of these

2. Pyrex glass mainly comprises of ______.
   A. celenium
   B. lead
   C. zinc
   D. boron

3. ________ is produced by the thermal pyrolysis of ethylene dichloride.
   A. ethylene oxide
   B. vinyl chloride
   C. trichloroethylene
   D. ethanol amine

4. The percentage alcohol content in freshly prepared natural and fortified wine may be respectively around ________ percent.
   A. 7-14 and 14-30
   B. 10 - 20 and 40 - 50
   C. 7-14 and 40-50
   D. 14 - 30 and 40-50

5. Is an endothermic reaction.
   A. absorption of SO_3 in 98% H_2SO_4.
   B. $C + H_2O = CO + H_2$.
   C. thermal dissociation of iron pyrites.
   D. both **B** and **C**

6. The setting of plaster of Paris involves
   A. hydration
   B. loss of CO_2
   C. dehydration
   D. hydrolysis

7. The production of penicillin in a fermenter is carried out under process conditions
   A. 25°C, 2 atm.
   B. 90°C, 45 atm.
   C. 50°C, 10 atm.
   D. 30°C, 200 mm Hg (absolute).

8. In _________ glass basic oxide is not present
   A. pyrex
   B. flint
   C. quartz
   D. all **A**, **B**&**C**

9. In comparison to oils, the fats have
   A. much higher reactivity to oxygen
   B. less unsaturated glycerides of fatty acids
   C. more unsaturated glycerides of fatty acids
   D. lower melting point

10. Which of the following statements is incorrect?
    A. ordinary glass is represented chemically by $Na_2O .CaO.6SiO_2$
    B. hard glass which is used for making laboratory glass wares is a mixture of sodium borosilicate and aluminium borosilicate
    C. glass isdecolorised during its manufacture by adding antimony oxide, mangnese dioxide or arsenic oxide
    D. red color is imparted to glass by addition of arsenic oxide

11. The conversion of caprolactum to nylon–6 is by condensation polymerisation in a reactor maintained at ________ pC
    A. 500-600
    B. 10-30
    C. <0
    D. 250-280

12. The purpose of fluorescent dyes being added in detergents is to
    A. act as fabric brightener

B. act as tarnish inhibitor for metals like German silver

C. attain distinctiveness from other brands

D. none of these

13. __________ is not an abrasive material.

A. bakelite

B. carborundum

C. pumice

D. corundum

14. Coals having the highest calorific value are

A. lignite

B. sub-bituminous

C. anthracite

D. peat

15. In the treatment of water alum is used as a coagulant to remove .

A. bacteria

B. colour

C. turbidity

D. all **A**, **B** and **C**

16. When the following is formed by finely ground Calcium aluminate and silicate.

A. explosive

B. cement

C. cermet

D. abrasive

17. In sulphite process of, paper manufacturing the cooking liquor is .

A. sodium sulphite and sodium bisulphite

B. magnesiumsulphate and magnesium bicarbonate

C. magnesiumsulphite and free SO_2 in acid medium

D. none of these

18. Coal produced by high temperature carbonisation produces.

A. larger quantity of tar compared to low temperature carbonization

B. inferior coke compared to low temperature carbonization

C. less of gases compared to liquid products

D. none of these

19. In which of the following reactions nickel is not used as a catalyst.

A. shift conversion

B. ammonia cracking/dissociation

C. oil hydrogenation

D. steaming reforming of naphtha

20. In which of the following molasses is the starting material

A. alcohol

B. essential oil

C. fatty acids

D. massecuite

21. The industrial production of BHC (Benzene hexachloride) taking place by the chlorination of benzene .

A. which is an addition reaction

B. in presence of sunlight

C. which is a substitution reaction

D. in absolute dark

22. The synthesis of alcohol from molasses is carried out in fermentation which operates at a temperature of ____ °C.

A. 150

B. 30

C. 300

D. 5

23. The ore concentrating metallurgical process which involves a chemical is known as _________.

A. froth floatation

B. electromagnetic separation

C. roasting

D. none of these

24. Which among the following is not likely to be a constituent of vegetable oils.

A. Citric acid

B. Glycerol

C. Oleic acid

D. Stearic acid

25. Vegetable oils containing high amount of saturated fatty acids is also known as

A. semi-drying

B. drying

C. non-drying

D. none of these

26. Most commonly used catalyst for the hydrogenation of oil is (A) ________ and (B) ________ is a catalyst poison.

A. (A) platinum (B) sulphur

B. (A) palladium (B) oxygen

C. (A) nickel (B) sulphur
D. (A) nickel (B) oxygen

27. The catalyst used for the production of ammonia from methane and air as raw materials are (i) __________ for steam reforming of methane and (ii) __________ for ammonia synthesis.
A. *i* - Fe/Al_2O_3; *ii* - Ni/Al_2O_3
B. *i* - Ni/Al_2O_3; *ii* - Cu - ZnO/Al_2O_3
C. *i* - Ni/Al_2O_3; *ii* - Fe/Al_2O_3
D. *i* - Fe/Al_2O_3; *ii* - Cu - ZnO/Al_2O_3

28. Which of the following statements is incorrect?
A. Both temperature and pressure in the digestor is less in case of the sulphite method as compared to that in the sulphate method
B. Kraft method of pulp manufacture can process all types of fibrous raw materials
C. Digestion time for bagasse is less than that for wood base materials
D. none of these

29. Fumigant insecticides posess the property to .
A. are absorbed throughout the plant
B. emit poisonous vapour
C. kill insects, when they eat it
D. none of these

30. Which of the following chemical compounds is used in the bleaching of paper pulp?
A. magnesium sulphite
B. activated clay
C. chlorine or chlorine dioxide
D. bromine

31. Polyvinyl chloride (PVC) falls under which of the following categories
A. a condensation polymerisation product
B. a thermosetting material
C. made by employing emulsion polymerisation
D. none of these

32. Natural rubber mainly comprises of
A. polyisoprene
B. polystyrene
C. polybutadiene
D. polychloroprene

33. The raw material which is used industrially in the manufacturing of viscose rayon is
A. bamboo
B. bagasse
C. recalyptious wood
D. fine teak wood

34. Which of the following mixtures is termed as Bordeaux?
A. insecticide
B. inorganic fungicide
C. explosive
D. fertilizer

35. The main object for roasting of metallurgical ores is to .
A. remove CO_2 & H_2O
B. dehydrate it
C. sinter the ore
D. remove arsenic & sulphur

36. Silicone belongs to which of the following categories?
A. monomer
B. inorganic polymer
C. thermoplastic
D. none of these

37. During the manufacture of sugar high magnesia lime is added to hot sugar cane juice to
A. flocculate the impurities
B. facilitate fast filtration
C. both **A** and **B**
D. neither **A** nor **B**

38. Dynamite, a controlled explosive, is manufactured by reacting nitroglycerine and _________.
A. sulphuric acid
B. phosphoric acid
C. hydrochloric acid
D. hydroflouric acid

39. Which of the following statements is wrong?
A. Mordant dyes are applied mainly to wools
B. Azoic dyes are mostly applied on cotton fabrics
C. Basic dyes (e.g. amino derivatives) are applied mostly to paper
D. none of these

40. In the manufacture of which of the following calcareous and argillaceous materials are used
A. carbon disulphide
B. cement
C. lead
D. none of these

41. In which of the following categories alkyl benzene sulfonate (ABS) falls ?

A. detergent
B. polyester
C. rubber
D. pesticide

42. Raw materials required for the commercial production of calcium carbide includes limestone and ________.
A. fuel oil
B. sand
C. coke
D. soda ash

43. Sea water contains about ____________ gm/ liter sodium chloride.
A. 10
B. 50
C. 25
D. 2

44. Unless the air used in aerobic fermentation is sterilized the
A. recovery of product will be difficult
B. contamination of pure culture would result
C. uniformity of product cannot be achieved
D. none of these

45. Platinum is used in several chemical industries as a catalyst. However, it is highly prone to poisoning by
A. carbon
B. arsenic
C. lead
D. sulphur

46. Alkylation of benzene with propylene produces isopropyl benzene which is known as .
A. mercaptans
B. cumene
C. neoprene
D. gelatin

47. The sulphonation reaction by which detergents are produced by dodecyl benzene is __________.
A. endothermic
B. exothermic
C. irreversible
D. both **B** and **C**

48. The converting of rubber into a thin sheet or coating on fabric is an example of _______
A. valcanisation
B. extrusion
C. calendering
D. mastication

49. In compared to mercury electrolytic cells the diaphragm electrolytic cell
A. produces70% NaOH solution.
B. requiresless specific power consumption for the production of chlorine.
C. requires lesser investment for similar capacity.
D. all **A**, **B** and **C**

50. Which of the Graphite possesses following properties?
A. moderator used in nuclear reactor.
B. allotrope of carbon.
C. electrical insulator.
D. both **A**and **B**

51. In the manufacture of soda ash by Solvay process which of the following may act as a catalyst.
A. NH_3
B. Coke
C. NaCl
D. CaO

52. For producing soda ash the process used is known as _______ process.
A. tromp
B. head chamber
C. chance
D. Solvay

53. In Solvay process for manufacture of soda ash the raw materials are limestone __________.
A. coke and caustic soda
B. brine and coal
C. coke and sand
D. none of these

54. Beer may contain about _________ percent alcohol.
A. 2-8
B. 18-23
C. 27-32
D. 1-4

55. Which of the following terms represent hydrolysis of sugar?
A. esterification
B. inversion
C. hydration
D. none of these

56. What is the object behind cooling the digest chip at the bottom portion of the digester by injecting cold black liquid to .

A. avoid mechanical weakening of fibre

B. remove lignin by way of crystallisation

C. increase the cellulose content

D. none of these

57. Coloured glass is produced by mixing of colored salts which is addition of oxide of__________ to impart greenish blue color to the glass.

A. manganese

B. chromium

C. copper

D. arsenic

58. Which of the following terms represent sulphuric acid completely saturated with sulphur trioxide?

A. concentratedsulphuric acid.

B. dilutesulphuric acid.

C. oleum.

D. sulphurous acid.

59. The most suitable dye for synthetic fibers is ______.

A. acid dye

B. mordant dye

C. azoic dye

D. pigment dye

60. __________ is an unsaturated fatty acid?

A. stearic acid

B. lauric acid

C. palmitic acid

D. oleic acid

61. High temperature carbonization of dry coal yields many valuable products including tar which is about __________ percent.

A. 3

B. 12

C. 22

D. 0.3

62. The commercial production of bleaching powder i.e. calcium chloro hypochlorite is doneby the reaction of chlorine and .

A. slaked lime

B. soda lime

C. calcium perchlorate

D. none of these

63. Separation of Penicillin from fermented broth is by _____.

A. extraction with amyl or butyl acetate.

B. Azeotropic distillation.

C. evaporation.

D. extractive.

64. Which of the following statements is wrong?

A. strongly calking coal should not be used in the Lurgi gasifier

B. gaseous fuels required less percentage of excess air for combustion as compared to liquid fuels

C. water gas is called blue gas because of the color of the flame, when it is burnt

D. acetylene gas cannot be used for illumination purpose

65. Iron ore may also be called as __________.

A. bauxite

B. galena

C. hematite

D. chalcopyrite

66. In the manufacture of sodium hydroxide from brine by electrolysis _________ is obtained as a by-product.

A. chlorine

B. ammonium chloride

C. sodium carbonate

D. sodium bi-carbonate

67. Which of the following is not achieved by hydrogenation of oil?

A. improve its resistance to oxidation.

B. remove double bonds.

C. raise its melting point.

D. none of these.

68. ___________ is chemically represented asCaCl(OCl).

A. hypo

B. bleaching powder

C. plaster of paris

D. aqua regia

69. ___________is produced by condensation polymerisation and is chemically known as Phenol formaldehyde

A. teflon

B. bakelite

C. polyester

D. nylon-66

70. Which of the following statements is wrong?
A. production of cold SBR employs lower pressure as compared to that of hot SBR.
B. cold rubber (SBR) is superior as compared to hot rubber (SBR).
C. polymerisation temperature can be modify the properties of SBR.
D. none of these.

71. Which is the most widely used coagulant for removing suspended impurities from water?
A. bleaching powder
B. calcium sulphate
C. chlorine
D. alum

72. Which of the following is the main constitute of Limestone?
A. $CaCO_3$
B. $CaSO_4$
C. $MgCO_3$
D. Na_2CO_3

73. Which of the following terminologies represent Nylon-6?
A. polyamide
B. polyester
C. thermosetting resin
D. none of these

74. __________ is the main constituent of cotton fiber?
A. gelatine
B. cellulose
C. lignin
D. starch

75. One of the uses of phthalic anhydride is .
A. for making nylon-6
B. as plasticisers
C. in making PVC
D. in insecticides manufacture

76. High temperature carbonization of coal mainly produces .
A. coke
B. phenol
C. ammonia
D. tar

77. What are Enzymes?
A. proteins with high molecular weight (around 10,000)
B. catalyst for temperature sensitive reactions
C. derived from living organisms
D. all **A**, **B** and **C**

78. Recovery of NH_3 present in Coke oven gas in an integrated steel plant done is in the form of .
A. $(NH_4)_2SO_4$
B. liquid NH_3
C. NH_4Cl
D. $(NH_4)_2\ NO_3$

79. The removal of permanent hardness of water can be achieved by.
A. passing it through sodium hexametaphosphate
B. addition of soda ash to it.
C. Treating it with zeolites.
D. all **A**, **B** and **C**

80. Which of the following industries produces glycerene as by product?
A. soap
B. pain
C. detergent
D. oil hydrogenation

81. In which of the following high purity is required?
A. firefighting purposes.
B. making protective gas (95% N_2 + 5% H_2) for annealing of cold rolled steel strip coils.
C. both **A** and **B**
D. neither **A** nor **B**

82. The catalyst usedfor splitting of fats is .
A. $CaCO_3$
B. ZnO
C. Al_2O_3
D. Fe

83. The __________ salt of fatty acids forms Metallic soap.
A. potassium
B. sodium
C. both potassium & sodium.
D. aluminum or calcium

84. In sulfuric acid manufacture by contact process SO_3 is absorbed in 97% H_2SO_4 instead of water, because
A. SO_3 gas is sparingly soluble in water.
B. water forms an acid mist, which is difficult to absorb.
C. the purity of acid is affected.
D. scale formation in the absorber is to be avoided.

85. Cation exchanger is regenerated usually with
A. alum solution
B. H_2SO_4
C. NaOH
D. hydrazine

86. Refining of cane sugar involves addition of hydrated lime to the sugar syrup followed by carbonation. The purpose of this refining is to
A. adjust the pH of the syrup.
B. improve the rate of crystallisation of sugar.
C. reduce the viscosity of the syrup.
D. remove the coloring matter from the syrup.

87. Which of the following term defines bittern?
A. starting material for the production of iodine.
B. by-product of chlor-alkali industry.
C. unsaturated fat.
D. none of these.

88. What is the bromine content in ppm of sea water?
A. 500
B. 70
C. 5
D. 1700

89. Which of the following is a thermosetting plastic?
A. polystyrene
B. epoxy polymer
C. polythene
D. P.V.C.

90. Which of the following metals is derived from the chalcopyrite ore?
A. copper
B. lead
C. iron
D. tin

91. Removal of naphthalene from coke oven gas is done by,
A. passing it through electrostatic precipitator.
B. adsorbing on palladium.
C. scrubbing with wash oil.
D. absorbing in ethanolamine.

92. Last trace of color bodies from the pulp is removed by using at the end bleaching agent.
A. chlorine dioxide (ClO_2)
B. mercaptans
C. SO_2 gas
D. MgO

93. __________is chemically termed as polymethyl methacrylate (PMMA).
A. nylon-6
B. bakelite
C. Perspex
D. teflon

94. What is the appropriate percent of oil in fish?
A. 5
B. 10
C. 20
D. 35

95. Bio-degradable detergents
A. can be readily oxidised.
B. have an isoparaffinic structure
C. pose problem in sewerage treatment
D. should not be used as it spoil the clothing

96. Like chalcopyrite __________ is also used as a flux in the smelting of copper.
A. coke breeze
B. lime powder
C. silica/quartz
D. dolomite

97. Which of the following mixtures form lubricating greases?
A. mineral oil, soap and additives
B. mineral oil and metallic soap
C. fatty oil and metallic soap
D. mineral oil and fatty oil

98. Which allotrope of sulphur is insoluble in carbon distribution ?
A. Monoclinic sulphur
B. Rhombic sulphur
C. Plastic sulphur
D. Milk of sulphur

99. The use of Multistage catalytic converter is not possible in
A. conversion of SO_2 to SO_3
B. NH_3 synthesis reaction
C. both **A**and**B**
D. neither **A** nor **B**

100. The making of mattresses or flexible foam is __________,
A. polyamides
B. PVC
C. polyurethanes
D. silicone

101. When water is added to calcium oxide it produces__________,

A. exothermic heat

B. hissing sound

C. slaked lime

D. all **A**, **B** and **C**

102. _________is a constituent of vinegar.

A. Around 10% alcohol

B. Around 1% acetic acid

C. Both **A** and **B**

D. Neither **A** nor **B**

103. The treatment of sewage by the use of chlorine _________.

A. grease separation

B. aids in flocculation

C. the biological oxygen demand (BOD)

D. causes bulking of activated sludge

104. Solvent extraction produces oil which.

A. is odourless.

B. has low free fatty acid content.

C. both **A** and **B**

D. neither **A** nor **B**

105. Coal tar is largely used on a

A. binder in making carbon electrodes

B. binding material for coal briquettes

C. fuel in rotary kiln

D. all **A**, **B** and **C**

106. __________of hard work is done by sodium bisulfite

A. dechlorination

B. deaeration

C. both **A** and **B**

D. neither **A** nor **B**

107. Coagulant are normally used _______ filtration.

A. before

B. to avoid

C. after

D. during

108. The research why during the manufacturing of sulphuric acid the temperature of molten sulphur is not increased beyond 160°C is that .

A. it decomposes on further increasing the temperature

B. its viscosity is not reduced on further heating (hence pressure drop on pumping it, cannot be further reduced)

C. it is very corrosive at elevated temperature

D. none of these

109. __________ is known by the trade name Litharge,

A. lead oxide

B. zinc sulphide

C. zinc oxide

D. cellulosic photographic material

110. Most of the vegetable oils have flash point around __________ °C.

A. 50

B. 100

C. 200

D. 300

111. Fabric stains caused due to blood, tea etc. can be removed by __________ when added in the detergents.

A. caustic soda

B. sodium borate

C. sodium silicate

D. sodium tripolyphosphate (STPP)

112. Chemical formula of the dense soda ash used in the manufacture of glass is

A. Na_2CO_3

B. $Na_2CO_3.10H_2O$

C. $Na_2CO_3.H_2O$

D. Na_2HCO_3

113. The use of reverse osmosis is practiced in

A. purification of oxygen

B. separation of helium from natural gas

C. desalination of brackish water to produce potable later

D. separation of isotopes of uranium from gaseous uranium hexaflouride

114. Sodium tripolyphosphate is present as an additive in synthetic detergent because it

A. facilitates its use even in hard water

B. does not allow redeposition of dirt on the cleaned surface.

C. inhibits its corrosive effects.

D. none of these.

115. Riboflavin falls under the category of .

A. vitamin

B. anti-malarial drug

C. analgesic drug

D. anaesthetics

116. The coal gasifier technology Kopper-Totzek
A. can give ammonia synthesis gas ($H_2 + N_2$)
B. operate at very high pressure
C. is a moving bed reactor
D. cannot use coking coal

117. Which of the following statements is not correct?
A. pasteurisation of milk involves moderate heating followed by cooling
B. enzyme is a complex nitrogenous compound
C. bakeries and breweries make use of yeasts
D. Oils and fats are alkaloids

118. In a fermentation chemical reaction if the concentration of enzymes is high, the rate of reaction will be __________ the concentration of reactants.
A. independent of
B. directly proportional to
C. inversely proportional to
D. proportional to the square of

119. Which of the following hydrazine is commonly used for?
A. in printing industry
B. as a starting material for 'hypo'
C. as rocket fuel
D. in photographic industry

120. Synthesis of organic compound used for the manufacturing of organic compounds is a mixture of
A. C & H_2
B. N_2 & H_2
C. CO & H_2
D. CO_2 & H_2

121. For manufacturing of paper a good quality pulp should contain high .
A. cellulose
B. lignin
C. both **A** and **B**
D. none of these

122. 10% oleum consists of 10% free
A. H_2SO_4
B. SO_2
C. SO_3
D. H_2SO_3

123. Dirt removal by soaps and/or detergent does not involve __________ of the dirt particles.
A. wetting
B. emulsification
C. precipitation
D. dispersion

124. What is the by-product produced during manufacture of soda ash by Solvay process,
A. $CaCl_2$
B. NH_4Cl
C. NH_3
D. NaOH

125. Subjecting ethyl benzene to dehydrogenation produces
A. styrene
B. benzoic acid
C. phenol
D. naphthalene

126. __________is the chemical name of analgesic drug aspirin.
A. acetylsalicylic acid
B. nictonic acid
C. calcium acetate
D. methyl salicylate

127. Which of the following may be attributed to Lindane?
A. a by-product of BHC manufacture.
B. BHC (Benzene Hexa Chloride) containing 99% γ-isomer.
C. not a fumigant.
D. both **B** and **C**

128. $Ca(OH)_2$ is also called
A. gypsum
B. slaked lime
C. quicklime
D. limestone

129. The reason why Solvay process is not used for the manufacturing of potassium carbonate is because ________.
A. has high water solubility and is unstable
B. is prone to thermal decomposition
C. is soluble in ammonium chloride and potassium chloride solution
D. all **A**, **B** and **C**

130. Aluminum metal is obtained from pure alumina in the presence of fused cryolite by
A. electrolytic oxidation
B. electrolytic reduction

C. electrolysis
D. none of these

131. For the purpose of water softening $Na_2CO_3$is added to water of low alkalinity to,
A. permit the use of alum as a coagulant
B. facilitate easy regeneration of zeolite
C. increase the softening capacity of zeolite
D. all A, B and C

132. What is the percentage of uranium in the uranium are carnotite found in Jadugoda mines (Jharkhand).
A. 0.1 to 0.5
B. 5 to 10
C. 15 to 25
D. 1 to 5

133. Phthalic anhydride is produced by
A. oxidation of naphthalene
B. dehydrogenation of ethyl benzene
C. oxidation of benzene
D. none of these

134. Which of the following chemicals can be produced by catalytic oxide of naphthalene?
A. phenol
B. styrene
C. phthalic anhydride
D. none of these

135. When ethylene glycol reacts with dimethyl terephthalate (DMT) it produces.
A. PVC
B. dacron
C. nylon-6
D. polyester

136. The explosive RDX is chemically called.
A. cyclotrimethylene trinitramine
B. trinitro resorcinol
C. cyclotetramethylenetetranitramine
D. trinitro benzene

137. SO_3 is best absorbent in __________.
A. 20% oleum
B. 65% oleum
C. 78%H_2SO_4
D. 98%H_2SO_4

138. Major constituent of the Portland cement is ___________.
A. Calcium carbonate
B. Calcium oxide
C. Tricalcium silicate
D. Calcium sulphate

139. The most common raw material for alcohol distilleries is.
A. molasses
B. methylated spirit
C. benzol
D. none of these

140. It is not possible to liquefy gases by .
A. adiabatic expansion through a throttle valve (Joule-Thomson expansion).
B. adiabatic expansion against a piston or in a turbine
C. merely compressing it beyond critical pressure
D. exchange of heat with colder stream

141. If an oil or fat has high acid value then this is an indication of .
A. storage under improper conditions
B. its smaller molecular weight
C. absence of unsaturation
D. none of these

142. Which of the following product, produces yeast
A. pasteurised milk
B. pencillin
C. wine
D. antibiotics

143. The co-product produced during the electrolytic production of caustic soda is __________.
A. Na_2CO_3
B. $NaHCO_3$
C. H_2
D. none of these

144. Chemical name of Bakelite is
A. polyurathane
B. polyvinyl chloride (PVC)
C. phenol formaldehyde
D. polybutadiene

145. Potassium preserved & transported under
A. alcohol
B. water
C. kerosene oil
D. liquid ammonia

146. Which of the following chemicals is usually used for anion exchanger?
A. NaOH
B. alum solution

C. H_2SO_4

D. hydrazine

147. Under which of the following categories parathion and malathion fall ?

A. pesticides

B. Tranquilisers

C. pain-relieving drugs (analgesic)

D. plasticisers for unsaturated polysters

148. From which of the following chemicals Caprolactum (a raw material for nylon-6 manufacture) is produced ?

A. phenol

B. pyridine

C. benzene

D. naphthalene

149. Which of the following statements is correct for pears like transparent soaps?

A. usually such soft soap which contain alcohol and cane sugar and washed with methylated spirit to achieve transparency

B. metallic soaps with frothing agent from which glycerene has not been recovered

C. metallic soaps with frothing agent and free stearic acid to achieve transparency

D. none of these

150. The pyrolysis of __________ produces perchloro-ethylene.

A. Ethylene dichloride

B. Chlorobenzene

C. Carbon tetrachloride

D. Chlorinated paraffin

151. __________ is an yellow pigment.

A. zinc sulphides

B. titanium dioxide

C. lead chromates

D. ferrous sulphate

152. Condensation of acetone which produces Bisphenol is

A. diethyl ether

B. formaldehyde

C. ethyl alcohol

D. phenol

153. The property of hypo (sodium thio sulphate) which makes it useful in photography is .

A. photochemical

B. oxidising

C. complex forming

D. reducing

154. Carbonaceous matter of the sewage __________

A. causes reduction in its dissolved oxygen content thereby endangering the life of aquatic creatures.

B. increases the quantity of chlorine used for its purification.

C. reducessulphate ions to sulphides causing obnoxious smell.

D. all A, B and C

155. __________ is produced by the catalytic oxidation of toluene.

A. tri-nitro-toluene

B. phenol

C. styrene

D. benzene

156. In the sulfuric acid manufacture the use of vanadium pentaoxide as catalyst is preferred as compared to platinum catalyst as it __________

A. gives higher conversion efficiency

B. has a longer life and is not poisoned by arsenic

C. handles lower SO_2 content gas (7 -10% SO_2), thus increasing the capital cost of the plant

D. all A, B and C

157. Which of the following substances is an ore of lead?

A. Chalcopyrite

B. Galena

C. Quartz

D. Siderite

158. Kraft pulp is manufactured by a/an __________ process.

A. alkaline

B. neutral

C. acidic

D. none of these

159. The reactor in which the hydrogenation of oil takes place is

A. plug flow

B. trickle bed

C. autothermal

D. none of these

160. Chloroform is industrially produced from acetone and__________.
   A. ammonium chloride
   B. calcium hypochlorite
   C. phosgene
   D. chlorine
161. Which of the following statements is wrong?
   A. the rate of chemical reaction is independent of the concentration of reactants at high concentration of enzymes
   B. wine, rum & Vodka are prepared by the formulation of fruit juice, sugar beet & rye respectively
   C. protein catalysts are called enzymes
   D. total sugar content in molasses is about 10%
162. Which of the following statements about antibiotic is correct?
   A. inhibits/destroys the growth of micro organisms
   B. is an anaesthetic
   C. is used as a pain reliever
   D. is an antimalarial.
163. The compound which does not fall under the category of an elastomer is
   A. Nitrile-butadiene
   B. Polyisoperene
   C. Neoprene
   D. None of these
164. The commercial production of Styrene (a monomer for the production of polystyrene) is done by.
   A. catalytic dehydrogenation of ethyl benzene
   B. fermentation of starch
   C. dehydration of ethyl alcohol followed by hydrogenation
   D. reacting ethylene oxide with acetal-dehyde
165. In which of the following categories Dacron falls?
   A. thermosetting material
   B. condensation product of dimethyl terephthalate and ethylene glycol
   C. condensation product of hexa-methylene diamine and adipic acid
   D. none of these
166. In general favorable conditions for the liquefaction of the gases are .
   A. high pressure & low temperature
   B. low pressure & low temperature
   C. low pressure & high temperature
   D. high pressure & high temperature
167. The earlier use of platinum catalyst for sulphuric acid manufacture by contact process has now been stopped because of its
   A. high cost
   B. fragile nature
   C. easy poisoning tendency
   D. all **A**, **B** and **C**
168. Butadiene is mainly used .
   A. as an anti-skinning agent in paint
   B. in the manufacture of synthetic rubber
   C. as a plasticizer for unsaturated polyester
   D. none of these
169. The base characteristics of thermosetting plastic is that it .
   A. is useful for spinning
   B. can be repeatedly melted
   C. cannot be melted after forming
   D. is useful for melt casting
170. The correct sequence of major operation in the manufacturing of sulphuric acid from elemental sulphur is
   A. furnace $\rightarrow$ converter $\rightarrow$ absorber
   B. converter $\rightarrow$ furnace $\rightarrow$ absorber
   C. furnace $\rightarrow$ evaporator $\rightarrow$ absorber
   D. furnace $\rightarrow$ converter $\rightarrow$ evaporator
171. The major function of bubbling SO_2 through hot sugar cane juice is to .
   A. act as an acidifying agent
   B. increase the crystal size
   C. increase its concentration
   D. increase the amount of molasses
172. Penicillin was discovered by
   A. Alexender
   B. Dr. Zivago
   C. Flaming
   D. none of these
173. What is the sugar content in sugarcane on cane basis expressed as percent by weight?
   A. 15 to 20
   B. 5 to 10
   C. 20 to 30
   D. 1 to 5

174. Sucrose belongs to the category of .
A. polysaccharide
B. disaccharide
C. monosaccharides
D. none of these

175. Activated carbon is used in water treatment primarily to control the __________.
A. turbidity
B. taste and odour
C. bacterial growth
D. none of these

176. Froth floatation process for ore concentration utilises the __________ of ore particles.
A. density difference
B. wetting characteristics
C. terminal velocities
D. none of these

177. The reduction of rancidity of the fatty oil can be achieved by its
A. purification
B. hydrogenation
C. decoloration
D. oxidation

178. When rubber is vulcanised its ______ decreases.
A. tensile strength
B. stickiness
C. maximum service temperature
D. resistance to the action of organic solvent

179. The sedimentation of sewage is speed up by the use of .
A. lime
B. chlorine
C. sulphuric acid
D. sodium bisulphate

180. In India starting raw material for the commercial production of ethyl alcohol is
A. rice
B. molasses
C. fruit of Mahua tree
D. maize

181. The production of phenol formaldehyde.
A. is an abrasive material.
B. employs condensation polymerisation.
C. employs addition polymerisation.
D. is a monomer.

182. Amongst the following __________ is not an insecticide.
A. hydrocyanic acid
B. sodium flouride
C. nicotine
D. hexane

183. For the oxidation of SO_2 to SO_3 the equilibrium yield for the reaction at any given temperature andpressure P is proportional to
A. $1/P$
B. P
C. $P^{1/2}$
D. P^2

184. Which of the following is produced when water containing temperature hardness is boiled?
A. CO_2
B. $CaHCO_3$
C. $CaSO_4$
D. none of these

185. During the setting of cement, shrinkage volume __________ does not depend on
A. sand to cement ratio
B. drying period
C. water to cement ratio
D. ambient temperature fluctuation

186. Which of the following glass is used for the manufacture of optical glass?
A. Crooke's
B. Pyrex
C. Flint
D. Soda

187. __________ is produced by the inversion of sucrose.
A. fructose
B. glucose
C. both **A** and **B**
D. neither **A** nor **B**

188. During paper manufacture by Kraft process white cooking liquor consists of a mixture is caustic soda__________
A. sodiumsulphide& sodium carbonate
B. sodiumsulphite& sodium sulphide
C. sodiumsulphite& sodium carbonate
D. none of these

189. A mineral may also be termed as'ore' if
A. a metal can be economically extracted from it

B. it contains e" 40% metal
C. the metal present in it is costly
D. all **A**, **B** and **C**

190. In __________ reactions enzymes are used as organic catalyst.
A. photochemical
B. biochemical
C. electrochemical
D. chemical

191. In soap making the fatty materials used is __________.
A. detergents
B. fatty acid
C. tallow
D. fatty alcohols

192. Which of the following is the starting material for the production of styrene butadiene rubber (SBR).
A. ethyl alcohol
B. ethylene
C. both **A** and **B**
D. neither **A** nor **B**

193. Permanent hardness of water is removed by __________.
A. boiling
B. filtration
C. distillation
D. none of these

194. In a strong solution of 20% oleum100 kg of oleum will contain 20 kg of,
A. SO_3 and 80kg of H_2SO_4
B. SO_3 for each 100 kg of H_2SO_4
C. H_2SO_4 and 80kg of SO_3
D. none of these

195. In water treatment chlorine is generally used for
A. disinfection and control of taste &odour
B. control of bacteria
C. corrosion control
D. removing turbidity

196. Fumes produced byoleum are of
A. $SO_2 + H_2SO_4$
B. H_2SO_4
C. SO_3
D. SO_2

197. The mechanism of gas liquefaction by Linde process employs
A. merely compressing the gas beyond its critical pressure
B. adiabatic expansion through a throttle valve (Joule-Thomson expansion)
C. exchange of heat with colder stream
D. adiabatic expansion against a piston or in a turbine

198. __________ does not require a filler during manufacture.
A. coloured paper
B. bond paper
C. blotting paper
D. writing paper

199. Bamboo and other good quality fibrous raw materials for the manufacture of paper contain __________ percent cellulose.
A. 10
B. 50
C. 80
D. 95

200. Recoveryof glycerine from lye is effected by
A. evaporation followed by vacuum distillation
B. extractive distillation technique
C. liquid extraction technique
D. none of these

201. The calcium carbide furnace is operated at a temperature ranging from _______ °C.
A. 200-300
B. 700-850
C. 2000-2200
D. 4000-4500

202. Which of the following statements is wrong?
A. bagasse fibre contains both lignin & cellulose
B. fibrillation of fibre during paper manufacture is done to develop the strength in paper
C. alkali consumption in digestion/cooking of bamboo is measured in terms of permangnate number
D. presence of sodium sulphate in pulp makes the pulp bleachability poor

203. Full from of DDT (an insectisede) is dichloro diphenyl
A. trichloromethane
B. trichloroethane
C. tetrachloroethane
D. tetrachloromethane

204. Glass manufacturing process does not involve .
A. sintering
B. melting
C. annealing
D. shaping or forming

205. Maleic anhydride is produce from .
A. *n*-butene
B. benzene
C. either **A** or **B**
D. neither **A** nor **B**

206. __________ is the raw material of commercial production of alum.
A. galena
B. gypsum
C. feldspar
D. bauxite

207. Which of the following represents chemical name of Grignard reagent?
A. ethyl magnesium chloride
B. mono chloro acetic acid
C. methyl magnesium chloride
D. dichloro phenol

208. Which of the following parameters differentiate between vegetable oils and fats?
A. density
B. composition
C. physical state
D. chemical properties

209. Which of the following substances is used as a flux in the extraction of iron ore (hematite) blast furnace?
A. manganese
B. limestone
C. bauxite
D. quartz

210. __________ is added as additive/ water proofing agent to lower the hydrophilic (moisture loving) characteristic of cement.
A. formic acid
B. xanthates
C. calcium &aluminum stearate
D. stearic acid

211. __________ is a substance produced by a living organism and capable of anti-microbial activity.
A. antibiotic
B. disinfectant
C. antiseptic
D. none of these

212. It is necessary to deaerated water in its treatment as it .
A. helps in controlling its taste and odour
B. minimises its turbidity
C. minimises its corrosiveness
D. none of these

213. In which of the following categories aryl benzene sulphonate falls?
A. detergent
B. coating ingredient for photographic film
C. plasticiser for unsaturated polyester
D. starting material for the synthesis of glycerine

214. Which method is used for the concentration of sulphide ores?
A. electro-magnetic separation
B. roasting
C. froth floatation
D. smelting

215. In the manufacture of which of the following compounds adipic acid is used as an intermediate
A. bakelite
B. nylon-66
C. perspex
D. polystyrene

216. In DCDA process of H_2SO_4 manufacture the conversion of SO_2 to SO_3 ($2SO_2 + O_2 \rightleftharpoons 2SO_3$) is carried out in multistage where the thereverse reaction becomes small at around 550° C. The percentage equilibrium conversion of SO_2 to SO_3 can be increased by,
A. maintaining low temperature & pressure in the converter
B. increasing the oxygen concentration
C. removing some quantity of SO_3 during intermediate stage
D. putting more quantity of V_2O_5 catalyst in the converter

217. The thermal pyrolysis of ethylene dichloride to produce vinyl chloride (CH_2 = CH.Cl) at a pressure and temperature of .
A. 4 kgf/cm^2 & 500°C
B. 10 kgf/cm^2 & 1000°C
C. 40 kg/cm^2 & 200°C
D. 100 kgi/cm^2 & 500°C

218. __________ is not a food additive?

A. invertage
B. citric acid
C. benzoyl peroxide
D. ammonium chloride

219. __________ is commonly known as hydrochloric acid.

A. green acid
B. muriatic acid
C. oil of vitriol
D. strong organic acid

220. Which of the following is also known as CaO?

A. quick lime
B. calcite
C. slaked lime
D. limestone

221. Which of the following is chemical formula of Glauber's salt?

A. $Na_2SO_4.10H_2O$
B. $(NH_4)_2SO_4$
C. CaCl(OCl)
D. $CaSO_4.H_2O$

222. Which of the following statements is incorrect regarding the properties of glasses.

A. compressive strength much greater than their tensile strengths
B. sharp melting points
C. high electrical insulation properties
D. low co-efficient of thermal expansion

223. The process by which styrene is produced from ethyl benzene is known as

A. dehydrogenation
B. dehydration
C. oxidation
D. alkylation

224. Which of the following chemicals is produced using molasses as the starting raw material?

A. dimethyl ether
B. ethyl alcohol
C. methyl alcohol
D. benzol

225. Constituents of gun powder include .

A. potassium nitrate
B. charcoal
C. sulphur
D. all **A**, **B** and **C**

226. Sea water contains bromine around __________ ppm

A. 70
B. 1875
C. 2500
D. 640

227. During cement manufacture the function of adding gypsum is to

A. increase the density of cement
B. decrease the cement setting rate
C. both **A** and **B**
D. neither **A** nor **B**

228. The process in which the maximum Corrosion problem in digestion and handling equipments is encountered is in process of paper pulp manufacture is

A. sulphite
B. mechanical
C. sulphate/kraft
D. neutral sulphite semi-chemical

229. Which of the following is a polysaccharide.

A. glucose
B. starch
C. maltose
D. sucrose

230. Which of the following compounds is also known as poly tetrafluoro ethylene (P.T.F.E.)?

A. decron
B. teflon
C. bakelite
D. celluloid

231. Dehydration of ethyl benzene produces

A. styrene
B. phenol
C. ethyl alcohol
D. cumene

232. Catalytic dehydrogenation of __________ produces acetone?

A. ethyl benzene
B. phenol
C. isopropanol
D. naphthalene

233. The production of calcium and magnesium __________ causes permanent hardness of water.

A. carbonate
B. sulphites & chlorides

C. bi-carbonates
D. none of these

234. In the manufacture of newsprint paper __________ is used
A. groundwood
B. wrapping
C. tissue
D. board

235. Which of the following is correct for viscose rayon?
A. cannot yield textile grade fibre
B. utilises H_2SO_4, NaOH and CS_2 during its manufacture
C. cannot be made from sulphite pulp
D. none of these

236. The third largest constituent of air is in argon whose percentage by volume in air is .
A. 0.14
B. 0.34
C. 0.94
D. 1.4

237. Dacron falls under which of the following categories?
A. polyester
B. polyamide
C. inorganic polymer
D. unsaturated polyester

238. Which of the following is the chemical name of soda ash?
A. potassium carbonate
B. sodium bicarbonate
C. Sodiumthiosuphate
D. none of these

239. Bakelite falls under which of the following categories?
A. an inorganic polymer
B. same as poly tetrafluoroethylene (PTFE)
C. same as thermosetting phenol-formaldehyde
D. not a polymer

240. The black liquor recovery produces tall oil as a byproduct. Which is .
A. used in the manufacture of greases, emulsions & soaps
B. composed mainly of rosin & fatty acids
C. a black, sticky & viscous liquid
D. all A, B and C

241. Which of the following has nitrogen as an essential component?
A. fats
B. mineral salts
C. carbohydrates
D. proteins

242. Grignard reagent is the chemical name of .
A. sodium bicarbonate
B. ethyl magnesium chloride
C. sodium thiosulphate
D. sodium sulphite

243. Which of the following does not use sodium carbonate (Soda ash) in its manufacture?
A. baking powder
B. fire extinguishers
C. sugar
D. detergents

244. How many tons of bamboo or wood is required for production of one ton of dry paper pulp?
A. 1
B. 2.5
C. 5
D. 10

245. Coagulant used in water treatment is .
A. hydrogen peroxide
B. chloramine
C. ferrous sulphate
D. chlorine

246. In comparison to nylon fibers Dacron (or terylene) fibers have
A. poorerdyeability.
B. better heat & acid resistant properties.
C. poorer resistance to alkalis.
D. all A, B and C

247. In which of the following low purity oxygen is used?
A. medicinal purposes
B. L.D. steel making
C. cutting and welding of metals
D. chemical oxidation processes

248. __________ is used for the manufacturing of nylon 6–6.
A. hexamethylene diamine andadipic acid
B. dimethyl terephthalate and ethylene glycol
C. hexamethylene diamine and maleic anhydride
D. caprolactum

249. Which of the following types of reactions is involved in hydrogenation of vegetable oil?

A. homogenous
B. endothermic
C. exothermic
D. autocatalytic

250. The mechanism of removal of dirt by soaps is by_________.

A. increasing the surface tension
B. supplying hydrophyllic group
C. decreasing wettability
D. none of these

Answer

1.	D	36.	B	71.	D	106.	C
2.	B	37.	C	72.	A	107.	A
3.	B	38.	A	73.	A	108.	B
4.	A	39.	D	74.	B	109.	A
5.	D	40.	B	75.	B	110.	C
6.	A	41.	A	76.	A	111.	B
7.	A	42.	B	77.	D	112.	C
8.	C	43.	C	78.	A	113.	C
9.	B	44.	B	79.	D	114.	A
10.	D	45.	B	80.	A	115.	A
11.	B	46.	B	81.	C	116.	A
12.	A	47.	D	82.	B	117.	D
13.	A	48.	C	83.	D	118.	A
14.	C	49.	C	84.	B	119.	C
15.	D	50.	D	85.	B	120.	C
16.	B	51.	A	86.	A	121.	A
17.	A	52.	D	87.	D	122.	C
18.	D	53.	B	88.	B	123.	C
19.	A	54.	A	89.	B	124.	A
20.	A	55.	B	90.	A	125.	A
21.	A	56.	A	91.	C	126.	A
22.	B	57.	C	92.	A	127.	B
23.	C	58.	A	93.	C	128.	B
24.	A	59.	A	94.	C	129.	C
25.	C	60.	D	95.	A	130.	B
26.	C	61.	A	96.	C	131.	A
27.	C	62.	A	97.	A	132.	A
28.	D	63.	A	98.	C	133.	A
29.	B	64.	B	99.	B	134.	C
30.	C	65.	C	100.	C	135.	B
31.	C	66.	A	101.	D	136.	A
32.	B	67.	D	102.	C	137.	D
33.	D	68.	B	103.	A	138.	C
34.	B	69.	B	104.	D	139.	A
35.	D	70.	D	105.	D	140.	C

141. A
142. C
143. C
144. C
145. C
146. A
147. A
148. C
149. A
150. C
151. C
152. D
153. C
154. D
155. B
156. D
157. B
158. A
159. B
160. B
161. D
162. A
163. D
164. A
165. D
166. A
167. D
168. B
169. C
170. A
171. A
172. A
173. B
174. B
175. B
176. B
177. B
178. B
179. A
180. B
181. B
182. D
183. B
184. A
185. A
186. C
187. B
188. A
189. A
190. B
191. C
192. C
193. C
194. A
195. A
196. C
197. B
198. C
199. B
200. A
201. C
202. D
203. B
204. A
205. C
206. D
207. A
208. C
209. B
210. C
211. A
212. C
213. A
214. C
215. B
216. C
217. A
218. D
219. B
220. A
221. A
222. B
223. A
224. B
225. D
226. A
227. B
228. D
229. B
230. B
231. A
232. C
233. B
234. A
235. B
236. C
237. A
238. D
239. C
240. D
241. D
242. B
243. D
244. B
245. C
246. D
247. D
248. A
249. C
250. D

GATE 2015-2000

(1 MARK QUESTION)

GATE 2015

1. Match the chemicals written on the left with the raw materials require to produce them mentioned on the right.

I. Single Superphosphate (SSP)	P. Rock Phosphate + Sulfuric Acide + Ammonia
II. Triple Superphosphate (TSP)	Q. Brine
III. Diammonium Phosphate (DAP)	R. Rock Phosphate + Sulfuric Acid
IV. Caustic soda	S. Rock Phosphate + Phosphoric Acid

Codes

A. I-Q, II-R, III-S, IV-P
B. I-S, II-P, III-Q, IV-R
C. I-R, II-S, III-P, IV-Q
D. I-S, II-R, III-P, IV-Q

Sol. **C.** Single Superphosphate (SSP) ⟶
Rock Phosphate + Sulfuric Acid (H_2SO_4)
Triple Superphosphate (TSP) ⟶
Rock Phosphate + Phosphoric Acid (H_3PO_4)
Diammonium Phosphate (DAP) ⟶
Sulfuric Acid (H_2SO_4) + Ammonia (NH_3)
Caustic soda ⟶ Brine

2. Match the polymer mentioned on the left with the catalyst used for it manufacture given on the right.

I. Low density polyethylene	P. Ziegler-Natta catalyst
II. High density polyethylene	Q. Traces of oxygen
III. Polyethylene terephthalate	R. Butyl lithium
IV. Polyvinyl chloride	S. Antimony

Codes

A. I-Q, II-R, III-S, IV-P
B. I-S, II-P, III-Q, IV-R
C. I-Q, II-P, III-S, IV-R
D. I-S, II-R, III-P, IV-Q

Sol. **C.** Low density polyethylene → Traces of oxygen
High density polyethylene → Ziegler – Natta catalyst
Polyethylene terephthalate → Antimony
Polyvinyl chloride → Butyl lithium

GATE 2014

3. Match the following.

Group I	*Group II*
P. $\left(\frac{\partial G}{\partial n_i}\right)_{T,P,n_{j\neq i}}$	I. Arrhenius equation
Q. $\left(\frac{\partial G}{\partial n_i}\right)_{S,V,n_{j\neq i}}$	II Reaction equilibrium constant
R. $\exp\left(\frac{-\Delta G^0_{\text{reaction}}}{\text{RT}}\right)$	III. Chemical potential
S. $\Sigma(n_i d\mu_i)_{T,P} = 0$	IV. Gibbs-Duhem equation

A. Q-III, R-I, S-II
B. Q-III, R-II, S-IV
C. P-III, R-II, S-IV
D. P-III, R-IV, S-I

Sol. **C.** P – III, R – II, S – IV

4. Catalytic cracking is

A. a hydrogen addition process
B. a carbonnm rejection process
C. an exothermic process
D. a cooking process

Sol. **C.** An exothermic process

5. Which one of the following statements is correct?

A. The major components of biodiesel are triglycerides
B. Biodiesel is essentially a mixture of ethyl esters
C. Biodiesel is highly aromatic.
D. Biodiesel has a very low aniline point

Sol. **B.** Biodiesel is essentially a mixture of ethyl esters.

GATE 2013

6. A wet solid is dried over a long period of time by unsaturated air of non-zero constant relative humidity. The moisture content eventually attained by the solid is termed as the
   A. unbound moisture content
   B. bond moisture content
   C. free moisture content
   D. equilibrium moisture content

Sol. D. Equilibrium moisture content

GATE 2012

7. In petroleum refining catalytic reforming is used to convert
   A. paraffin's and napthenes to aromatic
   B. paraffin's to hydrogen to aromatic
   C. gas oil diesel and gasoline
   D. light olefins to gasoline

Sol. A. Paraffins and napthenes to aromatic

8. The final boiling points of gasoline, diesel, Atmospheric Gas Oil (AGO) and lubricating oils vary as
   A. gasoline > diesel > AGO > lubricating oils
   B. lubricating oils > AGO > diesel > gasoline
   C. AGO > lubricating oils > diesel > gasoline
   D. lubricating oils > diesel > AGO > gasoline

Sol. B. lubricating oils > AGO > diesel > gasoline

9. The main unit processes used for the production of hydrogen from natural gas are Steam Reforming (SR), Pressure Swing Adsorption (PSA), Low Temperature Water Gas Shift (LTEGS) reaction and High Temperature Water Gas Shift (HTWGS) reaction. The correct sequence of these in the plant is
   A. SR, LTWGS, HTWGS, PSA
   B. PSA, SR, LTWGS, HTWGS
   C. SR, HTWGS, LTWGS, PSA
   D. PSA, HTWGS, LTWGS, SR

Sol. C. SR $\rightarrow$ HTWGS $\rightarrow$ LTWGS $\rightarrow$ PSA

GATE 2011

10. Match the polymerisation mechanism in Group I with the corresponding polymers in Group II.

Group I	*Group II*
P. Chain growth/ addition Polymerization Q. Step growth/ condensation	I. Polyethylene II. Polyvinyl chloride III. Polyethylene terephthalate

   A. P-III; Q- I, II
   B. P- I, III Q-III
   C. P- II, III;Q- I
   D. P-I; Q-II, III

Sol. B. Addition polymerisation – Polyethylene, Polyvinyl chloride

Condensation polymerisation – polyethylene terephthalate

11. Which one of the following sequences is arranged according to increasing calorific value?
   A. Producer gas, Natural gas, Water gas
   B. Natural gas, Producer gas, Water gas
   C. Producer gas, Water gas, Natural gas
   D. Water gas, Natural gas, Producer gas

Sol. C. Producer gas < Water gas < Natural gas

12. The correct sequence of processequipment used in the production of sulphuric acid from sulphur by contact process is
   A. burner, catalytic converter, 98% sulphuric acid absorption tower, oleum absorption column
   B. catalytic converter, oleum absorption column, 98% sulphuric acid absorption tower, burner
   C. burner, catalytic converter, oleum absorption column, 98% sulphuric acid adsorption tower
   D. burner, oleum absorption column, catalytic converter, 98% sulphuric acid absorption tower

Sol. C. The correct sequence of process equipment used in the production of sulphuric acid by contact process is burner, catalytic converter, oleum absorption column, 98% sulphuric acid absorption tower

13. Hydrotreating is used for
   A. removal of water from crude oil
   B. treatment of crude oil with water
   C. improving octane number of gasoline
   D. removal of sulphur and nitrogen from petroleum fractions

Sol. D. Hydrotreating is used for removal of sulphur and nitrogen from petroleum fractions.

14. Zeolite ZSM-5 is added to commercial FCC catalyst for
   A. promoting SO_2 reduction
   B. promoting CO oxidation
   C. improving tolerance to metal content in feed
   D. enhancing octane number

Sol. D. Zeolite ZSM-5 is added to commercial FCC catalyst for enhancing octane number.

GATE 2010

15. In the manufacturing of caustic soda from brine, which one of the following statements is true?

A. The membrane cell cannot produce concentrated NaOH solution, and cannot tolerate calcium andmagnesium ions in the feed brine

B. The membrane cell cannot produce concentrated NaOH solution, and can tolerate calcium and magnesium ions in the feed brine

C. The membrane cell can produce concentrated NaOH solution, andcannot tolerate calcium and magnesiumions in the feed brine

D. The membrane cell can producecon centrated NaOH solution, and can tolerate calcium and magnesium ions in the feed brine

Sol. D. The membrane cell can produce concentrated NaOH solution, and can tolerate calcium and magnesium ions in the feed brine

16. For making superphosphate by acidulation of phosphate rock, use of nitric acid is desirable because

A. nitric acid is less expensive than sulphuric acid

B. the availability of nitrogen enhances the value of the superphosphate as a fertilizer

C. the process produces non-hygroscopic 14. Superphosphate as a fertilizer

D. the process produces superphosphate having higher phosphorus content then the sulphuric acid

Sol. A. Nitric acid produces triple superphosphate which in turn has higher phosphorus content compared to that of sulphuric acid.

17. Match the processes in Group I with the products in Group II.

Group I	*Group II*
P. Claus process Q. Linde process R. Lurgi process	1. Syngas 2. Oxygen 3. Sulphur

Codes

	P	Q	R
A.	1	2	3
B.	2	1	3
C.	3	1	2
D.	3	2	1

Sol. D. Claus process → Sulphur

Linde process → Oxygen

Lurgi process → Syngas

GATE 2009

18. In petroleum refining operations, the process used for converting paraffins and naphthalenes to aromatics is

A. catalytic reforming

B. catalytic cracking

C. hydrocracking

D. alkylation

Sol. A. Catalytic reforming is used to convert low naphthas into aromatic compounds.

19. The active component of catalysts used in steam reforming of methane to produce synthesis gas is

A. Nickel

B. iron

C. platinum

D. palladium

Sol. A. The active component of catalysts used in steam reforming of methane to produce synthesis gas is nickel.

GATE 2008

20. Which one of the following is not a major constituent of crude oil?

A. Paraffins

B. Olefins

C. Naphthenes

D. Aromatics

Sol. C. Naphthene is converted into aromatic compounds and thus not a constituent of product, i e., crude oil.

GATE 2007

21. Which of the following is desirable in gasoline but undesirable in kerosene?

A. Aromatics

B. Mercaptans

C. Naphthenic acid

D. Paraffins

Sol. A. Smoke comes out while burning the aromatic compounds which is not desirable in kerosene as it is used in households.

22. In the sulphite process for paper manufacturing, the 'cooking liquor' is

A. magnesium bisulphite and sulphur dioxide in acid medium

B. magnesium sulphite and magnesium dicarbonate

C. sodium sulphite and magnesium sulphite

D. sodium sulphite andsodium bisulphite and sulphur dioxide

Sol. **A.** Magnesium bisulphite and sulphur dioxide in acid medium.

GATE 2006

23. The oxo-reaction is used for converting

A. alcohol to aldehyde

B. paraffins to olefin

C. olefin to aldehyde

D. aldehyde to alcohol

Sol. **C.** Oxo-reaction is used for converting olefin to aldehydes.

24. In a fluid catalytic cracking unit, the nature of the reactions occurring in the reactor and the regenerator is

A. Reactor - Exothermic, Regenerator - Exothermic

B. Reactor - Exothermic, Regenerator - Endothermic

C. Reactor - Endothermic, Regenerator - Exothermic

D. Reactor - Endothermic, Regenerator – Endothermic

Sol. **C.** In a FCC unit, cracking reaction is an endothermic process while regenerating the catalyst used is an exothemic process.

GATE 2005

25. Which of the following is not employed in the commercial production of linear polyvinyl chloride?

A. Emulsion polymerization

B. Suspension polymerization

C. Addition polymerization

D. Condensation polymerization

Sol. **C.** Addition polymerization.

26. Which of the following is a detergent?

A. Benzene hexachloride

B. Cellulose nitrate

C. Polyvinyl chloride

D. Alkyl benzene sulphonate

Sol. **D.** Alkyl benzene sulphonate is a detergent.

27. The chief raw material for the commercial production for methanol is

A. Synthesis gas

B. Formaldehyde

C. Acetic acid

D. Ethanol

Sol. **A.** Synthesis gas (CO and H_2 mixture) is the chief raw material for commercially producing methanol as given below

$$CO + 2H_2 \xrightarrow[300-400°C, 200-300 atm]{CuO, ZnO, Cr_2O_3} CH_3OH$$

GATE 2004

28. Prilling tower is found in the flow sheet for the manufacturing of

A. Ammonia

B. urea

C. superphosphate

D. triple superphosphate

Sol. **B.** Prilling tower is used the manufacturing of urea.

29. The proper arrangement of the petroleum fractions in the order of their boiling points is

A. lubricating oil > diesel > petrol > LPG

B. lubricating oil > petrol > diesel > LPG

C. petrol > lubricating oil > diesel > LPG

D. petrol> diesel >LPG >lubricating oil

Sol. **A.** The various petroleum products in the decreasing order of their respective boiling points are - lubricating oil > diesel > petrol > LPG

GATE 2003

30. 'Nylon 66'is so named because

A. the average degree of polymerization of the polymer is 1966

B. the number of carbon atoms between two nitrogen atoms are 6

C. the number of nitrogen atomsbetween two carbon atoms are 6

D. the polymer was first synthesized in 1966

Sol. **B.** Six carbon atoms are present in between two nitrogen atoms in 'Nylon-66.

31. The catalytic converter for conversion of SO_2 to $SO_3$by contact process should have a feed with SO_2 content between

A. 2-5%
B. 7-10% 31
C. 12-15%
D. 20-25%

Sol. **B.** Feed should have 7-10% SO_2 content in contact process for converting SO_2 to SO_3.

32. The composition of fresh feed to the high temperature high pressure urea autoclave is
A. excess liquid ammonia and liquified CO_2
B. excess liquid ammonia and compressed CO_2 gas
C. excess liquid ammonia and excess compressed CO_2 Gas
D. compressed NH_3 gas and excess compressed CO_2

Sol. **B.** Excess amount of liquid ammonia is used with and compressed CO_2 gas in high pressure-high temperature urea autoclave.

GATE 2002

33. Among the following, choose one that is not an exothermic process.
A. Methanol synthesis
B. Catalytic cracking
C. Ammonia synthesis
D. Oxidation of sulphur

Sol. **B.** Catalytic cracking is an endothermic process.

34. The commonly used solvent in supercritical extraction is
A. methyl ethyl ketone
B. water
C. carbon tetrachloride
D. carbon dioxide

Sol. **D.** CO_2 is used as a solvent in supercritical extraction.

35. The average boiling point of aviation turbine fuelis closest to that of
A. lubricating oils
B. LPG
C. Diesel
D. Kerosene

Sol. **D.** Kerosene

GATE 2001

36. One of the steps during refining of cane sugar consists of addition of hydrated lime to the sugar syrup followed by carbonation ofthe resulting solution.The purpose of this step is to
A. adjust the pH of the syrup
B. remove the colouring matter from the syrup
C. reduce the viscosity of the syrup
D. improve the rate of crystallization of sugar

Sol. **A.** This step is used for pH adjustment.

37. Styreneis produced from ethylbenzene by the process of
A. Dehydrogenation
B. oxidation
C. alkylation
D. dehydration

Sol. **D.** Catalytic in FCC is endothermic process while regeneration is exothermic process.

38. In the Fluid Catalytic Cracker (FCC), the cracking reaction is (A).......... and the regeneration is (B)........

	A	B
A.	exothermic	endothermic
B.	exothermic	exothermic
C.	endothermic	endothermic
D.	endothermic	exothermic

Sol. **D.** Catalytic in FCC is endothermic process while regeneration is exothermic process.

GATE 2000

39. In petroleum refining, the process used for conversion of hydrocarbon to aromatics is
A. catalytic cracking
B. catalytic reforming
C. hydrotreating
D. alkylation

Sol. **B.**Catalytic reforming is used for conversion of hydrocarbons to aromatics in petroleum refining

40. Commercially ethylene is produced from naphtha by
A. catalytic cracking
B. catalytic dehydrogenation
C. pyrolysis\`\`\`\`\`\`\`\`\`\`\`\`
D. hydrocracking

Sol. **D.** Commercially ethylene is produced from naphtha hydrocracking

41. Triple superphosphate manufactured by reacting
A. phosphate rock with phosphoric acid
B. phosphate rock with sulphuric acid
C. phosphate rock with nitric acid
D. ammonium phosphate with phosphoric acid

Sol. A. Triple superphosphate is manufactured by reacting phosphate rock with phosphoric acid.

(2 Marks Questions)

GATE 2015

42. A typical batch filtration cycle consists of filtration followed by washing, one such filtration unit operating at constant pressure difference first filters a slurry during which 5 L of filtrate is collected in 100 s. This is followed by washing. Which is done for t_w second and uses 1 L of was water. Assume the following relation to be applicable between the applied pressure drop Δp, cake thickness L at time t, and volume of liquid V collected in time t.

$\frac{\Delta p}{L} = k_1 \frac{dv}{dt}, L = k_2 V,$ if L is changing

k_1 and k_2 can be taken to be constant during filtration and washing. The was time t_w, (in s) (upto one decimal place) is

Sol. Given, $\frac{\Delta \rho}{L} = k_1 \frac{dv}{dt}$ and $L = K_2 V$

$[k_1 \text{ and } k_2 \text{ can be taken constant}]$

So, $\frac{dV}{dt} = \frac{1}{k_1} \cdot \frac{\Delta p}{k_2 V}$

$$or \int_0^5 V\,dV = \frac{\Delta p}{k_1 k_2} \int_0^{100} dt$$

$$or \quad \frac{25}{2} = \frac{\Delta p}{k_1 k_2} \times 100$$

$$\text{or} \quad \frac{\Delta p}{k_1 k_2} = \frac{1}{8}$$

Also, final rate of filtration,

$$\frac{dV}{dt} = \frac{\Delta p}{k_1 k_2} = \frac{1}{8 \times 5} = \frac{1}{40}$$

$$\text{Washing time, } t_w = \frac{\text{volume of wash water used}}{\text{final rate of filtration}}$$

$$t_w = \frac{1}{\left(\frac{1}{40}\right)} = 40\ sec$$

GATE 2014

43. Determine the correctness or otherwise of the following Assertion [A] and Reason [R].

Assertion (A) Significant combustion of coke takes place only if it is heated at higher temperature in presence of air.

Reason (R) $C + O_2 \longrightarrow CO_2$ is an exothermic reaction.

A. Both [A] ands [R] are true and [R] is the correct reason for [A]

B. Both [A] and [R] are true but [R] is not the correct reason for [A]

C. [A] is correct but [R] is false

D. Both [A] and [R] are false

Sol. B. Both (a) and (r) are true but (r) is not the correct reason for (a).

44. Match the raw materials of Group I and II with the final products of Group III

Group I	*Group II*	*Group III*
P_1. Ethylene P_2. Propylene P_3. Adipic acid P_4. Terephthalic acid	Q_1. Ammomia Q_2. 1-Butene Q_3. Ethylene glyco Q_4. Hexamethylene diamine	R_1. Synthetic fibre R_2. Nylon 66 R_3. LLDPE R_4. Acrylonitrile

A. $P_1 + Q_2 \longrightarrow R_3, P_2 + Q_1 \longrightarrow R_4, P_3 + Q_4 \longrightarrow R_2, P_4 + Q_3 \longrightarrow R_1$

B. $P_1 + Q_1 \longrightarrow R_3, P_2 + Q_3 \longrightarrow R_1, P_3 + Q_4 \longrightarrow R_4, P_4 + Q_2 \longrightarrow R_2$

C. $P_1 + Q_2 \longrightarrow R_2, P_2 + Q_3 \longrightarrow R_1, P_3 + Q_4 \longrightarrow R_3, P_4 + Q_1 \longrightarrow R_4$

D. $P_1 + Q_1 \longrightarrow R_4, P_2 + Q_2 \longrightarrow R_3, P_3 + Q_4 \longrightarrow R_2, P_4 + Q_3 \longrightarrow R_1$

Sol. A. Ethylene + 1 – Butene → LLDPE

Propylene + Ammonia → Acrylonitrile

Adipic acid + Hexamethylene diamine → Nylon 66

Terephthalic acid + Ethylene glycol → Synthetic fibre

GATE 2013

45. Match the reactant-product combination in Group 1 with the unit process in Group 2

Group 1	*Group 2*
P. Propylene – butanol Q. Cumene – phenol R. Butane – butadiene S. Ethylene dichloride-vinyl chloride	1. Pyrolysis 2. Dehydrogenation 3. Hydroformylation 4. Peroxidation

A. P-3, Q-2, R-4, S-1

B. P-2, Q-4, R-3, S-1

C. P-1, Q-3, R-2, S-4

D. P-3, Q-4, R-2, S-1

Sol. D. Propylene – butanol : Hydroformylation

Cusmene – phenol ; Peroxidation

Butane – butadiene ; Dehydrogenation

Ethylene dichloride – vinyl chloride : Pyrolysis

46. Identify which of the following statements are false.

A. Oils with an oleic radical (1 double bond) are more suitable than oils with a linolenic radical (3 double bonds) as film forming vehicles for

B. Production of synthesis gas from coal and steam is an endothermic process.

C. Use of chlorine for bleaching of wood pulp results in the release of dioxins.

D. In the manufacture of urea from ammonia, the main intermediate product formed is ammonium bicarbonate.

A. P and Q

B. R and S

C. Q and R

D. P and S

Sol. **D.** P and S

GATE 2012

47. Match the processes in Group I with catalysts in Group II and find the correct answer using the codes given below the Groups.

Group I	*Group II*
P. Fischer - Tropsch synthesis	1. Nickel
Q. Formaldehyde from Methanol	2. Fe_2O_3
R. Hydrogenation of Vegetable oils	3. Silver
S. Dehydrogenation of ethyl benzene	4. Cobalt

Codes

	P	Q	R	S
A.	3	4	1	2
B.	4	2	1	3
C.	4	3	1	2
D.	3	4	2	1

Sol. **C.** Fischer - Tropsch synthesis → Co

Formaldehyde from Methanol → Ag

Hydrogenation of vegetable oils → Ni

Dehydrogenation of ethyl benzene → Fe_2O_3

48. Match polymers in Group I With polymer characteristics in Group II.

Group I	Group II
P. Polyethylene	1. Elastomer
Q. Phenol-formaldehyde Polymer	2. Fibre
R. Polyiosoprene	3. Thermoplastic
S. Polyester	4. Thermosetting polymer

Codes

	P	Q	R	S
A.	3	4	1	2
B.	4	2	3	1
C.	3	2	1	4
D.	4	3	1	2

Sol. **A.** Polyethylene → Thermoplastic

Phenol-formaldehyde → polymer Thermosetting polymer

Polyiosoprene → Elastomer

Polyester → Fibre

GATE 2011

49. Match the reactions in Group I with the products in Group II.

Group I	*GroupII*
P. Ammoxidation Q. Nitration R. Dehydrogenation S. Oxidation	I. Aniline from benzene II. Benzoic acid from toluene III. Acrylonitrile from propylene IV. Styrene from ethylbenzene

A. P-III, Q-II, R-IV, S-II

B. P-IV, Q-I, R-III, S-II

C. P-I, Q-III,R-IV, S-II

D. P-I, Q-II, R-III, S-IV

Sol. **A.** Ammoxidation - Acrylonitrile from propylene

Nitration - Aniline from from propylene

Dehydrogenation - Styrene from ethylbenzene

Oxidation - Benzoic acid from toluene

GATE 2010

50. Match each of the following techniques of polymerization in Group I, with the corresponding process characteristics in Group II.

Group I	*Group II*
P. Bulk Q. Solution R. Suspension S. Emulsion	1. Polymer with very high molecular weight can be obtained 2. Heat removal is crucial but very difficult 3. Small amount of undesired low molecular weight polymer is formed 4. Polymer concentration in the product stream is low

Codes

	P	Q	R	S
A.	1	2	3	4
B.	2	4	3	1
C.	2	3	4	1
D.	2	3	1	4

***Sol.* B.** Bulk - Heat removal is crucial but very difficult

Solution - Polymer concentration in the product stream is low

Suspension - small amount of undersirable low molecular weight polymer is formed

Emulsion - Polymer with very high molecular weight can be obtained

51. Match each of the polymers in Group I with the raw material in GroupII, from which they are made.

Group I	*Group II*
P. Polyester Q. Polyamide R. Viscose rayon S. Epoxy resin	1. Ethylene glycol 2. Adipic acid 3. Cellulose 4. Bisphenol

Codes

	P	Q	R	S
A.	1	2	3	4
B.	2	1	3	4
C.	1	2	4	3
D.	3	2	4	1

***Sol.* A.** Polyester - Ethylene glycol

Polyamide - Adipic acid

Viscose rayon - Cellulose

Epoxy resin – Bisphenol

GATE 2009

52. Match the products in Group I with the name of processes in Group II.

Group I	*Group II*
P. Sodium carbonate Q. Ammonia R. Sulphuric acid	1. Haber 2. Solvay 3. Fischer-Tropsch 4. Contact

Codes

	P	Q	R
A.	2	1	4
B.	4	1	2
C.	3	4	2
D.	2	1	3

***Sol.* A.** Sodium carbonate - Solvay

Ammonia – Haber

Sulphuric Acid- Contact

53. Match the products in Group I with the material in Group II

Group I	*Group II*
P. Ethylene Q. Methanol R. Phthalic anhydride	1. Natural gas 2. Synthesis gas 3. Naphtha 4. Naphthalene

Code

	P	Q	R
A.	1	2	3
B.	2	1	4
C.	3	1	4
D.	3	2	4

***Sol.* D.** Ethylene - Naphtha

Methanol - Synthesis gas

Phthalic anhydride – Naphthalene

54. Match the unit processes in Group I with the industries in Group II.

Group I	*Group II*
P Steam cracking Q. Hydrocracking. R. Condensation	1. Petroleum refining 2. Petrochemicals 3. Polymers 4. Soaps and detergents

Code

	P	Q	R
A.	1	2	3
B.	2	3	3
C.	1	2	4
D.	2	1	3

***Sol.* D.** Steam cracking – Petrochemical

Hydrocracking – Petroleum refining

Condensation – Polymer

GATE 2008

55. Matchthe unit processes in Group I with industries in Group II.

Group 1	*Group II*
P. Saponification Q. Calcination R. Alkylation	1. Petroleum refining 2. Synthetic fibres 3. Cement 4. Soaps and detergents

Codes

	P	Q	R
A.	1	3	4
B.	2	3	4
C.	4	2	1
D.	4	3	1

Sol. **D.** Saponification - Soaps and detergents
Calcination - Cement
Alkylation - Petroleum refining

56. Which one of the following process sequences is used in the production of synthesis gas?
 A. Desulphurization → Steam reforming → Hot K_2CO_3 cycle
 B. Steam reforming → Desulphurization → Hot K_2CO_3 cycle
 C. Hot K_2CO_3 cycle → Steam reforming → Desulphurization
 D. Hot K_2CO_3 cycle → Desulphurization → Steam reforming

Sol. **D.** Sequence used in the production of synthesis gas: Hot K_2CO_3 cycle → Desulphurization → Steam reforming

57. Which one of the following process sequences is used in the sugar industry?
 A. Ca_2HPO_4/Lime treatment → Crystallization → Crushing
 B. Ca_2HPO_4/Lime treatment → Multiple stage evaporation → Crystallization
 C. Crushing → Crystallization → Ca_2HPO_4/ Lime treatment
 D. Multiple stage evaporation → crystallization → Ca_2HPO_4/Lime treatment

Sol. **B.**Process sequence in sugar industry: Ca_2HPO_4/ Lime treatment → Multiple stage! evaporation → Crystallization

GATE 2007

58. Match the chemicals in Group I with their functions in Group II

Group I	*Group II*
P. Styrene Q.Tert-dodecyl mercapton R.Potassium pyrophosphate	1. Buffer 2. Catalyst 3. Modifier 4. Monomer

Codes

	P	Q	R
A.	1	4	4
B.	4	1	2
C.	4	1	3
D.	4	3	1

Sol. **D.** Styrene - Monomer
Tert-dodecyl mercapton – Modifier
Postassium pyrophosphate - Buffer

59. Match the products in Group I with its applications in the industries of Group II.

Group I	*Group II*
P. Lithium stearate Q. Magnesium stearate R. Aluminium sulphate	1. Cosmetics 2. Paper 3. Dry cleaning 4. Greases

Codes

	P	Q	R
A.	2	3	1
B.	3	4	2
C.	4	2	3
D.	4	1	2

Sol. **D.** Lithium stearate - Grease
Magnesium – Cosmetics
Aluminum sulphate - Paper

60. Match the synthetic fibres in GroupI with their classifications in Group II

Group I	*Group II*
P. Rayon Q. Orlon R. Dacron	1. Polyamide 2. Polyester 3. Cellulose 4. Acrylic

Codes

	P	Q	R
A.	2	3	1
B.	3	4	2
C.	3	1	3
D.	3	3	4

Sol. **C.** Rayon - Cellulose
Orlon - Acrylic
Dacron - Polyester

61. Match the petrochemical derivatives in Group I with the raw materials in Group II

Group I	*Group II*
P. Acrylonitrile Q. Ammonia R. Dodecene	1. Methane 2. Ethane 3. Ethylene 4. Propylene

Codes

	P	Q	R
A.	1	2	1
B.	2	1	2
C.	3	4	3
D.	4	1	4

Sol. **D.** Acrylonitrile - Propylene
Ammonia – Methane
Dodecene – Propylene

GATE 2006

62. Pair the following industrial processes in Group with the catalysts used in Group II

Group I	*Group II*
P. Oxidation of o-xylene to phthalic anhydride Q. Oxidation of ethanol to acetaldehyde R. Oxidation of ammonia to oxides of nitrogen	1. V_2O_5 2. Pd 3. Ag 4. Pt

A. P - 1, Q - 3, R - 4
B. P - 2, Q - 1, R - 4
C. P - 4, Q - 3, R - 1
D. P - 1, Q - 3, R – 2

Sol. **A.** O-xylene + $O_2 \xrightarrow{V_2O_5}$ Phthalic anhydride

$$2(CH_3CH_2OH) + O_2 \xrightarrow{Ag} 2(CH_3CHO) + 2H_2O$$

$NH_3 \xrightarrow{Pt}$ Nitrogen oxide

63. Pair the following reactorsin Group I with their products in Group II.

Group I	*Group II*
P. Arc furnance Q. Fermenter R. Hydrogenator	1. Citric acid 2. Calcium carbide 3. Saturated fats 4. Alum

A. P - 4, Q - 1, R - 3
B. P - 2, Q - 1, R - 3
C. P - 4, Q - 3, R - 1
D. P - 2, Q - 3, R – 1

Sol. **B.** Arc furnace - Calcium
Fermenter - Citric acid
Hydrogenator - Saturated fats

64. Pair the following polymers in Group l with their chain characteristics in Group ll

Group I	*Group II*
P. HDPE Q. LDPE R. LLDPE	1. Very few branches 2. Short and regular branches 3. High branching with both short and long chain branches

Codes

A. P - 1, Q - 2, R - 3
B. P - 2, Q - 1, R - 3
C. P - 1, Q - 3, R - 2
D. P - 2, Q - 3, R - 1

Sol. **A.** HDPE - Very few branches
LDPE - Short and regular branches
LLDPE - High branching (both short chain)

65. Choose the most appropriate pairs from Group I and Group II

Group I	*Group II*
P. Nitration Q. Sulphonation R. Carbonation	1. Detergent 2. Pulp and paper 3. Explosives 4. Sugar

Codes

A. P - 1, Q - 2, R - 4
B. P - 3, Q - 1, R - 2
C. P - 3, Q - 1, R - 4
D. P - 3, Q - 2, R – 4

Sol. **B.** Nitration - Explosive
Sulphonation – Detergent
Carbonation - Pulp and paper

66. Match the appropriate pairs from Group I and Group II. Group II

Group I	*Group II*
P. Carbon disulphide Q. Caprolactum R. Gypsum	1. Nylon-6 2. Nylon-66 3. Phosphoric acid 4. Viscose rayon

A. P - 4, Q - 2, R - 3
B. P - 2, Q - 1, R - 3
C. P - 3, Q - 1, R - 4
D. P - 4, Q - 1, R- 3

Sol. **D.** Carbon disulphide - Viscose rayon
Caprolactum - Nylon-6
Gypsum - Phosphoric acid

GATE 2005

67. Match the products in Group I, with the raw materials in Group II.

Group I	*Group II*
P. Nylon-66 Q. Terylene	1. Chlorodifluoro methane 2. Dimethyl terephthalate and ethylene glycol 3. Acetylene and hydrogen cyanide 4. Hexamethylene diamine and adipic acid

A. P - 3, Q – 4
B. P - 4, Q - 3
C. P - 4, Q – 2
D. P - 1, Q - 2

Sol. **C.** Nylon-66 - Hexamethylene diamine and adipic acid Terylene - Dimethyl terephthalate and ethylene glycol Catalyst

68. Match the products in Group I with the catalysts used for its production in Group II.

Group I	*Group II*
P. Nitric acid Q. Formaldehyde	1. Silver oxide 2. Raney nickel 3. Activated carbon 4. Platinum-Rhodium

A. P - 1, Q – 2
B. P - 2, Q - 3
C. P - 3, Q – 4
D. P - 4, Q - 1

Sol. **D.** Nitric acid - Platinum-Rhodium
Formaldehyde - Silver oxide

69. Match the products in Group I with the raw materials in Group II.

Group 1	*Group II*
P. Caustic soda Q. Soda ash	1. Ammonia and sulphuric acid 2. Sodium carbonate and slaked lime 3. Salt and limestone 4. Salt and sulphuric acid

A. P - 2, Q – 3
B. P - 1, Q - 2
C. P - 3, Q – 4
D. P - 4, Q - 1

Sol. **A.** Caustic soda - Sodium carbonate and slaked lime
Soda ash - Salt and limestone

70. Match the feeds in Group I with the processes in Group II.

Group I	*Group II*
P. Gas oil Q. Residual crude	1. Acetylene production 2. Ethylene production 3. Coking 4. Cracking

A. P - 3, Q – 4
B. P - 4, Q - 3
C. P - 2, Q – 3
D. P - 1, Q - 4

Sol. **B.** Gas oil - Cracking Residual
Residual crude - Coking

GATE 2004

71. Multiple effect evaporators are commonly used in the manufacturing of

P. paper Q. superphosphate
R. sugar S. fats

A. P and Q
B. P and R
C. P and S
D. R and S

Sol. **C.** Multiple effect evaporators are used in both paper and sugar industries.

72. Match the processes in Group l with the products in Group II.

Group I	*Group II*
P. DCDA process Q. Mercury cell	1. Sodium hydroxide 2. Sulphuric acid 3. Sodium carbonate 4. Nitric acid

A. P - 1, Q – 4
B. P - 1, Q - 3
C. P - 2, Q – 3
D. P - 2, Q - 1

Sol. **D.** DCDA process - H_2SO_4
Mercury cell - NaOH

73. Match the products in Group I with the raw materials in Group II.

Group I	*Group II*
P. Urea Q. Polyester	1. Ammonia and carbon dioxide 2. Dimethyl terephthalate and ethylene glycol 3. Ammonia and carbon monoxide 4. Hexamethylene diamine and adipic acid

A. P - 1, Q – 4
B. P - 3, Q - 2
C. P - 3, Q – 2
D. P - 1, Q - 2

Sol. **D.** Urea - NH_3 and CO_2
Polyester - Dimethyl terephthalate and ethylene glycol

74. Match the products in Group I with the nature of reactions in Group II.

Group I	*Group II*
P. Polyethylene Q. Nylon R. Polystyrene	1. Condensation polymerization 2. Addition polymerization

A. P - 1, Q - 1, R – 2
B. P - 2, Q - 2, R - 1
C. P - 1, Q - 2, R – 1
D. P - 2, Q - 1, R – 2

***Sol.* B.** Polyethylene and - Condensation
Nylon - polymerization
Polystyrene - Addition polymerization

75. Match the process in Group I with the catalysts used in Group II.

Group I	*Group II*
P. Sulphuric acid manufacture Q. Vegetable oil hydrogenation	1. Platinum 2. Vanadium pentoxide 3. Iron 4. Raney nickel

A. P - 3, Q – 1
B. P - 2, Q - 1
C. P - 2, Q – 4
D. P - 4, Q – 2

***Sol.* C.** H_2SO_4 manufacturing - V_20_5
Vegetable oil - Raney nickel hydrogenation

GATE 2003

76. Identify the group in which all the polymers mentioned can be used to make fibres.
A. Butadiene copolymers, Polyamides, Urea aldehydes
B. Cellulose derivatives, Polyisoprene, Polyethylene
C. Cellulose derivatives, Polyamides, Polyethanes
D. Polypropylenes, Polyvinylchloride, Silicones

***Sol.* B.** Cellulose derivatives, polyisoprene and polyethylene can be used to manufacture fibres.

77. The preferred reacting system for oxidation of o-xylene to phthalic anhydride is
A. jacketed liquid phase CSTR
B. jacketed steam heated multitubular reactor
C. multitubular reactor with cooling multistage
D. multitubular reactor with interstage cooling

***Sol.* A.** Jacketed liquid phase CSTR is used for oxidation of O-xylene to manufacture phthalic anhydride.

78. Match the items in Group I with the items in Group II.

Group I	*Group II*
P. Calcium ammonium nitrate Q. $CaCl_2$– NaCl liquor	1. Fertilizer industry 2. Paper and pulp 3. Soda ash industry

A. P -1, Q -3
B. P - 1, Q - 2
C. P - 3, Q – 1
D. P - 2, Q – 3

***Sol.* B.** Calcium ammonium - Fertilizer nitrate
$CaCl_2$, NaCl liquor - Pulp and paper

79. Match the items in Group I with the items in Group II. Group II

Group I	*Group II*
P. Black liquor Q. Activated silica alumina R. Press mud	1. Petroleum refining 2. Sugar factory

A. P - 1, R – 2
B. P - 2, Q - 1
C. Q - 1, R – 2
D. Q - 2, R - 1

***Sol.* B.** Petroleum refinery - Black liquor
Sugar - Activated silica alumina

80. In a refinery, petroleum crude is fractionated into gas fraction, light ends, intermediate distillates, heavy distillates, residues and by products. The group of products including gas oil, diesel oil and heavy fuel oil belongs to the fraction
A. heavy distillates
B. intermediate distillates
C. light ends
D. residues

***Sol.* A.** Gas oil, diesel oil and heavy fuel oil all belong to heavy distillates.

81. The order of preference for feedstock to a catalytic reformer is
A. catalytic naphtha - coking naphtha - virgin naphtha
B. coking naphtha - virgin naphtha - catalytic naphtha
C. virgin naphtha - catalytic naphtha - coking naphtha
D. virgin naphtha - coking naphtha - catalytic naphtha

***Sol.* B.** The preferential order is
Coking naphtha → Virgin naphtha → Catalytic naphtha

GATE 2002

82. In the manufacturing of sulphuric acid from elemental sulphur,which of the following sequences of major operations is followed?

A. Furnace → Converter → Absorber
B. Furnace → Evaporator → Absorber
C. Furnace → Converter → Evaporator
D. Converter → Furnace →Absorber

Sol. **A.** The operational sequence in the manufacturing of H_2SO_4 from elemental sulphur is
Furnace → convertor → absorber

83. Consider the production of ammonia from methane and air as raw materials. The catalysts used are ....(A).... for steam reforming of methane and ....(B).... for ammonia synthesis.

	A	B
A.	Ni/Al_20_3	$Cu\text{-}ZnO/Al_2O_3$
B.	Fe/Al_2O_3	$Cu\text{-}ZnO/Al_2O_3$
C.	Ni/Al_2O_3	Fe/Al_2O_3
D.	Fe/Al_2O_3	Ni/Al_2O_3

Sol. **C.** Steam reforming of methane - $NiAl_2O_3$
Ammonia synthesis - Fe/Al_2O_3

GATE 2001

84. In the converter of the contact process for the manufacturing of H_2SO_4, the equilibrium conversion of SO_2 ....(A).... with increase in the temperature and ....(B).... with increase in mole ratio of SO_2, to air

	A	B
A.	increases	decreases
B.	decreases	increases
C.	increases	increases
D.	decreases	decreases

Sol. **B.** In contact process, the equilibrium conversion of to SO_2 decreases with increase in temperature and increases with increase in mole ratio of SO_2 to air.

85. For the hydrogenation of oils, ...(A).... is commonly used as catalyst, and ....(B).... is a catalyst poison.

	A	B
A.	platinum	Sulphur
B.	palladium	oxygen
C.	nickel	sulphur
D.	nickel	oxygen

Sol. **C.**

5. Fluid Mechanics and Mechanical Operations

1.1 Fluid Mechanics: Fluid mechanics may be broadly divided into fluid statics and fluid dynamics

- Classification of fluids: The fluids can be classified into the following

(a) Ideal Fluid: Such a fluid which is incompressible and is having no viscosity is known as an ideal fluid. Ideal fluids do not exist in practice as all the fluids which exist have some viscosity.

Newtonian fluids are those in which the viscous stresses arising from its flow, at every point, are linearly proportional to the local strain rate i.e. the rate of change of its deformation over time.

Non – Newtonian fluids are those in which the viscosity is variable based on applied stress or force.

- **Fluid Statics:** In a stationary column of static fluid, the pressure at any point is the same in all directions. Further, the pressure will also be constant in any cross section parallel with the earth's surface but will vary from height to height.
- **Manometers:** A manometer is an important device for measurement of pressure difference. A simple manometer is shown is fig. 1.1

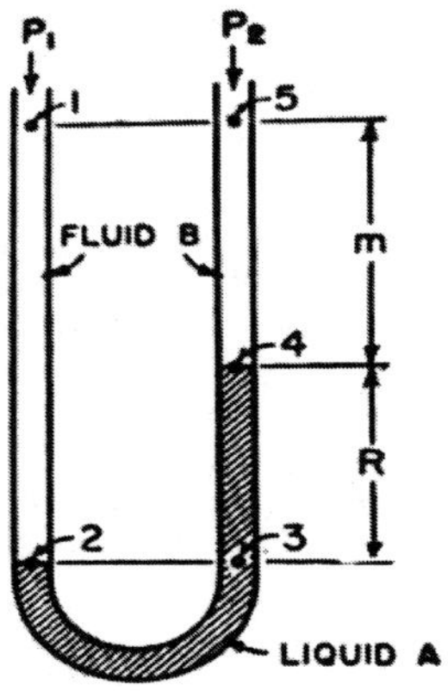

Figure 1.1

The arms of the U tube is filled with a liquid A of density ρ_A gm/sec. The arms of the U tube above liquid A are filled with fluid B which is immiscible with liquid A and has a density ρ_B gm/sec. A pressure P_1 in kg/cm^2 is exerted in one arm of the U tube and a pressure P_2 on the other. As a result of the difference in pressure $P_1 - P_2$ the meniscus in one branch of the U tube will be higher than in the other. The vertical distance between these two surfaces is R cm. Writing the pressure balance in the two limbs of the manometer.

$$P_1 + \rho_B\ (\text{m} + \text{R}) \left(\frac{g}{g_c}\right) - \text{R}.\rho_A \left(\frac{g}{g_c}\right) - \text{m}\ \rho_B \left(\frac{g}{g_c}\right) = P_2$$

or

$$\Delta\text{P} = P_1 - P_2 = \text{R}\ (\rho_A - \rho_B) \left(\frac{g}{g_c}\right) \quad \text{.... (1.1)}$$

We can see that this relationship is independent of the distance m and also of the dimensions of the U tube, provided that P_1 and P_2 are measured in this same horizontal plane.

- **Inclined Manometer:** For measuring small differences in pressure, the inclined manometer shown in

Fig. 1.2 may be used

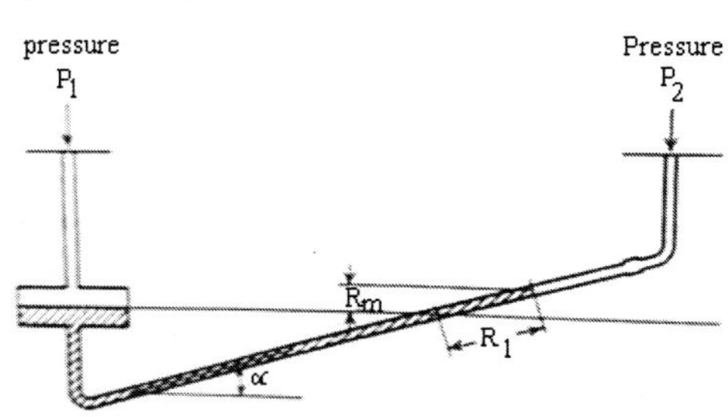

Figure 1.2

In this type of manometer, one leg of the manometer is inclined in such a manner that for a small magnitude of Rm, the meniscus in the inclined tube must move a considerable distance along the tube. This distance is Rm divided by the Sine of ∝ i.e. the angle of inclination. By making ∝ small, the magnitude of Rm is multiplied into a long distance R_1 and a large reading becomes equivalent to a small pressure difference. Writing pressure balance equation

$$P_1 - P_2 = \left(\frac{g}{g_c}\right) R\, (\rho_A - \rho_B) \text{ Sin} \propto \qquad .... (1.2)$$

In this type of manometer it is necessary to provide an enlargement in the vertical leg so that the movement of the meniscus in the enlargement is negligible within the operating range of the instrument.

- **Distribution of Velocities:** In a circular pipe the point velocities at different points in the cross section is not constant but changes from point to point with respect to the centre. The point velocity is maximum at the centre and goes on decreasing toward the pipe wall. By measuring the velocities in a circular pipe at different distances from the centre at a reasonable distance from the entrance to the pipe a plot can be prepared as shown in the fig. 1.3

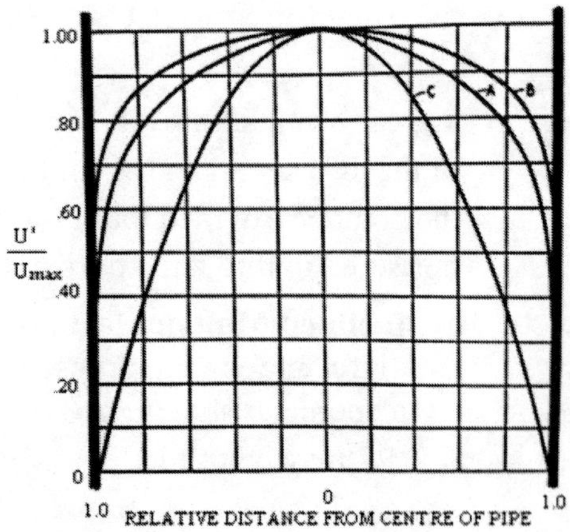

Figure 1.3

If local velocity u' is plotted against distance from the walls, different curves are obtained as shown in equation 1.4. In viscous flow (curve c) the curve is a true parabola. In this case the average velocity over the whole cross section is 0.5 times the maximum. In turbulent flow, however, the curve is somewhat flattened in the middle, and the average velocity is about 0.8 middle, and the average velocity is about 0.8 times the maximum. Changes in roughness, direction, or temperature etc. distort the shape and proportion of the velocity curves.

- **Bernoull's Theorem:** Applying the principle of conservation of energy to a fluid flow system an important law known as Bernoull's Theorem may be derived.

  The Bernoull's equation without friction can be derived by applying the manometer balance to the steady flow of a fluid in potential flow. The manometer balance provides the following equation.

$$\frac{dp}{\rho} + \left(\frac{g}{gc}\right) Z + \frac{1}{gc}\, d\left(\frac{U^2}{2}\right) = 0 \qquad ....(1.3)$$

Integrating between two stations A and B and since ρ is constant

$$\frac{p_A}{\rho} + \left(\frac{g}{gc}\right) Z_A + \frac{U_A^2}{2gc\ D} = \frac{p_B}{\rho} + \left(\frac{g}{gc}\right) Z_B + \frac{U_B^2}{2gc} \qquad ....(1.4)$$

All the terms in equation (1.4) is scalar and have the dimensions of energy per unit mass. It can be seen that with no friction if the velocity is reduced either the height above the datum Z or pressure or both must increase.

- **Equation of Continuity:** this also originates from the principle of conservation of mass. For continuous flow of a mass of fluid.

$$\rho_1 A_1 V_1 = \rho_2 A_2 V_2 = \text{Constant}$$

Since $\rho_1 = \rho_2$

∴ $Q = A_1 V_1 = A_2 V_2$

Where A_1, V_1 are the cross sectional area and velocity at section 1 and similarly A_2 , V_2 at section 2 respectively.

- **Friction Losses:** Friction loss of a fluid flowing through a pipe is a special case of general law of resistance between a solid and fluid in relative motion. These friction losses may be of many kinds. Consider a solid body immersed in a stream of fluid. Let the length of body be D and area of contact between the solid and the fluid be A. Based on dimensional analysis it has been shown that

$$\frac{F}{A} = \frac{\rho u^2}{gc} \emptyset \left(\frac{D\, u\rho}{\mu}\right)$$

where,

F = total resisting force

u = velocity of fluid past the body

ρ = density of fluid

$\emptyset$ = some function whose precise form must be determined for each specific case

In a specific case of a fluid flowing through a circular pipe of length L, the pressure drop is given by the Fanning equation as given below

$$\Delta P_f = \frac{2 f u^2 L \rho}{g_c D} \quad ....(1.5)$$

A plot of f as a function of $\left(\frac{D u \rho}{\mu}\right)$ is available for calculation of the pressure drop.

- **Measurement of Fluids:** Methods of measuring fluids may be classified as follows

1. Hydrodynamics methods
   a. Orifice meter
   b. Venturimeter
   c. Pitot tube
   d. Weirs
   e. Rotameter
2. Direct weighing or measurement orifice meters: An orifice meter is a most commonly and cheap flow measuring devise which is constructed of a thin plate containing an aperture through which a fluid passes. It may be placed in the side or bottom of a container.

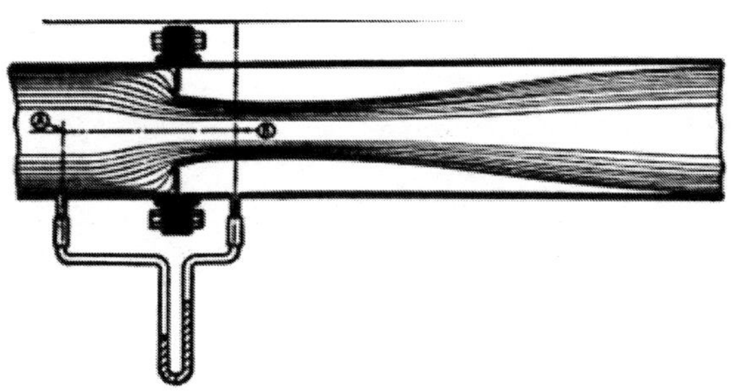

Figure 1.4 Orifice meter

Applying Bernoulli's equation between point A and B

$$X_A + \frac{U_A^2}{2\,gc} + \frac{P_A}{\rho_A} - F + w = \frac{U_B^2}{2\,gc} + \frac{P_B}{\rho_B} + X_B \quad ....(1.6)$$

In case of a horizontal pipe section

$X_A = X_B$ and assuming no friction losses

F = 0.

Further since the same fluid is passing through the orifice

$\rho_A = \rho_B = \rho$ and w = 0.

Thus simplifying equation 1.6

$$U_B^2 - U_A^2 = \frac{2gc}{\rho}(P_A - P_B)$$

$$= 2\text{gc}\left(\frac{P_A - P_B}{\rho}\right)$$

$$= 2\text{gc}\,(\Delta H)$$

or $\sqrt{U_B^2 - U_A^2} = \sqrt{2\text{gc}\,(\Delta H)}$ ....(1.7)

If the fluid is allowed to issue as a jet by removing the pipe section immediately after the orifice plate, than the minimum diameter of the stream would be less than the diameter of the orifice. This point of minimum cross section is known as ***Vena contracta.***

- **Venturimeter:** A venturimeter overcomes the most important disadvantage of an orifice meter of high pressure loss due to sudden contraction and expansion. This is done by making the entry and exit sections tapered and smooth. Thus, there is minimum of energy losses.

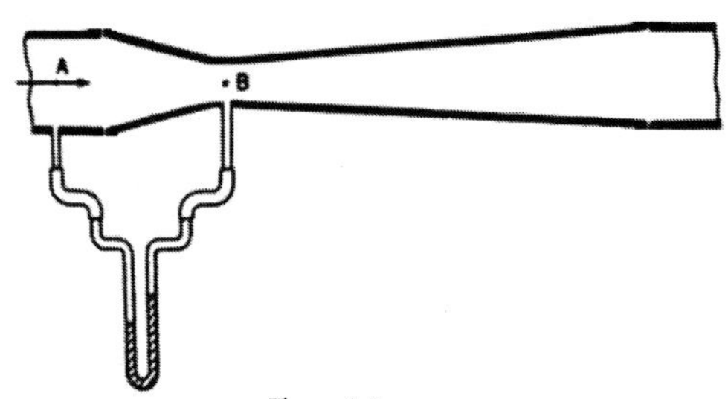

Figure 1.5

$\sqrt{U_B^2 - U_A^2} = C_v\sqrt{2g_c(\Delta H)}$ ....(1.8)

Where U_B is the velocity at the throat of the venturi. In the case of the venturimeter the value of the coefficient of discharge C_v is 0.98 and hence the head permanently lost is from 1/8 to 1/10 ΔH

For both the venturi and the orifice if the diameter of the smaller section is 1/5th the pipe diameter or less U_A^2 is so small compared to U_B^2 that the former may be neglected and the equation becomes

$U_0 = C_0\sqrt{2g_c(\Delta H)}$ for orifice meter

and $U_v = C_v\sqrt{2g_c(\Delta H)}$ for venturimeter

While comparing the venturimeter with the orifice meter both the cost of installation and cost of operation must be considered. The orifice is cheap and easy to install while a venturimeter is expensive but its cost is more than recovered in operation. Thus, venturimeters are installed as a permanent device.

- **Pilot tube:** Whereas a venturimeter and orificemeter measures average velocity of the whole stream, a pilot tube is used to measure the velocity at one point only. A pilot tube is a simple device having two tubes, one connected horizontally in the direction of point where velocity is to be measured while the other is connected to the base of the pipe as shown in the figure 1.16

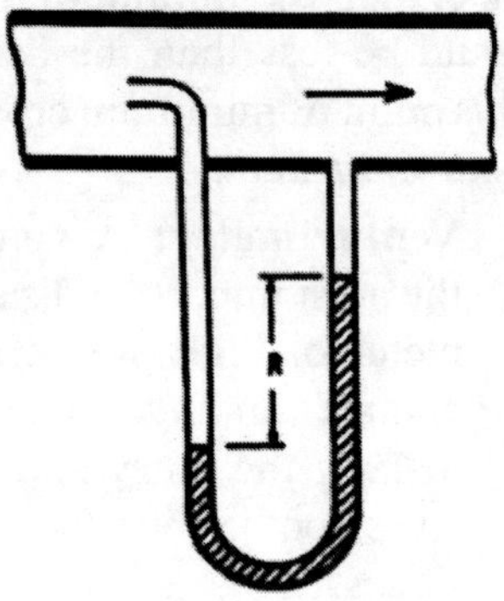

Figure 1.6 Pitot tube

The reading R of the manometer indicated the velocity head

$$\Delta H_p = \frac{u^2}{2g_c} \qquad \text{.... (1.9)}$$

- **Weirs:** Weirs are used only for liquids flowing in open channels and not in pipelines. A weir consists of a partition or obstruction with a notch cut in it to carry the flow of liquid. The amount of flow is determined by observing the level of the liquid above the edge of the opening in the weir, at a point far enough upstream to be free from the disturbances that occur in the neighborhood of the notch. Various forms of weirs are used with rectangular, V – shaped and curve – sided notches.

Multiple Choice Question

1. The _______________ equation is used to predict the pressure drop in a packed bed.
   A. Hagen-Poiseulle's
   B. Blake-Plummer
   C. Leva's
   D. Kozney-Karman
2. The gas capacity is a function of ___________ for a centrifugal fan or blower.
   A. speed
   B. $(\text{speed})^{0.5}$
   C. $(\text{speed})^2$
   D. $(\text{speed})^3$
3. The ratio of the speed of the _______________ is known as Mach number
   A. sound to that of the fluid.
   B. fluid to that of the light.
   C. fluid to that of the sound.
   D. light to that of the fluid.
4. Which of the following statements is wrong?
   A. In a static mass of liquid, the pressure at a point is the same for all liquids.
   B. As per Pascal's law, the pressure at a point in a static or uniformly moving fluid is equal in all directions.
   C. Pressure decreases exponentially with elevation in an isothermal atmosphere.
   D. Atmospheric pressure = absolute pressure-gage pressure.
5. Which of the following statements is wrong?
   A. Rise of water in capillary tubes reduces with the increasing diameter of capillary tubes.
   B. Newton's law of viscosity is not applicable to the turbulent flow of fluid with linear velocity distribution.
   C. A fluid mass is free from shearing forces, when it is made to rotate with a uniform velocity.
   D. Laminar flow of viscous liquids is involved in the lubrication of various types of bearings.
6. For fluid flowing in a circular pipe the fanning friction is
   A. inversely proportional to Reynolds number.
   B. not a function of the roughness of pipe wall.
   C. both (a) & (b).
   D. neither (a) nor (b).
7. The ____________ fluid pressure is measured of a piezometer opening
   A. total
   B. undisturbed
   C. static
   D. dynamic
8. For flow of fluid through beds of solids having large NRe Black-Plummer equation is
   A. $\frac{\Delta p}{\rho} = 4f.\frac{L}{D}.\frac{V^2}{2gc}$
   B. $\frac{-\Delta p.gc.D^2p.\varepsilon^2}{L.\overline{V_0}.\mu\ (1-\varepsilon)^2} = 150$
   C. $f_p = \frac{150(1-\varepsilon)}{N_{Re}} + 1.75$
   D. $\frac{-\Delta p}{\delta.L}.\frac{gc}{V_0^2}.\frac{D_p.\varepsilon^3}{1-\varepsilon} = 1.75$
9. For the impeller diameter of a centrifugal pump at constant speed____________.
   A. capacity varies directly with
   B. head varies as the square of
   C. horsepower varies as the cube of
   D. all (a), (b) and (c)
10. What is the value of n in Stoke's law regime, if the terminal velocity of a particle moving through fluid varies as dp^n?
    A. 1
    B. 1.5

C. 2

D. 0.5

11. In which of the following cases Dean number is applicable?

A. psychrometry.

B. fluid flow through helical pipes.

C. fluid-particle interaction.

D. power consumption in agitated vessels.

12. The equation $f^{0.5} = 4.07 \log_e (N_{Re}f)^{-0.6}$ is known as

A. Fanning equation

B. Von-Karman equation

C. Colebrook formula

D. none of these

13. The conversion factor of poise to stoke is

A. dividing by specific gravity.

B. dividing by density (gm/c.c).

C. multiplying with density (gm/c.c).

D. multiplying with specific gravity.

14. For fluids flowing in turbulent flow the pressure drop is a function of _____________.

A. $V^{2.7}$

B. $V^{1.8}$

C. $V^{-0.2}$

D. V^2

15. Which of the following statements is not correct?

A. A hydrometer used for the determination of specific gravities of liquids works on the principle of buoyant forces.

B. Turbulent fluid flow is characterised by the rapid fluctuation of instantaneous pressure & velocity at a point.

C. In case of unsteady fluid flow, the velocity at any given point does not change with time.

D. Surface tension of a liquid is because of the difference in magnitude of adhesive & cohesive forces.

16. In SI unit the dynamic viscosity has the unit of _____________.

A. kg/m . sec

B. N/m^2

C. m /N.sec.

D. m^2/sec.

17. In case of steady state laminar flow at constant pressure drop the flow at velocity varies as the square root of the cylindrical pipe diameter for a _____________ fluid.

A. dilatant

B. Newtonion

C. pseudoplastic

D. Bingham plastic

18. _____________ is known as the pressure coefficient.

A. viscosity

B. surface tension

C. gravity

D. inertial

19. If Q is the flow rate then the drag force of the float of a rotameter is

A. $\propto Q^{0.5}$

B. $\propto Q$

C. $\propto Q^2$

D. constant

20. For a liquid flowing through an orifice meter the vena-contracta form has _____________.

A. minimum liquid cross-section.

B. minimum velocity of fluid stream.

C. more diameter compared to orifice diameter.

D. none of these.

21. The velocity head which is represented by the kinetic energy of fluid per unit weight is given by

A. $\rho v^2/g_c$

B. $v^2/2g_c$

C. $2v^2/g_c$

D. $\rho.v^2/2g_c$

22. The shear stress for a given rate of angular deformation of fluid according to Newton's law of viscosity is ___________ where μ is fluid viscosity.

A. $1/\mu^2$

B. μ

C. $1/\mu$

D. μ^2

23. For turbulent fluid flow in a circular pipe the equation for pressure drop is.

A. $\frac{32\,\mu L V}{g_c\,.\,D^2}$

B. 64/Re

C. $4f.\frac{L}{D}.\frac{v^2}{2g_c}.\rho$

D. $f.\frac{L}{D}.\rho.\frac{V^2}{2g_c}$

24. In a centrifugal gas compressor constant volumetric flow conditions ______________ in the gas discharge pressure will result in horsepower increase.
   A. increase
   B. decrease
   C. not change
   D. exponentially decrease
25. If a bed of solids is continuously fluidized then
   A. solids are completely entrained.
   B. velocity of the fluid is very small.
   C. there is no entrainment of solids.
   D. the pressure drop is less than that for batch fluidization.
26. The unit of absolute viscosity (Poise) is
   A. m^2/sec
   B. gm/cm. sec
   C. cm^2/ sec
   D. gm/cm.sec
27. For laminar flow in a pipe a Bingham plastic fluid will follow a velocity profile ________________.
   A. flat
   B. parabolic
   C. flat near the wall and parabolic in the middle
   D. parabolic near the wall and flat in the middle
28. The minimum compressibility is exhibited by
   A. Water at room temperature
   B. Nitrogen at room temperature
   C. Air at room temperature
   D. Oxygen at room temperature
29. The relationship between friction factor to Reynold number $f^{0.5} = 4\ \log_e(N_{Re}/f)^{-0.4}$, is called the __________ equation
   A. Nikuradse
   B. Blausius
   C. Colebrook
   D. Von-Karman
30. If M is the metacentre for an unstable equilibrium of a floating body then
   A. *M* & *G* coincide
   B. *M* is below *G*
   C. *M* is above *G*
   D. none of these
31. If a fluid jet discharging from a 5 cm diameter orifice has a 4 cm diameter at its vena-contractors what would be the coefficient of contractor
   A. 0.64
   B. 0.32
   C. 0.94
   D. 1.65
32. In rough pipes under complete turbulence conditions the
   A. friction factor depends upon NRe only.
   B. rough and smooth pipes have the same friction factor.
   C. laminar film covers the roughness projections.
   D. friction factor is independent of the relative roughness.
33. The device which can be used for direct measurement of volumetric flow rate of a slurry is
   A. Pitot tube
   B. Venturimeter
   C. Rotameter
   D. Orificemeter
34. Which of the following parameters is responsible for fluid resistance to shear?
   A. cohesion
   B. rate of transfer of molecular momentum
   C. both (a) and (b)
   D. neither (a) nor (b)
35. A hydraulically smooth pipe is the one for which the friction factor
   A. is zero irrespective of the Reynolds number.
   B. for a given Reynolds number remains constant even on further smoothening of the pipe.
   C. is not a function of Reynolds number
   D. none of these
36. When a fluid flow through a circular pipe, the
   A. velocity profile becomes flatter and flatter with increasing Reynolds number.
   B. point of maximum instability exists at a distance of 2r/3 from the pipe wall (r = pipe radius).
   C. mean flow velocity is about 0.5 times the maximum velocity.
   D. skin friction drag, shear stresses, random orientation of fluid particles and slope of velocity profile at the wall are more.
37. A relief valve is required to be put in a reciprocating pump with the purpose to
   A. protect the pump against developing excessive pressure.
   B. control the rate of discharge.

C. facilitate unidirectional flow of liquid.

D. reduce the discharge pressure.

38. The main function of a hydraulic accumulator is to

A. increase the pressure and store/ accumulate it.

B. accumulate pressure to increase force.

C. ensureintermittent supply of hydraulic pressure.

D. generate high pressure to operate hydraulic machines like cranes, lifts, presses etc

39. The energy of flow downstream of the machine is considerably decreased in case of a ______________.

A. centrifugal fan

B. turbine

C. blower

D. centrifugal pump

40. The apparent viscosity of pseudo plastic fluids ____________ with increase in the shear rate.

A. remains same

B. decreases

C. increases

D. may increase or decrease; depends on the magnitude of shear rate

41. The ratio of actual discharge to theoretical discharge known as coefficient of discharge for an orifice is approximately ______________.

A. 0.42

B. 0.62

C. 0.82

D. 0.98

42. Which of the following statements is correct?

A. The discharge through a venturimeter depends upon Δp only and is independent of orientation of the meter.

B. For a given pressure difference, the discharge of gas is greater through a venturimeter, when compressibility is taken into account, than when it is neglected.

C. The overall pressure loss is the same in a given pipe line, whether a venturimeter or a nozzle with the same throat dia is used.

D. A venturimeter with a given gage difference discharges at a greater rate, when the flow is vertically downward through it, than when the flow is vertically upward.

43. Which of the following conditions should be satisfied with respect to the derivative of flow variables for steady flow to occur?

A. $\frac{\partial}{\partial t} = \text{constant}$

B. $\frac{\partial}{\partial t} = 0$

C. $\frac{\partial}{\partial s} = \text{constant}$

D. $\frac{\partial}{\partial s} = 0$

44. The reason for the delivery of insufficient quality of liquid by a pump is

A. low rpm

B. too high a lift

C. air leak in the inlet

D. all (a), (b) and (c)

45. The non-dimensional number $(\Delta P/pu^2)$ is author form of Konzney-Carman equation and is proportional to __________.

A. $\frac{L/D_p}{R_e}$

B. $\frac{Re^2}{D_p/L}$

C. $\frac{Re}{D_p/L}$

D. $\frac{L/D_p}{Re^2}$

46. The pressure drop in a fully opened globe value as compared to that in a gate value for the same terminal conditions and value size is

A. more

B. equal

C. less

D. either (a) or (b); depends on the viscosity of the fluid

47. What in the velocity of air in m/sec for a pitot tube being used for measuring velocity of air?

A. 14.1

B. 5

C. 56.22

D. 28.2

48. A reciprocating pump may develop into a maximum of ______________ theoretical suction lift in meters.

A. 5

B. 10

C. 100
D. 50

49. In a reciprocating pump an air vessel is provided to
A. reduce discharge fluctuation.
B. increase the pump efficiency.
C. reduce the danger of cavitation.
D. avoid the necessity of priming.

50. Which of the following measurements is done by a Prandtle pilot tube?
A. velocity at a point in the flow.
B. pressure difference in pipe flow.
C. pressure at a point.
D. average flow velocity.

51. The ratio of pressure forces to ____________ forces is known as pressure coefficient.
A. inertial
B. viscous
C. gravity
D. none of these

52. For large water discharge rate from a river or from an open channel the weir used is not ___________ shape.
A. circular
B. triangular
C. trapezoidal
D. rectangular

53. For an incompressible fluid flow, the fluid
A. is frictionless.
B. temperature remains constant.
C. density does not change with pressure & temperature.
D. compressibility is greater than zero.

54. For laminar flow between parallel plates, the maximum to average velocity is ________________.
A. 1
B. 1.5
C. 2
D. 2.5

55. Laminar flow region exists during agitation, when Reynolds number is
A. more then 2500
B. more then 1000
C. more then 50
D. less then 10

56. If a centrifugal pump being used to pump water is employed to pump a more viscous oil then in the later case the pump.
A. capacity is reduced.
B. requires more power.
C. develops a lower head.
D. all (a), (b) and (c).

57. For fluid flow in pipes the critical velocity _______________.
A. independent of fluid density.
B. increases as pipe diameter increases.
C. increases as fluid viscosity increases.
D. none of these.

58. If lift force is to be developed then?
A. The body should be stream lined.
B. The main stream velocity must approach the velocity of sound in that fluid medium.
C. Circulation around the body is essentially required.
D. The body should be bluff body.

59. For an orifice meter the discharge coefficient is a function of ________________.
A. ratio of orifice dia to pipe dia.
B. Reynolds number at the orifice.
C. both (a) and (b).
D. none of the above parameters, and has a constant value of 0.61.

60. For continuous flow of ideal fluids the continuity equation is based on principal of conservation of ________________.
A. energy
B. mass
C. momentum
D. both (b) & (c)

61. Which of the following statements is correct for the blades of a centrifugal impeller to be curved forward?
A. outlet tip of a blade curves in a direction opposite to that
B. inlet tip of a blade curves in a direction opposite to that
C. inlet tip of a blade is towards the direction
D. outlet tip of a blade is towards the direction

62. Which of the following is not dimensionless?
A. Weber number
B. Froude number
C. Kinematic viscosity

D. Pressure-co-efficient

63. It is possible for flow to occur from low pressure to high pressure when.

A. Impossible in a constant cross-section conduit.

B. Flow through a converging section.

C. Flow of liquid upward in a vertical pipe.

D. Flow of air downward in a pipe.

64. The applicable of Stoke's law is valid only when the particle Reynolds number is _______________.

A. < 1

B. < 5

C. > 1

D. none of these

65. If the terminal conditions and filling sizes are the same the least friction loss is incurred in a/an _______________.

A. T-joint

B. union

C. 90°bend

D. 45° elbow

66. Which of the following statements is correct for the blades of a centrifugal impeller to be curved forward?

A. outlet tip of a blade curves in a direction opposite to that

B. inlet tip of a blade curves in a direction opposite to that

C. inlet tip of a blade is towards the direction

D. outlet tip of a blade is towards the direction

67.

If Q_1 and Q_2 are the actual and theoretical discharges respectively then the percentage slip in a reciprocating pump set is given by the percentage of.

A. Q_2/Q_1

B. Q_1/Q_2

C. $\frac{Q_2 \quad Q_1}{Q_1}$

D. $\frac{Q_2 \quad Q_1}{Q_2}$

68. Which of the following may be used for very accurate measurement of flow of gas at low velocity?

A. Rotameter

B. Segmental Orificemeter

C. Pitot tube

D. Hot wire annemometer

69. What will be the pressure head of 320 m of water in m of CCl_4 (sp.gr = 1.6)

A. 100

B. 200

C. 320

D. 160

70. In comparison to a single acting reciprocating pump a double acting pump of almost same size and working under same pressure levels would give almost double_______________.

A. efficiency

B. discharge

C. head

D. none of these

71. The direct measurement of which of the following cannot be done by the use of a venturimeter

A. friction loss in pipe flow.

B. datum difference in the stretch of pipe flow.

C. pressure difference in the flow through pipeline

D. all (a), (b) and (c).

72. For fluid flow through a packed bed of solid which of the following equation is called the Kozney-Karman equation?

A. $\frac{p}{p} = 4f.\frac{L}{D}.\frac{V^2}{2gc}$

B. $fp = \frac{150\,(1- \ )}{N_{Re}} \quad 1.75$

C. $\frac{-\ p\,.\,gc\,.\,D^2p\,.\quad ^3}{L.\,\bar{V}o\,.\quad (1 - \ )^2.} \quad 150$

D. $\frac{-\ p}{.\,L}.\frac{gc}{V^2o}.\frac{-Dp\,.\quad ^3}{1 -} \quad 1.75$

73. Which of the following is accomplished by agitation of liquids in agitators?

A. Dispersing gas in liquid

B. Dispersing immiscible liquid in form of emulsion

C. Suspending solid particles

D. All of these

74. The applicability of Pascal's law is valid while

the fluid is

A. at rest and when the frictionless fluid is in motion.

B. at rest.

C. frictionless and at rest.

D. none of these.

75. Which of the following is equivalent to a pressure of 10 m head of water in kN/m_2.

A. 98

B. 49

C. 147

D. 196

76. In a stationary fluid and flow in Newton's law range the drag coefficient for the motion of spherical particles is

A. 0.44

B. 0.044

C. 4.4

D. 44

77. For a flow past a sphere if the velocity V and the Reynolds number increase from very low value, the drag force for Re<<1

A. increases linearly with V.

B. decreases as V^2.

C. decreases linearly with V.

D. none of these.

78. Which of the following measuring devices is not based on the Bernoll's equation.

A. orificem

B. pitot tube

C. venturimeter

D. none of these

79. The following specifications pertain to a centrifugal pump which is used to develop.

Flow 1000 liters/minutes of water.

Power = 4 H.P.; Speed = 800 rpm

Head = 8 meters.

If its speed is halved, what will be the new discharge___________ liters/minute?

A. 500

B. 750

C. 200

D. 1000

80. The design of pipes having diameter 14 inches or more are based on ________________.

A. outside diameter

B. schedule number

C. inside diameter

D. none of these

81. When water flowing in a pipe is suddenly brought to rest by closing the valve water hammer occurs. The extent of pressure thus produced due to water hammer depends on the

A. Time taken to close the valve

B. fluid velocity in the pipe

C. pipe length

D. all (a), (b) and (c)

82. For ________________ fluid, the rate of shear versus the shear stress curves are time dependent.

A. rheopectic

B. thixotropic

C. both (a) & (b)

D. neither (a) nor (b)

83. For which of following flow meters the pressure head remains constant?

A. pitot tube

B. venturimeter

C. rotameter

D. orificemeter

84. Highly viscous liquids and pastes are agitated by.

A. Propellers

B. turbine agitators

C. multiple blade paddles

D. none of these

85.

In case of Newtonion fluids flowing in laminar flow in a pipe of diameter 'D' and length 'L' incurring a pressure drop 'ΔP' with average velocity 'V_{avg}', what is the shear rate at the pipe wall.

A. $4 . V_{avg}/D$

B. $D \Delta p/4L$

C. $8 . V_{avg}/D$

D. $D \Delta p/8L$

86. Name the device which is the most efficient and best for measuring very small flow rate of gases.

A. Orifice meter

B. Venturimeter

C. Rotameter

D. Flow nozzle

87. The maximum head loss will incur in the pipe bend with ________________.

A. U-bend

B. 30° bend

C. 45° bend

D. 90° bend

88. Which of the following parameters is a major factor for speed of sound in an ideal gas?

A. temperature

B. specific volume

C. pressure

D. none of these

89. Among the characteristics of a perfect gas is that it

A. does not satisfy $PV = nRT$.

B. can't develop shear stresses.

C. has constant specific heat.

D. is incompressible and has zero viscosity.

90. The ratio N_{Re}^2/N_{Fr} is termed ________ number.

A. Euler

B. Galileo

C. Brinkman

D. Archimedes

91. Water is pumped by a centrifugal pump if the same pump is used for pumping a thicker oil $(\mu = 35\ cp)$, it will develop ____________.

A. less head & capacity

B. less head & more capacity

C. more head

D. more capacity

92. For Newtonion fluid flowing in laminar flow in circular pipe if the velocity distribution is a function of the distance'd' measured from the centre line of the pipe then if followers a __________ relationship.

A. linear.

B. parabolic

C. hyperbolic

D. logarithmic

93. What would be the hydraulic radius for flow in an open channel of large width where depth of flow is x?

A. x

B. x/2

C. x/3

D. 2x/3

94. Tubes having lower BWG will have.

A. outer diameter

B. lower cross-section

C. lower thickness

D. inner diameter

95. Water is flowing through a main pipe of I.D. 4 m at a velocity of 5 m/sec. If this main pipe is bifurcated into two pipes of I.D 2 each. What would be the average velocity through the bifurcated pipes?

A. 5 m/sec

B. 10 m/sec

C. 20 m/sec

D. 52 m/sec

96. In which of the following case a self-priming centrifugal pump can be used?

A. booster service.

B. pumping industrial wastes.

C. pumping liquid fertilisers (e.g. liquid NH_3).

D. all (a), (b) and (c).

97. The kinematic viscosity has a dimension of________________.

A. $ML^{-2}T^{-2}$

B. L^2T^{-1}

C. $ML^{-2}T^{-1}$

D. none of these

98. In which of the following ranges the viscosity of water at 40°C lies?

A. 1 to 2 kg/m.s

B. 0.5×10^{-3} to 1×10^{-3} kg/m.s

C. 0.5 to 1 kg/m.s

D. 1×10^{-3} to 2×10^{-3} kg/m.s

99. The Prandtl mixing length is

A. independent of the shear stress.

B. independent of radial distance from the pipe axis.

C. zero at the pipe wall and is a universal constant.

D. useful for computing laminar flow problems.

100. Paddle agitator

A. is suitable for mixing low viscosity liquids.

B. produce axial flow.

C. moves at very high speed.

D. none of these.

101. If a fluid under turbulent flow conditions is entering

into a pipe then the transition length would be around ___________ times the pipe diameter.

A. 5
B. 50
C. 500
D. 1000

102. In which of the following device maximum head loss will occur for a given fluid flow rate?

A. Orifice meter
B. Flow nozzle
C. Venturimeter
D. All of them incur the same head loss

103. In which of the following cases Pascal law is not applicable?

A. static
B. uniformly moving
C. accelerating frictionless
D. none of these

104. Which of the following statements is correct pertaining to transition or entrance length in fluid flow?

A. Transition length for turbulent flow of Newtonion fluids in a smooth pipe of diameter'd' is equal to 50 D.
B. Transition length for laminar flow of Newtonion fluids in a pipe of diameter'd' is equal to 0.05. D.N_{Re}.
C. The length of entrance region of pipe, in which full development of fluid flow takes place such that velocity profile does not change downstream, is called the transition length.
D. all (a), (b) and (c).

105. It is not possible for the boundary layer separation to occur during fluid flow

A. at points of abrupt changes in the flow directions.
B. in case of boundaries experiencing form drag.
C. in laminar flow.
D. none of these.

106. For a ______________ fluid the apparent viscosity increases with shear rate.

A. viscous
B. Newtonion
C. dilatant
D. non-viscous

107. Variation of shear stress (τ) with velocity gradient $\frac{dv}{dy}$, constant pressure temperature is shown as followers. The flow of Bingham plastic is represented by the curve

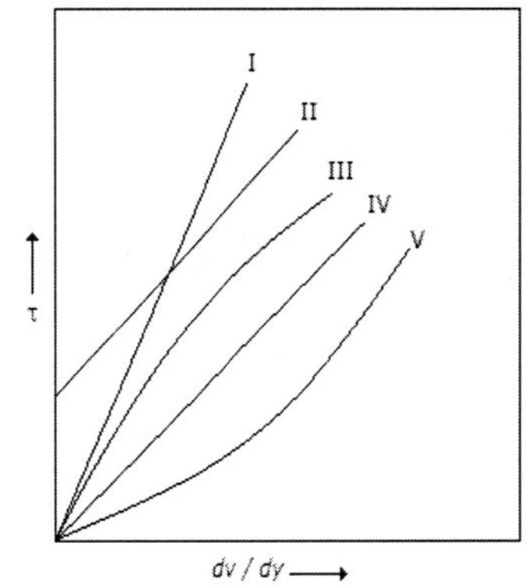

A. I
B. II
C. III
D. V

108. The principle of measuring the difference between the impact and the static pressure forms the principle of a ___________.

A. rotameter
B. pitot tube
C. flow nozzle
D. venturimeter

109. For pumping raw sewage or sludge the most common pump is

A. Gear pump
B. Centrifugal pump
C. Reciprocating pump
D. Electromagnetic pump

110. Which of the following is correct for a pipe exit fitted with a nozzle

A. power transmitted through the nozzle is maximum, when the head lost due to friction in the pipe is equal to one third of the total supply head.
B. conversion of pressure head to kinetic head is facilitated.
C. conversion of kinetic head to pressure head is facilitated.
D. both (a) and (b)

111. For constant rate fluid flow through a tapering pipe the flow is ___________.

A. unsteady uniform
B. steady non uniform

C. unsteady non uniform
D. steady uniform

112. Such a case where turbulent flow may occur is
A. very slow motion
B. highly viscous fluid
C. very narrow passages
D. none of these

113. If an incompressible fluid under turbulent flow conditions is flowing through a pipe then the flow rate Q is proportional to $(\Delta p)^n$ where Δp is the pressure drop the value of exponent n is _______________.
A. 1
B. 0
C. < 1
D. > 1

114. For the Euler's equation of motion to be integrated which of the following assumptions is to be made?
A. The fluid is incompressible.
B. The flow is rotational and incompressible.
C. The fluid is non-viscous.
D. The continuity equation is satisfied.

115. For impeller diameter D (inches), speed of pump (N rpm) and head developed (H) in ft the correct relationship for a centrifugal pump is _______________.
A. $D = 1840\ H^{0.5}/N$
B. $H = 1840\ D^{0.5}/N$
C. $D = 1840\ N/H^{0.5}$
D. $D = 1840\ H/N$

116. During the operation of a fluidized bed reactor the
A. hot spots are formed at very high tempture.
B. temperature is more or less uniform.
C. segregation of the solids occurs.
D. temperature gradients are

117. For industrial uses centrifugal pumps are more popular in comparison to reciprocating pumps mainly because of?
A. very low speeds.
B. high vibrations.
C. higher capital &maintenance cost.
D. smaller discharge.

118. The ratio of _______________ to the product of velocity head and density for flow past immersed body is known as the drag coefficient.
A. shear force
B. shear stress
C. average drag per unit projected area
D. none of these

119. The parameter which is not dimension – less is
A. Pressure co-efficient
B. Kinematic viscosity
C. Froude number
D. None of these

120. A diffuser pump is a special type of liquid transporting device where _______________ are minimum.
A. cavitation
B. bearing losses
C. shock losses
D. disk friction

121. Certain conditions in which a pulsating discharge flow is most likely to happen is in ___________.
A. Reciprocating pump
B. Axial flow pump
C. Gear pump
D. Centrifugal pump

122. A streamline is
A. defined for uniform flow only.
B. always the path of a particle.
C. drawn normal to the velocity vector at every point.
D. the line connecting the mid-points of flow cross-sections.

123. When the equilibrium position of the float in a rotameter is attained the following relationship holds good. D_f is drag force, B_f is buoyant force on the float and W_f is the weight of the float.
A. $D_f + B_f = W_f$
B. $D_f + B_f + W_f = 0$
C. $D_f = B_f + W_f$
D. none of these

124. The stagnation point in fluid flow represents a point where the _______________ is zero.
A. flow velocity
B. total energy
C. pressure
D. all (a), (b) and (c)

125. When a plane surface of in-regular shape is immersed vertically in a state mass of fluid the centre of gravity

A. is different for different fluids.
B. is independent of the specific weight of the fluid.
C. is at the centroid of the plane surface.
D. lies above the centroid of the plane surface.

126. When the ______________ a free vortex is foromed.
A. velocity changes linearly with radial distance.
B. radial component of velocity is same everywhere.
C. stream lines are not circular.
D. flow is necessarily rotational.

127. When particle Reynolds number is __________ around a sphere the flow is known as creeping flow.
A. <2100
B. <0.1
C. >2.5
D. <500

128. The most convenient and popular measurement of liquid flow through an open channel is
A. segmental orifice
B. notch
C. hot wire anemometer
D. rotameter

129. Bernoulli's equation may be written in the form $\frac{p}{\rho}+\frac{g}{g_c}\cdot Z+\frac{V^2}{2g_c}=\text{constan}$. Each term of this equation represented the total energy per unit
A. mass
B. specific weight
C. volume
D. none of these

130. A fluid element is undergoing changes from one position to another position such that the angle between the two sides change. What would be the type of motion?
A. Linear deformation
B. Rotation
C. Translation
D. Angular deformation

131. Which of the following statements is not correct?
A. The viscosity of gas increases with increase in temperature.
B. The eddy viscosity is a fluid property.
C. The viscosity of a liquid increases with decrease in temperature.
D. The eddy viscosity is a function of the type of turbulence involved.

132. When a flow is taking place in plane Poiseuille region the velocity distribution in a direction normal to the direction of flow is ______________.
A. linear
B. parabolic
C. hyperbolic
D. none of these

133. When a valve is opened in a pipeline the nature of fluid flow is
A. uniform
B. unsteady
C. laminar
D. steady

134. For pressure drop calculation the equivalent diameter for a duct of square cross section is given by
A. x
B. πx
C. 2x
D. x/2

135. If the velocity of flow of fluid changes from point to point in the fluid at any instant what would be the type of flow?
A. Unsteady
B. Turbulent
C. Rotational
D. Non-uniform

136. The force exerted by the ________________ is defined as drag.
A. solid on the fluid.
B. fluid on the solid in the direction of flow.
C. fluid on the solid in a direction opposite to flow.
D. none of these.

137. When the flow is ______________ the Bernoulli's equation is not applicable
A. viscous.
B. irrotational.
C. incompressible.
D. all (a), (b) & (c).

138. In which of the following categories toothpaste falls?

A. Newtonion liquid
B. Bingham plastic
C. pseudoplastic
D. dilatant

139. Water is filled in a conical tank with a bottom opening of cross sectional are A and is placed as shown in the figure. If density of water is ρ_L and ρ_{air} is negligible then what is the quantum of force F with which plate X must be pushed up to prevent water from falling?

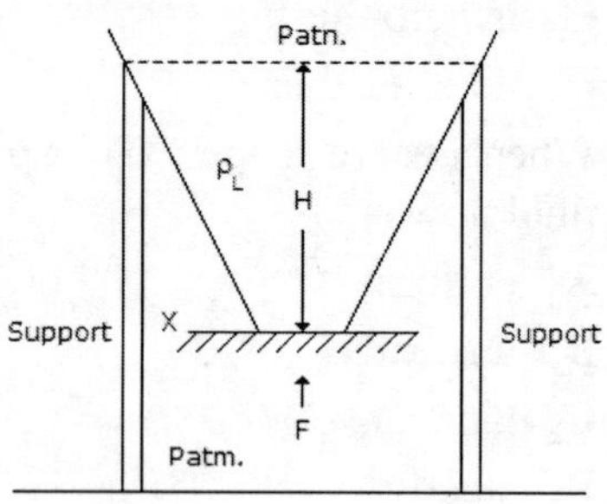

A. $\rho_L .V_g/3$
B. $\rho_L .A.H_g$
C. $\rho_L . V_g$
D. $\rho_L . V_g/2$

140. The fluid velocity in the boundary layer is that part of a moving fluid which is
A. constant.
B. affected by the fluid flow pressure.
C. affected by the presence of a solid boundary.
D. all (a), (b) and (c).

141. The discharge for a fluid flow through a round tube varies
A. inversely as the pressure drop.
B. as the square of the radius.
C. inversely as the viscosity.
D. linearly as the viscosity.

142. The most convenient pumping of slurries is by a _________ pump.
A. reciprocating
B. gear
C. screw
D. centrifugal

143. In a centrifugal pump volute type of casing is provided to
A. reduce the discharge fluctuation.
B. convert velocity head to pressure head.
C. convert pressure head to velocity head.
D. increase the discharge.

144. In comparison to centrifugal pumps reciprocating pumps
A. do not need priming.
B. run at a lower speed for the same discharge.
C. deliver fluid with pulsating/fluctuating discharge.
D. can be run with discharge line valve closed for a short

145. Which of the following parameters is responsible for the actual lift of a pump to be less than the theoretical lift
A. frictional resistance through pipes, fittings & passages.
B. leakage& pressure decreasing at higher elevations.
C. specific gravity & temperature of the liquid.
D. all (a), (b) and (c).

146. In a barometer the ideal fluid to be used is mercury mainly due to its
A. high density.
B. low capillary action.
C. low compressibility.
D. very low vapor pressure.

147. If the velocity of flow of fluid changes from point to point in the fluid at any instant what would be the type of flow?
A. Unsteady
B. Turbulent
C. Rotational
D. Non-uniform

148. A gas
A. is incompressible.
B. signifies absence of density.
C. can resist shearing action.
D. is a supercritical vapor.

149. Which of the following is analogous to the speed of a sound wave in a gas?
A. an elementary wave in an open channel.
B. a disturbance travelling upstream in moving fluid.
C. flow in an open channel.
D. none of these.

150. The function of an air vessel fitted in a reciprocating pump is that if
A. results in non-uniform discharge.
B. decreases the work done.

C. increases the work done.
D. causes cavitation.

151. In an open channel flow through a circular channel the ratio of the depth of flow to the diameter of the circulation channel under maximum discharge conditions will be
A. 0.1
B. 0.55
C. 0.95
D. 1.85

152. If the density difference of the liquid phase to be separated is very small (as in milk cream separator), the most suitable separator is
A. disc bowl centrifuge
B. Sharples super centrifuge
C. hatch basket centrifuge
D. sparkler filter

153. The ratio of viscosity of water to that of air at room temperature is approximately
A. 15
B. 55
C. 155
D. 1050

154. The measurement of _______________ of a fluid is done by a pilot tube.
A. average flow rate
B. Pressure
C. average velocity
D. point velocity

155. Which of the following schedule number will be invalent to the minimum pipe wall thickness?
A. 160
B. 120
C. 80
D. 40

156. What would be the Reynolds number for flow of water at room temperature through 2cm dia pipe at an average velocity of 5 cm/sec?
A. 100
B. 10
C. 2000
D. 1000

157. For a centrifugal pump to operate the net positive suction head (NPSH) must be
A. greater than the vapour pressure of the liquid.
B. less than barometric pressure.
C. less than the vapour pressure of the liquid.
D. equal to the vapour pressure of the liquid.

158. Most suitable conveyor for transportation of sticky material is
A. apron conveyor
B. belt conveyor
C. screw conveyor
D. pneumatic conveyor.

159. Which of the following is an example of unsteady fluid flow for the fluid flow conditions and flow pattern change with the passage of time at a position?
A. Low velocity flow of a highly viscous liquid through a hydraulically smooth pipe.
B. Water flow in the suction and discharge pipe of a reciprocating pump.
C. Discharge of water by a centrifugal pump being run at a constant rpm.
D. Water discharge from a vertical vessel in which constant level is maintained.

160. When a valve is opened in a pipeline the nature of fluid flow is
A. uniform
B. unsteady
C. laminar
D. steady

161. Comparing the performance of a centrifugal pump with another pump the term specific speed relates to
A. dynamic similarity
B. same speed
C. geometrical similarity
D. same efficiency

162. Which of the following represents the function of relief value?
A. unloads a pump.
B. provides back pressure for a cylinder.
C. is a directional control valve.
D. none of these.

163. Traces of liquid for present in Coke oven gas is separated using
A. electrostatic precipitator
B. cyclone separator
C. strainer
D. none of these

164. For a fluid flowing in a circular tube what would be the distribution of shear stress?

A. linear with radius for laminar flow only.

B. linear with radius for turbulent flow only.

C. linear with radius for both laminar & turbulent flow.

D. parabolic with radius for both laminar & turbulent flow.

165. In a stationary fluid the drag co-efficient in hindered settling in comparison to free setting for motion of spherical particles in

A. more

B. equal

C. less

D. more or less, depending on the type of particle

166. If a particle is falling slowly in viscous liquid with Reynolds number less than 1. Then the correct statement is?

A. Drag force and gravitational forces are important.

B. Drag, gravitational and buoyancy forces are important.

C. Inertial and drag forces are important.

D. None of the above.

167. The discharge for a fluid flow through a round tube varies

A. inversely as the pressure drop.

B. as the square of the radius.

C. inversely as the viscosity.

D. linearly as the viscosity.

168. Which of the following will happen as far as the bulk modulus of elasticity of a liquid is concerned

A. is independent of temperature & pressure.

B. is zero for incompressible liquid.

C. decreases with pressure.

D. increases with pressure.

169. When two pipe systems become equivalent the two quantities which are same are

A. Head & discharge

B. Length & diameter

C. Length & discharge

D. Friction factor & diameter.

170. Which of the following statements is not valid pertaining to flow of an incompressible fluid through a venturimeter.

A. Discharge of fluid through a venturimeter depends upon the gage difference irrespective of the orientation of venturimeter.

B. Venturimeter incurs less power loss compared to an equivalent orifice meter.

C. Venturimeter occupies less space than an orifice meter.

D. For frictionless flow, the fluid pressure entering the venturi meter will be exactly equal to that leaving the venturimeter.

171. When water is lifted from a reservoir to a greater height than the initial level it is called a ______________.

A. Tunnel

B. Siphon

C. Penstock

D. Pressure pipeline

D. fluid friction is zero

172. Two pipes have to be connected which have to be dismantled at intervals for maintenance what will be the type of connector?

A. union

B. reducer

C. elbow

D. tee

173. Which of the following parameters functionally influence the absolute viscosity of a fluid.

A. motion

B. pressure & temperature

C. shearing stress

D. both (b)&(c)

174. Which of the following statements correctly define a fluid?

A. cannot remain at rest under the action of shear force.

B. continuously expands till it fills any container.

C. permanently resists distortion.

D. is incompressible.

175. Air at room temperature and atmospheric pressure is flowing through a rotameter giving a certain reading for a flow rate of 100 cc/sec. In the same rotameter if helium (molecular weight 4) is used and rotameter shows the same reading the flow rate now in cc/sec is ______________.

A. 26

B. 42

C. 269

D. 325

176. What will be the proportion of capillary rise of mercury in a small diameter tube if σ = surface tension of mercury and d is diameter?

A. $1/d$
B. $1/\sigma$
C. σ
D. d

177. For a flow through a converging-diverging tube the correct statement as given below is.
A. Throughout the converging portion of the tube, the density increases in the downstream direction.
B. No shock wave develops in the tube when the Mach number at exit is greater than unity.
C. The value of Mach number is always unity at the throat.
D. none of these.

178. Cavitation is harmful for the pumps and can be prevented by
A. maintaining suction head lower than the vapour pressure.
B. maintaining the suction head sufficiently greater than the vapour pressure.
C. suitably designing the pump.
D. maintaining suction head = developed head.

179. What will be the effect on the fraction of the orifice pressure differential that is permanently lost when the ratio of orifice diameter to pipe diameter is increase?
A. increases
B. decreases
C. increases exponentially
D. remains unchanged

180. The equivalent of one horsepower is
A. 550 lbf.ft/second
B. 550 kgf.m/second
C. 550 lbf.ft./hr
D. both (a) and (b)

181. The purpose of doing priming in a centrifugal pump is to
A. develop effective pressure rise by the pump.
B. increase the mass flow rate of fluid.
C. avoid chances of separation inside the impeller.
D. none of these.

182. The continuity equation for one dimensional flow of an incompressible fluid in unsteady state in x-direction is
A. $\frac{\partial u}{\partial x} = 0$
B. $\frac{\partial u}{\partial x} = -\frac{\partial \rho}{\partial t}$
C. $\frac{\partial \rho}{\partial t} = 0$
D. $\frac{\partial (\rho u)}{\partial x} = 0$

183. When the flow is laminar the kinetic energy correction factor for velocity distribution is
A. 0.5
B. 1.66
C. 1
D. 2

184. What will be the bed pressure in an air fluidized bed of catalyst particles with dp = 0.05 cm and $\rho_p = 200$ kg/m^3, bed height 60 cm and bed porosity 0.5 expressed in cm of water.
A. 90
B. 60
C. 45
D. 30

185. At critical flow conditions ($R_e = 2100$) a fluid having $\left(\frac{\mu}{\rho} = 0.01 \text{ cm}^2/\text{sec}\right)$ moving through a pipe of diameter 3 cms. What will be the velocity of flow in cm/sec?
A. 7
B. 630
C. 700
D. 7000

186. Two fluid X and Y are flowing one by one through a pipe. Y is having density and specific gravity just double of that of fluid X at the same flow rate/average velocity, if the pressure drop per unit length of pipe incurred by fluid X is Δp then the pressure for fluid Y will be equal to
A. Δp^2
B. $2\Delta p$
C. $\Delta p/2$
D. Δp

187. A flow is termed steady flow when
A. conditions are the same at the adjacent points at any instant

B. rate of the velocity change is constant.

C. conditions do not change with time at any point.

D. conditions change steadily with time.

188. An example of pressure flow is fluid flow in___________.

A. river.

B. pipe

C. open channel

D. partially filled pipeline

189. For a fluidization operation water (density 1000 kg/m^3) is the fluidizing medium which is flowing through a 0.5 m bed made up of a 1 mm dia glass spherical particles (density 2500 kg/m^3). If at the onset of fluidization the bed voidage is 40 % the pressure drop across the bed is ___________.

A. 4.4 KPa

B. 3.7 KPa

C. 2.94 KPa

D. none of these

190. A piezometric opening measures the ______________.

A. static pressure

B. point velocity

C. total pressure

D. dynamic pressure

191. Which of the following statements is correct for the formation of fluidized beds?

A. pressure forces equal gravity forces.

B. gravity force is less than the fluid friction.

C. sum of the fluid friction and pressure forces is equal and opposite to gravity forces.

D. fluid friction is zero

192. For water flowing through a tube of I.D. 5cm the Reynolds numbers is 1500. Through the same pipe if a liquid of 5 centipoises viscosity and 0.8 specific gravity starts flowing at the same velocity then the pressure drop will

A. increase

B. data insufficient to predict pressure drop

C. decrease

D. remain same

193. Which of the following types of devices is used for measurement of small pressure differences?

A. pitot tube.

B. inclined tube manometer.

C. U-tube manometer.

D. none of these.

194. What is the effect on NPSH when the pump speed is increased?

A. can either increase or decrease; depends on other factors

B. increases

C. decreases

D. remains unaltered

195. The crushing energy required to create new surface is given by

A. Rittinger's law

B. Taggart's rule

C. Fick's law

D. none of these

196. ______________ number is the ratio of pressure and inertia force.

A. Froude

B. Weber

C. Euler

D. Mach

197. The terminal velocity of a solid spherical particle flowing in the Newton's law range through a stationary fluid mass is ______________ to the fluid viscosity.

A. independent of

B. inversely proportional to

C. directly proportional to

D. inversely proportional to the square root of

198. Separation of materials of the same density based on their sizes by using their different rates of flow is called

A. sizing

B. sorting

C. flocculation

D. elutriation

199. In screen analysis, notation +5 mm/-10 mm means passing through

A. 5 mm screen and retained on 5mm screen

B. 10 mm screen and retained on 10 mm screen

C. both 5 mm and 10 mm screen

D. neither 5 mm nor 10 mm screen

200. The centrifugal pumps used in chemical industries are mostly _______________ driven.

A. steam

B. gas turbine

C. electric motor
D. diesel engine

201. A Redwood viscometer measures the time taken for gravity flow of a fixed volume of liquid which is directly proportional to
A. density.
B. ratio of absolute viscosity to density.
C. Reynolds number.
D. absolute viscosity.

202. What would be the horse power requirement for a fluid being pumped at the rate of 10 lb/sec.
A. 1
B. 1/55
C. 5.5
D. 10/55

203. Which of the following pumps is ideal for pumping a medium viscosity lubricating oil?
A. vane
B. piston
C. plunger
D. centrifugal

204. Gear pump
A. is a positive displacement pump.
B. can be started with delivery valve closed.
C. is a non-positive displacement pump.
D. is a centrifugal pump.

205. In a centrifugal pump the discharge line is usually fitted with a check value for the purpose of controlling the
A. back flow of fluid in the event of stoppage of pump.
B. flow of liquid during operation of the pump.
C. discharge pressure.
D. all (a), (b) and (c).

206. In a venture meter the exit cone angle is ______________ the entrance cone angle.
A. smaller than
B. greater than
C. equal to
D. either (a) or (b)

207. In a fast breeder reactor the molten sodium is transported by ________________ pump.
A. gear
B. plunger
C. electromagnetic
D. reciprocating

208. In which category of non-Newtonion fluid sewage sludge falls?
A. Pseudo plastic
B. Bingham plastic
C. dilatant
D. none of these

209. In a venturimeter the permanent loss is about ______________ percent of the pressure drop in the upstream cone.
A. 1
B. 10
C. 40
D. 70

210. If δ is the boundary layer thickness and x is the distance from the leading edge in the direction of flow, the ratio (δ/x) varies as ___________. Assume laminar boundary layer flow is taking place over a flat plate.
A. $Re^{0.5}$
B. Re
C. 1/Re
D. $Re^{-1/2}$

211. An incompressible fluid is flowing in fully turbulent flow ($Re>10^5$) in a pipe of diameter under a constant pressure gradient. On which of the following parameter the volumetric flow will depend?
A. d
B. d^2
C. $d^{2.5}$
D. d^4

212. A floating body displaces volume of liquid equivalent to its
A. own weight
B. own volume
C. submerged volume
D. submerged weight

213. Which of the following statements is wrong?
A. Manometric liquid should have high surface tension.
B. The vacuum pressure is always the negative gauge pressure.
C. The pressure of the liquid measured by a piezometer tube is the gauge pressure.
D. The point at which the resultant pressure on immersed surface acts, is known as the centre of gravity.

214. The ____________ diameter of a pipe is indicated by its nominal size

A. outer
B. inner
C. approximate
D. none of these

215. The flow through a/an ______________ is known as steady uniform flow.

A. long pipe at constant rate.
B. expanding tube at constant rate.
C. long pipe at decreasing rate.
D. none of these.

216. For solid spherical particles falling through a stationary fluid mass in the Stoke's law range the terminal velocity is proportional to be

A. difference in the densities of the particle & fluid.
B. inverse of fluid viscosity.
C. square of particle size.
D. all (a), (b) and (c).

217. What would be the effect on pressure drop if the diameter of the pipe is halved assuming laminar flow.

A. increase
B. remain same
C. be quadrupled
D. decrease

218. What should be the position of the of the deliver valve of a centrifugal pump while it is being started? it should be

A. either opened or closed; it does not make any deference
B. closed.
C. either opened or closed; depending on the fluid viscosity.
D. opened.

219. The maximum height to which water can be lifted by a pump at any temperature largely depends on _________.

A. barometric pressure
B. diameter of the impeller
C. speed of the impeller
D. both (b) and (c)

220. What would be the coefficient of contraction of a fluid jet which in discharging from a 2 inch diameter orifice and at vena – contracta has a diameter of 1.75 inch.

A. 0.87
B. 0.766
C. 1.3
D. none of these.

221. In which of the following region of flow the fanning equation $\frac{\Delta p}{\rho} = \left[(4f).\left(\frac{L}{D}\right).\left(\frac{v^2}{2g_c}\right)\right]$ Is applicable?

A. transition
B. laminar
C. turbulent
D. both (b) and (c)

222. The viscosity of water _________ with increase in pressure.

A. decreases
B. also increases
C. remains constant
D. First decreases, and then increases

223. Which of the following law/equations forms the basis for capillary tube method of viscosity measurement?

A. Hagen-Poiseuille's equation
B. Navier-stokes equation
C. Stoke's law
D. none of these

224. Which of the following is an undesirable effect created by cavitation?

A. Develops noise
B. Increase in thrust
C. Decrease in effect
D. Develops high pressure

225. What would be the pressure drop required to fluidize a bed of particles of density 2000kg/m^3which a porosity of 0.6 and particles density 2000 kg/m^3

A. 14.86 kPa
B. 11.77 kPa
C. 25.61 kPa
D. 21.13 kPa

226. Multistage centrifugal pumps are generally used for

A. high head.
B. highly viscous liquid.
C. low head but high discharge.
D. slurries of high solid concentration.

227. The density of the float material in case of a rotameter should be__________ that of the flowing liquid.
A. more than
B. equal to
C. less than
D. either (a) or (b)

228. In which of the following categories sewage sludge belongs to.
A. Bingham plastic
B. pseudoplastic
C. Newtonion
D. dilatent

229. The boundary layer thickness for a flow past a flat plate will be proportional to _________ where x is the distance along the plate in the direction of flow.
A. x
B. 1/x
C. x
D. 1/x

230. The implication that a pump is a high speed pump is that it is a/an _________ pump.
A. mixed flow
B. centrifugal
C. axial flow
D. none of these.

231. If Δp in the pressure drop for a length of tube L, and diameter 'D' in which a Newtonion liquid (density P, viscosity μ) is flowing with a velocity v in laminar region then Δp is proportional to
A. $L\,\rho v^2/D$
B. $\mu\,V/L$
C. $L\,v^2/D^2$
D. $D\,\rho v^2/L$

232. For a fluid flow operation which of the following statements is wrong?
A. the ratio of average velocity to the maximum velocity for turbulent flow of Newtonion fluid in circular pipes is 0.5.
B. the Newtonion fluid velocity in a circular pipe flow is maximum at the center of the pipe.
C. Hagen-Poiseuille equation is applicable to the laminar flow of Newtonion fluids.
D. Navier-Stokes equation is applicable to the analysis of viscous flows.

233. What would be the value of Fanning friction factor when Darcey friction factor is f_1?
A. $f_1/4$
B. $4f_2$
C. $2f_1$
D. $f_1/2$

234. If the schedule number of a pipe of a particular nominal size is increased the
A. I.D. of the pipe decreases.
B. wall thickness also increases.
C. O.D. of the pipe remains constant.
D. all (a), (b) and (c)

235. Which of the following devices may be integrated to an automatic flow controller?
A. venturimeter
B. flow nozzle
C. rotameter
D. none of these

236. For flow taking place in streamline regime the velocity at a certain point is
A. independent of time
B. constant
C. both (a) & (b)
D. neither (a) nor (b)

237. The variation of power requirement and pressure for specific centrifugal force air blower operating at constant speed and capacity is proportional as
A. inversely as gas density.
B. directly as gas density.
C. directly as square of gas density.
D. directly as square root of gas density.

238. During the flow regime where Reynold's number is low the
A. gravity forces control.
B. viscous forces control.
C. viscous forces control and inertial forces are unimportant.
D. viscous forces are unimportant.

239. Which of the following statements is correct?
A. Mach number in an incompressible fluid is always unity.
B. Fanning friction factor is inversely proportional to Reynolds number.
C. The property of a randomly packed bed (with raschig rings) is given by the ratio of the total volume to the volume of voids in the bed.

D. Mach number is given by the ratio of the speed of the fluid to that of sound in the fluid under conditions of flow.

240. For laminar flow in a pipe average velocity υ is equal to
A. Umax
B. 0.5 Umax
C. 0.25 Umax
D. 2 Umax

241. The friction factor according to the definition of friction is _____.
A. always inversely proportional to the Reynolds number.
B. not dependent on the roughness of the pipe.
C. not dimensionless.
D. none of these.

242. Bernoulli's equation cannot be applied when the flow is
A. rotational.
B. turbulent.
C. Unsteady.
D. all of the above.

243. When is a closed conduit flow occurs in the laminar region the
A. $V_{max} = 2V_{av}$
B. $V_{max} = V_{av}$
C. $V_{max} = 1.5V_{av}$
D. $V_{max} = 0.5V_{av}$

244. Fluid flow through a V– notch is represented by ___________.
A. $H^{3/2}$
B. $H^{1/2}$
C. $H^{5/2}$
D. $H^{2/3}$

245. For an orificemeter the point where vena contracta occurs does not depend upon the
A. type of orifice.
B. pipe roughness.
C. ratio of pipe diameter to orifice diameter.
D. density, viscosity & compressibility of the fluid.

246. Gas A having a density =1.5 kg/m^3, viscosity = 2×10^{-5} kg/m.sec. is flowing at a superficial velocity of 2 m/s through a packed bed incurring a pressure drop of 8400 Pa/m. the particle size is 0.5 cm and porosity of bed 0.5. If gas A is replaced by gas B with density 1.5kg/m^3 and viscosity 3×10^{-5}kg/m.sec which flows through the bed at 3m/s what would be the pressure drop will?
A. 8400 Pa/m
B. 12600 Pa/m
C. 16800 Pa/m
D. 18900 Pa/m

247. Which of the following represents the unsteady uniform flow in a pipe.
A. long pipe at constant rate.
B. long pipe at decreasing rate.
C. expanding tube at constant rate.
D. expanding tube at increasing rate.

248. For fluid flowing between any two points then flow is _____.
A. steady, incompressible, irrotational.
B. unsteady, compressible, irrotational.
C. steady, compressible and along a streamline.
D. steady, compressible, irrotational.

249. In winch of the following categories of fluids rubber latex may be put?
A. pseudoplastic
B. dilatent
C. Newtonion
D. Bingham plastic

250. A _________ pump has working characteristics signifying mixed flow.
A. turbine
B. diaphragm
C. piston
D. none of these

251. Flow of ___________ may be controlled by the use of cocks.
A. solids
B. any liquid
C. water
D. none of these

252. If 'x' and 'y' are the sides of rectangular cross-section channels the equivalent diameter for ΔP calculation for a fluid flowing through it will be equal to
A. $\frac{2xy}{x+y}$
B. $\frac{x+y}{2xy}$
C. $\frac{x+y}{xy}$
D. $\frac{xy}{x+y}$

253. Which of the following statements is with respect to a turbine impeller?
   A. does not produce tangential current.
   B. produces only radial current.
   C. is effective over wide range of viscosities.
   D. produces only tangential current.
254. Which of the following parameters represents the rate of change of moment of momentum with a flowing fluid?
   A. torque applied
   B. work done
   C. force exerted
   D. power developed
255. Which of the following ranges represent the coefficient of discharge (C_d) for an orifice plate?
   A. 0.58 to 0.8
   B. 0.93 to 0.98
   C. 0.2 to 0.3
   D. 0.02 to 0.03
256. The most convenient and economical measurement of water flow rate in a pipe of approximately 3.5 m diameter may be done by ____________.
   A. Pitot tube
   B. rotameter
   C. Orificemeter
   D. venturimeter
257. Surface tension is represented in following units _____.
   A. ML^{-2}
   B. MT^{-2}
   C. MLT^{-2}
   D. $ML^{-2}T$
258. ____________ is a phenomenon which may occur during pumping of a liquid solution containing dissolved gases, which may come out of the solution forming gas pockets.
   A. sublimation
   B. cavitation
   C. stripping
   D. evaporation
259. The pressures drop in a pipeline does not depend on _____.
   A. size of pipe
   B. velocity of fluid
   C. length of pipe and number of bends
   D. none of these
260. A Flow in which each liquid particle has a definite path and their path do not cross each other is called
   A. steady flow
   B. streamline flow
   C. uniform flow
   D. turbulent flow
261. The speed of a pump of an unit size that _____________ is called the specific speed of the pump.
   A. delivers unit discharge at unit head.
   B. delivers unit discharge at unit power.
   C. requires unit power for unit head.
   D. none of these.
262. Which of the following friction is responsible for producing formation of wakes?
   A. skin
   B. disk
   C. form
   D. none of these
263. For a centrifugal pump with a constant diameter impeller the
   A. head varies as the speed.
   B. head varies as the square of speed.
   C. horsepower input varies as the square of speed.
   D. its capacity varies directly as the square of speed.
264. Because of sudden expansion the head loss may be expressed as
   A. $\frac{V_1-V_2}{2gc}$
   B. $\frac{(V_1-V_2)^2}{2gc}$
   C. $\frac{V_1^2-V_2^2}{gc}$
   D. $\frac{V_1^2-V_2^2}{2gc}$
265. The friction losses of about _________ percent of maximum velocity occurs in a Venturimeter.
   A. 2
   B. 8
   C. 20
   D. 12
266. Which type flow occurs when the pipe Reynolds's member is 6000?
   A. transition
   B. viscous

C. turbulent
D. laminar

267. When solid spherical particles fall through a stationary fluid mass in Newton's law regime the terminal velocity varies as the ________ of its diameter.

A. inverse
B. square root
C. first power
D. second power

268. Which of the following can be measured by a hot wire anemometer?

A. temperature of liquids.
B. pressure of liquids.
C. velocity of gases.
D. velocity of liquids.

269. Water is being pumped by a centrifugal pump upto a certain height. If the same pump is used to pump an oil of specific gravity 0.8 at the same rate and the same height. The power consumption will now ________.

A. data insufficient to predict
B. decrease
C. increase
D. remain same

270. Stoke's law is valid, when particle is less than ________mm.

A. 2
B. 100
C. 2100
D. 700

271. In which of the following cases steady non-uniform flow will occur?

A. expanding tube at increasing rate.
B. long pipe at decreasing rate.
C. long pipe at constant rate.
D. expanding tube at constant rate.

272. Which of the following parameters is responsible for momentum transfer in laminar flow of fluid?

A. density
B. viscosity
C. velocity gradient
D. none of these

273. The depth of flow in the channel________ is the normal depth in open channel.

A. corresponding to uniform flow.
B. corresponding to steady flow.
C. measured normal to the channel bed.
D. none of these.

274. ________ fluids can be handled by the use of diaphragm valves.

A. corrosive
B. solid suspended
C. viscous
D. non-Newtonion

275. For ________ fluid in steady state laminar flow condition, the fluid velocity varies as the square of the cylindrical pipe diameter.

A. Newtonion
B. pseudoplastic
C. dilatant
D. non-Newtonion

276. For transportation of ________ diaphragm pumps are best.

A. fluids
B. solids
C. liquids
D. slurries

277. Which of the following parameters governs the theoretical head developed for a centrifugal pump?

A. fluid velocity leaving
B. diameter of
C. speed of
D. all (a), (b) and (c)

278. In which of the following glass tube diameters (in mm) the capillary rise of mercury would be maximum?

A. 0.5
B. 1
C. 2
D. 5

279. Operations which require ________ are best done by a propeller type centrifugal pump.

A. low capacity at high heads.
B. high capacity at low heads.
C. High capacity at high heads.
D. low capacity at low heads.

280. The variation of discharge through a rectangular weir may be expressed as________.

A. $H^{5/2}$
B. $H^{2/5}$
C. $H^{1/2}$
D. $H^{3/2}$

281. Which of the following statements represents the Bernoulli's equation for steady ideal fluid flow?
   A. energy is constant throughout the fluid
   B. velocity is constant along a stream line.
   C. energy is constant along a stream line, but may vary across stream lines
   D. none of these
282. Which of the following sets of pressure intensities are equivalent?
   A. 4.33 psi, 10 ft. of water, 8.83 inches of Hg.
   B. 4.33 psi, 10 ft. of water, 20.7 inches of Hg.
   C. 10 psi, 19.7 ft. of water, 23.3 inches of Hg.
   D. 10 psi, 19.7 ft. of water, 5.3 inches of Hg.
283. What would be the hydraulic mean radius for flow through packed bed of spherical particles of size D_p and porosity ε?
   A. $\frac{D_p}{6}\left(\frac{\varepsilon}{1-\varepsilon}\right)$
   B. $\frac{2}{3}D_p\left(\frac{1-\varepsilon}{\varepsilon}\right)$
   C. $\frac{2}{3}D_p\left(\frac{\varepsilon}{1-\varepsilon}\right)$
   D. $\frac{D_p}{6}\left(\frac{1-\varepsilon}{\varepsilon}\right)$
284. A large Reynold number is indication of
   A. laminar flow
   B. high turbulent flow
   C. steady flow
   D. smooth and streamline flow
285. Pressure drop in a packed bed for laminar flow is given by the __________ equation.
   A. Kozney-Karman
   B. Leva's
   C. Fanning friction factor
   D. Blake-Plummer
286. In which of the following categories of fluid paper pulp falls?
   A. pseudoplastic
   B. Bingham plastic
   C. Newtonion
   D. dilatant
287. Which of the following expressions is valid for the flow taking place through a system of three pipes of different diameter, length and friction factors connected in parallel given Q = flow rate V = fluid velocity, f = friction factor?
   A. $Q = Q_1 + Q_2 + Q_3$
   B. $f = f_1 + f_2 + f_3$
   C. $V_1 = V_2 = V_3$
   D. $Q_1 = Q_2 = Q_3$
288. The Mach number for a supersonic flow of fluid through a pipeline is _________.
   A. 0
   B. 1
   C. < 1
   D. > 1
289. The dirt is removed from flowing fluid by
   A. strains
   B. clarifier
   C. coagulant
   D. gravity setter
290. Water flowing at a linear velocity of 10 in ft/sec through a 24 inch pipe will have discharge in ft^3/sec
   A. 7.65
   B. 32.36
   C. 48.22
   D. 125.6
291. Which of the following can be identified as the most serious disadvantage of an Orifice meter?
   A. it is very costly.
   B. it is not very accurate.
   C. most of the pressure drop is not recoverable.
   D. it is not suitable for measuring gas flow.
292. Which of the following represents the main use for Plunger pumps?
   A. higher pressure
   B. viscous mass
   C. slurries
   D. none of these
293. An indication of_________ is by the schedule number of the pipe.
   A. roughness
   B. material density
   C. size
   D. wall thickness
294. What will be the transition length for liquid flow in turbulent region through a smooth pipe in terms of multiples of pipe diameter?
   A. 0.5
   B. 5
   C. 50

D. 100

295. In which of the following situations Froude number is not significant?
A. for unbaffled tank.
B. when there is no vortex formation.
C. for Reynolds number greater than 300.
D. none of these.

296. When the pump capacity is expressed in horsepower the main parameters are
A. suction lift
B. discharge head
C. specific gravity of the liquid
D. all (a), (b) and (c)

297. Large diameter of pipelines carrying fuel gas are provided with disc compensators to________.
A. account for the pressure variation side the pipeline.
B. keep the pipe in proper orientation.
C. account for contraction/expansion of pipe due to temperature changes of the surroundings.
D. make the pipe joint leak-proof.

298. Which of the following represents an axial flow impeller?
A. turbine
B. paddle
C. propeller
D. all (a), (b) and(c)

299. A centrifugal pump has to be designed so as to minimise the________.
A. loss of static head
B. loss of kinetic head
C. back flow through impeller
D. none of these

300. For a flow taking place on a flat plate what is the ratio of displacement thickness to nominal thickness for a linear distribution of velocity in the boundary layer?
A. 0.5
B. 1
C. 1.5
D. 2

301. The main considerations for Bernoulli's equation are
A. various masses
B. various momentums
C. different forms of mechanical energy
D. none of these

302. If Stoke's law is applicable for motion of spherical particles in a stationary fluid which of the following expressions holds good?
A. $\frac{24}{N_{Re,p}}$
B. $\frac{48}{N_{Re,p}}$
C. $\frac{16}{N_{Re,p}}$
D. $\frac{64}{N_{Re,p}}$

303. Flow of fluid in turbulent regime occurs when,
A. momentum transfer is on molecular scale only.
B. shear stress is caused more effectively by cohesion than momentum transfer.
C. fluid particles move in an orderly manner.
D. shear stresses are generally larger than in a similar laminar flow.

304. The reason for less energy loss in flow through venturimeter in comparison to flow nozzle is because in venturimeter________.
A. throat diameter is more.
B. length is shorter.
C. sudden expansion of flow in the downstream occurs.
D. distance between the throat and the inlet is more.

305. For pipe flow at constant diameter capacity is proportional to
A. head
B. $head^2$
C. $head^{1/2}$
D. $head^{3/2}$

306. The energy loss over a length of pipeline according to Darcy-Weisbach equation for pipe flow is ________ the mean velocity of flow.
A. inversely proportional to square of
B. directly proportional to square of
C. inversely proportional to
D. directly proportional to

307. Reynolds number for non-circular cross section is
A. $\frac{V*2P}{4*v}$

B. $\frac{V*P}{v}$

C. $\frac{V*4P}{v}$

D. $\frac{V*P}{4v}$

Where, V = Mean Velocity

v = Kinematic Viscosity

P = ratio of cross sectional area to the wetted perimeter

308. The continuity equation
   - **A.** relates mass flow rate along a stream tube.
   - **B.** stipulates that Newton's second law of motion must be satisfied at every point in the fluid.
   - **C.** relates work and energy.
   - **D.** none of these.

309. A 4 HP centrifugal pump running at a speed of 800 rpm delivers water at a rate of 1000 liter/min at a head of 8 meter. What would be the new power consumption in HP when it starts running at half of the above speed?
   - **A.** 0.5
   - **B.** 1
   - **C.** 2
   - **D.** 4

310. The operating range of fluidization velocity _________ with decrease in partial pressure for the size to be fluidised.
   - **A.** unpredictable from the data
   - **B.** widens
   - **C.** does not change
   - **D.** squeezes

311. Pumping of crude oil from oil well is recommended to be done by_________.
   - **A.** Gear pump
   - **B.** Screw pump
   - **C.** Single stage centrifugal pump
   - **D.** Duplex/triplex reciprocating pump

312. Which of the following ratio is known as Froude number?
   - **A.** drag stress to shear stress.
   - **B.** inertial stress to shear stress.
   - **C.** shear stress to gravitational stress
   - **D.** inertial stress to gravitational stress.

313. Which of the following statements is wrong as far as a streamline flow is concerned?
   - **A.** it is always parallel to the main direction of the fluid flow.
   - **B.** the fluid lying between any two streamlines can be considered to be in isolation and the streamline spacing varies inversely as the velocity.
   - **C.** it is a line across which there is no flow and it is equivalent to a rigid boundary.
   - **D.** streamlines intersect at isolated point of zero velocity and infintie velocity.

314. Which of the following statements define the minimum porosity for fluidization?
   - **A.** that corresponding to completely fluidised bed.
   - **B.** less than that of the' static bed.
   - **C.** the porosity of the bed when true fluidisation begins.
   - **D.** that corresponding to static bed.

315. A centrifugal pump can be used to draw water from upto a maximum depth of
   - **A.** 150 feet.
   - **B.** dependent on the speed of the pump.
   - **C.** 34 feet.
   - **D.** dependent on the power of the pump.

316. The flow measuring device which is an area meter is _________.
   - **A.** Orifice meter
   - **B.** Anemometer
   - **C.** Venturimeter
   - **D.** Rotameter

317. The fluid velocity varies as the cube of the cylindrical pipe diameter in case of steady state laminar flow at constant pressure drop for _______ fluid.
   - **A.** dilatent
   - **B.** pseudoplastic
   - **C.** Newtonion
   - **D.** Bingham plastic

Answer

1.	B	36.	D	71.	D	106.	C
2.	A	37.	A	72.	C	107.	B
3.	C	38.	D	73.	B	108.	B
4.	A	39.	B	74.	B	109.	C
5.	B	40.	B	75.	A	110.	D
6.	C	41.	B	76.	A	111.	B
7.	B	42.	A	77.	A	112.	D
8.	D	43.	B	78.	D	113.	C
9.	D	44.	D	79.	A	114.	A
10.	C	45.	A	80.	A	115.	A
11.	B	46.	A	81.	D	116.	B
12.	B	47.	D	82.	C	117.	C
13.	B	48.	B	83.	C	118.	C
14.	D	49.	A	84.	B	119.	B
15.	C	50.	A	85.	C	120.	C
16.	A	51.	B	86.	C	121.	C
17.	A	52.	A	87.	A	122.	C
18.	A	53.	C	88.	A	123.	A
19.	D	54.	B	89.	C	124.	A
20.	A	55.	D	90.	B	125.	B
21.	B	56.	D	91.	A	126.	A
22.	B	57.	B	92.	B	127.	B
23.	C	58.	C	93.	A	128.	B
24.	A	59.	C	94.	B	129.	A
25.	A	60.	B	95.	B	130.	D
26.	B	61.	D	96.	D	131.	B
27.	D	62.	C	97.	B	132.	B
28.	A	63.	B	98.	B	133.	B
29.	A	64.	A	99.	D	134.	A
30.	B	65.	B	100.	B	135.	D
31.	A	66.	C	101.	B	136.	B
32.	D	67.	D	102.	A	137.	D
33.	C	68.	D	103.	D	138.	D
34.	C	69.	B	104.	D	139.	B
35.	B	70.	B	105.	D	140.	C

141. C
142. D
143. D
144. D
145. D
146. D
147. A
148. D
149. A
150. B
151. C
152. B
153. B
154. D
155. D
156. D
157. A
158. A
159. B
160. B
161. A
162. C
163. C
164. C
165. A
166. B
167. C
168. D
169. A
170. C
171. B
172. A
173. D
174. A
175. C
176. C
177. B
178. B
179. B
180. A
181. C
182. A
183. B
184. B
185. A
186. B
187. C
188. B
189. A
190. C
191. B
192. A
193. B
194. B
195. A
196. C
197. B
198. A
199. B
200. C
201. B
202. A
203. A
204. A
205. A
206. A
207. C
208. B
209. B
210. D
211. C
212. A
213. D
214. C
215. A
216. D
217. A
218. B
219. A
220. B
221. D
222. D
223. A
224. D
225. B
226. A
227. A
228. A
229. A
230. C
231. A
232. A
233. A
234. D
235. C
236. C
237. B
238. C
239. D
240. B
241. D
242. B
243. A
244. C
245. A
246. B
247. B
248. A
249. A
250. A
251. B
252. A
253. C
254. A
255. A
256. A
257. B
258. B
259. D
260. B
261. A
262. C
263. B
264. B
265. A
266. C
267. B
268. C
269. B
270. A
271. D
272. C
273. A
274. A
275. A
276. D
277. D
278. A
279. B
280. D
281. C
282. A
283. A
284. B
285. A
286. B
287. A
288. D
289. A
290. D
291. C
292. A
293. D
294. C
295. B
296. D
297. C
298. C
299. B
300. A
301. C
302. A
303. D
304. C
305. D
306. B
307. A
308. A
309. A
310. C
311. D
312. D
313. A
314. C
315. C
316. D
317. B

GATE 2015-2000

(1 Mark Question)

GATE 2015

1. Two different liquids are flowing through two different pipes of the same diameter. In the first pipe, the flow is laminar with centerline velocity, $v_{\max 1}$, whereas in the second pipe, the flow is turbulent. For turbulent flow, the average velocity is 0.82 times the centerline velocity, $v_{\max 2}$, For equal volumetric flow rates in both the pipes, the ratio $v_{\max 1}/v_{\max 2}$ (upto two decimal places) is____________.

Sol. For laminar flow, the centerline velocity is

$$v_{avg\ 1} = 0.5\ v_{\max\ 1}$$

For turbulent flow, the centerline velocity is

$$v_{avg\ 2} = 0.82\ v_{\max\ 2} \qquad \text{[given]}$$

Since, volumetric flow rates are same in both pipes

$$Q_1 = Q_2$$

$$A_1 v_{avg\ 1} = A_2 v_{avg\ 2}$$

Given, $A_1 = A_2$

So, $v_{avg\ 1} = v_{avg\ 2}$

$\Rightarrow 0.5\ v_{\max\ 1} = 0.82 v_{\max\ 2}$

$\therefore \quad \frac{v_{\max 1}}{v_{\max 2}} = 1.64$

2. A cylindrical packed bed of height 1 m is filled with equal sized spherical particles. The particles are non-porous and have a density of 1500 kg/ m^3. The void fraction of the bed is 0.45. The bed is fluidized using air (density 1 kg/m^3). If the acceleration due to gravity is 9.8 m/s^2, the pressure drop (in Pa) across the bed at incipient fluidization (up to one decimal place) is__________.

Sol. According to the question,

$$\frac{\Delta p}{L} = g(1-\varepsilon)(\rho_p - \rho)$$

where, $\varepsilon = 0.45$,

$$\rho_p = 1500\ \text{kg/m}^3 \text{and } \rho = 1\text{kg/m}^3$$

and $L = 1$m (height of the bed)

$$\Delta p = 9.8 \times (1 - 0.45)(1500 - 1) \times 1$$
$$= 8079.61\text{Pa}$$

3. For uniform laminar flows over a flat plate, the thickness of the boundary layer δ, at a distance x from the leading edge of the plate follows the relation

A. $\delta\ (x) \propto x^{-1}$
B. $\delta\ (x) \propto x$
C. $\delta\ (x) \propto x^{1/2}$
D. $\delta\ (x) \propto x^{-1/2}$

Sol. **A.** For uniform laminar flow, boundary layer thickness,

$$\delta\ (x) = \frac{4.99x}{\sqrt{\text{Re}}} = \frac{4.99x}{\sqrt{\frac{\rho V x}{\mu}}}$$

or $\quad \delta\ (x) \propto x^{1/2}$

GATE 2014

4. In case of a pressure driven laminar flow of a Newtonian fluid of viscosity (μ) through a horizontal circular pipe, the velocity of the fluid is proportional to

A. μ
B. $\mu^{0.5}$
C. μ^{-1}
D. $\mu^{-0.5}$

Sol. **C.** For pressure driven flow of Newtonian fluid,

$$\frac{\Delta p}{l} = \frac{8\mu v}{R^2}$$

$$\therefore \quad v \propto \frac{1}{\mu}$$

5. Which of the following statements is correct?

P. For a rheopectic fluid, the apparent viscosity increases withtime under a constant applied shear stress

Q. For a pseudo plastic fluid, the apparent viscosity decreaseswith time under a constant applied shear stress

R. For a Bingham plastic, the apparent viscosity increases exponentially with the deformation rate

S. For a dilatants fluid, the apparent viscosity increases with increasing deformation rate

A. P and Q

B. Q and R

C. R and S

D. P and S

Sol. **D.** For rheopectic fluid, viscosity increases with time under dilatants shear stress. For dilated fluid, viscosity increases with increase in rate of shear strain

$$\text{i.e., } \tau = \mu\left(\frac{du}{dy}\right)^n \text{ for } n > 1$$

6. In order to produce fine solid particles between 5 and 10 μm, the appropriate size reducing equipment is

A. fluid energy mil

B. hammer mil

C. jaw crusher

D. smooth roll crusher

Sol. **A.** Fluid energy mill is used to produce fine solid particles between 5 and 10 μm.

7. Slurries are most conveniently pumped by a

A. syringe pump

B. diaphragm pump

C. vacuum pump

D. gear pump

Sol. **C.** The diffusive flux will be maximum at the vent.

GATE 2013

8. An open tank contains two immiscible liquids of densities (800 kg/m^3 and 1000 kg/m^3) as shown in the figure. If g = 10 m/s^2, under static conditions, the gauge pressure at the bottom of the tank in Pa is

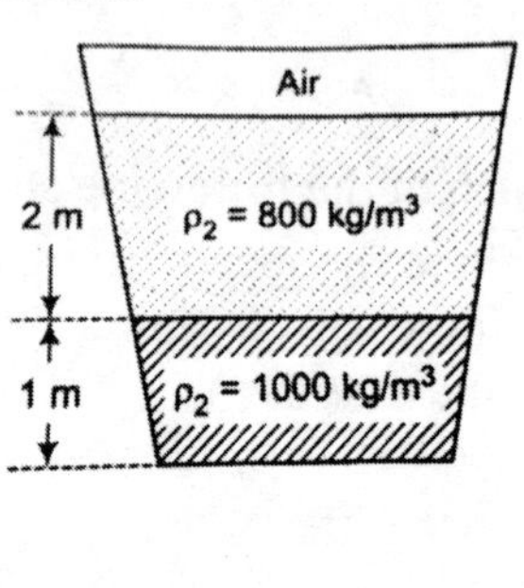

A. 23000

B. 22000

C. 24000

D. 26000

Sol. **D.** The gauge pressure at bottom

$$p_g = \rho_1 g h_1 + \rho_2 g h_2$$

$$= 800\times10\times2+1000\times10\times1 = 26000 \text{ Pa}$$

9. The apparent viscosity of a fluid is given by $0.007\left|\frac{dv}{dy}\right|^{0.3}$, where $\left(\frac{dv}{dy}\right)$ is the velocity gradient. The fluid is

A. bingham plastic

B. dilatants

C. pseudoplastic

D. thixotropic

Sol. **B.** Given, $\mu = 0.007\left(\frac{dv}{dy}\right)^{0.3}$

The curve between apparent viscosity and velocity gradient will be

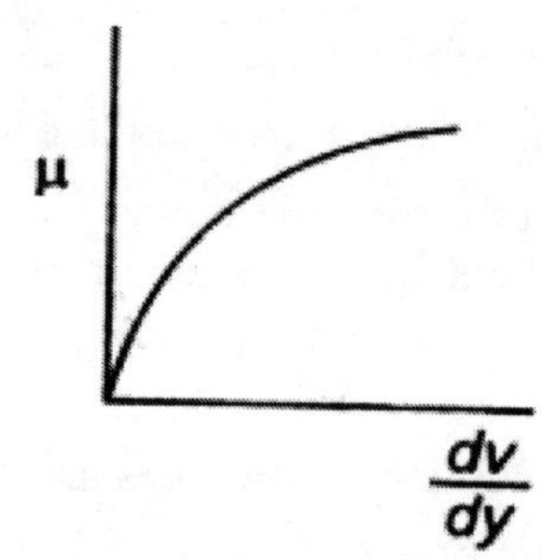

It shows dilatant behavior of the fluid.

10. The mass balance for a fluid with density (ρ) and velocity vector(v) is

A. $\frac{\partial p}{\partial t} + \nabla.(\text{pv}) = 0$

B. $\frac{\partial p}{\partial t} + \text{v}.(\nabla \text{p}) = 0$

C. $\frac{\partial p}{\partial t} + \text{p}(\nabla.\text{v}) = 0$

D. $\frac{\partial p}{\partial t} - \text{v}.(\nabla \text{p}) = 0$

Sol. **A.** The continuity equation in three dimension coordinates is, $\frac{\partial p}{\partial t} + \frac{\partial}{\partial x}(\text{pu}) +$

$$\frac{\partial}{\partial y}(\text{pv}) + \frac{\partial}{\partial z}(\text{pw}) = 0 \qquad \text{or}$$

$$\frac{\partial p}{\partial t} + \left(\hat{\mathbf{i}}\frac{\partial}{\partial x} + \hat{\mathbf{j}}\frac{\partial}{\partial y} + \hat{\mathbf{k}}\frac{\partial}{\partial z}\right) \blacksquare$$

$(pu\hat{\mathbf{i}} + pv\hat{\mathbf{j}} + pw\hat{\mathbf{k}}) = 0$

or $\frac{\partial p}{\partial t} + \nabla (p\,v) = 0$

11. An incompressible Newtonian fluid, filled in an annular gap between two concentric cylinders of radii R_1 and R_2 as shown in the figure, is flowing under steady state conditions. The outer cylinder is rotating with an angular velocity of Ω while the inner cylinder is stationary. Given that $(R_2 - R_1) << R_1$, The profile of the θ – component of the velocity v_θ can be approximated by.

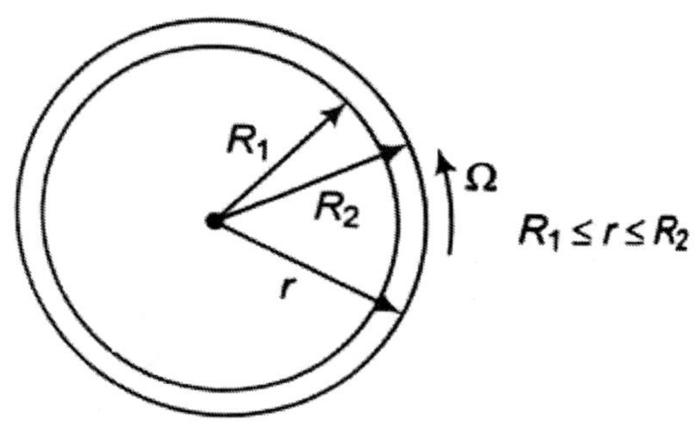

A. $R_2\Omega$

B. $\frac{(r-R_2)}{(R_2-R_1)} r$

C. $\frac{(r+R_1)}{(R_2+R_1)} R_1\Omega$

D. $\frac{(r-R_1)}{(R_2-R_1)} R_2\Omega$

Sol. **D.** Applying equation of continuity for this flow

$$\frac{\partial}{\partial r}\left(\frac{1}{r}\frac{\partial}{\partial r}(r\,v_\theta)\right) = 0$$

Integrating $\frac{\partial}{\partial r}(r\,v_\theta) = c_1 r,$

Integrating again, $r\,v_\theta = c_1\frac{r^2}{2} + c_2$ ....(i)

at $r = R_1, v_\theta = 0$

$\therefore\ 0 = c_1\frac{R_1^2}{2} + c_2$

$c_2 = -\frac{c_1}{2}R_1^2$

at $r = R_2, v_\theta = R_2\Omega$

$\therefore\ R_2^2\Omega = c_1\frac{R_2^2}{2} + c_2 = c_1\frac{R_2^2}{2} - \frac{c_1}{2}R_1^2$

$R_2^2\Omega = \frac{c_1}{2}(R_2^2 - R_1^2)$

From Eq. (i),

$$r\,v_\theta = c_1\frac{r^2}{2} - \frac{C_1}{2}R_1^2 = \frac{C_1}{2}(r^2 - R_1^2)$$

$$r\,v_\theta = \frac{R_2^2\Omega(r^2-R_1^2)}{(R_2^2-R_1^2)} = \frac{R_2^2\Omega(r-R_1)(r+R_2)}{(R_2-R_1)(R_2+R_1)}$$

The approximate expression,

$$v_\theta = \frac{R_2^2\Omega(r-R_1)}{(R_2-R_1)}\left\{\begin{matrix}\therefore r + R_1 \approx R_2 + R \\ \frac{R_2}{r} \approx 1\end{matrix}\right\}$$

12. For a Newtonian fluid flowing in a circular pipe under steady state conditions in fully developed laminar flow, the Fanning friction factor is

A. $0.046\ \text{Re}^{-0.2}$

B. $0.0014+\frac{0.125}{\text{Re}^{0.32}}$

C. $\frac{16}{\text{Re}}$

D. $\frac{24}{\text{Re}}$

Sol. **C.** Fanning friction factor,

$$f = \frac{16}{\text{Re}}$$

13. In the Tyler standard screen scale series, when the mesh number increases from 3 mesh to 10 mesh, then

A. the clear opening decreases

B. the clear opening increase

C. the clear opening is unchanged

D. the wire diameter increases

Sol. **C.** When mesh number increases in Tyler standard screen scale series. The clear opening decreases.

14. Taking the acceleration due to gravity to be 10 10m/s² the separation factor of a cyclone 0.5 m in diameter and having a tangential velocity of 20 m/s near the will is

A. 120

B. 130

C. 160

D. 100

Sol. **C.** Separation factor, $s = \frac{r\omega^2}{g}$

Here $\omega = \frac{v}{r} = \frac{20}{\left(\frac{0.5}{2}\right)} = 80$

$\therefore\ s = 0.25 \times \frac{80\times 80}{10} = 160$

GATE 2012

15. Water is flowing under laminar conditions in a pipe of length L. if the diameter of the pipe is

doubled, for a constant volumetric flow rate, the pressure drop across the pipe

A. decreases 2 times

B. decrease 16 times

C. increase 2 times

D. increases 16 times

***Sol.* B.** By Hagen-Poiseuille's equation for laminar flow,

$$\frac{\Delta p}{L} = \frac{32\mu v}{D^2} \qquad \text{....(i)}$$

Also, volumetric flow rate,

$$Q = Av = \frac{\pi D^2}{4}.v$$

$$\therefore \qquad v = \frac{4Q}{\pi D^2}$$

Using Eqs. (i) and (ii),

$$\Delta p \propto \frac{1}{D^4}$$

$$\frac{\Delta p_2}{\Delta p_1} = \left(\frac{D_1}{D_2}\right)^4 = \left(\frac{1}{2}\right)^4$$

$$= \frac{1}{16}$$

16. The local velocity of a fluid along a streamline can be measured by

A. Pitot tube

B. Venturi meter

C. Rotameter

D. Orifice meter

***Sol.* A.** Only Pitot tube measures the local velocity of a fluid along a streamline while all others measure average velocity of the fluid.

17. For uniform laminar flow (in the *X*-direction) past a flat plate at high Reynolds number, the local boundary layer thickness (δ) varies with the distance along the plate (x) as

A. $\delta \propto x^{1/4}$

B. $\delta \propto x^{1/3}$

C. $\delta \propto x^{1/2}$

D. $\delta \propto x$

***Sol.* C.** For uniform laminar flow, the boundary layer thickness is given by,

$$\frac{\delta}{x} \propto \frac{1}{\sqrt{N_{Re}}}$$

or $$\frac{\delta}{x} \propto \frac{1}{\sqrt{x}}$$

or $$\delta \propto \sqrt{x}$$

18. In a mixing tank operating at Reynolds number ($> 10^4$), if the diameter of the impeller is doubled (other conditions remaining constant), the power requirement increases by a factor of.

A. $\frac{1}{32}$

B. $\frac{1}{4}$

C. 4

D. 32

***Sol.* D.** Power number, $N_p = \frac{P}{\rho N^3 D^5}$

where, *P* is power input and *D* is diameter of impeller. For all conditions remaining constant,

$$P \propto D^5$$

$$\therefore \qquad \frac{P_2}{P_1} = \left(\frac{D^2}{D_1}\right)^5 = (2)^5$$

$$= 32$$

GATE 2011

19. Consider the following two cases of movement of particles. In case I, the particle move along the positive Y-direction and in case II, the particle move along the negative Y-direction. Which one of the following option corresponds to the correct direction of buoyancy acting on the particles?

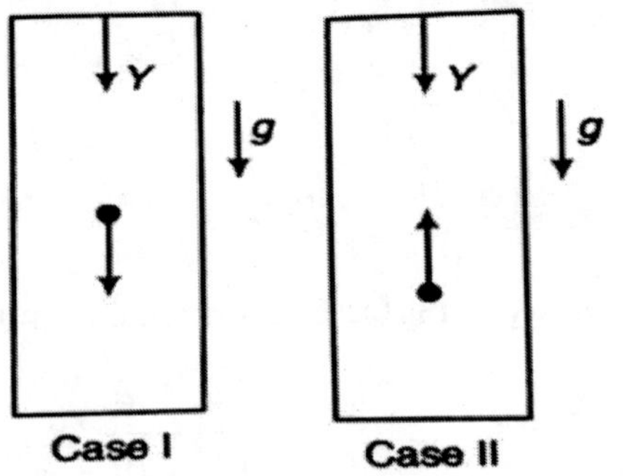

A. Positive Y-direction for both the cases

B. Negative Y-direction for case I, positive Y-direction for case II

C. Negative Y-direction for both the cases

D. Positive Y-direction for case I, negative Y-direction for case II

***Sol.* C.** Negative Y – direction for the cases corresponds to the correct direction of buoyancy acting on the particles.

20. Match the pumps in Group I with the corresponding fluid in Group II.

Group I	*Group II*
P. Gear pump Q. Peristaltic pump	I. Highly viscous liquid II. Aqueous sterile liquid III. Slurry

A. P-III, Q-I
B. P-II, Q-I
C. P-III, Q-II
D. P-I, Q-II

Sol. **D.** Gear pump – Highly viscous liquid
Peristaltic pump – Aqueous sterile liquid

21. In an orifice meter, if the pressure drop across the orifice is over estimated by 5%, then the percentage error in the measured flow rate is

A. +2.47
B. +5
C. – 2.47
D. – 5

Sol. **A.** for an orifice meter,

$$Q \propto \sqrt{\Delta p}$$

$$\text{or} \quad \frac{Q_2}{Q_1} = \sqrt{\frac{\Delta p_2}{\Delta p_1}}$$

Here, $\Delta p_1 = 100$ and $\Delta p_2 = 105$ (over estimated)

$$\therefore \quad \frac{Q_2}{Q_1} = \sqrt{\frac{105}{100}} = 1.0247$$

Percentage error in the flow rate

$$=\left(\frac{1.0247-1}{1}\right) \times 100$$

$=2.47\%$

GATE 2010

22. The critical speed (revolution per unit time) of a ball mill of radius *R,* which used balls of radius r is

A. $\frac{1}{2\pi}\sqrt{\frac{g}{\sqrt{Rr}}}$

B. $\frac{1}{2\pi}\sqrt{\frac{g}{R}}$

C. $\frac{1}{2\pi}\sqrt{\frac{g}{\sqrt{r}}}$

D. $\frac{1}{2\pi}\sqrt{\frac{g}{R-r}}$

Sol. **D.** Critical speed of the ball mill is $n_C = \frac{\omega_C}{2\pi}$

Where, ω_C = critical angular speed
Assume, m= mass of the ball
at critical speed,
Gravitational force = Centrifugal force

$$mg = m(R-r)\,(2\,\pi n_C)^2$$

$$\text{or,} \quad n_C = \frac{1}{2\pi}\sqrt{\frac{g}{R-r}}$$

23. The stream function in a xy-plane is given below $\psi = \frac{1}{2}x^2y^3$. The velocity for this stream function is

A. $xy^3\hat{\imath} - \frac{3}{2}x^2y^2\hat{\jmath}$
B. $\frac{3}{2}x^2y^2\hat{\imath} - xy^3\hat{\jmath}$
C. $\frac{3}{2}x^2y^2\hat{\imath} + xy\hat{\jmath}$
D. $xy^3\hat{\imath} + \frac{3}{2}x^2y^2\hat{\jmath}$

Sol. **B.** Stream function in a xy – plane is –

$$\psi = \frac{1}{2}x^2y^3$$

Velocity vector is –

$$\vec{V} = \frac{\partial\psi}{\partial y}\hat{\imath} - \frac{\partial\psi}{\partial x}\hat{\jmath}$$

$$= \frac{3}{2}x^2y^2\hat{\imath} - xy^3\hat{\jmath}$$

24. The height of a fluidized bed at incipient fluidization is 0.075 m and the corresponding voidage is 0.38. If the voidage of the bed increases to 0.5, then the height of the bed would be

A. 0.058 m
B. 0.061 m
C. 0.075 m
D. 0.093 m

Sol. **D.** We know that

$$h_1A(1-\epsilon_1) = h_2A(1-\epsilon_2)$$

$$h_2 = \left(\frac{1-\epsilon_1}{1-\epsilon_2}\right)$$

given, $h_1 = 0.075$, $\epsilon_1 = 0.38$
$h_2 = ?$, $\epsilon_2 = 0.5$

putting values –

$$h_2 = 0.075 \times \left(\frac{1-0.38}{1-0.5}\right)$$

$h_2 = 0.093$ m.

25. A storage vessel exposed to atmosphere (absolute pressure = 10.3 m of water) has a diameter of 3 m and is initially filed with water to a height of 2 m. The pump draws water from the vessel and is located at an elevation of 5 m above the bottom of the vessel. The frictional head loss in the suction pipe is 2 m of water. If the vapour pressure of the liquid at the temperature of operation is 3 m of water, then the available NPSH is

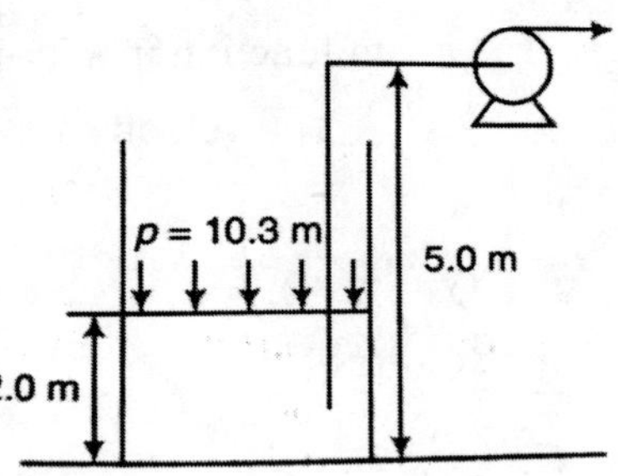

A. 2.3
B. 5.3
C. 6.3
D. 8.3

***Sol.* A.** NPSH $= \frac{p}{\rho g} - h_f - h_s - \frac{p_{vap}}{\rho g}$

Putting values,

Absolute pressure $\frac{p}{\rho g}$ = 10.3

Frictional head loss h_f = 2 m

Suction head h_s = 3 m

Vapour pressure $\frac{p_{vap}}{\rho g}$ = 3 m

NPSH = 10.3 – 2 – 3 –
= 2.3 m

26. In Hagen-Poiseuille flow through a cylindrical tube , the radial profiles of shear stress is
A. constant
B. cubic
C. parabolic
D. liner

***Sol.* D.** In Hagen – poiseuille flow through a cylindrical tube, the radial profile of shear stress would be linear.

GATE 2009

27. Under fully turbulent flow condition, the frictional pressure drop across a packed bed varies with the superficial velocity (v) of the fluid as
A. v^{-1}
B. v
C. $v^{3/2}$
D. v^2

***Sol.* D.** For fully turbulent flow condition, the frictional pressure drop across a packed bed can be written as

$$\frac{\Delta p}{l} = 1.75 \frac{(1-e)}{e^3} \cdot \frac{\rho V^2}{d}$$

Or pressure drop ∝ (Superficial velocity)^2

28. For a mixing tank operating in the laminar regime, the power number varies with the Reynolds number (Re) as
A. $Re^{-½}$
B. $Re^{1/2}$
C. Re
D. Re^{-1}

***Sol.* D.** For mixing tank operating in laminar regime,

Power number ∝ $\frac{1}{\text{Reynolds number (Re)}}$

GATE 2008

29. Three containers are filled with water up to the same height as shown. The pressure at the bottom of the containers is denoted as p_1, p_2 and p_3 which one of the following relationships is true?

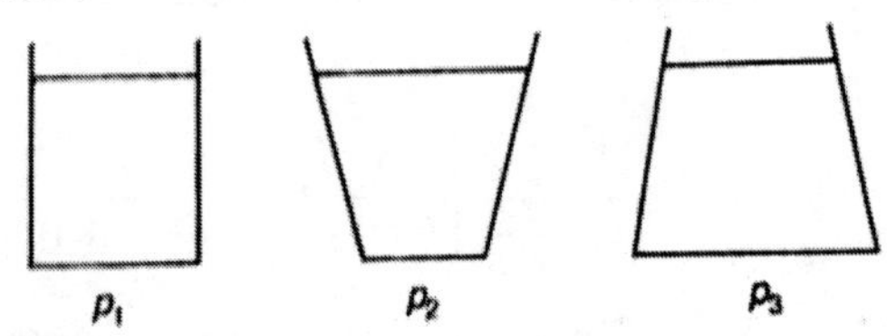

A. $p_3 > p_1 > p_2$
B. $p_3 > p_1 > p_3$
C. $p_1 > p_2 = p_3$
D. $p_1 = p_2 = p_3$

***Sol.* D.** Pressure, p $= \rho g h$

All containers are filled with water therefore, ρ is constant. The heights of all three tank are same,

$$h_1 = h_2 = h_3$$
$$p_1 = p_2 = p_3$$

30. Losses for flow through valves and fitting are expressed in terms of
A. drag coefficient
B. equivalent length of a straight pipe
C. shape factor
D. roughness factor

Sol. **B.** Losses through valve and fitting are expressed in terms of equivalent length of a straight pipe.

31. Given a pipe of diameter D, the entrance length necessary; to achieve fully developed laminar flow is proportional to (N_{Re} is Reynolds number).

A. DN_{Re}

B. $\frac{D}{N_{Re}}$

C. DN_{Re}^2

D. $\frac{D}{DN_{Re}^2}$

Sol. **A.** If x_t is length required to achieve fully developed laminar flow, then,

$$\frac{X_t}{D} = 0.05\ N_{Re}$$

$$X_t \propto DN_{Re}$$

32. For laminar flow conditions, the relationship between the pressure drop (δp_c) across in incompressible filter cake and the specific surface area (S_0) of the particles being filtered is given by which one of the following?

A. δp_c is a proportional to sS_0

B. δp_c is a proportional $1/S_0$

C. δp_c is a proportional S_0^2

D. δp_c is a proportional $1/S_0^2$

Sol. **C.** For laminar flow, the pressure drop across an incompressible filter cake,

$$\Delta p_C \propto S_0^2$$

33. The power required for size reduction in crushing is

A. proportional to $\frac{1}{\text{surface energy of the maerial}}$

B. proportional to $\sqrt{\frac{1}{\text{surface energy of the material}}}$

C. proportional to surface energy of the material

D. independent of the surface energy of the material

Sol. **C.** The power required for size reduction in crushing is proportional to surface energy of the material.

GATE 2007

34. Consider a soap film bubble of diameter D. If the external pressure is P_0 and the surface tension of the soap film is σ, the expression for the pressure inside the bubble is

A. P_0

B. $P_0 + \frac{2\sigma}{D}$

C. $P_0 + \frac{4\sigma}{D}$

D. $P_0 + \frac{8\sigma}{D}$

Sol. **D.** Let p be the pressure inside the bubble.

$$p \times \frac{\pi D^2}{4} = (2-\sigma) \times \pi D + p_0 \times \frac{\pi D^2}{4}$$

$$P = p_0 + \frac{8\,\sigma}{D}$$

35. In Tyler series, the ratio of the aperture size of a screen to that of the next smaller screen is

A. $\frac{1}{\sqrt{2}}$

B. $\sqrt{2}$

C. 1.5

D. 2

Sol. **B.** The ratio of the aperture size of a screen to that of the next smaller screen is $\sqrt{2}$.

36. Size reduction of coarse hard solids using a crusher is accomplished by

A. Attrition

B. compression

C. cutting

D. impact

Sol. **C.** Size reduction of coarse hard solid using a crusher is accomplished by cutting.

37. In constant pressure filtration, the rate of filtration follows the relation

(V = filtrate volume, t = time, k and c = constant).

A. $\frac{DV}{dt} = kV + c$

B. $\frac{DV}{dt} = \frac{1}{kV+c}$

C. $\frac{DV}{dt} = kV$

D. $\frac{DV}{dt} = kV^2$

Sol. **B.** In constant pressure filtration,

The rate of filtration $\frac{dV}{dt} = \frac{1}{kV+C}$

38. Sticky materials are transported by

A. apron conveyor

B. screw conveyor

C. belt conveyor

D. hydraulic conveyor

Sol. **B.** Sticky materials are transported by screw conveyor.

GATE 2006

39. Match the following for a centrifugal pump with impeller speed n.

P. Capacity Q. Head	1. Proportional to *nl* 2. Proportional to n^2 3. Proportional n^3

A. P-2, Q-1
B. P-1, Q-3
C. P-2, Q-3
D. P-1, Q-2

***Sol.* B.** for a centrifugal pump,

Capacity $\propto$ n

Head $\propto$ n^2

40. The magnitude of the force (in newton) required to hold a body of volume 0.05 m^3 and mass 40 kg in water (density 1000 kg/m3) at a depth of 0.1 m is (g = 9.18 m/s²).

A. zero
B. 98.1
C. 490.5
D. 882.9

***Sol.* B.** We know, $V = 0.05\ m^3$, $m = 40$ kg

$\rho = 1000\ kg/m^3, g = 9.81\ m/s^2$

weight + applied force = Buoyancy force

$mg + F = \rho gV$

$= \rho gV - mg$

$= 1000 \times 9.81 \times 0.05 - 40 \times 9.81$

= 98.1 N

GATE 2005

41. Match the following type of fluid (in group) their respective constitutive relation (in Group II), where, τ is the stress and ⍰ is the strain rate.

Group 1	*Group II*
P. Pseudo plastic Q. Bingham plastic	1. $\tau = \mu\dot{\gamma}$ 2. $\tau = \tau_0 + K\dot{\gamma}$ 3. $\tau = K\lvert\dot{\gamma}\rvert^n$; n < 1 4. $\tau = K\lvert\dot{\gamma}\rvert^n$; n > 1

A. P-1, Q-4
B. P-4, Q-1
C. P-2, Q-3
D. P-3, Q-2

***Sol.* D.** Pseudoplastic → $\tau = K\,\lvert\dot{\gamma}\rvert^n$ for n < 1

Bingham plastic → $\tau = \tau_0 + K\dot{\gamma}$

42. The critical speed of the ball mill of radius *R*, which contain balls of radius r is proportional to

A. $(R-r)^{-0.5}$
B. $(R-r)^{-1}$
C. (c) $(R - r)$
D. $(R - r)^2$

***Sol.* A.** The critical speed of the ball mill,

$$n_C = \frac{1}{\sqrt{2}}\sqrt{\frac{g}{R-r}}$$

$$\therefore n_C \propto (R-r)^{-0.5}$$

43. If the frequency of the stirrer in a mixing tank is increased by a factor of 2 while all other parameter are kept constant, by what factor is the power requirement increased at high Reynolds number?

A. 4
B. 8
C. 16
D. 32

***Sol.* B.** At high Reynolds number

Power $P \propto N^3$

$$\therefore \frac{P_2}{P_1} = \left(\frac{N_2}{N_1}\right)^3$$

$$P_2 = P_1(2)^3 = 8P_1$$

GATE 2004

44. The equivalent diameter for flow through a rectangular duct of width *B* and height *H* is

A. $\frac{HB}{2(H+B)}$

B. $\frac{HB}{(H+B)}$

C. $\frac{2HB}{(H+B)}$

D. $\frac{4HB}{(H+B)}$

***Sol.* C.** By definition equivalent diameter

$$d = 4\,r_H \quad \text{....(1)}$$

Where r_H = hydraulic radius

$$\therefore r_H = \frac{Cross-Sectional\ area}{Watted\ parameer} = \frac{BH}{2B+2H}$$

Comparing equivalent (I) and (II)

$$d = 4\frac{BH}{2B+2H}$$

$$= \frac{2BH}{B+H}$$

45. What is the force required (in newton) to hold a spherical balloon stationary in water at a depth of *H* from the air-water interface? The balloon is of radius 0.1m and is filled with air.

A. $\frac{4\pi g}{3}$

B. $\frac{0.1\pi g h}{4}$

C. $\frac{0.1\pi g h}{8}$

D. $\frac{0.04\pi gH}{3}$

Sol. **A.** Force required to hold the balloon = Buoyancy force

$$= \rho g V$$

$$= 1000 \times g \times \frac{4}{3}\pi(0.1)^3$$

$$= \frac{4\pi g}{3}$$

46. Match the system in Group I with equipment used to separate them in Group II.

Group I	*Group II*
P. Gas – solid Q. Liquid – liquid	1. Filter press 2. Cyclone 3. Decanter 4.Thickener

A. P-1, Q-2

B. P-2, Q-3

C. P-3, Q-4

D. P-4, Q-1

Sol. **B.** Gas – Solid – Cyclone

Liquid – Liquid – Decanter

47. For a cyclone of diameter 0.2 m with a tangential velocity of 15 m/s at the wall, the separation factor is

A. 2250

B. 1125

C. 460

D. 230

Sol. **D.** By definition, separation factor

$$S = \frac{V_0^2}{gr} = \frac{2V_0^2}{gd} \qquad \begin{Bmatrix} d = 0.2\, m, \\ v_0 = 15\, m/s \end{Bmatrix}$$

$$= \frac{2\times(15)^2}{9.81\times0.2}$$

$$= 230$$

48. For a partial settling in water at its terminal settling velocity, which of the following is true?

A. Buoyancy = weight + Drag

B. Weight = Buoyancy + Drag

C. Drag = Buoyancy + Weight

D. Drag = Weight

Sol. **D.** At terminal settling velocity,

Weight = Buoyancy + Drag

49. In constant pressure filtration,

A. resistance decreases with time

B. rate of filtration is constant

C. rate of filtration increases with time

D. rate of filtration decreases with time

Sol. **D.** In constant pressure filtration, rate of filtration decreases with time.

GATE 2003

50. A lubricant 100 times more viscous then water would have a viscosity (in Pa-s)

A. 0.01

B. 0.1

C. 1

D. 10

Sol. **B.** Velocity of water = 1 CP

Viscosity of lubricant = 100 CP

= 100 × 0.001 Pa-s

= 0.1 Pa-s

51. The velocity profile for a Bingham plastic fluid flowing (under laminar condition) in a pipe is

A. parabolic

B. flat

C. flat near the wall and parabolic in the middle

D. parabolic near the wall and flat in the middle

Sol. **D.** The velocity profile for a Bingham plastic fluid in laminar flow condition is parabolic near the wall and flat in the middle.

52. Energy requirement (per unit mass of material crushed / ground) is highest for

A. Jaw crusher

B. rod mill

C. ball mill

D. fluid energy mill

Sol. **D.** Energy requirement is highest for fluid energy mill out of given options.

GATE 2002

53. With increasing flow rate, the hydraulic efficient of a centrifugal pump

A. monotonically decreases

B. decreases and then increases

C. remains constant

D. increases and then decreases

Sol. **A.** With increasing flow rate, the hydraulic efficiency of a centrifugal pump monotonically decreases.

54. For flow past a flat plate, if x is the distance along the plate in the direction of flow, the boundary layer thickness is proportional to

A. $\sqrt{x}$
B. $1/\sqrt{x}$
C. X
D. 1/x

Sol. **A.** Boundary layer thickness $\propto \sqrt{x}$.

55. For turbulent flow of an incompressible fluid through a pipe, the flow rate Q is proportional to $(\delta p)^n$,where δp is the pressure drop. The value of exponent n is

A. 1
B. 0
C. < 1
D. > 1.

Sol. **C.** For turbulent flow through a pipe

$$Q \propto (\Delta p)^n$$

Here, n< 1

56. If d_p is the equivalent diameter of a non-spherical particle, V_p its volume and S_p its surface area, then its sphericity ϕ_s is defined by

A. $\emptyset_s = \frac{6\,V_p}{d_p S_p}$
B. $\emptyset_s = \frac{V_p}{d_p S_p}$
C. $\emptyset_s = \frac{6\,d_p S_p}{V_p}$
D. $\emptyset_s = \frac{d_p S_p}{V_p}$

Sol. **A.**

$$\text{Sphericity} = \frac{\text{Surface area of sphere of same volume as of particle}}{\text{Surface area of particle}}$$

$$\text{Here, } \frac{4}{3}\pi r^3 = V_P$$

$$\text{or } \frac{4}{3}\pi\left(\frac{d_p}{2}\right)^3 = V_P$$

$$\pi d_p^3 = 6V_P$$

$$\text{Surface area of sphere} = 4\,\pi r^2$$

$$S_P = 4\,\pi\frac{d_p^2}{4} = \pi d_p^2$$

$$= \frac{6V_p}{d_p}$$

$$\text{Sphericity} = \frac{6V_p}{d_p S_p}$$

57. Arrange the following size reduction equipment in the decreasing order of the average particle size produced by each of them.

A. jaw crusher, ball mills, fluid energy mills
B. ball mills, jaw crusher, fluid energy mills
C. fluid energy mils, jaw crusher, ball mills
D. fluid energy mills, ball mills, jaw crusher

Sol. **A.** Size reduction equipment in decreasing order of the average particle size4 produced by each of them is jaw crusher, ball mill, and fluid energy mill.

GATE 2001

58. The power number for a stirred tank becomes constant at high Reynolds number. In this limit, the variation of power input with impeller rotational speed (N) is proportional to

A. N^0
B. N^1
C. N^2
D. N^3

Sol. **D.** Power number $= \frac{P}{\rho N^3 D^5}$

For constant power number

Power input (P) $\propto N^3$.

59. The operational of a rotameter is based on

A. variable flow area
B. rotation of a turbine
C. pressure drop across a nozzle
D. pressure at a stagnation point

Sol. **A.** The operation of rotameter is based on variable flow area.

60. Applying a pressure drop across a capillary results in a volumetric flow rate Q under laminar flow conditions. The flow rate for the same pressure drop, in a capillary of the same length but half the radius is

A. $\frac{Q}{2}$
B. $\frac{Q}{4}$
C. $\frac{Q}{8}$
D. $\frac{Q}{16}$

Sol. **D.** Flow rate Q $\propto r^4$.

$$\frac{Q_2}{Q} = \left(\frac{r_2}{r}\right)^4$$

$$\frac{Q_2}{Q} = \left(\frac{r/2}{r}\right)^4$$

$$Q_2 = \frac{Q}{16}$$

GATE 2000

61. For a sphere falling in the constant drag coefficient regime, its terminal velocity depends on its diameter d as

A. D

B. $\sqrt{d}$

C. d^2

D. $\frac{1}{d}$

Sol. **C.**Terminal velocity u ∝ d^2.

62. In a fully turbulent flow (Re > 10^5) in a pipe of diameter *d* for a constant pressure gradient, the dependence of volumetric flow rate of an incompressible fluid is

A. D

B. d^2

C. $d^{2.5}$

D. d^4

Sol. **C.** Q ∝ $d^{2.5}$

63. In the laminar boundary layer flow over a flat plate, the ratio (δ/x) varies as

A. Re

B. $\sqrt{Re}$

C. $\frac{1}{Re}$

D. $\frac{1}{\sqrt{Re}}$

where, δ is the boundary layer thickness and x is the distance from the leading edge in the direction of flow.

Sol. **D.**

$$\frac{\delta}{x} \propto \frac{1}{\sqrt{Re}}$$

64. For laminar flow of a shear thinning liquid a pipe, if the volumetric flow rate is doubled, the pressure gradient will increase by a factor of

A. 2

B. < 2

C. > 2

D. $\frac{1}{2}$

Sol. **A.** If $Q_2 = 2Q_1$

$$\Delta\rho_2 = 2\Delta\rho_1$$

(2 Mark Quotations)

GATE 2015

65. A centrifugal pump delivers water at the rate of $0.22\ m^3/s$ from a reservoir at ground level to another reservoir at a height *H,* through a vertical pipe of 0.2 m diameter. Both the reservoirs are open to atmosphere. The power input to the pump is 90 kW and it operates with an efficiency of75%[given : fanning friction factor for pipe flow, $f = 0.004$. neglect other head losses gravitational acceleration, g = 9.8 m/s^2 and density of water is 1000 kg/m^3].The height *H* (in meter), to which the water can be delivered (upto one decimal place) is____.

Sol. Given, $Q = 0.22 m^2/s$

$$v_2 = \frac{0.22}{\frac{\pi}{4}\times(0.2)^2} = 7\ m/s$$

$$W = 90\ kW$$

W in head unit = $\frac{90\times10^3}{\rho Q g}$

Applying Bernoulli's equation between reservoirs (1) and (2)

$$\frac{\rho_1}{pg} + z_1 + \frac{v_1^2}{2g} + hf + \eta w = \frac{\rho_2}{pg} + z_2 + \frac{v_2^2}{2g} \ ....(i)$$

and $Z_2 = H$

$$h_f = \frac{4fHv^2}{2gD}$$

$$= \frac{4\times0.004\times H\times7^2}{2\times9.8\times0.2} = 0.2\ Hm$$

$$or\ 0.2\ H + 0.75 \times \frac{90\times10^3}{0.22\times1000\times9.8}$$

$$= H + \frac{49}{2\times9.8}$$

or 0.2 *H*+ 31.308 = *H* + 2.5

or 0.8 *H* = 28.808

H = 36.01 m

so, *H* = 36m

66. A spherical particle of 1 mm diameter is falling with a downward of 1.7 mm/s through a liquid (viscosity 0.04 Pa-s) at a low Reynolds number (Strokes regime). The liquid is flowing upward at a velocity of 1 mm/s. All liquid is flowing upward at a velocity of 1 mm/s. All velocities are with respect to a stationary reference frame.Neglecting the wall effects . The wall effects, the drag force

per unit projected area of the particle (in Pa) (upto two dedcimal place) is ___________.

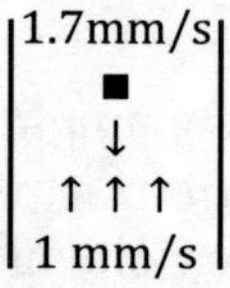

Sol. Given, $d = 1\text{mm} = 10^{-3}\text{m}$

Velocity of particle relative to the fluid,

$v = 1.7 + 1 = 2.7$ mm/s

$= 2.7 \times 10^{-3}\text{m/s}$

Drag force per unit projected area,

$$\rho = \frac{F_\rho}{A_\rho} = \frac{3\pi\eta dv}{\left(\pi\frac{d^2}{4}\right)} = \frac{12\eta v}{d}$$

$$\rho = \frac{12\times0.04\times2.7\times10^{-3}}{10^{-3}}$$

$$\rho = 1.296 \text{ Pa}$$

67. For fanning friction factor f (for flow in pipes) and drag coefficient C_D (for flow over immersed bodies), which of the following statement is correct?

P: f account only for the skin friction

Q: C_D account only for the form friction

R: C_D account for both skin friction and form friction

S: Both f and C_D depends on the Reynolds number

T: For laminar flow through a pipe, f doubles on doubling the volumetric flow rate

A. R, S, T

B. P, Q, S

C. P, R, S

D. P, Q, S, T

Sol. **C. P:** f accounts only for the skin friction

R: C_D accounts for both skin friction and form friction

S: Both f and C_D depends on the Reynolds number

GATE 2014

68. An incompressible fluid is flowing through a contraction section of length L and has a 1-D (x-direction) steady state velocity distribution, u $= u_0\left(1+\frac{2x}{L}\right)$. If $u_0 = 2$ m/s and $L = 3$ m, the convective acceleration (in m/s^2) of the fluid at L is ________.

Sol. Given, $u = u_0\left(1+\frac{2x}{L}\right)$

Differentiating with respect to t,

$$\frac{du}{dt} = a = u_0\left(0+\frac{2x}{L}\right)$$

$$a = \frac{2u_0}{L}u$$

$$= \frac{2u_0^2}{L}\left(1+\frac{2x}{L}\right)$$

at $x = L$,

$$a = \frac{2u_0^2}{L}\left(1+\frac{2L}{L}\right) = \frac{6u_0^2}{L}$$

$$a = \frac{6\times(2)^2}{3} = 8\text{ m/s}^2$$

69. Match the following.

Group I	*Group II*
P. Turbulence	I. Reciprocating pump
Q. NPSH	II. Packed bed
R. Ergun equation	III. Fluctuating velocity
S. Rotameter	IV. Impeller
T. Power number	V. Vena contracta

A. P-III, R-II, T-IV

B. Q-V, R-II, S-III

C. P-III, R-IV, T-II

D. Q-III, S-V, T-IV

Sol. **A.** Turbulence Fluctuating velocity,

Ergun equation is applied for packed bed

Power number is applicable for impeller.

70. In a steady and incompressible flow of a fluid (density = 1.25 kg m^{-3}), the difference between stagnation and static pressures at the same location in the flow is 30 mm of mercury (density = 13600 kg m^{-3}). Considering gravitational acceleration as 10 ms^{-2}, the fluid speed (in ms^{-1}) is ___________.

Sol. $\rho_1 - \rho_2 = \frac{1}{2}\text{p}v^2$

$$p'gh = \frac{1}{2}\text{p}v^2$$

$$13600\times(30\times10^{-3}) = \frac{1}{2}\times1.25\times v^2$$

v = 80.8 m/s

GATE 2012

71. A bed of spherical glass beads (density 3000 kg/m^3, diameter 1 mm, bed porosity 0.5) is to be fluidized by a liquid of density 1000kg/m^3 and viscosity 0.1 Pa-s. Assume that the Reynolds number based on particle diameter is small compared to one. If g = 10 m/s^2, then the minimum velocity (in m/s) required to fluidize the bed is

A. 3.33×10^{-4}
B. 3.33×10^{-1}
C. 3
D. 30

Sol. **A.** For packed bed, Kozney-Carman equation is

$$\frac{\Delta p}{L} = \frac{150\,\mu v}{\phi_s^2 D_p^2} \cdot \frac{(1-e)^2}{e^3} \quad \text{....(i)}$$

For minimum fluidization,

$$\frac{\Delta p}{L} = (1-e)(\rho_p - \rho)g \quad \text{....(ii)}$$

From Eqs. (i) and (ii),

$$v = \frac{\phi_s^2 D_p^2 (\rho_p - \rho)g}{150\,\mu} \cdot \frac{e^3}{(1-e)}$$

Here, $\phi_s = 1$(for sphere), $D_p = 1\text{mm} = 10^{-3}\text{m}$,

$e = 0.5$, $\rho_p = 3000\ \text{kg/m}^3$, $\rho = 1000\text{kg/m}^3$,

$g = 10\ \text{m/s}^2$, $\mu = 0.1\ \text{Pa} - \text{s}$

Putting values, we get

$v = 3.33 \times 10^{-4}\text{m/s}$

72. The steady, laminar velocity profile in the X-direction is

A. $v\left(\frac{y}{b}\right)$
B. $v\left[\left(\frac{y}{b}\right)^2 - 1\right]$
C. $v\left[1 - \left(\frac{y}{b}\right)^2\right]$
D. $v\left[1 - \left(\frac{y}{b}\right)\right]$

Sol. **D.** When the top plate is stationary, the velocity profile is linear, $v_x = v\left(1 - \frac{y}{b}\right)$

73. The force per unit area (in the X-direction) that must be exerted on the bottom plate to maintain the flow is

A. $\frac{\mu v}{b}$
B. $-\frac{\mu v}{b}$
C. $\frac{2\mu v}{b}$
D. $-\frac{2\mu v}{b}$

Sol. **A.** Force per unit area,

$$\tau = -\mu \frac{dv_x}{dy}$$

$$= -\mu \frac{d}{dy}\left[v\left(1 - \frac{y}{b}\right)\right]$$

$$= -\mu v\left(-\frac{1}{b}\right) = \frac{\mu v}{b}$$

GATE 2011

74. Two liquid (P and Q) having same viscosity are flowing through a double pipe heat exchanger as shown in the schematic below.

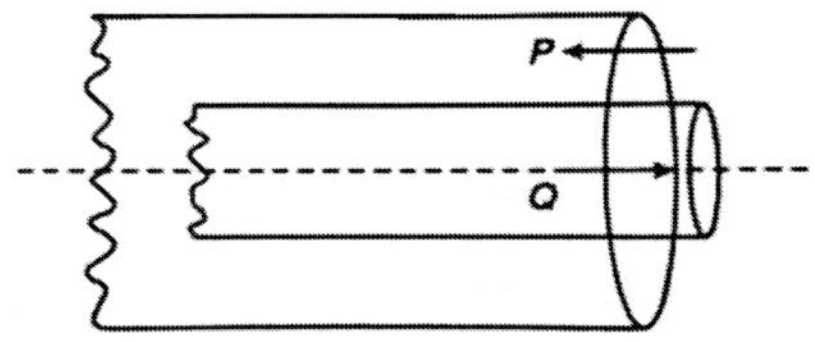

Densities of P and Q are 1000 kg/m^3 and 800 kg/m^3 respectively. The average velocities of the liquids P and Q are 1 m/s and 2.5 m/s respectively. The inner diameters of the pipes are 0.31 m and 0.1 m. Both pipes are 5 mm thick. The ratio of the Reynolds numbers Re_P to Re_Q is

A. 2.5
B. 1.55
C. 1
D. 4

Sol. **C.** Ratio of Reynolds numbers is

$$\frac{Re_P}{Re_Q} = \frac{\rho_P V_P D_P}{\mu_P} \times \frac{\mu_Q}{\rho_Q \times V_Q \times d_Q}$$

Putting values,

$\rho_P = 1000\ \text{kg/m}^3$, $V_P = 1$ m/s

$\rho_Q = 800\ \text{kg/m}^3$, $V_Q = 2.5$ m/s

$\mu_p = \mu_Q$

$d_Q = 0.1$ m

$$d_p = 0.31 - 0.1 - 2 \times \frac{5}{1000} = 0.2\ \text{m}$$

$$\frac{Re_P}{Re_Q} = \frac{1000}{800} \times \frac{1}{2.5} \times \frac{0.2}{0.1}$$

75. The partial size distribution of the feed and collected solids (sampled for same duration) for a gas cyclone are given below.

Size range (μm)	*Weight of feed in The size range (g)*	*Weight of collected solids in the size range (g)*
1-5	2.0	0.1
5-10	3.0	0.7
10-15	5.0	3.6
15-20	6.0	5.5
20-25	3.0	2.9
25-30	1.0	1.0

What is the collection efficiency (in percentage) of the gas cyclone?

A. 31

B. 60

C. 65

D. 69

***Sol.* C.** Collection efficiency

$$= \frac{\left(\frac{0.1}{2}+\frac{0.7}{3}+\frac{3.6}{5}+\frac{5.5}{6}+\frac{2.9}{3}+\frac{1}{1}\right)}{6} \times 100$$

= 65%

76. A liquid following through the following piping network. The length of pipe section *P*, *Q*, *R* and *S* shown in the schematic are equal. The diameter of the section *P* and *R* are equal and the diameter of the section *Q* is twice that of *S*. The flow is steady and laminar. Neglecting curvature and entrance effects, the ratio of the volumetric flow rate in the pipe section *Q* to that in *S* is

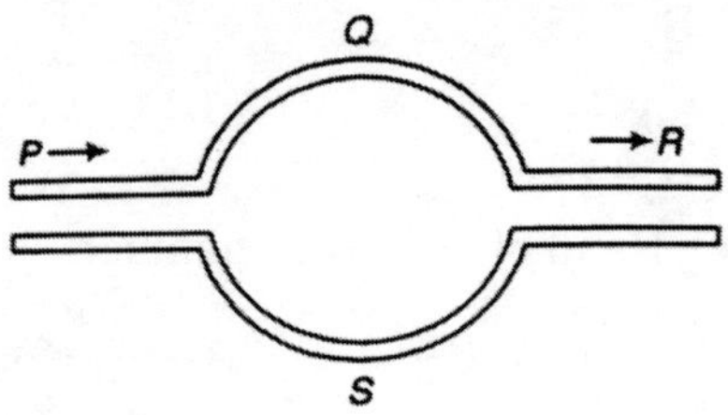

A.

B. 8

C. 2

D. 1

***Sol.* A.** By definition Hagen – Poiseuille's equation

$$\frac{\Delta P}{L} = \frac{32\mu\, V}{D^2} \quad \ldots\text{(i)}$$

Volumetric flow rate

$$Q = Av \quad \ldots\text{(ii)}$$

Comparing

$$Q = \frac{\pi D^2}{4} \times \frac{\Delta P}{L} \times \frac{D^2}{32\,\mu}$$

$$= \frac{\Delta P}{L} \times \frac{\pi}{128\mu} \times D^4$$

$$= Q \propto D^4$$

Given,

$$D_Q = 2D_S$$

$$\frac{Q_Q}{D_S} = \frac{D_Q^4}{D_S^4} = 2^4 = 16$$

Common Data for Questions 50 and 51

For a liquid following through a packed bed, the press drop per unit length of the bed $\frac{\Delta p}{L}$ is

$$\frac{\Delta p}{L} = \frac{150\mu_i \bar{V}_0}{(\varphi_s \bar{d}_p)^2} \times \frac{(1-\varepsilon)^2}{\varepsilon^3} + \frac{1.75 p_i \bar{V}_0^2 (1-\varepsilon)}{\varphi_s\, d_p \varepsilon^3}$$

Where, $\bar{V}_0$ is the superficial liquid velocity, ε is the bad porosity, $\bar{d}_p$ is average partial size, $\emptyset_S$ is partial sphericity, ρ_f is liquid density and μ_f is a liquid viscosity.

Given, $\bar{d}_p = 1 \times 10^{-3}$ m, $\emptyset_S = 0.8$, $\rho_f = 1000$ kg/m^3

$\mu_f = 10^{-3}$ kg/m-s

Particle density ρ_P = 2500 kg/m^3 and acceleration due to gravity, g = 9.8 m/s^2.

77. When $\bar{V}_0$is 0.005 m/s and ε = 0.5, which one of the following is the correct value for the ratio of the viscous loss to the kinetic energy loss?

A. 0.09

B. 1.07

C. 10.71

D. 93

***Sol.* C.** The ratio of the visous loss to the kinetic energy loss

$$= \frac{150\mu_f \overline{V_0}}{(\emptyset_S \overline{d_p})^2}, \frac{(1-\epsilon)^2}{\epsilon^3}$$

$$= \frac{1.75\rho_f \overline{V_0^2}(1-\epsilon)}{\emptyset_S \overline{d_p} \epsilon^3}$$

$$= \frac{150}{1.75} \times \frac{\mu_f}{\rho_f} \times \frac{(1-\epsilon)}{\emptyset_S \bar{d}_p \bar{V}_0}$$

$$= \frac{150}{1.75} \times \frac{1\times10^{-3}}{1000} \times \frac{(1-0.5)}{0.8\times 1\times10^{-3} \times 0.005}$$

$$= 10.71$$

78. On further increasing $\bar{V}_0$, incipient fluidisation is achieved. Assuming that the porosity of the bed remains unaltered, the pressure drop per unit length (in Pa/m) under incipient fluidization condition is

A. 3675

B. 7350

C. 14700

D. 72501

Sol. **B.** By definition incipient fluidization is

$$\frac{\Delta P}{L} = (1-\epsilon)(\rho_p - \rho_f)\, g$$

$$= (1 - 0.5) \times (2500 - 1000) \times 9.8$$

$$= 7350 \text{ Pa/m}$$

GATE 2010

79. A hydrometer with stem cross-sectional area of $2.82 \times 10^{-5} \text{m}^2$ is immersed in a very large vessel containing water as shown in the figure. The immersed volume is 15×10^{-6} m^3 and the length of the stem about water surface is L_W. If the entire volume of water is replaced by a liquid with specific gravity 1.5 and if the length of the stem above the liquid surface is L_l, then the difference, $L_l - L_W$ is

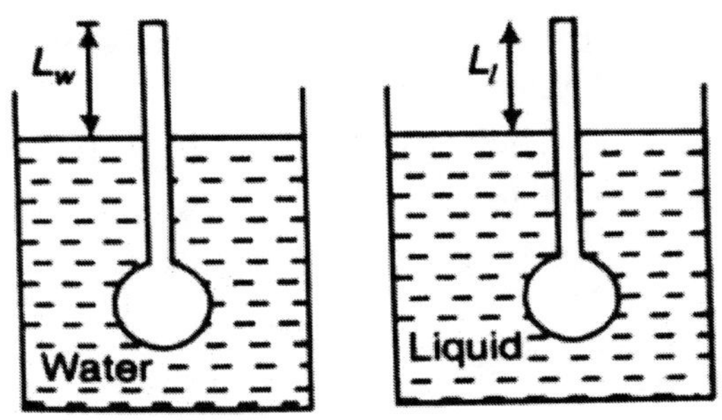

A. – 177 mm
B. 177 mm
C. – 266 mm
D. 266 mm

Sol. **B.** Assume, L be the total length of the hydrometer

For water,

Weight of the bulb = weight of the displayed water

Or $W = \rho V = 1000 \times 15 \times 10^{-6}$

$W = \rho\,(L - L_W)\cdot A = 15 \times 10^{-3}$ ....(i)

For liquid,

$W' = \rho' V'$

$W' = \rho'(L - L_L)\cdot A$

Weight of the bulb remains constant

$\therefore$ W = W'

Or $\rho'(L - L_L)\cdot A = 15 \times 10^{-3}$

$1500 \times (L - L_L) \times 2.82 \times 10^{-5}$

$= 15 \times 10^{-3}$ ....(ii)

Form equation (i)

$1000 \times (L - L_W) \times 2.82 \times 10^{-5} = 15 \times 10^{-3}$

$L - L_W = 0.532$ ....(iii)

From eqs. (ii) and (iii)

$L_l - L_W = 0.177$ m= 177 mm

80. The diameter of a drop of a liquid fuel changes with time due to combustion according to the relationship,

$$D = D_0\left(1 - \frac{t}{t_b}\right)$$

While burning, the drop falls at its terminal velocity under stockes flow regime. The distance it will travel before complete combustion is given by

A. $\frac{D_0^2\,\Delta\rho\, t_b\, g}{18_\mu}$

B. $\frac{D_0^2\,\Delta\rho\, t_b\, g}{36_\mu}$

C. $\frac{D_0^2\,\Delta\rho\, t_b\, g}{54_\mu}$

D. $\frac{D_0^2\,\Delta\rho\, t_b\, g}{108_\mu}$

Sol. **C.** the diameter of a drop of liquid fuel changes with time due to combustion according to the relationship,

$$D = D_0\left(1 - \frac{t}{t_b}\right)$$

At complete combustion,

D = 0

$\therefore$ $t = t_b$

And at

$t = 0, \quad D = D_0$

Thus, the initial diameter of the drop is D_0 and changes according to given equation and the time for complete combustion is t_b

From Stokes law, the velocity of drop,

$$v_P = \frac{D^2(\rho_p - p)g}{18\mu} = \frac{\Delta\rho g}{18\mu}\,.\,D^2$$

$$v = \frac{\Delta\rho g}{18\mu}.\,D_0^2\left(1 - \frac{t}{t_b}\right)^2$$

Also $\quad v = \frac{ds}{dt}$

Thus,

$$v\int_0^s ds = \frac{D_0^2\Delta\rho g}{18\mu}\left[t - \frac{2}{t_b}.\frac{t^2}{2} + \frac{1}{t_b^2}.\frac{t^3}{3}\right]_0^{t_b}$$

$$s = \frac{D_0^2\Delta\rho g}{18\mu}\left(t_b - t_b + \frac{t_b}{3}\right)$$

$$s = \frac{D_0^2\Delta\rho g.t_b}{54\mu}$$

GATE 2009

81. Two identical reservoirs, open at the top are drained through pipes attached to the bottom of the tanks as shown below. The two drain pipes are of the same length but of different diameters ($D_1 > D_2$)

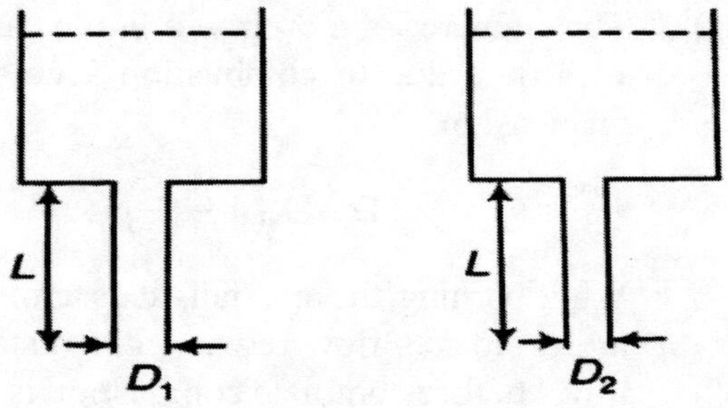

Assuming the flow to be steady and laminar in both drain pipes, if the volumetric flow rate in the larger pipe is 16 time of that in the smaller pipe, the ratio D_1/D_2 is

A. 2

B. 4

C. 8

D. 16

***Sol.* B.** We know that $Q = Av$

or $\frac{Q_2}{Q_1} = \frac{A_2}{A_1}, \frac{V_2}{V_1}$

And lengths of both drain pipes are same

Thus, $v_1 = v_2 = \sqrt{2gL}$

$\therefore \quad \frac{Q_2}{Q_1} = \frac{V_2}{V_1} = \left(\frac{D_2}{D_1}\right)^2$

$\frac{1}{16} = \left(\frac{D_2}{D_1}\right)^2$

or $\frac{D_1}{D_2} = 4$

82. The terminal settling velocity of a 6 mm diameter glass sphere (density = 2500 kg/m³) is a viscous Newtonian liquid (density = 1500 kg/m³) is 100 ìm/s. If the partial Reynolds number is small and the value of acceleration due to gravity is 9.81 m/s², then the viscosity of the liquid (in Pa-s) is

A. 100

B. 196.2

C. 245.3

D. 490.5

***Sol.* B.** According to Stokes law

$$V_P = \frac{D^2(\rho_P - \rho)g}{18\mu}$$

Putting value,

ρ_P = 2500kg/m³, ρ = 1500 kg/m³,

D = 6mm = 6×10^{-3}m,

v_P = 100 μm/s = 10^{-4} m/s

G = 9.81 m/s²

$$\mu = \frac{(6\times10^{-3})^2\times(2500-1500)\times9.81}{18\times10^{-4}}$$

= 196.2 pa.s

Statements for Linked Answer Questions 56 and 57

A free jet of water is emerging from a nozzle (diameter 75 mm) attached to a pipe (diameter 225 mm) as shown below

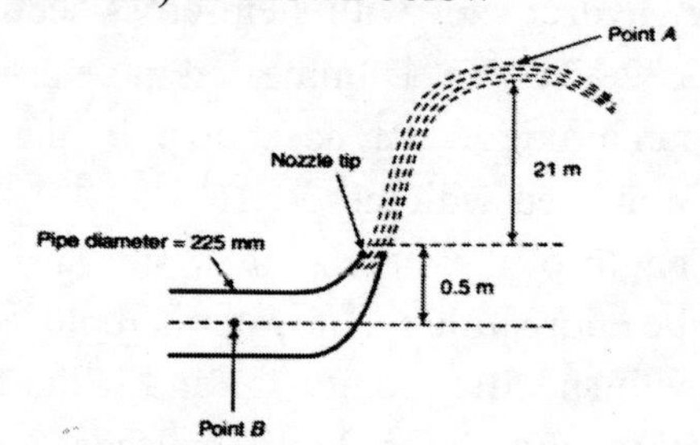

The velocity of water at point A is 18 m/s. Neglect friction in the pipe and nozzle. Use g = 9.81 m/s² . and density of water = 1000 kg/m³.

83. The velocity of water at the trip of the nozzle (in m/s) is

A. 13.4

B. 18.0

C. 23.2

D. 27.1

***Sol.* B.** Taking energy balance between point a and nozzle tip (T),

$$\frac{1}{2}mv_A^2 + mgh = \frac{1}{2}mv_T^2$$

$$v_T^2 = v_A^2 + 2gh \quad \begin{Bmatrix} v_A = 18\frac{m}{s} \\ h = 21\ m \end{Bmatrix}$$

$= (18)^2 + 2 \times 9.81 \times 21$

V_T = 27.1 m/s

84. The gauge pressure (in kPa) at point B is

A. 80.0

B. 100.0

C. 239.3

D. 367.6

***Sol.* B.** Let point nozzle tip is T.

Between point B and T

$A_B V_B = A_T V_T$

$V_B = V_T\left(\frac{D_T}{D_B}\right)^2$

Putting,

V_T = 27.1 m/s

D_T = 75 mm = 75 × 10^{-3} m

D_B = 225 mm = 225 × 10^{-3} m

$$\therefore \quad V_B = 27.1 \times \left(\frac{75\times10^{-3}}{225\times10^{-3}}\right)^2 = 3 \text{ m/s}$$

Applying Bernoulli's equation for point B and T,

$$\frac{P_B}{\rho} + g\, z_B + \frac{V_B^2}{2} = \frac{P_T}{\rho} + g\, z_T + \frac{P_B - P_T}{\rho}$$

$$= g\,(z_T - z_B) + \frac{1}{2}(V_T^2 - V_B^2)$$

Putting values, $z_T - z_B = 0.5$ m, $\rho = 1000$ kg.m^3

$$\therefore \frac{P_B - P_T}{1000} = 9.81 \times 0.5 + \frac{1}{2} \times [(27.1)^2 - (3)^2]$$

$$P_B - P_T = 367.6 \times \text{kPa}$$

∴ Point T is at 1 atm pressure.

∴ Gauge pressure at point

B $= P_B - P_T = 367.6$ kPa

GATE 2008

85. A pump draws oil (specific gravity 0.8) from a storage tank and discharges it into an overhead tank. The mechanical energy delivered by the pump to the fluid is 50 j/kg. The velocities at the section and the discharge points of the pump are 1 m/s and 7 m/s, respectively. neglecting friction losses and assuming kinetic energy correction factor to be unity, the pressure developed by the pump (in kN/m^2) is

A. 19.2
B. 20.8
C. 40
D. 80

***Sol.* B.** According to Bernoulli's equation

$$\frac{P_1}{\rho} + \frac{V_1^2}{2} + g\, z_1 + h_m = \frac{P_2}{\rho} + \frac{V_2^2}{2} + g\, z_2$$

Putting value,

$\rho = 0.8 \times 1000 = 800$ kg/m^3

$z_1 = z_2$, $v_1 = 1$ m/s, $v_2 = 7$ m/s

$h_m = 50$ j/kg

$$\frac{P_2 - P_1}{\rho} = \frac{1}{2}(V_1^2 - V_2^2) + h_m$$

$$\frac{P_2 - P_1}{800} = \frac{1}{2}(1^2 - 7^2) + 50$$

$$P_2 - P_1 = 20800 \text{ N/m}^2 = 20.8 \text{ kN/m}^2$$

86. Match the following Group I with Group II and select the correct answer using the codes given below the list

Group I	*Group II*
P. Euler number	1. Viscous force/Inertial force
Q. Froude number	2. Pressure force/Inertial force
R. Weder number	3. Inertial force/Gravitional number
	4. Inertial force/Surface tension force

A. P-1, Q-2, R-3
B. P-2, Q-3, R-4
C. P-3, Q-2, R-1
D. P-4, Q-3, R-2

***Sol.* B.** Eular number -pressure force / inertial force
Froude number - inertial force / Gravitation force
Weber number - inertial force / Surface tension

87. A steady flow field on an incompressible fluid is given by $\vec{V}$ = (Ax +By) $\hat{\imath}$ – Ay $\hat{\jmath}$, where A = $1s^{-1}$, B = $1s^{-1}$ and x, y are in metre. The magnitude of the acceleration (in m/s^2) of a fluid particle at (1, 2) is

A. 1
B. $\sqrt{2}$
C. $\sqrt{5}$
D. $\sqrt{10}$

***Sol.* B.** Given,

$\vec{V}$ = (Ax – By) $\hat{\imath}$ – Ay $\hat{\jmath}$

Compare it with,

$\vec{V}$ = u$\hat{\imath}$ + v$\hat{\jmath}$

∴ u = Ax + By, v = – Ay

Given,

A = 1, B = 1

∴ u = x + y, v = – y

Now, $\vec{a} = a_x\hat{\imath} + a_y\hat{\jmath}$

Thus, a_x = x and a_y = y

Therefore,

$$|a| = \sqrt{a_x^2 + a_y^2} = \sqrt{x^2 + y^2}$$

Given, x = 1, y = 2

$$\therefore |a| = \sqrt{5}$$

88. Two identically sized spherical particles A and B having densities p. Assuming free settling under turbulent flow condition, the ratio of the terminal settling velocities of partial A to that of particle B is given by

A. $\sqrt{\frac{(\rho_A - \rho)}{(\rho_B - \rho)}}$

B. $\sqrt{\frac{(\rho_B - \rho)}{(\rho_A - \rho)}}$

C. $\frac{(\rho_A - \rho)}{(\rho_B - \rho)}$

D. $\frac{(\rho_B - \rho)}{(\rho_A - \rho)}$

***Sol.* A.** Terminal settling velocity of the particle

$$V = \sqrt{\frac{4d_p(\rho_p-\rho)g}{3C_D\rho}}$$

here, $d_1 = d_2$

thus, $\frac{u_1}{u_2} = \sqrt{\frac{(\rho_A-\rho)}{(\rho_B-\rho)}}$

89. Consider the scale-up of a cylindrical baffled vessel configured to have the standard geometry (i.e., height = diameter). In a order to maintain an equal rate of mass transfer under turbulent condition for a Newtonian fluid, the ratio of the agitator speeds should be (Given N_1, D_1 are agitator speed and vessel diameter before scale-up; N_2, D_2 agitator speed and vessel diameter after scale-up)

A. $\frac{N_1}{N_2} = \frac{D_1}{D_2}$

B. $\frac{N_1}{N_2} = \frac{D_2}{D_1}$

C. $\frac{N_1}{N_2} = \left(\frac{D_1}{D_2}\right)^{\frac{2}{3}}$

D. $\frac{N_1}{N_2} = \left(\frac{D_2}{D_1}\right)^{\frac{2}{3}}$

***Sol.* D.** According to definition the power number is

$$N_P = \frac{P}{N^3 D_a^5 \rho}$$

where, P = power input

D_a = diameter

N = Speed of agitation

For equal rate of mass transfer, power input per unit volume would be constant.

$$\frac{P}{V} = \frac{N_p N^3 D_a^2 \rho}{\pi}$$

$$v = \pi r^2 h = \frac{\pi D^2}{4} \cdot D$$

Height = diameter

$$\frac{P}{V} = \frac{4N_p N^3 D_a^2 \rho}{\pi}$$

Here, N_p and ρ are Constant,

$$\therefore \frac{P}{V} \cdot \frac{\pi}{4N_p\rho} = N^3 D_a^2 = \text{constant}$$

Therefore, for two cases,

$$N_1^3 D_1^2 = N_2^3 D_2^2$$

$$\frac{N_1}{N_2} = \left(\frac{D_2}{D_1}\right)^{2/3}$$

Statement for Linked Answer Questions 63 and 64

A siphon tube having a diameter of 2 cm draws water from a large open reservoir and discharges into the open atmosphere as shown in the figure. Assume incompressible fluid and neglect frictional losses.(g = 9.8 m/s²)

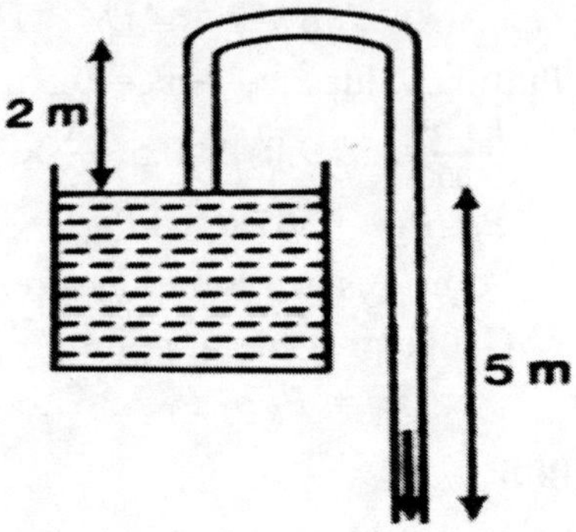

90. The velocity (in m/s) at the discharges point is

A. 9.9

B. 11.7

C. 98

D. 136.9

***Sol.* A.** According to question

The velocity at discharge point $v = \sqrt{2\,gh}$

$h = z_1 - z_2 = 5$m

$v = \sqrt{2 \times 9.8 \times 5}$

$v = 9.9$ m/s

91. The volumetric flow rate (in L/s) water at the discharges is

A. 3.11

B. 3.67

C. 30.77

D. 42.99

***Sol.* A.** According to question Q = Av

Where, A = area of tube = $\frac{\pi D^2}{4}$

Putting value, D = 2cm = 2×10^{-2} m

$Q = \frac{\pi}{4} \times (2 \times 10^{-2})^2 \times 9.9$

$= 31.1\times 10^{-2}$ m³/s

$= 31.1\times 10^{-4}\times 10^3$ L/s = 3.11 L/s

GATE 2007

92. A pipeline system carries crude oil of density 800 kg/m³. The volumetric flow rate at point 1 is 0.28 m³/s. The cross sectional areas of the branches are 1, 2 and 3 are 0.012, 0.008 and 0.004 m² respectively. All the three branches are in a horizontal plane and the friction is negligible. If

the pressure at the points 1 and 2 are 270 kPa and 240 kPa respectively, then the pressure at point 2 is

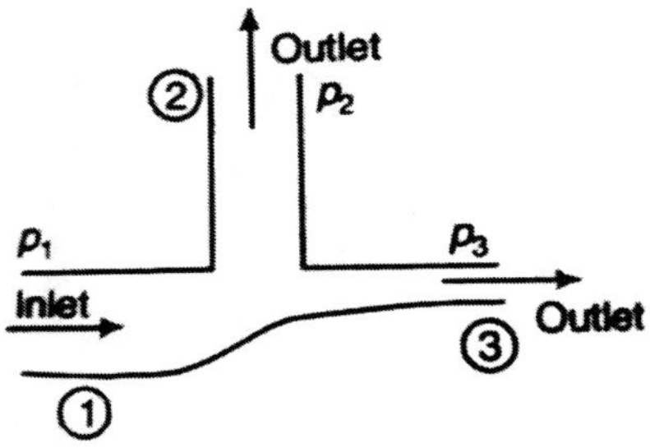

A. 202 kPa
B. 240 kPa
C. 284 kPa
D. 355 kPa

***Sol.* A.** we know, $Q_1 = A_1V_1$

$A_1 = 0.012\ m^2, \quad A_2 = 0.08\ m^2,$
$A_3 = 0.04\ m^2$
$P_1 = 270\ kPa = 270 \times 10^3 Pa$
$P_2 = 240\ kPa = 240 \times 10^3 Pa$
$\rho = 800\ kg/m^3$
$Q_1 = 0.28 m^3/s$
$Q_1 = A_1V_1$
$\therefore \quad V_1 = 23.33\ m/s$

Now, $Q_1 = Q_2 + Q_3$
$Q_1 = A_2V_2 + A_3V_3$
So, $0.28 = 0.008\ V_2 + 0.004\ V_3$
$2V_2 + V_3 = 70$

Applying Bornoulli's equation for points 1 and 3,

$$\frac{p_1}{\rho}+\frac{V_1^2}{2} = \frac{p_3}{\rho}+\frac{V_3^2}{2}$$

$$V_3^2 = \frac{2(p_1-p_3)}{\rho} + V_1^2$$

Putting values,

$$V_3^2 = \frac{2\times(270\times10^3-240\times10^3)}{800} + (23.33)^2$$

$V_3 = 24.88\ m/s$

Now from equation (i)

$2V_2 + 24.88 = 70$
$V_2 = 22.56\ m/s$

Again applying Bernoulli's equation for points (i) and (ii),

$$\frac{p_1}{\rho}+\frac{V_1^2}{2} = \frac{p_2}{\rho}+\frac{V_2^2}{2}$$

$$P_2 = P_1 + \frac{\rho}{2}(V_1^2 - V_2^2)$$

$$P_2 = 270 \times 10^3 + \frac{800}{2} \times [(23.33)^2 - (22.56)^2]$$

$P_2 = 284130\ Pa$
$P_2 = 284\ kPa$

93. The figure shows the idealized view of a return elbow or U bend, which is connected to two pipes by flexible hoses that transmit no force. Water with density 1000kg/m³ flows at velocity of 10 m/s through the pipe, which has a uniform ID of 0.1m. The gauge pressure at points 1 and 2 are 304 kPa and 253 kPa respectively. The horizontal force F required to keep the elbow in position is

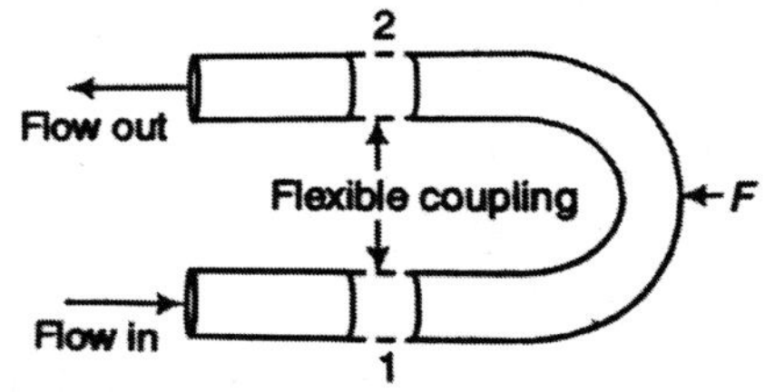

A. 1574 N
B. 1970 N
C. 5942 N
D. 7533 N

***Sol.* B.** we know volumetric flow rate through the pipe,

$$Q = Av = \frac{\pi d^2 V}{4}$$

$$\left\{ \begin{array}{c} given: \\ \rho = 1000\frac{kg}{m^3} \\ v = 10\frac{m}{s}, d = 0.1\ m \\ p_1 = 304 \times 10^3\ Pa \\ p_2 = 253 \times 10^3\ Pa \end{array} \right\}$$

$$Q = \frac{\pi}{4} \times (0.1)^2 \times 10$$
$$= 0.0785\ m^2$$

Taking force balance

Net force = rate of change of momentum

$p_1A_1 - p_2A_2 - F = \rho Q\,[(-V_2) - V_1]$

$F = (p_1 - p_2)A + 2\rho Q\,v$

$[A_1 = A_2 = A]$

And $[V_1 = V_2 = V]$

$F = (304 - 253) \times 10^3 \times \frac{\pi}{4} \times (0.1)^2$
$+ 2 \times 1000 \times 0.0785 \times 10$

$F = 1970\ N$

94. The figure shows a series-parallel configuration of three identical centrifugal pumps. The head increases ΔH across a single such pump varies with flow rate Q according to $\Delta H = a - bQ^2$. The expression for the total head increases $\Delta H = H_2 - H_1$ in terms of a and b the total flow rate Q_1 for this configuration is given by

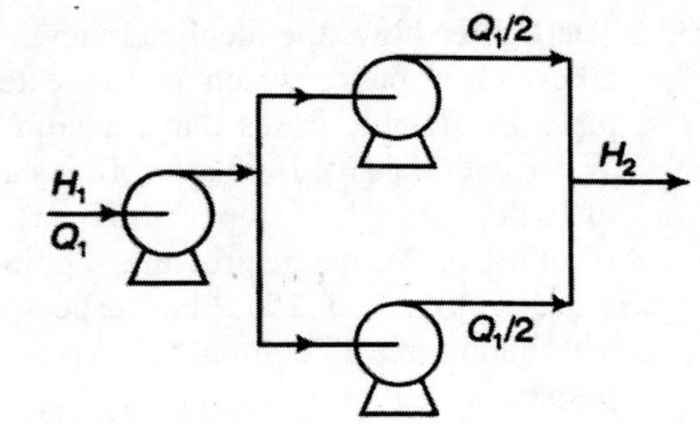

A. $2a - \frac{5}{4}bQ_2^1$
B. $2a - bQ_1^2$
C. $2a - 2bQ_1^2$
D. $a - bQ_1^2$

Sol. **A.** Given, head increase across a single pump is

$\Delta H = a - bQ^2$

For pump 1,

$H_1' - H_1 = a - bQ_1^2$

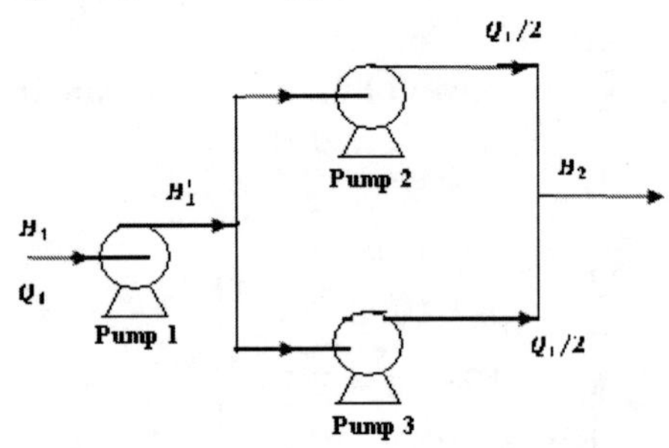

Since, two identical centrifugal pumps are in parallel, the head increase will be equal for two pumps.

$\therefore H_2 - H_1' = a - b\left(\frac{Q_1}{2}\right)^2$

Putting value of H_1' from Eq. (i),

$$H_2 - (H_1 + a - bQ_1^2) = a - \frac{1}{4}bQ_1^2$$

$$H_2 - H_1 = 2a - \frac{5}{4}bQ_1^2$$

95. The pressure differential across a venturimeter, inclined at 45^0 to the vertical (as shown in the figure) is measured with the help of a manometer to estimate the flowrate of a fluid flowing through it. If the density of the flowing fluid is ρ and the density of the manometer fluid is ρ_m, the velocity of the fluid at the throat can be obtain from the expression

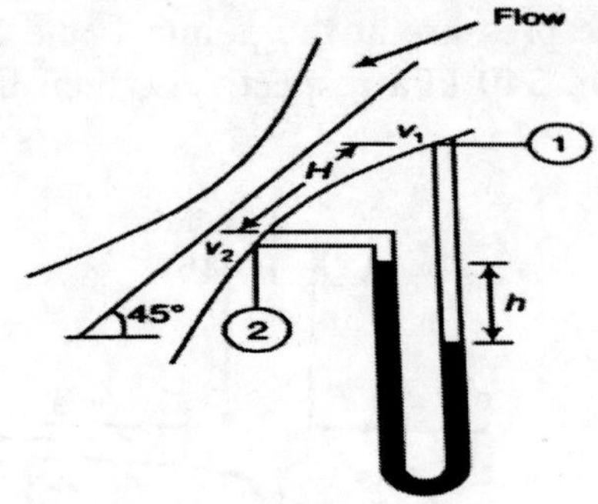

A. $\frac{v_2^2 - v_1^2}{2g} = \frac{(h\rho_m - \rho)}{\rho} + H \sin 45$

B. $\frac{v_2^2 - v_1^2}{2g} = \frac{h\rho_m}{\rho} + H \sin 45$

C. $\frac{v_2^2 - v_1^2}{2g} = \frac{h\rho_m}{\rho}$

D. $\frac{v_2^2 - v_1^2}{2g} = \frac{(h\rho_m - \rho)}{\rho}$

Sol. **D.** According to Bernoulli's equation for points 1 and 2,

$$\frac{p_1}{\rho} + \frac{V_1^2}{2} = \frac{p_2}{\rho} + \frac{V_2^2}{2}$$

$$\frac{V_2^2 - V_1^2}{2} = \frac{(p_1 - p_2)}{\rho} = \frac{\rho_m gh - \rho gh}{\rho} = \frac{h(\rho_m - \rho)}{\rho}$$

96. In the stocks regime, the terminal velocity of particles for centrifugal sedimentation is given by

$u_t = \omega^2 r\,(\rho_p - \rho)\, d_p^2 / 18\,\mu$

Where, ω: angular velocity; r : distance of the particle from the axis of rotation ;

ρ_p : density of the particle; ρ : density of the fluid;

d_p : diameter of the particle and μ : viscosity of the fluid.

In a bowl centrifugal classifier operating at 60 rpm with water (μ = 0.001 kg/m-s), the time taken for a particle (d_p = 0.0001m, specific gravity = 2.5) in second to traverse a distance of 0.05 m from the liquid surface is

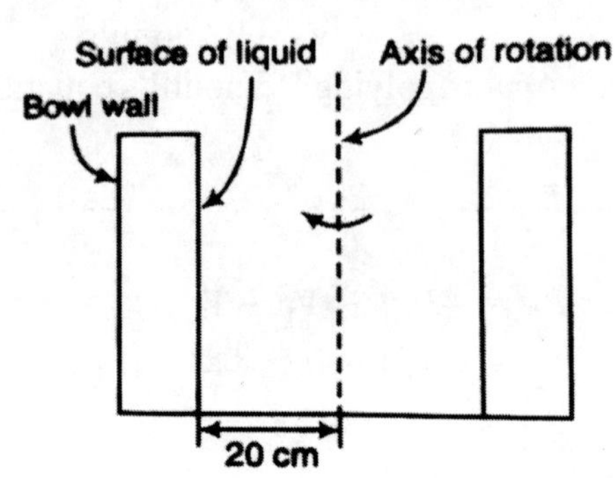

A. 4.8
B. 5.8
C. 6.8
D. 7.8

***Sol.* D.** According to question the terminal velocity of particles for centrifugal sedimentation is given by

$$u_t = \frac{\omega^2 r(\rho_p - \rho)d_p^2}{18\mu}$$

given:

$$\left\{\begin{array}{l} \omega = 60rpm = \frac{2\pi rad}{s} \\ \rho = 1000\ kg/m^3 \\ \mu = \frac{0.001kg}{ms} \\ d_p = 0.0001\ m \\ \rho_p = 2500\frac{kg}{m^3} \\ r = 0.2m \end{array}\right\}$$

Putting given values

$$u_t = \frac{(2\pi)^2 \times 0.2 \times (2500-1000) \times (0.0001)^2}{18 \times 0.001}$$

$= 6.57 \times 10^{-3}$ m/s

Time required to traverse a distance of 0.05 m

$$= \frac{0.05}{6.57 \times 10^{-3}}$$

$= 7.6$ s

Statement of linke Answer Questions 70 and 71

97. A fluidized bed (0.5 m dia, 0.5 m high) of spherical particles (diameter = 2000 μm), specific gravity = 2.5) uses water as the medium. The porosity of the bed is 0.4 .The Ergun equation for the system is

$$\Delta p/L = 4 \times 10^5\, v_{mf} + 1 \times 10^7 v_{mf}^2$$

(SI unit, v_{mf} in m/s)

$\Delta p/L$ (SI unit) at minimum fluidization condition is (g = 9.8 m/s^2).

A. 900
B. 8820
C. 12400
D. 17640

***Sol.* B.** At minimum fluidization condition,

$$\frac{\Delta p}{L} = (1 - e)(\rho_p - \rho)\,g$$

Putting value

e = 0.4, $\rho_p = 2.5 \times 1000 = 2500$ kg/m^3

ρ = 1000 kg/m^3

So, $= \frac{\Delta p}{L} = (1 - 0.4) \times (2500 - 1000) \times 9.8$

$= 8820$

98. The minimum fluidization velocity (mm/s) is

A. 12.8
B. 15.8
C. 24.8
D. 28.8

***Sol.* B.** We known that $\frac{\Delta p}{L} = 4 \times 10^5 V_{mf} + 1 + 10^7 V_{mf}^2$

At minimum fluidization condition,

$8820 = 4 \times 10^5 V_{mf} + 1 + 10^7 V_{mf}^2$

$V_{mf} = 1.58 \times 10^{-2}$ m/s

$= 15.8$ mm/s

99. A tube diameter D and length L is initially filled with a liquid of density ρ and viscosity μ. It is then pushed out by the application of a constant force F to the plunger as shown in the figure. Assuming laminar flow and pseudo steady state, the time required to expel one-half of the liquid out of the tube is

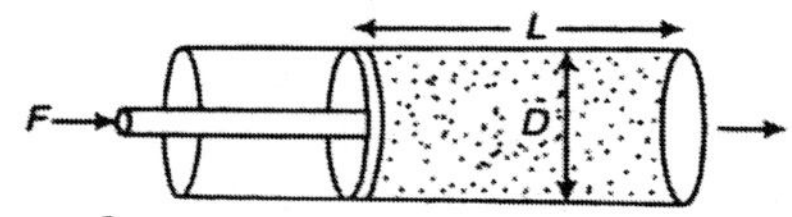

A. $\frac{3\pi\mu L^2}{F}$

B. $\frac{3\pi\mu D^2}{F}$

C. $\frac{6\pi\mu L^2}{F}$

D. $\frac{6\pi\mu DL}{F}$

***Sol.* A.** We need to push liquid from length L to $\frac{L}{2}$.

By Hagen-Poiseuille's equation

$$\Delta p = \frac{32\mu vl}{D^2}$$

$$\frac{F}{A} = \frac{32\mu vl}{D^2}$$

$$\frac{F}{\left(\frac{\pi D^2}{4}\right)} = \frac{32\mu vl}{D^2}$$

$$F = 8\pi\mu L v$$

$$F = 8\pi\mu L\frac{dL}{dt}$$

$$\int_0^t dt = \frac{8\pi\mu}{F}\int_{L/2}^{L} L\,dL$$

$$t = \frac{8\pi\mu}{F}\left[\frac{L^2}{2} - \frac{1}{2} \times \left(\frac{L}{2}\right)^2\right]$$

$$t = \frac{3\pi\mu L}{F} \times \frac{3}{8}L^2$$

$$t = \frac{3\pi\mu L^2}{F}$$

GATE 2006

100. A liquid is pumped at the flow rate Q through a pipe of length L. The pressure drop of the fluid across the pipe is Δp. Now a leak develops at the mid-point of the length of the pipe and the fluid leaks at the rate of Q/2. Assuming that the friction factor in the pipe remains unchanged, the new pressure drop across the pipe for the same inlet flow rate (Q) will be

A. (1/2) Δp
B. (5/8) Δp
C. (3/4) Δp
D. Δp

***Sol.* B.** For Flow through a pipe is $\Delta p = \frac{2fLV^2\rho}{D}$

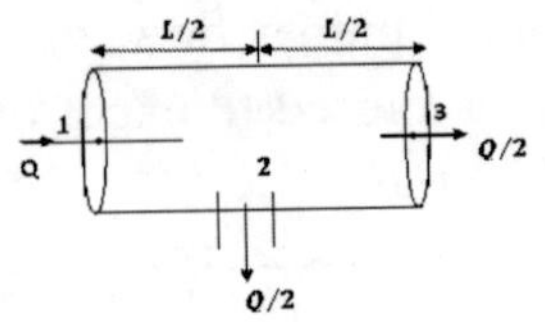

$$\text{or} \quad \Delta p = \frac{2fL\rho}{D}\left(\frac{4Q}{\pi D^2}\right)^2$$

$$Q = Av = \frac{\pi D^2}{4} v$$

$$\text{or} \quad v = \frac{4Q}{\pi D^2}$$

$$\Delta p = KLQ^2 \qquad \text{....(i)}$$

$$K = \frac{32 f \rho}{\pi^2 D^5}$$

Since all other parameters are constant

Now, leak develops at mid length at $L/_2$ and flow rate becomes half $Q/_2$ therefore, pressure drop across the pipe

$\Delta p' =$ pressure drop between points 1 and 2 + Pressure drop between point 2 and 3

$$= \frac{KLQ^2}{2} + \frac{KL}{2}\left(\frac{Q}{2}\right)^2$$

$$\Delta p' = \frac{5}{8} KLQ^2 \qquad \text{....(ii)}$$

Comparing equation (i) and (ii)

$$\Delta p' = \frac{5}{8}\Delta p$$

101. In a laminar flow through a pipe of the radius R, the fraction of the total fluid flowing through a circular cross-section of radius R/2 centred at the pipe axis is

A. 3/8
B. 7/16
C. 1
D. 3/4

***Sol.* B.** According to question laminar flow in pipe,

$$\frac{U}{u_{max}} = 1 - \left(\frac{r}{R}\right)^2$$

Now,

Flow rate through a differential cross section of radius dr,

$$dQ = u.2\pi r dr$$

Integrating this equation

Let Q = the total flow rate

$Q_{1/2}$ flow rate through circular cross section of radius $R/_2$

$$\int_0^{Q_{1/2}} dQ = 2\pi u_{max} \int_0^{R/2} \left[1 - \left(\frac{r}{R}\right)^2\right] r\, dr$$

$$Q_{1/2} = 2\pi u_{max}\left[\frac{\pi^2}{2} - \frac{1}{4R^2}\pi^4\right]_0^{R/2}$$

$$= 2\pi u_{max} \times \frac{7}{64} R^2$$

$$= \frac{7}{32} 2\pi u_{max} R^2$$

Total flow through the pipe $Q_{av} = u_{av} \times \pi R^2$

$$u_{av} = \frac{u_{max}}{2} \text{ (for laminar flow)}$$

Putting value,

$$Q_{av} = \frac{u_{max}}{2}\pi R^2 \frac{\pi}{2} u_{max} R^2$$

So fraction of the fluid through a cross section of radius

$$R/_2 = \frac{Q_{1/2}}{Q_{av}} = \frac{7}{16}$$

102. A bed fluidized by water is used for cleaning sand contaminated with salt. The particles of sand ant salt have the same shape and size but different densities (ρ_{sand} = 2500 kg/m^3 and ρ_{salt} = 2000 kg/m^3). If the initial volume fraction of the salt in the mixture is 0.3 and if the initial value of the minimum fluidization velocity (V_{mf}) is 0.9 m/s, find the final value of

the V_{mf} in (m/s) when the sand is washed free of the salt. Assume that the bed characteristics (bed porosity and solid surface area per unit volume) do not change during the operation and that the pressure drop per unit length is directly proportional to the fluid velocity

A. 0.70
B. 0.90
C. 1.00
D. 1.46

***Sol.* C.** Taking force balance

$\Delta p \times A$ = Weight of the bed – Buoyancy force

$\Delta p \times A = LA(1-e)\,[\rho_{sand} \times 0.7 + \rho_{salt} \times 0.3]\, g$
$LA\,(1-e)\,\rho_{water}\, g$

Putting given value,

$\rho_{sand} = 2500$ kg/m^3
$\rho_{salt} = 2000$ kg/m^3

And initially, sand : salt = 0.7 : 0.3

$\therefore \frac{\Delta p}{L} = (1-e)\,g$
$[2500 \times 0.7 + 2000 \times 0.3 - 1000]\, g$

$\frac{\Delta p}{L} = 1350\,(1-e)\,g$ ....(i)

At final condition, all salt is washed away.

So, $\frac{\Delta p}{L} = (1-e)\,g \;\; [\rho_{sand} - \rho_{water}]\, g$

$\frac{\Delta p'}{L} = 1500\,(1-e)\,g$ .... (ii)

Comparing equation (i) and (ii) we get

$\frac{\Delta p'}{\Delta L} = 1.11$

$\frac{\Delta p}{L} \propto V$

$\therefore \frac{V'_{mf}}{V_{mf}} = 1.11$

Hence,

$V_{mf} = 0.9$ m/s

$\therefore \quad V'_{mf} = 1$ m/s

103. Two spherical particles have the same outer diameter but are made of different materials. The first one (with material density ρ_1) is solid, whereas the second (with material density ρ^2) is a hollow sphere with the inner shell diameter equal to half the outer diameter. If both the sphere have the same terminal velocity in any fluid, then the ratio of the their material density ρ_2/ρ_1 is

A. 1
B. 8/7
C. 2
D. 8

***Sol.* B.** For solid sphere,

$m_1 = \rho_1 V_1$

$m_1 = \rho_1 \times \frac{\pi D^3}{6}$ ....(i)

For hollow sphere

$m_2 = \rho_2 \times \frac{\pi}{6}\left[D^3 - \left(\frac{D}{2}\right)^3\right]$

$m_2 = \rho_1 \times \frac{\pi D^3}{6} \times \frac{7}{8}$ ....(ii)

Given, terminal velocity of two spheres are same

$u_1 = u_2$ ....(iii)

We have,

$u = \sqrt{\frac{4d_p(\rho_p - \rho)g}{3C_D\rho}}$

But, $d_p = \frac{6m}{A\rho_p}$

$u = \sqrt{\frac{8g}{AC_D\rho}\left(1 - \frac{\rho}{\rho_p}\right)m}$

since, $\frac{\rho}{\rho_p} << 1$ and A is same for two spheres.

$\therefore u \propto \sqrt{m}$

From eq. (iii),

$m_1 = m_2$

Putting values from eqs. (i) and (ii),

$\rho_1 \times \frac{\pi D^3}{6} = \rho_2 \times \frac{\pi D^3}{6} \times \frac{7}{8}$

$\frac{\rho_2}{\rho_1} = \frac{8}{7}$

104. A filtration is conducted at constant pressure to recover solids from dilute slurry. To reduce the time of filtration, the solid concentration in the feed slurry in increased by evaporating half the solvent. If the resistance of the filter medium is negligible, the filtration time will be reduce by a factor of

A. 1
B. 2
C. 4
D. 8

***Sol.* C.** Relationship for constant filtration t $\propto V^2$

Or $\frac{t_1}{t_2} = \left(\frac{V_2}{V_1}\right)^2 = \left(\frac{1}{2}\right)^2$

$\therefore \quad t_2 = \frac{t_1}{4}$

Statement for Linked Answer Questions 78 and 79

A continuous grinder obeying the Bond crushing law grinds a solid at the rate of 1000 kg/h from the initial diameter of 10 mm to the final diameter of 1 mm.

105. If the market now demands particle of size 0.5 mm, the output rate of the grinder (in kg/h) for the same power input would be reduced to

A. 227
B. 474
C. 623
D. 856

Sol. **C.** According to question, energy required to crush the particles from size L_1 to L_2 is

$$E = E_i\sqrt{\frac{100}{L_2}\left(1-\frac{\sqrt{1}}{\sqrt{q}}\right)}, q = \frac{L_1}{L_2}$$

Or $E = 10E_i\left(\frac{1}{\sqrt{L_2}}-\frac{1}{\sqrt{L_1}}\right) = 10E_i\left(\frac{1}{\sqrt{1}}-\frac{1}{\sqrt{10}}\right)$

$= L_1 = 10$ m

$= 10E_i \times 0.684$

And

$$E_2 = 10E_i \times \left(\frac{1}{\sqrt{0.5}}-\frac{1}{\sqrt{10}}\right) = 10E_i \times 0.684$$

Since, $\frac{E_1}{E_2}$ =0.623

M = feed rate and P = power required

$E = \frac{p}{m}$

Then,

$$\frac{(p_1/m_1)}{(p_2/m_2)} = 0.623$$

Putting value, $p_1 = p_2$ and $m_1 = 1000$kg/h

$\therefore\ m_2 = m_1 \times 0.623$

= 623 kg/h

106. In order to restore the output back to 1000 kg/h, an additional grinder was installed. The two grinders can be operated either in series (configuration-1) or parallel (configuration-2).
Compare the two configuration in terms of the additional power consumption over the case above.

A. Configuration-1 consumes less power than configuration-2
B. Configuration-2 consumes less power than configuration-1
C. Both configuration consume the same power
D. Configuration-2 consume less or more power then configuration-1 depending on how the feed in distributed between the two grinders in configuration-2 (the parallel configuration)

Sol. **A.** Two grinders in series (configuration – 1) will consume less power compared to two grinders in parallel (configuration – 2).

107. The liquid surface in a cylindrical bucket of radius R rotating about is axis acquires a parabolic profile given by the equation $y = a + br^2$, Where y is height of the liquid surface from the bottom of the bucket at a radial distance r from the bucket axis. If the liquid has density ρ, then the mass of the liquid in the bucket is

A. $\pi\rho R^2\left(\frac{a+bR^2}{2}\right)$
B. $\pi\rho R^2\left(a+\frac{bR^2}{2}\right)$
C. $\pi\rho R^2 a$
D. $\pi\rho R^2(a+bR^2)$

Sol. **B.** The differential volume of strip is

$dV = 2\pi r dr.y$

Mass of liquid in this differential volume

$dm = \rho$ dv

$dm = \rho. 2\pi r dr. y$

Integrating this equation

$$\int_0^m dm = 2\pi\rho\int_0^R r.(a+br^2)dr$$

$$(\because y = a + br^2)$$

$$m = 2\pi\rho\left[a\frac{R^2}{2}+b\frac{R^4}{4}\right]$$

$$m = \pi\rho R^2\left(a+\frac{bR^2}{2}\right)$$

GATE 2005

108. A dam of width 50 m is used to hold water in a reservoir. If the water height is 10 m from the bottom of the dam, what is the total force F acting on the dam due to the water? Assume $g = 10$ m/s^2, and the fluid density is 1000 kg/m^3.

A. $F = 12.5 \times 10^6\ N$
B. $F = 25 \times 10^6\ N$
C. $F = 50 \times 10^6\ N$
D. $F = 5 \times 10^6\ N$

Sol. **C.** Area of dam = 50 × 10 = 500 m^2

Total force of the dam

$$F = \rho gh.\ A = 1000 \times 10 \times 10 \times 500 = 50 \times 10^6\ N$$

109. The relation between the stress τ and the strain rate (du_x/d_y) for the rapid flow of a granular material is

given by $\tau = B\left(\frac{du_x}{dy}\right)^2$, where B is constant. If M, L and T are the mass, length and time dimension respectively, what is the dimension of the constant B?

A. $[ML^{-1}T^{-1}]$
B. $[ML^{-1}T^{-2}]$
C. $[MT^{-1}]$
D. $[ML^{-1}]$

Sol. **D.** According to question $\tau = B\left(\frac{du_x}{dy}\right)^2$

or $\frac{F}{A} = B\left(\frac{du_x}{dy}\right)^2$

Substituting dimensions for each unit.

$$\frac{[MLT^{-2}]}{[L^2]} = B \times \left(\frac{[LT^{-1}]}{[L]}\right)^2$$

$$[ML^{-1}T^{-2}] = B \times [T^{-2}]$$

$$B = [ML^{-1}]$$

Statement for Linked Answer and Questions 83 and 84

Two tanks A and B of cross-sectional area 1 m^2 each, contain a fluid of density 1000 kg/m^3 and viscosity 1kg/(ms.). The tanks are connected by a pipe of diameter 0.02 m and length 1m, and a check valve at the bottom. Assume that the flow is a laminar and there is no friction in the check valve. In the initial state, the height of the fluid in the tank is A is 6 m and the height of the fluid in tank B is 2 m (as shown in the figure below). The check valve in the opened and the fluid flows from tank A to tank B till the levels in the two tanks are equal in the final state. Assume g = 10 m/s^2 in the calculations.

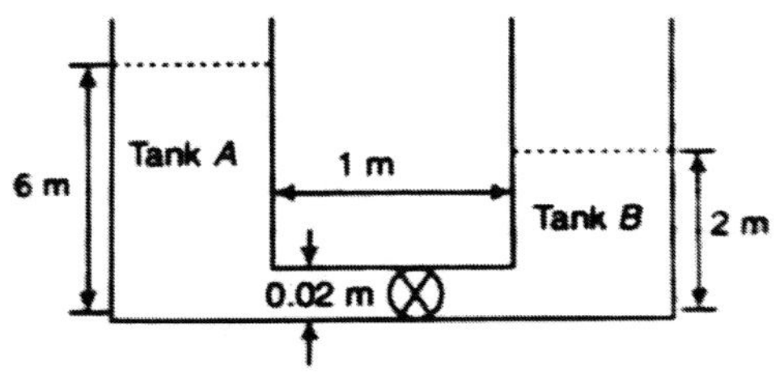

110. What is the average fluid velocity in the pipe as soon as the valve is opened?

A. 0.25 m/s
B. 0.5 m/s
C. 1 m/s
D. 2 m/s

Sol. **B.** According to Hagen – Poiseuille equation,

$$\Delta p = \frac{32\ L\mu v}{D^2} \quad \text{or} \quad v = \frac{\Delta p\ .\ D^2}{32\ L\mu v}$$

here, $\Delta p = \rho g(z_1 - z_2) = 1000 \times 10 \times (6 - 2)$

$\Delta p = 4 \times 10^4$pa, D = 0.02m

L = 1m, μ = 1kg / ms

Putting value we get

$$v = \frac{4 \times 10^4 \times (0.02)^2}{32 \times 1 \times 1} = 0.5\ ms$$

111. What is the total energy loss between in the initial and final states due to the fluid flow?

A. 2 × 10^4 J
B. 16 × 10^4 J
C. 8 × 10^4 J
D. 4 × 10^4 J

Sol. **D.** Initial level in tank A, z_1 = 6 m

Final level in tank A, $z_2 = \frac{6+2}{2} = 4$ m

Thus mechanical energy loss

$$E = \frac{\Delta p}{\rho} = \frac{\rho g\,(z_1 - z_2)}{\rho}$$

$= 1 \times (6 - 4) \times 1000 = 2000$ kg

So total energy loss $= 20 \times 2000 = 4 \times 10^4$ J

112. A centrifuge of diameter 0.2 m in a pilot plant rotates at a speed of 50 Hz in order to achieve effective separation. If the centrifuge is scaled up to a diameter of 1 m in the chemical plant and the same separation factor is to be achieved, what is the rotational speed of the scaled up centrifuge?

A. 15 Hz
B. 22.36 Hz
C. 30 Hz
D. 44.72Hz

Sol. **B.** Relationship for separation factor as

$$S = \frac{r\omega^2}{g}$$

Given that $s_1 = s_2$

Or $r_1\omega_1^2 = r_2\omega_2^2$

$$\omega_2^2 = \frac{0.2}{1} \times (50)^2$$

$$\omega = 22.36 \text{ Hz}$$

113. What is the terminal velocity in m/s , calculated from stokes law for a particle of diameter 0.1×10^{-3} m, density 2800 kg/m³ settling in water of density 1000 kg/m³ and viscosity 10^{-3} kg/ms? (Assume g = 10m/s²)

A. 2×10^{-2}

B. 4×10^{-3}

C. 10^{-2}

D. 8×10^{-3}

***Sol.* C.** Stockes law for terminal velocity can be written as

$$u_p = \frac{d_p^2(\rho_p - \rho)g}{18\mu}$$

Putting value,

$d_p = 0.1\times 10^{-3}$ m, $\rho_p = 2800$ kg/m^3,

$\rho = 1000$ kg/m^3, $\mu = 10^3$ klg/ms, g = 10 m/s^2

We get

$$u_p = \frac{(0.1\times10^{-3})^2(2800-1000)\times10}{18\times10^{-3}} = 10^{-2} \text{ m/s}$$

GATE 2004

114. Viscosity of water at 40°C in the range of

A. $1 \times 10^{-3} - 2 \times 10^{-3}$ kg/m-s

B. $0.5 \times 10^{-3} - 1 \times 10^{-3}$kg/m-s

C. 1 – 2 kg/m-s

D. 0.5 – 1 kg/m-s

***Sol.* B.** Viscosity of water of room temperature is 1 CP

$= 10^{-2}$ poise $= 10^{-2}\frac{g}{cm-s} = 10^{-3}$ kg/m – s

Increase in temperature, viscosity will also decrease

$0.5\times 10^{-3} - 1 \times 10^{-3}$ kg / m – s

115. For the manometer setup shown in the figure, the pressure difference $P_A - P_B$ is given by

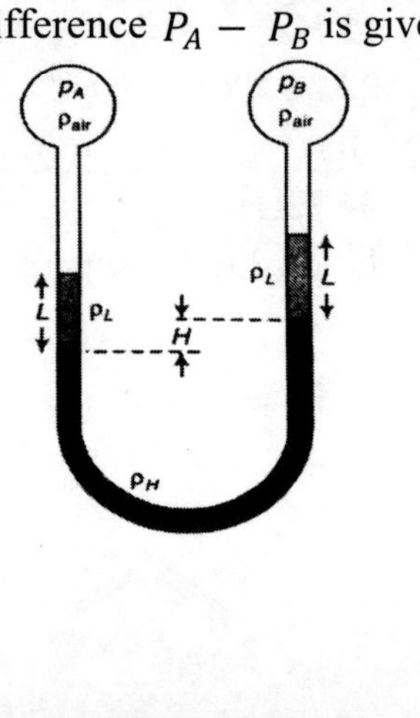

A. $(\rho_H - \rho_{air})gH$

B. $(\rho_H - \rho_L)gH$

C. $(\rho_H - \rho_L)gH + (\rho_L - \rho_{air})g(L - H)$

D. $(\rho_H - \rho_L)gL + (\rho_L - \rho_{air})gH$

***Sol.* A.** consider PQ as datum line,

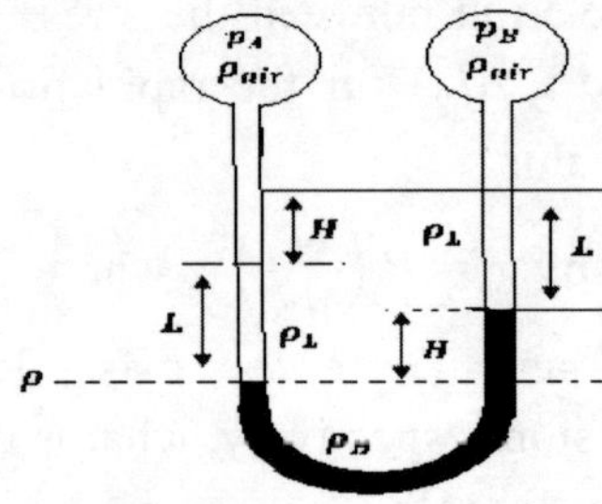

Then we have

$$p_A + \rho_L \times gL + \rho_{air} \times gH = p_B + \rho_H \times gH + \rho_L \times gL$$

$$p_A - p_B = (\rho_H - \rho_{air})\,gH$$

116. A conical tank with a bottom opening of cross-sectional area A is filled with water and is mounted on supports as shown in the figure What is the force F with which plate X must be pushed up to prevent water from leaking? Assuming that the density of air is negligible as compared to the density water ρ_L.

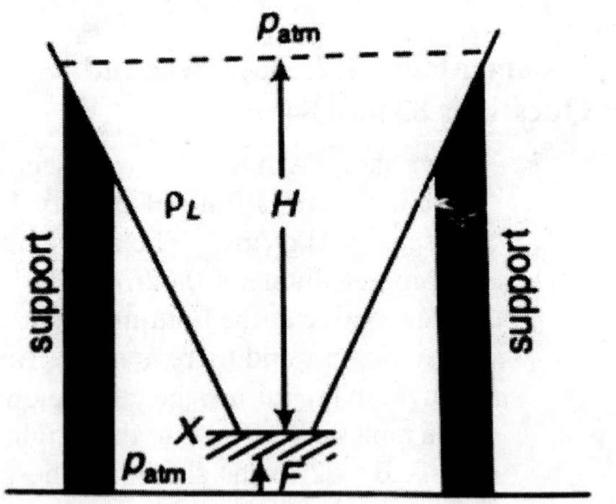

A. ρ_LVg

B. ρ_{LL}AHg

C. ρ_LVg/2

D. ρ_LVg/3

***Sol.* B.** force required to push the plate

$$F = \rho_L \text{ gH . A}$$

117. Three piping networks as shown in the figure are placed horizontally. They are made using identical pipe segments and are subjected to the same

pressure drop across them. Assuming no pressure losses at junctions, the flow rates across the three network are related as

$Q_1 : Q_2 : Q_3$ ::

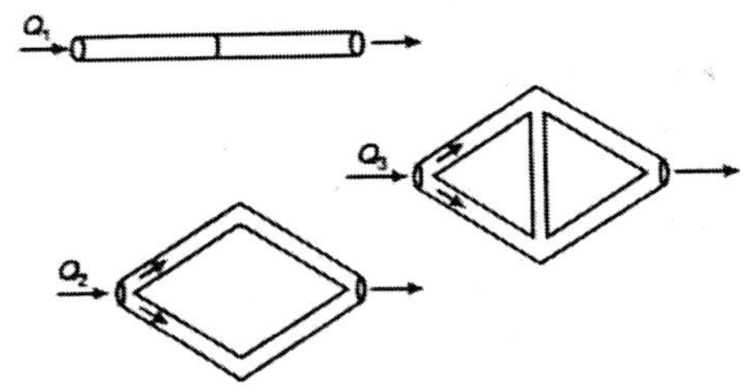

A. $1 : \sqrt{3} : 2$

B. 1 : 2 : 3

C. 1 : 2 : 2

D. $1 : \sqrt{2} : \sqrt{2}$

Sol. **C.** We know that volumetric flow rate is

$$Q = Av = \frac{\pi D^2}{4} v$$

Hagen – Poiseuille's equation can be written as ______

$$v = \frac{\Delta p D^2 . V}{32 \mu L}$$

Substituting values

$$Q = \frac{\pi D^2}{4} \cdot \frac{\Delta p D^2 . V}{32 \mu L}$$

Thus, $Q \propto \frac{\Delta p}{L}$ [D, μ are constant]

Pressure drop is constant across three networks but flow rate in 2nd and 3rd network will be $\frac{Q2}{Q}$ and $\frac{Q3}{Q}$ respectively.

So, $Q_1 : Q_2 : Q_3 = 1 : 2 : 2$

118. To keep the power input constant for a stirred vessel operating under fully developed turbulent flow condition (constant power number), if the impeller diameter is increases by 20% the impeller speed should be decreased by a factor of

A. $(1.2)^{3/2}$

B. $(1.2)^{3/5}$

C. $(1.2)^{\frac{2}{3}}$

D. $(1.2)^{5/3}$

Sol. **B.** by definition power number $N_p = \frac{p}{N^3 D^5 \rho}$

$$\therefore \quad N_1^3 D_1^5 = N_2^3 D_2^5$$

$$[N_p,\ p \text{ and } \rho \text{ are constants}]$$

$$\frac{N_2}{N_1} = \left(\frac{D_1}{D_2}\right)^{5/3} \qquad N_2 = \frac{N_1}{(1.2)^{5/3}}$$

$$[D_2 = 1.2 D_1]$$

119. A centrifugal filtration unit operating at a rotational speed of ω has inner surface of the liquid (density ρ_L) located at a radial distance R from the axis of rotation. The thickness of the liquid film is δ and no cake is formed. The initial pressure drop during filtration is

A. $\left(\frac{1}{2}\right) \omega^2 R^2 \rho_L$

B. $\left(\frac{1}{2}\right) \omega^2 \delta^2 \rho_L$

C. $\left(\frac{1}{2}\right) \omega^2 \delta \rho_L (2R + \delta)$

D. $\left(\frac{1}{2}\right) \omega^2 R \rho_L (R + 2\delta)$

Sol. **C.** Pressure drop is

$$\Delta p = \frac{1}{2} \rho_L \omega^2 [(R + \delta)^2 - R^2]$$

$$= \frac{1}{2} \rho_L \omega^2 [\delta^2 + 2R\delta]$$

$$= \frac{1}{2} \rho_L \omega^2 \delta (\delta + 2R)$$

120. V_{mf} is the minimum fluidization velocity for a bed of particles. An increases in the superficial gas velocity from $2V_{mf}$ to $2.5V_{mf}$ results in (all velocities are smaller than the entrainment velocity of the particles) no change in

A. drag on particles

B. drag on column walls

C. the bed height

D. the bed voidage

Sol. **C.** The height of the fluidized bed does not change with the increase in superficial gas velocity.

121. The kozney-Carman equation, rewritten in terms of non-dimensional numbers give $\left(\frac{\Delta\rho}{\rho u^2}\right)$ proportional to

A. $\frac{(L/D_p)}{R_e}$

B. $\frac{R_e}{D_p/L}$

C. $\frac{\left(\frac{L}{D_p}\right)}{R_e^2}$

D. $\frac{R_e^2}{D_p/L}$

Sol. **A.** Kozney – Carman equation can be written as

$$\frac{\Delta p}{L} = \frac{150\ \mu v\ (1-e)^2}{\emptyset_s^2 D_p^2 \quad e^3}$$

$$\frac{\Delta p}{\rho v^2} = \frac{150 L (1-e)^2}{\emptyset_s^2 e^3} \cdot \frac{\mu v}{D_p^2} \times \frac{1}{\rho v^2}$$

$$\frac{\Delta p}{\rho u^2} \propto \frac{\mu}{\rho v D_p} \times \frac{L}{D_p}$$

$$\propto \frac{1}{R_e} \times \frac{L}{D_p}$$

$$\propto \frac{(L/D_p)}{R_e}$$

122. The cumulative mass fraction of particles smaller than size d_j for a collection of N_i particles of diameters d_i and massm_i $(i = 1, 2, 3 \ldots\ldots, \infty)$ is given by

A. $\dfrac{\sum_{i=1}^{j} N_i d_i^3}{\sum_{i=1}^{\infty} N_i d_i^3}$

B. $\dfrac{\sum_{i=1}^{j} N_i m_i d_i^3}{\sum_{i=1}^{\infty} N_i m_i d_i^3}$

C. $\dfrac{\sum_{i=1}^{j} N_i m_i d_i^2}{\sum_{i=1}^{\infty} N_i m_i d_i^2}$

D. $\dfrac{\sum_{i=1}^{j} N_i m_i d_i}{\sum_{i=1}^{\infty} N_i m_i d_i}$

Sol. **B.** Required cumulative mass fraction

$$= \frac{\sum_{i=1}^{j} N_i m_i d_i^3}{\sum_{n=1}^{\infty} N_i m_i d_i^3}, \text{ weight fraction}$$

GATE 2003

123. A pipe has a porous section of length L as shown in the figure. Velocity at the start of this section is V_0. If fluid leaks into the pipe through the porous section at a volumetric rate per unit area q $(x/L)^2$, what will be the axial velocity in the pipe at any x? Assume incompressible one dimensional flow, i.e., no gradients in the radial direction.

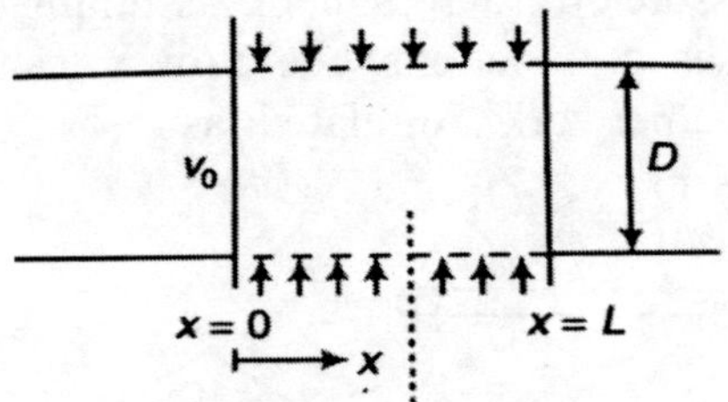

A. $V_x = V_0 + q\dfrac{x^3}{L^2 D}$

B. $V_x = V_0 + \dfrac{1}{3}q\dfrac{x^3}{L^2}$

C. $V_x = V_0 + 2q\dfrac{x^3}{LD}$

D. $V_x = V_0 + \dfrac{4}{3}q\dfrac{x^3}{L^2 D}$

Sol. **D.** Flow rate is $Q_0 = A \,.\, V_0 = \dfrac{\pi D^2}{4} \,.\, V_0$ ....(i)

[V_0 = velocity at the start]

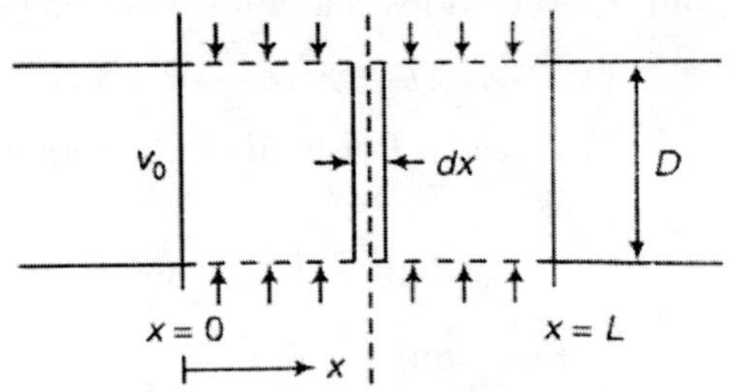

Let the small elemental section of width dx at the distance of x from the start.

Thus, $\dfrac{dQ}{\pi D.dx} = q\left(\dfrac{X}{L}\right)^2$

$$\int_{Q_0}^{Q_x} dQ = \frac{\pi D q}{L^2}\int_0^x x^2 dx$$

Integrating this equation we get

$$Q_x - Q_0 = \frac{\pi D q}{3L^2} \,.\, x^3$$

$$\frac{Q_x}{(\pi D^2/4)} - \frac{Q_0}{(\pi D^2/4)} = \frac{\pi D q}{3L^2} \,.\, x^3 \,.\, \frac{1}{(\pi D^2/4)}$$

[Divided by $\dfrac{\pi D^2}{4}$ both side]

$$V_x - V_0 = \frac{4}{3} q \frac{X^3}{DL^2}$$

124. A centrifugal pump is used to pump water through a horizontal distance of 150 m and then raised to an overhead tank 10 m above. The pipe is smooth with an ID of 50 mm. What head (of water) must the pump generate at its exit (E) to deliver water at a flow rate of 0.001 m³/s? The fanning friction factor f is 0.0062

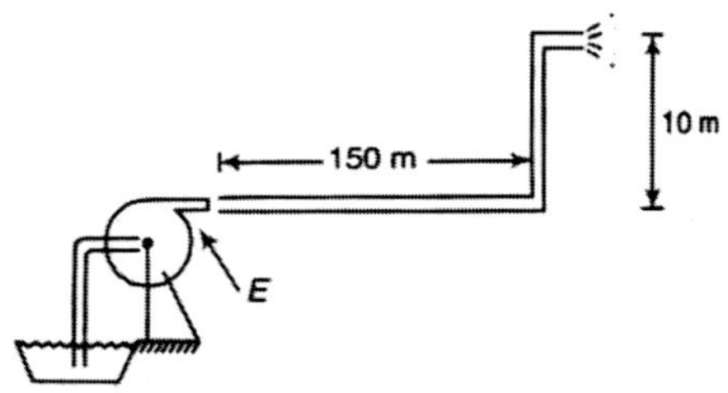

A. 10 m
B. 11 m
C. 12 m
D. 20 m

Sol. **B.** According to mass flow rate Q = Av = $\frac{\pi D^2}{4}$. v

Putting value

L = 150+10 = 160 m

D = 50 mm = 0.05 m

Q = 0.001 m³/s, f = 0.0062

$0.001 = \frac{\pi \times (0.05)^2}{4} \times V$

V = 0.51 m/s

And friction loss can be written as

$$h_f = \frac{fLv^2}{2gD}$$

$$= \frac{4 \times 0.0062 \times 160 \times (0.51)^2}{2 \times 9.8 \times 0.05}$$

= 1.05 m

Now velocity head $= \frac{v^2}{2g} = \frac{(0.51)^2}{2 \times 9.8} = 0.013$

Total head = 1.05 + 0.013 + 10 = 11 m

125. The pressure differential across a vertical venturimeter (shown in figure) is measured with the help of a mercury manometer to estimate flow rate of water flowing through it. The expression for the velocity of water at the throat is

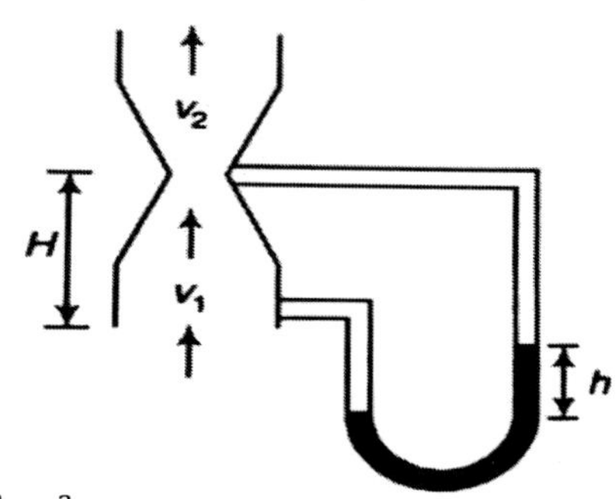

A. $\frac{V_2^2 - V_1^2}{2g} = h\frac{\rho_m}{\rho_w}$

B. $\frac{V_2^2 - V_1^2}{2g} = h\frac{(\rho_m - \rho_w)}{\rho_w}$

C. $\frac{V_2^2}{2g} = H + h\frac{(\rho_m - \rho_w)}{\rho_w}$

D. $\frac{V_2^2 - V_1^2}{2g} = h\frac{(\rho_m - \rho_w)}{\rho_m}$

Sol. **B.** According to Bernoulli's equation,

$$\frac{p_1}{\rho} + gz_1 + \frac{1}{2}V_1^2 = \frac{p_2}{\rho} + gz_2 + \frac{1}{2}V_2^2$$

$$\frac{p_1 - p_2}{\rho g} + (z_1 - z_2) = \frac{1}{2g}(V_2^2 - V_1^2) \quad \text{....(i)}$$

or $\left(\frac{p_1}{\rho} + z_1\right) - \left(\frac{p_2}{\rho} + z_2\right) = h\left(\frac{\rho_m}{\rho_w} - 1\right)$ ....(ii)

Comparing equation (i) and (ii)

$$\frac{V_2^2 - V_1^2}{2g} = \frac{h(\rho_m - \rho_w)}{\rho_w}$$

Statement for Linked Answer and question 99 and 100

A bed of spherical particle (specific gravity 2.5) of uniform size 1500 μm is 0.5 in diameter and 0.5 m high. In the packed bed state, the porosity may be taken as 0.4 Ergun's equation for the above particle –fluid system (in SI unit) is given below

$\Delta\rho/L = 375 \times 10^3 V_{OM} + 10.94 \times 10^6 V_{OM}^2$ (SI units)

126. If water is to be used as the fluidization medium, the minimum fluidization velocity, V_{OM} is

A. 12 mm/s
B. 16 mm/s
C. 24 mm/s
D. 28 mm/s

Sol. **B.** Given that

$$\frac{\Delta p}{L} = 375 \times 10^3 V_{om} + 10.94 \times 10^6 V_{om}^2 \quad \text{....(i)}$$

For fluidization,

$$\frac{\Delta p}{L} = (1 - e)(\rho_p - \rho)g \quad \text{....(ii)}$$

Comparing equation (i) and (ii)

$375 \times 10^3 V_{om} + 10.94 \times 10^6 V_{om}^2$

$= (1 - e)(\rho_p - \rho)g$

Putting value

e = 0.4, ρ_p = 2.5×1000 = 2500 kg/m³

ρ = 1000 kg/m³, g = 9.8 m/s²

We get $10.94 \times 10^6 V_{om}^2 + 375 \times 10^3 V_{om} - 8820 = 0$

Solving equation

$V_{om} = 16 \times 10^3$ m/s

= 16 mm/s

127. In actual operation, the above bed has a height = 1m. What is the porosity of the fluidized bed?

A. 0.2
B. 0.5
C. 0.7
D. 0.8

Sol. **C.** for fluidization $\frac{\Delta p}{L} = (1-e)(\rho_p - \rho)g$

or $L \propto \frac{1}{(1-e)} = \frac{L_1}{L_2} = \frac{1-e_2}{1-e_1}$

Putting value $L_1 = 0.5$ m, $L_2 = 1$ m, $e_1 = 0.4$

We get $\frac{0.5}{1} = \frac{1-e_2}{1-0.4}$

Solving $1 - e_2 = 0.3$

$e_2 = 0.7$

128. The basic filtration equation is given as $\frac{dt}{dv} = \frac{\mu}{A\Delta p}\left(\frac{\alpha\, CV}{A} + R_m\right)$, Where V is volume of the filtrate, A is the filtration area; α is specific cake resistance, μ is the viscosity of the filtrate, and C is the concentration of solids in the feed slurry. In a 20 min constant rate filtration, 5 m^3 of filtrate was obtained. If this is followed by a constant pressure filtration, how much more time in minutes will it take for another 5 m^3 of filtrate to be produced? Neglect filter medium resistance, R_m assume incompressible cake.

A. 10
B. 20
C. 25
D. 30

Sol. **B.** Given, $\frac{dt}{dV} = \frac{\mu}{A\,\Delta p}\left(\frac{\propto CV}{A} + R_m\right)$

For constant filtration rate, t = 20 min, v = 5 m^3
Afterwards for constant filtration, another 5 m^3 volume is collected as filtrate.

For collecting V=5 m^3

Time required =20 min

GATE 2002

129. The drag coefficient for a bacterium moving in water at 1 mm/s will be of the following order of magnitude (assume size of the bacterium to be one micron and kinematic viscosity of water to be $10^{-6} m^2/s$)

A. 24000
B. 24
C. 0.24
D. 0.44

Sol. **B.** Reynols number of the particle can be written as

$$(N_{Re})_p = \frac{\rho_p u_p d_p}{\mu}$$

Putting values,

u_p = 1 mm/s = 10^{-3} m/s

d_p = 1 micron/s = 10^{-6} m/s

μ = 10^{-6} m^2/s

$$(N_{Re})_p = \frac{\rho_p \times 10^{-3} \times 10^{-6}}{10^{-6}}$$

Density of the particle will be of the order of 10^3

or ρ_p = x × 10^3 kg/m^3

$(N_{Re})_p$ = x

Coefficient of drag coefficient is

$$C_d = \frac{24}{(N_{Re})_p}\left[1 + 0.15(N_{Re})_p^{\,0.687}\right]$$

Thus, order of magnitude of drag coefficient = 24

130. A gas (density = 1.5 kg/m^3, viscosity = 2 × 10^{-5}kg/m-s) flowing through a packed bed (particle size = 0.5 cm, porosity = 0.5) at a superficial velocity of 2 m/s causes a pressure drop of 8400 pa/m. The pressure drop for another gas with density of 1.5 kg/m^3 and viscosity of 3 × 10^{-5}kg/ms, following 3 m/s will be

A. 8400 pa/m
B. 18900 pa/m
C. 12600 pa/m
D. 16800 pa/m

Sol. **B.** For gas A, it is given that

ρ = 1.5 kg/m^3, μ = 2 × 10^{-5} kg/ms, d_p = 0.5 cm = 5 × 10^{-3} m

e = 0.5, u = 2 m/s, $\frac{\Delta p}{l}$ = 8400 pa/m

We know that flow through packing is

$$\frac{\Delta p}{l} = 150\frac{(1-e)^2}{e^3}\frac{\mu u}{d^2} + 1.75\frac{(1-e)}{e^3}\frac{\rho u^2}{d} = A + B$$

Now $B = 1.75\frac{(1-e)}{e^3}\frac{\rho u^2}{d}$

Putting value

B $= \frac{1.75\times(1-0.5)\times1.5(2)^2}{(0.5)^3\times5\times10^{-3}}$

= 8400

Or B $= \frac{\Delta p}{L}$

Thus, the satisfying equation for pressure drop in this care is

$$\frac{\Delta p}{L} = 1.75\frac{(1-e)}{e^3}\frac{\rho u^2}{d}$$

Now, for gas B,

ρ = 1.5 kg/m^3, u = 3 m/s

$\therefore \quad \frac{\Delta p}{L} = 1.75 \times \frac{(1-0.5)}{(0.5)^3} \times \frac{1.5\times(3)^2}{5\times10^{-3}}$

= 18900 pa/m

131. What is the critical rotational speed in rev/s. for a ball mill of 1.2 m diameter charged with 70 mm diameter balls?

A. 0.5
B. 1.0
C. 2.76
D. 0.62

***Sol.* A.** Given, d = 1.2 m

d_p = 70 mm = 0.07 m

Critical rotation speed, ω_C = 0.5 rev/s

132. A sand mixture was screened through a standard 10-mesh screen. The mass fraction of the oversize material in feed, overflow and underflow were found to be 0.38, 0.79 and 0.22 respectively. The screen effectiveness based on the oversize is

A. 0.50
B. 0.58
C. 0.68
D. 0.62

***Sol.* B.** Overall balance

A = B + C ....(i)

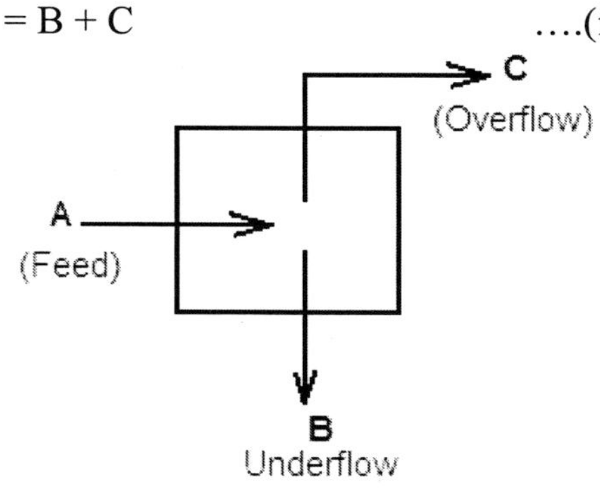

Taking material balance

0.38 A = 0.79 C + 0.22 B

Since A = B + C

0.38 A = 0.79 (A – B) + 0.22 B

41 A = 57 B ....(ii)

C = A – B

$= A - \frac{41}{57}A = \frac{16}{57}A$

So screen effectiveness can be written as

$= \frac{0.79\,C}{0.38\,A} = \frac{0.79}{0.38} \times \frac{16}{57} = 0.58$

GATE 2001

133. A Bingham fluid viscosity μ = 10 Pa-s and yield stress τ_0 = 10 kpa is sheared between flat parallel plates separated by a distance 10^{-3}m. The top plate is moving with a velocity of 1 m/s. The shear stress on the plate is

A. 10 Kpa
B. 20 Kpa
C. 30 Kpa
D. 40 Kpa

***Sol.* B.** Bingham fluid for shear stress can be written as

$$\tau = \tau_0 + \mu\frac{dv}{dy}$$

Putting value

μ = 10 Pa–s

τ_0 = 10 kPa = 10^4 Pa

$\therefore \quad \tau = 10^4 + 10 \times 10^3$

$\left[\frac{dv}{dy} = \frac{1}{10^{-3}} = 10^3\ s\right]$s

$\tau = 2 \times 10^4$ Pa

τ = 20 kPa

134. The energy required per unit mass to grind limestone particle of very large size to 100 μm is 12.7kWh/ton. An estimate (using bond's law) of the energy to grind the particles from a very large size to 50 μm is

A. 6.25 kW-h/ton
B. 9.0 kW-h/ton
C. 18 kW-h/ton
D. 25.4 kW-h/ton

***Sol.* C.** According to Bond's law the energy required to grind particle from an infinite size to particle size D_p can be written as__

$$E = \frac{K}{\sqrt{D_p}}$$

$$\frac{E_1}{E_2} = \sqrt{\frac{(D_p)_2}{(D_p)_1}}$$

Here,

$$E_1 = 12.7 \text{ kW} - \text{h/ton}$$

$$(D_p)_1 = 100\ \mu m,\ (D_p)_2 = 50\ \mu m$$

Putting value

$$\frac{12.7}{E_2} = \sqrt{\frac{50}{100}}$$

$$E_2 = 18 \text{ kW} - \text{h/ton}$$

GATE 2000

135. The sphericity of a solid particles of cubicle shape is

A. Π

B. $\left(\frac{\pi}{6}\right)^{1/3}$

C. $\left(\frac{\pi}{6}\right)^{1/2}$

D. $\frac{\pi}{3}$

***Sol.* B.** According to question sphericity

$$\Psi = \frac{\textit{Surface area of sphere of same volume as that of particle}}{\textit{Surface area of particle}}$$

Now,

$$\frac{4}{3}\pi r^3 = a^3 \text{ (r = radius of sphere)}$$

$$r = \left(\frac{3}{4\pi}\right)^{1/3} a$$

$$\therefore \quad \Psi = \frac{4\pi r^2}{6a^2}$$

$$\Psi = \frac{2\pi}{3} \times \left(\frac{3}{4\pi}\right)^{2/3}$$

$$\Psi = \left(\frac{\pi}{6}\right)^{1/3}$$

136. A 30% (by volume) suspension of spherical sand particles in a viscous oil has a hindered settling velocity of 4.44 μm/s. If the Richardson-Zaki hindered settling index is 4.5, then the terminal velocity of a send grain is

A. 0.9 μm/s

B. 1 μm/s

C. 22.1 μm/s

D. 0.02 μm/s

***Sol.* B.** Terminal velocity of the sand grain = 1 $\mu m/s$

137. A 1 m high bed made up of 2 mm particles is to be fluidized by an oil (density 900 kg/m³, viscosity 0.01 pa-s). If at the point of incipient fluidization. the bed voidage is 39% and the pressure drop across the bed is 10 kpa, then the density of particle is

A. 2571 kg/m³

B. 3514 kg/m³

C. 4000 kg/m³

D. 4350 kg/m³

***Sol.* A.** for incipient fluidization,

$$\frac{\Delta p}{L} = (1 - e)\left(\rho_p - \rho_f\right)g$$

Given,

$$\Delta p = 10 \text{ kPa} = 10^4 \text{ Pa}$$

$$L = 1 \text{ m}$$

$$\epsilon = 0.39,\ \rho_f = 900 \text{kg/m}^3$$

$$g = 9.8 \text{ m/s}^2$$

Putting values,

$$\frac{10^4}{1} = (1 - 0.39) \times \left(\rho_p - 900\right) \times 9.8$$

$$\rho_p = 2572.8 \text{ kg/m}^3$$

138. A free jet of water of cross-section area 0.01 m² and velocity of 20 m/s strikes a plate and then flows in the plane parallel to the plate as shown in the figure below. The horizontal component of the force on the support is

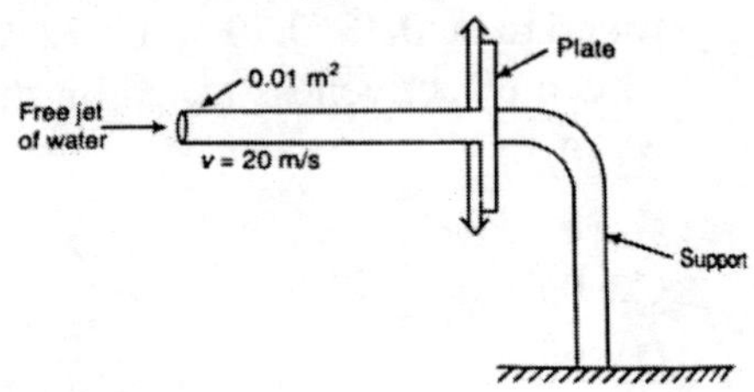

A. 200 N

B. 400 N

C. 2000 N

D. 4000 N

***Sol.* D.** By definition, Horizontal component of force on the support

$$F = \rho \cdot Qv$$

$$F = \rho \cdot Av \cdot V \qquad \left\{\begin{array}{c} given: \\ A = 0.01\ m^2 \\ v = 20\ m/s \end{array}\right\}$$

$$F = 1000 \times 0.01 \times 20 \times 20$$

$$F = 4000 \text{ N}$$

6. HEAT TRANSFER

Basic concepts of Heat Transfer

Heat Transfer is one of the most important unit operations prevalent in chemical process and allied industries.

There are three basis modes of heat transfer in which heat transfer in which heat may be transferred from a hot to a cold body.

– Conduction

– Convection

– Radiation

- Conduction: conduction is a transfer of heat energy at microscopic level by collision of particle within a body due to temperature gradient.
- Fourier's law of heat conduction the rate of heat transfer by conduction is governed by Fourier's law which states that the rate of heat flow through a uniform material is directly proportional to the area of heat transfer and temperature gradient in the direction of heat flow and inversely proportional to the length of path of heat flow.

$$Q \propto A\left[-\frac{dT}{dx}\right]$$

Or $$Q = -KA\frac{dT}{dx}$$

Where,

Q = rate of heat transfer by conduction

K = proportionally constant which is known as thermal conductivity of the material it is the characteristic proportionally of the material

A = cross section area perpendicular to the direction of heat flow

dT = temperature difference between the two faces of the material of thickness dx.

$\frac{dT}{dx}$ = Temperature gradient in the direction of heat flow

The negative sign in the Fourier's law equation has a specific signification in that this indicates the decrease in temperature with increase in x. for a heat flow to occur this gradient must be negative.

- Heat flow through a plane wall: consider a plane wall having uniform thickness of x made of a material having thermal conductivity $K$and cross section area A. The Hot and Cold surface are at T1 and T2 respectively. The system Cs under study state where input and output are equal and there is accumulation.

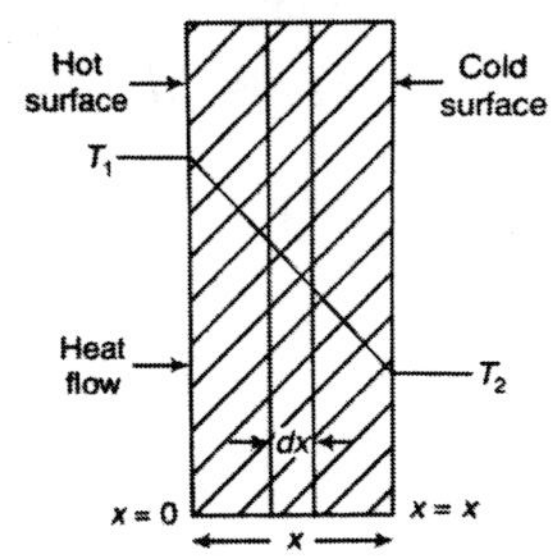

Under steady state condition if Q is the rate of heat transfer then

$$Q = -kA\left[\frac{dT}{dx}\right]$$

$$or \quad Q\int_0^x dx = -KA\int_{T_1}^{T_2} dT$$

$$or \quad Q = \frac{KA(T_1 - T_2)}{x}$$

$$or \quad Q = -\frac{kA}{x}\Delta T = \frac{\Delta T}{x/KA}$$

$$or \quad Q = \frac{\Delta T}{R}$$

Where, R = Thermal resistance = $\frac{x}{KA}$

- Heat flow through composite wall: A Composite wall may consist of several layers of different materials….

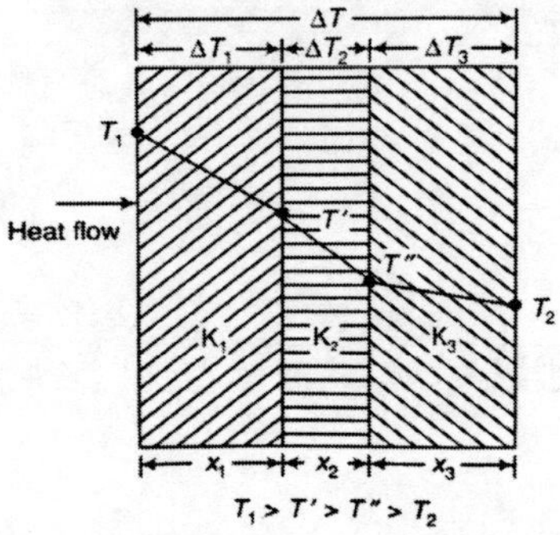

Let three layer of insulation of material having thermal conductivity is given k_1, k_2, k_3 and of thickness x_1, x_2, x_3 respectively have face temperature T_1, T_2, T_3 and T_4.

Applying Fourier's law, the rate of heat flow through the layers under study is state are

$$Q = \frac{K_1 A}{x_1}(T_1 - T_2)$$
$$= \frac{K_2 A}{x_2}(T_2 - T_3)$$
$$= \frac{K_3 A}{x_3}(T_3 - T_4)$$

Understudy state you u Q will be identical through all the layers

Since $\Delta T = \Delta T_1 + \Delta T_2 + \Delta T_3$

$$Q\left[\frac{1}{\frac{K_1 A}{x_1}} + \frac{1}{\frac{K_2 A}{x_2}} + \frac{1}{\frac{K_3 A}{x_3}}\right] = \Delta T$$

Or $Q[R_1 + R_2 + R_3] = \Delta T$

Where, $R_1 = \frac{K_1 A}{x_1}$; $R_2 = \frac{K_2 A}{x_2}$: $R_3 = \frac{K_3 A}{x_3}$

Or $Q = \frac{\Delta T}{\sum R}$ where $\sum R = R_1 + R_2 + R_3$

$$= \frac{Temperature\ driving\ force}{Thermal\ resistance}$$

- Heat flow through cylinder there are numerous application of heat flow through pipe tube and cylinder in process industries let us consider it there was a hollow cylinder having inner and Outer radius R1 and R2 are two respectively the cylinder is of length L and the thermal conductivity of the material of cylinder in k

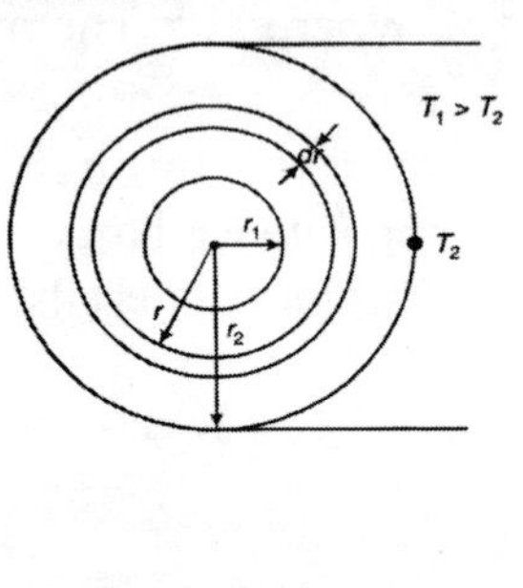

Analyzing and elemental error of dr having radius are concentric with men cylinder by Fourier's law

$$Q = -K(2\pi r L)\left[\frac{dT}{dr}\right]$$

Or $\int_{r_1}^{r_2} \frac{dr}{r} = \frac{-K(2\pi r L)}{Q}\int_{T_1}^{T_2} dT$

Or $ln\left(\frac{r_2}{r_1}\right) = \frac{2\pi LK(T_1 - T_2)}{Q}$

Defining log mean radius $r_m = \frac{r_2 - r_1}{\ln\frac{r_2}{r_1}}$

$$Q = \frac{K(2\pi r_m L)(T_1 - T_2)}{r_2 - r_1}$$

This heat flow equation is analysis to back off the felt wall is area term is calculated by taking log mean radius.

- Heat flow through a sphere : Analysing the case of heat flow through is there let us assume the inner and Outer radius R R1 and R2 be maintained air temperatures T1 and t2 respectively further taking an elemental thickness of dr it a radius R are formed the concentric centre by Fourier's law

Where $r_m = \sqrt{r_1 r_2} = Geometricmeanradius$

- Critical Radius of Insulation: Insulation is provided to reduce the heat lost to the atmosphere which may occur by conduction, convection and radiation. In a pipe as the thickness of insulation is increase to reduce the heat loss, with the increase in thickness the outs surface area through which is lost also increase and beyond a certain point a situation may occur when the net total heat loss may increase more rapidly than the resistance. The point thickness of insulation, Resistance of insulation per unit length of pipe

$$R_c = \frac{\ln {}^{r}/{r_i}}{2\pi K}$$

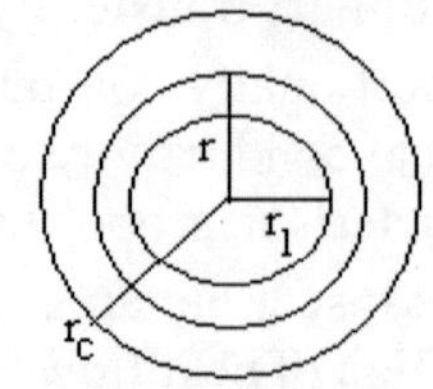

Heat may also be lost from
the outside surface by convection through the air film.
The resistance will be represented by

$$R_a = \frac{1}{(h_a)(2\pi r)} \text{ for unit length}$$

Where, h_a = film coefficient of air

$$\sum R = R_e + R_a = \frac{\ln {}^{r}/{}_{r_i}}{2\pi K} + \frac{1}{(h_a)(2\pi r)}$$

At critical radius, the resistance should be maximum

$$\therefore \frac{dR}{dr} = \frac{1}{2\pi k_b} d \ln\left(\frac{r}{r_i}\right) + \frac{1}{(h_a)(2\pi r)} d\left(\frac{1}{r}\right)$$

$$= \frac{1}{(2\pi k_b)^r} - \frac{1}{(h_a)(2\pi r)r^2}$$

At critical radius $\frac{dR}{dr} = 0, r = r_c$

$$\therefore r_c = \frac{K_b}{h_a}$$

It should be valid that the critical radius should be as small possible so that application of insulation will result in a reduction and hot an increase in the heat loss from pipe.

- Convection: Heat transfer by convection is due to maeroscopie fluid motion caused by difference in density. The effectiveness of heat transfer by convection depends largly upon the extent of mixing of the fluid . The physical properties of the fluid like density, viscosity as well as the thermal properties like thermal conductivity, specific heat etc. control the process of convective heat transfer.

  Convection can be material or free when the temperature difference produces secondary density difference resulting in mass moment there is no external force applied within the system in natural convection. On the other hand when the mass motion of the fluid is caused by an external devics like a pump or blows the process is called the forced convection.

- Dimension less numbers in Heat Transfer: The following dimensionless numbers are among the controlling parameter of heat transfer by convection.

$$Nussult\ Number(Nu) = \frac{Rate\ of\ heat\ transfers\ by\ convection}{Rate\ of\ heat\ transfer\ by\ conduction}$$

$$Nu = \frac{hA\Delta T}{KA\frac{\Delta T}{x}}$$

$$Nu = \frac{hx}{K}$$

For pipe $Nu = \frac{hD}{K}$

Larger the value of Nu, larges will be the rate of heat transfer.

$$\text{Reynolds Number (Ra)} = \frac{Inertial\ force}{Viscous\ force} = \frac{Du\rho}{\mu}$$

Where, ρ and μ are density and viscosity of the fluid flowing in a pipe of diameter D, for circular pipes

- Re < 2300 – laminar flow
- 2300 < Re < 10000 – Transient flow
- Re > 10000 – turbulent flow

However, the above classification may changes depending upon local condition affecting the flow. Reynolds number is the most important group which controls the nature of flow and consequently the rate of convective heat transfer.

- Prandtl number: It provides a measure of effect of momentum and energy transfer by diffusion in velocity and thermal boundary layer respectively.

$$Pr = \frac{Momentum\ diffusivity\ through\ the\ fluid}{Thermal\ diffusivity\ through\ the\ fluid} = \frac{C_p M}{K}$$

  For air and gases Pr may be around for wats Pr may be around 10 to 20 for viscous oil and greases Pr $>10^3$ or 10^4 and for liquid metals Pr may less than 0.01.

- Grash of number: It measure the ratio of heat transferred into a fluid to that of the thermal capacity of fluid

$$St = \frac{Heat\ transfer\ coefficient}{Heat\ flow\ for\ unit\ temperature\ difference} = \frac{Nu}{(Re)(Pr)}$$

- Newton's Law of cooling: According to the Newton's law of cooling the rate of heat transfer during cooling is proportional to the temperature difference between the system and surroundings.

- Convective heat Transfer coefficient:

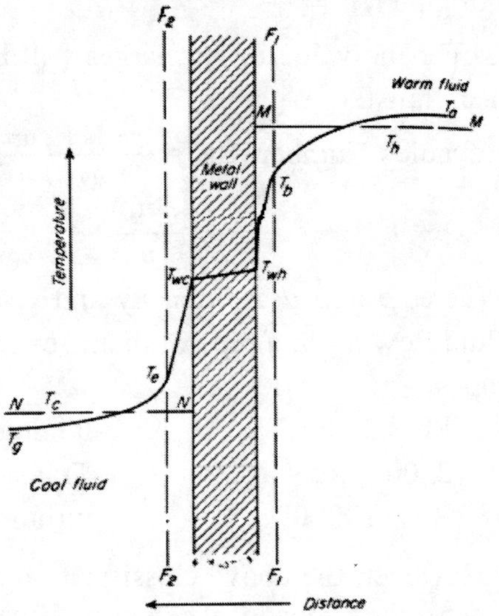

T_a, T_b, T_c -Temperature gradient from bulk fluid to metal wall

T_d, T_e, T_f – Temperature gradient from metal wall to bulk fluid

- Analysing the case of heat transfer by convection hot to cold fluid separated by a metal wall, the temperature gradient has shown to be T_a........ to T_f.

It is customary to and convenient to use as the temperature of the fluid neither the maximum temperature t_a nor t_b but rather the average temperature of the fluid such as around be found by completely mixing the fluid and taking its temperature (T_1).

According to the two film theory a fluid film is postulated to be present adhering to the pipe surface and ferias the fluid film is assumed to contain the entire resistance to heat flow or either side.

On this postulation a film coefficient h is defined as

$$h_i = \frac{Q}{A_1(T_1 - T_c)}$$

h_i is analogous to $\left(\frac{k}{x}\right)$ of conductive resistance

$$= \frac{1}{h_1 A_1}$$

Similarly for the other side $h_2 = \frac{Q}{A_2(T_d - T_2)}$

Resistance of the metal wall $= \frac{x}{KA_m}$

Where, A_1, A_2 and A_m are heat transfer area pertaining to either side of metal wall and the area respectively.

$$\therefore Q = \frac{\Delta T}{\Sigma R} = \frac{\Delta T}{(1/h_1A_1) + \frac{x}{KA_m} + \frac{1}{h_2A_2}}$$

or Multiplying by A_1 on both numerator and denominator

$$Q = \frac{A_1 \Delta T}{(1/h_1) + \frac{x}{K}\left(\frac{A_1}{A_m}\right) + \frac{1}{h_2}\left(\frac{A_1}{A_2}\right)}$$

Defining

$U_1 =$

Overall heat transfer coefficient based or

$$= \frac{1}{(1/h_1) + \frac{x}{K}\left(\frac{A_1}{A_m}\right) + \frac{1}{h_2}\left(\frac{A_1}{A_2}\right)}$$

$$\therefore Q = U_1 A_1 \Delta T$$

Similarly$Q = U_2 A_2 \Delta T$

Where U_2 is overall heat transfer coefficient based on outside surface area

For tubular surface suiec heat transfer area is equal to πDL, the area ratios $\frac{A_1}{A_2} = \frac{D_1}{D_2}$

Hence,

$$U_1 = \frac{1}{(1/h_1) + \frac{x}{K}\left(\frac{A_1}{A_m}\right) + \frac{1}{h_2}\left(\frac{D_1}{D_2}\right)}$$

Concept of controlling resistance: For this walled cylinder of small thickness $D_1 \approx D_2$ andD_1/D_2, D_1/D_metc. will be nearly Unity. Further if $h_2 \gg h_1$ then h_1 will be small in comparison toh_2 and $\left(\frac{1}{h_1}\right)$ or $h_1$will be controlling under these circumstance

$$U_1 \approx h_1$$

- Empirical Equations for that Transfer coefficient: dimensional analysis method for developing a relationship between a heat transfer coefficient and the process variable affecting this phenomena dimensional analysis approach is a adopted. Thus, for heat flow to a fluid in turbulent flow in a long straight pipe the dimensional analysis approach given

$$\frac{hD}{K} = \emptyset\left[\left(\frac{Du\rho}{\mu}\right), \left(\frac{C_p\mu}{K}\right)\right]$$

The product $(u\rho)$ is known as mass velocity G

$$\therefore \qquad \frac{hD}{K} = \emptyset\left[\left(\frac{DG}{\mu}\right), \left(\frac{C_p\mu}{K}\right)\right]$$

The exact nature of this function has been is evaluated by testing it on very large number of conditions. The most common equation is known as Dittus - Buells equation.

$$\frac{hD}{K} = 0.023\left(\frac{DG}{\mu}\right)^{0.8}\left(\frac{C_p\mu}{K}\right)^{0.4}$$

It has been cheeked experimentally onair and other gases and on wats, hydrocarbon, oils and various organic liquids. It is applicable over a wide range of Reynolds number from 10,000 to 50,000 and Prandtl numbers from 0.73 to 95. The modified form of this equation known as Sieder – Tate equation is

$$\frac{hD}{K} = 0.023\left(\frac{DG}{\mu}\right)^{0.8}\left(\frac{C_p\mu}{K}\right)^{0.4}\left(\frac{\mu}{\mu_w}\right)^{0.14}$$

- Heat transfer from condensing vapors: The process of condensation of a pure substance is isothermal. However the condensation of mixed vapors at constant pressure mayoccur there a temperature range. As the vapor condenser latent heat is liberated and there is flow of heat to the surface depending upon the condition of the cold surfers condensation may occur in either of the two ways film condensation and drop wise condensation Nusselt has derived the equation for the rate of heat transfer in film type condensation thus, for a vertical tube the average heat transfer coefficient for a condensate film in laminar region is 4/3 times the local coefficient at the bottom of the tube.

$$h = 0.943\left[\frac{K_f^3\rho_f^2 g\lambda}{\Delta T_0 L\mu_f}\right]^{1/4}$$

It is assumed that that temperature gradient is constant across the film and that $1/\mu$ varies linearly with temperature. The reference temperature for evaluatingμ_f, K_f and ρ_f (at film temperature) is given by the equation

$$T_f = T_h - \frac{3(T_h - T_w)}{4}$$
$$= T_h - \frac{34T_0}{4}$$

Where, T_f is reference temperature

T_h is temperature of condensing vapor

And $T_w$is temperature of outside surface of tube wall

For horizontal tubes the corresponding is

$$h = 0.725\left[\frac{K_f^3\rho_f^2 g\lambda}{\Delta T_0 D_0\mu_f}\right]^{1/4}$$

For a stack of tubes placed horizontally if N is number of tube in stack

$$h_N = 0.725\left[\frac{K_f^3\rho_f^2 g\lambda}{N\Delta T_0 D_0\mu_f}\right]^{1/4}$$

Where h_N is average coefficient for entire stack.

- Heat transfer to boiling liquids: heat transfer to boiling liquids find wide applications in evaporation distillation steam generation and petroleum refining this phenomenon is known as a pool boiling of saturated liquids. Consider horizontal element immersed in a vessel containing a boiling liquid. The graph between heat flux various temperature drop can be represented as

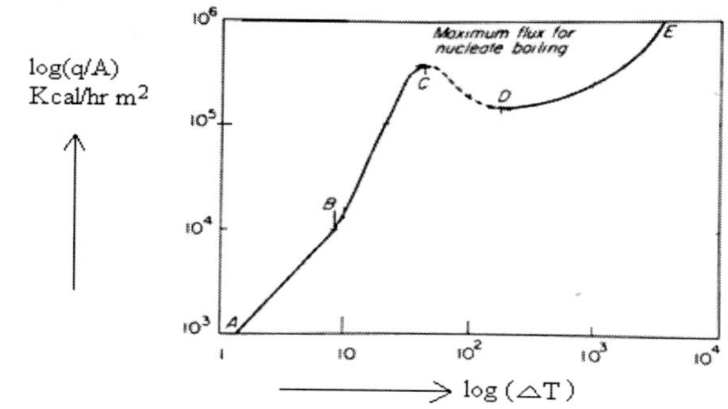

Defining, $\frac{q}{A}$ = the heat flux

ΔT = temperature difference between the temperature of elements that the boiling liquid

Four distinct zone of this plot are evident viz, AB, the natural convection: BC, the nucleate boiling, CD, the transition boiling and DE, film boiling. The first zone is at low temperature with straight slope of about 1.25, the second zone is also straight but with greater slope between 3 and 4 indicating higher heat transfer rate. The end of this zone is at a definite point of maximum flux with this temperature drop known as critical temperature drop corresponding to'pean flux'. In the third zone, the heat flux decreases but the temperature drop increases and reach a minimum point D which is known "Leiden frost point" finally in the last zone DE, the heat flux rises again with temperature drop surpassing the previous maximum at C. Each of these four zones represents a definite mechanism of boiling. Whereas in the first zone the mechanism in that of heat transfer is natural convection, at a great temperature drop, in zone BC, the rate of bubble generation is large enough to increase the velocity of circulation thereby increasing coefficient below the critical point at nucleate boiling tiny bubbles occupy small portion of heating surface. However, still most of the surface is in direct contact with liquid. Eventually, bubble generation become so rapid that they tend to coalesce on the heating surface to form a lays of insulating vapor. This is transition boiling zone in which increasing the temperature drop increases the thickness of the vapor film. As a result the heat flux and the heat transfer coefficient both fall in this zone. Around the Leiden frost point, the hot surfers is convert with a quiescent film of vapor through which heat transfer is only by conduction. As the temperature drop increases further the bubbles detach there selves from the interface and rice through the liquid. During this zone, virtually all the resistance to heat transfer is offered by the vapor film covering the heating element. As the temperature drop increases still further, the heat flux rises first slowly and then more rapidly us radiation heat transfer becomes prominent. This boiling action is known a film boiling film boiling is not desired in not commercial liquipmenter.

- Radiation Heat Transfer: It is defined as the transfer of energy by means of electromagnetic where caused by temperature difference. Unlike conduction and convection it does not require any medium. The most common example of radiation heat transfer is the solar energy. other examples include furnaces combustion chamber etc. further, whereas in conduction and convection rate of heat transfer is directly proportional to the temperature difference, the radiant heat energy is proportional to the temperature difference to the fourth pows.
- Stefan Boltzmann law: This state that the total energy emitted by a black body is proportional to the fourth pows of the absolute temperature.

$$E = \sigma T^4$$

Where, E= energy radiated unit time per unit area $\frac{N}{m^2}$

σ = Stefen – Boltzmann constant

$= 5.67 \times 10^{-8}\ w/m^2K^4$

T = absolute temperature

- Black body:A body which absorbs all radiation incidents upon it without reflecting any part of it is known as black body.
- White body: If all the radiation falling on a body are reflected it is known as white body. Gaseous hydrogen, oxygen and nitrogen are examples of white body.
- Grey body: A grey body is defined as one whose absorptivity does not vary with temperature and wavelength of radiation.
- Emissivity ability of surface of a body to radiate heat is called Emissivity. It is also defined as the ratio of the emissive pows of anybody to the emissive pows of a black body at equal temperature.
- When radiation strikera material surface, part of the radiation is reflected part is absorbed and part is transmitted for a material surface reflectivity (ρ), absorptivity (α) and transmit-vity (τ) are defined as the fraction of the radiation incident reflected, absorbed and transmitted respectively.

  Thus, $T + \alpha + \tau \approx 1$
- Kirchhoff's law: It states that the ratio of the emissive pows of a body to the emissive pows of a black body at the same temperature is equal to the absorptivity of the body. Therefore emissivity and aborptivity of a body are equal.
- Specular and Diffused Reflection: There are two types of reflection phenomena. When the radiation strikera material surface, the reflection is called specular ifthe angle of incidence is equal to the angle of reflection. On the other hand when an incident beam of radiation is distributed uniformly in all directions after reflection, the reflection is called diffused.
- Heat exchanger: A heat exchanger may be defined as equipment which is used to transfer energy from a hot fluid to a cold fluid with minimum investment and running cost. The principal type of heat exchangers in common user are

  (i) Double pipe heat exchanger

  (ii) Shell and tube heat exchanger

  (iii) Plate heat exchanger

  (iv) Fin type heat exchanger

  (v) Agitated vessel
- Flow arrangement in a heat exchanger

  (i) Co-current or parallel floor

  (ii) Counter-current flow

  (iii) Cross flow
- Logarithm Team temperature difference

  In a heat exchanger both the hot and cold fluids continuously undergo temperature variation. Therefore, the actual temperature driving force is variable all along the heat exchanger. Under those conditions the logarithmic mean temperature difference (LMTD) is defined as the temperature difference which is presumed to be constant and would give the same rate of heat transfer as actually occurs under variable conditions of temperature differences.

LMTD for parallel flow Heat Exchanger

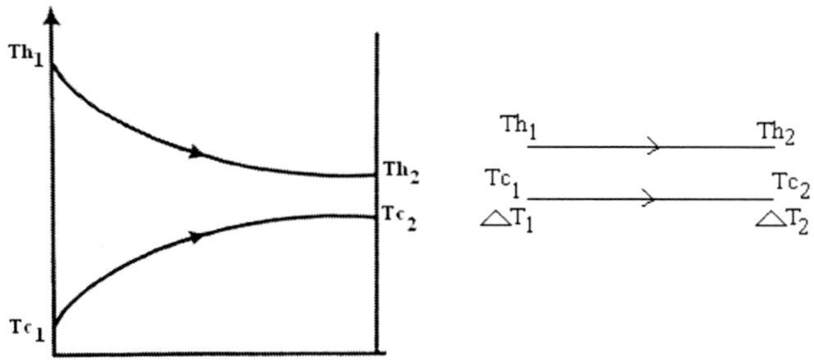

$$LMTD\ (\Delta T_m) = \frac{\Delta T_1 - \Delta T_2}{\ln \Delta T_1 / \Delta T_2}$$

Similarly LMTD for Counter-current flow

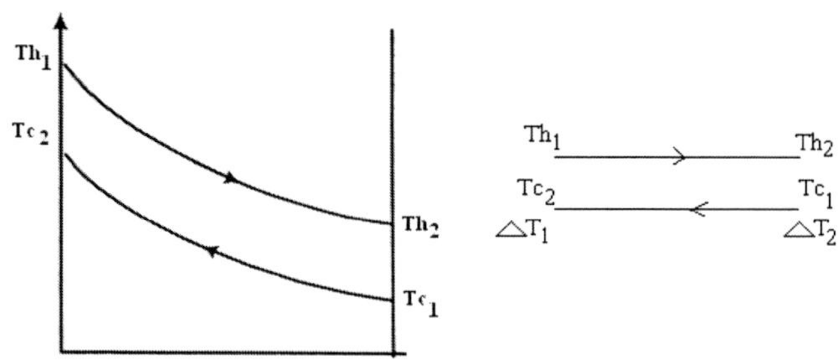

$$LMTD\ (\Delta T_m) = \frac{\Delta T_1 - \Delta T_2}{\ln \Delta T_1 / \Delta T_2}$$

- Temperature difference for cross flow – In a shell and tube heat exchanger having more than one pass, the direction of flow of either shell side or tube side or both may be reversed or criss – cross. Under their conditions the flow is neither truly parallel nor counter – current. This is known as cross flow. A connection factor is applied to the (ΔT_m) counter – current which depends upon the temperature conditions and number of tube or shell pass.
- Evaporator: An evaporator is equipment in which separation of a solvent is carried out from a solution of a volatile solvent and a nonvolatile solute. Evaporators are commonly used in the concentration and fruit juices industries. The main design parameters are determination of the heat transfer area and the steam consumption. Required to effect a specified separation.
- Evaporators can be of single effect or multiple effect evaporator:
- Design of a single effect evaporator

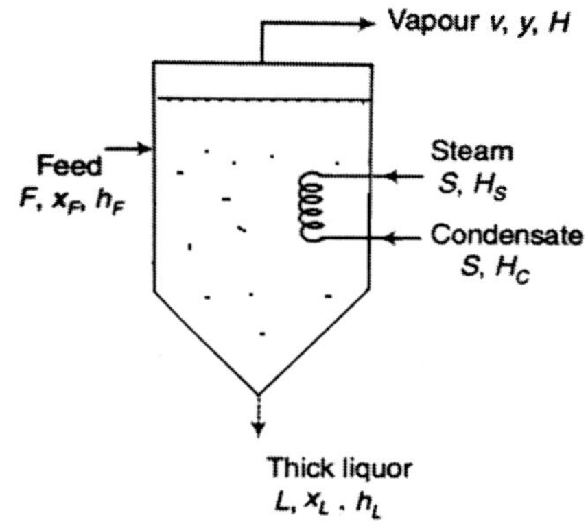

Where, F = feed rate, kg/hr
L = Thick liquor rate kg/hr
V = vapor rate kg/hr

x_f, x_L, y are mass fractions of solute in feed, their liqueur and vapor respectively.

The basic heat transfer equation is

$$Q = UA\Delta T$$

Q = Rate of heat transfer
U = Overall heat transfer coefficient
A = Area of heat transfer
ΔT = Temperature difference between steam and the boiling liquid

Overall material balance

F = L + V

Solute balance,

$$F \cdot x_f = L \cdot x_L + V \cdot y$$

Heat balance equation

Enthalpy of feed + enthalpy of steam = enthalpy of thick liquor + enthalpy of condensate

$$Fh_F + SH_s = VH + Lh_L + Sh_e$$
$$Fh_F + S(H_s - h_e) = VH + Lh_L$$

Or $$Fh_F + S\lambda = VH + Lh_L$$

Where,

h_F,$h_L$and h_e are enthalpy of feed, thick liquor and steam condense respectively

$$V = steam\ rate, \frac{kg}{hr}$$
$$H_s = enthalpy\ of\ steam$$
$$\lambda = latent\ heat\ of\ steam$$
$$= H_s - h_e$$

- Duhring's Rule: According to this rule, if the boiling point of a solution is plotted against the boiling point of pure wats at the same pressure, a straight line results for a given concentration of solution for all different liner are obtained.
- Boiling point Elevation: Presence of a solute in wats causes the boiling point elevation (BPE) is the difference between boiling point of a same temperature.

- Steam Economy: Steam economy of an evaporator is the kg of wats vaporized. It depends on the number of effects. Whereas in a single effect evaporator will always be less than one or at the most nearly equal to one, in a triple effect evaporator it may be 2.4 – 2.5 or even more.
- Evaporator capacity: It is defined as the number of kg of wats vaporized per hour
- Multiple Effect Evaporator: In a multiple effect evaporator.

Multiple Choice Questions

1. The term Leidenfrost point is associated with the.
   A. heat transfer between two highly viscous liquids.
   B. condensation of the saturated vapor on a cold surface.
   C. concentration of a corrosive solution by evaporation.
   D. boiling of a liquid on a hot surface.
2. Tube length/diameter ratio does not affect the tube side heat transfer coefficient when the flow is in the __________ zone.
   A. laminar
   B. transition
   C. both **A** and **B**
   D. highly turbulent
3. In shell and tube heat exchanger the use of baffles is to
   A. decrease the shell side heat transfer co-efficient.
   B. increase the cross-section of the shell side liquid.
   C. increase the shell side heat transfer co-efficient.
   D. force the liquid to flow parallel to the bank.
4. In an evaporator the actual temperature drop largely depend on the
   A. depth of liquid over heating surface.
   B. pressure difference between steam chest and vapour space.
   C. feed
   D. all **A**, **B** and **C**
5. The ratio of quantity of water vaporised to that of steam fed to the evaporator is known as
   A. rate of vaporisation
   B. capacity
   C. economy
   D. rate of evaporation
6. If a constant electrical current is passed through an electrical conductor of resistance R is the amount of heat produced is
   A. I^2R^2
   B. I^2R
   C. IR
   D. IR^2
7. With the increases in temperature the j_H factor also known as col-burn factor .
   A. increases
   B. remains unchanged
   C. may increases or decreases; depends upon the temperature
   D. decreases
8. In a multiple effect evaporator the steam economy is independent of
   A. boiling point elevations
   B. ratio of the weight of the thin liquor to thick liquor
   C. temperature of the feed
   D. rate of heat transfer
9. Tube length/diameter ratio does not affect the tube side heat transfer coefficient when the flow is in the __________ zone.
   A. laminar
   B. transition
   C. both **A** and **B**
   D. highly turbulent
10. The film coefficient of heat transfer which offers__________ resistance is known as the controlling resistance.
    A. lower
    B. no
    C. the largest
    D. the least

11. During condensation of vapour the pressure of any non-condensation gas
   A. is desirable to increase the film co-efficient
   B. increases the rate of condensation
   C. decreases thermal resistance
   D. none of these

12. What will be the outside temperature of an insulating wall will thermal conductivity K = 0.04 W/m°K and thickness 0.16 m. when the heat flux across the wall is 10 W/m^2 and the cold inside wall temperature is -5°C?
   A. 40°C
   B. 25°C
   C. 35°C
   D. 30°C

13. One shell and tube and another finned tube heat exchanger is available for heating by condensing steam. The tube side heat transfer coefficients for both the exchangers are for same optimum operation the recommended configuration is
   A. shell and tube heat exchanger with air on shell side and steam inside tubes
   B. finned tube heat exchanger with air outside and steam inside
   C. shell and tube heat exchanger with air inside tubes and steam on shell side
   D. finned tube heat exchanger with air inside and steam outside

14. In an evaporator the increases in liquid level will result is
   A. decreased capacity
   B. increased true temperature drop
   C. increase in liquor film co-efficient
   D. decreased effect of hydrostatic head

15. In SI unit the heat transfer coefficient may be expressed as
   A. W/m°K
   B. W/m^2°K
   C. J/m°K
   D. J/m^2°K

16. The thermal conductivity of solids depends on its
   A. temperature
   B. porosity
   C. both A & B
   D. neither A nor B

17. Identify the wrong statements .
   A. distillation process is not the same as evaporation
   B. the equivalent diameter of heat transfer for a duct of square cross-section (having each side as 'x') is equal to 4x
   C. in case of heat transfer by purely forced convection, G_R/R_e^2 d" 1
   D. the effectiveness of nucleate boiling depends basically on the ease with which the bubbles are formed and detached from the heating surface

18. A 10 cm dia steam pipe, carrying steam at 180°C, is covered with an insulation (conductivity = 0.6 W/m.°C). It losses heat to the surroundings at 30°C. Assume a heat transfer co-efficient of 0.8 W/m^2.°C for heat transfer from surface to the surroundings. Neglect wall resistance of the pipe and film resistance of steam. If the insulation thickness is 2 cms, the rate of heat loss from this insulated pipe will be
   A. less than the steam pipe with 5 cms insulation
   B. less than that of the uninsulated steam pipe
   C. greater than that for uninsulated steam pipe
   D. equal to that of the uninsulated steam pipe

19. With increase in heat transfer area in a shell and tube heat exchanger the purchase cost per unit heat transfer area
   A. remain constant
   B. increase
   C. decrease
   D. passes through a maxima

20. Identify the wrong statement
   A. phase change in case of a pure fluid at a given pressure from liquid to vapor or vice-versa occurs at saturation temperature
   B. the log mean temperature difference (LMTD) for counter flow and parallel flow can be theoretically same when any one of the fluids (hot or cold fluid) passes through the heat exchanger at constant temperature
   C. in case of a 1 - 2 shell and tube heat exchanger, the LMTD correction factor value increases sharply, when a temperature cross occurs
   D. the controlling resistance in case of heating of air by condensing steam is in the air film

21. By definition the equivalent diameter for pressure drop calculations is that for corresponding heat transfer operation .
    A. smaller that
    B. equal to
    C. greater than
    D. not related with
22. For film wise condensation on vertical surface the average heat transfer coefficient is propositional to (ΔT = temperature drop across condensate film)
    A. $(\Delta T)^{3/2}$
    B. $(\Delta T)^{2}$
    C. $(\Delta T)^{1/4}$
    D. ΔT
23. In a continuous crystallizer the crystal size depends on which of the following.
    A. degree of turbulence
    B. rate of heat transfer
    C. degree of super saturation
    D. all **A**, **B** and **C**
24. Which of the following relates both heat and mass transfer ?
    A. Lewis relationship
    B. Froude number
    C. Nusselt number
    D. Kutateladze number
25. If on a heating surface directly vaporisation is taking places directly the phenomenon is called
    A. vapour binding
    B. nuclear binding
    C. film binding
    D. none of these
26. For concentrating cold feed a backward feed multiples effect is a better option because it provides
    A. high economy
    B. lower capacity
    C. both **A&B**
    D. lower economy
27. In heat transfer operation heat waves
    A. travel in straight line
    B. cannot pass through vacuum
    C. can be reflected by a mirror
    D. both **A** and **C**
28. For heat transfer by radiations which of the following may be considered as opaque
    A. gases
    B. liquids
    C. solids
    D. both **B&C**
29. The thermal conductivity non-metallic amorphous solid with increased in temperature
    A. Remain constant
    B. Increase
    C. Decrease
    D. First decrease upto certain temperature and then increase
30. What are the factor upon which the shell side pressure drop in a shell and tube heat exchanger dose not depends ?
    A. tube diameter and pitch
    B. baffle spacing and shell diameter
    C. viscosity, density and mass velocity of shell side fluid
    D. none of these
31. For heat transfer from a hot tube surface to the flowing fluid the bulk of heat conversion heat transfer resistance lies in
    A. uniformly distributed throughout the fluid
    B. in the central core of the fluid
    C. mainly confined to a thin film of fluid near the surface
    D. none of these
32. For an evaporator steam consumption may be expressed by
    A. C/E
    B. CE
    C. E/C
    D. 1/CE
    (Where the C & E are capacity and economy of the evaporator respectively)
33. The cause of liquid entrainment is an evaporator
    A. high heat transfer rate
    B. high vacuum in the evaporator
    C. foaming in the solution
    D. high evaporation rate
34. Following air by condensing steam which type of heat exchange is best ?

A. shell and tube type

B. plate type
C. double pipe type with fin on air side
D. double pipe type with fin on steam side

35. When porosity increases the thermal conductivity of a solid substance ?
A. Remain unchanged
B. Decrease
C. Increase
D. may increase of decrease; depends on the solid

36. An agitator is rotated by a slowly falling body weight 40 kg. through a height of 4 m in an insulated container counting 20 kg. of water initially at 25°C. The setup is used to conduct Joule's experiment and the process repeated 500 times. If the acceleration due to gravity is 9.8 m/sec^2 and the heat capacity of agitator is neglected, the temperature of water (in °C) is
A. 25
B. 34.4
C. 40.5
D. 26.8

37. For concentrated black liquor in paper industries which type of evaporated is used
A. multiple effect evaporators followed by a crystalliser
B. single effect evaporator
C. multiple effect evaporator
D. single effect evaporator followed by a crystalliser

38. With increase in liquid level in a vertical tube evaporator, the overall heat transfer coefficient
A. may increase or decrease; depends upon the feed
B. decrease
C. increase
D. is not affected

39. Which of the following expression is the correct statement of Reynold's analogy ?
A. N_{st} á f
B. N_{Re} á f
C. N_{st} á N_{Re}
D. N_{Nu} á f

40. A body is said to be an ideal black body when
A. absorptivity = 1
B. Transmissivity = 1
C. emissivity = 0
D. reflectivity = 1

41. In a counter current shell and tube heat exchanger hot water enters in the tube side at a flow rate of 0.01 m^3/min at 80°C and leaves at 50°C. 0.05 m^3/min of cooling oil (specific heat 2 KJ/kg.K and density 800 kg/m^3 enters at 200°C). For this system the log mean temperature difference is
A. 32
B. 50
C. 37
D. 45

42. Fourier's law is applicable to the heat transfer by
A. radiation
B. convection
C. conduction
D. all **A**, **B&C**

43. The term ΔT is the heat transfer equation $Q = UA\Delta T$ represents
A. arithmetic mean temperature difference
B. geometric mean temperature difference
C. logarithmic mean temperature difference
D. the difference of average bulk temperatures of hot and cold fluids

44. In heat transfer the fouling film or dirt factor
A. is a dimensionless quantity
B. does not provide a safety factor for design
C. accounts for additional resistances to heat flow
D. none of these

45. The thermal efficiency of a reversible heat engine operating between two given thermal reservoirs is 0.4. The device is used either as a refrigerator or as a heat pump between the same reservoirs. Then the coefficient of performance as a refrigerator $(COP)_R$ and the co-efficient of performance as a heat pump $(COP)_{HP}$ are
A. $(COP)_R = (COP)_{HP} = 0.6$
B. $(COP)_R = 2.5$; $(COP)_{HP} = 1.5$
C. $(COP)_R = 1.5$; $(COP)_{HP} = 2.5$
D. $(COP)_R = (COP)_{HP} = 2.5$

46. Floating heat is provided in a heat exchanger to
A. facilitate cleaning of the exchanger
B. increase the heat transfer area
C. relieve stresses caused by thermal expansion
D. increase log mean temperature gradient

47. In which of the following cases of radiation Kirchhoff's law is applicable?
   A. monochromatic
   B. total
   C. both **A&B**
   D. neither **A&B**

48. 10 kg/sec of a dilute aqueous solution is heated by steam condensation at 95°C at a flow rate of 10 kg/sec. Which is the best arrangement in a 1-1 shell and tube heat exchanger
   A. counter flow with process stream on shell side
   B. parallel flow with process stream on tube side
   C. parallel flow with process stream on shell side
   D. counter flow with process stream on tube side

49. Pick out the wrong statement.
   A. orifice baffles are never used in a shell and tube heat exchanger
   B. split ring type and pull through type floating heads are two commonly used floating heads is heat exchangers
   C. in a horizontal tube evaporator, surface blanketing by air is avoided
   D. pressure drop on the shell side of a heat exchanger depends upon tube pitch also

50. The function of a steam jet ejector is to
   A. superheat the steam
   B. create vacuum
   C. remove condensate from the steam pipelines
   D. none of these

51. A double pipe heat exchanges whose inner pipe has fins on the outside will have its annul diameter _________ than that of another exchanger of the same size without fins
   A. unpredictable
   B. less
   C. more
   D. same

52. The steadystate temperature distribution is a very large thin plane wall having uniform surface temperature follows _________ law (thermal conductivity of wall material reaming constant)
   A. parabolic
   B. logarithmic
   C. linear
   D. logarithmic

53. In a coke oven plant regenerative operation of an air preheater is by heat transfer
   A. by direct contact of hot gas with air
   B. through a metallic wall
   C. by heating ab intermediate material (like chequor bricks) and then heating the air from this hot material
   D. none of these

54. Value of overall heat transfer coefficient is highest in the process of heat exchange
   A. is stream condenses
   B. from air to CO_2
   C. from air to heavy tar
   D. from steam to oil

55. which of the following parameter influence the economy of an evaporator
   A. steam pressure
   B. number of effect
   C. feed temperature
   D. both **B&C**

56. with rise in temperature thermal conductivity of solid metals
   A. remises same
   B. increases
   C. decreases
   D. may increase of decreasedepending upon temperature

57. The value of Nusselt number for fully developed laminar, constant property flow in a pipe at uniform heat flux is around___________.
   A. 83
   B. 4.364
   C. 0.72
   D. 18

58. Which of the following types of evaporators is used for concentrating viscous and heat sensitive liquids?
   A. long tube
   B. open pan
   C. agitated film
   D. none of these

59. Which of the following force correctly define the Grashoff number as represented by
   $\beta\,(T_s - T_\infty)\,\rho^2.\,L^3/\mu^2$,

A. viscous
B. inertial
C. elastic
D. none of these

60. When the Prandtl number is __________ the thickness of thermal boundary layer is more in comparison to that of hydrodynamic boundary layer.
A. >5
B. <1
C. 1
D. >1

61. A lower heat transfer coefficient fluid flowing is an extended surface heat exchanger
A. can flow either inside or outside the tubes
B. flows outside the tubes
C. should not be used as it gives very high pressure drop
D. flows through the tube

62. Which of the following is valid for a perfectly transparent surface (like gases)?
A. reflectivity = 0
B. transmissivity = 1
C. reflectivity = 0
D. all (a), (b) & (c)

63. The mathematical from of Graetz number is __________.
A. $\frac{mC_p}{kL}$
B. $\frac{kL}{mC_p}$
C. $\frac{mC_p}{k\mu}$
D. $\frac{kL}{mC_p}$

64. By defection the wetted perimeter for heat transfer and pressure drop for a fluid flowing in an annulus space are
A. never different
B. different
C. linearly related
D. same

65. The condition that formation of crystals can occur during crystallization is that the solution should be __________
A. under saturated
B. supersaturated
C. saturated
D. all A, B and C

66. In an evaporator, handling a cold dilute feed to produce thick viscous liquor, backward feeding as compared to forward feeding result in
A. increased economy
B. lower capacity
C. no effect on economy
D. decreased economy

67. Which of the following is valid for backward feed multiple effect evaporator?
A. no pumps are required between successive effects
B. feed flows from low pressure to high pressure
C. feed is introduced in the first effect
D. none of these

68. Identity the statement which is wrong
A. in industrial practice, sub-cooling of condensate is required, when the condensate is a volatile liquid and is to be transferred for storage
B. the condensing film co-efficient is about 3 times: lower for vertical condenser as compared to the equivalent horizontal condenser for identical situation
C. film co-efficient for vaporization decreases as a result of vapor binding
D. overall heat transfer co-efficient in a heat exchanger is controlled by the value of the film co-efficient, which is higher

69. The conversion factor from BTU/ hr.ft°.F to kcal/hr.m.°C is __________
A. 1.49
B. 4.88
C. 1
D. none of these

70. Which of the following dimensionless group represents the ratio of momentum diffusivity to thermal diffusivity?
A. Prandtl
B. Stanton

C. Grashoff
D. Nusselt

71. The ratio of is known as Grashoff number
A. inertial to viscous forces
B. buoyancy to viscous forces
C. inertial to viscous forces
D. buoyancy to surface tension forces

72. An aqueous solution is concentrated from 10 to 40% solids evaporating 30000 kg of water in an evaporator. The amount of solids handled by the system in kg is
A. 4000
B. 4600
C. 3000
D. 9000

73. When mixing of cold and warm liquids take place, most of the heat transfer takes place by
A. conduction
B. convection
C. radiation
D. both **A** & **C**

74. In natural convection the rate of heat transfer is increased in such system having
A. low temperature gradients
B. high co-efficient of thermal expansion
C. low density change with temperature
D. high viscosity

75. Which of the following condition apply for an insulator?
A. low thermal conductivity
B. less resistance to heat flow
C. a porous structure
D. high thermal conductivity

76. In the nucleate boiling regimes the heat flux is proportional to ___________ if ΔT is excess temperature
A. $(\Delta T)^4$
B. (ΔT)
C. $(\Delta T)^3$
D. $(\Delta T)^2$

77. If the tube side liquid outlet temperature is equal to the shell side fluid outlet temperature in a 1 – 1 concurrent heat exchanger, then the LMTD is
A. equal to the difference between hot fluid inlet temperature and cold fluid outlet temperature
B. 0
C. ∞
D. equal to the difference between hot and cold fluids inlet temperature

78. The monochromatic emissivity of a grey body is
A. unity
B. different for all wavelengths
C. same for all wavelengths
D. zero

79. If the bubbles formed on a submerged hot surface get absorbed in the pool of liquid the phenomena is collect___________ boiling.
A. nucleate
B. pool
C. low pressure
D. none of these

80. For a perfect black body the absorptivity and reflectivity are respectively __________
A. 1 and 0
B. 1 and ∞
C. 0 and 0.5
D. 0 and 1

81. In a condenser provision is made for vent pipe to
A. remove non-condensable gases
B. facilitate easy cleaning of tubes
C. purge the condenser
D. none of these

82. When the ambient temperature is T_a°C, the heat transfer by radiation between two bodies at temperature T_1 and T_2 depends on
A. $T_2 - T_a$
B. $T_1 - T_2$
C. $T_1 - T_a$
D. none of these

83. Which of the following products of dimensionless groups represent Nusselt number in forced convection?
A. Re and Pr
B. Re and Sc
C. Pr and Gr
D. Re and Gr

84. In a heat exchanger the heat transfer efficiency can be improved by
A. keeping the heat transfer surface clean
B. increasing the tube length

C. enhancing the fluid pumping rate
D. none of these

85. For that transfer by radiation for a black body __________ where ε is emissivity and α is absorptivity.
A. $\alpha \neq 1, \varepsilon = 1$
B. $\alpha \neq 1, \varepsilon \neq 1$
C. $\alpha = 1; \varepsilon \neq 1$
D. $\alpha = 1; \varepsilon = 1$

86. Applying the principle of Reynolds analogy Stanton number may be expressed as __________ where f is fanning friction factor.
A. f
B. $f/4$
C. $f/2$
D. $2f$

87. When the emissivity and absorptivity of a body become equal
A. for circular bodies
B. at a particular temperature
C. under thermal equilibrium
D. none of these

88. Identify the wrong statement.
A. superheated steam is preferably not used for process heating because of its low heat transfer film co-efficient
B. in a shell and tube heat exchanger, the shell pressure drop is maximum for orifice baffles
C. S.I. unit of fouling factor is Watt/m^2.°K
D. longitudinal fins are used in extended surface heat exchangers, when the direction of fluid flow is parallel to the axis of the tube

89. When Newton's law of cooling is applied for small temperature difference, the heat transfer is proportional to (Δt temperature difference)
A. Δt
B. Δt^2
C. Δt^3
D. Δt

90. A composite flat wall of a furnace is made f two materials 'A' and 'B' The thermal conductivity of 'A' is twice of tat of material 'B' while the thickness of layer of 'A' is half that of B. lf the temperature at e two sides of the wall are 400 and 1200pK, then the temperature drop (inpK) across the layer of material 'A' is
A. 133
B. 150
C. 125
D. 160

91. Three objects of different slopes is, a cube, sphereand thin circular plate made of same material and having same mass are all maintained at 300°C and allowed to cool in natural air. Which will cool the slowest?
A. all will cool at the same rate
B. cube
C. sphere
D. plate

92. Tube pitch may be defined as the _______ of tube diameters and the clearances.
A. sum
B. ratio
C. difference
D. none of these

93. Which of the following statements is wrong?
A. steam condensate is corrosive because of the presence of dissolved oxygen in it
B. by increasing the number of shell passes, the temperature cross in a shell and tube heat exchanger can be prevented
C. an underdesigned steam trap will back up the condensate instead of discharging it out
D. film baling is desirable in commercial equipments due to high heat transfer rate at low temperature drop

94. The radiation heat transfer _________ with increase in the distance between the heat source and object receiving the heat.
A. decreases
B. increases exponentially
C. remains unaffected
D. increases

95. If a mass of solid takes 10 minutes for cooling down from 75°C to 70 °C the time required to cooled down from 75°C to 65 °C will be minutes?
A. <10
B. >10
C. 10
D. Either (b) or (c), depends on the mass of the body

96. The hydrodynamic and thermal boundary layers become identical when the Prandtl number

_________.

A. 0.5

B. 1

C. 1.5

D. 80

97. High pressure steam for concentrating a dilute aqueous solution in a multiple effect evaporator system. Which of the following statements is most correct?

A. boiling point elevation in a single effect system is much higher than that in any effect in a multi effect system

B. total amount of vapor produced per Kg of feed system is much higher than in any effect in a multieffect system

C. heat transfer co-efficient in a single effect is much lower than that in any effect in a multi effect system

D. total heart transfer area of all the effects is – less than that in a single effect evaporator system

98. Which of the following groups characterise natural convection?

A. Grashoff number

B. Prandtl number

C. Peclet number

D. Reynolds number

99. If h_i is the heat transfer coefficient inside of the pipe the critical radius of insulation for the cylindrical pipe is _________.

A. K/h_0

B. $2K/h_0$

C. $2h_i/K$

D. $2K/h_0$

100. Steam pipe lines invariably are a trap to remove _____________.

A. none – condensable

B. condensate

C. steam

D. none of these

101. With increase in pressure and at constant temperature the thermal conductivities of gases will _________.

A. decreases

B. may increase or decrease; depends upon the pressure

C. increase

D. remain same

102. Calculation of the rate of heat transfer may be done by multiplying overall heat transfer coefficient, the temperature differences and the .

A. Nusselt number

B. heat transfer area

C. heating volume

D. none of these

103. At which value of Prandtl number the heat transfer by condition in the turbulent core of a fluid flowing through a heated pipe is negligible?

A. 0.2

B. 0.4

C. 0.6

D. 0.8

104. Which of the following equation is correct?

A. $j_H = (St)(Pr)^{2/3} = f/2$

B. $j_H = (St)^{1/3}(Pr) = f/2$

C. $j_H = (St)(Pr)^{1/3} = f/2$

D. $j_H = (St)^{2/3}(Pr) = f/2$

105. If K is thermal conductivity of insulating material and ho is the heat transfer coefficient at the outer surface, what will be the critical radius of insulation for a spherical shell?

A. $h_o/2K$

B. $2K/h_o$

C. K/h_o

D. h_o/K

106. Heat transfer by conduction is taking place from a droplet or a spherical particle to a surrounding stagnant film. What value of Nusselt number (Nu = hD/k) is expressed?

A. 0.5

B. 2

C. 10

D. 100

107. Under fully developed (thermally and hydro dynamically) laminar flow through a circular pipe, if surface temperature is constant what would be the value of the Nusselt number?

A. 1.66

B. 88.66

C. 3.66

D. Dependent on N_{Re} only

108. The heat transfer equipment used for small scale evaporator is .

A. multiple effect evaporator
B. heat exchanger
C. condenser
D. steam jacketed kettle

109. What is the reason for low thermal conductivity of heat insulating materials?

A. high specific heat
B. high proportion of air space
C. dense structure
D. none of these

110. Dropwise condensation takes place most conveniently on surface.

A. coated
B. oily
C. glazed & polished
D. smooth

111. Which of the following types of heat exchanger is used in the radiator of an automobile engine?

A. counter-current
B. cross-current
C. co-current
D. direct current

112. Which of the following cannot recover the sensible heat of hot industrial flue gases?

A. ceramic recuperator
B. rectangular
C. economiser
D. none of these

113. Transfer of heat from pipe wall to the liquid takes place in pipe flow by

A. condition only
B. forced convection only
C. forced conversion and conduction
D. free and forced convection

114. The equilibrium state corresponds to equality of temperature whenever an interphase heat transfer takes place. Which of the following parameters will be the controlling condition in interphase mass transfer process?

A. concentrations
B. mass transfer coefficient
C. chemical potential
D. active co-efficients

115. Which of the following range of values will be applicable for Colburn analogy?

A. 120 to 400
B. 0.6 to 120
C. 0.5 to 5
D. 0.001 to 1

116. By what factor the heat transfer coefficient will change when heat transfer is taking place in a tube under turbulent flow condition?

A. 1.74
B. 1
C. 6.1
D. 37

117. In shell and tube heat exchanger vibrations may be induced due to

A. vibrations transmitted through piping and/or supports due to external reasons
B. oscillations in the flow of shell/tube side's fluid
C. flow of fluid on the tube and shell sides
D. all **A**, **B** and **C**

118. In comparison to single effect evaporators multiple effect evaporators are prepared to .

A. decrease the steam economy & increase the capacity
B. increase the steam economy & the capacity
C. increase the steam economy & decrease the capacity
D. decrease the steam economy & the capacity

119. The composite wall of a furnace consists of two insulating materials 10 and 20 cm thick with thermal conductivity 0.05 and 0.1 $Wm^{-1}°C$ respectively. The inner wall is at 700°C and the ambient air is at 30°C. The heat transfer coefficient between the outer surface of wall and air is 20 $W/m^{-2}°C$. What would be the heat loss from outer surface in W/m^2?

A. 165.4
B. 167.5
C. 172.5
D. 175

120. The dimensionless Prandtl number is expressed as____________.

A. $\frac{\mu}{h \cdot C_p}$
B. $\frac{C_p \cdot \mu}{a}$

C. $\frac{C_p \cdot \mu}{k}$

D. $\frac{hD}{k}$

121. For the overall heat transfer coefficient which is the following is the lowest?

A. molten sodium

B. dowtherm

C. water

D. air

122. Which type of evaporator is preferred for the concentration of foamy and frothy liquors?

A. agitated film evaporator

B. long tube vertical evaporator

C. open pan evaporator

D. none of these

123. Which of the following is the unit of heat transfer coefficient?

A. BTU/hr. ft^2°F

B. BTU/hr. ft

C. BTU/hr. °F. ft.

D. BTU/hr. °F

124. When heat exchange start operating in the asymptotic range of temperature it

A. provide very large heat transfer co-efficient

B. results in making part of the heating surface inactive

C. results in abruptly increased velocity

D. none of these

125. A one dimensional slab of thickness 0.2m has a left maintained at 80°C whereas the right face is at 30°C. The thermal conductivity and heat transfer coefficient from right face in 1.2 W/m.K and10 W/m.K respectively. Under steady state, the temperature of the fight face in °C will be .

A. 77.2

B. 71.2

C. 63.8

D. 48.7

126. In SI unit the numerical value of Stefan-Boltzman constant is

A. 5.6697 × 10^{-8}W/m^2.°K^4

B. 0.1714 × 10^{-8}W/m^2.°K^4

C. 5.6697 × 10^{-8}kcal/m^2. °K^4

D. 0.1714 × 10^{-8}kcal/m^2. °K^4

127. Which of the following materials of construction is rarely used for construction of heat exchanger tubes?

A. copper

B. plain carbon steel

C. lead

D. stainless steel

128. Heat is flowing through a composite well consisting of two plate A and B placed in series of thermal conductivities K_A and K_B and specific heat capacities Cp_A and Cp_B respectively. Under the condition that plate B has twice the thickness of plate A, the temperature difference across plate A will be greater than that of B. When

A. $C_{PA} > C_{PB}$

B. $C_{PA} < C_{PB}$

C. $k_A < 0.5k_B$

D. $k_A > 2\ k_B$

129. The Dittus-Boelter equation for heat transfer coefficient for fluid flowing inside a clean pipe is represented as

$$h = 0.023\ K/D \left(\frac{DV\rho}{\mu}\right)^{0.8} \cdot \left(\frac{C_p.\mu}{k}\right)^{0.4}.$$

thus equation is valid for Reynolds number.

A. <2100

B. 2100 – 4000

C. >4000

D. > 10000

130. The heat radiation of ———— wavelength will be emitted in case of a hot black body.

A. one fixed

B. small

C. all

D. large

131. Which type of mean area is used to calculate the rate of heat transfer when the heat transfer mode is by conduction in a hollow sphere?

A. geometric

B. arithmetic

C. logarithmic

D. either **A**, **B** or **C**

132. What is the range of steam side heat transfer coefficient in an evaporator in Kcal/hr.m^2.°C?

A. 10 – 50

B. 100 – 500

C. 1000 – 1500
D. 5000 – 15000

133. 1 Kg of water at 0°C is heated and changed to superheated steam at 1 atm pressure and 200°C. The major heat consumption in the process will be to
A. heat of water from 0°C to 100°C
B. evaporate the water
C. to superheat the steam
D. data insufficient, can't be predicate

134. A heating element at a temperature of 800°C is located in a furnace being maintained at 300°C. At a radiation flux 8 KW/m^2, the elements temperature is
A. 11.2KW/m^2
B. 12.0KW/m^2
C. 14.6KW/m^2
D. 16.5KW/m^2

135. If the heat conduction is negligible in the buffer zone the corresponding value of Prandtl number is
A. no
B. low
C. high
D. extremely low

136. Minimum absorptivity will occur when radiation is encountered, from
A. aluminum foil
B. iron plates
C. coal dust
D. refractory bricks

137. For an ideal solution the boiling point elevation.
A. increases rapidly with temperature rise
B. decreases rapidly with temperature rise
C. in independent of pressure
D. both **B** and **C**

138. A correction (FT) is applied for LMTD for flow.
A. counter
B. parallel
C. cross
D. none of these

139. What is the limiting value of the minimum and maximum baffle spacing respectively in a shell and tube heat exchanger (D = inside diameter of shell)?
A. D/5 and D
B. D/2 and 2D
C. D/4 and 2D
D. D and 2D

140. It is desired to reduce the tube side pressure drop keeping the flowrate the same. Which of the following types of heat exchanger is recommended?
A. 1-2 heat exchanger
B. 1-1 heat exchanger
C. 3-2 heat exchanger
D. 2-4 heat exchanger

141. The shell side fluid velocity in a shell and tube heat exchange cannot be changed by changing the .
A. no. of baffles
B. tube diameter
C. tube layout
D. tube pitch

142. The cooling of air in cooled internal combustion engine the mode of heat transfer is
A. conduction
B. natural convection
C. forced convection
D. none of these

143. What will be the range of overall heat transfer coefficient if the inner and outer flow coefficient are h_1 and h_2 respectively?
A. dependent on metal resistance
B. always between h_1 and h_2
C. always higher than h_2
D. always less than h_1

144. The rate of heat transfer through a hollow cylinder of r_1& r_2.
A. difference of radii (r_2-r_1)
B. ratio of radii (r_2/r_1)
C. product of radii (r_2*r_1)
D. sum of radii (r_1+r_2)

145. Which of the following type of boiling occurs during boiling of milk in an open vessel?
A. film
B. saturated nucleate
C. sub-cooled
D. none of these

146. From a well-insulated high pressure boiler the radiation heat losses may be approximately __________ percent.

A. 1
B. 7
C. 18
D. 26

147. Fourier's law of steady state one dimensional heat conduction does not assume that
A. constant temperature gradient exists
B. there is no internal heat generation
C. material is anisotropic
D. boundary surfaces are isothermal

148. Which of the following statements is correct in a gas liquid shell and tube heat exchanger ?
A. gases to be heated/cooled is normally routed through the shell side, because the corrosion caused by the cooling water or steam condensate remain localised to the tubes
B. gases under high pressure are routed through the tube side, because high pressure gases are corrosive in nature
C. presence of a non-condensable gas decreases the condensing film co-efficient
D. All **A**, **B** or **C**

149. Which dimensionless group is the controlling parameter in natural convective heat transfer?
A. Graetz number
B. Eckert number
C. Grashoff number
D. Bond number

150. Which type of evaporator employs an annular downlake?
A. basket type
B. horizontal
C. long tube vertical
D. none of these

151. Which is the range of Prandtl number for the Colburn analogy to be valid?
A. 0.06 to 120
B. 0.6 to 120
C. 1 to 103
D. 1 to 50

152. What is the effect of boiling point elevation in multiple effect evaporator?
A. reduce the capacity
B. increase the economy
C. reduce the economy
D. none of these

153. It is known that the total emissivity of a perfect black body is unity. What will be the value for a real body?
A. 0
B. 1
C. >1
D. between 0 and 1

154. For a multiple effect evaporator system with forward feed arrangement the pressure wall be
A. highest at the outlet of the last effect
B. least at the outlet of the last effect
C. least at the inlet of the first effect
D. highest at the inlet of the last effect

155. Which of the following is a wrong statement?
A. Swenson-Walker crystalliser is a batch crystalliser
B. the first stage of crystal formation is called nucleation
C. super saturation of the solution is the driving potential for a crystal growth
D. the liquor left after the removal of crystals from a solution is called mother liquor

156. A shell and tube heat exchanger is used for condensing steam with water as the cooling medium. Heat transfer coefficient for water whether on shell or tube heat side is the same. The best arrangement is
A. vertical heat exchanger with steam on tube side
B. vertical heat exchanger with steam on shell side
C. horizontal heat exchanger with steam on tube side
D. horizontal heat exchanger with steam on shell side

157. Which of the following statements is wrong?
A. unit of thermal conductance is W/°K
B. reciprocal of the resistance to heat flow is called thermal conductance
C. thermal conductance of a wall of thickness 'L', thermal conductivity 'k' and heat flow area 'A' is kL/A
D. none of these

158. Fin efficiency is defined as the ratio of heat transfer across the fin surface to the theoretical heat transfer across an equal area held at the
A. surrounding temperature
B. temperature of the fin end

C. average temperature of the fin
D. constant temperature equal to that of the base

159. The measurment of higher BWG for a definite tube outside diameter is higher
A. weight per unit length
B. cross-sectional area
C. tube thickness
D. none of these

160. At _________ pressure film boiling starts occurring.
A. sub-atmospheric
B. negative
C. atmospheric
D. very high

161. What is the range of velocity when electro-magnetic waves propagate in vacuum?
A. 3×10^5
B. 3×10^8
C. 3×10^{10}
D. 3×10^{12}

162. The time rate of heat transfer per unit _________ is known as heat flux.
A. volume
B. area
C. length
D. none of these

163. For heavy heat loads the recommended type of heat exchanger is ________ type.
A. plate film
B. double pipe
C. series and parallel set of shell and tube
D. none of these

164. In which of the following cases double pipe heat exchanger is recommended for use?
A. because it is less costly
B. when heat transfer area required is very low, i.e (100-200 ft^2)
C. when heat transfer area required is very high
D. because it occupies less floor area

165. If the liquid level is increased in a vertical tube evaporator then it will result is an increase in
A. liquor-film co-efficient
B. velocity of circulation
C. both **A** and **B**
D. neither **A** nor **B**

166. In a double pass shell and tube heat exchanger the tube side pressure loss is equal to _________ the velocity head.
A. square root of
B. four times
C. square of
D. twice

167. Which of the following dimensionless group is empirically related to Nusselt number for forced convection?
A. Reynolds
B. Prandtl
C. both **A&B**
D. neither **A** nor**B**

168. The rate of heat transfer through a pipe wall is given by $q = \frac{2\pi K(T_i - T_o)}{\ln(r_i/r_o)}$, for a thin walled cylinder this equation can be approximated by

A. $q = \frac{2\pi K[(T_i + T_o)/2]}{\ln(r_o/r_i)}$

B. $q = \frac{2\pi r_i K(T_i - T_o)}{(r_o - r_i)}$

C. $q = \frac{2\pi K(T_i - T_o)}{(r_o - r_i)}$

D. $q = \frac{2\pi K(T_i - T_o)}{[(r_o + r_i)/2]}$

169. Which of the following parameters represent the heat transfer rate per unit area?
A. thermal diffusivity
B. heat flux
C. thermal conductivity
D. heat transfer co-efficient

170. Which of the following represents the phenomena of convective heat transfer?
A. Fourier's law
B. Newton's law
C. Fick's law
D. none of these

171. The purpose of maintening vacuum in the vapour space of an evaporator is to
A. get economical temperature difference by using moderate pressure steam

B. facilitate forward feeding in multiple effect evaporation

C. achieve very high concentration of the final product

D. concentrate heat sensitive materials

172. Heat flux through several resistances in series analogous to the current flowing through several __________.

A. capacitors in series

B. resistances in parallel

C. resistances in series

D. none of these

173. Which of the following types of surfaces are suitable for dropwise condensation to occur?

A. Polished

B. clean and dirt free

C. contaminated cooling

D. smooth clean

174. Which of the following dimensionless groups is unimportant in forced convection?

A. Prandtl number

B. Reynolds number

C. Grashoff number

D. none of these

175. Which of following statements is wrong?

A. heat transfer by radiation cannot occur across an absolute volume

B. the amount of heat involved in the condensation or vaporisation of 1 kg of a fluid is the same.

C. propagation velocity for travel of heat radiation through vacuum is equal to the velocity of the light

D. in case of a shell and tube heat exchanger, the pressure drop through the shell is proportional to the number of times the fluid crosses the bundle between baffles

176. In such cases where the Prandtl number is greater than the Schmidt number the

A. thermal boundary layer lies inside the concentration boundary layer

B. thermal & concentration boundary layers are of equal thickness

C. concentration boundary layer lies inside the thermal boundary layer

D. hydrodynamic (i.e., momentum) boundary layer is thicker than the other two

177. Very common uses of multiple effect evaporators are in the manufacture of

P. Paper Q. Superphosphate

R. Sugar S. Fats

A. R and S

B. P and Q

C. P and R

D. P and S

178. Which of the following has the highest thermal conductivity?

A. water

B. brick

C. air

D. silver

179. The function of a steam traps in steam carrying pipelines in to

A. release excess steam pressure by bleeding steam

B. condense steam

C. remove condensate and inert gases

D. none of these

180. It vaporisation from pools of water is to be carried out the rate of heat transfer will be affected by the

A. viscosity of water

B. surface tension of water

C. nature of heating surface and distribution of bubbles

D. all **A**, **B**and **C**

181. The boiling heat transfer coefficient __________ as the difference between the wall temperature and bulk temperature increases.

A. continues to increase

B. continues to decrease

C. goes through a minimum

D. goes through a maximum

182. Identify the correct statement

A. Rate = Resistance/Driving force

B. Driving force = Rate × Resistance

C. Rate = Driving force × Resistance

D. Resistance = Driving force × Rate

183. According to the Kirchhoff's law the ratio of the total radiating power to the absorptivity of a body depends on the

A. temperature of the body only

B. wavelength of monochromatic radiation

C. both **A** and **B**

D. nature of material of body

184. In case the outlet temperature of hot and cold fluids are the same in parallel flow heat exchanger then the log mean temperature difference is

A. minimum

B. maximum

C. zero

D. infinity

185. With increase in the thickness of insulation of a hot cylindrical pipe at given ambient temperature the rate of heat loss would

A. decrease

B. increase

C. first decrease and then increase

D. first increase and then decrease

186. If the logarithm of the numerical value of the temperature difference between the liquid and the room is plotted against time the nature of plot will be of the form

A. ellipse

B. straight line

C. parabola

D. circular arc

187. Which of the following color paints has the maximum absorption coefficient?

A. black

B. grey

C. yellow

D. white

188. In which of the following cases the log mean temperature difference con not be used?

A. heat transfer co-efficient over the entire heat exchanger is not constant

B. the heat capacity is not constant and there is a phase change

C. there exists an unsteady state

D. none of these

189. Which of the following represents Nusselt number?

A. $\frac{C_P \mu}{K}$

B. $\frac{hD}{K}$

C. $\frac{h\, C_P}{\mu}$

D. $\frac{C_P \mu}{h}$

190. It is required to reduce the emissivity of a surface by reducing the emissivity of the surface. Which of the following operation will give best results?

A. painting the surface black

B. painting the surface white (with aluminum paint)

C. giving the surface a mirror finish

D. roughening the surface

191. The rate of heat transfer by radiation dose not depend upon the

A. distance between the heat source and the object receiving the heat

B. surface area and temperature of the heat source

C. type of absorbing surface

D. none of these

192. If the heat transfer area and other conditions are same then the radio of the capacity of single effect to that of a triple effect evaporator is

A. 3

B. 0.33

C. 1

D. 1.33

193. The heat transfer equation Dittus-Boelter

$$\left[(i.e.)\ h = 0.023\frac{K}{D}(Re)^{0.8}(Pr)^{0.4}\right]$$

This equation is not applicable when there is

A. low Reynold's number

B. very low Grashoff number

C. molten metals

D. all **A**, **B** and **C**

194. The reason because of film boiling in avoided in commercial equipment's is that

A. the heat transfer rate is low in view of the large temperature drop

B. it is not economic

C. it is difficult to maintain

D. none of these

195. Which of the following expression represent the critical radius 'r' of insulation on pipe?

A. $r = h/k$

B. $r = k/h$

C. $r = 2k/h$

D. $r = k/2h$

196. Metals way have variable thermal conductivity with respect to temperature as represented by $K = K_0 + at$. Where K is the thermal conductivity and T is the temperature (°K). The unit of 'a' in SI system will be

A. W/mk

B. W/m

C. W/mk^2

D. None, 'a' is just a number

197. The reason why Dittus-Boelter equation is not applicable for molten metals is due to its low __________.

A. Prandtl number

B. Grashoff number

C. thermal conductivity

D. viscosity

198. Such surfaces which are ________ promoted the film condensation.

A. dirty

B. coated

C. clean and smooth

D. oily

199. Which of the following accessories is provided in the vapor line of an evaporator for removing the entrained liquid?

A. bleed point

B. vent

C. catchall

D. baffle

200. The Reynold's number of the shell side fluid in a shell and tube heat exchanger _________ if the baffle spacing increases.

A. remain unchanged

B. increase

C. increase or decrease depending on number or shell passes

D. decrease

201. Sometimes a heat exchanger may have a floating head to

A. avoid buckling of tubes

B. decrease the pressure drop

C. facilitate its lengthening, if needed

D. provide support for tubes

202. In a heat exchanger the outlet temperature of cooling water is not allowed to exceed 50°C mainly to avoid __________.

A. uneconomic LMTD

B. excessive corrosion

C. decrease in heat exchanger efficiency

D. its evaporation loss

203. The absorptivity of black body is equal to _________.

A. 1

B. 0.78

C. 0

D. 0.95

204. Concentration of a highly concentrated solution is to be done. Which of the following is the most suitable evaporator?

A. open pan evaporation

B. agitated film evaporator

C. long tube vertical evaporator

D. none of these

205. If feed temperature is decreased is an evaporator it

A. reduces the economy

B. increases the heating area required

C. both **A** and **B**

D. decreases the heating area required

206. In heat transfer to boiling liquids the most controlling factor for the rate of bubble detachment from the hot solid surface is ___________

A. liquid viscosity

B. liquid density

C. hot surface temperature

D. interfacial l tension

207. If D is inside diameter of the shall in a shall and tube heat exchanger, a 25 percent cut in its segmental baffles is equal to

A. 0.50 *D*

B. 0.75 *D*

C. 0.25 *D*

D. none of these

208. In which of the following situations multiple effect evaporation is recommended?

A. large scale evaporation of liquor is needed

B. fuel is cheaply available

C. evaporation on small scale is to be done

D. corrosive liquids are to be concentrated

209. When Prandtl number value is ________ Prandtl and Reynolds's analogies are identical.
   A. 0.5
   B. 5
   C. 1
   D. 0
210. A wall is made of two different solids I and II. At steady state the temperature variation in this plane wall is shown below. The thermal conducting of material I

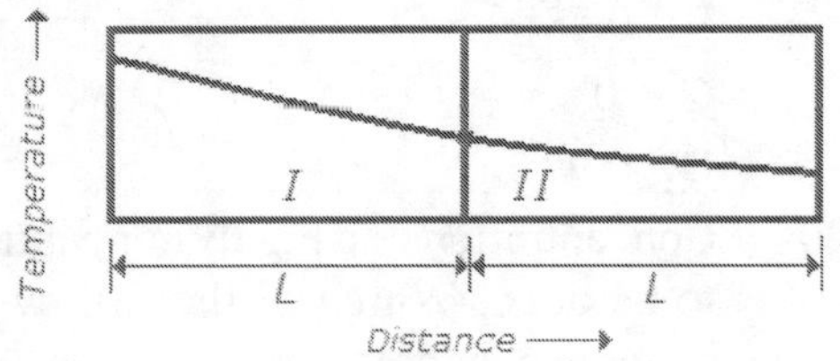

   A. is smaller than that of II
   B. is greater than that of II
   C. can be greater than or smaller than that of II
   D. is greater than that of II
211. In a shall and tube heat exchanger the fluid characteristic deciding the routing of a fluid is _______.
   A. corrosiveness
   B. viscosity
   C. fouling characteristic
   D. none of these
212. The interpretation of one transfer unit for a heat exchanger is.
   A. the size of the exchanger in which heat transfer rate is 1 kcal/hr.
   B. a section of the exchanger in which change in temperature of one stream equals the average driving force in the section.
   C. both **A** and **B**
   D. none of these
213. In which type of the following evaporator tube cleaning is conveniently possible.
   A. basket type
   B. horizontal tube
   C. calendria
   D. none of these
214. Sometimes a beck word feed multiple effect evaporators may be preferred over forward feed units due to following reason.
   A. heat sensitive material can be handled
   B. equal heat transfer coefficients exist in various effects
   C. most concentrated liquid is at highest temperature
   D. there is no additional cost of pumping
215. A metal ball having density and specific heat equal to 300 kg/ m^3 and 0.4 KJ/kg.K respecting left in air at 30°C the ball has a radians of 0.1 m and initially at a temperature of 90°C.If the heat transfer coefficient is 50 W/m^2. What will be the time taken in horns for ball to cool to 60°C?
   A. 55.5
   B. 0.55
   C. 555
   D. 0.15
216. The presence of air in a condensing vapor will result is _______ in the rate of condensation of vapor.
   A. increase
   B. decreases
   C. no affect
   D. increase the condensing film coefficient as well as
217. Which of the following may be called baffle spacing?
   A. should be less than one fifth the diameter of the shell
   B. is not the same as baffle pitch
   C. should be less than the inside diameter of the shell
   D. none of these
218. When a fluid moves under the influence of changes in fluid pressure produced by external work it is known as ___________.
   A. conduction
   B. natural convection
   C. forced convection
   D. radiation
219. A stirred vessel is equipped with immersed coils carrying condensing steam at 150 °C has a coil area of 1.2 m^2. The vessel is changed with 1000 Kg of liquid at 30°C which is to be heated to 120°C. The overall heat transfer coefficient for the liquid is 1500 W/m^2. °C and the specific heat to be 4 KJ/Kg. °C. The time taken for the liquid reach the desired temperature is ____________.

A. 22 min
B. 44 min
C. 15 min
D. 51 min

220. In a case of heat transfer by conduction through a cylinder arithmetic mean area can be used for calculation is
A. thin wall having the value of $A_0A_i/2$
B. having the value $A_0/A_i>2$
C. thick walled
D. both **B** and **C**

221. What is the mass of moisture removed in kg if 1000 kg of wet solids are dried from 60% to 20% moisture (by weight)?
A. 200
B. 500
C. 400
D. 520

222. In the condensation of a vapor under saturated conditions it is required to remove ________ heat.
A. super
B. sensible
C. latent
D. both **B&C**

223. A cylinder (length = diameter) a sphere and a cube are of equal mass and of same material are identially at 500°C. All of them are dropped in a quenching both containing a large volume of cooling each attaining the bath temperature eventually. The time required for 90% change in temperature is the smallest for
A. cube
B. sphere
C. cylinder
D. equal for all three

224. In which of the following situations double pipe heat exchangers are recommended for use.
A. overall heat transfer coefficient is very high
B. requirement of heat transfer area is low
C. a corrosive liquid is to be heated
D. high viscosity liquid is to be cooled

225. For heart transfer by conduction in a series of flat resistances the overall thermal resistance may be equal to the.
A. average of all resistances
B. sum of all resistances
C. minimum resistances presents in the series
D. maximum resistances in the series

226. Thermal diffusivity may be exposed as
A. $\frac{k}{\rho\, C_P}$
B. $\frac{\rho\, C_P}{k}$
C. $\frac{C_P\, \mu}{a}$
D. $\frac{\mu}{h\, C_P}$

227. For most of the liquids the thermal conductivities ______________ with rise in temperature.
A. remains unchanged
B. decreases
C. may increase or decrease; depends on the liquid
D. increases

228. Which of the following conditions must be fulfilled for achieving asymptotic conditions for a fluid flow in laminar flow through a long tube?
A. exit-fluid temperature > wall temperature
B. Graetz number > 100
C. exit fluid temperature = wall temperature
D. exit-fluid temperature < wall temperature

229. For cooling of hydrocarbon by water the overall heat transfer coefficient is approximately
A. 50 -100 Kcal/hr.m^2.°C
B. 50 -100 W/m^2°K
C. 50 -100 BTU/hr. ft.^{2}°F
D. 1000 - 1500 BTU/hr. ft.^{2}°F

230. For a fluid flowing over a heated plate at what numerical value Prandtl number and thermal boundary layers will be equal?
A. 1
B. <1
C. >1
D. none of these

231. Which of the following conditions must be fulfilled for a heat conduction to occur when unsteady condition?
A. three dimensional heat flow is concerned
B. temperature distribution is dependent on time

C. heat flows in one direction only

D. temperature distribution is independent of time

232. ________ **chart is known as transient heat conduction chart.**

233. On which of the following types of surfaces Fourier's law of heat conduction applies?

A. non-isothermal

B. isothermal

C. both**A** and **B**

D. nether **A** nor**B**

234. If the heat transfer coefficients in film wise and drops wise condensation are represented as h_f and h_d respectively for condensation of pure vapors than

A. $h_f > h_d$

B. h_f could be greater or smaller than h_d

C. $h_f < h_d$

D. $h_f = h_d$

235. Heat transfer by natural convection is taking place from a vertical plate. If the temperature differences between wall and onside bulk is ΔT than the rate of heat transfer is proportional to ________

A. $\Delta T^{1/2}$

B. $(\Delta T)^{5/4}$

C. $(\Delta T)^{1/4}$

D. $(\Delta T)^{3/4}$

236. In an evaporator a circulation pump is located to

A. avoid frequent priming

B. avoid cavitation

C. create more suction head

D. none of these

237. Which of the following has maximum absorptivity?

A. aluminum foil

B. refractory bricks

C. iron plate

D. coke breeze

238. Identify the most important factor on which the heat conducted through a wall in a unit time will depend on.

A. material of the wall

B. area of the wall perpendicular to heat flow

C. thickness of the wall

D. temperature difference between the two surfaces of the wall

239. Which of the following expressions represent harmonic mean temperature difference?

A. $\frac{\Delta T_1 - \Delta T_2}{\Delta T_1 \, . \, \Delta T_2}$

B. $\frac{2 \, . \, \Delta T_1 \, . \, \Delta T_2}{(\Delta T_1 + \Delta T_2)}$

C. $\frac{2 \, . \, \Delta T_1 \, . \, \Delta T_2}{(\Delta T_1 - \Delta T_2)}$

D. $\Delta T_1 \Delta T_2$

240. Identify the wrong statement

A. steam jet ejector is used for vapor compression in a thermal recompression evaporator

B. the capacity of an evaporator is reduced by the boiling point elevation

C. corrosive liquid is normally passed through the tubes in shell and tube heat exchanger

D. heat sensitive material should be concentrated in high pressure evaporators

241. Identity the material which has the minimum thermal conductivity.

A. nitrogen

B. steel

C. carbon black

D. tar

242. The reason that steam condensate is recovered by steam traps and recycled for use as boiler feed water is become of its low

A. dissolved solids content

B. hardness

C. suspended solids content

D. all **A** and **C**

243. Certain aqueous solutions like anhydrous Na_2SO_4 whose solubility decreases with rise in temperature

A. backward feed

B. vacuum

C. high pressure

D. none of these

244. If D is inside diameter of the shall in a shall and tube heat exchanger, a 25 percent out in its segmental baffles is equal to

A. 0.50 *D*

B. 0.75 *D*

C. 0.25 *D*

D. none of these

245. By what factor the inside heat transfer coefficient is increased if in a 1-2 shell and tube heat exchanger operating under turbulent flow conditions the number of tube passes is increased to 8?

A. $2^{0.8}$

B. $4^{0.8}$

C. $4^{0.4}$

D. $2^{0.4}$

246. Concentration of a solution of NaOH is accompanied with boiling point elevation which ___________.

A. increases rapidly with temperature rise

B. is almost independent of pressure

C. is almost independent of temperature

D. both **A** and **C**

247. Which type of heat exchanger is the most suitable, when the temperature of shell side fluid is much higher than that of tube side.

A. single pass, fixed tube sheet

B. three pass, fixed tube sheet

C. U-tube

D. none of these

248. Which law of radiation states that maximum wavelength of radiation is inversely proportional to the temperature?

A. Planck's

B. Stefan-Boltzmann's

C. Wine's displacement

D. none of these

249. Which of following color paints has minimum absorption coefficient?

A. grey

B. white lead

C. light cream

D. black

250. The application of Duhring's rule is in

A. humidification

B. distillation

C. evaporation

D. humidification

Answer

1.	D	37.	C	73.	B	109.	B
2.	D	38.	B	74.	B	110.	B
3.	C	39.	A	75.	A	111.	D
4.	D	40.	A	76.	C	112.	D
5.	C	41.	A	77.	B	113.	C
6.	B	42.	C	78.	C	114.	B
7.	A	43.	C	79.	B	115.	B
8.	A	44.	C	80.	A	116.	C
9.	B	45.	C	81.	A	117.	D
10.	C	46.	C	82.	D	118.	B
11.	D	47.	C	83.	A	119.	A
12.	C	48.	A	84.	A	120.	C
13.	B	49.	A	85.	D	121.	D
14.	A	50.	B	86.	C	122.	B
15.	B	51.	B	87.	C	123.	A
16.	C	52.	C	88.	C	124.	B
17.	B	53.	C	89.	A	125.	D
18.	B	54.	A	90.	D	126.	A
19.	D	55.	D	91.	C	127.	C
20.	C	56.	C	92.	A	128.	C
21.	A	57.	B	93.	D	129.	D
22.	C	58.	C	94.	A	130.	C
23.	D	59.	A	95.	B	131.	C
24.	A	60.	B	96.	B	132.	D
25.	B	61.	B	97.	B	133.	B
26.	A	62.	B	98.	A	134.	D
27.	D	63.	A	99.	A	135.	C
28.	D	64.	B	100.	B	136.	A
29.	B	65.	B	101.	A	137.	A
30.	D	66.	A	102.	B	138.	C
31.	C	67.	B	103.	C	139.	A
32.	A	68.	D	104.	A	140.	B
33.	C	69.	A	105.	B	141.	B
34.	C	70.	A	106.	B	142.	B
35.	B	71.	B	107.	C	143.	A
36.	B	72.	A	108.	D	144.	B

145. A
146. B
147. C
148. D
149. C
150. A
151. B
152. A
153. D
154. B
155. A
156. B
157. C
158. D
159. B
160. D
161. B
162. B
163. C
164. B
165. D
166. B
167. D
168. C
169. B
170. B
171. A
172. C
173. C
174. C
175. A
176. A
177. C
178. D
179. C
180. D
181. C
182. B
183. A
184. D
185. C
186. B
187. A
188. D
189. B
190. A
191. D
192. C
193. D
194. A
195. B
196. C
197. A
198. C
199. C
200. D
201. A
202. B
203. A
204. A
205. C
206. D
207. A
208. A
209. C
210. A
211. D
212. A
213. A
214. C
215. D
216. B
217. C
218. C
219. D
220. A
221. C
222. B
223. A
224. B
225. B
226. A
227. B
228. C
229. C
230. A
231. B
232. A
233. C
234. C
235. D
236. C
237. D
238. D
239. B
240. D
241. A
242. D
243. B
244. B
245. B
246. D
247. B
248. C
249. B
250. C

GATE 2015-2000

(1 Mark Question)

GATE 2015

1. Two infinitely large parallel plates (I and II) are held at temperatures T_{I} and T_{II}, $(T_{\text{I}} > T_{\text{II}})$ respectively and placed at a distance 2d apart in vacuum. An infinitely large flat radiation shield (III) is placed in parallel in between I and II. The emissivities of all the plates are equal. The ratio of the steady state radiative heat fluxes with and without the shield is

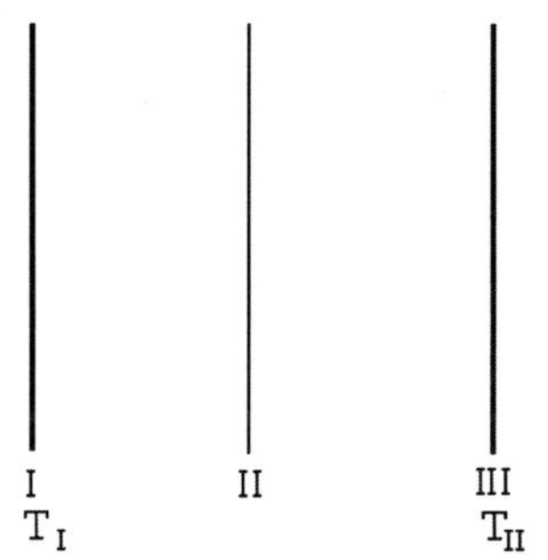

A. 0.5
B. 0.75
C. 0.25
D. 0

Sol. **A.** Heat exchange between shield I and II,

$$\frac{Q_{12}}{A} = \frac{\sigma(T_1^4 - T_2^4)}{\frac{1}{\varepsilon}+\frac{1}{\varepsilon}-1}$$

Where, ε = emissivity

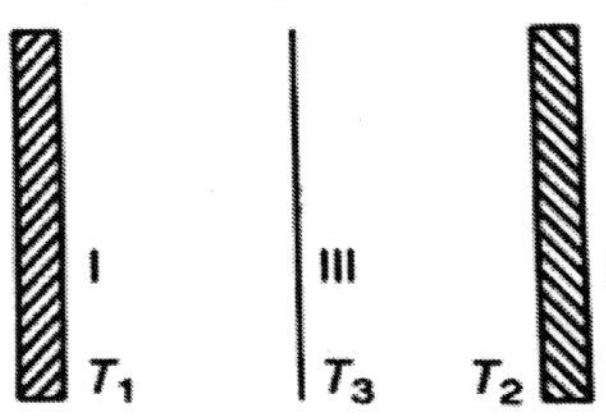

$$\frac{Q_{12}}{A} = \frac{\sigma(T_1^4 - T_2^4)}{\frac{2}{\varepsilon} - 1}$$

In presence of radiation shield,
Heat exchange between shield I and III = heat exchange between shield II and III

$$\frac{Q_{12}}{A} = \frac{Q_{13}}{A} = \frac{\sigma(T_1^4 - T_3^4)}{\frac{2}{\varepsilon}-1} = \frac{\sigma(T_3^4 - T_2^4)}{\frac{2}{\varepsilon}-1}$$

or $T_3^4 = \dfrac{T_1^4 - T_2^4}{2}$

so, $\dfrac{Q_{13}}{A} = \dfrac{\sigma(T_1^4 - T_3^4)}{\frac{2}{\varepsilon}-1} = \sigma\dfrac{\left(T_1^4 - \frac{(T_1^4 - T_2^4)}{2}\right)}{\frac{2}{\varepsilon}-1} = \dfrac{1}{2}\dfrac{Q_{12}}{A}$

$\dfrac{Q_{13}}{Q_{12}} = \dfrac{1}{2}$ or 0.5

2. Consider a solid block of unit thickness for which the thermal conductivity decreases with an increase in temperature. The opposite faces of the block are maintained at constant but different temperatures, T(x = 0) > T (x = 1). Heat transfer is by steady state conduction in x-direction only. There is no source or sink of heat inside the block. In the figure below, identify the correct temperature profile in the block.

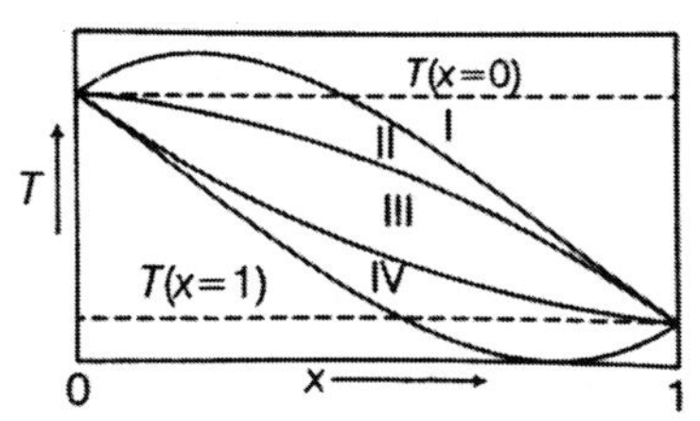

A. I
B. II
C. III
D. IV

Sol. **C.**

Let T (x = 0) = T_1 and T (x = 0) = T_2

From Fourier's law, we have

$$\frac{q}{A} = -k\frac{dT}{dx} \quad \text{....(i)}$$

At steady state conduction, $T_1 > T_2$

$$-\frac{dT}{dx} = \text{constant} = \frac{q}{A}$$

or $\frac{q}{A}$ dx = – KdT

Where, K = decreasing with temperature

Let K = $K_0(1 - \alpha T)$

$$\frac{q}{A}\int_0^x dx = -\int_{T_1}^{T} K_0(1-\alpha)dT$$

$$or\ \frac{q}{A}x = -K_0(T - T_1) + \frac{\alpha}{2}(T^2 - T_1^2) \quad ....(ii)$$

So, integrating Eq. (i) between x = 0 and x = 1

$$\frac{q}{A}\int_{x=0}^{x=1} dx = \int_{T_1}^{T_2} -K_0(1-\alpha T)dT$$

$$or\ \frac{q}{A}(1) = -K_0(T_2 - T_1) + \frac{\alpha}{2}(T_2^2 - T_1^2) \quad ....(iii)$$

Dividing Eq. (ii) by Eq. (iii),

$$\alpha(T^2 - T_1^2) - 2K_0(T - T_1)$$
$$= x(\alpha T_2^2 - \alpha T_1^2) - 2K_0(T_2 - T_1)$$

From the above expression for temperature profile and the temperature pass through a minimum.

GATE 2014

3. Steam economy of a multiple effect evaporator system is defined as

A. kilogram of steam used per hour

B. kilogram of steam consumed in all the effects for each kilogram of steam fed

C. kilogram of steam used in all the effects for each kilogram of water vaporized per hour

D. kilogram of water vaporized from all the effects for each kilogram of steam fed to the first effect

Sol. **C.**

$$steam\ economy = \frac{Amount\ of\ water\ vaporised}{Amount\ of\ stem\ used\ in\ first\ effect}$$

GATE 2013

4. The effectiveness of a heat exchanger in the ε-NTU method is defined as

A. $\frac{increase\ in\ temperature\ of\ the\ cold\ fluid}{decrease\ in\ temperature\ of\ the\ hot\ fluid}$

B. $\frac{actual\ exit\ temperature\ attained\ by\ the\ cold\ fluid}{maximum\ exit\ temperature\ attained\ by\ the\ cold\ fluid}$

C. $\frac{actual\ exit\ temperature\ attained\ by\ the\ hot\ fluid}{minimum\ exit\ temperature\ attainable\ by\ the\ cold\ fluid}$

D. $\frac{actual\ heat\ transfer\ rate}{maximum\ possible\ heat\ transfer\ rate\ from\ hot\ fluid\ to\ cold\ fluid}$

Sol. **D.** effectiveness of a heat exchanger

$$= \frac{actual\ heat\ transfer\ rate}{maximum\ possible\ heat\ transfer\ rate\ from\ hot\ fluid\ to\ cold\ fluid}$$

5. In a pool boiling experiment, the following phenomena were observed.

P. Natural convection

Q. Film boiling

R. Transition boiling

S. Nucleate boiling

What was the correct sequence of their occurrence?

A. P, Q, R, S

B. S, R, Q, P

C. Q, R, P, S

D. P, S, R, Q

Sol. **D.** Correct sequence is

Natural convection → nucleate boiling → Transition boiling → Film boiling

GATE 2012

6. For heat transfer across a solid fluid interface, which one of the following statements is not true when the Biot number is very small compared to 1

A. conduction resistance in the solid is very small compared to convection resistance in the fluid

B. temperature profile within the solid is nearly uniform

C. temperature drop in the fluid is significant

D. temperature drop in the solid is significant

Sol. **D.**

$$biot\ number\ (Bi) = \frac{Conductive\ resistance}{convective\ resistance}$$

When $Bi \ll 1$, then the conductive resistance is very small therefore, temperature drop in the solid is not significant.

7. A solid sphere with an initial temperature T_i is immersed in a large thermal reservoir of temperature T_o. The sphere reaches a steady temperature after a certain time t_1. If the radius of the sphere is doubled, the time required to reach the steady state will be

A. $\frac{t_1}{4}$

B. $\frac{t_1}{2}$

C. $2t_1$

D. $4t_1$

Sol. **C.** we know that

$$\frac{T - T_o}{T_i - T_o} = \exp\left(-\frac{hA}{mC_p}t\right)$$

For all conditions remaining same,

$$\frac{A_1 t_1}{m_1} = \frac{A_2 t_2}{m_2}$$

$$\frac{\pi r_1^2 t_1}{\rho \cdot \frac{4}{3}\pi r_1^3} = \frac{\pi r_2^2 t_2}{\rho \cdot \frac{4}{3}\pi r_2^3}$$

or $\frac{r_2}{r_1} t_1 = t_2$

or $t_2 = 2t_1$

8. If the Nusselt number (Nu) for heat transfer in a pipe varies with Reynolds number (Re) as Nu $\propto Re^{0.8}$, then for constant average velocity in the pipe, the heat transfer coefficient varies with the pipe diameter D as

A. $D^{-1.8}$

B. $D^{-0.2}$

C. $D^{0.2}$

D. $D^{1.8}$

Sol. **B.** $Nu \propto Re^{0.8}$

$$\Rightarrow \quad \frac{hD}{k} \propto \left(\frac{\rho v D}{\mu}\right)^{0.8}$$

$$\Rightarrow \quad h \propto D^{-0.2}$$

GATE 2011

9. Consider two black bodies with surfaces S_1 (area = 1 m^2) and S_2 (area = 4 m^2). They exchange heat only by radiation, 40% of the energy emitted by S_1 is received by S_2. The fraction of energy emitted by S_2 that is received by S_1 is

A. 0.05

B. 0.1

C. 0.4

D. 0.6

Sol. **B.** forradiation heat transfer the objects will receive heat energy in proportion to their surface areas Putting $s_1 = 1\ \text{m}^2$: $s_2 = 4\text{m}^2$and $f_{12} = 0.4$

$$\frac{s_1}{s_2} = \frac{f_{21}}{f_{12}}$$

$$f_{21} = \left(\frac{1}{4}\right) \times 0.4 = 0.1$$

Same answer may be deduced by logical analysis also. Since s_2 is having four times greater surface area than s_1, therefore, the smaller object s_1 will receive 4 time lesser energy since its exposed surface area is four times smaller. Hence 0.4/4 = 0.1 is correct answer

10. In film type condensation over a vertical tube, local heat transfer coefficient is

A. inversely proportional to local film thickness

B. directly proportional to local film thickness

C. equal to local film thickness

D. independent of local film thickness

Sol. **A.** a logical analysis will indicate that as the film thickness increases the resistances to heat transfer will increase thereby, heat transfer coefficient will decrease. Film thickness and heat transfer coefficient are inversely proportional

GATE 2010

11. The ratio Nusselt number to Biot number is

A. conductive resistance of fluid/conductive resistance of solid

B. conductive resistance of fluid/convective resistance of fluid

C. conductive resistance of fluid/conductive resistance fluid

D. unity

Sol. **A.** $N_u = \frac{hD}{k}$ and Biot number $Bi = \frac{hx}{k}$ where L is characteristic length Biot number is a less commonly used dimension less group of heat transfer. It indicates the extent of internal heat resistance of any solid body also exchanging heat from outside.For such systems having less than one Biot number the internal heat equilibrium is fast. It is sort of conductive resistance ofthe solid.

Hence correct answer is **A.**

12. Which one of the following statements about baffles in a shell and tube heat exchanger is false? Baffles

A. act as a support to the tube bundle

B. reduce the pressure drop on the shell-side

C. alter the shell-side flow pattern

D. help in increasing the shell-side flow pattern

Sol. **B.** The obvious answer is **B.** because the baffles inside a shell and tube heat exchanger create obstruction and increase turbulence thereby increasing the pressure drops instead of decreasing. **B.** is false

GATE 2009

13. During the transient convective cooling of a solid object, Biotnumber $\rightarrow 0$ indicates

A. uniform temperature throughout the object

B. negligible convection at the surface of the object

C. significant thermal resistance within the object

D. significant temperature gradient within the object

Sol. **A.**biot umber physically signifies fast internal heat equilibrium and is a measure of internal heat resistance, Biot number → O will therefore indicate a quick interval equilibrium inside the object is uniform temperature throughout the object.

GATE 2008

14. Transient three-dimensional heat conduction is governed by one of the following differential equations

    (α = thermal diffusivity, K = thermal conductivity and

    ψ = volumetric rate of heat generation)

    A. $\frac{1}{\alpha}\frac{\partial T}{\partial t} = \nabla T + \psi K$

    B. $\frac{1}{\alpha}\frac{\partial T}{\partial t} = \nabla T + \frac{\psi}{K}$

    C. $\frac{1}{\alpha}\frac{\partial T}{\partial t} = \nabla^2 T + \psi K$

    D. $\frac{1}{\alpha}\frac{\partial T}{\partial t} = \nabla^2 T + \frac{\psi}{K}$

Sol. **D.**writing the energy balance equation in transient 3-dimensional heat conduction

$$\rho C_p \frac{\partial T}{\partial t} = \nabla . K\nabla T + \psi$$

$$\frac{\rho C_p}{K}\frac{\partial T}{\partial t} = \nabla^2 T + \frac{\psi}{K}$$

$$\frac{1}{\alpha}\frac{\partial T}{\partial t} = \nabla^2 T + \frac{\psi}{K}$$

GATE 2006

15. A stagnant liquid film of 0.4 mm thickness is held between two parallel plates. The top plate is maintained at 40⁰ C and the bottom plate is maintained at 30⁰ C. If the thermal conductivity of the liquid is 0.14 W/(m-K),then the steady state heat flux (in W/m²) assuming one-dimensional heat transfer, is

    A. 3.5

    B. 350

    C. 3500

    D. 7000

Sol. **C.** applying Fourier's law of heat conduction

x = 0.4 mm = $(4 \times 10^{-4} m)$, ΔT = 10⁰K

K = 0.14 W/(m – K)

$q/A = k\frac{\Delta T}{\Delta x} = \frac{0.14\times10}{4\times10^{-4}} = 3500 W/m^2$

GATE 2005

16. An electrically heated element is submerged in a pool of water at its saturation temperature. As the temperature of the element increases, the maximum heat transfer coefficient is observed

    A. in the free convection regime

    B. between the nucleate boiling and partial nucleate boiling mixed with unstable film boiling regimes

    C. in the incipient nucleate boiling regime

    D. in the stable film boiling regime without significant radiation effects

Sol. **B.** nucleate boiling is a type of boiling that takes place when the surface temperature of the element is hotter than the saturated liquid temperature but below the critical heat flux. Therefore, the maximum heat transfer coefficient is observed in the nucleate boiling and partial boiling at peak flux.

17. Baffles are used in heat exchangers in order to

    A. increase the tube side fluid's heat transfer coefficient

    B. promote vibration in the heat exchanger

    C. promote cross flow and turbulence in the shell side fluid

    D. prevent shell expansion due to thermal effects

Sol. **C.** baffles are mechanical devices used in the shell side of shell and tube heat exchanger to promote turbulence and cross flow in the shell of such exchanges. In addition to this they also provide mechanical support to the tube bundle.

18. In film type condensation of liquid along a vertical tube, the thickness of the condensate layer increases towards the bottom. This implies that the local heat transfer coefficient

    A. increases from top to bottom

    B. decreases from top to bottom

    C. remains constant from top to bottom

    D. first increases and then decreases from top to bottom

Sol. **B.** in film type condensation on vertical tubes the thickness of the condensate layer grows progressively and increases from top to bottom. It is known that the local heat transfer coefficient will decrease with increasing thickness because the increased film thickness will offer more resistance to heat transfer.

GATE 2004

19. In forced convection, the Nusselt number Nu is a function of
   A. Re and Pr
   B. Re and Gr
   C. Pr and Gr
   D. Re and Sc

Sol. **A.** according to the Dittus – Boelter equation in forced convection Nu = F (Re, Pr) valid between Re 10000 – 500000 Hence nusselt number Nu is a function of Re and Pr.

20. For an ideal black body,
   A. absorptivity = 1
   B. reflectivity = 1
   C. emissivity = zero
   D. transmissivity = 1

Sol. **A.** by definition an ideal black body absorbs all the radiation incident upon it. Therefore, abortively is 1 for such a black body. The reflectivity, emissivity and transitivity are all equal to zero.

GATE 2003

21. Three solid objects of the same material and of equal masses –a sphere, a cylinder (length = diameter) and a cube are at 500°C initially. These are dropped in a quenching bath containing a large volume of cooling oil each attaining the bath temperature eventually. The time required for 90% change of temperature is smallest for
   A. cube
   B. cylinder
   C. sphere
   D. equal for all the three

Sol. **A.** since the material and mass of all the three solid objects is sphere, cylinder (length = diameter) and cube are same, the rate of cooling will depend upon heat transfer surface area which is the outer and exposed surface area of the three object –

Surface area of a sphere = πd^2

Surface area of cylinder = $\pi dL = \pi d^2$

Surface area of a cube = $6\ d^2$

Therefore time required for 90% change of temperature will be smallest for a cube since the cube has the maximum surface area.

22. A dilute aqueous solution is to be concentrated in an evaporator system. High pressure steam is available. Multiple effect evaporator system is employed because
   A. total heat transfer area of all the effects is less than that in a single effect evaporator system
   B. total amount of vapour produced per kg of feed steam in a multi-effect system is much higher than in a single effect
   C. boiling point elevation in a single effect system is much higher than that in any effect in a multi-effect system
   D. heat transfer coefficient in a single effect is much lower than that in any effect in a multi-effect system

Sol. **B.** the advantage of a multiple effect evaporator system over a single effect is that each kg of high pressure steam introduced in the multiple effect system evaporates progressively slightly less than 1 kg of vapour in each effect. Steam economy which is defined as the kg of vapour produced in the multiple effects per kg of steam fed is much higher than a single effect system. For triple effect system the steam economy be as high as 2.7 – 2.8.

23. The unit of resistance to heat transfer is
   A. $J/(m^2\text{-}K)$
   B. J/(m-k)
   C. $W/(m^2\text{-}K)$
   D. $m^2 - K/W$

Sol. **D.** the unit of resistance to heat transfer is denoted by $\frac{x}{K}$ (Ref. Fourier's law)

Which is $\frac{m}{(w/m-K)} = \frac{m^2-K}{w}$

GATE 2002

24. If the baffle spacing in a shell and tube heat exchanger increases, then the Reynolds number of the shell side fluid
   A. remains unchanged
   B. increases
   C. increases or decreases depending on number of shell passes
   D. decreases

Sol. **D.** the function of baffles is to create flow disturbance and increase the turbulence. This increase in turbulence increases the Reynolds number and subsequently the heat transfer coefficient on shell side. If the baffles spacing is increase the degree of turbulence created will reduce thereby decreasing the Reynolds number.

GATE 2001

25. The heat transfer by radiation from a mild steel surface is to be reduced by reducing the emissivity of the surface. This can be best achieved by

<u>**A.**</u> painting the surface black
<u>**B.**</u> painting the surface white
<u>**C.**</u> giving the surface a mirror finish
<u>**D.**</u> roughening the surface

Sol. <u>**C.**</u> the heat transfer by radiation from the mild steel surface may be reduced by giving the surface a mirror finish which will minimize the heat transfer. A mirror finish will reflect the entire radiation whereas in all the remaining three cases input radiation will be absorbed to various extents.

26. Heat transfer by natural convection is enhanced in systems with

<u>**A.**</u> high viscosity
<u>**B.**</u> high coefficient of thermal expansion
<u>**C.**</u> low temperature gradients
<u>**D.**</u> low density change with temperature

Sol. <u>**A.**</u> heat transfer by natural convection is enhanced in systems with high viscosity because according to the Dittus-Boelter equation Nu = f $(Re)^a (\text{Pr})^b$ In natural convection Re will be less than 10,000, therefore, (Pr) group will be more prominent. Pr = $\frac{C_p \mu}{k}$ will improve at high viscosity (μ) and low thermal conductivity (k). Therefore, systems with high viscosity will enhance the heat transfer rate in natural convection.

GATE 2000

27. A sphere of radius R_1 is enclosed in a sphere of radius R_2. The view (or shape) factor for radiative heat transfer of the outer sphere with respect to the inner sphere is

<u>**A.**</u> zero
<u>**B.**</u> $\frac{R_2}{R_1 + R_2}$
<u>**C.**</u> 1
<u>**D.**</u> $\left(\frac{R_1}{R_2}\right)^2$

Sol. <u>**D.**</u> given radii are R_1 and R_2

If $F_{12} = 1$ then

$$F_{21} = \frac{A_1}{A_2} \times F_{12}$$

$$= \frac{4\pi R_1^2}{4\pi R_2^2} \times 1$$

$$\therefore \; F_{21} = \left[\frac{R_1}{R_2}\right]^2$$

(2 Mark Question)

GATE 2015

28. In the figure, the temperature profiles of cold and hot fluids in counter current double pipe heat exchanges (in different mode of operation) are shown on the left. For each case, match the heat exchange process for the fluid represented by the bold curve with the options given on the right.

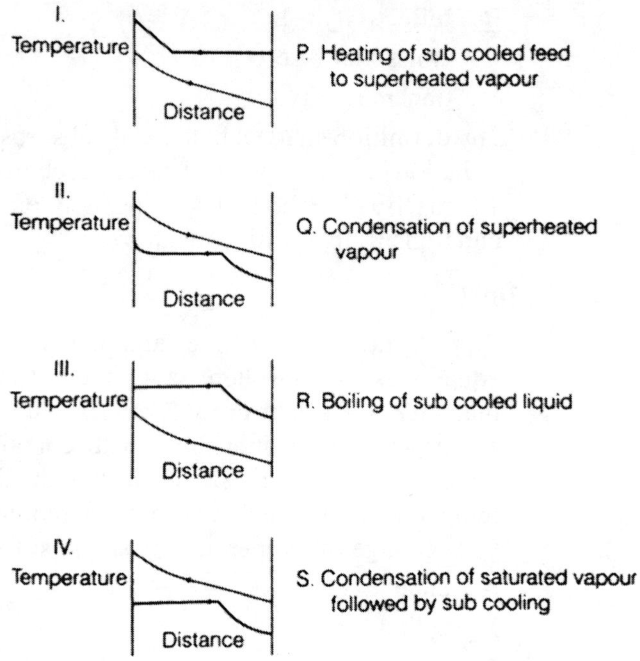

Codes

<u>**A.**</u> I-P, II-Q, III-R, IV-S
<u>**B.**</u> I-P, II-Q, III-S, IV-R
<u>**C.**</u> I-Q, II-P, III-S, IV-R
<u>**D.**</u> I-Q, II-S, III-P, IV-R

Sol. <u>**C.**</u>(P) heating of sub cooled feed to superheated vapour:

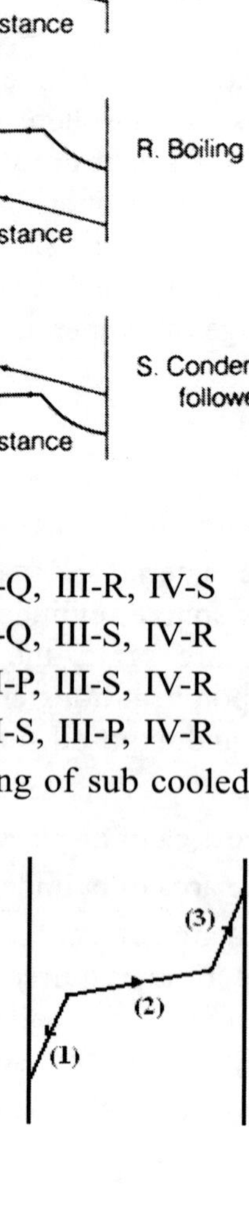

(1) Sub cooled to saturated
(2) Saturated upto boiling point
(3) Saturated to superheat

(Q) Condensation of superheated vapour:

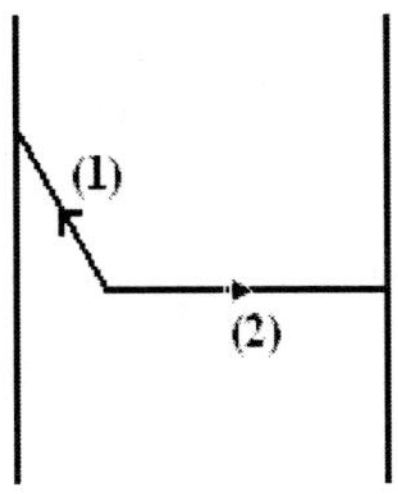

(1) Cooling of superheated vapour to saturated vapour
(2) Codensation of saturated vapour to liquid

(R) Boilling of sub cooling liquid:

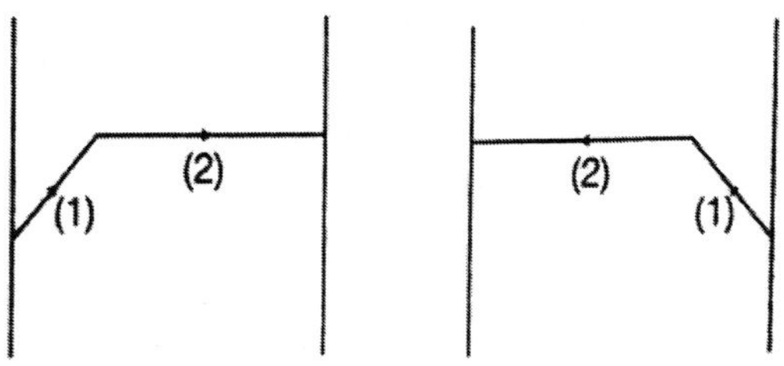

(1) Heating of sub cooled liquid to saturated liquid
(2) Boiling of saturated liquid

(S) Condensation of saturated vapour followed by sub cooling:

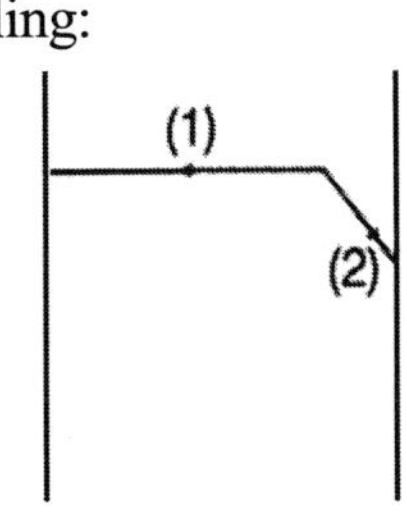

(1) Condensation of saturated vapour to saturated liquid
(2) Sub cooling of saturated liquid

29. A heated solid copper sphere of surface area A and volume V is immersed in a large body of cold fluid. Assume the resistance to heat transfer inside the sphere to be negligible and heat transfer coefficient (h), density (ρ), heat capacity (C) and thermal conductivity (k) to be constant. Then, at time t the temperature difference between the sphere and the fluid is proportional to

A. $\exp\left(-\frac{hA}{\rho VC}t\right)$

B. $\exp\left(-\frac{\rho VC}{hA}t\right)$

C. $\exp\left(-\frac{4\pi k}{\rho CA}t\right)$

D. $\exp\left(-\frac{\rho CA}{4\pi k}t\right)$

Sol. **A.**time $T = f$

Applying first law of thermodynamics, if body temperature is T then

$$hA(T_\infty - T) = \rho C_p V\frac{dT}{dt}$$

where, C_p= heat capacity of body

$$\int_{T_0}^{T}\frac{dT}{T_\infty - T} = \int_0^t \frac{hA}{\rho V C_p}dt$$

or $$ln\frac{T-T_\infty}{T_0-T_\infty} = -\frac{hA}{\rho V C_p}t$$

or $$T - T_\infty = e\left[-\frac{hA}{\rho VC}t\right]$$

30. Air is flowing at a velocity of 3 m/s perpendicular to a long pipe as shown in the figure below. The outer diameter of the pipe is d = 6 cm and temperature at the outside surface of the pipe is maintained at 100°C. The temperature of the air far from the tube is 30°C.

Given data,

V = 18 × $10^{-6} m^2/s$, thermal conductivity, k = 0.03 W/(mK)

according nusselt number correlation,

Nu = $\frac{hd}{k}$ = 0.024 × $Re^{0.8}$, the rate of heat loss per unit

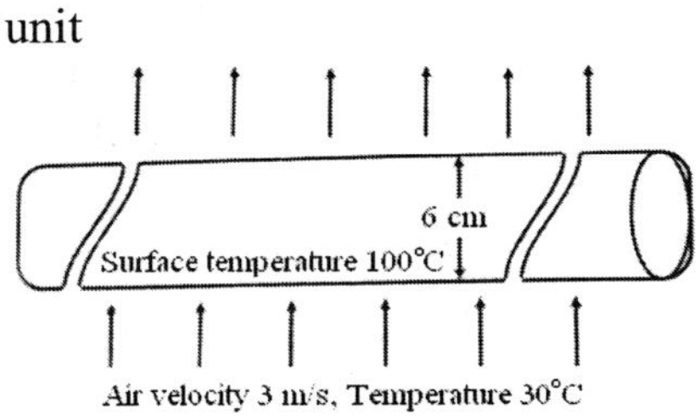

Length (W/m) from the pipe to air (upto one decimal place) is______ .

Sol. given, velocity, $v = 3\ m/s$

Diameter, $d = 6 \times 10^{-2}\ m$

Viscosity, v = $18 \times 10^{-6} m^2/s$

Thermal conductivity, $K = 0.03\ W/mK$

$$Nu = \frac{hd}{k} = 0.024\ Re^{0.8}$$

$$Re = \frac{\rho Vd}{\mu} = \frac{vd}{V} = \frac{3\times6\times10^{-2}}{18\times10^{-6}} = 10000$$

So, $\frac{hd}{k} = 0.024 \times (10000)^{0.8} = 38.037$

or $h = \frac{0.03 \times 38.037}{6 \times 10^{-2}} = 19.0185\ W/m^2K$

Rate of heat loss/length

$= h \cdot (\pi d)(T_s - T_a)$

$= 19.0185 \times \pi \times 6 \times 10^{-2} \times (100 - 30)$

$= 250.943\ W/m$

GATE 2014

31. Abrick wall of 20 cm thickness has thermal conductivity of 0.7 W $m^{-1}K^{-1}$. An insulation of thermal conductivity 0.2 W $m^{-1}K^{-1}$ is to be applied on one side of the wall, so that the heat transfer through the wall is reduced by 75%. The same temperature difference is maintained across the wall before and after applying the insulation. The required thickness (in cm) of the insulation is __________.

Sol. Case – I

$$Q = \frac{k_1 A \Delta T}{x_1}$$

$$Q = \frac{A \Delta T}{\left(\frac{x_1}{k_1}\right)}$$

Case – II

$$0.25\ Q = \frac{A\ \Delta T}{\frac{x_1}{k_1} + \frac{x_2}{k_2}}$$

$$\Rightarrow \quad \frac{1}{4} \times \frac{A \Delta T}{\left(\frac{x_1}{k_1}\right)} = \frac{A\ \Delta T}{\frac{x_1}{k_1} + \frac{x_2}{k_2}}$$

$$\Rightarrow \quad \frac{4\ x_1}{k_1} = \frac{x_1}{k_1} + \frac{x_2}{k_2}$$

$$\Rightarrow \quad x_2 = \frac{3\ x_1}{k_1} \times K_2 = \frac{3 \times 20 \times 0.2}{0.7}$$

$$\Rightarrow \quad x_2 = 17.1\ cm$$

32. Oil with a flow rate of a 1000 kg/h is to be cooled using water in a double-pipe counter-flow heat exchanger from a temperature of 70°C to 40°C. Water enters the exchanger at 25°C and leaves at 40°C. The specific heats of oil and water are 2 kJ $kg^{-1}K^{-1}$, respectively. The overall heat transfer coefficient is 0.2 kW $m^{-2}K^{-1}$. The minimum heat exchanger area (in m^2) required for this operation is __________.

Sol. log mean temperature difference (LMTD)

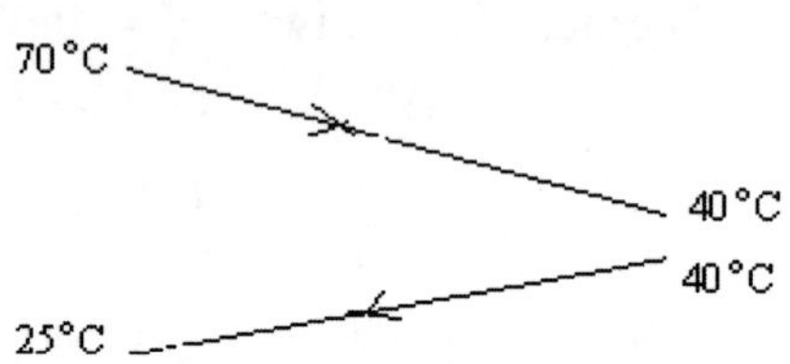

$$\Delta T_m = \frac{(T_{n2} - T_{c1}) - (T_{n1} - T_{c2})}{ln\left(\frac{T_{n2} - T_{c1}}{T_{n1} - T_{c2}}\right)}$$

$$= \frac{(40-25)-(70-40)}{\ln\left(\frac{40-25}{70-40}\right)} = \frac{15-30}{\ln\left(\frac{15}{30}\right)} = \frac{15}{\ln 2}$$

Heat balance,

$$Q = UA\Delta T_m = (ms\Delta T)_{oil}$$

$$U = 0.2\ \text{kW}/m^2\text{-K} = 200\text{W}/m^2\text{-K}$$

$$m = 1000\ \text{kg/h} = \frac{10}{36}\text{kg/s}$$

$$S_{oil} = 2\text{kJ/kg-K} \quad = 2000\text{J/kg-K}$$

$$\Delta T_{oil} = 70 - 40 \quad = 30\ K$$

$$\therefore \quad A = \frac{ms\,\Delta T}{U \Delta T_m} = \frac{10}{36} \times \frac{2000 \times 30 \times \ln 2}{200 \times 15} = 3.85\ m^2$$

33. The bottom face of a horizontal slab of thickness 6 mm is maintained at 300°C. The top face is exposed to a flowing gas at 30°C. The thermal conductivity of the slab is 1.5 W $m^{-1}K^{-1}$ and the convective heat transfer coefficient is 30 W $m^{-2}K^{-1}$. At steady state, the temperature (in °C) of the top face is ________.

Sol. let the temperature of the top face is T(°C)

$$Q_{conduction} = Q_{convection}$$

$$kA\frac{\Delta T}{x} = UA\Delta T'$$

Putting values,

$$\Rightarrow \quad \frac{1.5 \times (300 - T)}{6 \times 10^{-3}} = 30(T - 30)$$

$$\Rightarrow \quad T = 271°\text{C}$$

GATE 2013

34. Calculate the heat required (in kJ, up to 1 digit after the decimal point) to raise the temperature of 1 mole of a solid material from 100 °C to 1000°C. The specific heat C_p of the material (in J/mol-K) is expressed as C_p = 20 + 0.005T, where T is in K. Assume no phase change.

A. 25.7 kJ

B. 20.8 kJ

C. 21.7 kJ

D. 28.3 kJ

Sol. **C.** we have,

$$dQ = nC_p dT$$

$$1 \times (20 + 0.005T)dT$$

Heat required increasing temperature from to , T_1 to T_2,

$$Q = \int dQ$$

$$= \int_{T_1}^{T_2} (20 + 0.005T)dT$$

$$= 20(T_2 - T_1) + \frac{0.005}{2}(T_2^2 - T_1^2)$$

$$T_1 = 373\ K, T_2 = 1273\ K$$

$$Q = 20(1273 - 373) + \frac{0.005}{2}[(1273)^2 - (373)^2]$$

Or $Q = 18000 + 3703 = 21703$ J

Or $Q = 21.7$ kJ

35. In a double pipe counter-current heat exchanger, the temperature profiles shown in the figure were observed. During operation, due to fouling inside the pipe, the heat transfer rate reduces to half of the original value. Assuming that the flow rates and the physical properties of the fluids do not change, the LMTD (in °C) in the new situation is

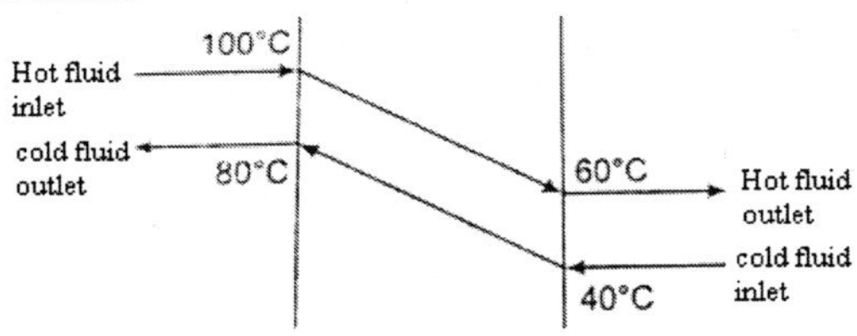

A. 0

B. 20

C. 40

D. indeterminate

***Sol.* C.**

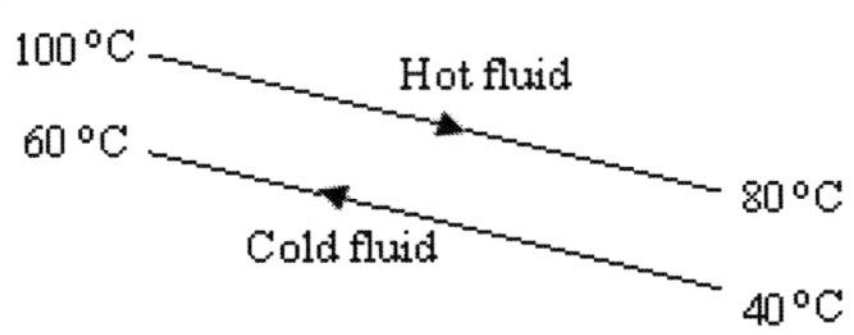

Heat lost by hot fluid,

$$q = (m\,C_p)_h(100 - 60)$$

$$= 40 \times (m\,C_p)_h$$

Heat gained by cold fluid,

$$q = (m\,C_p)_c(80 - 40) = 40 \times (m\,C_p)_c$$

After fouling, let the outer temperature of hot and could fluid becomes T_1 and T_1 respectively heat lost by hot fluid,

$$\frac{q}{2} = (m\,C_p)_h \times (100 - T_1)$$

From Eqs. (i) and (iii),

$$\frac{40}{2} = 100 - T_1$$

$$T_1 = 80°\text{C}$$

Similarly, heat gained by cold fluid,

$$\frac{q}{2} = (m\,C_p)_c \times (T_2 - 40)$$

From Eqs. (ii) and (iv),

$$\frac{40}{2} = T_2 - 40$$

$$T_2 = 60°\text{C}$$

$LMTD = 100 - 60°\text{C}$ or $80 - 40 = 40°\text{C}$

GATE 2012

36. In a 1-1 pass floating head type shell and tube heat exchanger, the tubes (od = 25 mm, id = 21 mm) and arranged in a square pitch. The tube pitch is 32 mm. The thermal conductivity of the shell side fluid is 0.19 W/m-K and the Nusselt number is 200. The shell side heat transfer coefficient in $W/m^2 - K$ rounded off to the nearest integer is

A. 1100

B. 1400

C. 1800

D. 2100

***Sol.* C.** $Nu = \frac{h \times D}{k}$

$$\Rightarrow 200 = \frac{h \times 0.021}{0.19}$$

$$\Rightarrow \quad h \approx 1800\ W/m^2 - K$$

37. The one-dimensional unsteady state heat conduction equation in a hollow cylinder with a constant heat source q is

$$\frac{\partial T}{\partial t} = \frac{1}{r}\frac{\partial}{\partial r}\left(r\frac{\partial T}{\partial t}\right) + q$$

If A and B are arbitrary constants, then the steady state solution to the above equation is

A. $T(r) = -\frac{qr^2}{2} = \frac{A}{r} + B$

B. $T(r) = -\frac{qr^2}{4} + A\,In\,r + B$

C. $T(r) = A\,In\,r + B$

D. $T(r) = -\frac{qr^2}{4} + A\,In\,r + B$

***Sol.* B.** one dimensional heat conduction equation as

$$\frac{\partial T}{\partial t} = \frac{1}{r}\frac{\partial}{\partial r}\left(r\frac{\partial T}{\partial r}\right) + q$$

Steady state,

$$\frac{\partial T}{\partial r} = 0$$

$$\Rightarrow \quad \frac{1}{r}\frac{\partial}{\partial r}\left(r\frac{\partial T}{\partial r}\right) + q = 0$$

$$\Rightarrow \quad \partial\left(r\frac{\partial T}{\partial r}\right) = -qr\partial r$$

Integrating,

$$r\frac{\partial T}{\partial r} = -\frac{qr^2}{2} + A$$

$$\partial T = \left(-\frac{qr}{2} + \frac{A}{r}\right)\partial r$$

Integrating,

$$T(r) = -\frac{qr^2}{4} + A\ln r \cdot$$

GATE 2011

38. Oil at 120° C is used to heat water at 30° C in a 1-1 concurrent shell and tube heat exchanger. The available heat exchange area is S_1. The exit temperature of the oil and the water streams are 90°C and 60° C respectively. The cocurrent heat exchanger is replaced by a 1-1 countercurrent heat exchanger having heat exchange area S_2. If the exit temperature and the overall heat transfer coefficients are same, the ratio of S_1 to S_2 is

A. infinite
B. 1.1
C. 0.91
D. zero

***Sol.* B.**

$120^0C \xrightarrow{Oil} 90^0C$

$30^0C \longrightarrow 60^0C$

For cocurrent case

$$\Delta T \text{ (LMTD)} = \frac{(120-30)-(90-60)}{ln\left(\frac{120-30}{90-60}\right)}$$

$= 54.61^0C$

$Q = (UA_1)(54.61)$ ....(i)

For countercurrent case

$120^0C \longrightarrow 90^0C$

$60^0C \longleftarrow 30^0C$

Since ΔT_1 and ΔT_2 are both equal to 60^0C the LMTD which is logarithm mean temperature difference will also be equal to 60^0C

Now since the OHTC and exit temperature are same in both the cases = Q will also be same

$\therefore \quad Q = (UA_2)(60)$ ....(ii)

Combining equation (i) and (ii)

$$\frac{A_1}{A_2} = \frac{60}{54.61} = 1.1$$

39. Heat is a generated uniformly within a solid slab. The slab separates fluid 1 from fluid 2. The heat transfer coefficients between the solid slab and the fluid are h_1 and h_2 ($h_2 > h_1$) respectively. The steady state temperature profile (T vs x) for one-dimensional heat transfer is correctly shown by,

A. 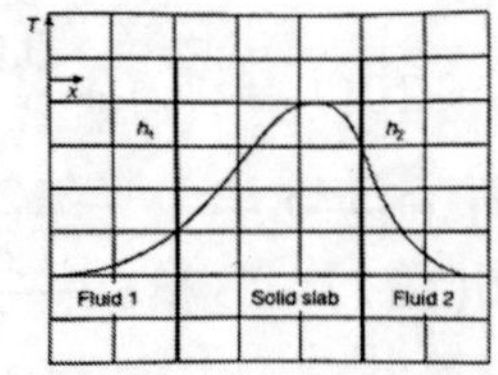

B. 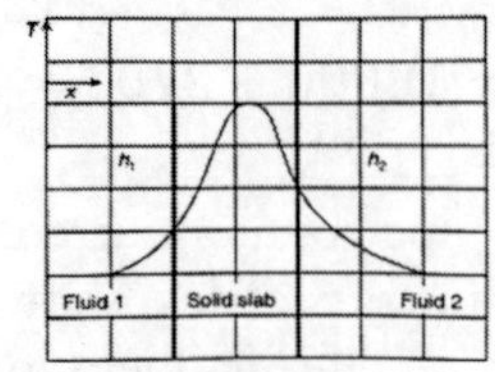

C. 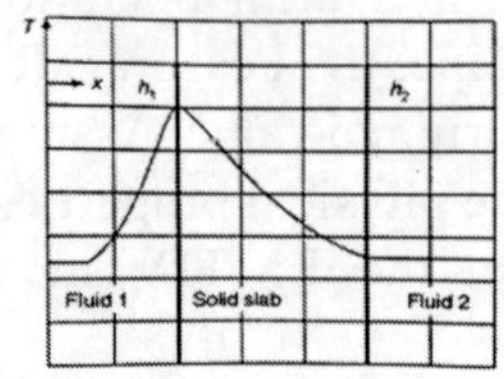

D. 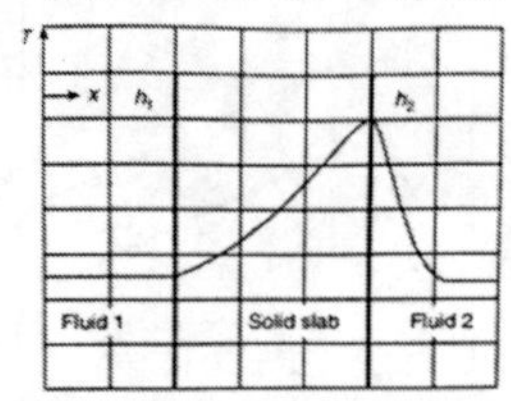

***Sol.* D.** heat is generated within the slab and on the outside faces fluid 1 and 2 are the heat transfer coefficients h_1 and h_2 respectively **($h_2 > h_1$).**

A higher value of h_2 will mean heat is dissipated in fluid 2 at a faster rate than fluid 1. This will result in a temperature profile falling sharply on fluid 2 sides whereas on fluid 1 side the temperature profile will fall slowly. This temperature profile trend is evident in **D.**

GATE 2010

40. The figure below shows steady state temperature profiles for one dimensional heat transfer within a solid slab for the following cases:

P. Uniform heat generation with left surface perfectly insulated

Q. Uniform heat generation with right surface perfectly insulated

R. Uniform heat consumption with left surface perfectly insulated

S. Uniform heat consumption with right surface perfectly insulated

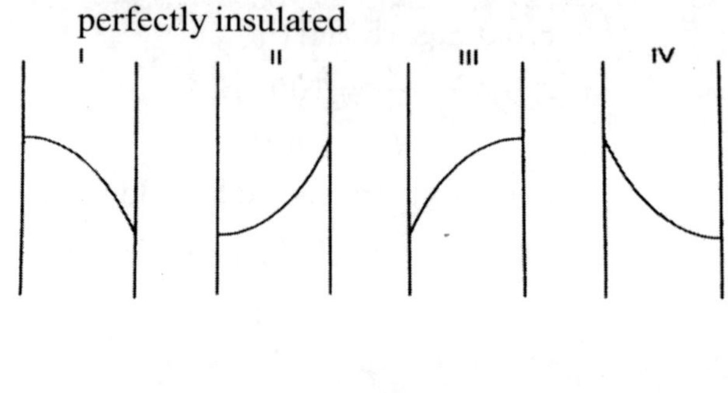

Match the profiles with appropriate cases.

A. P- I, Q-III, R-II, S-IV

B. P-II, Q-III, R-I, S-IV

C. P-I, Q- IV, R- II, S-III

D. P-II, Q- IV, R- I, S-III

Sol. **A.** for the case when there is uniform heat generation with left surface perfectly insulated the temperature profile will show very small decrease in temperature near the left face and a sharp fall thereafter towards right face as there will be heat loss from right face. This temperature profile is shown in profile I

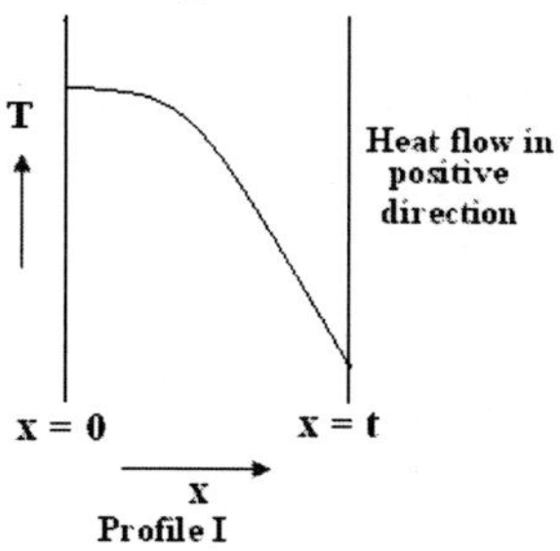

For the case when the left face is insulated and there is uniform heat consumption, similarly, there will be near constant temperature condition near the left face and the temperature will rise sharply thereafter as there is uniform heat consumption within the solid slab. This profile is evident in profile II

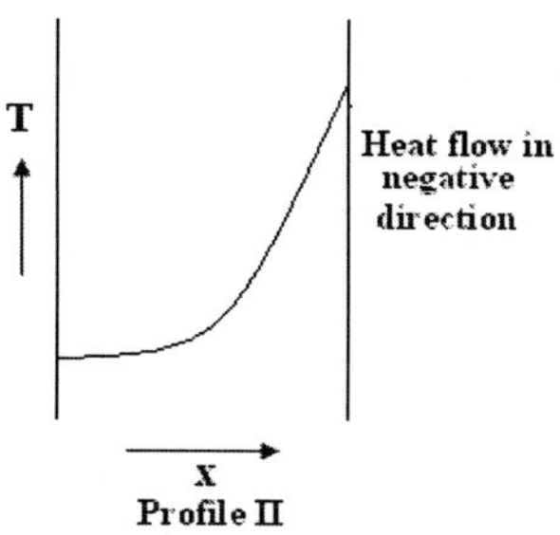

Similarly for heat generation with right surface perfectly insulated, there will be almost flat temperature profile near right surface and sharp fall towards left surface. This temperature profile is shown in III

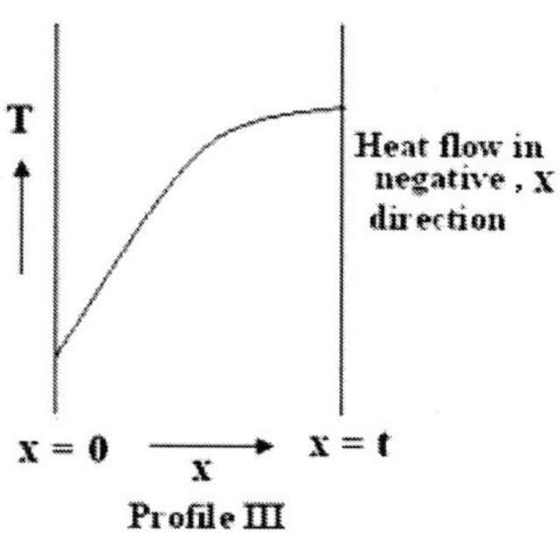

Applying the same logic for uniform heat consumption with right surface perfectly insulated, the temperature profile will be as in (iv) wherein near right surface the temperature change is flat and because of heat consumption the temperature rises sharply towards left surface with heat flow in positive x direction.

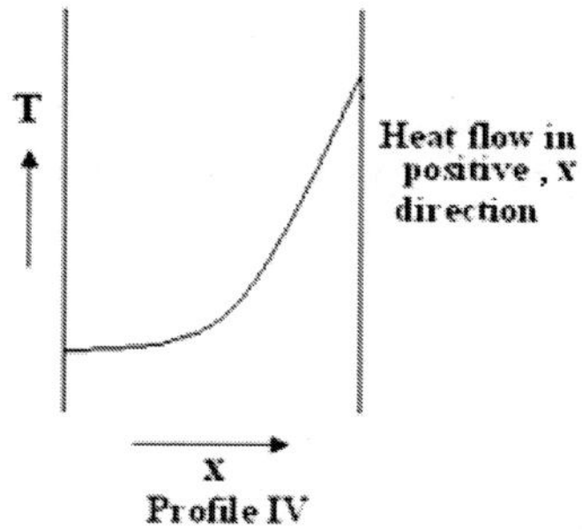

41. The view factor matrix for two infinitely long coaxial cylinders, shown in the figure below, is

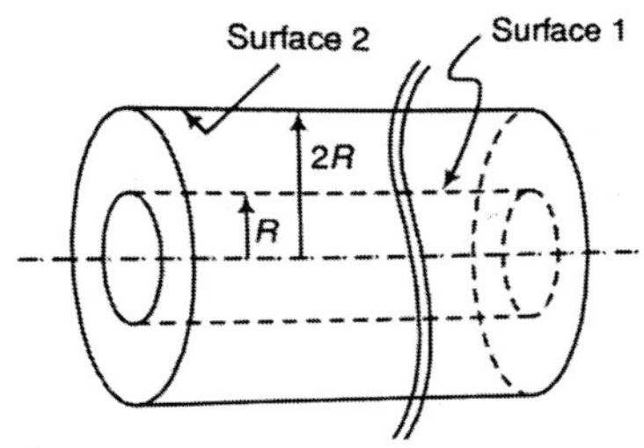

A. $\begin{bmatrix} 0 & 1 \\ 0.5 & 0.5 \end{bmatrix}$

B. $\begin{bmatrix} 0 & 1 \\ 1 & 0 \end{bmatrix}$

C. $\begin{bmatrix} 1 & 0 \\ 0 & 1 \end{bmatrix}$

D. $\begin{bmatrix} 0.5 & 0.5 \\ 0 & 1 \end{bmatrix}$

Sol. **A.**

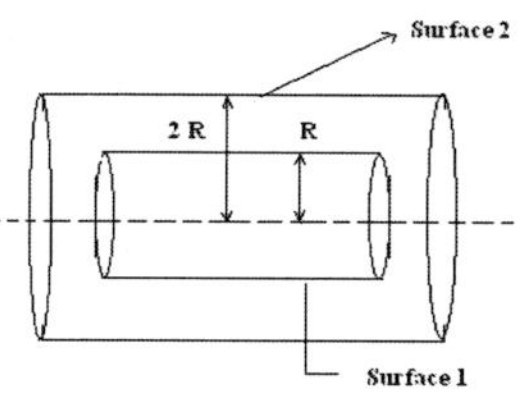

The view factor matrix may be written as

$$\begin{bmatrix} F_{11} F_{12} \\ F_{21} F_{22} \end{bmatrix}$$

Since surface 2 is receiving radiation falling upon it from surface 1

$$F_{12} = 1$$

$$\text{Also } \frac{F_{12}}{F_{21}} = \frac{Area\ of\ surface\ 2}{Area\ of\ surface\ 1} = \frac{(2\pi)(2R)h}{(2\pi)(R)h}$$

$$\therefore \quad F_{21} = \frac{1}{2} = 0.5$$

Since $F_{21} + F_{22} = 1$

$$F_{22} = 1 - 0.5 = 0.5$$

$\therefore$ The view factor matrix is $\begin{bmatrix} 0 & 1 \\ 0.5 & 0.5 \end{bmatrix}$

Common data for questions 24 and 25

Hot oil at 150° C is used to pre-heat a cold fluid at 30° C in a 1:1 shell and tube heat exchanger. The exit temperature of the hot oil is 110°C. Heat capacities (product of mass flow rate and specific heat capacity) of both the streams are equal. The heat duty is 2kW.

42. Under co-current flow conditions, the overall heat transfer resistance ($1/UA$) is

A. 0.4° C/W
B. 0.04° C/W
C. 0.36°C/W
D. 0.036°C/W

Sol. **D.** 150 → 110

30 → 70

By energy balance

$$M_1 C_{p1} (150 - 110) = M_1 C_{p2} (T - 30)$$

Given that $M_1 C_{p1} = M_2 C_{p2}$

$\therefore$ T = 70°C

Now, LMTD

$$(\Delta T) \text{ concurrent} = \frac{(150-30)-(110-70)}{lu\left(\frac{150-30}{110-70}\right)}$$

$$= 72.82^0\text{C}$$

$$Q = UA\Delta T$$

2kw = 2000 , w = (UA) 72.82

or $\frac{1}{UA} =$ 0.036°C/W

43. Under counter-current flow conditions, the overall heat transfer resistance ($1/UA$) is

A. 0.4° C/W
B. 0.04° C/W
C. 0.36°C/W
D. 0.036°C/W

Sol. **B.** For counter current flow

150 → 110

70 ← 30

Both ΔT_1 and ΔT_2 = 80°C

$\therefore$ LMTD (ΔT) = 80°C

Now, Q = UAΔT

2000 = (UA) 80

$$\therefore \frac{1}{UA} = \frac{80}{2000} = 0.04\ ^0\text{C/W}$$

GATE 2009

44. A well –insulated hemispherical furnace (radius = 1m) is shown below

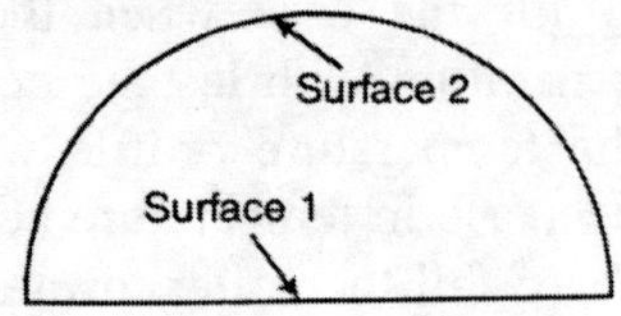

The self-view factor of radiation for the curved surface 2 is

A. 1/4
B. 1/2
C. 2/3
D. 3/4

Sol. **B.** Applying $\frac{F_{12}}{F_{21}} = \frac{A_2}{A_1}$

Or $\frac{1}{F_{21}} = \frac{2\pi r^2}{\pi r^2}$ for hemispherical furnace

$$\therefore F_{21} = \frac{1}{2}$$

45. A double – pipe heat exchanger is to be designed to heat 4 kg/s of a cold feed from 20 to 40° C using a hot stream available at 160° C and a flow rate of 1 kg/s. The two streams have equal specific heat capacities and the overall heat transfer coefficient of the heat exchanger is 640 W/(m²-K). Then the ratio of the heat transfer areas required for the co-current to counter-current modes of operations is

A. 0.73
B. 0.92
C. 1.085
D. 1.25

Sol. **C.** by energy balance

$$4 (Cp_1) (40 - 20) = 1(Cp_2) (160 - T)$$

Where T is outlet temperature of hot fluid

Since $Cp_1 = Cp_2$

$\therefore$ $T = 80^0\text{C}$

$$\text{LMTD } (\Delta T) \text{ concurrent} = \frac{(160-20)-(80-40)}{ln\left(\frac{160-20}{80-40}\right)} = 79.82^0\text{C}$$

$$\text{LMTD } (\Delta T) \text{ countercurrent} = \frac{(160-40)-(80-20)}{ln\left(\frac{160-20}{80-20}\right)} = 86.56^0\text{C}$$

Since Q and U are same for both countercurrent and concurrent flow

By Q = UA ΔT

$$\frac{(Area)\ cocurrent}{Area\ (countercurrent\ )} = \frac{86.56}{79.82} = 1.085$$

46. For the composite wall shown below (Case 1), the steady state interface temperature is 180°C. If the thickness of layer P is doubled (Case 2), then the rate of heat transfer (Assuming 1-diamansional conduction) is reduced by

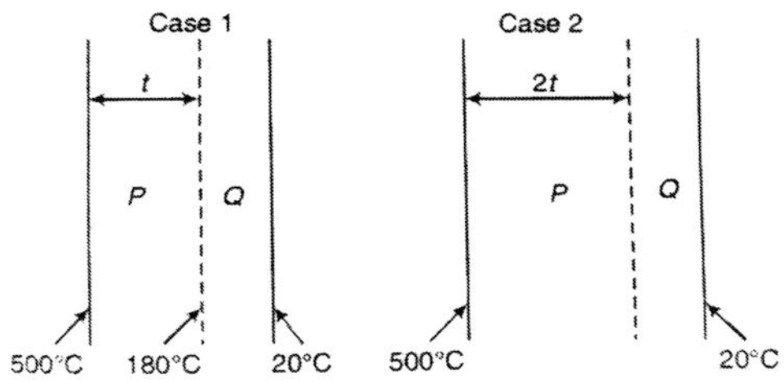

A. 20%
B. 40%
C. 50%
D. 70%

Sol. **B.** by Fourier's law

$$\frac{q}{A} = \frac{\Delta T}{x/k} = \frac{\Delta T}{R} \quad \text{where R = Resistance} = \frac{x}{k}$$

Under steady state same heat will flow through both

For Case 1 $\therefore \frac{q}{A} = \frac{500-180}{R_1} = \frac{180-20}{R_2}$

$\therefore$ $R_1 = 2R_2$

Hence $\frac{q}{A} = \frac{160}{R_2}$

Case 2

When the thickness of layer is doubled then correspondingly resistance will also be doubled to $2R_1$ the total resistance being $(2R_1 + R_2)$

Now $\frac{q_2}{A} = \frac{500-20}{(2R_1+R_2)} = \frac{480}{(4R_1+R_2)}$

$= \frac{480}{5R_2}$

Under steady state

$$\frac{q_2}{A} = \frac{480}{5R_2} = \frac{96}{R_2}$$

Hence percentage reduction in rate of heat transfer

$$= \frac{q_1/A - q_2/A}{q_1/A} \times 100$$

$$= \frac{160/R_2 - 96/R_2}{160/R_2}$$

$= 40\%$

Common data for questions 47 and 48

A slab of thickness *t* with one side (x = 0) insulated and the other side (x = *t)* maintained at a constant temperature T_0 is shown below.

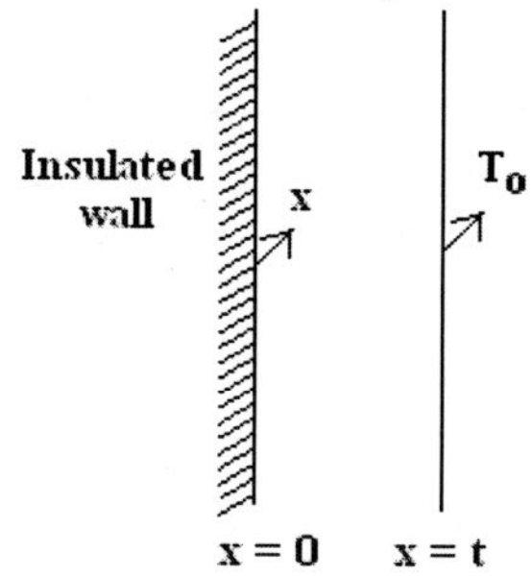

A uniformly distributed internal heat source produces heat in the slab at the rate of *S* W/m³. Assume the heat conduction to be steady and 1-dimansional along the x-direction.

47. The maximum temperature in the slab occurs at x equal to

A. zero
B. t/4
C. t/2
D. t

Sol. **A.**

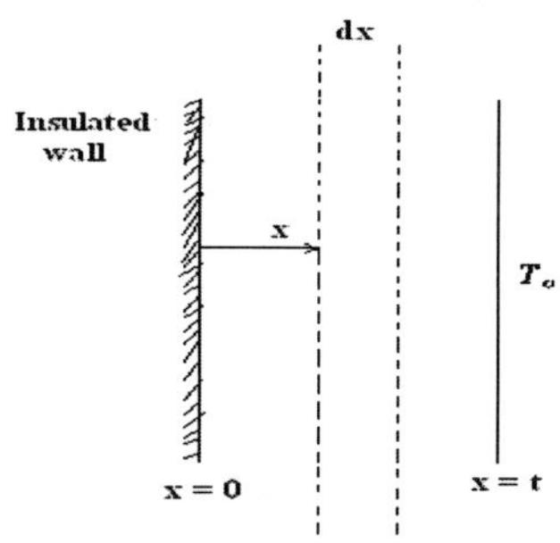

By logical analysis it can be deduced that since at x = 0 the slab is insulated, heat will flow in x direction and a temperature gradient will exist will x = 0 being the point of maximum temperature and decreasing progressively toward x = t. Hence, the maximum temperature in the slab will obviously at x = 0

However, a mathematical derivation also indicates the same conclusion. Writing energy balance over a differential element of thickness dx at x for unit surface area

$$q_{x+dx} - q_x = (S)\,dx$$

or $S = \frac{q_{x+dx} - q_x}{dx} = \frac{d(q_x)}{dx}$

or $q_x = k\frac{dT}{dx}$

$\therefore \frac{d}{dx}\left(-k\frac{dT}{dx}\right) = S$

Integrating and evaluating the constant of integration

At x = 0, heat flux = 0 constant of integration = 0

Hence, $-k\frac{dT}{dx} = Sx$

Maximum temperature will occur when the first derivation is zero

$\frac{dT}{dx} = 0$ and $Sx = 0$

Hence x = 0

If the maximum temperature of the slab will occur at x = 0

48. The heat flux at x = t is

A. zero

B. St/4

C. St/2

D. St

Sol. **D.** refer the equation above

$$q = -k\frac{dT}{dx} = Sx$$

Hence heat flux at x = t

$$q = St$$

GATE 2008

49. Two plats of equal thickness (t) and cross-sectional area are joined together to form a composite as shown in the figure. If the thermal conductivities of the plates are *k* and *2k*, then the effective thermal conductivity of the composite is

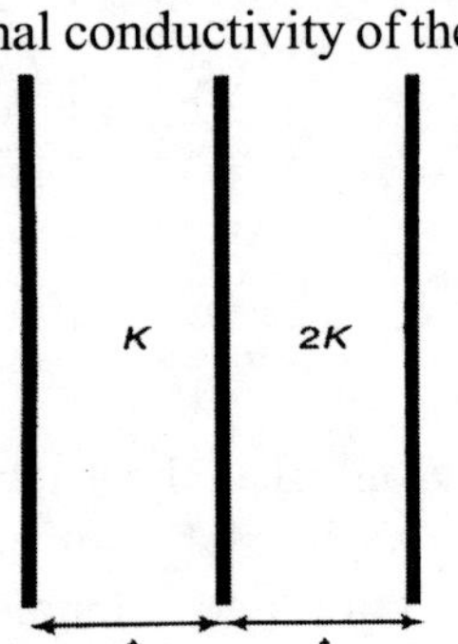

A. 3*k*/2

B. 4*k*/3

C. 3*k*/4

D. 2*k*/3

Sol. **B.** By Fourier's law $Q = -kA\frac{dT}{dx}$

Hence the resistance term is $\frac{x}{k}$

If K_{ef} is the overall thermal conductivity of the composite then

$$R = R_1 + R_2$$

$$\frac{2t}{K_{ef}.A} = \frac{t}{kA} + \frac{t}{(2k)A}$$

or $\frac{2t}{K_{ef}} = \frac{t}{k} + \frac{t}{2k} = \left(\frac{3}{2}k\right)$

Hence $K_{ef} = \frac{4}{3}k$

50. A metallic ball [ρ = 2700 kg/m³ and C_p = 0.9 kJ / (kg-°C)] of diameter 7.5 cm is allowed to cool in air at 25°C. When the temperature of the ball is 125°C, it is found to cool at the rate of 4°C /min. If thermal gradients inside the ball are neglected, the heat transfer coefficient (in W/((m²-°C)) is

A. 2.034

B. 20.34

C. 81.36

D. 203.4

Sol. **B.** for metallic ball = 2700 kg/m³

C_p = 0.9 kJ/kg°C

d = 7.5 cm T_b = 125°C, T_{air} = 25°C

rate of cooling $\frac{dT}{dt}$ = 4°C/min

$= \frac{1}{15}$°C/sec

By energy balance

$$hA(T_b - T_{air}) = (\text{vol. of ball})\ \rho C_p \frac{dT}{dt}$$

$$h\,\pi d^2(T_b - T_{air}) = \left(\frac{\pi d^3}{6}\right)\rho C_p\frac{dT}{dt}$$

$$\text{or} \quad h = \left(\frac{d}{6}\right)\left(\frac{\rho C_p}{T_b - T_{air}}\right)\frac{dT}{dt}$$

Putting the numerical value in consistent units

$$h = \left(\frac{0.075}{6}\right)\left(\frac{2700\times900}{125-25}\right)\frac{1}{15}$$

$$h = 20.25\ \text{W/m}^{2\,0}\text{C}$$

51. Hot liquid is following at a velocity of 2 m/s through a metallic pipe having an inner diameter of 3.5 cm and length of 20 m. The temperature at the inlet of the pipe is 90°C. Following data is given for liquid at 90°C.

Density = 950 kg/m³;

Specific heat = 4.23 kJ /(kg-°C)

Viscosity = 2.55 × 10⁻⁴ kg/(m-s);

Thermal conductivity = 0.685 W/(m-°C)

The heat transfer coefficient [in kW/(m²-°C)] inside the tube is

A. 222.22

B. 111.11

C. 22.22

D. 11.11

Sol. **D.** Following data u = 2 m/s d = 0.035 m

ρ = 950 kg/m³ L = 20 m

C_p = 4.23 kJ/kg°C = 4230 J/kg°C

$\mu = 2.55\times 10^{-4}$ kg/(m/sec)

K = 0.685 W/m^{0}C

h has to be calculated by Dittus – Boelter equation

$$\frac{hd}{k} = 0.023\ (R_e)^{0.8}(P_r)^{0.33}$$

$$\text{Hence Re} = \frac{du\rho}{\mu} = \frac{950\times2\times0.035}{2.55\times10^{-4}}$$

$\therefore$ Re = 260784

$$\text{Pr} = \frac{Cp_{\mu}}{\mu} = 1.57$$

Substituting the values of Re and Pr

$$h = (0.023)\left(\frac{k}{D}\right)(260784)^{0.8}(1.57)^{0.33}$$

= 11256 W/m^2 ^{0}C

= 11.256 kw/m^2 ^{0}C

52. The temperature profile for heat transfer from one fluid to another separated by a solid wall is

A.

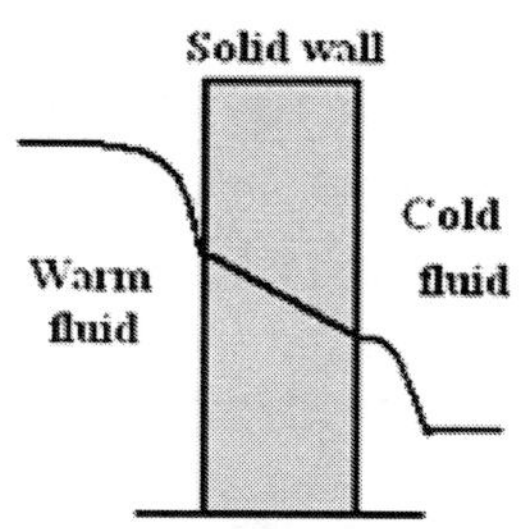

B.

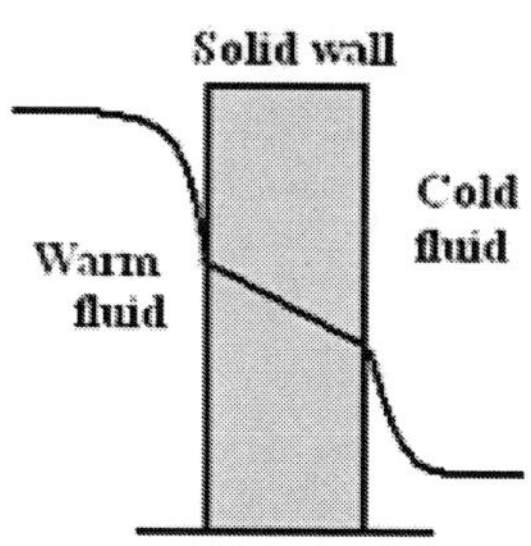

C.

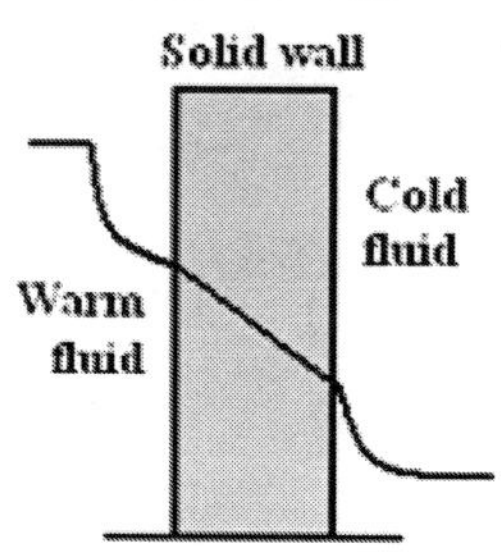

D.

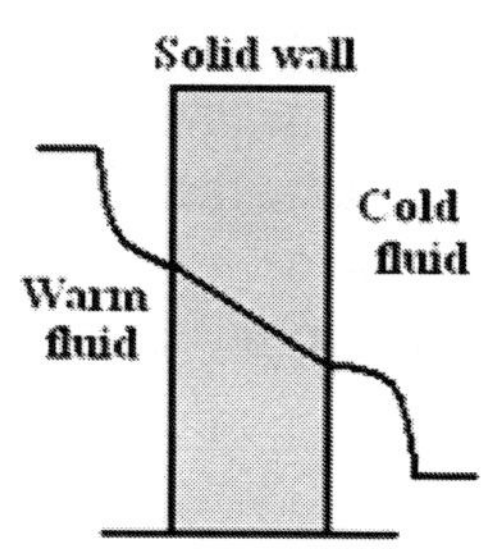

Sol. **D.** A careful steady of the temperature profiles given in four options reveal that in both **A.** and **B.** on the warm sides the temperature is falling sharply near the wall which should not happen as near the wall with the wall under steady state condition the fluid temperature should reach near equilibrium with only minor variation. This is correct for both sides. Temperature profile near the wall should not change sharply. The temperature profile which fulfils this criteria in option **D.**

53. In a shell and tube heat exchanger, if the shell length is L_S, the baffle spacing is L_B and the thickness of baffle is t_b, the number of baffles on the shell side, N_B is

A. $\frac{L_s}{L_B+t_b}$

B. $\frac{L_s}{L_B+t_b} - 1$

C. $\frac{L_s}{L_B+t_b} + 1$

D. $\frac{L_s}{L_B+t_b} + 2$

Sol. **B.** Number of baffles on the shell side is determined on the basis of interval construction specifications and is given by

$$N_B + 1 = \frac{L_S}{L_B+t_B}$$

$$\text{or} \quad N_B = \frac{L_S}{L_B+t_B} - 1$$

GATE 2007

4. For the two long concentric cylinders with surface areas A_1 and A_2, the view factor F_{22} is given by

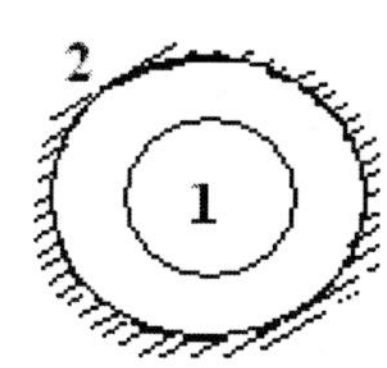

A. Zero

B. 1

C. $1 - \frac{A_1}{A_2}$

D. $\frac{A_1}{A_2}$

Sol. **C.** Since $\frac{F_{12}}{F_{21}} = \frac{A_2}{A_1}$ .... (i)

and $F_{11} + F_{12} = 1$

Given that $F_{11} = 0$

Hence $F_{12} = 1$

Further $F_{21} + F_{22} = 1$

$F_{22} = 1 - F_{21}$

Form equation (1) $F_{21} = 1 - F_{12}\left(\frac{A_1}{A_2}\right)$

$$F_{22} = 1 - \frac{A_1}{A_2}$$

55. The composite wall of an oven consists of three materials *A*,*B* and *C*. Under steady state operating conditions , the outer surface temperature T_{so} is 20°C, the inner surface temperature T_{si} is 600°C and the oven air temperature is T_{∞} = 800°C. For the following data:

Thermal conductivities K_A = 20 W/(m-k) and K_c = 50 W/(m-k);

Thickness t_A = 0.3 m, t_B = 0.15 m and t_c = 0.15 m

Inner-wall heat transfer coefficient h = 25 W/(m²-K),

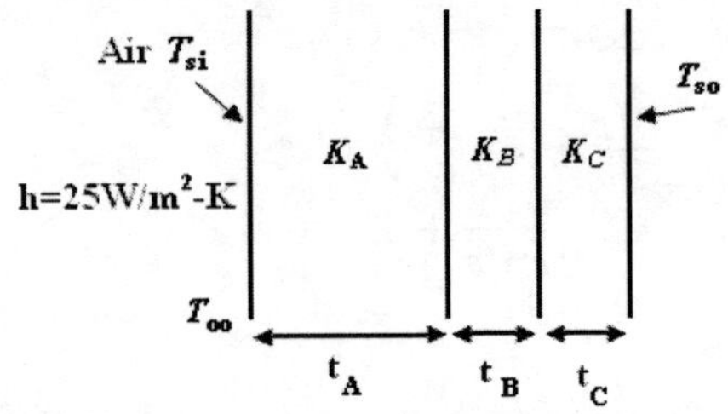

The thermal conductivity K_B in W/(m-k) of the material *B*, is calculated as

A. 35

B. 1.53

C. 0.66

D. 0.03

Sol. **B.** Data given

x_A = 0.3 m, x_B = 0.15 m, x_C = 0.15 m

K_A = 20 W/m⁰K K_C = 50 W/m⁰K

h = 25 W/m⁰K and K_B =?

T_{si} = 600⁰C, T_{∞} = 800⁰C

T_{so} = 20⁰C

Under steady state condition same heat will flow through all insulations and through outer surface

$$\therefore \quad \frac{T_{si} - T_{so}}{\left[\frac{L_A}{K_A} + \frac{L_B}{K_B} + \frac{L_C}{K_C}\right]} = h\,(T_{\infty} - T_{si})$$

$$\therefore \frac{L_A}{K_A} + \frac{L_B}{K_B} + \frac{L_C}{K_C} = \frac{1}{h}\left[\frac{T_{si} - T_{so}}{T_{\infty} - T_{si}}\right]$$

Now putting numerical value

$$\frac{0.3}{20} + \frac{0.15}{K_B} + \frac{0.15}{50} = \frac{1}{25}\left[\frac{600-20}{800-600}\right]$$

or $\qquad \frac{0.15}{K_B} = 0.098$

$\therefore \qquad K_B = 1.53$ W/m⁰K

56. Water enters a thin walled tube (*L = 1 m, D = 3mm)* at an inlet temperature of 97⁰C and mass flow rate 0.015 kg/s.The tube wall is maintained at a constant temperature of 27⁰C. Given the following data for water,

Density ρ = 1000 kg/m³

Viscosity μ= 489 × 10⁻⁶ N-s/m²

Specific heat Cp = 4184 J / (kg-K)

Inside heat transfer coefficient

h = 12978 W/(m²-K).

The outlet temperature of water in ⁰C is

A. 28

B. 37

C. 62

D. 96

Sol. **B.** Data given is

m = 0.015 kg/sec, ρ = 1000 kg/m³

$\mu = 489 \times 10^{-6}$ N-s /m²

C_p = 4184 J/(kg⁰K), h = 12978 W/(m² ⁰K)

L = 1m, D = 3mm = 3×10^{-3} m

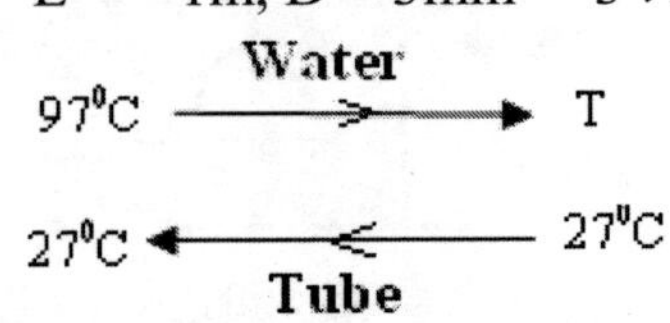

Let T be the outlet temperature of water

By energy balance

$$m\,C_p d_{dT} = h_A(\Delta T)_{LMTD}$$

Putting numerical value

$$(0.015)\,(4184)\,(97 - T) = (12978)\,(3.14\times 3\times 10^{-3}) \times 1 \times \left[\frac{(97-27)-(T-27)}{ln\left(\frac{97-27}{T-27}\right)}\right]$$

Simplifying

$$ln\left(\frac{97-27}{T-27}\right) = 1.948$$

$$\therefore \quad \frac{70}{T-27} = 7.07$$

$$\therefore \quad T = 37^0C$$

57. A hot fluid entering a well-stirred vessel cooled by feeding cold water through a jacket around the vessel. Assume the jacket is well–mixed. For the following data:

Mass flow rate of the hot fluid = 0.25 kg/s

Mass flow rate of cold water = 0.4 kg/s

Specific heat of oil = 6000 J/(kg-K)

Specific heat of cold water = 4184 J/(kg-K)

The inlet and exit temperature of the hot fluid are 150°C and 100°C respectively,

Inlet temperature of cold water = 20°C

The overall heat transfer coefficient is 500 W/(m²-k).

The heat transfer area in m², is

A. 1.82

B. 2.1

C. 3

D. 4.26

Sol. **A.** Terminal temperature of oil is 150°C and 100°C whereas that of water is entering 20°C at inlet. Let T be the temperature of water leaving

By energy balance

$M_w C_{pw}$ (T – 20) = $M_F C_{pF}$ (150 – 100) ....(i)

Given M_w = 0.4 kg/sec,

M_F = 0.25 kg/sec. , C_{pw} = 4184 J/(kg-°K)

C_{pF} = 6000 J/(kg-°K)

Putting these values in equation (i)

$$0.4 \times 4184 \times (T - 20) = 0.25 \times 6000 \times 50$$

or $T = 64.8^0\text{C}$

$$(\Delta T)_{LMTD} = \frac{(150-64.8)-(100-20)}{ln\left(\frac{150-64.8}{100-20}\right)}$$

$= 82.57^0\text{C}$

Now Q = $hA(\Delta T)_{LMTD}$

$M_w C_{pw} (T - 20) = hA(\Delta T)_{LMTD}$

or (0.4) ×(4184) × (64.8 – 20)

= 500 × A × 82.57

A = 1.82 m²

58. Consider a liquid stored in a container exposed to its saturated vapour at constant temperature T_{sat}. The bottom surface of the container is maintained at a constant temperature $T_s<T_{sat}$ while its side walls are insulted. The thermal conductivity is K_l of the liquid, its latent heat of vaporisation λ and density ρ_l are known. Assuming a linear temperature distribution in the liquid, the expression for the growth of the liquid layer δ as a function of time t is given by

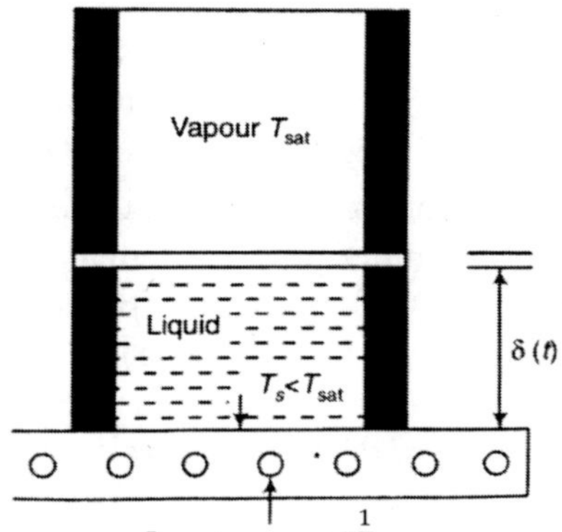

A. $\delta(t) = \left[\frac{4K_l(T_{sat}-T_s)}{\rho_l\lambda}\right]^{\frac{1}{2}}$

B. $\delta(t) = \left[\frac{K_l(T_{sat}-T_s)}{2\rho_l\lambda}\right]^{\frac{1}{2}}$

C. $\delta(t) = \left[\frac{2K_l(T_{sat}-T_s)}{\rho_l\lambda}\right]^{\frac{1}{2}}$

D. $\delta(t) = \left[\frac{K_l(T_{sat}-T_s)}{\rho_l\lambda}\right]^{\frac{1}{2}}$

Sol. **D.** Writing energy balance equation for the system. Since side walls are insulated heat loss is zero

Since heat transferred through conduction = Heat transferred to condense liquid

Hence $K_l A\left[\frac{(T_{sat}-T_s)}{\delta}\right] = m\lambda$

Also m = $(A\,\delta\rho_l)$

Substituting

$$\frac{(K_l-At)}{\delta}(T_{sat}-T_s) = (A\delta\rho_l)\lambda$$

$$\delta(t) = \left[\frac{K_l(T_{sat}-T_s)}{\rho_l\,\lambda}t\right]^{1/2}$$

59. The following list of options P, Q, Rand S are some of the important considerations in the design of a shall and tube heat exchanger.

P. Square pitch permits the use of more tube in a given shell diameter

Q. The tube side clearance should not be less than one-fourth of the tube diameter

R. Baffle spacing is not greater than the diameter of the shell or less than one-fifth of the shell diameter

S. The pressure drop on the tube side is less than 10 psi

Pick out correct combination of true statements from the following

A. P, Q and R

B. Q,R and S

C. R, S and P

D. P, Q,R and S

Sol. **B.** Q, R and S combination are correct. P in not correct since on the basis of design consideration triangular pitch accommodates more number of tubes in a given shell diameter.

60. In a double pipe heat exchanger the ID and OD of the inner pipe are 4cm and 5 cm respectively. The ID of the outer pipe is 10 cm with a wall thickness of 1 cm. Then the equivalent diameters (in cm) of the annulus for heat transfer and pressure drop respectively are

A. 15, 5

B. 21, 6

C. 6, 19

D. 15, 21

Sol. **A.** For

Inner pipe i.d. = 4cm,

o.d. = 5cm

Outer pipe i.d. = 10 cm

The equation for equivalent diameter for annulus differs only in wetted perimeter. For heat transfer the wetted perimeter is whereas for pressure drop it is π (di + do)

Hence $(D_{eq})_{HT} = 4\pi \frac{\left[\left(\frac{10}{2}\right)^2 - \left(\frac{5}{2}\right)^2\right]}{\pi \times 5}$

= 15 cm

And $(D_{eq})_p = 4\pi \frac{\left[\left(\frac{10}{2}\right)^2 - \left(\frac{5}{2}\right)^2\right]}{\pi \times (10+5)}$

D_p = 5 cm

GATE 2006

61. One dimensional steady state heat transfer occurs from, a flat vertical wall of length 0.1 m into the adjacent fluid. The heat flux into this fluid is 21 W/m^2. The wall thermal conductivity is 1.73 W/(m-k). If the heat transfer coefficient is 30 W/(m^2-k) and the Nusselt number based on the wall length is 20, then the magnitude of the temperature gradient at the wall on the fluid side (in K/m) is

A. 0.7

B. 12.14

C. 120

D. 140

Sol. **D.** Since the Nusselt number can be expressed as

$$\text{Nu} = \frac{(dT/dy)_{y=0}}{(\Delta T/L)}$$

Where $\left(\frac{dT}{dy}\right)_{y=0}$ represents the temperature gradient at the wall

Hence $\left(\frac{dT}{dy}\right)_{y=0} = \text{Nu}\left(\frac{\Delta T}{L}\right)$ ....(i)

Further $q = hA\Delta T$

or $\Delta T = \left(\frac{q}{A}\right)\left(\frac{1}{h}\right)$ ....(ii)

From equations (i) and (ii)

$$\left(\frac{dT}{dy}\right)_{y=0} = \text{Nu}\left(\frac{q}{A}\right)\left(\frac{1}{L}\right)\left(\frac{1}{h}\right)$$

It is given that

Nu = 20, and $\frac{q}{A}$ = 21 W/m^2

L = 0.1 m, h = 30 W/(m^2 – K)

Substituting these values wc gct

$$\left(\frac{dT}{dy}\right)_{y=0} = 20 \times 21 \times \frac{1}{0.1} \times \frac{1}{30}$$

= 140 K/m

62. An insulated cylindrical pipe of 0.2 m diameter has a surface temperature of 45°C. It is exposed to black body surroundings at 25°C. The emissivity and absorptivity of the insulation surface are 0.96 and 0.93, respectively. The convective heat transfer coefficient outside the insulation surface is 3.25 W/(m^2-k). The Stefan-Boltzmann constant is 5.67 ×10^{-8} W/(m^2-k^4).

The surrounding fluid may be assumed to be transparent.

Find the percentage contribution from radiation to the total heat transfer rate of the surrounding.

A. 30.9

B. 50.0

C. 57.6

D. 68.4

Sol. **D.** Data given in the problems are

$\sigma = 5.67 \times 10^{-8}$ W(m^{20}K^4)

$\epsilon = 0.96$, $\propto = 0.93$

h = 3.25 W/(m^{2} ^{0}K)

T_1 = 45 + 273 = 318^{0}K

T_2 = 25 + 273 = 298^{0}K

Radiation heat transfer is expressed by

$$q_r = \sigma\, A(\epsilon T_1^4 - \propto T_2^4)$$

Putting the values given

or $q_r = 5.67 \times 10^{-8} \times A \times [0.96 \times (318)^4 - 0.93 \times (298)^4]$

= (140.78)A

Now convection heat transfer is

$q_C = hA\Delta T$

= (3.25) A (318 – 298)

= (65) A

Hence, percentage contribution from radiation to total heat transfer rate

$= \frac{q_r}{q_r+q_c} \times 100$

$140.78 = \frac{(140.78\,A)}{148.78+65\,A} \times 100$

$= 68.41\ \%$

63. A process fluid has to be cooled from 22°C to 2°C using brine in a 2-4 shell and tube heat exchanger shown below. The brine enters at -3°C and leaves at 7°C. The overall heat transfer coefficient is 500 W/(m²-K). The design heat load is 30 kW. The brine flows on the tube side and the process fluid on the shell side. The heat transfer area in m² is

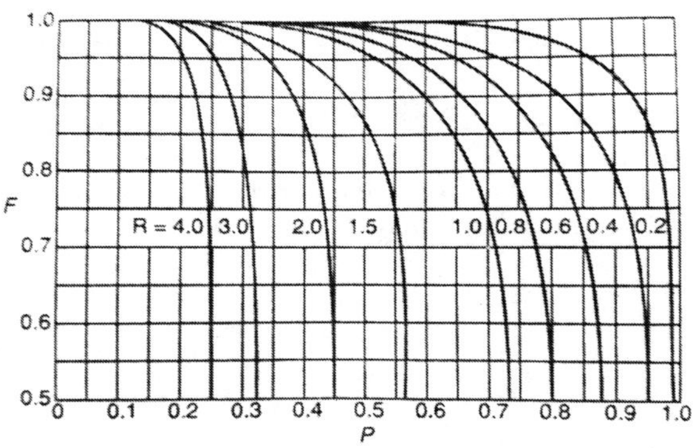

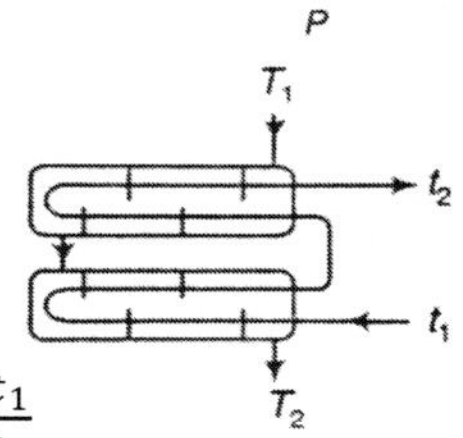

$P = \frac{t_2-t_1}{T_1-t_1}$

$R = \frac{T_2-T_1}{t_2-t_1}$

A. 1.1

B. 5.77

C. 6.59

D. 7.53

***Sol.* D.** According to given data

$T_1 = 22^0C, \quad T_2 = 2^0C$

$t_1 = -3^0C, \quad t_2 = 7^0C$

$P = \frac{t_2-t_1}{T_1-t_1} = \frac{7-(-3)}{22-(-3)} = 0.4$

and $R = \frac{T_1-T_2}{t_2-t_1} = \frac{22-2}{7-(-3)} = 2$

From the given graph corresponding to P = 0.4 and R = 2 the value of F is 0.87

Now, $(\Delta T)_{LMTD} = \frac{(22-7)-[2-(-3)]}{in\left(\frac{22-7}{2-(-3)}\right)}$

$= 9.1^0C$

Hence actual $(\Delta T) = (F)\ (\Delta T)_{LMTD}$

$= (0.87)\ (9.1)$

$= 7.92^0C$

Further, heat transferred

$Q = UA(\Delta T)_{actual}$

Given $U = 500\ W/m^{2\,0}K)$

$Q = 30\ kW = 30{,}000\ W$

$\therefore 30000 = (500)(A)(7.92)$

$\therefore\ A = 7.53\ m^2$

GATE 2005

64. A black body at a higher temperature T_H transfers energy by radiation to a black body at a lower temperature T_L. Initially $T_H = 1850°C$, $T_L = 500°C$ and the net rate of energy transfer is 25 W. After some time, when T_H=1500°C and T_L = 750°C, what is the net rate of energy transfer?

A. 8.73 W

B. 9.60 W

C. 13.89 W

D. 11.01 W

***Sol.* D.** At initial conditions

q= 25 W, T_H =1850⁰C = 2123 K

T_L= 500⁰C, = 773 K

and $q = \sigma A(T_H^4 - T_L^4)$

Putting the numerical value

$25 = \sigma A[(2123)^4 - (773)^4]$

or $\sigma A = 1.253 \times 10^{-12}\ W/K^4$

After sometime

T_H= 1550⁰C = 1773 K

T_L= 750⁰C = 1023 K

$q = \sigma A(T_H^4 - T_L^4)$

$q = 1.253 \times 10^{-12} \times [(1773)^4 - (1023)^4$

$q = 11.01\ W$

65. A circular tube of outer diameter 5 cm and inner diameter 4 cm is used to convey hot fluid. The inner surface of the wall of the tube is at a temperature of 80°C while the outer surface of the wall of the tube is at 25°C. What is the rate of heat transport across the tube wall per meter length of the tube at steady state, if the thermal conductivity of the tube wall is 10 W/(m-K)?

A. 13823 W/m

B. 15487 W/m

C. 17279 W/m

D. 27646 W/m

***Sol.* B.** Data given

$d_1 = 4$ cm, $d_2 =$ 5 cm

$T_1 = 80^0$C, $T_2 = 25^0$C

K = 10 W/m^{0}K

For heat transfer through a cylinder

$$q = \frac{\Delta T}{R}$$

Where, $R = \frac{ln\ (r_2/r_1)}{2\pi LK}$

$$= \frac{ln\ (2.5/2)}{(2)\times(3.14)(L)(10)}$$

$$= \frac{3.55\times10^{-3}}{L}$$

$$\therefore\ q = \frac{\Delta T}{R} = \frac{80-25}{\left(\frac{3.55\times10^{-3}}{L}\right)}$$

or $\frac{q}{L} = 15479$ W/m

66. Consider the flow of gas with density 1 kg/m^3 viscosity 1.5×10^{-5} kg/(m-s), specific heat C_p = 846 J / (kg-K) and thermal conductivity K = 0.01665 W/(m-K) in a pipe of diameter D = 0.01 m and length L = 1 m and assume the viscosity does not change with temperature. The Nusselt number for a pipe with (L/D) ratio greater than 10 and Reynolds number greater than 20000 is given by Nu = 0.026 Re$^{0.8}$Pr$^{1/3}$.

While the Nusselt number for a laminar flow for Reynolds number less than 2100 and (Re Pr D/L) <10 is Nu = 1.86 [Re Pr (D/L)]$^{1/3}$. If the gas flows through the pipe with an average velocity of 0.1 m/s , the heat transfer coefficient is

A. 0.68 W/(m^2 – K)
B. 1.14 W/(m^2 – K)
C. 2.47 W/(m^2 – K)
D. 24.7 W/(m^2 – K)

***Sol.* C.** Data given $\rho = 1$ kg/m^3,

$\mu = 1.5\times 10^{-5}$ kg/(m – s),

$C_p = 846$ J/(kg – K),

K = 0.01665 W/(m – K)

D = 0.01 m, L = 1 m, v = 0.1 m/s

Reynolds number

$$Re = \frac{\rho\ v\ D}{\mu} = \frac{1\times0.1\times0.01}{1.5\times10^{-5}} = 66.67$$

Prandtl number

$$Pr = \frac{C_p\ \mu}{K} = \frac{846\times1.5\times10^{-5}}{0.01665} = 0.762$$

Now

$$Re.Pr\left(\frac{D}{L}\right) = 66.67\times0.762\times\frac{0.01}{1}$$

$$= 0.51 < 10$$

Based on this condition the applicable expression

$$Nu = 1.86\left[Re.Pr\left(\frac{D}{L}\right)\right]^{1/3}$$

$$= 1.8\times(0.51)^{1/3}$$

$$= 1.487$$

Further, $Nu = \frac{hD}{K}$

$$h = \frac{NuK}{D}$$

$$h = \frac{1.487\times0.01665}{0.01}$$

h = 2.47 W/m^2 – K

67. A semi-infinite slab occupying the region x = 0 and x = ∞ is at an initial temperature T_0. At time t = 0, the surface of the slab at x = 0 is brought into contact with a heat bath at a temperature T_H. The temperature T (x, t) of the slab rises according to the equation $\frac{T_H - T(x,t)}{T_H - T_0}$ $= \frac{2}{\sqrt{\pi}}\int_0^{\frac{x}{\sqrt{4\alpha t}}} e^{-\acute{\eta}^2}\,d\acute{\eta}$, where, x is position and t is time. The heat flux at the surface x = 0 is proportional to

A. $t^{-1/2}$
B. $t^{1/2}$
C. t
D. $t^{3/2}$

***Sol.* A.**

$$\frac{T_H - T(x,t)}{T_H - T_0} = \frac{2}{\sqrt{\pi}}\int_0^{\eta} e^{-\eta^2}\,d\eta \quad \text{....(i)}$$

$$\eta = \frac{x}{\sqrt{4\alpha t}} \quad \text{....(ii)}$$

$$\frac{dT}{dx} = -(T_H - T_0).\frac{2}{\sqrt{\pi}}\frac{d}{dx}\left(\int_0^{\eta} e^{-\eta^2}\,d\eta\right)$$

$$\frac{dT}{dx} = \frac{-2(T_H - T_0)}{\sqrt{\pi}}\frac{d}{d\eta}\left(\int_0^{\eta} e^{-\eta^2}\,d\eta\right).\frac{d\eta}{dx}$$

$$\frac{dT}{dx} = \frac{2(T_H - T_0)}{\sqrt{\pi}}.e^{-\eta^2}\frac{d\eta}{dx} \quad \text{....(iii)}$$

Now since

$$\frac{d\eta}{dx} = \frac{1}{\sqrt{4\alpha t}} = \frac{2}{2\sqrt{\alpha t}}$$

Substituting this equation

$$\frac{dT}{dx} = -\frac{2(T_H - T_0)}{\sqrt{\pi}}.e^{-\eta^2}\frac{1}{2\sqrt{\alpha t}}$$

$$\frac{dT}{dx} = -\frac{(T_H - T_0)}{\sqrt{\pi\alpha t}}.e^{-\eta^2}$$

$$\left.\frac{dT}{dx}\right|_{x=0} = \frac{-(T_H - T_0)}{\sqrt{\pi\alpha t}}$$

Applying boundary condition at x = 0, η = o, and $e^{-\eta^2} = 1$

Further heat flux at the surface

$$\frac{q}{A} = -K\left.\frac{dT}{dx}\right|_{x=0}$$

$$\frac{q}{A} = \frac{K(T_H - T_0)}{\sqrt{\pi\alpha t}}$$

$$\frac{q}{A} \propto t^{-1/2}$$

68. A counter-current flow double pipe heat exchanger is used to heat water flowing at 1 kg/s from 40°C to 80°C. Oil is used for heating and its temperature changes from 100°C to 70°C.The overall heat transfer coefficient is 300 W/(m²-°C). If it is replaced by a 1-2 shell and tube heat exchanger with counter-current flow configuration with water flowing in shell and oil flowing in the tube, what is the excess area required with respect to the double pipe heat exchanger? The correction factor F_t for LMTD (Log Mean Temperature Difference) based on the above double pipe heat exchanger is 0.5. The heat transfer coefficient remains unchanged, and the same inlet and outlet conditions are maintained. C_{pwater} = 4180 J /(kg -°C), $C_{p,oil}$ = 2000 J/(kg-°C)

A. Zero

B. – 20.15 m²

C. 22.6 m²

D. 9.69 m²

Sol. **C.** Case I

100°C ——→ Oil —— 70°C

40°C ——←—— Water 80°C

$$(\Delta T)_{LMTD} = \frac{(100-80)-(70-40)}{ln\left(\frac{100-80}{70-40}\right)}$$

or $(\Delta T)_{LMTD}$ = 24.66°C

From the heat transfer equation

$$Q = UA(\Delta T)_{LMTD} = m_w C_{pw} \Delta T_w$$

Given that

U = 300 W/(m2-°C), m_w = 1 kg/s,

C_{pw} = 4180 J/(kg-°C)

ΔT_w = 80 – 40 = 40°C

Putting these value

300×A×24.66 = 1×4180×40

A = 22.59 m²

Case II

Temperature correction factor F_t = 0.5

$$\therefore (\Delta T)_{TRUE} = (F_t)(\Delta T)_{LMTD} = (0.5)(24.66) = 12.33°C$$

Now $Q = UA'(\Delta T)_{TRUE} = m_w C_{pw} \Delta T_w$

or $$A' = \frac{(1)(4180)(40)}{(300)(12.33)} = 45.19 \text{ m}^2$$

Hence, excess area required

= 45.19 – 22.59

= 22.6 m²

69. Fluid flows in an annulus of inner diameter 0.8 m and outer diameter 1 m. Heat is transferred to the fluid from inner tube surface of the annulus. What is equivalent diameter for heat transfer is metre?

A. 0.45

B. 0.20

C. 1.64

D. 0.90

Sol. **A.** For heat transfer the equivalent diameter

$$d_{eq} = \frac{4(flow area)}{(wetted\ perimeter)_{heat\ transfer}}$$

Given

d_i = 0.8 m, d_o = 1 m

$$d_{eq} = \frac{4\pi\left[\left(\frac{d_o}{2}\right)^2-\left(\frac{d_i}{2}\right)^2\right]}{\pi d_i} = \frac{d_o^2-d_i^2}{d_i} = \frac{1-0.64}{0.8}$$

$$d_{eq} = 0.45$$

Statement for linked answer question 70 and 71

A liquid of mass 7 kg and specific heat 4 kJ/(kg – °C) is contained in a cylindrical heater of diameter 0.15 m and height 0.40 m. The cylindrical surface of the heater is exposed to air at 25°C while the end caps are insulated, so that the heat transfer takes place only through the cylindrical surface.

The thickness of the wall of the heater = 2 mm

The wall thermal conductivity = 10 W/(m-K)

The heat transfer coefficient in the liquid = 100 W/(m²-K)

The heat transfer coefficient in air = 10 W/(m²-K)

The liquid in initially maintained at a temperature of 75°C. At time t = 0, the heater is switched off and the temperature of the liquid in the heater decreases due to heat loss across the cylindrical surface.

70. What is the overall heat transfer coefficient in W/(m²-K)?

A. 1

B. 4.04

C. 9.07

D. 10

Sol. **C.** The overall heat transfer coefficient will depend upon the sum of all there resistances

$$\frac{1}{U} = \frac{1}{h_{air}} + \frac{(r_o - r_i)}{K} ln\left(\frac{d_o}{d_i}\right) + \frac{d_o}{d_i} \cdot \frac{1}{h_l}$$

Data given x = 2 mm

d_o = 0.15+2×0.002 = 0.154 m

K = 10 W/(m-K)

h_l = 100 W/(m²-K)

h_{air} = 10 W/(m²-K)

Putting these value

$$\frac{1}{U} = \frac{1}{10} + \frac{(0.002)}{10} ln\left(\frac{0.154}{0.15}\right) + \frac{0.154}{0.15} \cdot \left(\frac{1}{100}\right)$$

U = 9.07 W/(m²-K)

71. What is the time required for the temperature of the liquid to reduce to 50°C after the heater is switched off, assuming lumped system analysis is valid?

A. 7.874 × 10³ s

B. 11.346 × 10³s

C. 16.828 × 10³ s

D. 23.213 × 10³ s

***Sol.* B.** Under unsteady state heat transfer conditions in a cylindrical vessel

$$\frac{T - T_\infty}{T_o - T_\infty} = e^{-(UA/mC_p)t} \quad ....(i)$$

Here T = 50°C, T_∞ = 25°C

T_o = 75°C, U = 9.07 W/(m²-K)

m = 7 kg, C_p = 4 kJ/ (kg-°C)

= 4000 J/(kg-°C), L = 0.4 m

Area A = $\pi\, d_i\, L$

A = 3.14 × 0.15 × 0.4

A = 0.1884 m²

Putting these values in equation (i)

$$\frac{50-25}{75-25} = e^{-(9.07 \times 0.1884/7 \times 4000)t}$$

or 0.6932 = (6.1028)(10⁻⁵)(t)

∴ t = 11.359 × 10³ s

GATE 2004

72. The left face of a one dimensional slab of thickness 0.2 m is maintained at 80°C and the right face is exposed to air at 30°C. The thermal conductivity of the slab is 1.2 W/(m-K) and the heat transfer coefficient from the right face is 10 W/(m²-K). At steady state, the temperature of the right face in °C is

A. 77.2

B. 71.2

C. 63.8

D. 48.7

***Sol.* D.** Data given is K = 1.2W/(m²-K), x = 0.2 m

Equation for heat flux through a rectangular slab is

$$\frac{q}{A} = \frac{K\Delta T}{x} = \frac{(1.2)(80-T)}{0.2}$$

Where T is temperature of the right face under steady state conditions

$$\therefore \quad \frac{q}{A} = (6)(80 - T)$$

Combining the resistance offered by the film on right face

$$\frac{q_1}{A} = \frac{\Delta T'}{\left(\frac{x}{K} + \frac{1}{h}\right)} = \frac{(1.2)(80-T)}{0.2}$$

$\Delta T'$ = 80 – 30 = 50°C, h = 10 W/(m²-K)

$$\therefore \quad \frac{q}{A} = \frac{50}{\left(\frac{0.2}{1.2} + \frac{1}{10}\right)} = 187.5$$

At steady state

$$\frac{q_1}{A} = \frac{q}{A} = 187.5 = (6)(80 - T)$$

∴ T = 48.75°C

73. A metal ball of radius 0.1 m at a uniform temperature of 90°C is left in air at 30°C. The density and the specific heat of the metal are 3000 kg/m³ and 0.4 kJ/(kg-K), respectively. The heat transfer coefficient is 50 W/(m²-K). Neglecting the temperature gradient inside the ball, the time taken (in hour) for the ball to cool to 60°C is

A. 555

B. 55.5

C. 0.55

D. 0.15

***Sol.* D.** For unsteady state heat transfer from metal ball

$$\frac{T - T_\infty}{T_b - T_\infty} = e^{-(hA/mC_p)t}$$

$$\frac{hA}{\rho V C_p} t = ln\left(\frac{T - T_\infty}{T_b - T_\infty}\right)$$

$$t = \left[\frac{\rho V C_p}{hA}\right] ln\left(\frac{T_b - T_\infty}{T - T_\infty}\right)$$

Here r = 0.1 m, T_b = 90°C, T_∞ = 30°C,

ρ = 3000 kg/m³,

C_p = 0.4 kJ/(kg-K) = 400 J/(kg°K)

h = 50 W/(m²-K), T = 60°C

Volume of metal ball $= \frac{4}{3}\pi r^3$ = V

Surface area $= 4\,\pi r^2$ = A

$$\therefore \quad \frac{V}{A} = \frac{0.1}{3} = 0.0333$$

Putting these data

$$t = \left(\frac{3000}{50}\right)(0.0333)(400)\ln\left(\frac{90-30}{60-30}\right)$$

t = 554.52 sec.

t = 0.15 hr.

74. It is desire to concentrate a 20% salt solution (20 kg of salt in 100 kg of solution) to a 30% salt solution in an evaporator. Consider a feed of 300 kg/min at 30⁰C. The boiling point of the solution is 110⁰C, the latent heat vaporization is 2100 kJ/kg and the specific heat of the solution is 4 kJ/(kg-K).The rate at which heat has to be supplied (in kJ/min) to the evaporator is

A. 3.06×10^5

B. 6.12×10^5

C. 7.24×10^5

D. 9.08×10^5

Sol. **A.**

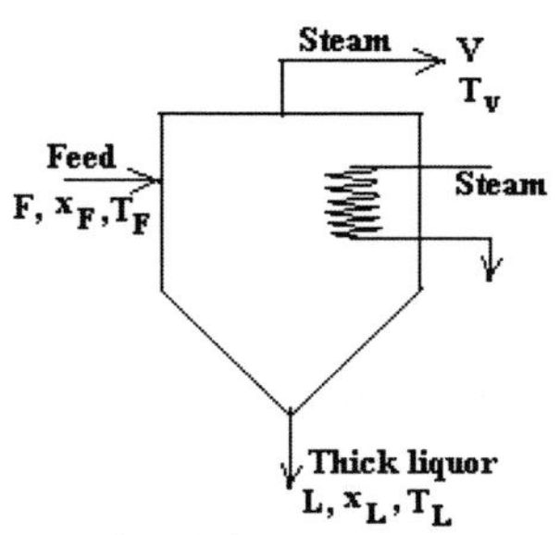

Data given is

$F = 300$ kg/min, $x_F = 0.2$,

$X_L = 0.3$

$T_F = 30^0$C, $T_L = T_V = 110$C

Material balance for solute

$(F)(x_F) = (L)(x_L) + Vy$

Since y = 0

(300)(0.2) =(L) (0.3)

or L = 200 kg/min.

Overall material balance

F = L + V

∴ V = F – L = 100 kg/min

∴ Rate of heat supplied

$Q = FC_p\Delta T + V\lambda$

$\lambda = 2100$ kJ/kg

$C_p = 4$ kJ/kg-K

$\Delta T = T_L - T_F = 110 - 30 = 80^0$C

Hence Q = (300)(4)(80) + (100)(2100)

= 306000

∴ $Q = 3.06 \times 10^5$ kJ/min

75. Hot water (0.01 m³ /min) enters the tube side of a co-current shell and tube heat exchanger at 80⁰C and leaves at 50⁰C. Cold oil (0.05 m³/min) of density 800 kg/m³ and specific heat of 2 kJ/(kg-k) enters at 20⁰C.The long mean temperature difference in⁰C is approximately

A. 32

B. 37

C. 45

D. 50

Sol. **A.** Data given

$T_{i,w} = 80^0$C, $T_{o,w} = 50^0$C, $V_w = 0.1$ m³/min

$V_{air} = 0.5$ m³/min, $\rho_{oil} = 800$ kg/m³

$C_{p,oil} = 2$ kJ/(kg-K), $C_{p,water} = 4.184$,

$T_{i,oil} = 20^0$C, $T_{o,oil} = ?$

By energy balance

$$(M_w)(C_p)_w(\Delta T)_w = (M_{oil})(C_p)_{oil}(\Delta T)_{oil}$$

or $$(V_w\,\rho_w)(C_p)_w(\Delta T)_w = (V_{oil}\,C_{p,oil})(C_p)_{oil}(\Delta T)_{oil}$$

or (1000)(0.01)(4.184)(80 – 50)

= (800)(0.05)(2)($T_{oil,}$ – 20)

$T_{oil,0} = 35.675^0$C

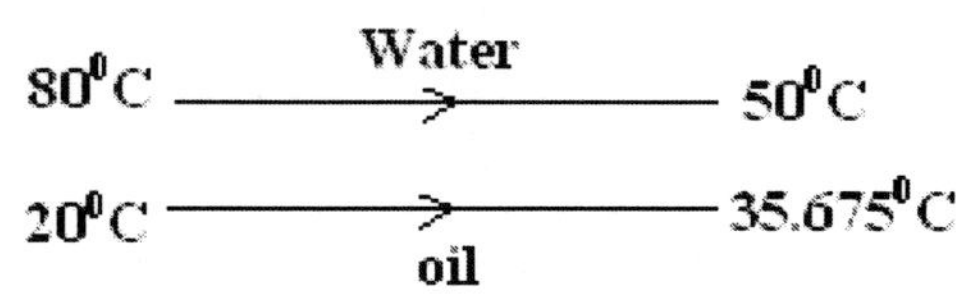

$$(\Delta T)_{LMTD} = \frac{(80-20)-(50-35.675)}{\ln\left(\frac{80-20}{50-35.675}\right)}$$

$\therefore (\Delta T)_{LMTD} = 31.9^0$C

GATE 2003

76. A process stream of dilute aqueous solution flowing at the rate of 10 kg/s is to be heated. Steam condensate at 95⁰C is available for heating purpose, also at a rate of 10 kg/s. A 1-1 shell and tube heat exchanger is available.The best arrangement is

A. counter flow with purpose stream on shell side

B. counter flow with purpose stream on tube side

C. parallel flow with purpose stream on shell side

D. parallel flow with purpose stream on tube side

Sol. A. Obviously a countercurrent flow arrangement will give a greater value of ΔT in comparison to parallel flow. Since, flow rates are identical, on the therefore basis of temperature conditions the high temperature stream should be in tube side to avoid heat loss to the surroundings. Hence, the process stream should be in the shell side.

77. The inner wall of a furnace is at a temperature of 700°C. The composite wall is made of two substances, 10 and 20 cm thickwith thermal conductivities of 0.05 and 0.1 W/(m-°C) respectively. The ambient air is at 30°C and the heat transfer coefficient between the outer surface of wall side air is 20 W/(m²-°C). The rate of heat loss from the outer surface in W/m²is

A. 165.4
B. 167.5
C. 172.5
D. 175

Sol. A.

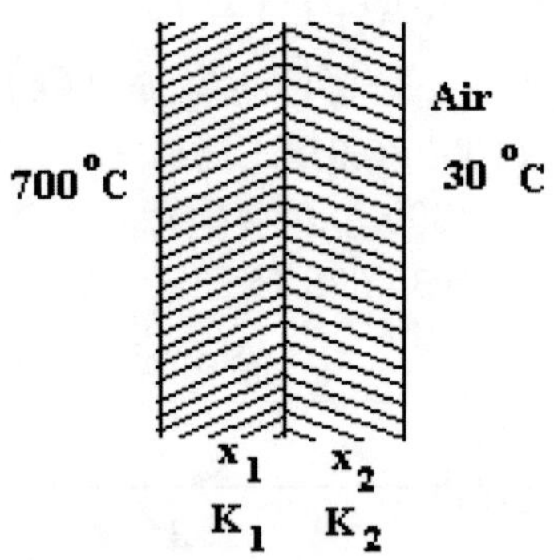

Data given

$T_1 = 700^0\text{C}, T_2 = 30^0\text{C}$
$x_1 = 10 \text{ cm} = 0.1 \text{ m}$
$x_2 = 20 \text{ cm} = 0.2 \text{ m}$
$K_1 = 0.05 \text{ W/(m-}^0\text{C)},$
$K_2 = 0.1 \text{ W/(m-}^0\text{C)}$
$\text{H} = 20 \text{ W/(m-}^0\text{C)}$

The rate of heat loss from the outer wall is

$$\frac{q}{A} = \frac{T_1 - T_2}{\left(\frac{X_1}{K_1} + \frac{X_2}{K_2} + \frac{1}{h}\right)}$$

or $$\frac{q}{A} = \frac{700-30}{\left[\left(\frac{0.1}{0.05}\right) + \frac{0.2}{0.1} + \frac{1}{20}\right]}$$

$\therefore$ $$\frac{q}{A} = 165.4 \text{ W/m}^2$$

78. Steam is to be condensed in a shell and tube heat exchanger 5 m long with a shell diameter of 1 m. Cooling water is to be used for removing the heat. Heat transfer coefficient for the cooling water whether on shell side or tube side is same.The best arrangement is

A. vertical heat exchanger with steam on tube side
B. vertical heat exchanger with steam on shell side
C. horizontal heat exchanger with steam on tube side
D. horizontal heat exchanger with steam on shell side

Sol. B. Condensation of steam on the outer surface of vertical tubes is known to produce higher heat transfer coefficient for condensation in comparison to horizontal tubes. Hence the best arrangement would be to use vertical heat exchanger with steam on shell side.

79. A fluid is following inside the inner tube of a double pipe heat exchanger with diameter '*D*'. For a fixed mass flow rate, the tube side heat transfer coefficient for turbulent flow conditions is proportional to

A. $D^{0.8}$
B. $D^{-0.2}$
C. D^{-1}
D. $D^{-1.8}$

Sol. A. Dittus-Boelter equation for heat transfer coef is

$$\frac{hD}{K} = 0.023\left(\frac{Dv\rho}{\mu}\right)^{0.8}\left(\frac{C_p\mu}{K}\right)^{0.4}$$

or $$\text{h} = (0.023)\left(\frac{K}{D}\right)\left(\frac{Dv\rho}{\mu}\right)^{0.8}\left(\frac{C_p\mu}{K}\right)^{0.4}$$

$$\text{h} \propto \left(\frac{K}{D}\right)\left(\frac{Dv\rho}{\mu}\right)^{0.8}$$

$\therefore$ $$\text{h} \propto (\text{K})\left(\frac{D^{0.8}}{D}\right)$$

or $$\text{h} \propto D^{0.2}$$

80. Air is to be heated by condensing steam. Two heat exchangers are available

(i) a shell and tube heat exchanger, and
(ii) a finned tube heat exchanger.Tube side heat transfer area is equal in both cases. The recommended arrangement is

A. finned tube heat exchanger with air inside and steam outside
B. finned tube heat exchanger with air outside and steam outside
C. shell and tube heat exchanger with air inside tubes and steam on shell side
D. shell and tube heat exchanger with air on shell side and steam inside tubes

Sol. **B.** Condensing steam and air being the two fluids exchanging heat, it is known that steam condensation produces substantially high value of heat transfer coefficient. The gas film coefficient is nonconductive with low film coefficient. Further fin tube offers high heat transfer area on the outside ie. on fin side. Hence air should be on the outside of the fin tube heat exchanger so that low heat transfer coefficient is compensated by high heat transfer area.

81. For a given ambient air temperature with increase in the thickness of insulation of a hot cylindrical pipe,the rate of heat loss from the surface would

A. decrease

B. first decrease and then increase

C. increase

D. first increase and then decrease

Sol. **B.** With increase in the thickness of insulation of a hot cylindrical pipe the rate of heat loss from the outer surface will decrease but simultaneously the increased thickness will also increase the outermost heat transfer area form which heat is being lost. At the point of critical thickness of insulation any further increase in thickness of insulation will result in increase in the heat loss instead of decrease. Therefore, it will first decrease and there increase.

GATE 2002

82. A 10 cm diameter steam pipe, carrying steam at 180°C is covered with an insulation (conductivity = 0.6 W/(m-°C)). It loses heat to the surroundings at 30°C. Assume a heat transfer coefficient of 0.8 W/(m²-°C) for heat transfer from surface to the surroundings. Neglect wall resistance of the pipe and film resistance of steam. If the insulation thickness is 2 cm the rate of heat loss from this insulated pipe will be

A. greater than that of the uninsulated steam pipe

B. less than that of the uninsulated steam pipe

C. equal to that of the uninsulated steam pipe

D. less than the steam pipe with 5 cm insulation

Sol. **B.** Rate of heat loss from insulated pipe

$$\frac{Q_1}{A} = \frac{180-30}{\frac{x}{K}+\frac{1}{h}} = \frac{150}{\frac{0.02}{0.6}+\frac{1}{0.8}}$$

$$= 0.78 \times 150 \qquad \text{....(i)}$$

Rate of heat loss from uninsulated pipe is

$$\frac{Q_2}{A} = \frac{180-30}{\left(\frac{1}{0.8}\right)} = 0.8 \times 150 \qquad \text{....(ii)}$$

Since $\frac{Q_1}{A} < \frac{Q_2}{A}$

∴ Heat loss will be lets than the uninsulated pipe.

83. 1000 kg of liquid at 30°C in a well-stirred vessel has to be heated to 120°C using immersed coils carrying condensing steam at 150°C. The area of the steam coils is 1.2 m² and overall heat transfer coefficient to the liquid is 1500 W/(m² -°C). Assuming negligible heat loss to surrounding and specific heat capacity of the liquid to be 4 kJ/(kg-°C), the time taken for the liquid to reach desired temperature will be

A. 15 min

B. 22 min

C. 44 min

D. 51 min

Sol. **D.** Data given

m = 1000 kg, T_l = 30°C

T = 120°C, T_∞ = 150°C

A = 1.2 m², h = 1500 W/m²-°C

C_p = 4 kJ/(kg-°C) = 4000 J(kg-°C)

For the unsteady state heat transfer

$$\frac{T-T_\infty}{T_l-T_\infty} = \exp\left[-\frac{(h)(A)}{(m)(C_p)}\right]t$$

or $$t = -\frac{mC_p}{hA}\ln\left(\frac{T-T_\infty}{T_l-T_\infty}\right)$$

Substituting the given values

$$t = \left[-\frac{(1000)(4000)}{(1500)(1.2)}\right]\ln\left(\frac{120-150}{30-150}\right)$$

t = 3080.6 sec.

t = 51 min

84. A composite wall consists of two plates *A* and *B* placed in series normal to the flow of heat. The thermal conductivity are K_A and K_B and the specific heat capacities are C_{pA} and C_{pB} for plates A and B respectively. Plate *B* has twice the thickness of plate *A*. At steady state, the temperature difference across plate *A* is greater than that across plate *B*, when

A. $C_{pA} > C_{pB}$

B. $C_{pA} < C_{pB}$

C. $K_A < 0.5\ K_B$

D. $K_A > 2\ K_B$

***Sol.* C.** Applying Fourier's law

$$Q_A = K_A \frac{\Delta T_A}{x} \text{ or } \Delta T_A = \frac{(Q_A)(x)}{K_A}$$

Similarly for B

$$Q_B = K_B \frac{\Delta T_B}{2x} \text{ or } \Delta T_B = \frac{(Q_B)(2x)}{K_B}$$

Under steady state conditions both the plates which are in series will have the same heat flux

Hence, $Q_A = Q_B = Q$

It is given that ΔT_A is greater than

$$\Delta T_A > \Delta T_B$$

Substituting $\frac{(Q_A)x}{K_A} > \frac{(Q_B)(2x)}{K_B}$

or $\frac{1}{K_A} > \frac{2}{K_B}$

or $K_A > 0.5\, K_B$

GATE 2001

85. The Sieder–Tate correlation for heat transfer in turbulent flow in a pipe gives Nu $\propto Re^{0.8}$, where Nu is the Nusselt number and Re is Reynolds number for the flow. Assuming that this relation is valid, the heat transfer coefficient varies with pipe diameter D as

A. $D^{-1.8}$
B. $D^{-0.2}$
C. $D^{0.2}$
D. $D^{1.8}$

***Sol.* B.** Given Nu $\propto Re^{0.8}$

or $\frac{hD}{K} \propto \left(\frac{\rho\, v\, D}{\mu}\right)^{0.8}$

or $h \propto \left(\frac{K}{D}\right)\left(\frac{\rho\, v\, D}{\mu}\right)^{0.8}$

or $h \propto \frac{D^{0.8}}{D}$

or $h \propto D^{-0.2}$

86. The overall heat transfer coefficient for a shell and tube heat exchanger for clean surface are U_0 = 400 W/(m² – K). The fouling factor after one year of operation is found to be h_{do} = 2000 W/(m² – K). The overall heat transfer coefficient at this time is

A. 1200 W/(m² – K)
B. 894 W/(m² – K)
C. 333 W/(m² – K)
D. 287 W/(m² – K)

***Sol.* C.** Given U_o = 400 W/(m²-K)

h_{do} = 2000 W/(m²-K)

Hence after one year of operation, U will consist of fouling factor also

$$\frac{1}{U} = \frac{1}{U_o} + \frac{1}{h_{do}}$$

$$\frac{1}{U} = \frac{1}{400} + \frac{1}{2000}$$

U = 333 W/(m²-K)

87. The heat flux (from outside to inside) across an insulating wall with thermal conductivity K = 0.04 W/(m-k) and thickness 0.16 m is 10 W/m². The temperature of the inside wall is –5°C. The outside wall temperature is

A. 25°C
B. 30°C
C. 35°C
D. 400°C

***Sol.* C.** Applying Fourier's law

$$\frac{q}{A} = K\frac{\Delta T}{(\Delta x)}$$

Given

$\frac{q}{A}$ = 10 W/m²

K = 0.04 W/(m-K), Δx = 0.16 m

Substituting these values

$$10 = (0.04)\left(\frac{\Delta T}{0.16}\right)$$

or $\Delta T = 40$

Since inside wall temperature is – 5°C

$\Delta T = T_o - T_i = 40$

or $T_o = 40 + T_i$

$= 40 + (-5)$

$T_o = 35°C$

GATE 2000

88. A steel sphere of radius 0.1 m at 400 K is immersed in oil at 300 K. If the center of the sphere reaches 350 K in 20 min, how long will it take for a 0.05 m radius steel sphere to reach the same temperature (at the centre) under identical conditions? Assume that the convective heat transfer coefficient in infinitely large.

A. 5 min
B. 10 in
C. 20 min
D. 40 min

***Sol.* A.** For second sphere having 0.05 m radius assuming t sec is taken to reach 350°K then

$$\frac{t}{20} = \frac{(4)(\pi)(0.05)^2}{(4)(\pi)(0.1)^2}$$

or t = 5 min

89. A composite flat wall of a furnace is made of two materials A and B. The thermal conductivity of A is twice of that of material B, while thickness oflayer of A is half of that of B. If the temperatures at the two sides of the wall are 400 K and 1200 K, then the temperature drop (in kelvin) across the layer of material A is

A. 125
B. 133
C. 150
D. 160

Sol. **D.**

Let T be the temperature of interface then under steady state condition

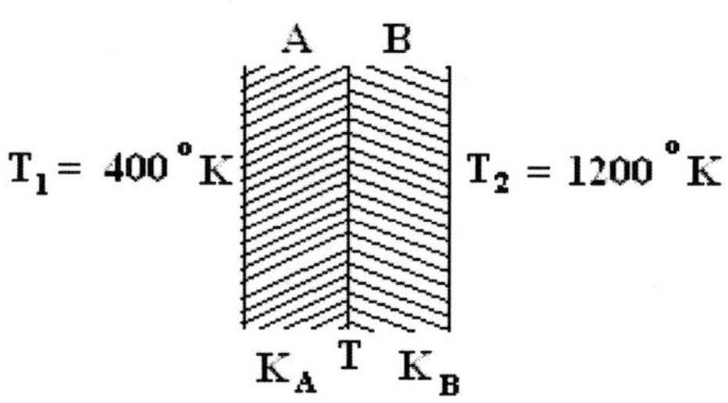

$$Q = \frac{T_1 - T}{R} = \frac{T - T_2}{R_2}$$

Where $R_1 = \frac{\Delta x_A}{(K_A)A}$ and $R_2 = \frac{\Delta x_B}{(K_B)A}$

Hence $(T_1 - T)\frac{(K_A)(A)}{\Delta x_A} = (T - T_2)\frac{(K_B)(A)}{\Delta x_B}$

Since $\Delta x_A = \frac{\Delta x_B}{2}$ and $K_A = 2K_B$

$T_1 = 400^0$K, $T_2 = 1200^0$K

$(400 - T)$ (4) = (T – 1200)

or T = 560⁰K

Hence, temperature drop across layer of material A is

$$\Delta T_A = 560 - 400 = 160^0\text{K}$$

90. For turbulent flow in a tube, the heat transfer coefficient is obtained from the Dittus-Boelter equation. If the tube diameter is halved and the flow rate is doubled, then the heat transfer coefficient will change by a factor of

A. 1
B. 1.74
C. 6.1
D. 37

Sol. **C.** Data given is

$$Q_2 = 2Q_1 \; and \; D_2 = \frac{1}{2}D_1$$

$$Re = \frac{Dv\rho}{\mu} = \left(\frac{\rho}{\mu}\right)\left(\frac{D_1}{D_2}\right) a$$

or $Re \propto \frac{Q}{a}$

or $\frac{Re_2}{Re_1} = \left(\frac{Q_2}{Q_1}\right)\left(\frac{D_1}{D_2}\right) = 2 \times 2$

or $\frac{Re_2}{Re_1} = 4$

Dittus – Boelter equation is

$$\frac{hD}{K} = (0.023)\,(Re)^{0.8}(\rho r)^{0.4}$$

$$h \propto \frac{Re^{0.8}}{D}$$

Hence, $\frac{h_2}{h_1} = \left(\frac{Re_2}{Re_1}\right)^{0.8}\left(\frac{D_2}{D_1}\right)$

$= (4)^{0.8}(2)$

∴ $\frac{h_2}{h_1} = 6.1$

91. A fluid flows through a cylindrical pipe under fully developed, steady state laminar flow conditions. The tube wall is maintained at constant temperature. Assuming constant physical properties and negligible viscous heat dissipation,the governing equation for the temperature profiles is (z-axial direction; r-radial direction)d

A. $U_{max}\left[1 - \left(\frac{r}{R}\right)^2\right]\left(\frac{\partial T}{\partial z}\right) = \frac{k}{\rho C_p}\left[\frac{1}{r}\frac{\partial}{\partial r}\left(r\frac{\partial T}{\partial r}\right) + \frac{\partial^2 T}{\partial z^2}\right]$

B. $U_{max}\left[1 - \left(\frac{r}{R}\right)^2\right]\left(\frac{\partial T}{\partial z}\right) = \frac{k}{\rho C_p}\left[\frac{1}{r}\frac{\partial}{\partial r}\left(r\frac{\partial T}{\partial z}\right) + \frac{\partial^2 T}{\partial r^2}\right]$

C. $U_{max}\left[1 - \left(\frac{r}{R}\right)^2\right]\left(\frac{\partial^2 T}{\partial z^2}\right) = \frac{k}{\rho C_p}\left[\frac{1}{r}\frac{\partial}{\partial r}\left(r\frac{\partial T}{\partial r}\right) + \frac{\partial^2 T}{\partial r^2}\right]$

D. $U_{max}\left[1 - \left(\frac{r}{R}\right)^2\right]\left(\frac{\partial T}{\partial z}\right) = \frac{k}{\rho C_p}\left[\frac{1}{z}\frac{\partial}{\partial z}\left(z\frac{\partial T}{\partial z}\right) + \frac{\partial^2 T}{\partial r^2}\right]$

Sol. **A.** For steady state laminar flow the governing eq. for temperature profile is given (z – axial direction, r – radial direction) at the condition

Here, $\frac{\partial T}{\delta r} = 0$

The governing equation can be written as

$$\rho C_p\left(u_z\frac{\partial T}{\partial z}\right) = K\left[\frac{1}{r}.\frac{\partial}{\partial r}\left(r\frac{\partial T}{\delta r}\right) + \frac{\partial^2 T}{\partial z^2}\right]$$

or $u_z\frac{\partial T}{\partial z} = \frac{K}{\rho C_p}\left[\frac{1}{r}.\frac{\partial}{\partial r}\left(r\frac{\partial T}{\delta r}\right) + \frac{\partial^2 T}{\partial z^2}\right]$ ....(i)

For laminar flow,

$$u_z = u_{max}\left[1 - \left(\frac{r}{R}\right)^2\right]$$

Putting this value in equation (i),

$$u_{max}\left[1 - \left(\frac{r}{R}\right)^2\right]\frac{\partial T}{\partial z} = \frac{K}{\rho C_p}\left[\frac{1}{r}.\frac{\partial}{\partial r}\left(r\frac{\partial T}{\delta r}\right) + \frac{\partial^2 T}{\partial z^2}\right]$$

7. Mass Transfer

1. Mass transfer by operation may be defined as the net movement of mass from one stream, phase or component to another with or without chemical reaction. It is one of the most common unit operation and may occur in diffusion, Absorption, Distillation, Extraction, Drying, Humidification – dehumidification Adsorption. The driving force for all the mass transfer operations is the concentration difference. Mass transfer occur due to contact between a solution to be separated with other insoluble phase. Such a contact results is transfer of a component from one phase to the other. Mass transfer is the most important parameter which given as the rate of mass transfer. It is based on the two film theory. The gas phase to the interface and the liquid face mass transfer coefficient deals with transfer from the interface to the bulk of the liquid.

1.1. Rate equation for gas – phase mass transfer: The rate equation for the mass transfer of component A through a gas phase consisting of components A and B may be written as

$$\frac{dNA}{dt} = (K_G)A(P_{A1} - P_{A2})$$

Where N_A = Moles of A transferred (gm mole)

K_G = Gas phase mass transfer coefficient gm mole/(sec)(cm^2)(atm)

A = Area normal to direction of mass transfer cm^2

P_A = Partial pressure of component A, atm 1 and 2 refer to values at 1 and 2 is the phase

Normally it is presumed. That no resistance to transits occurs at the interface and that the gas face. The mass transfer coefficient is similar to a heat transfer coefficient. It is a function of the physical properties of the gas phase, the characteristics of the equipment and the conditions of flow.

1.2. Molecular diffusion: in most of the application the gas phase will be in turbulent flow and the transfer process within the gas phase will consist of transfer by mixing within the turbulent portion and molecular diffusion the region near the interface. Where component A is being transferred through a stagnant layer of a non-diffusing component is so that the process is true molecular diffusion. The equation for molecular diffusion is

$$N_A = \frac{(D_V)(P)(A)}{RTBP_{BM}}$$

Where D_v = Diffusion coefficient for component A through B, cm^2/sec.

P = Total pressure, atm

T = Absolute temperature, °K

R = Gas law constant (1.987 cal/mole °K)

P_{BM} = Log mean value of partial pressure of component B over the path of diffusion.

The governing equation for molecular diffusion is given by Fick's law of diffusion which state that the diffusion flux

$$J_{AZ} \propto \frac{dC_A}{dZ}$$

Or

$$J_{AZ} = -D_{AB}\frac{dC_A}{dZ}$$

This is analogous to Fourier's law of heat the negative sign signifies conduction that diffusion occurs in the direction of concentration gradient.

J_{AZ} is diffusion flux.

1.3. Diffusion coefficient for Gases: Diffusion coefficients are determined experimentally and for different systems the values are different. However, in many cases it may be required to be estimated. Gilliland proposed the following equation for estimating the diffusion coefficient for a binary system.

$$D_{AB} = \frac{0.0043T^{1.5}}{P({V_A}^{1/2} + {V_B}^{1/2})} = \sqrt{\frac{1}{M_A} + \frac{1}{M_B}}$$

Where,

$D_{AB} = Diffusivity\,(cm^2/sec.)$

T = Temperature, °K

P = Absolute pressure, atm

M_A,M_B = Molecular weight of component A and B

V_A,V_B = Molecular volume at normal boiling points, cc/g mole

The coefficient of diffusion in binary mixtures varies only slightly with composition and is usually considered independent of composition. Fir given system their equation

$$D_{AB} = \frac{C\,T^{1.5}}{P}$$

Where, C is constant.

This states that the diffusion coefficient is inversely proportional to the pressure and directly proportional to the absolute temperature raised to the power 1.5.

Multiple Choice Question

1. Determination of humidity can be done by the measurement of the _______________ of a fiber.
   A. electrical resistance
   B. strength
   C. thermal conductivity
   D. none of these.

2. When $N_{pr} = N_{sc} =$ ____________ then the terms mass diffusivity, the thermal diffusivity and the momentum diffusivity are the same.
   A. 1
   B. 0.5
   C. 10
   D. 0

3. When the equilibrium and operating lines are both straight lines for a packed adsorption tower the ratio HETP/HYU$_{OG}$ _____________ the absorption factor.
   A. is one at unity value of
   B. increases with increase in
   C. both (a) and (b)
   D. neither (a) nor (b)

4. In a distillation column increase in the reflux ratio from the minimum
   A. number of plates decreases very slowly first and then more and more rapidly.
   B. slope of the operating line in stripping section decreases.
   C. total cost first decreases and then increases.
   D. liquid flow increases while the vapor flow decreases for a system.

5. According to the definition of relative humidity it is the ratio of the
   A. actual humidity to saturation humidity.
   B. partial pressure of the vapour to the vapour pressure of the liquid at gas temperature.
   C. partial pressure of the vapour to the vapour pressure of the liquid at room temperature.
   D. none of these

6. At constant reflux conditions in a batch distillation the overhead product composition ___________ with time
   A. does not vary
   B decreases
   C increases
   D. may increase on decrease, depends on the system

7. During sensible cooling process which of the following parameters is constant?
   A. Partial pressure of vapour
   B. Specific humidity
   C. Both (a) and (b)
   D. Neither (a) nor (b)

8. Problem relating to simultaneous heat and mass transfer fall under the domain of ___________ member.
   A. Lewis number
   B. Sherwood number
   C. Schmidt number
   D. Prandtl number

9. The operating ____________ in case of an absorber
   A. line always lies above the equilibrium curve.
   B. velocity is more than the loading velocity.
   C. line can be either above or below the equilibrium curve.
   D. line always lies below the equilibrium curve.

10. The point at which the tie line reduces to a point on the solubility curve is called a plait point. What is the number of plait point for a ternary system containing two pairs of partially miscible liquids?
    A. 0
    B. 1

C. 2

D. 3

11. A____________ provides maximum contact surface area among the various liquid-vapor contacting devices listed below?

A. Bubble cap column

B. Sieve plate column

C. Randomly packed column

D. Wetted wall column

12. Which of the following statements is correct for mass transfer coefficient of liquid?

A. affected more by temperature than that for gases.

B. affected much less by temperature than that for gases.

C. not affected by the temperature.

D. none of these.

13. According to film theory the relation between mass transfer coefficient (K) and diffusivity (D) is

A. $K \propto D$

B. $K \propto D^{0.5}$

C. $K \propto D^{1.5}$

D. $K \propto D^2$

14. Which of the following statements is correct for a limitation on the rate of solid-liquid extraction.

A. diffusion of the solute from the solution in contact with the particle to the main bulk of the solution.

B. phase change of the solute as it dissolves in the solvent.

C. diffusion of the solute through the solvent in the pores to the outside of the particle.

D. all 'a', 'b' & 'c'.

15. If S_1 and S_2 are the slopes of operating line and equilibrium curves respectively the absorption factor is.

A. $S_2 x\ S_1$

B. S_1/S_2

C. S_1 / S_2

D. $S_1 - S_2$

16. In the design of a co-current absorber which of the following quantities need not be fixed?

A. Flow rate of gas

B. Flow rate of entering liquid

C. Heat gain or loss

D. None of these

17. An example of absorption accompanied with chemical reaction is in the absorption of

A. benzol present in coke oven gas by wash oil

B. ammonia in water.

C. SO_2 in alkaline solution.

D. all (a), (b) and (c).

18. Which of the following equipments is frequently used for adiabatic humidification-cooling operation with recirculating liquid.

A. induced draft cooling tower

B. natural draft cooling tower

C. spray chamber

D. none of these

19. By definition chemical absorption also known as Chemisorption is

A. a reversible phenomenon.

B. same as "Van der Waals" adsorption.

C. an irreversible phenomenon.

D. characterized by adsorption of heat.

20. The relative volatility for a binary system ___________ with increase is pressure?

A. remains same.

B. decreases.

C. increases.

D. either (a) or (b), depends on the system.

21. If extraction is required to be done for a system having very low density difference, which of the following will be the most suitable?

A. Pulsed extractor

B. Centrifugal extractor

C. Packed extraction tower

D. Mixer-settler extractor

22. In a distillation column operating at total reflux which of the following parameter is at minimum value?

A. condenser load

B. number of plates

C. reboiler load

D. all (a), (b) and (c)

23. Between two fluid phases the rate of mass transfer does not essentially depend on they ___________ of the two phase.

A. chemical properties

B. physical properties

C. interfacial area

D. degree of turbulence

24. The primary parameter which is considered for determination of number of bubble caps per tray in a distillation column is
    A. feed composition.
    B. allowable gas velocity.
    C. allowable gas and liquid velocities.
    D. allowable liquid velocity.
25. The overhead product in a distillation column contains
    A. only saturated liquid
    B. only one component
    C. any number of components
    D. two components
26. What will be the molality of a solution containing 0.30 kg mole of solute and 600 kg of solvent?
    A. 0.50
    B. 0.60
    C. 2
    D. 1
27. For a gas – vapor mixture the dew point.
    A. increases with increase in pressure.
    B. increases with temperature rise.
    C. decreases with decrease in pressure.
    D. decreases with temperature rise.
28. ____________ may be determined by the use of Fenske equation
    A. optimum reflux ratio.
    B. maximum number of ideal plates.
    C. minimum number of theoretical plates.
    D. height of the distillation column.
29. For a system of unsaturated mixture the difference of wet bulb temperature and adiabatic saturation temperature will be ______________.
    A. +ve
    B. zero
    C. –ve
    D. none of these
30. In a distillation column the theoretical number of plates may be calculated by Fenske's equation subject to the condition that
    A. relative volatility is reasonably constant.
    B. mixture (to be separated) shows positive deviation from ideality.
    C. multicomponent distillation is involved.
    D. mixture (to be separated) shows negative deviation from ideality.
31. Which of the following is represented by the term "cooling range" in a cooling tower.
    A. hot water entering the tower and the wet bulb temperature of the surrounding air.
    B. cold water leaving the tower and the wet bulb temperature of the surrounding air.
    C. hot water entering the tower and the cooled water leaving the tower.
    D. none of these
32. When a drum drier is in operation which of the following is the controlling parameter?
    A. Diffusion
    B. Heat transfer
    C. Neither (a) nor (b)
    D. Both (a) and (b)
33. A ____________ draft cooling tower may be used when the temperature and humidity of air is low.
    A. induced
    B. natural
    C. forced
    D. none of these
34. In case of steam distillation the amount of steam required per unit quantity of distillation will be reduced by
    A. lowering the total pressure.
    B. raising the temperature.
    C. both (a) and (b).
    D. neither (a) nor (b).
35. The application of Raoult's law is valid for.
    A. non-volatile solute.
    B. all liquid solutions
    C. non-ideal solution only.
    D. solvents.
36. Such absorbents which has how viscosity are preferred for the reasons of.
    A. improved flooding characteristics.
    B. low pressure drop on pumping.
    C. rapid absorption rates and good heat transfer characteristics.
    D. all (a), (b) and (c).
37. The operation of a drier under vacuum conditions is done to
    A. Dry materials having high bound moisture content.
    B. Reduce drying temperature.

C. Dry those materials which have very high unbound moisture content.

D. Increase drying temperature.

38. For liquid dispersion in a continuous gas phase which of the following column may be used?

A. Packed

B. Pulse

C. Sieve plate

D. Bubble cap

39. Air is to be cooled at constant pressure to 30°C from the initial condition of relative humidity of 50% pressure 101.3 kPa and temperature 40°. The cooled air will have.

A. higher wet bulb temperature.

B. higher dew point.

C. higher relative humidity

D. higher absolute (specific) humidity.

40. Thiele-Geddes equation is applicable in which of the following mass transfer operation?

A. Liquid-liquid extraction

B. Solid-liquid extraction

C. Multicomponent absorption with chemical reaction

D. Multicomponent distillation.

41. Which of the following statements is correct?

A. The selectivity of solvent used in solvent extraction is unity at the plate point.

B. In case of liquid-liquid extraction, no separation is possible, if the selectivity of the solvent used is unity.

C. with increase in temperature, the selectivity of solvent of the solvent used in solvent extraction decreases.

D. All (a), (b) and (c).

42. For a liquid solution having two or more components the separation of the components cannot be effected by

A. liquid extraction

B. fractional crystallization

C. absorption

D. evaporation

43. When the mass transfer is by __________ Stefan's law is applicable.

A. diffusion

B. bulk flow

C. both 'a' & 'b'

D. neither 'a' nor 'b'

44. A decrease in the __________ of the solution may occur when a solute is added in the solvent.

A. boiling point

B. vapour pressure

C. freezing point

D. both (b) and (c)

45. By definition of physical absorption it is

A. an irreversible phenomenon.

B. accompanied by evolution of heat.

C. a reversible phenomenon.

D. both (b) and (c)

46. The amount of heat in kcal required to increase the temperature of 1 kg of gas and its accompanying vapor by 1°C is known as

A. sensible heat

B. latent heat

C. specific heat

D. humid heat

47. If G is the gas flow rate and S is the solvent flow rate then the absorption factor may be increased if we.

A. decrease both '*G*' and '*S*'.

B. increase both '*G*' and '*S*'.

C. increase '*S*' and decrease '*G*'.

D. increase '*G*' and decrease '*S*'

48. A defined separation is to be effected at minimum reflux ratio then

A. separation is most efficient.

B. number of plates is infinity.

C. number of plates is zero.

D. minimum number of the theoretical plates is required.

49. In a cooling tower makeup water is required from time to time to replace the water lost by __________.

A. drift

B. evaporation

C. blow down and leakage

D. all (a), (b) and (c)

50. For a distillation column the overall efficiency may be represented by

A. the ratio of number of ideal plates to actual plates.

B. same as the Murphree efficiency.

C. always more than the point efficiency

D. the ratio of number of actual plates to ideal plates.

51. In a distillation column the phenomena of weeping
    A. provides large interfacial surface for mass transfer.
    B. due to very high gas velocity.
    C. results due to very high gas velocity.
    D. results due to very low gas velocity.

52. Which of the following systems will form minimum boiling azeotrope at 1 atm.
    A. hydrochloric acid-water
    B. ethyl alcohol-water
    C. benzene-toluene
    D. all (a), (b) and (c)

53. The total vapor pressure varies __________ with the composition (as mole fraction) for an ideal solution.
    A. exponentially
    B. linearly
    C. inversely
    D. none of these

54. In large diameter tower if the path of liquid across the plate is very long Murphree plate efficiency may be ___________ percent.
    A. <100
    B. >100
    C. 100
    D. none of these

55. Diffusion in its simple form is also called __________ diffusion.
    A. pressure
    B. thermal
    C. concentration
    D. forced.

56. For a given set of operating conditions which of the following gas liquid contacting devices will incur the least pressure drop?
    A. wetted wall
    B. perforated tray
    C. packed
    D. bubble cap

57. Which of the following statements is wrong?
    A. Even a forced draft cooling tower cannot cool water below the wet bulb temperature.
    B. Different solids have different equilibrium moisture curves.
    C. Heat removal load on cooler remains constant, even with increase of the reflux ratio in a distillation column.
    D. Total condenser or reboiler is considered equivalent to one theoretical plate in McCabe-Thiele method of theoretical plate calculation for distillation column.

58. The most suitable drier for free flowing granular materials is a __________ drier.
    A. rotary
    B. drum
    C. freeze
    D. cylinder

59. Which of the following statements regarding rotary dryer is not correct?
    A. Hold up of a rotary drier is defined as the fraction of the dryer volume occupied by the solid at any instant. The best performance for rotary drier is obtained, when the holds up is in the range of 0.05 to 0.15.
    B. Flights (located in the inside shell of rotary dryer) lift the material being dried and shower it down through the current of hot air/gases. It extends from the wall to a distance which is about 8-12% of the inside diameter of shell.
    C. Rotary dryer is suitable for drying sticky material.
    D. Recommended peripherial speed of a rotary drier is in the range of 10 to 30 meters/ minute.

60. Pick out the correct statement. Which of the following statement is correct.
    A. Electrodylsis is used for the refin-ning/ purification of petrol.
    B. Electrodylsis is used in the manufacture of NaOH.
    C. Dialysis process is used for the production of fresh water from brine.
    D. both (b) and (c).

61. Increasing the reflux reflux ratio above optimum for a binary mixture does not result in the increase of
    A. area between operating line and 45° diagonal x-y diagram.
    B. tower cross-section.
    C. condenser and reboiler surfaces
    D. none of these.

62. By the use of McCabe-Thiele method at infinite ratio

A. the overhead product is minimum.
B. both the operating lines coincide with diagonal.
C. neither (a) nor (b)
D. both (a) and (b).

63. During the operation of a bubble cap distillation column if the throughput is decreased in comparison to the design capacity its efficiency
A. increases
B. remains same
C. decreases
D. may increase or decrease; depends on individual design

64. If large temperature changes are involved plate towers are preferred to packed towers because of reason that the
A. packing may be crushed due to thermal expansion/contraction of the components of the equipments.
B. viscosity variation of the fluids may cause channeling/flooding
C. pressure drop in the equipment will be very high.
D. none of these.

65. In which of the following cases the Raoult's law is applicable?
A. ideal solutions
B. mixture of water and alcohol
C. non-ideal gases
D. real solutions

66. Isotopes of methane may be separated by using the principle of diffusion.
A. Thermal
B. Pressure
C. Concentration
D. Force

67. What will be the humid heat in KJ/°K a gas if the specific heat is 0.2KJ/°K?
A. 0.017
B. 0.107
C. 0.31
D. 0.215

68. ________ may be explained as an operation when the soluble material in largely on the surface of an insoluble solid and is merely washed off the solvent.
A. percolation
B. decoction
C. elutriation or elution
D. none of these

69. For a batch drying operation the rate of drying curve in the constant rate period is absorbed, where
A. cracks develop on the surface of the solid.
B. rate of drying decreases abruptly.
C. surface evaporation of unbound moisture occurs.
D. none of these.

70. Which of the following statement is correct?
A. Ponchan-Savarit method is more accurate than McCabe-Thiele method to determine the number of theoretical stages required for the separation of non-ideal binary system by distillation.
B. Solvent used in extractive distillation should be of high volatility.
C. A sharp decrease in pressure drop is an indication of flooding in a distillation column.
D. Flash distillation is suitable for separating components which have very close boiling temperature.

71. It takes 6 hours to dry a wet solid from 50% moisture content to the critical moisture content of 15%. How much longer it will take to dry the solid to 10% moisture content, under the same drying conditions? (The equilibrium moisture content of the solid is 5%).
A. 51 min
B. 15 min
C. 71 min
D. 94 min

72. Among tie binary systems listed below ________ is an example of a maximum boiling azeotrope.
A. Water-hydrochloric acid.
B. Water-ethyl alcohol.
C. n-heptane-n-octane.
D. Acetone-carbon disulphide.

73. Such packed towers which are very tall are usually divided into series of beds to that may .
A. reduce liquid hold-up
B. avoid channeling
C. avoid flooding
D. reduce the overall pressure drop

74. Variation of binary diffusivity in gases and liquids may be represented respectively as
   A. $T^{3/2}$ and T
   B. T and $T^{3/2}$
   C. $T^{3/2}$ and T
   D. T and $T^{3/2}$

75. __________ Moisture is the amount of moisture contained by substance in excess of the equilibrium moisture.
   A. Critical
   B. Free
   C. Bound
   D. Unbound

76. The number of gm moles of solute per _________ of solvent is knows as molality
   A. litre
   B. gm
   C. c.c
   D. kg

77. At plant point for a ternary system
   A. density difference between the two equilibrium phases become zero.
   B. distribution coefficient
   C. for solute will be unity selectively of solvent will be unity
   D. none of these

78. For air – water vapor system the value of Lewis number (Le = Sc/Pr) is approximately ________
   A. 1
   B. 600
   C. 3.97
   D. 0.24

79. From among the following T – H diagrams which is valid for dehumidification operation?

A.

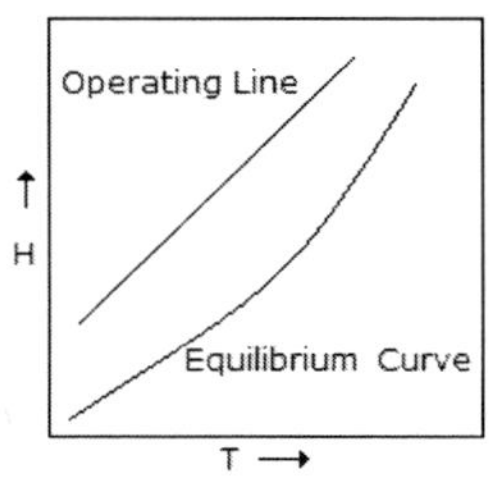

B.

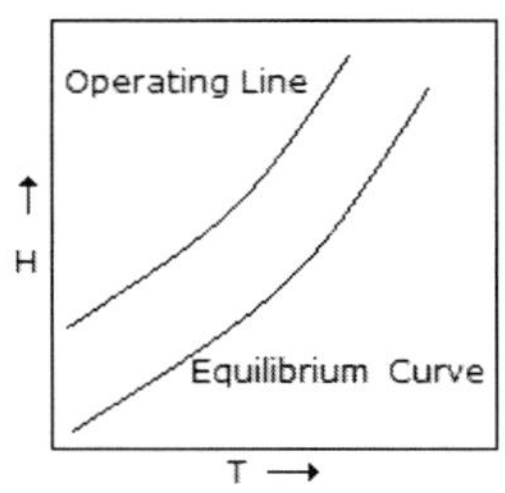

C.

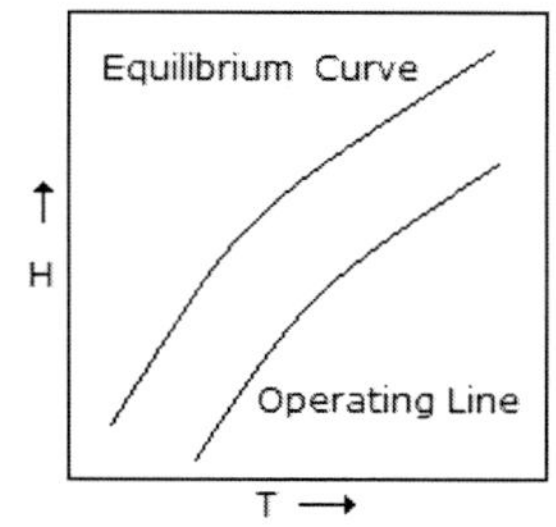

D.

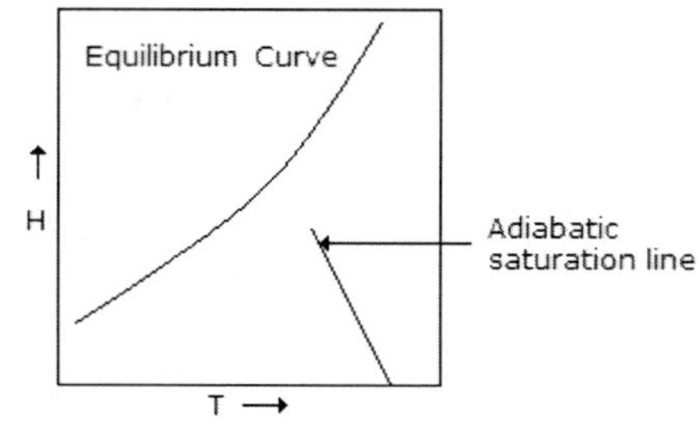

80. Distillation of an aqueous solution of methanol in a tray column using light pressure steak as a source of heat for a given reflex ratio and overhead composition two options are available (i) a reboiler is used and (ii) no reboiler is used but steam is fed directly to the bottom of the column. As compared to option in (ii)
   A. less number of trays are required.
   B. more number of trays are required but the residue composition remains unchanged.
   C. composition of the residue remains unchanged
   D. more number of trays are required and the residue composition is more dilute in methanol.

81. In distillation practice packed towers are prefered over plate towers because of
   A. high pressure drop and low hold up.
   B. low pressure drop and high hold up
   C. low pressure drop and low hold up.
   D. high pressure drop and high hold up.

82. The solubility of gases in liquids, at fixed pressure __________ with increase in temperature.
   A. remains same
   B. decreases
   C. increases
   D. either (a) or (b), depends on the system

83. Bollman extractor is a special purpose equipment which
   A. is a centrifugal extractor.
   B. is used for extraction of oil from oilseed.
   C. is a static bed leaching equipment.
   D. employs only counter-current extraction.

84. Which of the following relationships is valid for partial pressure distribution of on ideal gas diffusing through another stagnant ideal gas at steady state?

A. Exponential
B. linear
C. Cubic
D. parabolic

85. Which of the following statements is correct?

A. Free moisture is removed during falling rate drying period.
B. The sum of free moisture and equilibrium moisture in a solid is called the critical moisture.
C. Bound moisture is removed during constant rate drying period.
D. The driving force for mass transfer in a wet solid is its free moisture content.

86. In a liquid-liquid extraction column of mixture settler type the functions of a coalescer is that it

A. helps in increasing the settling rate of the bubbles.
B. comprises of thin bed of substances of extended surface having high porosity.
C. helps in increasing the bubble size entering the settler.
D. all 'a', 'b', & 'c'.

87. Which of the following dimensionless group corresponds to the Nusselt number in heat transfer?

A. Sherwood
B. Peclet
C. Stanton
D. Schmidt

88. Creation of vacuum in a vacuum crystalliser is done by

A. compressed air jet
B. suction pump
C. steam jet ejector with a barometric condenser
D. none of these

89. Which of the following dimensionless groups is equal to (N_{RE}. N_{SC}) as used in mass transfer operations.

A. Sherwood
B. Stanton
C. Peclet
D. none of these

90. Which of the following statement is correct?

A. Kremsor – Brown – Souder's equation is used to calculate the efficiency of a plate tower.
B. For a given separation in a binary distillation column, with increase in reflux ratio (above the minimum reflux ratio), the fixed cost of the distillation column first increases and then decreases.
C. In Ponchon-Savarit method of no. of theoretical plate determination, the stripping and rectifying operating lines are vertical at total rcflux.
D. The slope of the stripping line is always less than unity.

91. The pressure drop in packed tower being irrigated with liquid ________ that in dry packed tower at the same gas flow rate.

A. is greater than
B. is same as
C. is lower than
D. cannot be predicted as data are insufficient

92. Which the mechanism of transfer is the same only then the relation $S_c = P_r = 1$ is valid.

A. heat and mass
B. mass and momentum
C. heat and momentum
D. heat, momentum and mass

93. Liquid diffusivity is of the order of __________ cm^2 /second. Which of the following ranges in cm^2/second correspond to diffusivity of liquid?

A. 0.1
B. 0.01
C. 10^{-5} to 10^{-6}
D. > 1

94. Solvents having high latent heat of vaporization are not recommended to be used in liquid extraction because

A. it cannot be recovered by distillation.
B. it will decompose while recovering by distillation.
C. its recovery cost by distillation may be prohibitively high.
D. the pressure drop and hence the pumping cost will be very high

95. For azeotrpic mixtures the equilibrium liquid composition compared to the vapor composition is
   A. less
   B. more
   C. same
   D. either more or less; depends on the system
96. Experiment carried out in a wetted wall tower determines the
   A. volumetric co-efficient
   B. molal diffusivity
   C. mass transfer co-efficient
   D. none of these.
97. The maximum percentage extraction achieved in the unit is. The minimum liquid rate to be used is an absorber corresponds to an operating line
   A. f slope equal to I
   B. tangential to the equilibrium curve
   C. slope equal to Op 1
   D. vane of these
98. Under the condition that Raoult's law and Dalton's law are applicable for liquid and vapour phase respectively of a binary system, the relative volatility may be define as the ratio of
   A. vapour pressure of component A to that of component B.
   B. partial pressure of component A to the total pressure.
   C. vapour pressure of component A to the total pressure.
   D. vapour pressure of component A to the partial pressure of A.
99. The purpose of providing an agitator in crystalliser is
   A. formation of nucle
   B. crystal growth
   C. avoiding deposition on cooler surfaces.
   D. all (a), (b) and (c).
100. In a distillation column if the feed composition, feed quality and separation is the same number of plates will increase with total pressure.
   A. increase
   B. remain same
   C. decrease
   D. data insufficient, can't be predicted
101. In a Vanaspati ghee manufacturing plant the process of hydrogenation of oil is carried out in a/an_______________
   A. agitated vessel
   B. packed tower
   C. sieve tray column
   D. bubble cap column
102. The lesser the solubility of a given solute in a liquid in an interphase mass transfer the higher are the chances that the transfer process will be
   A. liquid phase resistance-controlled.
   B. impossible.
   C. driven by a non-linear driving force.
   D. gas phase resistance controlled.
103. In a triangular graph paper representing a ternary liquid system the sides of equilateral-triangular represent.
   A. a binary mixture.
   B. a pure component.
   C. a ternary mixture.
   D. partially miscible ternary system.
104. For rigorous and accurate design of a distillation column , McCabe – Thiele method is inadequate and Ponchan-Savarit procedure is needed , when
   A. a total condenser is used.
   B. saturated feed is not used.
   C. the latent heat of vaporization of the more and less volatile components are greatly different.
   D. an azeotrope forms.
105. An enhancement in the cooling effect is possible by
   A. increasing the air velocity and area of exposed wet surface.
   B. reducing the humidity of the air.
   C. reducing the barometric pressure.
   D. all (a), (b) and (c).
106. The primary parameter to decide the capacity of a rotary dryer is
   A. r.p.m
   B. inclination with ground surface
   C. neither (a) nor (b)
   D. both (a) and (b)
107. The extraction efficiency of a counter current extractor __________ with increase in axial mixing
   A. increases
   B. remains unchanged

C. decreases
D. depends on the pressure of the system

108. A decreases in reflux ratio means
A. separation becomes more efficient.
B. column diameter increases.
C. number of plates increases.
D. none of these.

109. The basic equation of diffusion is applicable to
A. steady state equimolar counter diffusion
B. steady state diffusion for stagnant case
C. Fick's first law of diffusion
D. liquid M.T.C. by penetration theory.

110. Which of the following statements is not correct?
A. in the McCabe –Thiele diagram for binary distillation, vertical feed line represents saturated liquid feed and horizontal feed line represents saturated vapour feed.
B. in small columns, local efficiency is larger than Murphree efficiency; in large columns local efficiency is smaller than Murphree efficiency.
C. for Laminar flow over a plate of length L, the local mass transfer co-efficient at a distance L from the leading edge is 1.5×10^{-2}m/s. Then the average mass transfer co-efficient for the plate is 2×10^{-2}m/s.
D. the concentration and hydrodynamic boundary layers over a flat plate are of equal thickness, if Schmidt number is equal to unity.

111. With reference to a catalyst, the Knudsen diffusivity of a catalyst is dependent on the
A. pore radius of the catalyst only.
B. molecular velocity only.
C. molecular mean free path only.
D. molecular velocity and por radius of the catalyst.

112. For an absorber the slope of the operating line is __________ where L = non-volatile solvent in liquid stream and G = insoluble gas in gas stream
A. L/G
B. G/L
C. always < 1
D. none of these

113. In extractive distillation the solvent used
A. is of low volatility.
B. forms a low boiling azeotrope.
C. forms a high boiling azeotrope.
D. does not alter the relative volatility of the original components.

114. For obtaining a given separation in distillation column, the minimum number of theoretical stage is obtained with.
A. minimum reflux ratio
B. optimum reflux ratio
C. total reflux
D. zero reflux ratio.

115. Among the following columns the __________ will incur the lowest pressure drop.
A. gas
B. liquid
C. neither (a) nor (b)
D. both (a) & (b)

116. Flash vaporisation is done at fixed T and P for one mole feed of a binary mixture. If Raoult's law applies to the system then changing the feed composition would effect.
A. the fraction vaporised but not the product composition.
B. the product composition as well as the fraction vaporised.
C. the product composition but not the fraction vaporised.
D. neither the product composition nor the fraction vaporised.

117. In a randomly counter current gas-liquid absorption column, with increase in the liquid flow rate at a fixed gas velocity the pressure drop __________.
A. remains unchanged
B. decreases
C. increases
D. decreases exponentially

118. During evaporative cooling process with re-circulated water supply which of the following remains constant?
A. partial pressure of vapour
B. relative humidity
C. wet bulb temperature
D. none of these

119. In concentrated solutions the diffusion is different than that in dilute solutions because of the changes in __________
A. viscosity
B. degree of ideality

C. both 'a '& 'b'
D. neither 'a' nor 'b'

120. In vegetable oil industry the most commonly used solvent is
A. furfurol
B. hexane
C. liquid SO_2
D. phenol

121. For a particular mass transfer operation a packed tower compared to a plate tower
A. is less costly when built in large sizes/ diameters.
B. allows installation of cooling coils.
C. incurs smaller pressure drop.
D. is more suitable, if suspended solids are present in fluid streams.

122. Relative volatility, for a binary system
A. decreases with increase in pressure
B. increases with increase in pressure
C. increases with increase in temperature at constant pressure
D. has no significance in distillation operation

123. In the given plot of saturated humidity (H_e) versus temperature (T) the constant enthalpy line, joining (H_1, H_2) and (H_2, T_2) has been shown. Which of the following statements is correct?

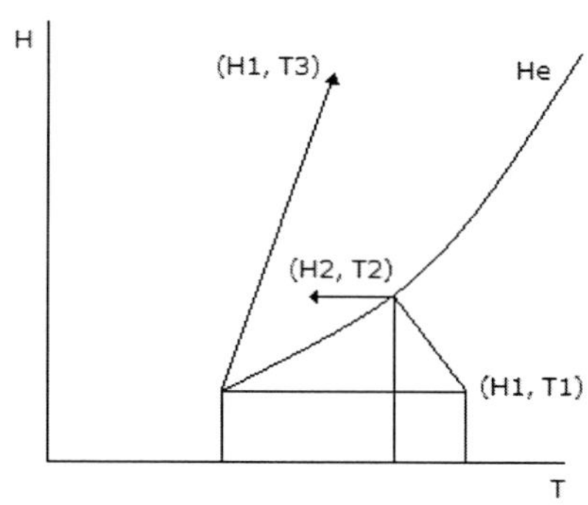

A. T1-Dew point temp; T2-Dry bulb temp; T3-Wet bulb temp.
B. T1-Dry bulb temp; T2-Wet bulb temp; T3-Dew point temp.
C. T1-Dew point temp; T2-Wet bulb temp; T3-Dry bulb temp.
D. T1-Wet bulb temp; T2-Dry bulb temp; T3-Dew point temp.

124. In a leaching process on which of the following parameters the stage efficiency depends?
A. rate of diffusion of the solute through the solid and into the liquid
B. time of contact between the solid and the solution
C. both (a) and (b)
D. vapour pressure of the solution

125. The deciding factor for minimum possible diameter of a packed column is
A. gas viscosity
B. flooding
C. liquid density
D. liquid hold up

126. The term solvent extraction is a special case of liquid-liquid extraction used for the separation of the components of liquid when
A. extracting solvent is cheaply & abundantly available.
B. viscosity of liquid components is very high.
C. one of the liquid components is heat sensitive.
D. one of the liquid components has very high affinity towards the solvent.

127. For the same system the diffusivity in concentrated solutions will differ from that in dilute solutions, because of the changes in
A. degree of ideality of the solution.
B. viscosity with concentration.
C. both (a) and (b).
D. neither (a) nor (b).

128. Occurrence of 'case hardening' during drying of high moisture solid cake will ________ the drying rap.
A. exponentially increase
B. increase
C. does not effect
D. decrease

129. During the operation of the distillation column, the
A. driving force for the vapour flow is the pressure drop.
B. pressure increases gradually from bottom to the top of the column.
C. liquids are not always at their bubble points.
D. none of these

130. Increasing the __________ would increase the rate of leaching
A. temperature
B. size of the solid

C. viscosity of solvent
D. pressure

131. When cooling takes place at constant total pressure out of contact with a liquid the temperature at which a gas-vapor mixture becomes saturated is known as
A. wet bulb temperature
B. dew point
C. dry bulb temperature
D. bubble point

132. The treatment of radioactive nuclear waste is in ___________.
A. Bollman extractor
B. rotating-disc contactor
C. mixer-settler extractor
D. pulsed column extractor

133. An azeotropic mixture is a __________ mixture.
A. binary
B. constant boiling point
C. ternary
D. none of these

134. The drying of sugar on commercial scale is done in a _________ dryer.
A. tray
B. spouted bed
C. rotary
D. truck

135. In distillation, overhead product contains
A. uncertain
B. any number of components
C. low components
D. only one components

136. For the same column diameter which of the following columns would give lowest pressure drop per unit height?
A. packed column (stacked)
B. bubble-cap column
C. sieve-plate column
D. randomly packed column

137. The solvent is __________ in extractive distillation
A. present in overhead stream.
B. of high volatility.
C. added to alter the relative volatility of the mixture.
D. of high viscosity to give high tray efficiency.

138. Which of the following relations indicate the internal reflux ratio (R_i) and the external reflux ratio(R_e) ?
A. $R_i = \dfrac{R_e}{1 - R_e}$
B. $R_i = \dfrac{R_e}{1 + R_e}$
C. $R_i = \dfrac{1 - R_e}{R_e}$
D. $R_i = R_e$

139. The steady state temperature attained by a small amount of liquid evaporating into a large quantity of unsaturated gas-vapor mixture is called ________ temperature.
A. wet bulb
B. dry bulb
C. dew point
D. none of these

140. In a counter current stripper, the operating line
A. lies below the equilibrium curve.
B. lies above the equilibrium curve.
C. can lie above or below the equilibrium curve.
D. is always parallel to the equilibrium curve.

141. Among the following columns __________ the incur the lower pressure drop.
A. bubble plate column
B. packed column (with stacked packing)
C. packed column (with dumped packing)
D. pulse column

142. Within the constant rate period of drying the drying time between fixed moisture content is proportional to __________ where T is thickness of the solid.
A. T
B. $T^{0.5}$
C. $T^{1.5}$
D. T^3

143. For the drying of large quantity of silica gel normally __________ is used.
A. freeze
B. through circulation
C. rotary vacuum
D. tray

144. In which type of dryer heat sensitive materials like certain pharmaceuticals can be dried?
   A. freeze
   B. indirect tray
   C. spray
   D. none of these
145. The regeneration of molecular sieves may be done by heating to____________°C.
   A. > 500
   B. < 150
   C. 200-330
   D. > 1000
146. What will be the effect on tower height if for the same system, the same liquid used in an absorber is decreased?
   A. decrease
   B. increase
   C. remain unaffected
   D. decrease or increase; depends on the type of liquid
147. The extraction efficiency of a counter current extractor __________ with increase in axial mixing
   A. increases
   B. remains unchanged
   C. decreases
   D. depends on the pressure of the system
148. Mass transfer between a pure liquid phase and a fixed gas which is ___________ happens during humidification.
   A. at a fixed temperature
   B. insoluble in the liquid
   C. non-ideal in nature
   D. soluble in the liquid
149. Which of the following is not a continuous drier?
   A. tunnel drier
   B. spray drier
   C. drum drier
   D. trey drier
150. The HETP does not vary with the __________ in an absorber.
   A. flow rate of liquid
   B. type and size of packing
   C. flow rate of gas
   D. none of these
151. Which of the following equations is correct if R_i and R_o represent internal and external reflux ratio respectively?
   A. $(1 + R_o) = R_o/R_i$
   B. $(1 + R_i) = R_o/R_i$
   C. $(1 - R_o) = R_o/R_i$
   D. $(1 - R_i) = R_o/R_i$
152. A mass transfer operation system has a flux of component A in a stagnant medium across a gas-liquid interface. Data available on overall mass transfer coefficient based on the liquid side for dilute system is 4×10^{-3} kg mol/m^2.s. and the equilibrium data is y = 2x. The flux across the interface in kg mol/m^2.s. for the bulk concentrations of A in gas phase and liquid phase as y = 0.4 and x = 0.01 respectively is
   A. 8.5×10^{-3}
   B. 8.5×10^{-4}
   C. 5.6×10^{-4}
   D. 5.6×10^{-3}
153. Which of the following statements is wrong?
   A. in case of an absorber, the operating line is always above the equilibrium curve.
   B. in case of a stripper, the equilibrium curve is always below the operating line.
   C. absorption factor is constant, when the equilibrium and operating lines are straight.
   D. in the absorption of low solubility gases, the liquid film is the controlling resistance.
154. A tall packed tower is usually divided into a series of small towers so as to
   A. reduce liquid hold up.
   B. minimize the overall pressure drop.
   C. avoid flooding.
   D. avoid channeling.
155. If the selectivity is unity in a liquid extraction process then.
   A. solvent flow rate should be very low.
   B. no separation will occur.
   C. separation of the constituents is the most effective.
   D. amount of solvent required will be minimum.
156. The principle of gaseous diffusion process is used during separation of __________.
   A. isotopes of helium

B. isotopes of uranium
C. SO_2 and air
D. N_2 and O_2 from air

157. Wettedwall column is effectively used for experimental determination of.
A. NTU
B. mass transfer co-efficient
C. diffusion co-efficient
D. none of these

158. Mass diffusivity and ________ has the same dimension.
A. momentum flux
B. thermal diffusivity
C. kinematic viscosity
D. Both (b) and (c)

159. The entire moisture content of a non-hygroscopic material in called the moisture.
A. equilibrium
B. bound
C. unbound
D. free

160. During cooling tower operation which of the following happens
A. effectiveness of forced draft cooling tower is reduced by the recirculation of the humid and hot discharged air.
B. in case of induced draft cooling tower, the fan is placed at the top of the tower.
C. a natural draft cooling tower is recommended to be used, when the air humidity is low and air temperature is also low.
D. effectiveness of a mechanical draft cooling tower is reduced with increase in air wet-bulb temperature.

161. Pick out the wrong statement.
A. decrease in Murphree plate efficiency is an indication of entrainment in distillation column.
B. in solvent extraction, the distribution co-efficient should be as large as possible for solvent economy.
C. with increase in pressure, the solubility of gases in solvent increases at a fixed temperature.
D. the capacity of a distillation column is maximum at total reflux.

162. What will be the pore size of a molecular sieve 5A is mm?
A. 0.05
B. 0.5
C. 5
D. 50

163. For K_L = 4.5 and K_G= 1.5. The slope of the equilibrium line is 3 for absorption of a solute from a gas mixture into a liquid solvent Which of the following resistance (s) is (are) controlling ?
A. interfacial
B. gas side
C. liquid side
D. both liquid and gas side

164. The relative volatity of a binary mixture may be defined as the ratio of vapor pressure of component 'A' to that of component 'B', when
A. liquid phase obeys Dalton's law and vapor phase obeys Raoult's law.
B. only vapor phase follows Raoult's law.
C. only liquid phase obeys Dalton's law.
D. vapor phase obeys Dalton's law and liquid phase obeys Raoult's law.

165. In comparison to isothermal absorption in a gas-liquid system the absorption with evolution of heat results in.
A. decreased solute solubility.
B. large minimum liquid to gas (L/G) ratio.
C. large number of trays.
D. all (a), (b) and (c).

166. Which type of distillation may be used to dehydrating alcohol?
A. molecular
B. azeotropic
C. extractive
D. steam

167. A distillation lower is to be used to separate equimolar binary mixture of components A and B (α_{AB} = 2) is into overhead fraction containing 99 mole % A and a bottom fraction containing 98 mole % B. According to Fenske equation, what will be the minimum number of plates required?
A. 28
B. 9
C. 12
D. 25

168. An evaporative cooling process with recirculated water supply would result is an increases of which of the following?

A. wet bulb temperature

B. partial pressure of vapour

C. relative humidity

D. both (b) and (c)

169. For solvent extraction perforated plate towers are recommended an they provide __________.

A. better contact of the phase.

B. higher contact area.

C. repeated coalescence & redispersion of the drop phase.

D. none of these.

170. Comparing the phenomena of heat and mass transfer operation which of the following statements is not correct?

A. Prandtl number in heat transfer is analogous to Schmidt number in mass transfer.

B. Reynolds number remains the same in both heat and mass transfer.

C. Reynolds number in mass transfer is analogous to Grashoff number in heat transfer.

D. Sherwood number in mass transfer is analogous to Nusselt number in heat transfer.

171. Which of the following properties is used in separation of two volatile liquids by distillation?

A. relative volatility

B. density difference

C. solubility

D. selectivity

172. In which type of dryer, the samples of finished product from various parts may exhibit non-uniformity in the moisture content?

A. freeze

B. tray

C. rotary

D. none of these

173. At azeotropic composition the relative volatility of a binary mixture is.

A. 1

B. > 1

C. 0

D. ∞

174. Which of the following statements is not correct?

A. 'Elution' means the desorption of the adsorbed solute by a solvent.

B. change in barometric pressure does not affect the relative humidity.

C. in case of a packed tower, the operating velocity is about half of the flooding velocity, generally.

D. the equilibrium moisture content of the solid can be reduced by increasing the absolute humidity.

175. Flash distillation operation is largely used in.

A. polymerization industry

B. soap manufacturing

C. ammonia synthesis plant

D. petroleum refining

176. What would be the humidity of a vapor free gas in percent of fixed pressure conditions if the humidity depends upon the partial pressure of vapor in the mixture?

A. 50

B. 0

C. 100

D. between 0 and 100

177. When equilibrium vapor pressure is __________ that of the pure liquid at the given temperature the liquid content is called bound moisture.

A. less than

B. equal to

C. more than

D. either (a) or (b); depends on the solid

178. Which of the following expressions represent the Lewis (Le) number?

A. $Sc \times Pr$

B. $St \times Sh$

C. $Pr \times St$

D. $Sh \times Pr$

179. The operation of a Swenson-Walker crystallizer is __________.

A. continuous

B. semi-batch

C. batch

D. cooling (adiabatic)-cum-evaporation

180. The difference between heat of adsorption and heat of normal condensation in physical adsorption phenomenon is

A. called integral heat of adsorption

B. equal to the heat of wetting.

C. equal to the heat of formation of surface compound.

D. zero

181. The mutual solubility of two liquids _______ with increase in temperature.

A. increases
B. remains unchanged
C. decreases
D. decreases exponentially

182. In the case of the diffusional resistance of ________ is zero the interface of the liquid & gas for absorption and stripping are present in equilibrium.

A. interface
B. liquid phase
C. gas phase
D. all 'a', 'b' & 'c'

183. For a single component co-current absorber the slope of the operating line of the plot in terms of mole ratio unit is.

A. +ve
B. 0
C. -ve
D. ∞

184. For optimum operation of an absorption column the liquid –gas ratio L/mG ranges from.

A. 2.5 to 3.5
B. 0.8 to 1.2
C. 1.25 to 2.0
D. 0.5 to 0.8

185. What would be the effect of maintaining high pressure at the bottom of a distillation column handling heat sensitive materials?

A. thermal decomposition of bottoms.
B. erosion of the tower.
C. very efficient operation.
D. increased relative volatility.

186. In mass transfer operation the interfacial area measurement may conveniently and accurately be done in a __________ column.

A. bubble cap plate
B. spray
C. packed
D. wetted wall

187. Which of the following driers is used in a paper manufacturing unit?

A. cylinder
B. fluidised bed
C. rotary
D. spray

188. At constant pressure conditions a sample of unsaturated air-water mixture is heated. Which of the following will happen?

A. the relative humidity increases.
B. the partial pressure of water vapour increases.
C. the specific humidity decreases.
D. the relative humidity decreases.

189. Theminimum reflux ratio for water-ethanol system.

A. is computed from the slope of the upper operating line that is tangent to the equilibrium curve.
B. is the optimum reflux ratio.
C. is computed from the intercept of the operating line.
D. cannot be computed.

190. If in a gas-liquid contactor, the fractional hold up of gas is 0.1 and gas bubble diameter is 0.5 mm, the interfacial area per unit volume of dispersion m^2/m^3 in is equal to.

A. 500
B. 900
C. 800
D. 1200

191. At a given temperature the amount of ad-sorbate on the absorbent ___________with the lowering of equilibrium pressure.

A. remains same.
B. decreases.
C. increases.
D. either (a) or (b), depends on the system

192. Sometimes the adsorption and desorption isotherms may not completely coincide due to the phenomena of adsorption hysteresis. In such case the equilibrium pressure is ___________that obtained for adsorption.

A. lower than
B. same as
C. higher than
D. either same or higher than

193. When coarse solid lumps is subjected to leaching the process may be called______________.

A. dissolution
B. decoction
C. percolation
D. agitation and settling

194. The cooling process in which only sensible heat is removed will also mean that the
  - A. relative humidity decreases
  - B. relative humidity increases
  - C. wet bulb temperature decreases
  - D. both (b) and (c)
195. A __________is not a continuous leaching equipment.
  - A. Pachuka tank
  - B. Kennedy extractor
  - C. Bollman extractor
  - D. Hydrocyclones
196. When a gas is to be dissolved in the liquid which is ________ usually co-current absorbers are used.
  - A. a mixture
  - B. sparingly soluble
  - C. a pure substance
  - D. highly soluble
197. Which of the following extractors is used for the concentration of radioactive nuclear waste?
  - A. Pulsed column
  - B. Mixer-settler
  - C. Ballman
  - D. Seive plate
198. An alcohol water azeotrobic mixture is formed of ___________percentage.
  - A. 90
  - B. 91.5
  - C. 95
  - D. 99
199. At a given partial pressure the moisture content of a substance at equilibrium is called the ___________ moisture.
  - A. bound
  - B. free
  - C. equilibrium
  - D. unbound
200. Which of the following expressions represent the definition of mass transfer co-efficient?
  - A. flux=concentration difference/co-efficient.
  - B. co-efficient = Flux/concentration difference.
  - C. flux = Co-efficient/concentration difference.
  - D. none of these.
201. For distillation column design for petroleum hydrocarbons Cox chart is used which is a plot of vapor pressure (P) and temperature (T) as represented by ____________.
  - A. log P *vs* T
  - B. P *vs* log T
  - C. log P *vs*. log T
  - D. T *vs* .P
202. Using steam table and Dalton's law of partial pressure which of the following can be calculated if the relative humidity and dry bulb temperature of moist air is knows.
  - A. humidity ratio
  - B. dew point
  - C. wet bulb temperature
  - D. none of these
203. For a gas-liquid contact operation, the Reynolds number of the liquid was increased 100 times for a laminar falling film. If penetration theory holds good how many times mass transfer coefficient (K_C) will increase?
  - A. 100
  - B. 10
  - C. 5
  - D. 1
204. A small amount of liquid evaporating into a large amount of unsaturated gas-vapor mixture under steady state temperature is known as ___________ temperature.
  - A. dew point
  - B. wet-bulb
  - C. dry-bulb
  - D. adiabatic saturation
205. Leaching and extraction involving a ternery mixture is represented on equilateral triangular coordinate in which a ________________.
  - A. binary mixture is represented by the apex
  - B. binary mixture is represented by any point inside
  - C. ternery mixture is represented by the sides
  - D. pure component is represented by the apex.
206. A decrease in tower diameter can be effected by ____________.
  - A. using higher reflux ratio.
  - B. use of increased tray spacing.
  - C. increasing the vapour flow rate.
  - D. increasing the liquid flow rate.

207. Which of the following conditions is to be fulfilled is steam distillation?
   A. temperature is more than 100° C.
   B. temperature is 100°C.
   C. product must be immiscible with water.
   D. temperature is higher than the boiling point of either component.

208. The slope of the operating line for rectifying section __________ as the reflux ratio increase.
   A. increases
   B. remains constant
   C. decreases
   D. data insufficient, cannot be predicted

209. Volumetric diffusivity has unit of __________.
   A. cm^2/sec
   B. cm^2/sec^2
   C. cm/sec
   D. cm^3/sec

210. In which of the following process steady state equimolar counter diffusion is involved?
   A. separation of a binary mixture by distillation.
   B. all liquid-liquid diffusion systems.
   C. absorption of NH_3 from air by water.
   D. all liquid-solid diffusion systems.

211. For a binary mixture at constant temperature the relative volatility __________ the total pressure.
   A. decreases with increase in
   B. increases with increase in
   C. is independent of
   D. none of these

212. Which of the following theories describes the mass transfer coefficient being directly proportional to D_{AB}.
   A. film
   B. surface-renewal
   C. penetration
   D. none of these

213. In cooling towers the temperature of water cooled in always __________ temperature of entering air.
   A. more than the wet bulb
   B. less than the wet bulb
   C. equal to the dry bulb
   D. equal to the wet bulb

214. In the salvage process of soda ash manufacture a __________ tower is used for production of sodium bicarbonate.
   A. plate
   B. baffle
   C. packed
   D. none of these

215. Which of the following dimension less group indicated the ratio of the heat and mass transfer in a cooling towers?
   A. Lewis
   B. Sherwood
   C. Grashoff
   D. none of these

216. Which of the following parameters does not affect the dew point of an unsaturated gas-vapor mixture?
   A. total pressure
   B. temperature
   C. composition
   D. all (a), (b) and (c)

217. Which of the following determines the slope of q-line?
   A. relative volatility
   B. reflux ratio
   C. thermal condition of the feed
   D. plate efficiency to be achieved

218. In extractive distillation the __________ is considered to be the most important factor in the choice of the separating agent.
   A. Toxicity
   B. Cost
   C. Availability
   D. Selectivity

219. Which of the following is the most efficient cooling tower
   A. induced draft
   B. forced draft
   C. natural draft
   D. atmospheric

220. In McCabeThiele's method ________ is considered equivalent to one theoretical stage.
   A. Partial condenser
   B. Reboiler
   C. Total condenser
   D. Both (b) and (c)

221. The process of moisture removal in case of freeze drying of food stuff is termed as __________.

A. adsorption
B. sublimation
C. evaporation
D. dehydration

222. In psychrometry the ___________ is directly involved.
A. Lewis relationship
B. Weber number
C. Galileo number
D. Dean number

223. In a distillation operation, the heat removed in condenser
A. decrease with increase in reflux ratio
B. increase with increase in reflux ratio
C. remains unaffected with change in reflux ratio
D. and the heat required in reboiler decrease with increase in reflux ratio.

224. Which of the following types of distillation methods is represented by Rayleigh's equation?
A. flash
B. continuous
C. differential
D. steam

225. During chemical dehumidification ________ remains constant.
A. dry bulb temperature
B. wet bulb temperature
C. partial pressure of vapour
D. none of these

226. During the constant rate period increasing the mass velocity of the gas, the rate of drying during the constant rate period ______________ if the conduction and radiation through the solid are neglected.
A. increases
B. decreases
C. increases linearly
D. remains same

227. ____________ cannot be used for both the absorption as well as the liquid-liquid extraction process?
A. packed towers
B. plate towers
C. spray towers
D. wetted wall columns

228. At minimum reflux ratio the operating cost of a distillation column is.
A. minimum
B. maximum
C. zero
D. infinite

229. Which of the following separation is an example of adsorption unit operation.
A. mustard oil from mustard seed
B. water from petrol
C. uranium from its ore
D. ammonia from coke oven gas

230. The caking tendency for powdery materials can be prevented by
A. having maximum possible points of contact.
B. providing irregular grain size.
C. providing minimum percentage of voids.
D. none of these.

231. What is the reflux ratio of a distillation column operating with 100 moles of reflux and 5 moles of overhead product?
A. 2
B. 0.5
C. 50
D. 150

232. In a distillation column plate efficiency is reduced due to the ____________ of liquid.
A. entrainment
B. foaming
C. weeping & dumping
D. all (a), (b) & (c)

233. What is the effect of temperature on molecular diffusivity?
A. increases with temperature.
B. may increase or decrease with temperature.
C. is independent of temperature.
D. decreases with temperature.

234. In _______ the process is followed by unsaturated air with dry bulb temperature 18°C and relative humidity 50% passing through water spray washer with constant water temperature at 45°C?
A. heating only
B. humidification only
C. both heating and humidification
D. evaporative cooling

235. Which of the following statements of Henry's law is correctly represented?

A. partial pressure of a component over a solution is proportional to its mole fraction in the liquid.

B. partial pressure is equal to the product of the mole fraction and total pressure.

C. partial pressure of a component over a solution is proportional to the mole fraction in the vapour.

D. vapour pressure is equal to the product of the mole fraction and total pressure.

236. In case of drying the critical moisture content is an indicates of

A. beginning of falling

B. end of falling

C. beginning of constant

D. none of these

237. A vapor-gas mixture has wet and dry bulb temperature of 25°C and 30°C respectively. At constant pressure when the mixture is heated to 45°C the wet bulb temperature will then be ___________.

A. 25

B. > 25

C. < 25

D. – 25

238. The height of a packed column distillation column is limited to ___________ meters.

A. height < 6

B. diameter < 0.6

C. both (a) & (b)

D. neither (a) nor (b).

239. Given that the degree of freedom for a binary mixture distillation process is 2, if the pressure is fixed then the number if independent variables in this process will be

A. 1

B. 2

C. 0

D. 3

240. Which of the following statements is not correct?

A. when water is cooled from 80 to 70°C by exposure to the air with a wet bulb temperature of 60°C, then both the approach & the range would be 10°C.

B. a horizontal line on the humidity chart indicates thetemperature changes at constant molal humidity.

C. for unsaturated air-water vapor mixture at atmospheric conditions, the wet bulb temperature & adiabatic cooling lines are the same.

D. relative saturation of unsaturated mixture of water vapor &air cannot be increased by either reducing the mixturetemperature or by increasing the total pressure.

241. Adsorption applied to gaseous separation is not present in_____________.

A. removal of objectionable odour and impurities from industrial gases.

B. recovery of valuable solvent vapors from dilute mixture with air and other gases.

C. dehumidification of air and gas.

D. decolouration of yellow glycerene.

242. A wet solid it required to be exposed to ________ air for removal of all the moisture.

A. perfectly dry

B. high temperature

C. highly humid

D. none of these

243. In distillation operation, the reflux ratio may vary between.

A. zero and one

B. zero and infinity

C. minimum and infinity

D. one and two

244. Low viscosity solvents are considered to be better for gas absorption because of their

A. rapid absorption rates.

B. better flow characteristics.

C. low pumping pressure drop.

D. all (a), (b) and (c).

245. Which of the following dimensionless groups does not directly affect the mass transfer?

A. Lewis relationship

B. Schmidt number

C. Sherwood number

D. Froude number

246. Ammonia content of the coke oven gas is removed by scrubbing with.

A. ethanolamine

B. dilute ammoniacal liquor

C. caustic solution

D. dilute HCl

247. Which of the following proportionality relationship correctly represent the permeability for gas flow though a capillary according to Poiseuille's law?
   A. μ
   B. $1/\mu$
   C. $\mu^{1/2}$
   D. μ^2
248. The leaching rate increases with increase in temperature due to
   A. increased diffusivity
   B. decreased liquid viscosity
   C. both (a) and (b)
   D. neither (a) nor (b)
249. Whether in a distillation column the plate efficiency can be greater than 100%?
   A. normally not; but is possible if infinite number of plates are put.
   B. yes; if the reflux ratio is maximum.
   C. never
   D. yes
250. The phenomena of drying of solids is accompanied by ______________ transfer.
   A. only mass
   B. only heat
   C. both heat and mass
   D. none of these
251. With increase in the ______________ gaseous diffusion coefficient increases.
   A. pressure
   B. temperature
   C. neither (a) nor (b)
   D. both (a) & (b)
252. What will be the boiling point of a mixture of 10% benzene vapor in air at 30°C and 700 mm Hg at 25°C and 750 mm Hg if it has a dew point of 20°C.
   A. 27.3
   B. 21.7
   C. 20
   D. 18.7
253. In the design of absorption column which of the following is not fixed by the process requirement?
   A. terminal concentrations of gas stream.
   B. flow rate of the entering gas.
   C. composition of the entering liquid.
   D. none of these.
254. The interpretation of an equilibrium for a two phase system is that.
   A. total pressure throughout the liquid phase is equal to that throughout the vapor phase.
   B. escaping tendency of each component from the liquid phase to the vapor phase is exactly equal to that from vapor phase to liquid phase.
   C. temperature of the liquid phase is equal to that of the vapor phase.
   D. all (a), (b) and (c).
255. When the effects of velocity, temperature and humidity of the gas and the thickness of the solid are the same the drying refers to ____________.
   A. constant rate
   B. unsaturated surface
   C. both (a) and (b)
   D. neither (a) nor (b).
256. Which of the following statements is wrong during solution of a binary distillation problem by McCabe-Thiele method.
   A. heat of mixing is negligible.
   B. sensible heat changes are negligible compared with latent heat changes.
   C. molar latent heats of all components are equal.
   D. none of these
257. When a solid is being dried under constant rate conditions then
   A. increasing the air temperature decreases the drying rate.
   B. increased air humidity decreases the rate of drying.
   C. surface evaporation of unbound moisture occurs.
   D. none of these.
258. The Hatta number is encountered in.
   A. multicomponent distillation.
   B. binary distillation.
   C. gas absorption without chemical reaction.
   D. gas absorption with chemical reaction.
259. High pressure drop will result in ____________ for absorbers.
   A. better gas liquid contact.
   B. increased efficiency.
   C. high operating cost.
   D. decreased efficiency.

260. The film theory was conceived and developed by.

A. Higbie
B. Fick
C. Ergun
D. Levenspeil

261. Which of the following should not be the characteristic of an ideal tower packing?

A. have large free cross-section
B. have low weight per unit volume
C. have large surface area per unit volume
D. hold up large amount of liquid in the packing

262. When unsaturated air is subjected to adiabatic saturation which of the following parameters remains almost constant?

A. dew point
B. dry bulb temperature
C. wet bulb temperature
D. none of these

263. The minimum number of theoretical plate in a distillation column may be calculated using Fenske-Underwood equation when there is __________.

A. constant relative volatility
B. constant molal overflow
C. total reflux
D. all (a), (b) and (c)

264. For a system of binary gas mixture the diffusivity does not depends upon the

A. nature of the components
B. pressure
C. temperature
D. none of these

265. Give the molal composition of overhead product (X_D) and reflux ratio (R_D), for the rectifying section of a distillation column the slope and intercept would be represented respectively by the

A. $\frac{x_D}{R_D+1}, \frac{R_D+1}{R_D}$
B. $\frac{R_D}{R_D+1}, \frac{x_D}{x_D+1}$
C. $\frac{x_D}{R_D+1}, \frac{R_D}{R_D+1}$
D. none of the above

266. The dew point of an unsaturated mixture of air and water vapor at constant pressure and temperature ____________ with increase in absolute humidity.

A. increases
B. decreases linearly
C. remains unchanged
D. decreases

267. ___________ distillation is best suited for concentrating water-ethyl alcohol solution.

A. high pressure
B. vacuum
C. atmospheric
D. none of these

268. The ratio L/mG is used to denote the ___________.

A. absorption factor
B. slope of operating line in absorber
C. slope of operating line in stripper.
D. Brinkman number

269. Which of the following statements does not apply when the Raoult's law is not valid?

A. attractive forces between like and unlike molecules in the solution are approximately equal.
B. size of component molecules are not equal.
C. component molecules are nonpolar.
D. chemical combination or molecular association between unlike molecules takes place in the formation of solution.

270. If P denotes the pressure of a gas the binary diffusivity is proportional to

A. P
B. $1/P$
C. $\sqrt{p}$
D. p^2

271. What would be the amount of crystal formed in kg of a saturated aqueous solution of 1 kg made by dissolving powdered sugar particles at 60°C and then cooling to 25°C. The solubility of sugar is 0.6 and 0.2 in kg of solute per kg of water at 60°C and 25°C respectively.

A. 0.175
B. 0.25
C. 0.2
D. 0.4

272. Which type of extractor is used for extraction of penicillin?
    A. centrifugal
    B. plate
    C. packed
    D. None of these.

273. In mass transfer the ratio $(N_{Sh}/N_{Re}.N_{sc})$ represents
    A. Stanton number
    B. momentum diffusivity
    C. Peclet number
    D. thermal diffusivity

274. Normally a co-current absorber will be used when a ____________ is required to be dissolved in liquid.
    A. sparingly soluble gas
    B. pure gas
    C. mixture of two gases
    D. ideal gas

275. In a distillation column priming
    A. reduces the overall pressure drop.
    B. results from very low gas velocity.
    C. is characterised by the presence of foam throughout the space between trays.
    D. is desirable from point efficiency consideration.

276. When minimum reflux rate is referred in connection with multistate tray tower it presumes.
    A. infinite trays and maximum reboiler heat load.
    B. minimum trays and minimum reboiler heat load.
    C. minimum trays and maximum reboiler heat load.
    D. infinite trays and minimum reboiler heat load.

277. What would be the moisture content of a solid on wet basis if X is that on the dry basis
    A. $\frac{X}{X+1}$
    B. $\frac{1+X}{X}$
    C. $\frac{1-X}{X}$
    D. $\frac{X}{1-X}$

278. Absorption of Co_2 can take place on
    A. alumina
    B. heated charcoal
    C. hot cupric oxide
    D. cold $Ca(OH)_2$

279. When the gas is saturated, the
    A. vapour is in equilibrium with the liquid at the gas temperature.
    B. partial pressure of vapour equals the vapour pressure of the liquid at room temperature.
    C. vapour is in equilibrium with the liquid at the room temperature.
    D. none of these.

280. When uniform concentration of liquid prevails over a plate then
    A. Murphree efficiency < point efficiency.
    B. Murphree efficiency > point efficiency.
    C. Murphree efficiency = point efficiency.
    D. Murphree efficiency "" point efficiency.

281. Which of the following statements is not correct?
    A. with decrease in the liquid flow rate for a fixed gas flow rate, the slope of the operating line decreases, in case of gas absorption process.
    B. when the concentration difference for the mass transfer becomes zero at the bottom of the gas absorption tower, then the upper end of the operating line touches the equilibrium curve.
    C. the operating line lies above the equilibrium curve in case of a gas desorber.
    D. the average distance between the operating line and equilibrium line is large, when the liquid-gas ratio is also large in case of a counter current gas absorption tower.

282. What would be the effect of weather conditions on the reflux ratio of an uninsulated distillation column being operated under constant pressure conditions?
    A. increases in summer
    B. increases in winter
    C. decreases in winter
    D. none of these

283. Separation is very efficient and convenient in a binary system when the relative volatility is
    A. <1
    B. > 1

C. 1
D. 0.5

284. ____________ moisture is the moisture in a solid when equilibrium vapor pressure is equal to that of the pure liquid at the same temperature.
A. unbound
B. bound
C. critical
D. free

285. In steam distillation if the quantity of the steam used is increased the temperature of distillation
A. depends on relative volatility
B. decreases
C. increases
D. remains unchanged

286. For optimum performance of an absorber the desirable value of absorption factor is
A. 0.5
B. 1
C. > 1
D. < 1

287. During the operation of a distillation column, the
A. vapors are not always at their dew points.
B. driving force for the liquid flow is its weight.
C. highest temperatures is near the feed plate.
D. all (a), (b) and (c)

288. At which of the following values of the psychometric ratio the adiabatic saturation temperature and wet bulb temperature becomes equal.
A. 1
B. > 1
C. < 1
D. ∞

289. What is the degree of freedom of an absorption process in which only one component is transferred between phases?
A. 3
B. 4
C. 2
D. 1

290. For drying of refractory bricks which of the following types of dryer is used.
A. conveyor
B. tunnel
C. tray
D. festoon

291. For which of the following operations Langmuir equation is associated
A. multicomponent absorption
B. adsorption
C. leaching
D. steam distillation

292. By analogy the dimensionless group in mass transfer corresponding to Prandtl number in heat transfer is __________ number.
A. Schmidt
B. Stanton
C. Sherwood
D. Peclet

293. Carbon berl saddle used for packing cannot be used for
A. H_2SO_4
B. alkalis
C. SO_2
D. oxidising atmosphere

294. The critical moisture content during drying of a solid increase with
A. increased drying rate
B. increased thickness of the solid
C. decreased drying rate
D. both (b) and (c).

295. For drying of sticky materials the recommended type of dryer is.
A. tray
B. fluidised bed
C. rotary
D. none of these

296. In supercritical extraction the commonly used solvent is
A. water.
B. methyl ethyl ketone.
C. carbon tetrachloride.
D. carbon dioxide

297. Flash distillation is
A. same as differential distillation.
B. used for multi component systems like crude refining.
C. same as simple distillation.
D. most useful for handling binary systems

298. At constant pressure conditions humidity depends only on the
A. partial pressure of vapour in the mixture.
B. density of the vapour.
C. heat capacity of the vapour.
D. none of these.

299. Preparation of coffee from coffee beans is done by leaching with ______________.
A. naphtha
B. hot water
C. cold water
D. dilute hot caustic solution

300. According to Gilliland's equation the diffusivity is
A. inversely proportional to '*P*'.
B. directly proportional to '$T^{1.5}$'.
C. dependent on the molecular volume & molecular weight of components.
D. all 'a', 'b' & 'c'.

301. Which of the following conditions will be valid at minimum reflux ratio in a distillation column?
A. minimum number of trays.
B. optimum number of trays.
C. maximum condenser size.
D. minimum reboiler size.

302. At low relative volatility a binary mixture should
A. use molecular distillation.
B. use steam distillation.
C. use high pressure distillation.
D. an azeotrope may be formed during distillation.

303. If air is unsaturated then
A. wet bulb temperature < dry bulb temperature
B. dew point < wet bulb temperature
C. both (a) and (b)
D. neither (a) not (b)

304. Which of the following statements is not correct?
A. mass transfer co-efficient does not account for gase-liquid diffusion.
B. contacting cold liquid with warm vapor-gas mixture exemplifies dehumidification process.
C. diffusion co-efficient and mass transfer co-efficient are not same in any mass transfer operation.
D. contacting cold vapor-gas mixture with hot liquid exemplifies humidification process.

305. If the reflux ratio is increased is a distillation column with fixed number of ideal stages the difference between the top and bottom product streams
A. increases
B. decreases
C. passes through a maximum
D. remains unaffected

306. The area covered by binodal curve which is, the area of heterogeneity ____________ as the temperature of extraction increases.
A. decreases
B. remains unchanged
C. increases
D. none of these

307. The percentage saturation of an unsaturated mixture of gas and vapor is _________its relative saturation
A. lower than
B. higher than
C. equal to
D. either (a) or (b), depends on the mixture omposition

308. In the Chilton-Colburn analogy for mass transfer, the product $(N_{st} . N_{sc}^{2/3})$ is equal to
A. $2f$
B. $f/2$
C. f
D. $1/f$

309. When both the fluids flow concurrently in an absorber the slope of the operating line is
A. 1
B. positive
C. uncertain
D. negative

310. Which of the following statements is not correct?
A. the Lewis number of a mixture is unity, when the thermal diffusivity is equal to the mass diffusivity.
B. 1 Nm^3 of dry air is lighter than 1Nm^3 of humid air.
C. the mass diffusivity, the thermal diffusivity and the eddy momentum diffusivity are the same for $N_{Sc} = N_{Pr} = 1$.

D. an azeotropic mixture of two liquids has boiling point higher than either of them, when it shows negative deviation from Raoult's Law.

311. The rate of leaching is not a function of

A. particle size
B. agitation
C. temperature
D. none of these

312. When water is recirculated during evaporative cooling process which of the following parameters decrease?

A. relative humidity
B. partial pressure of vapour
C. wet bulb temperature
D. none of these

313. The percentage saturation of an unsaturated mixture of gas and vapor is _________its relative saturation

A. lower than
B. higher than
C. equal to
D. either (a) or (b), depends on the mixture composition

314. Operation of a distillation column under minimum reflex will result in

A. reflux ratio will be maximum.
B. concentration of liquid and vapour leaving a plate will be same.
C. number of plates required will be maximum.
D. none of these.

315. In case of an absorber with single component the operating line is straight only when plotted in terms of __________units.

A. mole fraction
B. mole ratio
C. partial pressure
D. none of these

316. Separation of such components ___________ is recommend by flash distillation.

A. which form minimum boiling azeotrope.
B. having very close boiling points.
C. having very wide boiling points.
D. which form maximum boiling azeotrope.

317. Operation of a distillation column at infinite reflux may by analyzed by McCabe-Thiele method if

A. overhead product is maximum.
B. number of plates is maximum.
C. both the operating lines coincide with the diagram.
D. none of these.

318. The dew point at constant pressure will _________ with increase in temperature.

A. increase
B. remain unchanged
C. decrease
D. increase/decrease; depends on the temperature

319. In humidification operation bubble cap towers are not used even after being very effective because of the

A. difficulty in its fabrication.
B. high pressure drop of the gas.
C. high evaporation losses of water.
D. none of these.

320. By definitions the plate efficiency of a distillation column

A. is a function of the mass transfer between liquid and vapor.
B. increases due to weeping and dumping of liquid.
C. increases due to foaming.
D. increases due to liquid entrainment.

321. Diffusivity of vapor may be determined experimentally by ___________ method.

A. Wike's
B. wetted wall column
C. Winkel mann's
D. Gilliland's

322. The use of packed towers for distillation may be only in

A. small size
B. vacuum distillation.
C. multicomponent distillation.
D. high pressure operation.

323. The shape of operating line in an absorber corresponding to the minimum liquid rate is.

A. of slope = 0.1
B. of slope = 1
C. tangential to the equilibrium curve
D. none of these

324. In which of the following cases steam distillation is not recommended.
    A. azeotropic mixture is to be separated and the final product is miscible with water.
    B. liquids decompose, if distilled directly at atmospheric pressure.
    C. material cannot be distilled by indirect heating even under low pressure, because of the high boiling temperature.
    D. material to be distilled is thermally unstable or has the tendency to react with other components associated with it, at the boiling temperature.
325. Which of the following statements is not correct?
    A. the binary diffusivity in liquids is of the order of 10^{-5} cm^2 /sec.
    B. molecular diffusion in solid is much faster than that in liquids.
    C. according to Poiseuille's law, the permeability decreases with increase in temperature for flow of a gas through a given capillary.
    D. particles movement from higher concentration to lower concentration takes place in diffusion process.
326. In which type of distillation Rayleigh equation is applicable.
    A. differential
    B. flash
    C. equilibrium
    D. molecular
327. Desalination of sea water is based on the principle of.
    A. adsorption
    B. reverse osmosis
    C. absorption
    D. thermal diffusion
328. An equal amount of drug is administered as a sphere and as a cube. If it is known that the shape and size do not influence the mass transfer what would be the ratio of rate of dissolution in water at t = 0 for the cubic to spherical tablet;
    A. 1.94
    B. 0.54
    C. 1.24
    D. 1.04
329. For obtaining maximum efficiency in a gas-liquid absorption column.
    A. liquid stream should be distributed uniformly.
    B. gas stream should be distributed uniformly.
    C. both gas as well as liquid streams should be distributed uniformly.
    D. by passing should be completely avoided.
330. In a distillation column at flooding condition a sharp ___________ is observed
    A. decrease in liquid hold up in the column.
    B. increase in Murphree plate efficiency.
    C. decrease in pressure drop.
    D. increase in pressure drop.
331. Accordingto the penetration theory mass transfer coefficient varies with mass diffusivity as.
    A. $D^{0.5}$
    B. $D^{1.5}$
    C. D
    D. $1/D$
332. Which of the following theories has been developed by Danckwerts?
    A. film
    B. surface renewal
    C. penetration
    D. none of these
333. A________ provides maximum contact surface for a liquid-vapour system?
    A. packed tower
    B. wetted wall column
    C. bubble-cap plate column
    D. seive-plate column
334. The liquid-liquid extraction system consisting of one pair partly soluble e.g. water (A) – benzene (B) and acetic acid (C) is represented by fig ___________ ?

A.

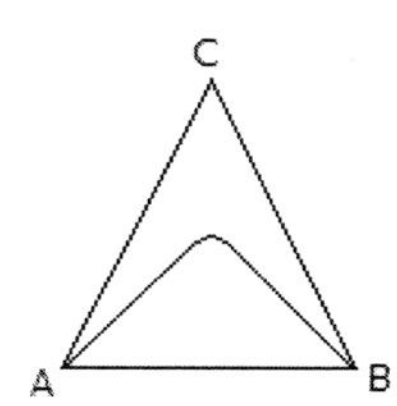

B.

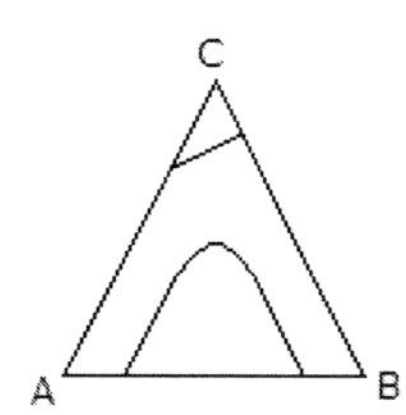

C. 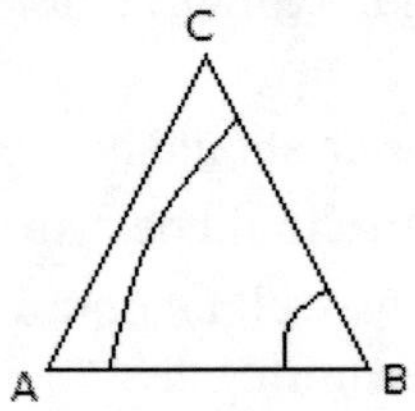

D. 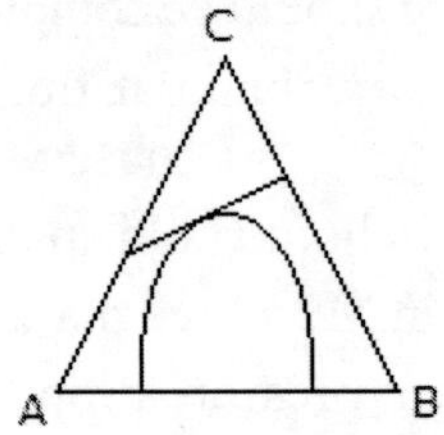

335. In a continuous distillation column in the stripping section.
A. vapour is stripped of low boiler.
B. liquid is enriched with high boiler.
C. liquid is stripped of high boiler.
D. none of these

336. The operation of a __________ is not under batch wise conditions.
A. Truck drier
B. Agitated pan drier
C. Fluidised bed drier
D. Vacuum shelf drier.

337. The operation of __________ is based on reverse osmosis.
A. dialysis
B. electrodilysis
C. diffusion
D. ultrafiltration.

338. A vacuum distillation unit operating at 60 mm Hg has a barometric condenser. The distillate will be collected through the vacuum drum located __________.
A. at ground level.
B. 10 meters above the ground.
C. 5 meters above ground.
D. 2 meters above ground.

339. In the selection of packing for absorbers the most important factor to be considered is the __________ of packing.
A. durability
B. size
C. porosity
D. cost

340. Which of the following is the correct statement for fractional solvent extraction.
A. results in low interfacial tension
B. employs two solvents
C. employs only one solvent
D. none of these

341. A wetted wall column has __________ mass transfer area for experimental determination of mass transfer coefficient.
A. unknown
B. calculated
C. known
D. not required

342. For a partially saturated mixture of vapour and gas the relative saturation can be increased by __________ of the mixture.
A. reducing the total pressure
B. reducing the temperature
C. increasing the total pressure
D. both (b) and (c)

343. The number of gm moles of solute per __________ of solvent is known as the molality.
A. gm
B. liter
C. Kg
D. m^3

344. The term adsorption equilibria in defined by.
A. Gibb's equation
B. Fick's law
C. Freundlich equation
D. none of these

345. The operation of Molecular distillation is.
A. high temperature distillation
B. very low pressure distillation
C. for heat-sensitive materials
D. both (b) and (c)

346. Which of the following solvents will be used for removal of H_2S present in naphtha reformed gas.
A. ethanolamine
B. HCl
C. K_2CO_3
D. vacuum gas oil

347. In a continuous distillation column in the rectifying section the
A. liquid is stripped of high boiling component.
B. vapour is enriched with high boiling component.

C. vapour is enriched with low boiling component.
D. none of these.

348. When the exposure of the wet solid is to the air of constant _________the drying will be under constant drying condition.
A. temperature
B. humidity
C. velocity
D. all (a), (b) & (c)

349. Porcelain raschig ring should not be used for treating concentrated_____________.
A. nitric acid
B. sulphuric acid
C. alkalis
D. hydrochloric acid

350. The rate of vaporization per unit of drying surface area during constant rate drying operation.
A. increases with time.
B. decreases with time.
C. does not change with time,
D. does not affect the moisture content of the wet solid.

351. What will be the approximate value of Schmidt number for air at 750°K and 1 atm
A. 1
B. 0.1
C. 0.01
D. 10

352. The most common method of estimation of mass transfer coefficient is by __________ theory.
A. surface renewal
B. film
C. penetration
D. none of these.

353. The expression for height of a transfer unit based on the gas phase for a gas absorption unit is given by which of the following? (G: superficial molar gas velocity, L : superficial molar liquid velocity, F_G: mass transfer coefficient, moles/m^2, a : interfacial area per unit volume of tower)

A. $\frac{G}{F_G \cdot a}$

B. $\frac{L}{F_G \cdot G}$

C. $\frac{G \cdot a}{F_G}$

D. $\frac{F_G}{G \cdot a}$

354. For the boundary layer development on a flat plate during convective mass transfers for velocity, concentration and temperature boundary will be same if
A. $N_{Sc} = N_{Le}$
B. $N_{Sc} = N_{Pr} = N_{Le}$
C. $N_{Sc} = 1$
D. $N_{Pr} = N_{Le}$

355. For the concentration of rubber latex which of the following evaporators is used.
A. Agitated film evaporator
B. Short tube evaporator
C. Long tube vertical evaporator
D. Calendria.

356. In a continuous counter current extractor as the reflux ratio is increased, the number of stages
A. increase
B. decrease
C. remain unchanged
D. can either increase or decrease, depends on the system

357. When composition is represented in triangular coordinates, the ternery compositions point falls ___________ of the triangle.
A. on the sides
B. inside
C. in the corners
D. none of these.

358. The mass transfer coefficient is expressed as ________ according to penetration theory (D_e is diffusivity and t is the timc)
A. $(D_e/t)^{1/2}$
B. $(D_e.t)^{1/2}$
C. $(4D_e/\pi t)^{1/2}$
D. $(4D_e/t)$

359. 10 kg of pure solvent S (solute zero) is mixed with 30 kg of feed F containing A at a mass fraction $x_f = 0.2$. In a single stage extraction process the mixture splits into an extract phase E and a raffinate phase R containing A at $x_B = 0.5$ and $X_R = 0.05$ respectively. What will be the total mass of the extract phase (in Kg).

A. 6.89

B. 8.89

C. 10

D. 8.25

360. The reason why reboiler is considered as one theoretical plate is because

A. of the assumption that vapour and liquid leaving the reboiler are in equilibrium.

B. reboiler itself contains one plate.

C. vapour is recycled to the column.

D. none of these.

361. With dry and wet bulb temperatures 38°C and 28°C respectively when unsaturated air is passed through water spray chamber maintained at 38°C the air will be

A. dehumidified

B. humidified

C. cooled

D. both (a) & (b)

362. When a sparingly soluble gas is dissolved in a liquid, the

A. liquid side co-efficient should be decreased.

B. liquid side coefficient should be increased.

C. gas side co-efficient should be increased.

D. gas side co-efficient should be decrease.

363. For a binary system increase in the total pressure of the system________ the boiling point of pure components.

A. remains unchanged

B. increases

C. decreases

D. may increase or decrease, depends on the system.

364. For efficient operation, a natural draft cooling tower should be filled with wooden grids covering about_________ present of tower height.

A. 10-15

B. 30-40

C. 70-80

D. 90-100

365. Adsorption of unit quantity of a gas on very large adsorbent on which a definite concentration of the adsorbed gas already exists will result in change enthalpy when is termed

A. differential heat of adsorption.

B. heat of wetting.

C. integral heat of adsorption.

D. none of these.

366. The relation of average mass transfer coefficient (K) with diffusivity (D) is explained by Penetration theory as follows

A. $K \propto D$

B. $K \propto D^{0.5}$

C. $K \propto D^{1.5}$

D. $K \propto D^2$

367. The equipment used for desulphurization of petrol is

A. Bollman extractor

B. rotating disc extractor

C. plate type extraction column

D. packed extraction column

368. An increase in rate of adsorption is effected as the

A. pressure decreases.

B. temperature decreases.

C. temperature increases.

D. size of adsorbent increases.

369. In comparison to physical absorption chemisorption has

A. lower adsorption rate.

B. lower capacity of the solvent for the solute gas.

C. increased utilisation of stagnant zones of the liquid phase.

D. none of these.

370. Which of the following systems form maximum boiling azeotrope at 1 atm.

A. acetone chloroform

B. benzene-ethyl alcohol

C. ethyl alcohol-water

D. none of these

Answer

1. A
2. A
3. C
4. C
5. B
6. B
7. C
8. A
9. A
10. A
11. D
12. A
13. A
14. D
15. B
16. B
17. C
18. C
19. C
20. B
21. B
22. B
23. A
24. C
25. C
26. A
27. C
28. C
29. A
30. A
31. C
32. B
33. A
34. C
35. D
36. D
37. B
38. A
39. C
40. D
41. D
42. C
43. C
44. D
45. D
46. C
47. C
48. B
49. D
50. A
51. D
52. B
53. B
54. B
55. C
56. A
57. C
58. A
59. C
60. B
61. A
62. A
63. A
64. A
65. A
66. A
67. D
68. C
69. C
70. A
71. C
72. A
73. B
74. A
75. B
76. A
77. C
78. A
79. A
80. A
81. C
82. B
83. B
84. A
85. D
86. D
87. A
88. C
89. C
90. C
91. A
92. A
93. C
94. C
95. C
96. C
97. B
98. A
99. D
100. A
101. A
102. A
103. C
104. C
105. D
106. B
107. C
108. A
109. A
110. A
111. D
112. A
113. A
114. D
115. B
116. B
117. C
118. C
119. C
120. B
121. B
122. B
123. A
124. C
125. C
126. A
127. C
128. C
129. A
130. A
131. C
132. B
133. A
134. C
135. B
136. D
137. B
138. A
139. C
140. D
141. D
142. C
143. C
144. D

145. D
146. C
147. A
148. B
149. D
150. A
151. A
152. B
153. B
154. D
155. B
156. B
157. B
158. D
159. C
160. D
161. D
162. B
163. B
164. D
165. D
166. B
167. C
168. D
169. C
170. C
171. B
172. B
173. A
174. D
175. D
176. B
177. A
178. A
179. A
180. B
181. A
182. A
183. C
184. C
185. A
186. D
187. A
188. D
189. A
190. A
191. B
192. A
193. C
194. D
195. A
196. C
197. A
198. C
199. C
200. B
201. A
202. C
203. B
204. B
205. D
206. B
207. C
208. A
209. D
210. A
211. A
212. A
213. A
214. B
215. A
216. B
217. C
218. D
219. A
220. D
221. B
222. A
223. A
224. C
225. A
226. A
227. D
228. A
229. B
230. D
231. A
232. D
233. A
234. C
235. A
236. A
237. B
238. C
239. A
240. D
241. D
242. A
243. A
244. D
245. D
246. B
247. B
248. C
249. C
250. C
251. B
252. D
253. D
254. D
255. C
256. D
257. C
258. D
259. C
260. A
261. D
262. C
263. D
264. D
265. B
266. A
267. B
268. A
269. C
270. B
271. C
272. A
273. A
274. B
275. C
276. A
277. A
278. B
279. A
280. C
281. C
282. B
283. B
284. A
285. B
286. C
287. B
288. A
289. A
290. B
291. B
292. A
293. D
294. D
295. A
296. D
297. A
298. B
299. B
300. D
301. C
302. D
303. C
304. A
305. A
306. A
307. C
308. B
309. D
310. B
311. D
312. D
313. A
314. C
315. B
316. C
317. C
318. A
319. B
320. A
321. C
322. A
323. C
324. A

325. B
326. A
327. B
328. C
329. C
330. D
331. A
332. B
333. A
334. A
335. B
336. C
337. D
338. A
339. C
340. B
341. C
342. D
343. C
344. C
345. D
346. A
347. B
348. D
349. C
350. C
351. B
352. B
353. A
354. B
355. A
356. B
357. B
358. C
359. B
360. A
361. B
362. B
363. B
364. A
365. A
366. B
367. B
368. B
369. C
370. A

GATE 2015-2000

(1 Mark Question)

GATE 2015

1. Benzene is removed from air by absorbing it in a non-volatile wash-oil at 100 kPa in a counter-current gas absorber. Gas flow rate is 100 mol/min, which includes 2 mol/min of benzene. The flow rate of wash-oil is 50 mol/min. Vapour pressure of benzene at the column conditions is 50 kPa. Benzene forms an ideal solution with the wash-oil and the column is operating at steady state. Gas phase can be assumed to follow ideal gas law. Neglect the change in molar flow rates of liquid and gas phases inside the column. For this process, the value of the absorption factor (up to two decimal places) is ____________.

Sol. absorption factor can be given as,

$$A = \frac{L_s}{mG_s}$$

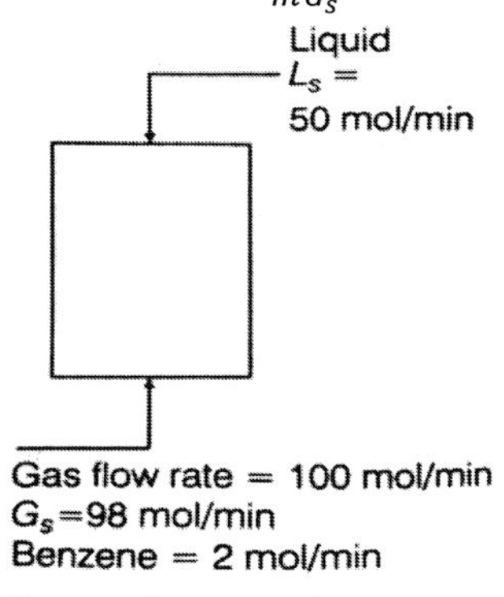

According to the question,

Raoult's law is applicable for gas phase

$$yp = x\, p_{sat}$$

$\Rightarrow$ $$y \times 100 = x \times 50$$

$\Rightarrow$ $$y = 0.5\, x$$

So, $$m = 0.5$$

$$A = \frac{50}{0.5 \times 98} = 1.02$$

2. A spherical naphthalene ball of 2 mm diameter is subliming very slowly in stagnant air at 25°C. The change in the size of the ball during the sublimation can be neglected. The diffusivity of naphthalene in air at 25°C is $1.1 \times 10^{-6} m^2/s$. The value of mass transfer coefficient is B $\times 10^{-3} m/s$, where B (up to one decimal place) is__________.

Sol. for spherical naphthalene ball,

Dimensionless Sherwood number = 2

$$\frac{K_C L}{D_{AB}} = 2$$

where, L = characteristic length

diameter of spherical ball, d = 2 mm

$$= 2 \times 10^{-3} m$$

$$K_C = \frac{2 \times D_{AB}}{d} = \frac{2 \times 1.1 \times 10^{-6}}{2 \times 10^{-3}}$$

$$K_C = 1.1 \times 10^{-3} \text{ m/s}$$

On comparing with the given value of mass transfer coefficient, we

get

$$B = 1.1$$

3. Identify the wrong statement amongst the following:

   A. steam distillation is used for mixture that are reimmiscible with water

   B. vacuum distillation is used for mixtures that are miscible with water

   C. steam distillation is used for mixtures that are immiscible with water

   D. vacuum distillation columns have larger diameters as compared to atmospheric columns for the same throughout

Sol. **D.** steam distillation is used for the mixture immiscible with water.

4. For a binary mixture of components A and B, N_A and N_B denote the total molar fluxes of components A and B, respectively. J_A and J_B are the corresponding molar diffusive fluxes. Which of the following is correct for equimolar counter-diffusion in the binary mixture?

A. $N_A + N_B = 0$ and $J_A + J_B \neq 0$
B. $N_A + N_B \neq 0$ and $J_A + J_B = 0$
C. $N_A + N_B \neq 0$ and $J_A + J_B \neq 0$
D. $N_A + N_B = 0$ and $J_A + J_B = 0$

Sol. A. for equimolar couner – diffusion in the binary mixture,

$$N_A + N_B = 0 \text{ and } J_A + J_B \neq 0$$

GATE 2014

5. In a completely opaque medium, if 50% of the incident monochromatic radiation is absorbed, then which of the following statements are correct?

P. 50% of the incident radiation is reflected
Q. 25% of the incident radiation is reflected
R. 25% of the incident radiation is transmitted
S. No incident radiation is transmitted

A. P and S
B. Q and R
C. P and Q
D. R and S

Sol. A. for opaque medium, transmittivity (τ) = 0

We know that

$$\alpha + \rho + \tau = 1$$

where α is absorptivity and ρ is reflectivity,

Thus, $\alpha + \rho + 0 = 1$

$$\rho = 1 - \alpha$$
$$= 1 - 0.5 = 0.5$$

∴ 50 % of incident radiation will be reflected and no radiation will be transmitted.

6. Packed towers are preferred for gas-liquid mass transfer operations with foaming liquids because

A. in packed towers, high liquid to gas ratios are best handled
B. in packed towers, continuous contact of gas and liquid takes place
C. packed towers are packed with random packings
D. in packed towers, the gas is not bubbled through the liquid pool

Sol. D. in packed towers, the gas is not bubbled through the liquid pool which is the reason that they are preferred for gas liquid mass transfer operations with foaming liquid

7. Assuming the mass transfer coefficients in the gas and the liquid phases are comparable, the absorption of CO_2 from reformer gas ($CO_2 + H_2$) in to an aqueous solution of diethanolamine is controlled by

A. gas phase resistance
B. liquid phase resistance
C. Both gas and liquid phase resistances
D. composition of the reformer gas

Sol. A. absorption of CO_2 from gas phase (CO_2+H_2) in to an aqueous solution of diethanolamine is controlled by gas phase resistance.

8. Which one of the following statements is correct for the surfacerenewal theory?

A. mass transfer takes place steady state
B. mass transfer takes place at unsteady state
C. contact time is same for all the liquid elements
D. mass transfer depends only on the film resistance

Sol. B. according to surface renewal theory, mass transfer takes place at unsteady state.

GATE 2013

9. The packing of an existing absorption tower is replaced with a new type of packing. The height of the packing and the inlet conditions are maintained the same as before. Tests reveal that the number of transfer units is lower than before. This indicates that the tower with the new packing, when compared to that with the old packing, will

A. have a higher rate of absorption of the solute from the gas stream
B. have a lower rate of absorption of the solute from the gas stream
C. have the same rate of absorption of the solute from the gas stream
D. have a lower height of transfer unit

Sol. B. when NTU is less than earlier, the new absorption tower will have low rate of absorption of solute from gas stream.

10. A wet solid is dried over a long period of time by unsaturated air of nom-zero constant relative humidity. The moisture content eventually attain by the solid is termed as the

A. unbound moisture content
B. bound moisture content
C. free moisture content
D. equilibrium moisture content

Sol. D. the moisture content attain by solid after this long drying period is known as equilibrium moisture content.

GATE 2012

11. In the Mc-Cabe-Thiele diagram, if the X-coordinate of the point of intersection of the q-line and the vapour-liquid equilibrium curve is greater than the X-coordinate of the feed point, then the quality of the feed is

A. superheated vapour
B. liquid below bubble point
C. saturated vapour
D. saturated liquid

***Sol.* B.** x_1 is X-coordinate of point of intersection of q-line and vapour-liquid equilibrium curve. For $x_1 >$ X- coordinate of feed point, then the quality of feed will be liquid below bubble point.

12. For which of the following combinations, does the absorption operation become gas-film controlled?

P. the solubility of gas in the liquid is very high.

Q. the solubility of gas in the liquid is very low.

R. the liquid-side mass transfer coefficient is much higher than the gas-side mass transfer coefficient.

S. the liquid-side mass transfer coefficient is much lower than the gas-side mass transfer coefficient.

A. P and Q
B. P and R
C. P and S
D. Q and R

***Sol.* B.** when the absorption became gas-film controlled, then thesolubility of gas in the liquid is very high and the liquid- side mass-transfer coefficient is higher than that of gas-side.

GATE 2011

13. Minimum input required to calculate the 'blank diameter' for a torispherical head is

A. crown radius
B. crown radius, knuckle radius and length of straight flange
C. knucle radius and length of straight flange
D. crown radius and knuckle radius

***Sol.* D.** calculation of blank diameter for a tori spherical head is required to be done by the use of crown radius and knuckle radius.

14. Ammonia (component 1) is evaporating from a partially filled bottle into surrounding air (component 2). The liquid level in the bottle and the concentration of ammonia at the top of the bottle are maintained constant. N_l is the molar flux relative to fixed location in space and J_l is the molar flux with respect to the average molar velocity of the constituent species in the gas phase. Assume the air in the bottle is stagnant. Which one of the following is correct?

A. N_1 = constant, $N_2 = 0$, $J_l + J_2 = 0$
B. $N_1 + N_2 = 0$, $J_l + J_2 = 0$
C. $N_1 + N_2 = 0$, J_l = constant $J_2 = 0$
D. N_1 = constant, $N_2 = 0$, J_1 = constant $J_2 = 0$

***Sol.* D.** N_1 = constant and $N_2 = 0$

J_1 = constant and $J_2 = 0$

15. Simultaneous heat and mass transfer are occurring in a fluid following over a flat plate. The flow is laminar. Te concentration boundary layer then, will coincide with the thermal boundary layer, when

A. Sc = Nu
B. Sh = Nu
C. Sh = Pr
D. Sc = Pr

***Sol.* D.** the concentration boundary layer will coincide with thermal boundary layer when Sc = Pr.

GATE 2010

16. The ratio of the thermal boundary layer thickness to the concentration boundary layer thickness is proportional to

A. Nu
B. Le
C. Sh
D. Pr

***Sol.* B.** the ratio of thermal boundary layer thickness to concentration layer thickness is proportional to lewi's number.

$$L_e = \frac{Termal\ \ diffusivity}{Mass\ \ diffusivity}$$

$$= \frac{Termal\ \ boundary\ \ layer\ \ thickness}{Thermal\ \ diffusivity}$$

17. The flooding velocity in a plate column, operating at 1 atm pressure is 3 m/s. If the column is operated at 2 atm pressure under otherwise identical conditions the flooding velocity will be

A. $\frac{3}{\sqrt{2}}$
B. $\frac{3}{2}$
C. 1
D. $\frac{3}{4}$

Sol. **A.** the expression for flooding velocity is

$$u = K\sqrt{\frac{\rho_{liq}}{\rho_{vap}}}$$

$$u = K\sqrt{\frac{\rho_{liq}}{p/RT}}$$

$$u \propto \frac{1}{\sqrt{p}}$$

Data given is

u_1 = 3 m/sec ; p_1 = 1 atm and p_2 = 2 atm

Hence, the ratios of flooding velocities

$$\frac{u_2}{u_1} = \sqrt{\frac{p_1}{p_2}}$$

Putting the values

$$u_2 = \frac{3}{\sqrt{2}} \text{ m/s}$$

GATE 2009

18. The Prandtl number of a fluid is the ratio of

A. thermal diffusivity to the momentum diffusivity
B. momentum diffusivity to thermal diffusivity
C. conductive resistance to convective resistance
D. thermal diffusivity to kinematic viscosity

Sol. **B.** the mathematical expression for Prandtl number is $\frac{C_p\,\mu}{K}$ and is defined as the ratio of momentum diffusivity to thermal diffusivity of a fluid

$$P_r = \frac{Momentum\ \ diffusivity}{Thermal\ \ diffusivity}$$

19. According to the penetration theory of mass transfer, the mass transfer coefficient (k) varies with diffusion coefficient (D) of the diffusing species as

A. D
B. $D^{-1/2}$
C. $D^{1/2}$
D. $D^{3/2}$

Sol. **C.** according to penetration theory of mass transfer, the mass transfer coefficient

$$k_C = 2\sqrt{\frac{D_{AB}}{\pi t}}$$

or

$$k_C \propto \sqrt{D_{AB}}$$

20. The ratio of the liquid to gas flow rate in a countercurrent gas absorption column is increased at otherwise identical condition. Which one of the following statements is true?

A. the operating lines shifts towards the equilibrium curve
B. the operating lines shifts away from the equilibrium curve
C. the concentration of the absorbed species increases in the exit liquid steam
D. the operating line does not shift

Sol. **B.** in a countercurrent gas absorption column, the equation for operating line is

$$G\,(Y_2 - Y_1) = L\,(X_2 - X_1)$$

or

$$Y_2 = Y_1 + \frac{L}{G}(X_2 - X_1)$$

In case the $\frac{L}{G}$ ratio increases, the slope of operating line will increase resulting in shifting the operating line away from the equilibrium curve.

GATE 2008

21. In a countercurrent gas absorber, both the operating and equilibrium relations are linear. The inlet liquid composition and the exit gas composition are maintained constant. In order to increase the absorption factor

A. the liquid flow rate should decrease
B. the gas flow rate should increase
C. the slope of the equilibrium line should increase
D. the slope of the equilibrium line should decrease

Sol. **C.** the absorption factor

$$A = \frac{L}{mG}$$

If L and G are maintained constant then for increasing the absorption factor A, the slope (m) of the equilibrium line should decrease

22. For a system containing species P, Q and R the composition at point K on the ternary plot is

A. *62.5 % P, 12.5% Q, 25%R*
B. 25 % P, 62.5% Q, 12.5%R
C. 12.5 % P, 62.5% Q, 25%R
D. 12.5 % P, 25% Q, 62.5%R

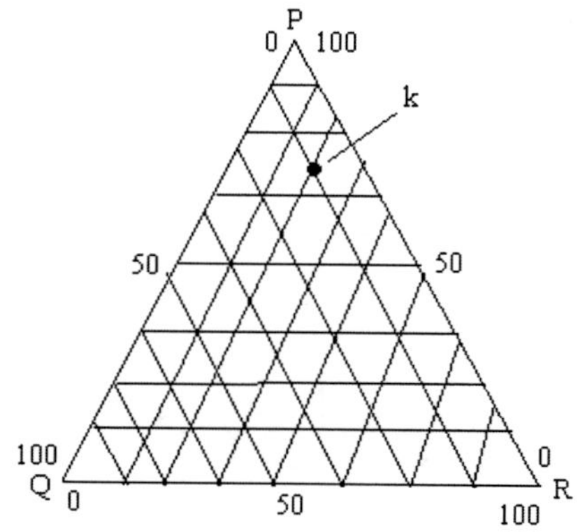

Sol. A. for the given three component ternary plot the point K represents the composition

P = 62.5% ; Q = 12.5% ; R = 25%

GATE 2007

23. The Grashof number is
   - A. thermal diffusivity/mass diffusivity
   - B. inertial force/surface tension force
   - C. sensible heat/latent heat
   - D. buoyancy force/viscous force

Sol. D. the Grashof number

$$G_r = \frac{g\,\beta\,(T_W - T_\infty)X^3}{V^2} = \frac{Buoyancy\ \ force}{Viscous\ \ force}$$

GATE 2006

24. Let d_h be the hydrodynamic entrance length for mercury in laminar flow in a pipe under isothermal conditions. Let d_t be its thermal entrance length under fully developed hydrodynamic conditions. Which one of the following is true?

Hence, $d_h = d_t$

25 . The Boussinesq approximation for the fluid density in the gravitational force term is given by which one of the following? (ρ_{ref} is the fluid density at the reference temperature T_{ref} and β is the thermal coefficient of volume expansion at T_{ref})
   - A. $\rho = \rho_{ref} + T_{ref}\ \beta(\rho - \rho_{ref})$
   - B. $\rho = \rho_{ref} - T_{ref}\ \beta(\rho - \rho_{ref})$
   - C. $\rho = \rho_{ref} - T_{ref}\ \beta(T - T_{ref})$
   - D. $\rho = \rho_{ref} - T_{ref}\ \beta(\rho - \rho_{ref})$

Sol. C. the Boussinesq approximation for fluid density in the gravitational force term is given by

$$\rho = \rho_{ref} - (\rho_{ref}).\,(\beta)(T - T_{ref})$$

26. An ideal single stage extraction process is used to treat 100 mol/s of an organic feed solution. The solute concentration in this solution is to be reduced from 0.5 mol % to 0.1 mol% .A pure solvent *S* is used. to reduced the solvent requirement by half for the same separation,
   - A. add one more ideal co-current stage
   - B. use another pure solvent S* whose partition coefficient is twice that 0f *S*
   - C. use solvent *S* containing 0.02 mole fraction of the solute
   - D. double the residence time of the solvent *S* in the contactor

Sol. B. to reduce the solvent requirement by half for the same separation, use another pure solvent S* whose partition coefficient is twice that of S.

GATE 2005

27. A distillation column at a pilot plant is scaled up by 3 times for industrial use at steady state. After scaling up
   - A. the number of theoretical trays increases by 3 times
   - B. the minimum reflux ratio increases by three times
   - C. the feed flow rate and product flow rates are increased by three times
   - D. the feed composition and product composition are increases by three times

Sol. C. after scaling up the distillation column by three times, the feed flow rate and product flowrates are increased by three times.

28. The thermal boundary layer is significantly thicker than the hydrodynamic boundary layer for
   - A. Newtonian liquid
   - B. polymeric liquids
   - C. liquid metals
   - D. gases

Sol. C. the thermal boundary layer is significantly thicker than the hydrodynamic boundary layer for liquid metals.

29. In a tray column, separating a binary mixture, with non-ideal stages, which one of the following statements is true?
   - A. point efficient can exceed 100%
   - B. murphree efficiency cannot exceed 100%
   - C. murphree efficiency can exceed 100%
   - D. both murphree and point efficiency can exceed 100%

Sol. **C.** in a tray column with non ideal stages separating a binary mixture, the Murphree plate efficiency can exceed 100%

30. The ratio of the diffusion coefficient in a gas to that in a liquid is of the order of
A. 10^5
B. 10^{-5}
C. 10^{-2}
D. 10^2

Sol. **A.** the ratio of diffusion coefficients in a gas to that in a liquid can be evaluated

$$D_{liq} = 10^{-9}, D_{gas} = 10^{-4}$$

Hence, the ratio is

$$\frac{D_{liq}}{D_{gas}} = 10^5$$

31. For turbulent flow past a flat plate ,when no from drug is present, the friction factor f and the Chilton-Colburn factor j_D are related as
A. f and j_D cannot be related
B. f is equal to j_D
C. f is greater than j_D
D. f is less than j_D

Sol. **C.** for the case of turbulent flow past a flat plate where no form drag is present as per Chilton – Colburn analogy.

$$J_D = \frac{f}{2}$$

$$\therefore \quad f > J_D$$

GATE 2004

32. A distillation column separates 10,000 kg/h of a benzene-toluene mixture as shown in the figure below. In the figure, x_F and x_D and x_W represent the weight fraction of benzene in the feeds distillate and residue respectively. The reflux ratio is

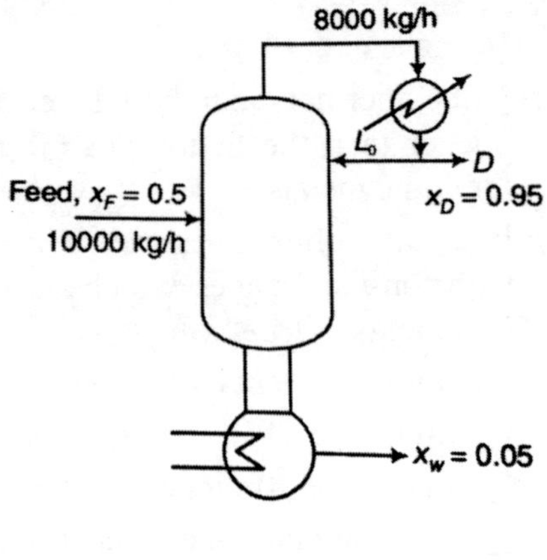

A. 0.5
B. 0.6
C. 1.0
D. 2.0

Sol. **B.**

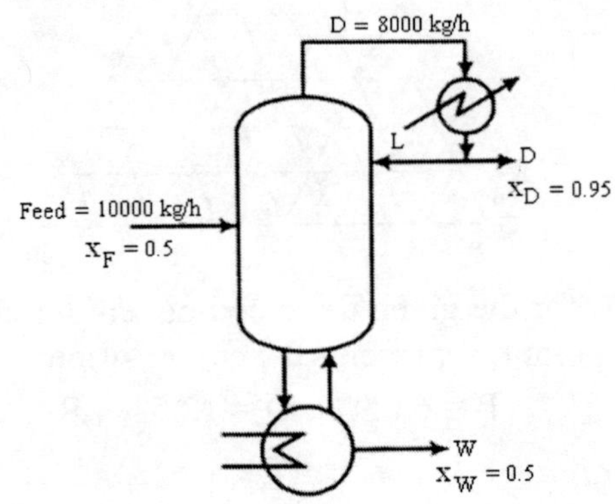

Writing the overall material balance

$$F = D + W$$

Writing the benzene balance

$$F \cdot X_F = DX_D + WX_W$$

Or (10000) (0.5) = 0.95 D + (10000 – D) 0.05

Solving D = 5000 kg/hr

Further, material balance around the overhead condenser

$$8000 = L + D$$

$$\therefore \quad L = 3000 \text{ kg/hr}$$

Hence, the reflux ratio

$$R = L/D = \frac{3000}{5000}$$

$$\therefore \quad R = 0.6$$

Statements for Linked Answer Questions 33 and 34

Pure aniline is evaporating through a stagnant air film of 1 mm thickness at 300 K and a total pressure of 100 kPa. The vapour pressure of aniline at 300 K is 0.1 kPa. The total concentration under these conditions is 40.1 mol/m^3. The diffusivity of aniline in air is $0.74 \times 10^{-5} m^2/s$.

33. The numerical value of the mass transfer coefficient is 7.4×10^{-3}. Its units are
A. m/s
B. cm/s
C. mol/(m²–s-Pa)
D. kmol/(m²-s-Pa)

Sol. **A.** given $D_{AB} = 0.74 \times 10^{-5} m^2/s$.

$$Z = 1 \text{ mm} = 10^{-3} m$$

$\therefore$Mass transfer coefficient $k_C = \frac{0.74 \times 10^{-5}}{10^{-3}}$

$\therefore \quad k_C = 7.4 \times 10^{-3}$ m/s

34. The rate of evaporation of aniline is 2.97×10^{-4}. Its units are

A. mol/s

B. mol/(m^2–s)

C. mol/(cm^2–s)

D. kmol/(cm^2–s)

Sol. **D.** given, $C_A = 4.01$ mol/m^3

Rate of evaporation

$$N_A = K_C \quad (C_A - C_{AS})$$

$$N_A \ (74 \times 10^{-3})\left(C_A \frac{m}{sec}\right)\left\{(40.1)\frac{mol}{m^3} - 0\right\}$$

$$= 296.74 \times 10^{-3} \frac{mol}{m^2 - S}$$

$$= 2.97 \times 10^{-4} \text{ kmol}/m^2 - S$$

Statement for Linked Answer and Questions 35 and 36

An air-water vapour mixture has a dry bulb temperature of 60^0 C and a dew point temperature of 40^0C. The total pressure is 101.3 kPa and the vapour pressure of water at 40^0C and 60^0C are 7.30 kPa and 19.91 kPa, respectively.

35. The humidity of air sample expressed as kg of water vapour/kg of dry air is

A. 0.048

B. 0.079

C. 0.1 22

D. 0.152

Sol. **A.** data given

$p = 101.3$ kPa

At dew point

$p_W = 7.3$ kPa

Humidity, $H = \left(\frac{P_W}{P - P_W}\right)\left(\frac{M_W}{M_{air}}\right)$

Molecular weight of water and air are 18 and 29 respectively

$$\therefore H = \left(\frac{7.3}{101.3 - 7.3}\right)\left(\frac{18}{29}\right)$$

H = 0.048 kg of water vapor per kg of dry air

36. The wet bulb temperature T_W for the above mixture would be

A. less than 40^0C

B. 40^0C

C. $40^0\text{C} < T_w < 60^0$ C

D. 60^0C

Sol. **C.** the wet bulb temperature for this mixture is

$$40^0\text{C} < T_w < 60^0 \text{ C}$$

37. Acetone is to be removed from air in an isothermal dilute absorber using pure water as solvent. The incoming air contains 5 mol% of acetone ($y_{in} = 0.05$). The design equation to be used for obtaining the number of trays (N) of the absorber is

$N + 2 = 6 \log\left(\frac{y_{in}}{y_{out}}\right)$. For 98% recovery of acetone, the number of trays required is/are

A. 1

B. 8

C. 9

D. 10

Sol. **B.**

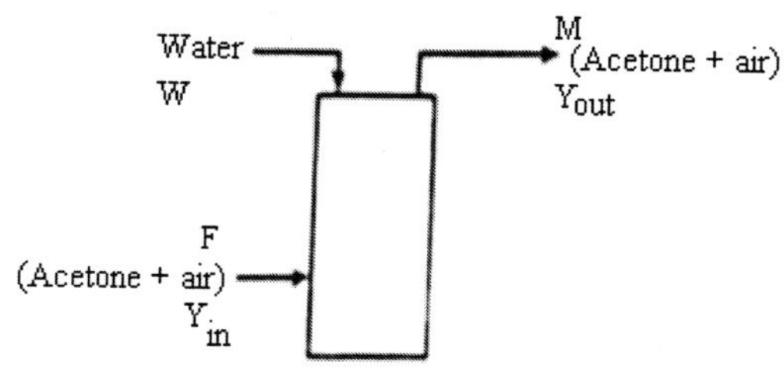

Mole fraction of acetone in M,

Mole fraction of Acetone in M

$$Y_{out} = \frac{0.1}{95 + 0.1} = 0.001$$

The design equation for obtaining the number of trays (N) for the absorber is

$$N + 2 = 6 \log\left(\frac{Y_{in}}{Y_{out}}\right)$$

$$= 6 \log\left(\frac{0.05}{0.001}\right)$$

$$\text{or} \quad N = 8.19$$

Hence number of trays required is 8.

GATE 2003

38. The diffusion coefficient, in m^2/s of acetic acid in benzene (liquid in liquid) is

A. 2.09×10^{-4}

B. 2.09×10^{-5}

C. 2.09×10^{-9}

D. 2.09×10^{-12}

Sol. **C.** the diffusion coefficient of acetic acid in benzene (liquid– in– liquid)

$$k_C = 2.09 \times 10^{-9} m^2/s$$

39. Component A is diffusing in a medium B. The flux N_A relative to a stationary point is equal to the flux due to molecular diffusion if

A. diffusion of A is in stagnant medium B
B. mass transfer is accompanied by reaction
C. molecular mean free path is high
D. there is equimolar counter-diffusion

Sol. D. the flux N_A relative to a stationary point is equal to the flux due to molecular diffusion when there is equimolar counter diffusion.

40. Minimum reflux ratio in a distillation column results
A. optimum number of trays
B. minimum reboiler size
C. maximum condenser size
D. minimum number of trays

Sol. C. minimum reflux ratio in a distillation column results in maximum condenser size, the reason being that at that time maximum vapour will be condensed in condenser as condensate i.e. L is minimum while D is maximum.

GATE 2002

41. The dimensionless group in mass transfer that is equivalent to Prandtl number in heat transfer is
A. Nusselt number
B. Sherwood number
C. Schmidt number
D. Stanton number

Sol. C. the dimensionless group in mass transfer that is equivalent to Prandtl number in heat transfer is

Schmidt number $\left(Sc = \dfrac{\mu}{\rho D_{AB}}\right)$

42. The Reynolds analogy for momentum, heat and mass transfer is best applicable for
A. gases in turbulent flow
B. gases in laminar flow
C. liquids in turbulent flow
D. liquids and gases in laminar flow

Sol. C. the Reynolds analogy for momentum, heat and mass transfer is best applicable for liquids in turbulent flow.

GATE 2001

43. The surface renewal frequency in Danckwerts' model of mass transfer is given by (k_L = mass transfer coefficient in m/s)
A. $\sqrt{k_L^2 D_A}$
B. $k_L^2 D_A$
C. $\dfrac{k_L^2}{D_A}$
D. $\dfrac{k_L}{D_A^2}$

Sol. C. according to Danckwerts' model of mass transfer the surface renewal frequency

$$S = \frac{k_L^2}{D_{AB}}$$

44. For gas absorption the height of a transfer unit, based on the gas phase, is given by (G = superficial molar gas velocity; L = superficial molar liquid velocity; F_G = mass transfer coefficient in mol/ m^2–s; a = interfacial area per unit volume of tower).
A. $\dfrac{G}{F_G a}$
B. $\dfrac{F_G}{G\, a}$
C. $\dfrac{G\, a}{F_G}$
D. $\dfrac{L}{F_G G}$

Sol. A. for gas absorption the height of a transfer unit based on the gas phase is given by

$$\frac{G}{F_G\, .a}$$

45. The Lewis relation for air-water humidification is given by (K_y = mass transfer coefficient of moisture in air; h_G = heat transfer coefficient; Cs = heat capacity of vapour gas mixture)
A. $\dfrac{h_G^2}{k_y C_s}$
B. $\dfrac{k_y C_s^2}{h_G} = 1$
C. $\dfrac{h_G}{k_y C_s}$
D. $\dfrac{k_y^2 h_G}{C_s} = 1$

Sol. C. the Lewis relation for air – water humidification is given by

$$\frac{h_G}{k_Y C_s} = 1$$

GATE 2000

46. The Grashof number is defined as the ratio of
A. buoyancy to inertial forces
B. buoyancy to viscous forces
C. inertial to viscous forces
D. buoyancy to surface tension forces

Sol. **B.** the Grashof number is defined as the ratio of buoyancy to viscous forces

$$Gr = \frac{Buoyancy\ force}{Viscous\ force}$$

47. The absorption factor is defined as

A. $\frac{L}{mG}$

B. $\frac{G}{mL}$

C. $\frac{m\ L}{G}$

D. $\frac{L\ G}{m}$

Where L = liquid flow rate, G = gas flow rate and m = slope of the equilibrium line

Sol. **A.** the absorption factor is defined as $\frac{L}{mG}$

48. At 750 K and 1 atm, the approximate value of the Schmidt number for air is

A. 0.01

B. 0. x1

C. 1

D. 10

Sol. **B.** at 750°K and 1 atm the approximate value of the Schmidt number for air is 0.1

49. For the nth tray (counted from the bottom of a distillation column), the murphree efficiency is given by

A. $\frac{Y_{n+1} - Y_n}{Y_n^* - Y_{n-1}}$

B. $\frac{Y_n - Y_{n-1}}{Y_n^* - Y_{n+1}}$

C. $\frac{Y_{n+1} - Y_n}{Y_{n+1} - Y_n}$

D. $\frac{Y_n^* - Y_{n-1}}{Y_n^* - Y_{n+1}}$

Sol. **B.** for the nth tray (counted from the bottom of a distillation column) the Murphree efficiency is given by

$$\frac{Y_n - Y_{n-1}}{Y_n^* - Y_{n-1}}$$

50. The McCabe ΔL law states that the

A. molar heat of vaporization of components are nearly equal

B. linear crystal growth rate depends on the degree of super saturation

C. linear crystal growth rate does not depend on the crystal size

D. linear crystal growth rate depends on the crystal size

Sol. **A.** as per the McCabe Thiele law the molar heats of vaporization of components are nearly equal.

(2 Mark Question)

GATE 2015

51. Consider a steady state mass transfer process between well-mixed liquid and vapor phase of a binary mixture comprising of components A and B. The mole fractions of component A in the bulk liquid (x_A) and bulk vapour (y_A) phases are 0.36 and 0.16, respectively. The mass transfer coefficients for component A in liquid and vapour phases are 0.1 mol/(m²s) and 0.05 mol/(m²s), respectively. The vapopur-liquid equilibrium can be approximated as $y_A^* = 2x_A$ for x_A less than 0.4. The mole fraction of A in the liquid at the interface (up to two decimal places) is ____________.

Sol. mole fraction in bulk liquid, $X_A = 0.36$

Mole fraction in bulk vapour, $Y_A = 0.16$

Mass transfer coefficients in liquid,

$k_L = 0.1 mol/m^2s$

Mass transfer coefficients in vapour,

$k_g = 0.05\ mol/m^2s$

Now use, $y_A^* = 2X_A$ for $X_A < 0.4$

So, $y_{Ai} = 2X_A$ ....(i)

At steady state, mass flux is constant

$$\therefore \quad k_L(X_A - X_{Ai}) = k_g(y_{Ai} - y_A)$$

$$0.1(0.36 - X_{Ai}) = 0.05\ (2X_A - 0.16)$$

$$0.11(0.36 - X_{Ai}) = 0.05\ (2\times 0.36 - 0.16)$$

$$\Rightarrow \quad X_{Ai} = 0.105$$

52. A multi-stage, counter-current liquid-liquid extractor is used to separate solute C from a binary mixture (F) of A and C using solvent B. Pure solvent B is recovered from the raffinate R by distillation, as shown in the schematic diagram below:

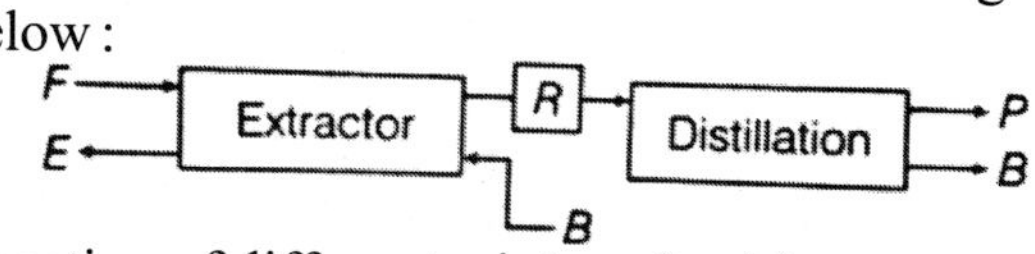

Location of different mixture for this process are indicated on the triangular diagram below. P is the solvent-free raffinate, E is the extract, F is the feed and Δ is the difference point from which the mass balance lines originate. The line PB intersects the bimodal curve at U and T. The lines P and FB intersect the bimodal at V and W, respectively.

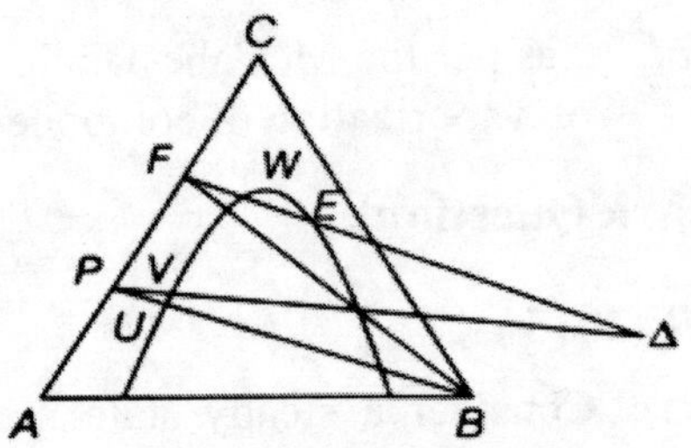

The raffinate coming out of extractor is represented in the diagram by the point

A. T
B. U
C. V
D. W

***Sol.* B.**

53. Adsorption on activated carbon is to be used for reducing phenol concentration in waste water from 0.04 mol/L to 0.008 mol/L. The adsorption isotherm at the operating temperature can be expressed as $q = 0.025C^{1/2}$, where q is the phenol concentration in solid (mol/g solid) and C is the phenol concentration in water (mol/L). The minimum amount of solid (in g) required per liter of waste water (up to one decimal place) is ____________.

Sol. Initial moles of phenol = 0.04 mol/L

Final moles of phenol in waste water = 0.008 mol/L

Number of moles of phenol in solid = 0.04 0.008 = 0.032

Let w is the weight of solid.

Given, Q = 0.025 $C^{1/3}$

$$\frac{0.032}{w} = 0.025 \times (0.008)^{1/3} \Rightarrow w = 6.4\ g$$

∴ Amount of solid required = 6.4 g/L of waste water.

54. A binary feed consisting of 25 mol% liquid and 75 mol% vapour is separated in a staged distillation column. The mole fraction of the more volatile component in the distillate product is 0.95. The molar flow rate of distillate is 50% of the feed flow rate and Mccabe-Thiele method can be used to analyse the column. The q-line intersects the operating line of the enriching section at (0.33, 0.5) on the x-y diagram. The slope of the stripping section operating line (up to one decimal place) is ________.

Sol. given, q = 0.25 and $X_D = 0.95$

Equation of operating line for rectifying section,

$$y_{n+1} - \frac{R}{R+1}X_n + \frac{X_D}{R+1}$$

Since, this line passes through (0.35, 0.5),

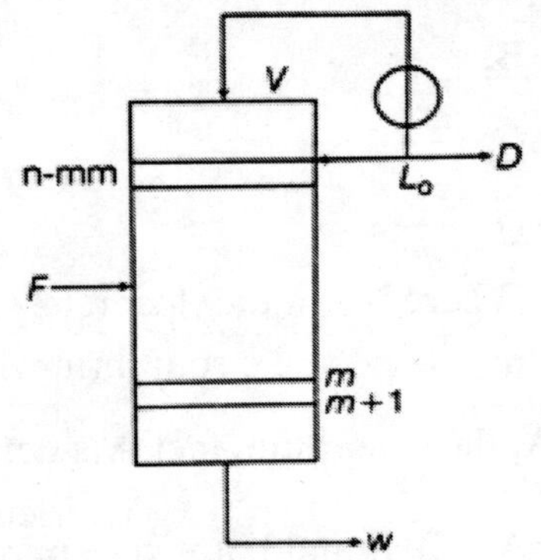

$$0.5 = \frac{R}{R+1} \times 0.35 + \frac{0.95}{R+1} = \frac{0.35R+0.95}{R+1}$$

$\Rightarrow$ 0.5R + 0.5 = 0.35R + 0.95

$\Rightarrow$ 0.15R = 0.45

$\Rightarrow$ R = 3

Reflux ratio $= \frac{L_0}{D} \Rightarrow L_0 = 3D$

A(∞) F = D + w

$\Rightarrow$ 2D = D + w

w = D and F = 2D

For stripping section,

$$y_{m+1} = \frac{L_m}{V_{m+1}}X_m - \frac{WX_w}{V_{m+1}}$$

In stripping section,

Lm = Ln + qF = 3D + 0.25 × 2D

Lm = 3.5D

and V_{m+1} = Lm – w = 3D – 0.5D = 2.5D

Slope of operating line in stripping section

$$= \frac{Lm}{V_{m+1}} = \frac{3.5}{2.5} = 1.4$$

GATE 2014

55. A spherical ball of benzoic acid (diameter = 1.5 cm) is submerged in a pool of still water. The solubility and diffusivity of benzoic acid in water are 0.03 kmol/m^3 and $1.25 \times 10^{-9} m^2/s$ respectively. Sherwood number is given as Sh = 2.0 + 0.6 Re$^{0.5}$ Se$^{0.5}$. The initialrate of dissolution (in kmol/s) of benzoic acid approximately is

A. 3.54×10^{-11}
B. 3.54×10^{-12}
C. 3.54×10^{-13}
D. 3.54×10^{-14}

Sol. **B.** Sherwood number,

$$Sh_0 = \frac{kL}{D},$$

where, D = diffusivity

L = characteristic length

K = mass transfer coefficient

$\therefore$ Rate of dissolution,

$$r_A = \text{kac}^*$$

$$= \frac{Sh_0.D}{L} \times 4\pi r^2 \times c^*$$

$$= \frac{Sh_0.D}{L} \times \pi \times L^2 \times c^*$$

$$= Sh_0 \times D \times \pi \times L^2 \times c^*$$

$$= 2 \times 1.25 \times 10^{-9} \times 3.14 \times 1.5 \times 10^{-2} \times 0.03$$

$$= 3.54 \times 10^{-12}\text{k} - \text{mol/s}$$

56. A wet solid of 100 kg is dried from a moisture content of 80 wt% to 10 wt%. The critical moisture content is 15 wt% and the equilibrium moisture content is negligible. All moisture contents are on dry basis. The falling rate is considered to be linear. It takes 5 h to dry the material in the constant rate period. The duration (in hour) of the falling rate period is _________.

Sol. for constant drying period,

$$t_1 = \frac{L_S}{A}\left(\frac{x_1 - x_c}{R_c}\right)$$

where, L_s = mass of dry solid

R_c = critical rate of during

x_i = initial moisture content

x_c = equilibrium moisture content

A = surface area.

For linear falling rate period,

$$t_2 = -\frac{L_S}{A}\int_{x_c}^{x_2}\frac{dx}{R}$$

Here, R = mx

At, x = $x_c, R_c = mx_c$

$\therefore$ $R = \frac{R_c}{x_c}x$

Putting values, $t_2 = -\frac{L_S}{A}\int_{x_c}^{x_2}\frac{x_c dx}{R_c x}$

$$= \frac{L_s x_c}{AR_c} ln\left(\frac{x_c}{x_2}\right)$$

Here, $t_1 = 5\text{h}, x_1 = 0.4, x_2 = 0.1, x_c = 0.15$

Putting values,

$$t_2 = \frac{5 \times 0.15}{(0.4 - 0.15)}\ln\left(\frac{0.15}{0.1}\right) = 1.22 \text{ h}$$

57. Two elemental gases (A and B) are reacting to form a liquid (C) in a steady state process as per the reaction A + B $\rightarrow$ C. The single pass conversion of the reaction is only 20% and hence recycle is used. The product is separated completely in pure form. The fresh feed has 439 mol% of A and B each along with 2 mol% impurities. The maximum allowable impurities in the recycle stream is 20 mol%. the amount of purge stream (in mole) per 100 moles of the fresh feed is ____________.

Sol. basis - 100 moles fresh feed (F)

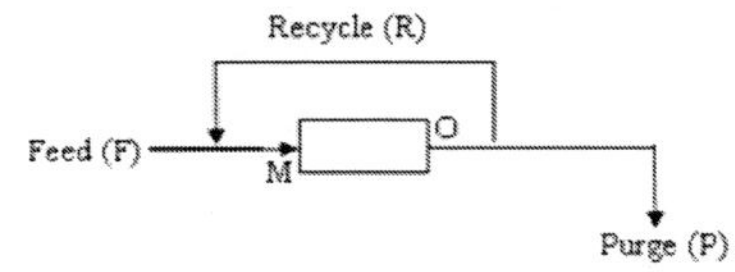

	Reactant A	*Reactant B*	*Impurities (I)*
Recycle (R)	0.4 R	0.4 R	0.2 R (given)
Feed (F)	49	49	2
Output (O)	0.8 (49 + 0.4 R)	0.8 (49 +0.4 R)	2 + 0.2 R

$\therefore$ Total output (O) = 8.4 + 0.84 R

Given, 20% impurities (I) in recycle (R)

$\therefore$ $\left(\frac{I}{O}\right)_{output} = 0.2$

$$\frac{2 + 0.2\,R}{80.4 + 0.84\,R} = 0.2$$

$$2 + 0.2\text{ R} = 16.08 + 0.168\text{ R}$$

$$0.032\text{ R} = 14.08$$

$$\text{R} = 440$$

$$\text{Purge, P} = \text{O} - \text{R}$$

$$= 80.4 + 0.84\text{ R} - \text{R}$$

$$= 80.4 - 0.16\text{ R}$$

$$= 80.4 - 0.16 \times 440$$

$$= 10 \text{ mole}$$

GATE 2013

58. The vapor-liquid equilibrium curve of a binary mixture A-B, may be approximated by a linear equation over a narrow range of liquid mole fractions ($0.2 < x_A < 0.3$) as follows

$$y_A^* = 10325\, x_A + 0.121$$

Here y_A^* is the mole fraction of A in the vapour. 100 moles of a feed ($x_{A,W}$= 0.28) is batch distilled to a final residue ($x_{A,W}$= 0.2). Using the Rayleigh behind in the distillation unit, up to 2 digits after the decimal point, is

A. 60.3 moles

B. 63.2 moles

C. 65.6 moles

D. 66.7 moles

Sol. **D.** applying Rayleigh equation for A,

$$\frac{dL}{L} = \frac{dx}{y-x}$$

$$\frac{dL}{L} = \frac{dx}{1.325x + 0.121 - x}$$

$$\int_{L_2}^{L_1} \frac{dL}{L} = \int_{x_2}^{x_1} \frac{dx}{0.121 + 0.325x}$$

$$\log\left(\frac{L_1}{L_2}\right) = \frac{1}{0.325}\log\left(\frac{0.121+0.325x_1}{0.121+0.325x_2}\right)$$

$$\log\frac{100}{L_2} = \frac{1}{0.325}\log\left(\frac{0.121+0.325\times 0.28}{0.121+0.325\times 0.2}\right)$$

$$= 3.077\log\left(\frac{212}{186}\right) \Rightarrow \frac{100}{L_2} = \left(\frac{212}{186}\right)^{3.077} = 1.496$$

$L_2 = 66.7$ moles

59. A crosscurrent cascade of N ideal stages is used to treat a feed stream of molar flow rate E. The feed stream contains a solute which is to be recovered by a pure solvent having a molar flow rate S. The solvent is divided equally between these N stages. The linear equilibrium curve relating the mole fractions x and y^* of the solute in the raffinate and the extract respectively, is given by $y^* = mx$. Assume dilute conditions. The ratio of the solute mole fraction in the original feed to that in the exit raffinate stream i.e., (x_0/x_N) is given by

A. $\left[1 + \left(\frac{mS}{NE}\right)\right]^N$

B. $\left[1 + \left(\frac{NS}{mE}\right)\right]^N$

C. $\left[1 + \left(\frac{NE}{mS}\right)\right]^N$

D. $\left[1 + \left(\frac{mE}{NS}\right)\right]^N$

Sol. **A.** the ratio of solute mol fraction in feed to that in exit raffinate stream is given by,

$$\frac{X_0}{X_N} = \left(1 + \frac{mS}{NE}\right)^N$$

60. A study was conducted in which water was pumped through cylindrical pipes made of a sparingly soluble solid. For a given pipe and certain flow conditions, the mass transfer coefficient k_c has been calculated as 1 mm/s using the correlation

$$Sh = 0.025\,Re^{0.6}Sc^{0.33}$$

If the velocity of the fluid and the diameter of the pipe are both doubled, what is the new value of k_c in mm/s, up to 2 digits after the decimal point?

A. 2.15 mm/s

B. 1.15 mm/s

C. 3.25 mm/s

D. 1.19 mm/s

Sol. **B.** Given, Sh = 0.025 $(Re)^{0.6}(Sc)^{0.33}$

or $\frac{K_c d}{D_{AB}} = 0.025\left(\frac{\rho v d}{\mu}\right)^{0.6}\left(\frac{\mu}{\rho D_{AB}}\right)^{0.33}$

$\therefore$ $\frac{K_{C_1}d_1}{K_2 d_2} = \left(\frac{v_1 d_1}{v_2 d_2}\right)^{0.6}$

Here, $d_2 = 2d_1$, and $v_2 = 2v_1$

$K_{C_1} = 1$ mm/s

Putting values,

$$\frac{1}{K_{C_2}\times 2} = \left(\frac{1}{2}\times\frac{1}{2}\right)^{0.6}$$

$$K_{C_2} = \frac{(4)^{0.6}}{2} = 1.15 \text{ mm/s}$$

Linked Answer Question 61 and 62

The vapor liquid equilibrium relation for an ideal binary system is given by

$$y_A^* = \frac{\alpha_{AB}X_A}{1+(\alpha_{AB}-1)X_A}$$

Here X_A and y_A^* are the mole fractions of species A in the liquid and vapor, respectively. The relative volatility (α_{AB}) is greater than unity.

61. The liquid mole fraction x_A at which the maximum difference between the equilibrium vapor mole fraction and liquid mole fraction occurs is

A. $\frac{1}{(1+\sqrt{\alpha_{AB}})}$

B. $\frac{0.75}{(1+\sqrt{\alpha_{AB}})}$

C. $\frac{0.5}{(\sqrt{\alpha_{AB}}+1)}$

D. $\frac{0.75}{(\sqrt{\alpha_{AB}}+1)}$

Sol. **A.** P = $y_A^* - X_A$

$$P = \frac{\alpha_{AB}X_A}{1+(\alpha_{AB}-1)X_A} - X_A$$

For P to be maximum, $\frac{dP}{dX_A} = 0$

$$\frac{\alpha_{AB}}{1+(\alpha_{AB}-1)X_A} - \frac{\alpha_{AB}X_A(\alpha_{AB}-1)}{[1+(\alpha_{AB}-1)X_A]^2} - 1 = 0$$

Or $\alpha_{AB}[1+(\alpha_{AB}-1)X_A] - \alpha_{AB}(\alpha_{AB}-1)X_A = [1+(\alpha_{AB}-1)X_A]^2$

$$\alpha_{AB} = 1 + (\alpha_{AB} - 1)^2 X_A^2 + 2(\alpha_{AB} - 1)X_A$$

Or $(\alpha_{AB} - 1)X_A + 2X_A - 1 = 0$

$$X_A = \frac{-2+\sqrt{4+4(\alpha_{AB}-1)}}{2(\alpha_{AB}-1)}$$

$$X_A = \frac{-1+\sqrt{\alpha_{AB}}}{\alpha_{AB}-1} = \frac{1}{1+\sqrt{\alpha_{AB}}}$$

62. A liquid having the composition found in the first part of the linked answer question, is flash distilled at a steady state to a final liquid mole fraction of 0.25. If α_{AB} is 2.5, the fraction of the feed vapourised is

A. 0.0
B. 0.20
C. 0.67
D. 0.74

***Sol.* C.** $X_F = \frac{1}{1+\sqrt{\alpha_{AB}}} = \frac{1}{1+\sqrt{2.5}} = 0.387$

$X_L = 0.25$

$y^* = \frac{2.5\times0.25}{1+(2.5-1)\times0.25} = 0.4545$

Applying material balance for A

$x_F \cdot F = V \cdot y^* + x_L \cdot L$

$0.387 \cdot F = V \cdot 0.4545 + 0.25(F - V)$

$0.137F = 0.2045\,V$

$\frac{V}{F} = 0.67$

GATE 2012

63. Consider the drying operation shown in the figure below for a solid loading (dry basis) of 50 kg/m^2 with a constant drying rate of 5 kg/m^2–h. The falling rate of drying is linear with moisture content.

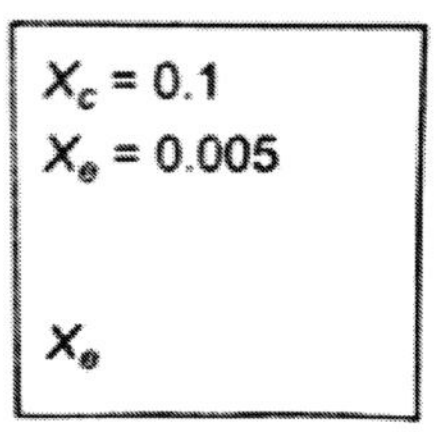

The drying time in hours required to reduce an initial moisture content of 25% to a final moisture content of 2% is

A. 1.5
B. 1.75
C. 3.25
D. 4.55

***Sol.* B.** for constant drying period,

$$t = \frac{L_s}{AR_c}(x_1 - x_2)$$

Here, $\frac{L_s}{A} = 50$ kg/m^2, $R_C = 5$ kg/m^2-h

$x_1 = 25\%$, $x_2 = x_c = 0.1$

Putting values, $t_1 = \frac{50}{5} \times (0.25 - 0.1) = 1.5$h

For falling rate period,

$$t = -\frac{L_s}{A}\int_{x_1}^{x_2} \frac{dx}{R}$$

Here, $R = \frac{5(x-0.005)}{(0.1-0.005)}$ (linear slope)

$= 52.63(x - 0.005)$

Thus, $t_2 = -50\int_{0.1}^{0.02} \frac{dx}{52.63(x-0.005)}$

$= -\frac{50}{52.63}\ln\left(\frac{0.02-0.005}{0.1-0.005}\right)$

$= \frac{50}{52.63}\ln\left(\frac{19}{3}\right)$

$\Rightarrow t_2 = 1.75$

Thus, total times of drying $= t_1 + t_2 = 3.25$h

Common Data for Questions 30 and 31

A counter – current extraction column is designed to remove 99% of solute C from a solution of solvent A and solute C using pure solvent B. The initial concentration of solute in the solution of A+C is 20 wt% and the total flow of solution is 1000 kg/h. If the equilibrium relationship is Y=2X, where Y = mass of C/mass of A and X = mass of C/mass of B.

64. The minimum flow rate of solvent B required in kg/h is

A. 1454
B. 1584
C. 1676
D. 1874

***Sol.* C.** given,

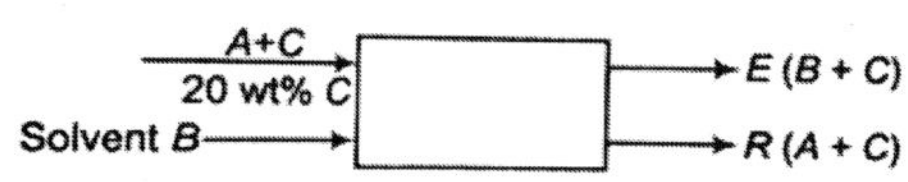

Given, F = 1000 kg/h

∴ $F_C = 0.2F = 200$ kg/h

Weight of C in R = 0.01 × 200 = 2 kg/h

Weight of C in E = 198 kg/h

Given, y = 2x

$$\frac{200}{800} = 2 \times \frac{198}{\text{Mass of B}}$$

⟹ Mass of B = 1584 kg/h

65. If the flow rate of B is 2400 kg/h, then the theoretical number of stages in the column, using Kremser's equation adjusted to the next integer is

A. 5
B. 9
C. 11
D. 13

Sol. **A.**

GATE 20011

66. An aqueous sodium chloride solution (10 wt%) is fed into a single effect evaporator at a rate of 10,000 kg/h. It is concentrated to a 20 wt% sodium chloride solution. The rate of consumption of steam in the evaporator is 8000 kg/h. The evaporator capacity (kg/h) and economy are, respectively

A. 5000, 0.625
B. 10,000, 0.625
C. 5000, 1.6
D. 10,000, 1.6

Sol. **A.**

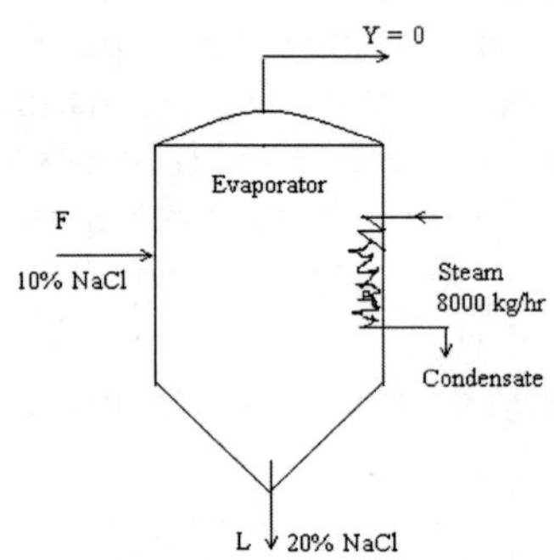

Given F = 10,000 kg/hr; S = 8000 kg/hr

Material balance for NaCl is

$$(0.1)\,F = (0.2)\,L$$

$$\therefore \quad L = 5000 \text{ kg/hr}$$

Capacity of evaporator = F – L = 5000 kg/hr

$$\text{Steam economy} = \frac{F-L}{S} = \frac{5000}{8000} = 0.625$$

67. A gas mixture is in contact with a liquid. Component P in the gas mixture is highly soluble in the liquid Possible concentration profiles during absorption of P are shown in the choices, where

x = mole fraction of P in a bulk liquid

y = mole fraction of P in bulk gas

x_i = mole fraction of P in the interface in liquid

y_i = mole fraction of P at the interface in gas

y^* = equilibrium gas phase mole fraction corresponding

To x_i

A.

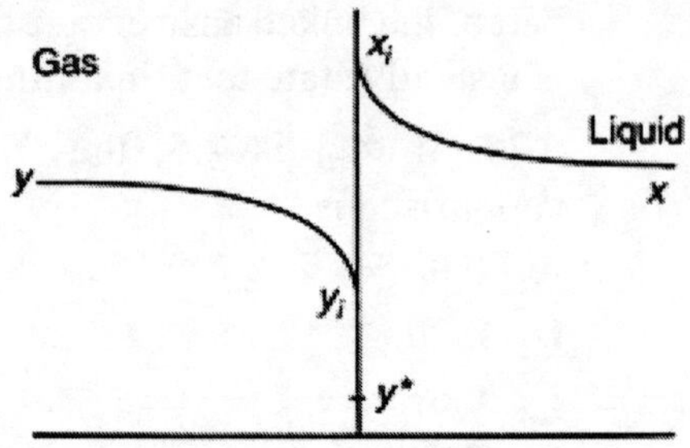

B.

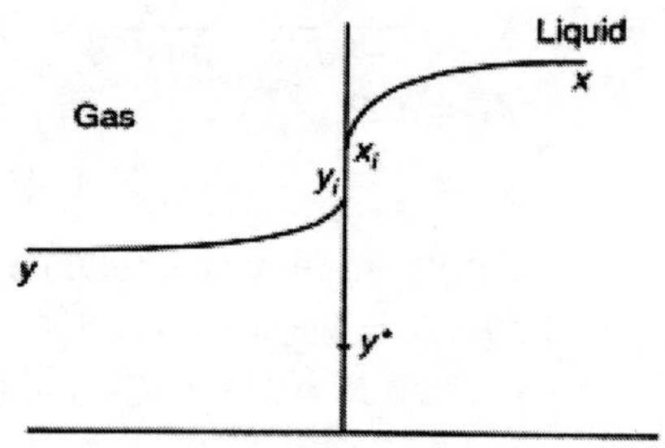

C.

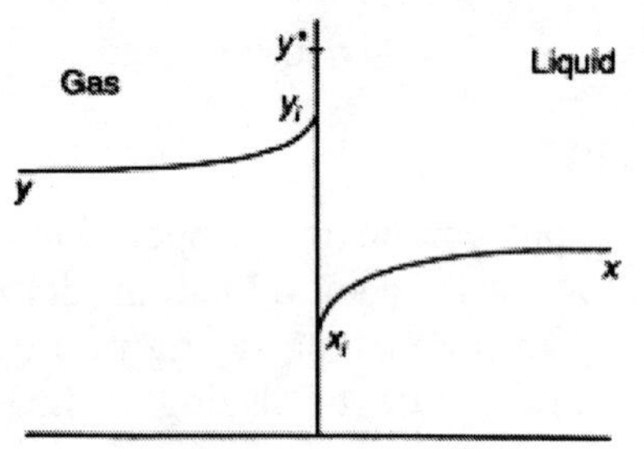

D.

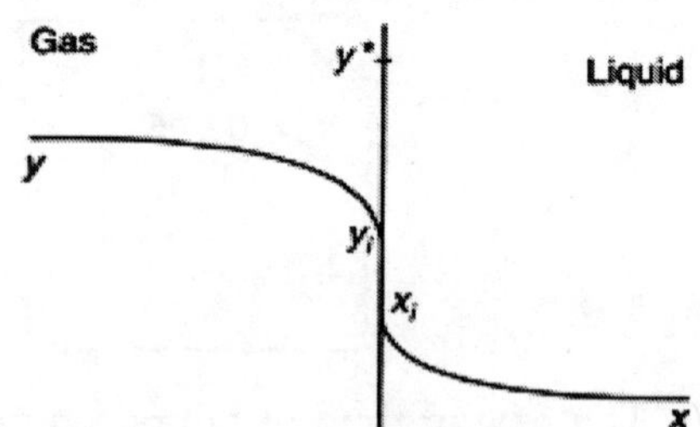

Sol. **A.** during absorption of P in the gas mixture which is highly soluble the possible concentration profile is as shown in fig (a). As can be observed from figure (a), y is falling due to absorption and tending towards y* but higher than it at because of resistance at the interface. Due to absorption of gas in liquid phase xi (the concentration of P in liquid phase at the interface) will be higher than yi and will show a decreasing trend as it moves towards bulk phase.

68. A batch of 120 kg wet solid has initial moisture content of 0.2 kg water /kg dry solid. The exposed area for drying is 0.05m²/kg dry solid. The rate of drying follows the curve given below

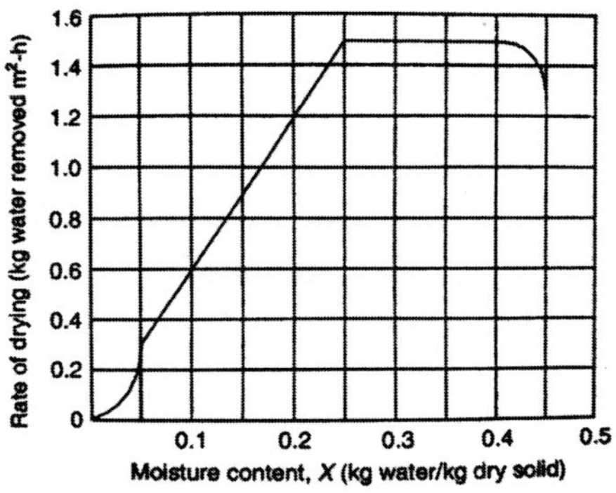

The time required (in hour) for drying this batch to a moisture content of 0.1 kg water/kg dry solid is

A. 0.033

B. 0.43

C. 0.6

D. 2.31

Sol. **D.** weight of wet solid = 120 kg

Initial moisture content $= 0.2 \frac{kg\ water}{kg\ dry\ solid}$

$= \frac{0.2}{1.2} \frac{kg\ water}{kg\ wet\ solid}$

∴ Total water in the solid batch $= 120 \left(\frac{0.2}{1.2}\right)$

= 20 kg

∴ Amount of dry solid (w)

= 120 – 20 = 100 kg

Referring to the given rate of drying curve

At x_1 = 0.2, Rate of drying

$= 1.2 \frac{kg\ water}{m^2 - hr}$

At x_2 = 0.1, Rate of drying

$= 0.6 \frac{kg\ water}{m^2 - hr}$

Rate of drying (R) = 6x

∴ Time required for drying

$$T = -\frac{W}{A}\int_{x_1}^{x_2} \frac{dX}{R}$$

Here, A = 0.05 m²/kg dry solid

∴ Total area = (0.05) (100) = 5m²

Hence, $t = -\frac{100}{5}\int_{0.2}^{0.1} \frac{dx}{6x}$

$= -\frac{20}{6} \ln\left(\frac{0.1}{0.2}\right)$

∴ t = 2.31 hr.

Common Data for Questions 73 and 74

A binary feed mixture containing equimolar quantities of components *S* and *T* is to be distilled in a fractionating tower at atmospheric pressure. The distilled contains 96 mol% *S*. The q-line (feed line) intersects the equilibrium line at $x' = 0.46$ and $y' = 0.66$ where x' and y' are mole fractions. Assume that the McCabe–Thiele method is applicable and the relative volatility is constant.

69. The minimum reflux ratio is

A. 1.6

B. 1.5

C. 0.66

D. 0.6

Sol. **B.** given $x_D = 0.96$, x' = 0.46, y' = 0.66

According to McCabe – Thiele method

$$\frac{R}{R+1} = \frac{0.96-0.66}{0.96-0.46} = \frac{0.3}{0.5}$$

∴ R = 1.5

70. The feed is

A. at dew point

B. at bubble point

C. superheated vapour

D. partially vapour

Sol. **D.** the feed is partially vapour

GATE 2010

71. At 25°C and 90% relative humidity, water evaporates from the surface of a lake at the rate of 1.0 kg/m². The relative humidity that will lead to an evaporation rate o 3.0 kg/m²/h, with other conditions remaining the same, is

A. 30%

B. 50%

C. 60%

D. 70%

Sol. **D.** rate of evaporation $N_A = 1 - \frac{p}{p_{sat}}$

or $N_A = 1$ – Relative humidity

Further, $\frac{N_{A2}}{N_{A1}} = \frac{1-(RH)_2}{1-(RH)_1}$

Data given is

$N_{A_1} = 1$ kg/$(m^2) - (h)$

$N_{A_2} = 3$ kg/$(m^2) - (h)$

$(RH)_1$ = 0.9

Putting the above values

$$\frac{3}{1} = \frac{1-RH_2}{1-0.9}$$

RH_2 = 70%

72. A liquid flows over a flat naphthalene plate of length L, at Reynolds number ($Re_L = L\rho u_\infty/\mu$) of 1500, as shown in the figure. The surface concentration of naphthalene is $C_{AS} > C_{A\infty}$ and the surface temperature is $T_S > T_{\infty}$, Assume Pr = Sc = 1.

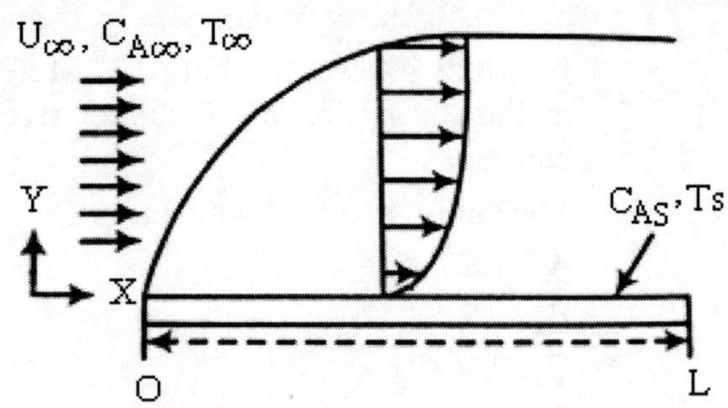

If at = L, $\left.\frac{\partial C_A^*}{\partial y^*}\right|_{y^*=0} = 10$, where $C^*_A = \frac{C_A - C_{AS}}{C_{A\infty} - C_{AS}}$ and y* = Y/L, then the Nusselt number and the friction coefficient x = L, are

A. 10, 1/75
B. 10, 10
C. 20, 10
D. 1/75, 5

Sol. **A.** according to Reynolds analogy

$$\frac{\partial}{\partial y}\left(\frac{C_A - C_{AS}}{C_{A\infty} - C_{AS}}\right) = \frac{\partial}{\partial y}\left(\frac{U_x}{U_\infty}\right) \quad ....(i)$$

Now, given

$$\left.\frac{\partial C_A^*}{\partial y^*}\right|_{y^*=0} = 10$$

or $$\left.\frac{\partial}{\partial y^*}\left(\frac{C_A - C_{AS}}{C_{A\infty} - C_{AS}}\right)\right|_{y^*=0} = 10$$

or $$\left.\frac{\partial}{\partial y^*}\left(\frac{U_x}{U_\infty}\right)\right|_{y^*=0} = 10 \quad ....(ii)$$

$$y^* = \frac{Y}{L}$$

$$dy^* = \frac{1}{L}dy \text{ and } y^* = 0$$

$\Rightarrow$ $$y = 0$$

$\therefore$ from equation (ii)

$$L.\left.\frac{\partial}{\partial y}\left(\frac{U_x}{U_\infty}\right)\right|_{y=0} = 10$$

or $$\left.\frac{L}{U_\infty}\frac{\partial U_x}{\partial y}\right|_{y=0} = 10$$

$$\left.\frac{\partial U_x}{\partial y}\right|_{y=0} = \frac{10U_\infty}{L} \quad ....(iii)$$

Now, fiction coefficient

$$f = \frac{\tau_0}{(\rho U_\infty^2/2)}$$

or $$f = \frac{\mu \left.\frac{\partial U_x}{\partial y}\right|_{y=0}}{(\rho U_\infty^2/2)}$$

or $$f = \frac{\mu.(10U_\infty/L)}{(\rho U_\infty^2/2)}$$

or $$f = \frac{20\,\mu}{\rho U_\infty L}$$

or $$f = \frac{20}{Re}$$

Given, $$Re = 1500$$

$\therefore$ $$f = \frac{20}{1500}$$

$$f = \frac{1}{75}$$

Nusselt number,

$$(Nu) = \left(\frac{f}{2}\right)(Re).(Pr)^{1/3}$$

Given $$Pr = 1$$

$\therefore$ $$Nu = \frac{1}{75} \times \frac{1}{2} \times 1500 \times 1$$

$$Nu = 10$$

Statements for Linked Answer Questions 73 and 74

Water is used to absorb ammonia from a gas mixture in a single separation stage contactor. The process is schematically represented in the figure below.

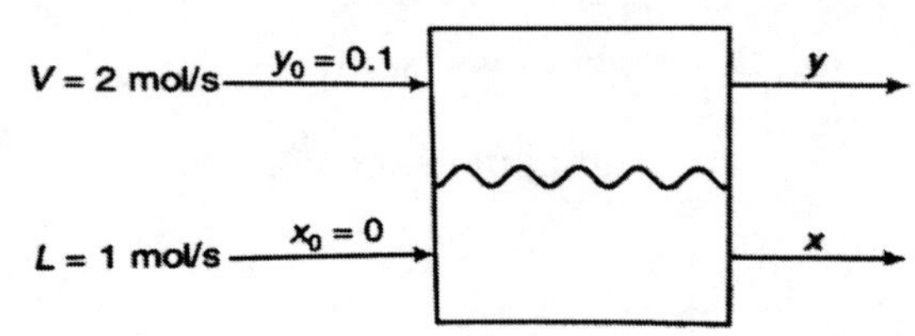

The molar gas and liquid flow rates and the inlet mole fractions are given in the figure. Both the liquid and the gas phases are well mixed and the equilibrium relation between y and x is given by y* = x.

73. If the stage is ideal, then the value of y is

A. 1/15
B. 1/10
C. 2/15
D. 1/6

Sol. **A.**

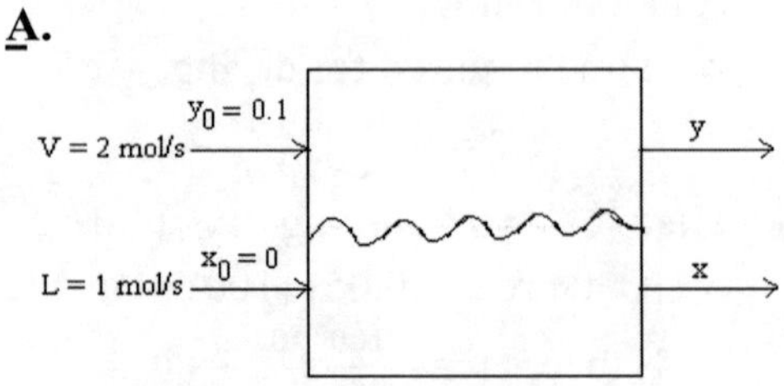

Writing the ammonia balance equation

$$(2)(0.1) + (1)(0) = (2)\,y^* + (1)\,x$$

$\therefore$ $$2y^* + x = 0.2$$

Given, $y^* = x$

$3y^* = 0.2$

$\therefore \quad y^* = \frac{0.2}{3} = \frac{1}{15}$

74. If the stage efficiency is 50% then the value of y is

A. 1/12

B. 1/6

C. 1/4

D. 1/3

Sol. **A.** the number of moles of ammonia absorbed from vapor to liquid phase at stage efficiency of 100%

$$n = (v)(y_0 - y^*) = 2\left(0.1 - \frac{1}{15}\right) = \frac{1}{15}$$

For stage efficiency equal to 50%

$$n' = \frac{1}{30}$$

or $\quad n' = v\,(y_0 - y')$

or $\quad \frac{1}{30} = 2\,(0.1 - y')$

$$y' = \frac{1}{12}$$

GATE 2009

75. For an incompressible flow, the x and y components of the velocity vector are

$$v_x = 2\,(x + y);\ v_y = 3\,(y + z);$$

Where x, y, z, are in meter and velocities are in m/s. Then the z component of the velocity vector (v_z) of the flow for the boundary condition $v_z = 0$ at $z = 0$ is

A. 5 z

B. –5 z

C. 2x + 3z

D. –2x – 3z

Sol. **B.** given $V_x = 2\,(x + y)$ and $V_y = 3\,(y + z)$

For an incompressible fluid,

$$\nabla . v = 0$$

$$\frac{\partial V_x}{\partial x} + \frac{\partial V_y}{\partial y} + \frac{\partial V_z}{\partial z} = 0$$

Here, $\frac{\partial V_x}{\partial x} = 2$ and $\frac{\partial V_y}{\partial y} = 3$

Putting the given values,

$$2 + 3 + \frac{\partial V_z}{\partial z} = 0$$

$$\frac{\partial V_z}{\partial z} = -5$$

or $\quad V_z = -5 + C$

Applying boundary condition,

$V_z = 0$ at $z = 0$

$\therefore \quad 0 = 0 + C$

$C = 0$

Therefore, $V_z = -5z$

76. Species A is diffusing at steady state from the surface of a sphere (radius = 1 cm) into a stagnant fluid. If the diffusive flux at a distance r = 3 cm from the centre of the sphere is 27 mol/cm^2 –s, the diffusive flux (in mol/cm^2 –s) at a distance r = 9 cm is

A. 1

B. 3

C. 9

D. 27

Sol. **B.** let N_r be the flux at distance x.

At steady state

$$4\pi\, r^2 N_r|_{r+\Delta r} - 4\pi\, r^2 N_r|_r = 0$$

or $\quad \lim_{\Delta r \to 0} \frac{r^2 (N_r)_{r+\Delta r} - r^2 (N_r)_r}{\Delta r} = 0$

$$\frac{d(r^2 Nr)}{dr} = 0$$

or $\quad r^2 Nr = \text{constant}$

Further, $\quad \frac{N_{r2}}{N_{r1}} = \frac{r_1^2}{r_2^2}$

Here, $\quad N_{r_1} = 27$ mol/cm^2-s

$r_1 = 3$ cm, $r_2 = 9$ cm

Putting values, $\frac{N_{r2}}{27} = \frac{3^2}{9^2}$

$N_r = 3$ mol/cm^2-s

77. The feed to a binary distillation column has 40 mol% vapour and 60 mol% liquid. Then, the slope of the q-line in the McCabe-Thiele plot is

A. –1.5

B. – 0.6

C. 0.6

D. 1.5

Sol. **A.** slope of the q-line is $-\frac{(1-f)}{f}$

Given $f = 0.4$

$\therefore$ Slope $= -\frac{(1-0.4)}{0.4} = -1.5$

78. The equilibrium moisture curve for a solid in the figure.

The total moisture contact of the solid is X and it is exposed to air of relative humidity H. In the table below, Group I lists the types of moisture, and Group II represents the regions in the graph.

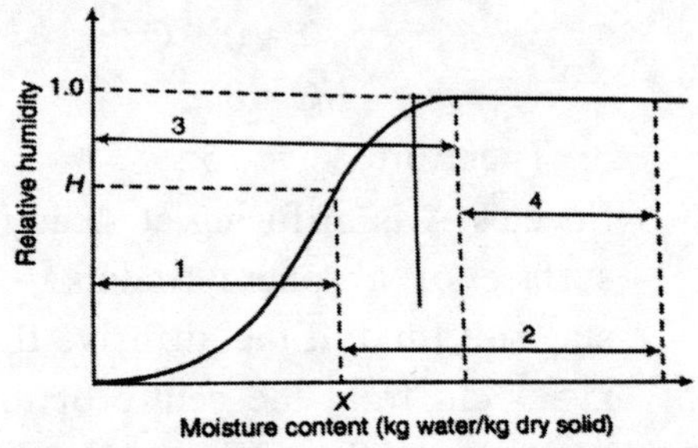

Group I	*Group II*
P. Equilibrium moisture	Region 1
Q. Bound moisture	Region 2
R. Unbound moisture	Region 3
S. Free moisture	Region 4

Which one of the following is the correct match?

A. P-1, Q-2, R-3, S-4

B. P-1, Q-3, R-4, S-2

C. P-1, Q-4, R-2, S-3

D. P-1, Q-2, R-4, S-3

***Sol.* B.** Equilibrium moisture - Region 1

Bound moisture - Region 3

Unbound moisture - Region 4

Free moisture - Region 2

79. An equimolar mixture for four hydrocarbons (1,2,3,4) is to be separated into high purity individual components using a sequence of simple distillation columns (one overhead and one bottom stream) Four possible schemes are shown below

Scheme P

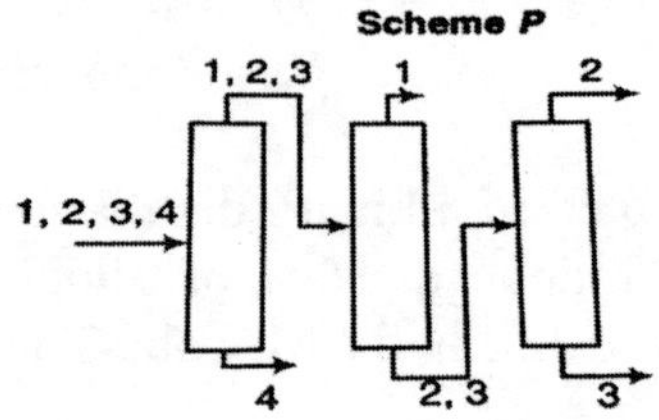

Scheme Q

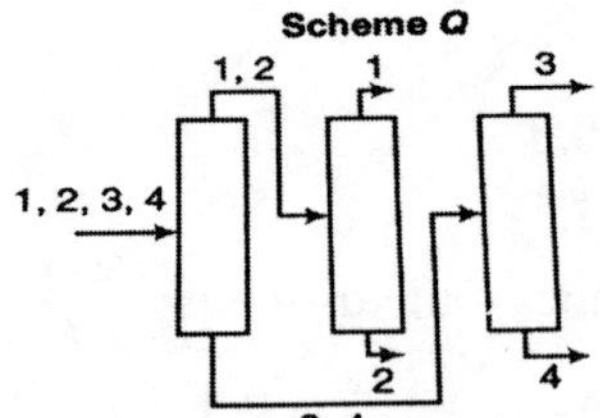

Scheme R

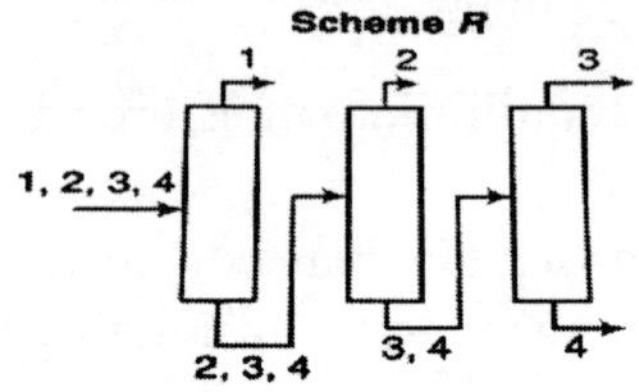

Scheme S

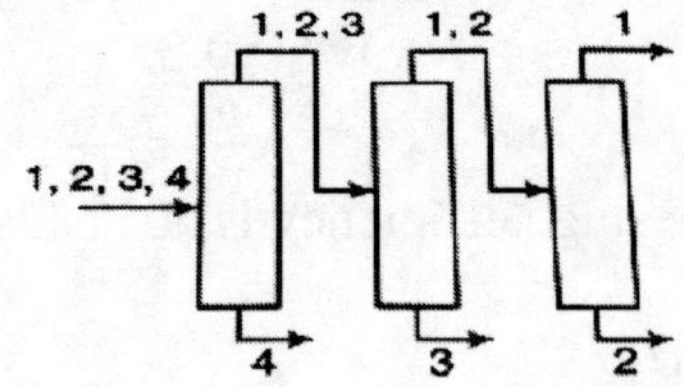

Component	K_i
1	6
2	3
3	2.5
4	1.1

Using the K_i ($= y_i^* / x_i$) values given above, the optimal scheme is

A. P

B. Q

C. R

D. S

***Sol.* D.** given

Component	$K_i(y_i^*/x_i)$
1	6
2	3
3	2.5
4	1.1

As can be observed from the table from the point of view of separation, component 4 is heaviest and component 1 is lightest. The heaviest component will go to the bottom of the first column while the lightest will come out as overhead product of the last column. Therefore, scheme S is the optimum scheme based on K_i values.

Common Data for Questions 84 and 85

A flash distillation drum (see figure below) is used to separate a methanol-water mixture. The mole fraction of methanol in the feed is 0.5 and the feed flow rate is 1000 kmol/h. The feed is preheated in a heater with heat duty Q_h and is subsequently flashed in the drum. The flash drum can be assumed to be an equilibrium stage, operating adiabatically. The equilibrium relation between the mole fraction of methanol in the vapour and liquid phases is $y^* = 4x$. The ratio distillate to feed flow rate is 0.5.

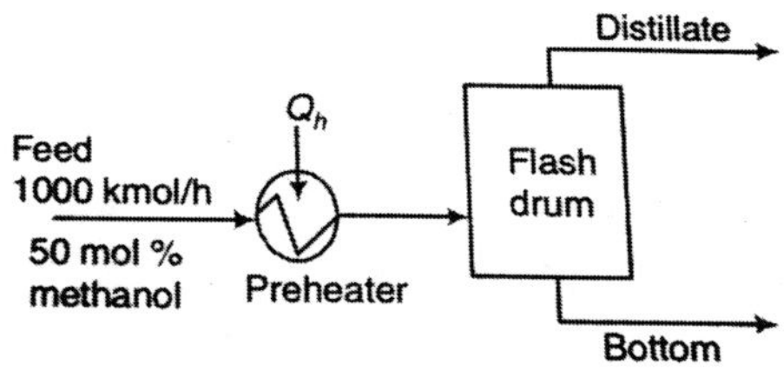

80. The mole fraction of methanol in the distillate is

A. 0.2

B. 0.7

C. 0.8

D. 0.9

Sol. **C.**

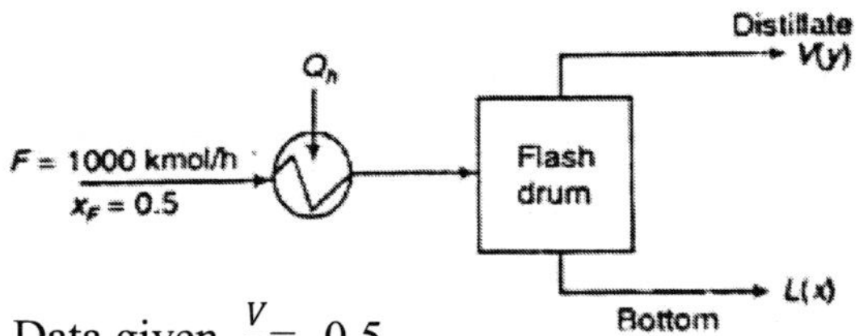

Data given $\frac{V}{F} = 0.5$

Writing overall material balance

$F = L + V$

$\therefore V = (1000)(0.5) = 500$ kmol/hr

$\therefore L = F - V = 500$ kmol/hr

Now, writing methanol balance

$F x_F = L x_L + Vy^*$

From the equation relationship given $y^* = 4x$

$\therefore$ Putting values

$(1000)(0.5) = (500)(x) + (4)(500)(x)$

or $x = 0.2$

Hence, mole fraction of methanol in distillate

$y^* = 4x$

$\therefore y^* = 0.8$

81. If the enthalpy of the distillate with reference to the feed is 3000 kJ/kmol and the enthalpy of the bottom with reference to the feed is –1000kJ/kmol, the heat duty of the preheater (Q_h in kJ/h) is

A. -2×10^6

B. -1×10^6

C. 1×10^6

D. 2×10^6

Sol. **C.** it is given that for enthalpy calculation the reference point is feed

Applying energy balance

$(F)(O)(Q_h) = (V)(3000) + (L)(-1000)$

$\therefore Q_h = (500)(3000) - (500)(1000)$

Hence, $Q_h = 1 \times 10^6$ kJ/hr

GATE 2008

82. A rectangular slab of thickness 2b along the X-axis and extending to infinity along the order direction is initially at concentration C_{A0}. At a time t = 0, both surface of the slab ($x = \pm b$) have their concentrations increased to C_{AW} and maintained at that value. Solute A diffuses into the solid. The dimensionless concentration C is defined as $= \frac{C_A - C_{A0}}{C_{Aw} - C_{A0}}$ .The diffusivity of A inside the solid is assumed constant. At certain time instant, which one of the following is the correct representation of the concentration profiles?

A. 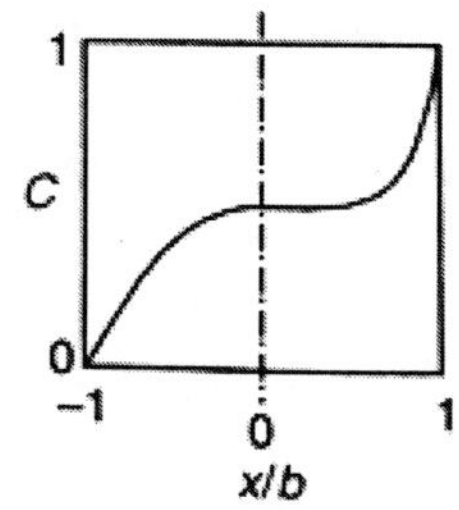

B. 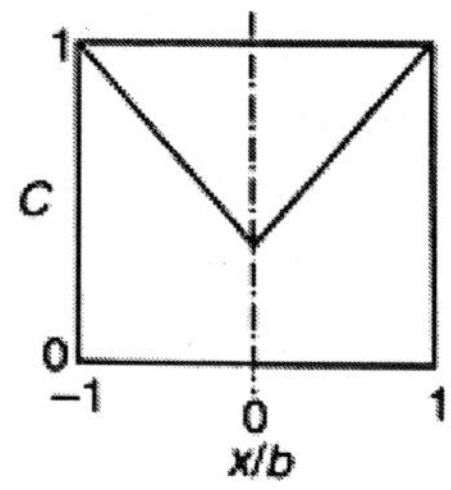

C. 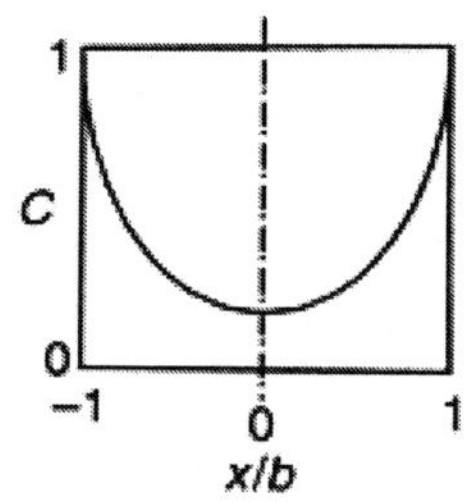

D. 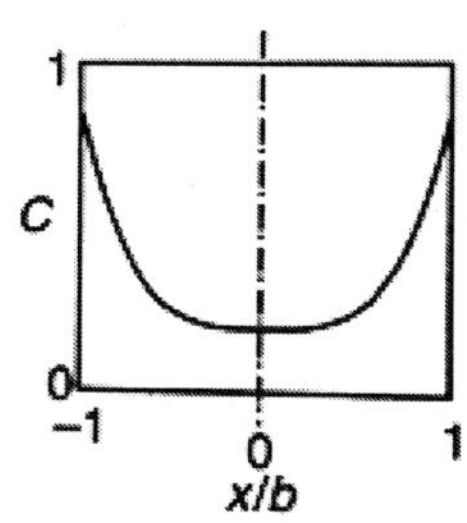

Sol. **C.** The correct representation of the concentration profile is as represented in option **C.**

83. In a binary mixture containing components A and B, the relative volatility of A with respect to B is 2.5 when mole fractions are used. The molecular

weights of A and B are 78 and 92 respectively. If the compositions are, however, expressed in mass fraction then the relative volatility will be

A. 1.18

B. 2.12

C. 2.5

D. 2.95

Sol. **C.** relative volatility remains same whether it is calculated on the basis of molefaction or mass fraction:

$$\alpha = \frac{y_A/x_A}{y_B/x_B} = 2.5$$

Option **C** is the correct answer.

84. An ideal flash vaporization is carried out with a binary mixture at a constant temperature and pressure. A process upset leads to an increase in the mole fraction of the heavy component in the feed. The flash vessel continues to operate at the previous temperature and pressure and still produces liquid and vapour. After the steady state is re-established.

A. the amount of vapour produced will increases

B. the amount of vapour produced will decrease

C. the new equilibrium compositions of the vapour and liquid products will be different

D. the new equilibrium compositions of the vapour and liquid products will remain as they were before the upset occurred

Sol. **D.** since after the process upset is restored after sometime, the unsteady state operation of the flash vessel will also return to the normal steady state condition.

Hence, after the steady state is re-established the new equilibrium compositions of the vapor and liquid products will also restored to the pre-upset condition and will remain the same.

Option is **D.** the correct answer.

85. A batch distillation operation is carried out to separate a feed containing 100 mol of a binary mixture A and B, The mole fraction of A in the feed is 0.7. The distillation progresses until the mole fraction of A in the residue decreases to 0.6. The equilibrium curve in this composition range may be linearized to y* = 0.7353 x + 0.3088. Here, x and y are the mole fractions of the more volatile component A in the liquid and vapour phases respectively. The number moles of residue is

A. 73.53

B. 48.02

C. 40

D. 30.24

Sol. **B.** for batch distillation

$$\text{In}\left(\frac{L_1}{L_2}\right) = \int_{x_2}^{x} \frac{dx}{y^*-x}$$

$$\text{or} \quad \text{In}\left(\frac{L_1}{L_2}\right) = \int_{X_2}^{X_1} \frac{dx}{(0.7353x+0.3088-x)}$$

$$\text{In}\left(\frac{L_1}{L_2}\right) = \int_{x_2}^{x_1} \frac{dx}{(0.3088-0.2647\,x-x)}$$

$$\text{In}\left(\frac{L_1}{L_2}\right) = \frac{1}{0.2647}\cdot \text{In}\left(\frac{0.308-0.2647\,x_1}{0.3088-0.2647\,x_2}\right)$$

Putting the data given

$$L_1 = 100 \text{ mol}$$

$$x_1 = 0.7, \quad x_2 = 0.6$$

$$\text{In}\left(\frac{100}{L_2}\right) = (0.37778)\ \text{In}\left(\frac{0.12351}{0.14998}\right)$$

$$\therefore \quad L_2 = 48.02 \text{ mole}$$

86. A packed tower containing Berl saddles is operated with a gas-liquid system in the countercurrent mode. Keeping the gas flow rate constant, if the liquid flow rate is continuously increased

A. the void fraction available for the gas to flow will decrease beyond the loading point

B. the gas pressure drop will decrease

C. liquid will continue to flow freely down the tower beyond the loading point

D. the entertainment of liquid in the gas will considerably decrease near the flooding point

Sol. **A.** if the liquid flow rate is continuously increased with constant gas flowrate, the void fraction available for the gas to flow will decrease beyond the loading point.

87. A sparingly soluble solute in the form of a circular disk is dissolved in an organic solvent as shown in the figure. The area available for mass transfer from the disk is A and the volume of the initially pure organic solvent is V. The disk is rotated along the horizontal plane at a fixed rpm to produce a uniform concentration of the dissolving solute in the liquid. The convective mass transfer coefficient under these conditions is K_c. The equilibrium concentration of the solute in the solvent is C*. The time required for the concentration to reach 1% of the saturation value is given by

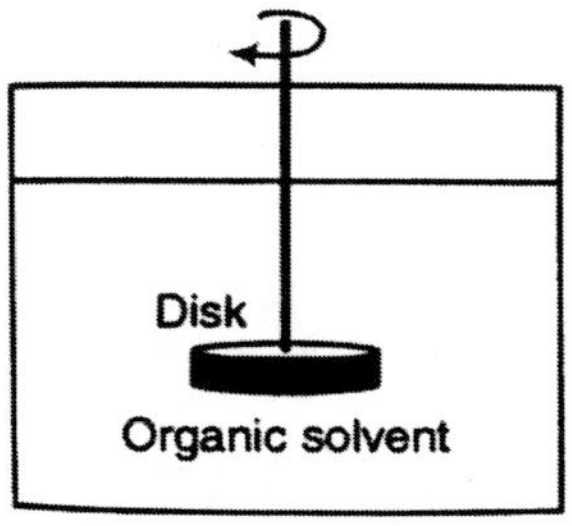

A. $\exp\left(-\frac{k_c A}{V}t\right) = 0.99$

B. $\exp\left(-\frac{k_c A}{V}t\right) = 0.01$

C. $\frac{V}{Ak_c}\exp(-0.99) = t$

D. $\frac{V}{Ak_c}\exp(0.01) = t$

Sol. **A.** A is the area of mass transfer and V is the volume of solvent.

Writing equation for solute balance

$$V\frac{dC}{dt} = k_C A(C^* - C)$$

Where C is the concentration at time t

$$\int \frac{dC}{C^*-C} = \int \frac{k_C A}{V}\,dt$$

or $\quad -\ln(C^* - C) = \frac{k_C A}{V}t + C_2$

Boundary condition is

$$C = 0 \text{ at } t = 0$$

$$-\ln C^* = 0 + C_2$$

Putting the value of the constant C_2

$$-\ln(C^* - C) = \left[\frac{k_C A}{V}\right]t - \ln C^*$$

or $\quad \frac{k_C A}{V}t = -\ln\left(\frac{C^*-C}{C^*}\right)$

Since, $\quad C = 0.01\,C^*$

Hence,

$$\left[\frac{k_C A}{V}\right]t = -\ln\left(\frac{C^*-0.01C^*}{C^*}\right)$$

$$\frac{k_C A}{V}t = -\ln(0.99)$$

$$\therefore \exp\left(-\frac{k_C A}{V}t\right) = 0.99$$

88. Air concentrated with solute P is brought in contact with water. At steady state, the bulk concentration of P in air and water are 0.3 and 0.02 respectively. The equilibrium equation relating the interface compositions is $Y_{p,\,i} = 0.25\,X_{p,\,i}$, I
Assume that the mass transfer coefficients F_G and F_L are identical. The gas phase mole fraction of P at the interface $(Y_{p,i})$ is

A. 0.0663
B. 0.075
C. 0.16
D. 0.3

Sol. **A.** data given $\quad y = 0.3;\ x = 0.02$

And $\quad F_L = F_C$

Writing

$$F_L(x_{p,i} - x) = F_C(y_{p,i} + y) \qquad \text{....(i)}$$

$$x_{p,i} - 0.02 = -y_{p,i} + 0.3$$

Equilibrium relationship given is

$$\therefore\ y_{p,i} = (0.25)\,x_{p,\,i} \qquad \text{....(ii)}$$

Combining equation (i) and (ii)

$$\frac{y_{p,i}}{0.25} + y_{p,i} = 0.32$$

or $\quad y_{p,i} = 0.064$

89. A feed (F) containing a solute is contacted with a solvent (S) in an ideal stage as shown in the figure below. Only the solute transfers into the solvent. The flow rates of all the streams are shown on a solute free basis and indicated by the subscript S. The compositions of the streams are expressed on a mole ratio basis. The extract leaving the contactor is divided into two equal parts, one part collected as the product (P) and the other stream is recycled to join the solvent. The equilibrium relationship is Y* = 2X.

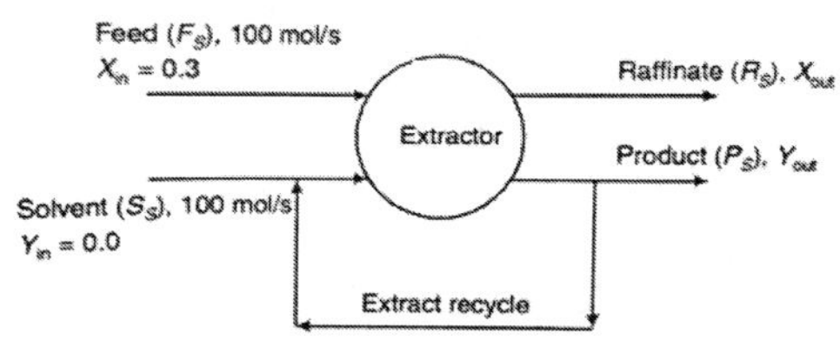

The product flow rate (P_s) and composition (Y_{out}) are

A. P_S = 50 mol/s, Y_{out} = 0.3

B. P_S = 100 mol/s, Y_{out} = 0.2

C. P_S = 200 mol/s, Y_{out} = 0.1

D. P_S = 100 mol/s, Y_{out} = 0.4

Sol. **B.**

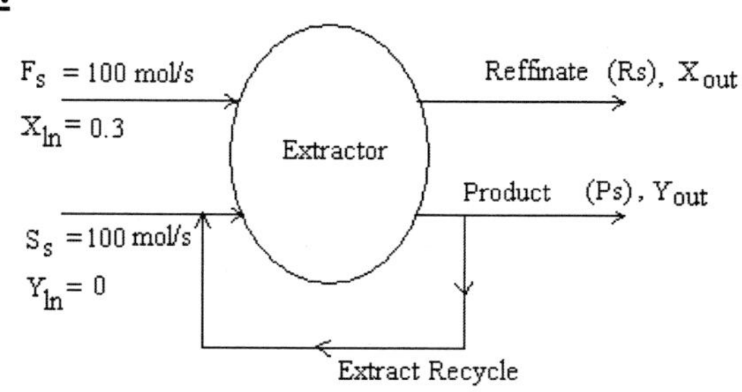

Writing the overall material balance equation

$$R_S + P_S = F_S + S_S$$

$$R_S + P_S = 200$$

$\therefore$ $R_S = P_S = 100$ mol/s

Now writing solute balance equation

$R_S \cdot X_{out} + P_S \cdot Y_{out} = F_S \cdot X_{in} + S_S \cdot Y_{in}$

Data given $Y_{out} = 2\, X_{out}$

Putting the values in solute balance equation

$(100)(X_{out}) + (100)(2\, X_{out})$

$= (100)(0.3) + (100)\,(0)$

$\therefore$ $X_{out} = 0.1$

Statements for Linked Answer Questions 90 and 91

A thin liquid film flows at steady state along a vertical surface as shown in the figure. The average velocity of the liquid film is 0.05 m/s. The viscosity of the liquid is 1 CP and its density is 1000 kg/m^3. The initially pure liquid absorbs a sparingly soluble gas from air as it flows down. The length of the wall is 2 m and its width is 0.5 m. The solubility of the gas in the liquid is 3.4×10^{-2} kmol/m^3 and isothermal condition may be assumed.

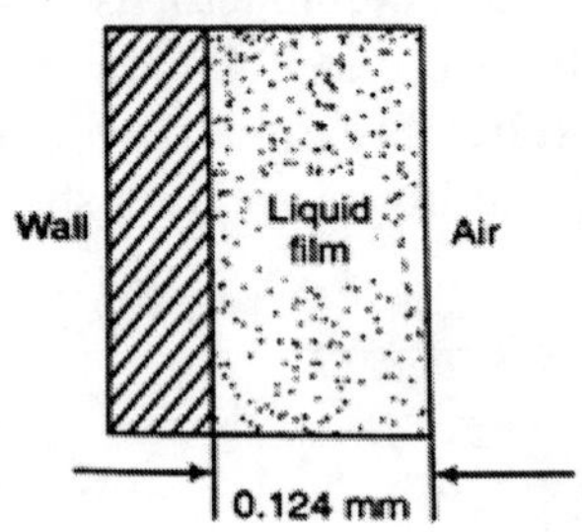

90. If the exit average concentration in the liquid is measured to be 1.4×10^{-2}kmol/m^3, the total mass transfer rate (in kmol/s) of the sparingly soluble gas into the liquid is

<u>A.</u> 0.133×10^{-4}

<u>B.</u> 0.434×10^{-7}

<u>C.</u> 3.4×10^{-2}

<u>D.</u> 17×10^{-2}

Sol. **<u>B.</u>** data given,

$\therefore\ \delta = 0.124$ mm1.24×10^{-4} m

v = 0.05 m/s; L = 2 m

Consider a differential elemental section of thickness Δx

$N_A = \text{v}\delta\text{b}\, dC_A = k_L \cdot \text{b.dx}\,(C_A^* - C_A)$

or $\quad \dfrac{dC_A}{dx} = \dfrac{k_L}{V\delta}(C_A^* - C_A)$

or $\quad \int \dfrac{dC_A}{(C_A^* - C_A)} = \int \dfrac{k_L}{V\delta}\, dx$

$$-\text{In}\,(C_A^* - C_A) = \frac{k_L\, x}{V\delta} + C_1 \quad \text{....(i)}$$

Applying boundary conditions

At x $= 0$, $C_A = 0$

$-\text{In}C_A^* = C_1$

Hence, putting the value of C_1

$$\frac{k_L\, x}{V\delta} = -\text{In}\left(\frac{C_A^* - C_A}{C_A^*}\right) \quad \text{....(ii)}$$

Given, $C_A = 1.4 \times 10^{-2}$ kmol / m^3

$C_A^* = 3.4 \times 10^{-2}$ kmol / m^3, x = L = 2 m

Substituting the values

$$k_L = \frac{(0.05)(1.24\times10^{-4})}{2} \times \left[In \frac{3.4\times10^{-2} - 1.4\times10^{-2}}{3.4\times10^{-2}}\right]$$

or $k_L = 1.65 \times 10^{-6}$ m/s

$\therefore$ Total mass transfer rate

$$N_A = k_L\, \text{A}\,(C_{Ai} - C_{Ao})_m \quad \text{....(iii)}$$

Since,

$$(C_{Ai} - C_{Ao})_m = \frac{(3.4\times10^{-2} - 0) - (3.4\times10^{-2} - 1.4\times10^{-2})}{\left(In \frac{3.4\times10^{-2} - 0}{3.4\times10^{-2} - 1.4\times10^{-2}}\right)}$$

$= 2.64\times 10^{-2}$ kmol / m^3

Substituting in equation (iii)

$N_A = (1.65\, \times 10^{-6})(2 \times 0.5)(2.64 \times 10^{-2})$

$= 0.434\times 10^{-7}$ kmol/sec.

91. The mass transfer coefficient k_c,av (in m/s), averaged along the length of the vertical surface is

<u>A.</u> 2.94×10^{-6}

<u>B.</u> 2.27×10^{-6}

<u>C.</u> 1.94×10^{-6}

<u>D.</u> 1.65×10^{-6}

Sol. **<u>D.</u>** the average mass transfer coefficient has already been calculate in the previous problem (Q. 90) equation (ii)

$$k_{C,}\, av = 1.65 \times 10^{-6}.$$

GATE 2007

92. The following figure depicts steady one–dimensional diffusion of water vapour from the surface of water taken in a conical flask at room temperature. Derive the governing equation for determining the concentration profile of water vapour in the gas medium. Neglect change of level of water due to condensation. The temperatures of the gas and the liquid media are identical and constant.

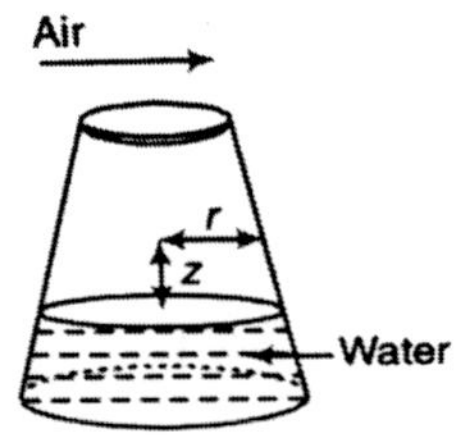

A. $\frac{d}{dz}\left[r^2 \frac{CD_{WA}}{1-X_W}\frac{dx_W}{dz}\right]=0$

B. $\frac{d}{dz}\left[\frac{CD_{WA}}{1-X_W}\frac{dx_W}{dz}\right]=0$

C. $\frac{d}{dz}\left[r \frac{CD_{WA}}{1-X_W}\frac{dx_W}{dz}\right]=0$

D. $\frac{d}{dz}\left[r\, CD_{WA}\frac{dx_W}{dz}\right]=0$

Sol. **A.** since $N_A = -D_{AB}\frac{dC_A}{dz}$

$$N_A = -D_{WA}\frac{d}{dz}[C_{A0}(1-x_W)]$$

$$= +(D_{WA})(C_{A0})\frac{dX_W}{dz}$$

Now, since $C_A = C_{A0}(1-x_A)$

$$N_A = -(D_{WA})\left(\frac{C}{1-X_W}\frac{dX_W}{dz}\right) \quad \ldots.(i)$$

Now, consider an elemental section at r having elemental thickness dr.

$$\frac{\pi(r+dr)^2 N_A|_z - \pi r^2 N_A|_{z+dz}}{dz} = 0$$

or $\frac{d}{dz}(r^2 N_A)=0 \quad \ldots.(ii)$

Combining equations (i) and (ii)

$$\frac{d}{dz}\left[r^2 \frac{CD_{WA}\cdot dX_W}{1-X_W \quad dz}\right]=0$$

93. In a distillation operation, it is desired to have a very high purity bottom product. Initially, a kettle-type reboiler is used at the bottom of the column and the following analytical equation is used to obtain the equilibrium trays in the exhausting section of the column

$$N_p - m + 1 = \frac{\log\left[\frac{x_m-(x_W/\alpha)}{x_W-(x_W/\alpha)}(1-\bar{A})+\bar{A}\right]}{\log(1/\bar{A})}$$

Where, x_m is the composition of the liquid leaving tray m. Tray m is the last equilibrium tray obtained by a McCabe-Thiele graph of the exhausting section. If the Kettle-type reboiler, is replaced by a thermo-syphon reboiler, the analytical equation for the exhausting section will be

A. $N_p - m + 1 = \frac{\log\left[\frac{x_m-(x_W/\alpha)}{x_W-(x_W/\alpha)}(1-\bar{A})+\bar{A}\right]}{\log(1/\bar{A})}$

B. $N_p + m + 1 = \frac{\log\left[\frac{x_m-(x_W/\alpha)}{x_W-(x_W/\alpha)}(1-\bar{A})+\bar{A}\right]}{\log(1/\bar{A})}$

C. $N_p - m = \frac{\log\left[\frac{x_m-(x_W/\alpha)}{x_W-(x_W/\alpha)}(1-\bar{A})+\bar{A}\right]}{\log(1/\bar{A})}$

D. $N_p - m + 2 = \frac{\log\left[\frac{x_m-(x_W/\alpha)}{x_W-(x_W/\alpha)}(1-\bar{A})+\bar{A}\right]}{\log(1/\bar{A})}$

Sol. **C.** in Kettle – type reboiler, the reboiler is considered to be equivalent to a tray so that the total number of trays is $N_p - m + 1$. However, in thermo – siphon reboiler, the number of trays will be $N_p - m$ only.

Hence, $N_p - m = \frac{\log\left[\frac{X_m-(X_W/\alpha)}{X_W-(X_W/\alpha)}(1-\bar{A})+\bar{A}\right]}{\log(1/\bar{A})}$

94. A 50 cm × 50 cm × 1 cm flat wet sheet weighing 2 kg, initially was dried from both the sides under constant drying rate period. It took 1000 s for the weight of the sheet to reduce to 1.75 kg. Another 1m × 1m ×1cm flat sheet is to be dried from one side only. Under the same drying rate and other conditions, time required for drying (in sec) from initial weight of 4 kg to 3 kg is

A. 1000

B. 1500

C. 2000

D. 2500

Sol. **C.** Case 1 Given,

$A = 2\times 50\text{ cm}\times 50\text{ cm} = 0.5\, m^2$

$X_1 - X_2 = 0.25/L_S$

$t = 1000$ s

Time required for drying,

$$t = \frac{L_S}{AR_C}(X_1 - X_2)$$

$$1000 = \frac{L_S}{0.5\times R_C}\times\frac{0.25}{L_S}$$

$$R_C = \frac{1}{2000}$$

Case 2

$$X_1 - X_2 = \frac{1}{L_S}$$

Now, $t = \frac{L_S}{AR_C}(X_1 - X_2)$

$$t = \frac{L_S}{1}\times 2000\times\frac{1}{L_S}$$

t = 2000 sec.

95. It is desired to reduce the concentration of pyridine in 500 kg of aqueous solution from 20 wt % to 5 wt % in a single batch extraction using chlorobenzene as solvent. Equilibrium compositions (end points of the tie line) in terms of weight per cent of pyridine-water-chlorobenzene are (5, 95, 0) and (11, 0 ,89).
The amount of pure solvent required in kg for the operation is

A. 607
B. 639
C. 931
D. 1501

***Sol.* B.**

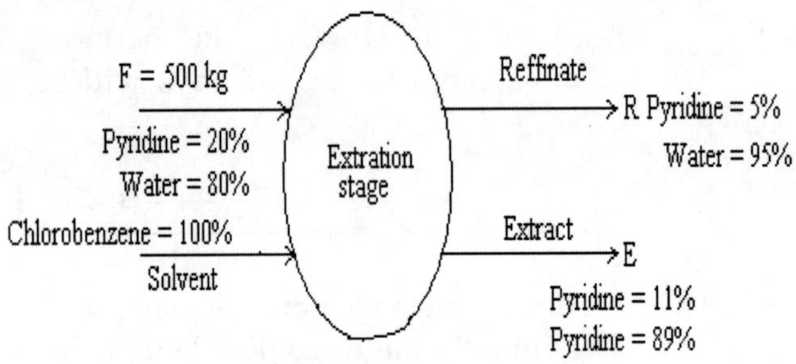

Writing the overall material balance

F + S = R + E

500 + S = R + E ....(i)

Writing water balance

(F) 0.8 = (R) 0.95

Since F = 500 kg

∴ R = 421 kg

Now writing pyridine balance

(F) 0.2 = (R) 0.05 + (E) 0.11

or 500 × 0.2 = 4210.5 + (E) 0.11

E = 718 kg

Substituting in equation (i)

S = R + E – 500

= 421 + 718 – 500

S = 639 kg

96. Benzene in an air-benzene mixture is to be reduced from 5.2 mol% in the feed to 0.5 mol% by contacting with wash oil in multistage countercurrent gas absorber. The inlet flow rate of air benzene mixture is 10 mol/s while benzene free wash oil comes in at 9.5mol/s. If the equilibrium mole ratios of benzene in air and benzene in oil, the number of equilibrium stages required to achieve the above separation will be

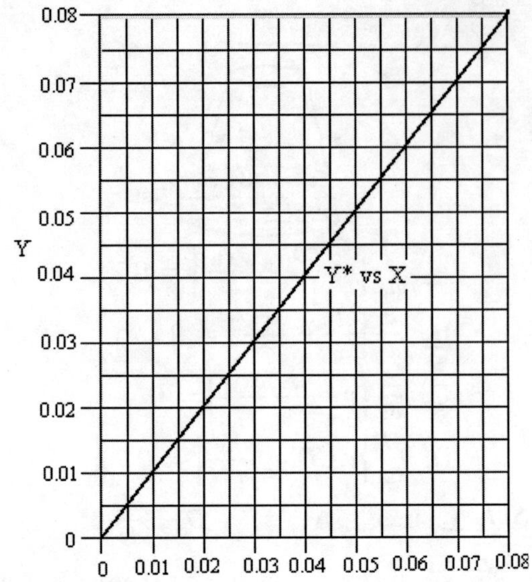

A. 12
B. 10
C. 8
D. 6

***Sol.* B.** data given

$V_{N+1} = 10$ mol/s

$y_{N+1} = 0.052$, $y_1 = 0.005$

$L_0 = 9.5$ mol/sec., $x_0 = 0$

Since $V_1 = V_{N+1} \quad (1 - y_{N+1})$

$= 10\ (1 - 0.052) = 9.48$ mol/sec

$V_1 = L_0(1 - X_0) = 9.5$ mol/sec.

Slope (m) $= \frac{9.5}{9.48} = 1$

Number of stages required

$$N = \frac{y_{N+1} - y_1}{y_1 - m\ x_0}$$

$$= \frac{0.052 - 0.005}{0.005 - 1 \times 0}$$

$N = 10$

97. Given the following statements listed from (P) to (T) select the correct combination of true statements from the choices that follow this list.

(P) Plate columns are preferred when the operation involves liquids containing suspended solids.

(Q) Packed towers are preferred if the liquids have a large foaming tendency.

(R) The pressure drop through packed towers is more than the pressure drop through plate columns designed for the same duty.

(S) Packed columns are preferred when large temperature changes are involved in distillation operations.

(T) Packed towers are cheaper than plate towers if highly corrosive fluids are handled

A. T, S, P

B. P, Q, T

C. S, R, T

D. R, Q, S

Sol. B. P, Q and T are true statement.

Statements for Linked Answer Questions 98 and 99

A simplified flowsheet is shown in the figure for production of ethanol from ethylene. The conversion of ethylene in the reactor completely separates ethylene (as top stream) and ethanol and water as bottoms. The last (distillation) column gives an ethanol-water azeotrope (90 mol% ethanol) as the final product and water as waste. The recycle to purge ratio is 34.

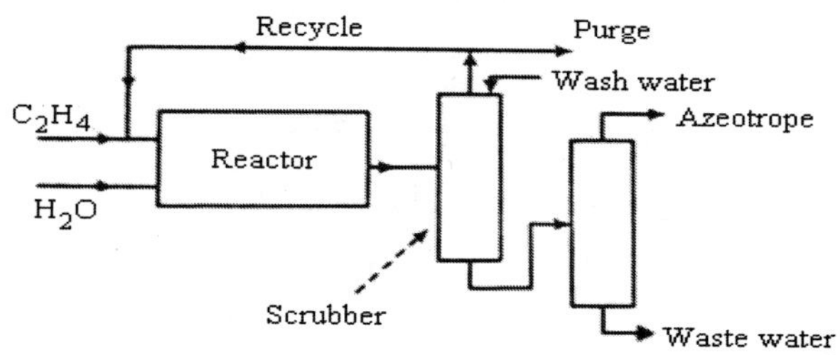

The reaction is $C_2H_4(g) + H_2O\,(g) \rightarrow C_2H_5\,OH\,(g)$

98. For an azeotropeproduct rate of 500 mol/h, the recycle gas flow rate in mol/h is

A. 30

B. 420

C. 1020

D. 1500

Sol. C.

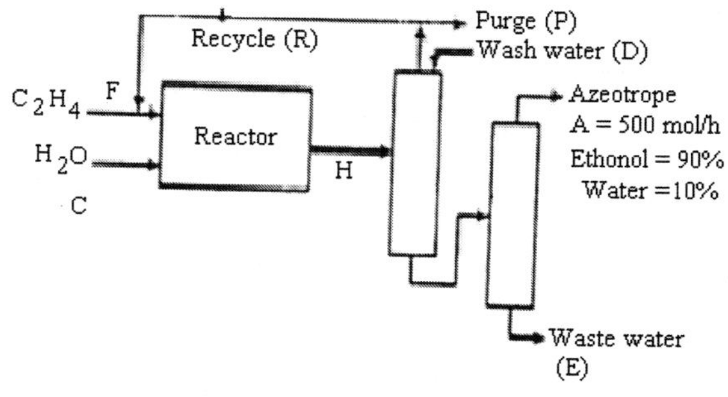

Azeotrope: A = 500 mol/hr

Ethanol = 90% = 450 mol/h

Conversion in the reactor = 30%

$\therefore \quad F = \frac{450}{0.3} = 1500$ mole/hr

$\therefore \quad P + R = 1500 - 450 = 1050$

Given $\frac{R}{P} = 34$ or $R = (34)\,P$

$\therefore \quad R = 34\,(1050 - R)$

Hence, $R = 1020$ mol/h

99. For the same process, if fresh H_2O feed to the reactor is 600 mol/h and wash water for scrubbing is 20% of the condensable coming out of the reactor, the water flow rate in mol/h from the distillation column as bottom is

A. 170

B. 220

C. 270

D. 430

Sol. B. given C = 600 mol/hr

For producing 450 mols of ethanol, 450 mols of water should reac with C_2H_4 in the reactor

Total condensable gases coming out of the reactor

= Ethanol + Unreacted water

= 450 + (600 – 450)

= 600 mol/hr

Wash water required for scrubbing

D = 0.2 × 600 = 120 mol/hr

Water in azeotrope (A) = 0.1 × 500 = 50 mol/hr

Now writing water balance around distillation column,

Water in H + Wash water (D) = Water is A + Waste water (E)

or 150 + 120 = 50 + E

E = 220 mols/hr.

100. A fluid obeying the constitutive equation

$\tau = \tau_0 + K\left(\frac{dv_x}{dy}\right)^{\frac{1}{2}}$; $\tau > \tau_0$ is held between two parallel plates at a distance d part. If the stress applied to the top plate is 3 τ_0, then the velocity with which the top plate moves the relative to the bottom plate would be

A. $2\left(\frac{\tau_0}{K}\right)^2$ d

B. $3\left(\frac{\tau_0}{K}\right)^2$ d

C. $4\left(\frac{\tau_0}{K}\right)^2$ d

D. $9\left(\frac{\tau_0}{K}\right)^2$ d

Sol. C. Given $\tau = \tau_0 + K\left(\frac{dV_x}{dy}\right)^{1/2}$

When $\tau > \tau_0$ for the top plate

$\tau = 3\tau_0$

Hence,

$$3\tau_0 = \tau_0 + K\left(\frac{dV_x}{dy}\right)^{1/2}$$

or $$2\tau_0 = \left(\frac{dV_x}{dy}\right)^{1/2}$$

$$\frac{dV_x}{dy} = \left(\frac{2\tau_0}{K}\right)^2$$

Now, after integration

$$V_X = 4\left(\frac{\tau_0}{K}\right)^2 y + Constt.$$

Boundary condition is

At y = 0, V_X = 0 at (bottom plate)

∴ 0 = 0 + Constt.

Constt. = 0

Hence, $V_X = 4\left(\frac{\tau_0}{K}\right)^2 Y$

∴ the velocity of top plate relative to bottom plate (y = d),

$$V = 4\left(\frac{\tau_0}{K}\right)^2 d$$

101. Experiments conducted with a sparingly dissolving cylinder wall in a following liquid yielded the following correlation for the Sherwood number Sh = 0.023 $(Re)^{0.83}(Sc)^{1/3}$. Assuming the applicability of the Chilton-Colburm analogy, the corresponding correlation for heat transfer is

A. St = 0.023 $(Gr)^{0.83}(Pr)^{1/3}$

B. Nu = 0.023 $(Re)^{0.83}(Pr)^{1/3}$

C. Jh = 0.023 $(Re)^{0.83}(Pr)^{2/3}$

D. Nu = 0.069 $(We)^{0.83}(Pr)^{\frac{4}{3}}$

***Sol.* B.** given Sh = 0.023 $(Re)^{0.8}(Sc)^{1/3}$

According to Chilton-Colburn analogy, the corresponding correlation for heat transfer is

$$N_u = 0.023\ (Re)^{0.8}(Sc)^{1/3}$$

Known as Dittus-Boelter equation

102. In a multistage countercurrent isothermal stripping column, the feed containing 0.05 mol of solute/mol of solute free oil is treated with steam. The absorption factor A = 0.65. The equilibrium relation is given by

Y* = 2X, where Y* and X refer to the equilibrium mole ratio in the steam and oil phases respectively. The Kremser equation is given as follows ('0' refer to liquid inlet at the top, N_p refers to the last stage at the bottom).

$$N_p = \frac{\log\left[\left[\frac{X_0 - \frac{Y_{N_p+1}}{m}}{X_{N_p} - \frac{Y_{N_p+1}}{m}}\right](1-A) + A\right]}{\log\left(\frac{1}{A}\right)}$$

If the steam is initially free of solute and its exit mole ratio (mol solute/mol steam) is 0.0624, then the number of equilibrium stage required is

A. 4.2

B. 5.2

C. 7.2

D. 8.2

***Sol.* B.** data given is

$$X_0 = 0.05, \quad Y_{N_P+1} = 0$$

$$Y_1 = 0.0624$$

Equilibrium relationship, Y* = 2X, m = 2

Absorption factor,

$$A = \frac{1}{mv} = 0.65$$

$$\therefore \quad \frac{L}{V} = 2\times 0.65 = 1.3$$

Writing the solute balance

$$L(X_0 - X_{N_P}) = V(Y_1 - Y_{N_P+1})$$

or $$\frac{L}{V}(0.05 - X_{N_P}) = 0.0624 - 0$$

or $$1.3(0.05 - X_{N_P}) = 0.0624$$

$$\therefore \quad X_{N_P} = 0.002$$

Given Kremser equation is

$$N_P = \frac{\left\{\left[\frac{\left(X_0 - \frac{Y_{N_P+1}}{m}\right)}{\left(X_{N_P} - \frac{Y_{N_P+1}}{m}\right)}\right](1-A)+A\right\}}{\log(1/A)}$$

Putting the values

$$N_P = \frac{\log\left(\frac{0.05-0}{0.002-0}\right)(1-0.65)+0.65}{\log(1/0.65)}$$

$$N_P = 5.2$$

103. 100 mol of a binary mixture F containing 60 mol% A (more volatile) and 40 mol% B is treated in a batch distillation still. After 1 h, 70 mol of the distillate D are collected leaving behind the residue W. Relative volatility α is 2. The governing equation is

$$\log\left(\frac{Fx_F}{Wx_W}\right) = \alpha \log\left[\frac{F(1-x_F)}{W(1-x_w)}\right]$$

The average mole fraction of A in the distillate is

A. 0.43

B. 0.61

C. 0.69

D. 0.73

***Sol.* C.** data given:

$$F = 100 \text{ mole}, \quad X_F = 0.6$$
$$D = 70 \text{ mole}, \quad \propto = 2$$

Writing overall material balance

$$F = D + W$$

or $$100 = 70 + W$$

$\therefore$ $$W = 30 \text{ mole}$$

The governing equation is given as

$$\text{Log} \left(\frac{FX_F}{WX_W}\right) = \propto \log\left[\frac{F(1-X_F)}{W(1-X_W)}\right]$$

$$\text{or } \log\left(\frac{(100)\times 0.6}{30\times X_W}\right) = 2\log\left[\frac{100(1-0.6)}{30(1-X_W)}\right]$$

$$\text{or } \log\left(\frac{2}{X_W}\right) = 2\log\left[\frac{4}{3(1-X_W)}\right]$$

$$\text{or } \log\frac{2}{X_W} = \left[\frac{4}{3(1-X_W)}\right]^2$$

$$2\times 9(1-X_W)^2 = 16\,X_W$$

$$9\,X_W^2 - 26\,X_W + 9 = 0$$

Now, writing material balance for A

$$FX_F = DX_D + WX_W$$

$$(100)\times 0.6 = 70\times X_D + 30\times 0.4$$

$$X_D = 0.69$$

104. A drug tablet of mass M_0 administered orally at time $t = 0$ reaches the intestine at time $t = \tau$ without losing any mass. From the intestine, the drug is absorbed into blood. The rate of absorption is found to be proportional to mass of the drug in the intestine with the proportionality constant k. Assuming no drug is lost from the blood, the total mass of the drug in the blood M_b, at a time $t \geq \tau$ is given by

A. $M_b = M_0[1 - exp\{-k(t-\tau)\}]$

B. $M_b = M_0[1 - exp\{-k\tau\}]$

C. $M_b = M_0[exp\{-k(t-\tau)\}]$

D. $M_b = M_0[1 - exp\{-k(t+\tau)\}]$

***Sol.* C.** for $t \leq \tau$, the mass of drug in the blood $M_B = M_0$

For $t > \tau$, Rate $\propto$ K,

Equivalent to first order reaction

$\therefore$ $M_B = M_0 e^{-K(t-\tau)}$

We can verify this relationship

since when $t = \tau$, $M_B = M_0$

Hence, for $t \geq \tau$

$$M_B = M_0 \exp\left[-k(t-\tau)\right]$$

105. A saturated vapour is fed to a distillation column at 180 kmol/h. Both the rectifying and stripping sections of the column operate at 60% of their respective flooding velocities. The flooding velocity of the rectifying section is twice that of the stripping section. The assumptions of constant molar vapour density throughout the column are valid. If the boil-up rate is 60 kmol/h then the relationship between the diameter of the rectifying section (d_r) and the stripping section (d_s) is

A. $d_r = \sqrt{2}d_s$

B. $d_r = \sqrt{3}d_s$

C. $d_r = 2\,d_s$

D. $d_r = 3\,d_s$

***Sol.* A.** for stripping section

$$\text{Feeding rate} = \left(\frac{Density}{Mol.\ Wt}\right)(\text{Area})(\text{Velocity})$$

$$\text{or} \quad 60 = \left(\frac{\rho}{M}\right)\left(\frac{\pi d_S^2}{4}\right)(0.6u_S) \quad \ldots\text{(i)}$$

Total vapour load in stripping section

$$\text{Rate} = 180 + 60 = 240$$

$$\therefore \quad 240 = \left(\frac{\rho}{M}\right)\left(\frac{\pi d_r^2}{4}\right)\times(0.6u_d) \quad \ldots\text{(ii)}$$

Combining equations (i) and (ii)

$$4 = \left(\frac{dr}{ds}\right)^2\left(\frac{u_d}{u_S}\right)$$

$$\text{Given} \quad u_d = 2\,u_S$$

$$\therefore \quad 4 = \left(\frac{dr}{ds}\right)^2(2)$$

$$\therefore \quad dr = \sqrt{2}ds$$

Common Data for Questions 106, 107 and 108

A binary distillation column separates 100 mol/h of a feed mixture into distillate D and residue W. The McCabe-Thiele diagram for this process is given below. The relative volatility for the binary system is constant at 2.4.

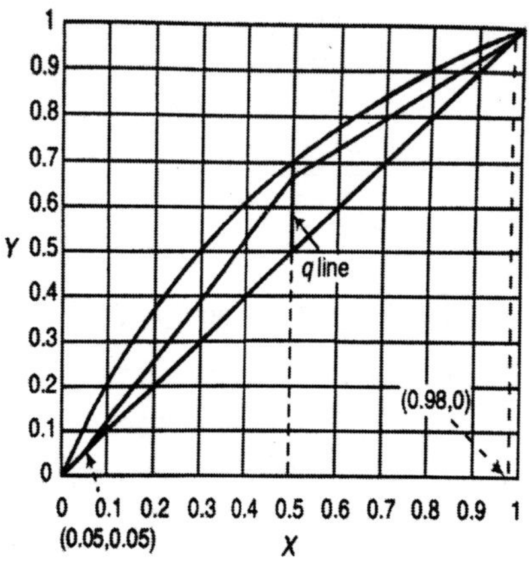

106. The distillate and residue flow rates (in mol/h) are

A. $D = 48.4$ $W = 51.6$

B. $D = 51.6$, $W = 48.4$

C. $D = 54.7$, $W = 45.3$

D. $D = 45.3$, $W = 54.7$

Sol. **A.** referring to the given figure

$$x_F = 0.5, x_D = 0.98, x_W = 0.05$$

Given, F = 100 mole/hr

Overall material balance

F = D + W

∴ 100 = D + W

Also, (100) (0.5) = (D) (0.98) + (100 – D) (0.05)

D = 48.4 mol/hr

And W = 100 – 48.4

= 51.6 mole/hr

107. The ratio of liquid to vapour molar flow rates in the rectifying section is

A. 0.64

B. 1.00

C. 1.55

D. 1.80

Sol. **A.** referring to the given figure. In rectifying section ratio of liquid to vapor molar flow rates in rectifying section

$$\frac{L}{V} = \text{Slope of upper line}$$

$$= \frac{0.98-.-0.68}{0.98-0.5}$$

$$\frac{L}{V} = 0.625$$

108. The minimum number of theoretical stages (inclusive of reboiler) for this process is

A. 5.2

B. 6.1

C. 7.8

D. infinite

Sol. **C.** the minimum number of theoretical stages,

$$N_m = \frac{\log\left[\left(\frac{x_D}{1-x_D}\right)\left(\frac{1-x_W}{x_W}\right)\right]}{\log(\propto)}$$

$$= \frac{\log\left[\left(\frac{0.98}{1-0.98}\right)\left(\frac{1-0.05}{0.05}\right)\right]}{\log(2.4)}$$

$$N_m = 7.8$$

Statements for Linked Answer Questions 109 and 110

Solvent C is used to extract solute B selectively from 100 kg/h feed mixture A + B in a steady state continuous process shown below. The solubility of C in the raffinate and the solubility of A in the extract are negligible. The extract is distilled to recover B in the bottom product. The overhead product is recycled to the extractor. The loss of solvent in the bottom in compensated by make-up solvent S_d. The total flow rate of the solvent stream S going to the extractor is 50 kg/h. The mass fraction ($X_i's$) of some selected stream are indicated in the figure below.

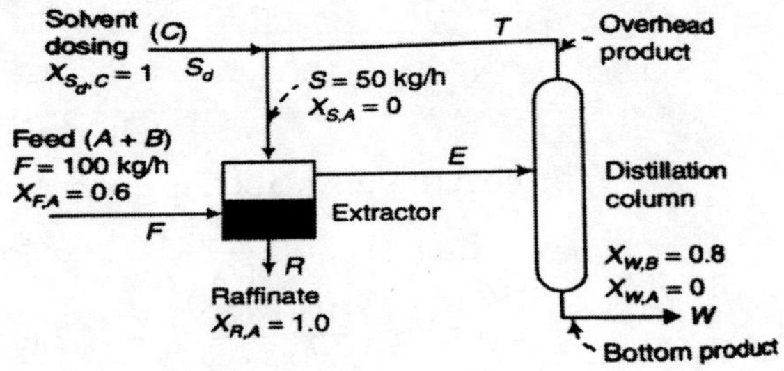

109. Distillation bottom flow rate W and solvent do rate S_d in kg/h are

A. $W = 50, S_d = 50$

B. $W = 100, S_d = 20$

C. $W = 10, S_d = 50$

D. $W = 50, S_d = 10$

Sol. **D.**

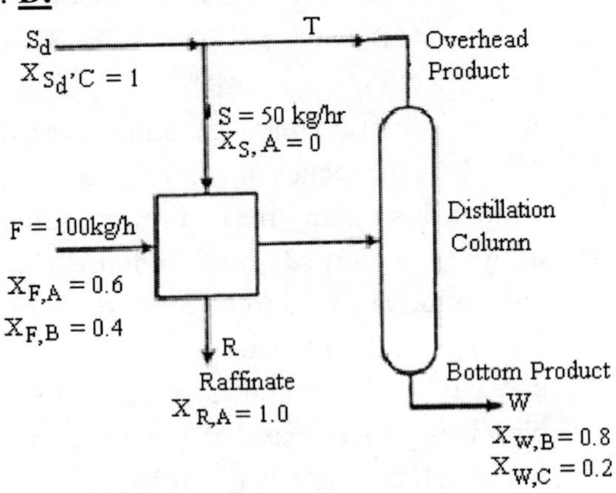

Referring to the figure

Writing the overall material balance

S_d + F = R + W

S_d + 100 = R + W

Writing material balance for A

(F) 0.06 = (R) 1

∴ R = 60 kg/h

Now, writing the material balance for B

(F) 0.4 = (W) 0.8

(100) 0.4 = (W) 0.8

∴ W = 50 kg/h

Again, writing material balance for C

$$(S_d)\,1 = (W)\,0.2$$

$$\therefore\ S_d = (50)\,0.2$$

$$\therefore\ S_d = 10 \text{ kg/hr}$$

110. Feed rate E to the distillation column and overhead product rate *T* in kg/h are

A. $E = 90, T = 40$

B. $E = 80, T = 20$

C. $E = 90, T = 50$

D. $E = 45, T = 20$

***Sol.* A.** If can be seen from the figure that

$$S_d + T = S$$
$$10 + T = 50$$
$$\therefore \quad T = 40 \text{ kg/hr}$$

Writing material balance around distillation column

$$E = T + W = 40 + 50$$
$$= 90 \text{ kg/hr}$$

Statements for Linked Answer questions 111 and 112

Consider the diffusion of a reactant A through a cylindrical catalyst pore of radius R and length *L>>R*. Reactant *A* undergoes a zero order reaction on the cylindrical surface of the pore. The following equation describes changes in the concentration of A within the pore due to the axial diffusion of A and the disappearance A due to reaction $\frac{d^2C_A}{dx^2} = k$, where C_A is the concentration of A at the distance x from the pore entrance and K is a constant.

111. If the concentration of A at the pore entrance (x=0) is C_{A0}, and x = L is dead and where no reaction occurs, the concentration profile of A in the pore is given by

A. $C_A(x) = \frac{Kx^2}{2} - KLx + C_{Ao}$

B. $C_A(x) = \frac{Kx(x-L)}{2} - C_{Ao}\frac{x}{L} + C_{A0}$

C. $C_A(x) = C_{Ao}$

D. $C_A(x) = C_{Ao}\left(\frac{L-x}{L}\right)$

***Sol.* B.** given $\frac{d^2C_A}{dx^2} = K$ ....(i)

Integrating ,

$$\frac{dC_A}{dx} = Kx + C_1 \quad \text{....(ii)}$$

C_1 is constant

Integrating again,

$$C_A = K\frac{x^2}{2} + C_1 x + C_2 \quad \text{....(iii)}$$

where C_2 is another constant

First boundary condition

When $x = 0,\ C_A = C_{A_0}$

Putting values in equation (iii)

$$C_{A_0} = 0 + 0 + C_2$$
$$\therefore \quad C_2 = C_{A_0} \quad \text{....(iv)}$$

Combining equations (iii) and (iv)

$$C_A = K\left(\frac{X^2}{2}\right) + C_1 x + C_{A_0} \quad \text{....(v)}$$

Now second boundary condition is

When $x = L,\ C_A = 0$

Putting values in equation (v)

$$0 = K\left(\frac{L^2}{2}\right) + C_1 L + C_{A_0}$$
$$\therefore \quad C_1 = -\frac{KL}{2} - \frac{C_{A0}}{L}$$

Now putting this value of C_1 in equation (v)

$$C_A = K\frac{X^2}{2} - \frac{KxL}{2} - \frac{C_{A0}\,x}{L} + C_{A_0}$$
$$C_A = \frac{Kx(x-L)}{2} - C_{A_0}\frac{x}{L} + C_{A_0}$$

112. The minimum pore length for A to be completely converted within the pore is

A. $\left(\frac{C_{A0}}{K}\right)^{1/2}$

B. $\left(\frac{C_{A0}}{2K}\right)^{1/2}$

C. $\left(\frac{2C_{A0}}{K}\right)$

D. $\left(\frac{2C_{A0}}{K}\right)^{1/2}$

***Sol.* D.** for minimum pore length,

$$\frac{dC_A}{dL} = 0$$
$$\text{or} \quad -\frac{Kx}{2} + C_{A_0}\left(\frac{x}{L^2}\right) = 0$$
$$\text{or} \quad \frac{K}{2} = \frac{C_{A0}}{L^2}$$
$$\therefore \quad L = \left[\frac{2C_{A0}}{K}\right]^{1/2}$$

Statements for Linked Answer Questions 113 and 114

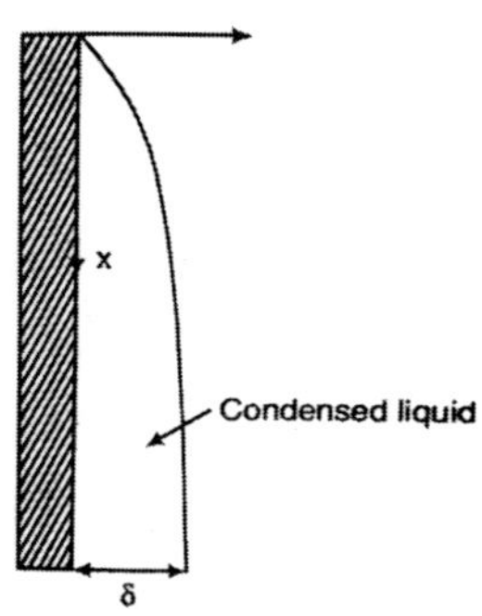

In film condensation on a vertical plane surface, the x directional velocity distribution is given by u(y) = $\frac{g(\rho_l-\rho_v)}{\mu_l}\left(\delta y-\frac{1}{2}y^2\right)$, where δ is the film thickness at any x.

113. The mass flow rate of the condensate m (x) through any axial position x per unit width of the plate is given by

A. $m(x) = \frac{g\,\rho_l(\rho_l-\rho_v)\delta^3}{3\mu_l}$

B. $m(x) = \frac{g\,(\rho_l-\rho_v)\delta^2}{3\mu_l}$

C. $m(x) = \frac{g\,\rho_v^2\delta^3}{\mu_l}$

D. $m(x) = \frac{g\,\rho_l\,\rho_v\,\delta^3}{3\mu_l}$

Sol. **A.** the x – directional velocity distribution is Given as

$$u(y) = \frac{g(\rho_l-\rho_v)}{\mu_l}\left[\delta y-\frac{y^2}{2}\right]$$

Hence, the mass flowrate of condensate per unit width of the plate is

$$\begin{aligned} m &= \int_0^\delta \rho_l\,u.dy \\ &= \int_0^\delta \rho_l\frac{g(\rho_l-\rho_v)}{\mu_l}\left[\delta y-\frac{y^2}{2}\right]dy \\ &= \frac{g\,\rho_l(\rho_l-\rho_v)}{\mu_l}\left[\delta\left(\frac{y^2}{2}\right)-\frac{1}{2}\left(\frac{y^3}{3}\right)\right]_0^\delta \\ &= \frac{g\,\rho_l(\rho_l-\rho_v)\delta^3}{3\mu_l} \end{aligned}$$

114. Differentiate m(x) with respect to δ to get the differential increases in condensate mass dm with film thickness ie, dm/dδ. Then obtain dm/dx assuming heat flux through the film to be due to conduction based on a linear temperature profiles between the vapour and wall. Hence, determine dδ/dx. Here μ_l is liquid viscosity, k_1 is thermal conductivity and λ is latent heat of condensation. T_v is the vapour temperature and T_w is the wall temperature.

A. $\frac{d\delta}{dx} = \frac{\mu_l\,K_l(T_v-T_W)}{g\,\rho_l(\rho_l-\rho_v)\,\lambda}\frac{1}{\delta^2}$

B. $\frac{d\delta}{dx} = \frac{\mu_l\,K_l(T_v-T_W)}{g\,\rho_v(\rho_l-\rho_v)\,\lambda}\frac{1}{\delta^3}$

C. $\frac{d\delta}{dx} = \frac{\mu_l(T_v-T_W)}{g\,\rho_l\,K_l(\rho_l-\rho_v)\lambda}\frac{1}{\delta^2}$

D. $\frac{d\delta}{dx} = \frac{\mu_l\,K_l(T_v-T_W)}{g\,\rho_l(\rho_l-\rho_v)\lambda}\frac{1}{\delta^3}$

Sol. **D.** it is already derived that

$$m = \frac{g\,\rho_l(\rho_l-\rho_v)\delta^3}{3\mu_l} \quad \text{..... (i)}$$

$$\therefore \frac{dm}{d\delta} = \frac{g\,\rho_l(\rho_l-\rho_v)\delta^2}{\mu_l} \quad \text{..... (ii)}$$

Writing energy balance equation

$$(dm)\lambda = K_l.dx.\frac{\Delta T}{\delta}$$

Putting the value of dm from equation (ii

$$\left[\frac{g\,\rho_l(\rho_l-\rho_v)\delta^2}{\mu_l}\right]d\delta.\lambda = K_l.dx.\frac{(T_v-T_v}{\delta}$$

$$\frac{d\delta}{dx} = \frac{\mu_l K_l(T_v-T_w)}{g\,\rho_l(\rho_l-\rho_v)\,\lambda\,\delta^3}$$

GATE 2005

115. In the triangular diagram represented below for a batch separation process, a stream F is mixed with a solvent B to produce R and E. Substance A is the carrier liquid and C is the solute to be extracted. The amounts of B and E are 1kg and 1.20 kg respectively. The length FM is 3.1 and length FB is 8.5 unit on the figure. The ratio R/E is estimated to be

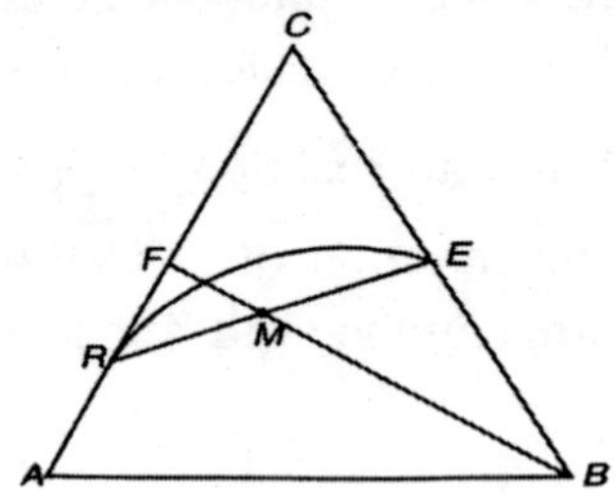

A. 1.285

B. 2

C. 0.751

D. 2.5

Sol. **A.** making a schematic block diagram

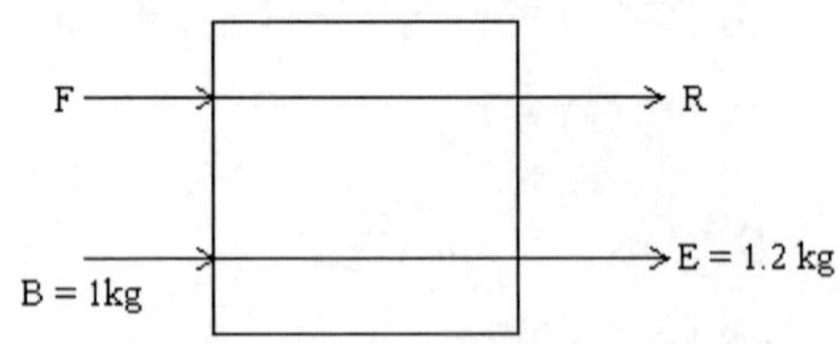

Referring to the triangular graph

$$\frac{F}{B} = \frac{Length\ Bm}{Length\ FM} = \frac{8.5-3.1}{3.1}$$

∴ F = 1.742 kg

Writing the material balance

F + B = R + E

1.742 + 1 = R + 1.2

∴ R = 1.542 kg

Again,

$$\frac{R}{E} = \frac{1.542}{1.2}$$

$$\therefore \quad \frac{R}{E} = 1.285$$

116. For a two phase feed, where 80% of the feed is vapourized under column conditions, the feed line slope in the McCabe-Thiele method for distillation column design, is

A. – 1/4
B. + 1/4
C. + 4
D. – 4

***Sol.* A.** slope of feed line, $m = \frac{q}{q-1}$

and $q = 1 - 0.8 = 0.2$

$\therefore \quad m = \frac{0.2}{0.2-1}$

or $m = -\frac{1}{4}$

117. Two solid discs of benzoic acid (molecular weight = 122) of equal dimensions are spinning separately in large volume of water and air at 300 K. the mass transfer coefficients for benzoic acid in water and air are 0.9×10^{-5} and 0.47×10^{-2} m/s respectively. The solubility of benzoic acid in water is 3 kg/m^3 and the equilibrium vapour pressure of benzoic acid in air is 0.04 kPa. Then the disc

A. dissolves faster in air than in water
B. dissolves faster in water than in air
C. dissolves at the same rate in both air and water
D. does not dissolve either in water on in air

***Sol.* B.** given,

$K_C = 0.9 \times 10^{-5}$ m/s, $C_S = 3$ kg/m^3

Mass transfer rate of disc in water

$$N_{Water} = K_C C_S = 0.95 \times 10^{-5} \times 3$$
$$= 2.7 \times 10^{-5} \text{ kg}/m^2\text{-sec}$$

Mass transfer rate of disc in air

$$N_{air} = K_p \frac{P_S M}{RT}$$

Given,

$K_p = 0.47 \times 10^{-2}$ m/s

$p_S = 0.04$ kPa ; M = 122, T = 300 K, R = 8.314

$$\therefore \quad N_{air} = (0.47 \times 10^{-2}) \frac{0.04 \times 122}{8.314 \times 300}$$
$$= 0.92 \times 10^{-5} \text{ kg}/m^2\text{-sec}$$

Comparing N_{Water} with N_{air}, it is evident that ($N_{Water} > N_{air}$) the disc will dissolve faster in water, than air.

118. Match the variation of mass transfer coefficient given by the theory in Group I with the appropriate variation in Group II.

Group I	*Group II*
P. Film theory	1. $\propto D_{AB}$
Q. Penetration theory	2. $\propto D_{AB}^{2/3}$
R. Boundary layer theory	3. $\propto D_{AB}^{1/2}$

A. P-1, Q-2, R-3
B. P-2, Q-1, R-3
C. P-1, Q-3, R-2
D. P-3, Q-2, R-1

***Sol.* C.** Film theory $K \propto D_{AB}$

Penetration theory : $K \propto D_{AB}^{\frac{1}{2}}$

Boundary layer theory: $K \propto D_{AB}^{2/3}$

119. A filter cake is dried with air at wet and dry bulb temperatures of 300 k and 323 k respectively. The heat transfer coefficient is 11W/m^2 –K and the latent heat of vaporization of water is 2500 kJ/kg. Mass transfer does not limit the process. Select the drying rate during constant rate period. Neglect conduction through the solid and radiation effects.

A. 1.32×10^{-2} kg/m^2-s
B. 0.71×10^{-2} kg/m^2-s
C. 4.53×10^{-2}kg/m^2-s
D. 0.10×10^{-3}kg/m^2-s

***Sol.* D.** for constant rate period,

$$R_C = \frac{h(T-T_W)}{\lambda_W}$$

Rate of drying

Given, h = 11 W/m^{2}K

T = 323 K, T_W = 300 K

λ_W = 2500 kJ/kg = 2500×10^3 J/kg

Hence, $R_C = \frac{11(323-300)}{2500 \times 10^3}$

$R_C = 0.1 \times 10^{-3}$ kg/m^2-s

120. A separation column for vapour-liquid contact processes 200 kmol/h of vapour. The flooding velocity is 3 m/s. If the column operates at 85% of flooding velocity and the down comer area is 10% of the total cross-sectional area, what is the

diameter of the column? Average density of vapour = 2 kg/m³ and its molecular weight = 44.

A. 0.82 m
B. 0.72 m
C. 0.78 m
D. 1 m

***Sol.* A.** mol. wt. of vapor =44, density of vapor

$= 2\ kg/m^2$

Vapor flow rate = 200 kmole/hr

$= (200)\left(\frac{44}{3600}\right)\left(\frac{1}{2}\right)$

$= 1.22 m^3/sec$

Given, flooding velocity = 3 m/sec

∴ Operating vapor velocity = 0.85 × 3

= 2.55 m/sec.

for vapor flow $= \frac{1.22}{2.55}$

Now, cross sectional area available

$= 0.48\ m^2$

Downcorner area occupied

= 10% of total area

Hence, total cross sectional area

$= \frac{0.48}{0.9} = 0.533$

If d is diameter of column

$\therefore \quad \frac{\pi d^2}{4} = 0.533$

$\therefore \quad d = 0.82\ m$

Statements for Linked Answer Questions 121 and 122

A binary gas mixture of a solute and a carrier gas is treated in a countercurrent gas absorption column, containing ideal trays, using a solvent. The compositions y and x (see figure below) are the mole fractions of the solute in the gas and in the liquid respectively. Also, V and L are the molar flow rates of the gas and liquid respectively. Assume that the carrier gas is insoluble in that the solvent and that vapour pressure of the solvent is very low at the given conditions of the column. Further, the gas and liquid streams are sufficiently dilute that L and V may be assumed to be constant throughout the column. The equilibrium relation is given by Y* = mx, where m is a positive constant.

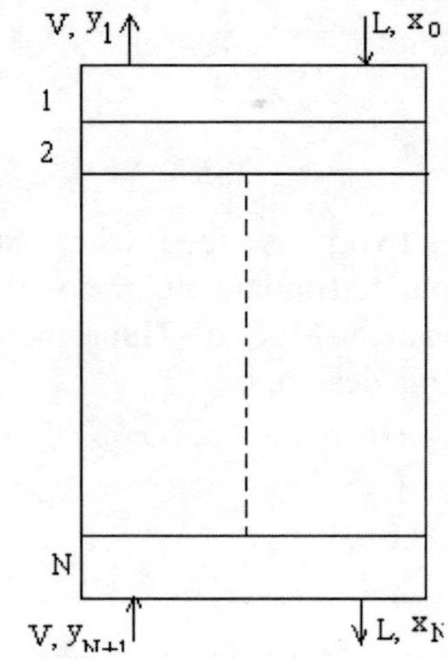

121. For any value of m, the change in liquid composition across a tray is independent of the trays location if

A. $L/V = 1$
B. $m\ L/V = 1$
C. $m/(LV) = 1$
D. $L/(mV) = 1$

***Sol.* D.** the change in liquid composition across a tray is independent of the tray location when

Slope of operating line = slope of equilibrium line

or $\frac{L}{V} = m$

or $\frac{L}{mv} = 1$

122. Under the correct condition corresponding to Q. 94, the number of ideal trays in the column is given by

A. $N = \frac{x_0 - x_N}{x_0 - \left(y_{N+\frac{1}{m}}\right)}$

B. $N = \frac{x_0 - (y_{N+1/m})}{x_0 - x_N}$

C. $N = \frac{x_N - x_0}{(y_{N+1/m}) - x_N}$

D. $N = \frac{(y_{N+1/m}) - x_N}{x_N - x_0}$

***Sol.* C.** the number of ideal trays in the column is given by

$$N = \frac{X_N - X_0}{(Y_{N+1}/m) - X_N}$$

GATE 2004

123. In a distillation operation, what is the effect of the temperature of the reflux stream (given below) on the condenser and reboiler loads?

Reflux conditions

1. Reflux stream is completely liquid and is at its bubble point
2. Reflux stream is below its bubble point

A. condenser and reboiler loads are the same in both the cases

B. reboiler load is the same in both the cases but condenser load is higher in case 2

C. condenser load is the same in both the cases but reboiler load is higher in case 2

D. both condenser and reboiler loads are higher in case 2 as compared to case 1

Sol. **D.** both condenser and reboiler loads are higher in case 2 as compared to case 1.

124. A long cylinder and a sphere both of 5 cm diameter are made from the same porous material. The flat ends of cylinder are sealed. Both the cylinder and sphere are saturated with the same solution of sodium chloride. Later both the objects are immersed for short and equal interval of time in a large tank of water which is well agitated. The fractions of salt remaining in the cylinder and the sphere are X_C and X_S respectively. Which of the following statement is correct?

A. $X_C > X_S$

B. $X_C = X_S$

C. $X_C < X_S$

D. X_C is greater / less than X_S depending on the length ofthe cylinder

Sol. **C.** Surface area available for salt diffusion in the cylinder,

$$A_C = 4\,\pi dl$$

Surface area available for salt diffusion in sphere,

$$A_S = 4\pi\left(\frac{d}{2}\right)^2 = \pi d^2$$

Since the cylinder is long l > d

$$\therefore \quad A_C > A_S$$

If can be concluded that due to larger surface area, the salt in cylinder will diffuse to a larger extent than that of sphere. Hence, lesser salt will remain in the cylinder compared to sphere.

125. In liquid-liquid extraction 10 kg of a solution containing 2 kg of solute C and 8 kg of solvent A is brought into contact with 10 kg of solvent B. Solvents A and B are completely immiscible in each other whereas solute C is soluble in both the solvents. The extraction process attains equilibrium. The equilibrium relationship between the two phases is Y* = 0.9X, where Y* is kg of C/kg of B and X is kg of C/kg of A. Choose the correct answer.

A. the entire amount of C is transferred to solvent B

B. less than 2 kg but more than 1 kg of C is transferred to solvent B

C. less than 1 kg of C is transferred to B

D. no amount of C is transferred to B

Sol. **B.** the problem is represented in a block diagram with the flowrates of streams for easy reference.

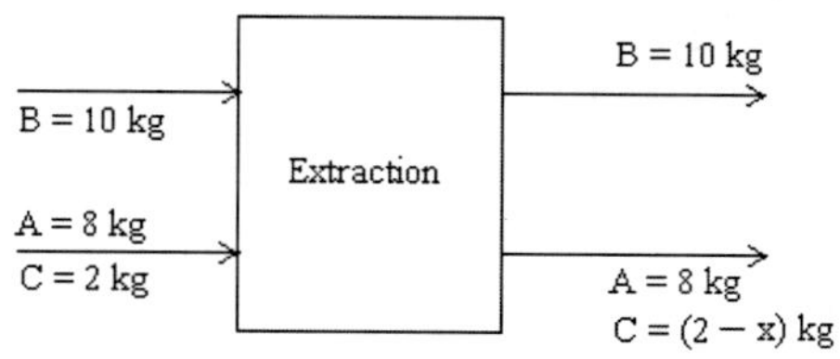

Given equilibrium relation is

$$Y^* = 0.9\,X$$

Substituting the values as per figure

$$\frac{X}{10} = (0.9)\left(\frac{(2-X)}{8}\right)$$

or $$X = \frac{18}{17}$$

Hence, less than 2kg but more than 1 kg of C is transferred to **B.**

126. At equilibrium the concentration of water in vapour phases (C*) in kg/m³ of air space and the amount of water (m) adsorbed per kg of dry silica gel are related by C* = 0.0667 m. To maintain dry conditions in a room of air space 100 containing 2.2 kg of water vapour initially, 10 kg of dry silica gel is kept in the room. The fraction of initial water remaining in the air space after a long time (during which the temperature in maintained constant as) is

A. 0.0

B. 0.2

C. 0.4

D. 1.0

Sol. **C.** assuming x is the amount of water adsorbed by dry silica.

The equilibrium relationship is given as

$$C^* = 0.0667\text{ m}$$

Since, $C^* = \left(\frac{(2.2-X)}{100}\right)$ and $m = \frac{x}{10}$

$$= \left(\frac{2.2-X}{100}\right) = 0.667\left(\frac{x}{10}\right)$$

or $$\text{x} = 1.32\text{ kg}$$

Hence, the fraction of initial water remaining in air space

$= \frac{2.2-X}{2.2}$

$= \frac{2.2-1.32}{2.2} = 0.4$

127. A 25 cm × 25 cm × 1 cm flat sheet weighing 1.2 kg initially was dried from both sides under constant drying rate conditions. It took 1500 sec for the weight of the sheet to reduce to 1.05 kg. Another 1 m × 1 m × 1 cm flat sheet of the same material is to be dried from one side only. Under the same constant drying rate conditions, the time required for drying (in sec) from its initial weight of 19.2 kg to 17.6 kg is

A. 1000
B. 1500
C. 2000
D. 2500

***Sol.* C.** initial weight = 1.2 kg

weight after drying = 1.05 kg

Case 1;

area of drying = (2) (25×25) cm^2 = 0.125 m^2

drying time = 1500 sec.

∴ Rate of drying , $R = \frac{1.2-1.05}{(0.125)1500}$

∴ $R = 8 \times 10^{-4}$ kg m^2-sec.

Case 2,

Area of drying for another sheet for one side only

= 1(m)×1(m) = 1m^2

drying time,t $= \frac{19.2-17.6}{(1)R}$

Since drying rate is same

∴ $t = \frac{1.6}{8 \times 10^{-4}}$

∴ t = 2000 sec.

128. A distillation column with N plates in being operated under normal conditions. At some point in time, the operation is shifted to total reflux condition (i,e., no product and residue are being withdrawn and feed to the column is stopped). At the new steady state,

A. composition of vapours and that of liquid do not vary throughout the column
B. reboiler load and condenser load are minimum
C. the top and bottom compositions are unchanged with and without total reflux
D. the top and bottom compositions correspond to the maximum enrichment achievable

***Sol.* D.** at new steady state of total reflux condition, the top and bottom compositions correspond to maximum enrichment achievable.

129. An aqueous solution of methanol is to be distilled in a tray column. High-pressure steam is available as a source of heat. For a given reflux ratio and overhead composition, two options are being explored:

1. a reboiler is used, and
2. no reboiler is used but steam is fed directly to the bottom of the column.

As compared to option 1 in option 2,

A. less number of trays are required,
B. composition of the residue remains unchanged
C. more number of trays are required but the residue composition remains unchanged
D. more number of trays are required and the residue composition is more dilute in methanol.

***Sol.* A.** in a distillation column a reboiler is considered as an extra tray. Hence, less number of trays are required in option 2 as compared to option 1.

GATE 2003

130. The equilibrium data of component A in the two phases B and C are given below:

x (mol of A/mol of B)	*y (mole of A/mole of C)*
1	0.5
2	4.125

The estimate of y for x = 4 by fitting a quadric expression of a from $y = mx^2$ for the above data is

A. 15.5
B. 16
C. 16.5
D. 17

***Sol.* C.** when the given data is fitted in the quadratic equation of the form

$y = mx^2$ the resulting expression is

$y = (1.03) x^2$

Hence, when x = 4, y = 16.48

∴ y = 16.5

131. Match the following dimensional numbers with the appropriate ratio of force.

P. Froude number Q. Reynolds number R. Friction number S. Nusselt number	1. Shear force/inertial force 2. Convective heat transfer/Conductive heat transfer 3. Gravitational force/viscous force 4. Inertial force/gravitational force

	P	Q	R	S
A.	1	2	5	3
B.	5	4	3	2
C.	5	4	1	2
D.	3	4	5	1

***Sol.* C.** Froude number - Inertial force/gravitational force

Reynolds number - Inertial force/viscous force

Friction factor - Shear force/inertial force

Nusselt - convective heat transfer/conductive heat transfer

132. Experiments were conducted to determine the flux of a species A in a stagnant medium across a gas-liquid interface. The overall mass transfer coefficient based on the liquid side for dilute systems for the above was estimated to be 4×10^{-3} kg-mol/m^2-s. The equilibrium data for the system is given as y = 2x. The flux across the interface (in kg-mol/m^2-s) for bulk concentrations of A in gas phases and liquid phases as y = 0.4 and x = 0.01, respectively is

A. 5.6×10^{-4}

B. 8.5×10^{-4}

C. 5.6×10^{-3}

D. 8.5×10^{-3}

***Sol.* B.** data given is is $X_{AL} = 0.01$, $Y_{AG} = 0.4$

The equilibrium relation is given as

$$Y = 2X$$

$$\therefore \quad X_{AG} = \frac{0.4}{2} = 0.2$$

$$(1 - X_A)_{LM} = \frac{(1-0.01)-(1-0.2)}{In\left(\frac{1-0.01}{1-0.2}\right)} = 0.89$$

$\therefore$ The flux across the interface,

$$N_A = k_C \frac{(X_{AG} - X_{AL})}{(1-X_A)_{LM}}$$

$$k_C = 4\times 10^{-3} \text{ kg mole/m}^2\text{-sec}$$

Substituting the values

$$N_A = \frac{(4\times10^{-3})(0.2-0.01)}{0.89}$$

$$\therefore \quad N_A = 8.5 \times 10^{-4} \text{ kg mole/m}^2\text{-sec}$$

133. H_2S is being absorbed in a gas absorber unit. The height of the transfer unit based on the overall mass transfer coefficient on the gas side is 0.4 m. The equilibrium data given by y = 1.5x. The bulk concentration of H_2S has to be reduced from 0.05 to 0.001 mole fraction in the gas side. The height of the tower (in metre) corresponding to an operating line given by y = 5x + 0.001 is

A. 2.0

B. 1.56

C. 1.0

D. 0.56

***Sol.* A.** Equilibrium relationship y = 1.5 x ....(i)

Operating line, y = 5x + 0.001 ....(ii)

Data given, $y_1 = 0.05$

From equation (ii),

$$0.05 = 5x_i + 0.001$$

$$\therefore \quad x_i = 0.0098$$

Substituting the value of x_i in equation (i)

$$y_{e_2} = (1.5)\ 0.0098 = 0.0147$$

Further given,

$$y_2 = 0.001$$

$$\therefore \quad (y - y_e)_m = \frac{(0.05-0.0147)-(0.001-0)}{In\left(\frac{0.05-0.0147}{0.001-0}\right)} = 0.00964$$

We know,

$$\text{NTU} = \frac{y_1 - y_2}{(Y-Y_e)_m}$$

$$\text{NTU} = \frac{0.05-0.001}{0.00964} = 5$$

Given, HTU = 0.4 m

$\therefore$ Height of tower

$$h = \text{HTU} \times \text{NTU}$$

$$h = 0.4 \times 5$$

$$h = 2 \text{ m}$$

134. The Reynolds number of the liquid was increased 100 fold for a laminar falling film used for gas-liquid contacting. Assuming the penetration theory is applicable, the fold-increase in the mass transfer coefficient (k_C) for the same system is

A. 100

B. 10

C. 5

D. 1

***Sol.* B.** according to the penetration theory,

$$k_C = 2\sqrt{\frac{D_{AB}}{\pi t}}$$

$$= 2\sqrt{\frac{D_{AB}\ .U}{\pi\ L}}$$

Hence,

$$\frac{k_C'}{k_C} = \sqrt{\frac{U'}{U}} \quad ....(i)$$

Now

$$Re = \frac{d\,u\,\rho}{\mu}$$

$$\therefore \quad \frac{Re'}{Re} = \frac{U'}{U} = 100 \qquad \text{....(ii)}$$

Hence, combining equations (i) and (ii)

$$\frac{k_C'}{k_C} = \sqrt{100} = 10$$

135. A pure drug is administered as a sphere and as a cube. The amount of drug is the same in the two tablets. Assuming that the shape and size do not influence the mass transfer the ratio of rate of dissolution in water at t = 0 for the cubic to spherical tablet is

A. 0.54

B. 1.04

C. 1.24

D. 1.94

Sol. **C.** we know that rate of dissolution will be proportional to the surface area

∴ Ratio of rate of dissolution for cubic to spherical tablet

$$\frac{r_C}{r_S} = \frac{A_C}{A_S}$$

$$= \frac{6\,l^2}{4\pi\left(\frac{d}{2}\right)^2}$$

$$= \frac{6}{\pi\left(\frac{l}{d}\right)^2} \qquad \text{....(i)}$$

Since,

$$V_C = V_S \quad \text{(volumes are same)}$$

$$l^3 = \frac{4}{3}\pi\left(\frac{d}{2}\right)^3 \text{ or } l^3 = \frac{\pi}{6}d^3$$

$$\therefore \quad \frac{l}{d} = \left(\frac{\pi}{6}\right)^{1/3} \qquad \text{....(ii)}$$

Combining equations (i) and (ii)

$$\frac{r_C}{r_S} = \frac{6}{\pi}\left(\frac{\pi}{6}\right)^{2/3}$$

$$= \left(\frac{6}{\pi}\right)^{1/3}$$

$$= 1.24$$

136. A solid is being dried in the linear drying rate regime from moisture constant k_C to X_0 to X_F. The drying rate is zero at X = 0 and the critical moisture content is the same as the initial moisture X_0. The drying time for m = (L_S/AR_C) is

A. m $(X_0 - X_F)$

B. m (X_0/X_F)

C. m In (X_0/X_F)

D. mX_0In (X_0/X_F)

Where, L_S = total mass of dry solids, A = total surface area for drying, R_C = constant maximum drying rate per unit area, and X = moisture content (in mass of water/mass of dry solids)

Sol. **D.** for lineare drying rate,

$$R = m\,X + C \qquad \text{....(i)}$$

Boundary conditions when X = 0, R = 0

$$\therefore \quad C_1 = 0$$

$$\text{Hence, } R = m\,X \qquad \text{....(ii)}$$

$$\text{Now at, } X = X_0, \quad R = R_C$$

$$\therefore \quad m = \frac{R_C}{X_0}$$

Combining equations (i) and (ii)

$$R = \left(\frac{R_C}{X_0}\right)X$$

Hence, for linear drying rate the drying time

$$t = -\frac{L_S}{A}\int_{X_0}^{X_F}\frac{dX}{R}$$

$$= -\frac{L_S}{A}\int_{X_0}^{X_F}\frac{X_0}{R_C\,X}.\,dX$$

$$= -\left(\frac{L_S}{A}\right)\frac{X_0}{R_C}\,\text{In}\left(\frac{X_F}{X_0}\right)$$

$$= -\frac{L_S\,X_0}{A\,R_C}\,\text{In}\left(\frac{X_0}{X_F}\right)$$

$$\therefore \quad t = \left(\frac{L_S}{A\,R_C}\right)X_0\,\text{In}\left(\frac{X_F}{X_0}\right)$$

$$\text{But } \frac{L_S}{A\,R_C} = M$$

$$\therefore \quad t = M\,X_0\,\text{In}\left(\frac{X_0}{X_F}\right)$$

137. The following plot gives the saturated humidity (H) versus temperature (T).

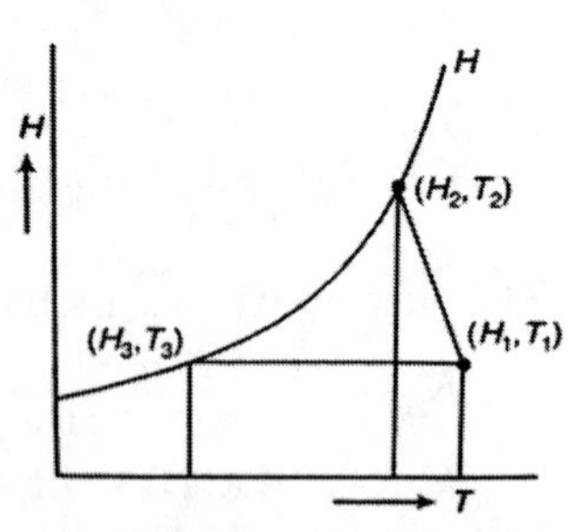

Line joining (H_1, T_1) and $(H_2$, T2) is the constant enthalpy line. Choose the correct one from the following options.

A. T_1 = Dew point temp; T_2 = Dry bulb temp; T_3 = Wet bulb temp

B. T_1 = Dew point temp; T_2 = Wet bulb temp; T_3 = Dry bulb temp

C. T_1 = Wet bulb temp; T_2 = Dry bulb temp; T_3 = Dew bulb temp

D. T_1 = Dry bulb temp; T_2 = Wet bulb temp; T_3 = Dew bulb temp;

***Sol.* D.** T_1 = Dry bulb temperature
T_2 = Wet bulb temperature
T_3 = Dew point temperature

138. Compound A is extracted from a solution of A + B into a pure solvent S. A co-current unit is used for the liquid-liquid extraction. The inlet rate of the solution containing A is 200 mol of B/h-m^2 and the solvent flow rate is 400 mol of S/h-m^2. The equilibrium data is represented by $Y = 3X^2$, where y is in mol of A/mol of B and X is in mol of A/mol of S. The maximum percentage extraction achieved in the unit is

A. 25%
B. 50%
C. 70%
D. 90%

***Sol.* B.** drawing a block diagram of the problem

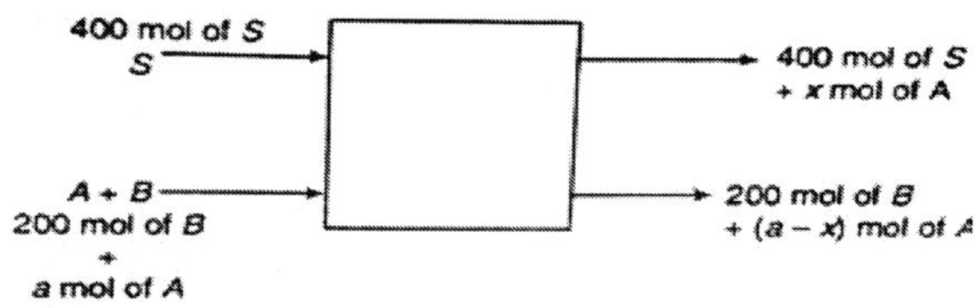

Data given,

$$Y = 3X^2$$

$$\text{or} \quad \frac{a-x}{200} = 3\left(\frac{x}{400}\right)^2$$

$$\text{or} \quad a - x = \left(\frac{3}{800}\right)x^2$$

The maximum percentage of extraction can be achieved when

$$a - x = x$$

$$\text{or} \quad a = 2x$$

Hence, maximum extraction

$$= \frac{x}{a} = \frac{x}{2x}$$

$$= 50\%$$

139. In distillation column sizing calculations, by short cut methods, match the following.

P. Underwood's equation	1. Number of real trays
Q. Fenske's equation	2. Column diameter
R. Gilliland's equation	3. Minimum number of ideal trays
S. Vapour velocity at flooding	4. Actual number of ideal trays
	5. Minimum reflux ratio
	6. Tray efficiency

A. P - 1, Q - 3, R - 4, S - 6
B. P - 2, Q - 5, R - 1, S - 3
C. P - 5, Q - 3, R - 6, S - 2
D. P - 5, Q - 3, R - 4, S - 2

***Sol.* C.** Underwood's equation – Minimum reflux ratio
Fenske's equation – Minimum number of ideal trays
Gilliland's equation – Tray efficiency
Vapour velocity at flooding – column diameter

GATE 2002

140. According to the Fenske equation, What will be the minimum number of plates required in a distillation column to separate an equimolar binary mixture of components A and B into an overhead fraction containing 99 mol % A and a bottom fraction containing 98 mol% B? (Assume that the relative volatility (α_{AB} = 2) does not change appreciably in the column)

A. 5
B. 9
C. 12
D. 28

***Sol.* C.** drawing the block diagram for easy reference

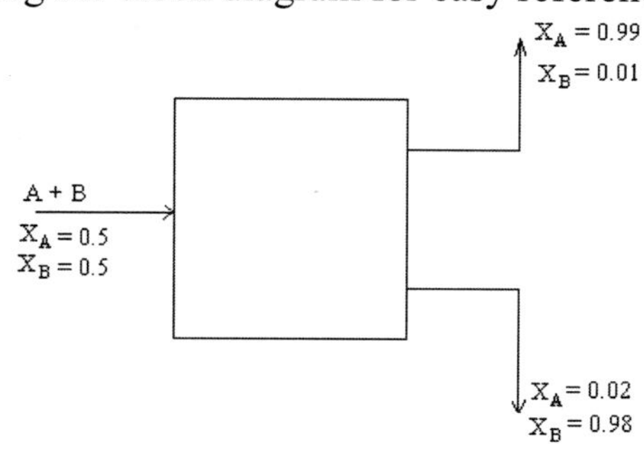

Fenske equation for minimum number of plates (total reflux)

$$N + 1 = \frac{\log\left[\left(\frac{x_D}{1-x_D}\right)\left(\frac{1-x_W}{x_W}\right)\right]}{\log \propto_{av}}$$

Given, x_D = 0.99; $x_W = 0.02$

∴ Substituting the values

$$N + 1 = \frac{\log\left(\frac{0.99}{1-0.99}\right)\left(\frac{1-0.02}{0.02}\right)}{\log \log 2}$$

$$\therefore \quad N + 1 = 12.24$$

$$\therefore \quad N = 11.24 = 12 \text{ plates}$$

141. It takes 6 h to dry to wet solid from 50% moisture content to the critical moisture content of 15%. How much longer will it take to dry the solid to 10 % moisture content under the same drying conditions (the equilibrium moisture content of the solid is 5%).

A. 15 min
B. 51 min

C. 71 min

D. 94 min

***Sol.* C.** given data

$t = 6hr = 6\times60$ min, $X_1 = 0.5$, $X_2 = 0.15$

$$t = \frac{L}{A\,R_C}(X_1 - X_2)$$

or $$360 = \frac{L}{A\,R_C}(0.5 - 0.15)$$

$$\therefore \quad \frac{L}{A\,R_C} = \frac{360}{0.35} \qquad ....(i)$$

Now, for falling rate period

$$R = mX + C$$

At equilibrium moisture content, X = 0.05, R = 0

$$\therefore \quad 0 = (0.05)\,m + C \qquad ....(ii)$$

$$\therefore \quad C = (-0.05)\,m$$

From equation (ii)

$$R = m\,(x - 0.05) \qquad ....(iii)$$

At critical moisture content, X = 0.15, R = R_C

$$\therefore \quad R_C = m\,(0.15 - 0.05)$$

or $$m = 10\,R_C$$

Now, from equation (iii)

$$R = 10\,R_C\,(x - 0.05)$$

Hence, drying time

$$t = -\frac{L_S}{A}\int_{X_1}^{X_2}\frac{dX}{R}$$

or $$t = -\frac{L_S}{A}\int_{0.15}^{0.10}\frac{dX}{(10\,R_C)(x-0.05)}$$

$$= -\left(\frac{L_S}{A\,R_C}\right)\left(\frac{1}{10}\right)\ln\left[\frac{0.1-0.05}{0.15-0.05}\right]$$

$$= -\left(\frac{360}{0.35}\right)\left(\frac{1}{10}\right)\ln\left(\frac{0.05}{0.10}\right)$$

$$\therefore \quad t = 71 \text{ min}$$

GATE 2001

142. The interfacial area per unit volume of dispersion, in a gas-liquid contactor, for fractional hold-up of gas = 0.1 and gas bubble diameter = 0.5 mm is given by (in m^2/m^3)

A. 500

B. 1200

C. 900

D. 800

***Sol.* B.** given data

$d = 0.5$ mm $= 0.5 \times 10^{-3}$ m, $\in = 0.1$

∴ Interfacial area per unit volume of dispersion,

$$a = \frac{6\varepsilon}{d} = \frac{(6)0.1}{0.5\times10^{-3}} = 1200\ m^2/m^3$$

143. 200 kg of solids (on dry basis) is subjected to a drying process for a period of 5000 s. The drying occurs in the constant rate period with the drying rate as $N_C = 0.5 \times 10^{-3}$ kg/m^2-s. The initial moisture content of the solid is 0.2 kg moisture kg of dry solid. The interfacial area available for drying is 4 m^2/100 kg of dry solid. The moisture content at the end of the drying period is (in kg moisture/kg dry solid)

A. 0.5

B. 0.05

C. 0.1

D. 0.15

***Sol.* C.** data given is

$N_C = 0.5 \times 10^{-3}$ kg/m^2-sec

$X_1 = 0.2$, $t = 5000$ sec, $X_2 = ?$

$A = 4\ m^2/100$ kg of dry solid

∴ Total area $= 8\ m^2$

$$N_C = \frac{L_S}{A}\frac{(X_1 - X_2)}{t}$$

Hence, $$0.5 \times 10^{-3} = \left(\frac{200}{8}\right)\frac{(0.2 - X_2)}{5000}$$

$$X_2 = 0.1$$

144. In a single stage extraction process, 10 kg of pure solvent S (containing no solute A) is mixed with 30 kg of feed F containing A at a mass fraction $X_F = 0.2$. The mixture splits into an extract phase E and a raffinate phase R containing A at $X_E = 0.5$ and $X_R = 0.05$ respectively. The total mass of the extract phase is (in kg)

A. 6.89

B. 8.89

C. 10

D. 8.25

***Sol.* B.** drawing a block diagram of the problem

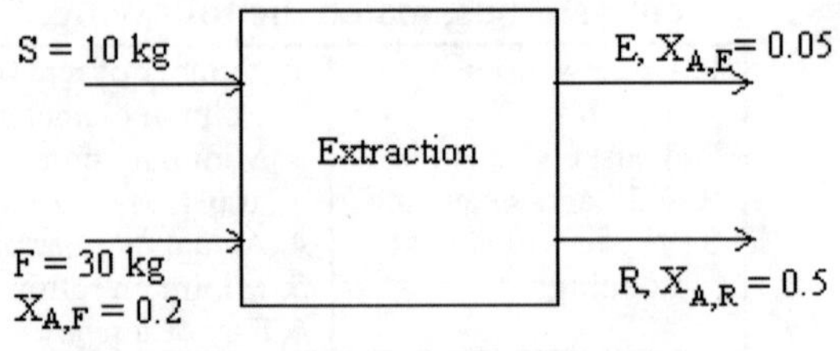

Writing overall material balance

$$F + S = R + E$$

or $$10 + 30 = R + E$$

or $$R + E = 40$$

Now writing material balance for A

$F.X_{AF} = E.X_{AE} + R.X_{AR}$

or $(30)0.2 = (E)0.5 + (40 - E)0.05$

$\therefore \quad E = 8.89$ kg

GATE 2000

145. The individual mass transfer coefficients (mol/m^2-s) for absorption of a solute from a gas mixture into a liquid solvent are $K_L = 4.5$ and $K_G = 1.5$. The slope of the equilibrium line is 3. Which of the following resistances is/are controlling?

A. liquid side
B. gas side
C. interfacial
D. both liquid and gas sides

Sol. **D.** both liquid and gas side resistances are controlling

Explanation: -

Since K_L is three time that of K_G so K_Gg should offer higher resistance but slope of equilibrium line is three. This means that the lesser value of K_G is compensated.

Hence, both liquid and gas side resistances are controlling.

146. In a laboratory test run, the rate of drying was found to be 0.5×10^{-3} kg/m^2-s when the moisture content reduced from 0.4 to 0.1 on a dry basis. The critical moisture content of the material is 0.08 on a dry basis. A tray dryer is used to dry 100 kg (dry basis) of the same material under identical conditions. The surface area of the material is 0.04 m^2/kg of dry solid. The time required (in second) to reduce the moisture content of the solids from 0.3 to 0.2 (dry basic) is

A. 2000
B. 4000
C. 5000
D. 6000

Sol. **C.** data given is

$R_C = 0.5 \times 10^{-3}$ kg m^2-sec

$L_S = 100$ kg

A = 0.04 m^2/kg of dry solid

$\therefore$ Total area $= 0.04 \times 100 = 4m^2$

Hence, time required

$$t = \left(\frac{L_S}{A}\right)$$

and $X_2 - X_1 = 0.3 - 0.2$

$$\therefore \quad t = \left(\frac{100}{4}\right)\left(\frac{0.1}{0.5\times10^{-3}}\right)$$

$\therefore \quad t = 5000$ sec.

8. Plant Design & Economics

Multiple Choice Question

1. A component of working capital investment is.
   A. depreciation
   B. maintenance and repair inventory
   C. utilities plants
   D. process equipment's

2. About __________ month supply of raw materials valued at delivery prices is included in the inventory of raw materials in the working capital.
   A. one
   B. three
   C. six
   D. twelve

3. The basis of __________ tax calculation is on gross earnings.
   A. capital gain
   B. property
   C. income
   D. excise

4. What will be the actual annual return on the principal expressed as present if the interest rate of 10% per period is compounded half yearly ?
   A. 10
   B. 20
   C. >20
   D. < 20

5. Which of the following is not included in the fixed charges for a chemical plant?
   A. property tax, insurance and depreciation
   B. interest on borrowed money
   C. rent of land and buildings
   D. repair and maintenance charges

6. Among the various methods the 'Six-tenth factor' rule is used for estimating the .
   A. utilities cost
   B. equipment cost by scaling
   C. equipment installation cost
   D. cost of piping

7. What will be the approximate electrical installation cost is in ordinary chemical plant?
   A. 10-15% of purchased equipment cost
   B. 3-10% of fixed capital investment
   C. either (a) or (b)
   D. neither (a) nor (b)

8. Which of the following statements is wrong?
   A. turn over = opening stock + production closing stock
   B. debt-equity ratio of a chemical company describes the lenders contribution for each rupee of owner's contribution i.e., debt-equity ratio = total debt/net worth
   C. working capital = current assets + current liability
   D. return on investment (ROI) is the ratio of profit before interest & tax and capital employed (i.e. net worth + total debt)

9. Which of the following statements is wrong?
   A. working capital turnover ratio = sales/net working capital
   B. net worth means paid up share capital and reserve & surplus (i.e. shareholders equity)
   C. return on equity = profit after tax/net worth
   D. total cost of production is more than net sales realisation (NSR) at breakeven point

10. What will be the depreciation for n[th] year by the sum of year digit method if 'P' is the investment made on an equipment 'S' which is its salvage value and 'n' the life of the equipment in years.
   A. $\frac{P-S}{n}$
   B. $1-\left(\frac{P}{S}\right)^{\frac{1}{m}}$
   C. $\frac{m}{n}.(P-S)$
   D. $\frac{2(n-m+1)}{n(n+1)}.(P-S)$

11. The payback method for the measurement of return on investment
    A. takes into account the cash inflows after the recovery of investments
    B. gives a correct picture of profitability
    C. does not measure the discounted rate of return
    D. underemphasises liquidity
12. What will be the annual depreciation cost by straight line method (in ₹) for a machine having an initial volue of ₹5000, service life of 5 years and final salvage value of ₹1000.
    A. 300
    B. 600
    C. 800
    D. 1000
13. When the ______ method of depreciation calculation is used annual depreciation cost are not constant.
    A. present worth
    B. straight line
    C. sinking fund
    D. declining balance
14. Which of the following cost components in a chemical plant falls under utilities cost?
    A. water supply
    B. medical services
    C. running a control laboratory
    D. property protection
15. What is the terminology (name) for a series if equal payments (e.g., deposit or cost) made at equal intervals of time?
    A. future worth
    B. perpetuity
    C. annuity
    D. capital charge factor
16. Which of the following pertains to Payback period?
    A. is affected by the variation in earnings after the recovery of the investment
    B. is the length of time over which the earnings on a project equals the investment
    C. and economic life of a project are the same
    D. all **A**, **B** and **C**
17. ______ is not a component of depreciation cost?
    A. repairs and maintenance cost
    B. loss due to accident/breakdown in the machinery
    C. loss due to obsolescence of the equipment
    D. loss due to decrease in the demand of product
18. The responsibility of the personnel working in the market research group is .
    A. cost estimation.
    B. product evaluation.
    C. equipment selection.
    D. equipment design
19. Gross earning is equal to the total income less ______________.
    A. total product cost
    B. income tax
    C. fixed cost
    D. none of these
20. Which of the following is the point of intersection of the plot for "break-even point"?
    A. fixed cost and sales revenue.
    B. total cost and sales revenue.
    C. fixed cost and total cost.
    D. none of these.
21. Which of the following statements is wrong?
    A. turnover ratio of a chemical plant is the ratio of gross annual sales to the fixed capital investment.
    B. longer tubes are less expensive per unit heat transfer area as compared to shorter tubes.
    C. a cost index is merely a number for a given year showing the cost at that time relative to a certain base year.
    D. plates with butt welded joints are less expensive compared to lap welded joints, because squaring of plates is not necessary.
22. Which of the following depreciation methods gives maximum value for the depreciation?
    A. diminishing balance
    B. sinking fund
    C. straight line
    D. sum of the year's digit
23. For the construction of pressure vessel the costliest material of construction is .
    A. high alloy steel
    B. low alloy steel
    C. titanium
    D. lead

24. Which of the following statements is wrong?
    A. consolidated income statement based on a given time period indicates surplus capital and shows the relationship among total income, costs & profit over the time interval
    B. net revenue is the total profit remaining after deducting all costs excluding taxes
    C. gross revenue is that total amount of capital received as a result of the sale of goods or service
    D. the ratio of immediately available cash to the total current liabilities is known as the cash ratio
25. Which of the following statements is wrong?
    A. insurance rates on annual basis in a chemical plant may be about 1% of the fixed capital investment
    B. the annual depreciation rate for machinery and equipments in a chemical process plant is about 10% of the fixed capital investment
    C. annual depreciation rate of buildings in a chemical plant is about 3% of its initial cost
    D. in a chemical industry, research and development cost amounts to about 15% of net sales realisation (NSR)
26. In which of the following depreciation calculations method the annual depreciation cost are constant?
    A. sum of the year's digit
    B. straight line
    C. declining balance
    D. none of these
27. What will be the equipment installation cost in a chemical process plant expressed as percent of purchased equipment cost?
    A. 10 to 20
    B. 35 to 45
    C. 55 to 65
    D. 70 to 80
28. What is roughly the construction expense expressed as percent of the total direct cost of the plant?
    A. 2
    B. 10
    C. 30
    D. 50
29. What is the percentage range of ratio of working capital to total capital investment for most chemical plant (except for non-seasonal based product)?
    A. 0.1 to 1
    B. 1 to 2
    C. 10 to 20
    D. 50 to 60
30. The scope of market analysis includes ________.
    A. product distribution
    B. opportunities
    C. competition from other manufactures
    D. economics
31. Which of the following includes the utilities cost in the operation of chemical process plant?
    A. general expenses
    B. plant overhead cost
    C. direct production cost
    D. fixed charges
32. Which of the following cost is included in the direct cost of the fixed capital?
    A. raw material costs
    B. onsite and offsite costs
    C. contingencies
    D. labor costs
33. Which of the following method for profitability evaluation of a project does not include
    A. rate of return on investment
    B. pay out period
    C. net present worth
    D. discounted cash flow
34. ₹10,000 is invested at 10% annual interest on compound basis. The gross amount at the end of 5 year will be .
    A. 15000
    B. 16105
    C. 18105
    D. 12500
35. A small scale project cost ₹10 lakhs for which the annual profit is likely to be 1.5 lakhs. If the project life is estimated as 10 years then the simple rate of return on investment is
    A. 15%
    B. 10%
    C. 1.5%
    D. 150%
36. When the interest is compounded nominal and

effective interest rates are equal?

A. in no case, they are equal
B. quarterly
C. annually
D. semi-annually

37. A breakeven point occurs in a manufacturing industry when the .

A. annual sales equals the fixed cost.t
B. total annual product cost equals the total annual sales.
C. total annual rate of production equals the assigned value
D. annual profit equals the expected value.

38. In calculating the depreciating by diminishing balance method, the depreciating amount during the nth year is calculated by multiplying by a fixed percentage 'N' to the

A. initial cost.
B. book value at the end of (n - 1)th year,
C. depreciation during the (n - 1)th year.
D. difference between initial cost and salvage value.

39. Which of the following depreciation calculation accounts for the interest on investment.

A. declining balance
B. straight line method
C. both (a) and (b)
D. neither (a) nor (b).

40. It is an usual practice that most chemical plants allocate an initial working capital amounting to 10-20% of the total capital investment. However this percentage may be increased to about in case of seasonal products manufacturing plants.

A. 30
B. 50
C. 75
D. 95

41. _________ is not a component of working capital?

A. semi-finished products in the process.
B. raw materials is stock.
C. transportation facilities.
D. finished products in stock.

42. Utilities are essential for a chemical plant which include steam, compressed air, water, oxygen etc. The fuel allocation for utilities in a normal chemical plant ranges from percentage of the total product cost.

A. 1 to 5
B. 10 to 20
C. 25 to 35
D. 35 to 45

43. As the pipe diameter increases the pumping costfor a given fluid?

A. decreases
B. remains the same
C. increases
D. may increase or decrease, depending upon whether the fluid is Newtonian or non-Newtonian

44. In compression to a small chemical process plant a large chemical plant has economic life.

A. almost equal
B. much more
C. only slightly more
D. slightly less

45. In case of chemical plant the following relationship is correct.

A. total product cost = manufacturing cost + general expenses
B. manufacturing cost = direct product cost + fixed charges + plant overhead costs
C. general expenses = administrative expenses + distribution & marketing expenses
D. total product cost = direct production cost + plant overhead cost.

46. Which of the following provides information about scheduling?

A. time of completion of job.
B. proper utilisation of machines.
C. means to minimise idle time for machines.
D. time of starting of job and also about how much work should be completed during a particular period.

47. What is the cost of instrumentation of a modern chemical plant expressed as a percent of the total plant cost?

A. 5 to 10
B. 20 to 30
C. 40 to 50
D. 60 to 70

48. In which of the following method declining balance calculation?

A. annual cost of depreciation is same every

year.

B. value of the asset decreases linearly with time.

C. annual depreciation is the fixed percentage of the property value at the beginning of the particular year.

D. none of these.

49. Which of the following is equal to the operating profit of a chemical plant?

A. profit before interest and tax i.e., net profit + interest + tax

B. profit after tax

C. profit after tax plus depreciation

D. net profit + tax

50. The operating cost is minimum in as far as the operating cost in concerned.

A. fast breeder reactor

B. thermal

C. hydroelectric

D. nuclear

51. Total capital investment for a chemical process plant = Fixed capital investment + .

A. direct production cost

B. working capital

C. overhead cost

D. indirect production cost

52. Which of the following is the key factor in deciding optimum number of effects in multiple effect evaporator?

A. cost benefit analysis.

B. evaporation capacity required.

C. floor area availability.

D. terminal parameters.

53. A balance sheet for a chemical plant shows its financial condition at any given time, however, it does not indicate the ___________of the plant.

A. long term debt

B. current asset

C. current liability

D. profit

54. Which of the following is not included is the manufacturing cost in a chemical company?

A. direct products cost.

B. fixed charges.

C. plant overheads.

D. administrative expenses.

55. What will be the present worth if an amount '*S*' available after '*n*' interest periods for an initial principal '*P*' with the discrete compound interest rate '*i*'?

A. $(1 + i)^n/S$

B. $S/(1 + i)^n$

C. $S/(1 + in)$

D. $S/(1 + n)^i$

56. ____________ is the costliest source of getting hydrogen on commercial scale for the manufacture of nitrogenous fertiliser?

A. coke oven gas

B. coal gasification

C. ectrolysis of water

D. Steam reforming of naphtha

57. Which of the following is the amount of simple interest during '*n*' interest period (i = interest rate based on one interest period, p = principal)

A. $p.i.n.$

B. $p(1 + i.n)$

C. $p(1 + i)n$

D. $p(1 - i.n)$

58. When the interest is compounded effective and nominal interest rates are equal.

A. annually

B. fortnightly

C. monthly

D. half-yearly

59. Which of the following can be obtained directly from the balance sheet of the company as the difference between current assets and current liabilities.

A. liquids assets

B. net working capital

C. cash ratio

D. current ratio

60. The cumulative cash flow is zero at the ________ for a typical project.

A. end of the design stage.

B. breakeven point.

C. end of the project life.

D. start up.

61. Which of the following components of fixed capital investment requires the minimum percentage of it? (Note; fixed capital investment of a chemical plant is the total amount of money needed to supply.)

A. Cost for piping
B. electrical installation cost.
C. equipment installation cost
D. equipment insulation cost

62. Which of the following statements is valid regarding depreciation?
A. does figure in the calculation of income tax liability on cash flows from an investment
B. costs (on annual basis) are constant when the straight line method is used for its determination
C. is the unavoidable loss in the value of the plant, equipment and materials with lapse in time
D. all (a), (b) and (c)

63. Construction of a plant required an investment of ₹100 lakhs which will take two years to start production. If the annual profit form the operation of the plant is ₹20 lakhs what will be the payback time?
A. 5 years
B. 7 years
C. 12 years
D. 10 years

64. It is a general is practice, that income taxes are based on the .
A. fixed cost
B. gross earning
C. total income
D. total product cost

65. What will be the purchased cost of equipment for a chemical plant in percent of the fixed capital investment?
A. 10 to 20
B. 20 to 40
C. 45 to 60
D. 65 to 75

66. ____________ is the costliest source of getting hydrogen on commercial scale for the manufacture of nitrogenous fertiliser?
A. coke oven gas
B. coal gasification
C. ectrolysis of water
D. mteam reforming of naphtha

67. Which of the following is the net present value of the annuity at an interest rate of i for an amount R at the end of every year for 'n' years?
A. $R\left[\frac{(1+i)^n-1}{i}\right]$
B. $\left[\frac{(1+i)^n-1}{i(1+i)^n}\right]$
C. $R(1+i)^n$
D. $R/(1+i)^n$

68. Which of the following is the ratio of gross annual sales to the fixed capital investment?
A. investment
B. cash reserve
C. turnover
D. capital

69. What will be the value of the investment at the end of five years for an investment of ₹1000 at an interest of 10% compounded quarterly.
A. $1000\left(1+\frac{0.1}{4}\right)^{20}$
B. $1000\,(1+0.1)^{20}$
C. $1000\left(1+\frac{0.1}{4}\right)^{5}$
D. $1000\left(1+\frac{0.1}{2}\right)^{5}$

70. ________ is not a component of the working capital.
A. minimum cash reserve.
B. product inventory.
C. in-process inventory.
D. storage facilities.

71. Indian chemical process plants are about ________ costlier than similar plants in U.S.A.
A. 15
B. 35
C. 55
D. 75

72. __________ is not a component of the fixed capital for a chemical plant facility?
A. raw materials inventory.
B. utilities plants.
C. process equipment.
D. emergency facilities.

73. The total product cost in a chemical process plant = manufacturing cost + .
A. general expenses
B. R & D cost
C. overhead cost
D. none of these

74. The measure of decrease in the value due to functional depreciation is due to.

A. breakdown or accident
B. ageing
C. obsolescence
D. wear and tear

75. The total product cost in a chemical process plant = manufacturing cost + .
A. general expenses
B. R & D cost
C. overhead cost
D. none of these

76. Chemical engineering plant cost index indicates The present cost of the plant = original cost

$$\text{x} \left(\frac{\text{index value at present}}{\text{index value at time original cost was obtained}}\right)$$

The major component of this cost index is .
A. fabricated equipment and machinery.
B. electrical equipment's and material.
C. process instruments and control.
D. pumps and compressor.

77. 'Lang factor' is defined as the ratio of the capital investment to the delivered cost of major equipment. The value of 'Langfactor' for fixed capital investment, for a solid-fluid processing chemical plant ranges from
A. 1.2 to 1.4
B. 2.5 to 2.7
C. 4.2 to 4.4
D. 6.2 to 6.4

78. Which of the following includes the expenditure on research and development (R & D) while making an estimate of the total product cost for a chemical plant?
A. direct production cost
B. overhead cost
C. general expenses
D. fixed expenses

79. __________ is not a current asset of a chemical company.
A. marketable securities
B. inventories
C. chemical equipment
D. none of these.

80. __________ does not come under the sales expenses for a product of a chemical plant?
A. customer service.
B. advertising
C. legal fees
D. warehousing

81. The costliest ceramic packing materials among the following is .
A. berl saddles
B. intalox saddles
C. raschig rings
D. pall rings

82. Among the following methods of depreciation calculation of the __________ in book valves greater than those obtained with straight line method.
A. sum of the years digit method
B. sinking fund method
C. multiple straight line method
D. declining balance method

83. In the affairs of company management a shareholder has __________ say in the company than a debenture holder.
A. more
B. less
C. same
D. no

84. A reactor is purchased for ₹80000 and has a service life of 10 years. The salvage value is estimated to be ₹10000. What will be the book value of the reactor after 5 year using sinking fund depreciation method?
A. 40096
B. 43196
C. 53196
D. 60196

85. Which of the following will include the effluent treatment cost in a chemical plant?
A. capital
B. fixed
C. utilities
D. overhead

86. Which of the following is helpful in Gantt chart (or Bar chart)?
A. inventory control.
B. preparing production schedule.
C. efficient utilisation of manpower and machines.
D. efficient despatching of products.

87. Profit = Revenue _________ (   ) ?

A. operating cost
B. total cost
C. book value
D. none of these

88. Making a chemical plant operational may incur about ___________ present of the fixed capital cost known as maximum production startup cost.

A. 1
B. 5
C. 10
D. 30

89. Which of the following relationship is invalid in 0financial accounting of a chemical plant?

A. assets = liabilities + net worth
B. total income = costs + profits
C. assets = equities
D. assets = capital.

90. The statement of six-tenths-factor rule, if the cost of a given unit at one capacity is known, then the cost of similar unit with "_______" times the capacity of the first unit is approximately equal to

A. n
B. $n^{0.6}$
C. $n^{0.4}$
D. n

91. Annuity is used in the calculation of the (Note; An annuity is a series of equal payments occurring at equal interval and this amount includes the sum of all payment plus interest. If allowed to accumulate at definite rate of interest – from the term of initial payment plus interest.)

A. cash ratio.
B. depreciation by sinking fund method.
C. manufacturing cost.
D. discrete compound interest.

92. Which of the following costs is not included in the total product cost of a chemical plant?

A. market survey
B. depreciation, property tax and insur-rance
C. operating labour, supervision and supplies
D. overhead and utilities

93. Which of the following is appropriate for depreciation?

A. decrease
B. increase
C. no change
D. none of these

94. What will be the slope of the straight line obtained from the log-log plot of capacities of the equipment vs. cost of the equipment following the sixed?

A. 0.1
B. 0.6
C. 0.2
D. 0.8

95. 'Factory manufacturing cost = direct production cost +.

A. fixed charges and plant overhead cost.
B. plant overhead cost and administrative expenses.
C. and plant overhead cost.
D. none of these.

96. According to straight line method of determination of depreciation, the value of a property decreases _________ with time.

A. linearly
B. exponentially
C. logarithmically
D. non-linearly

97. With 10% half yearly rate of interest a sum of ₹100 at the end of one year will be.

A. 121
B. 110
C. 97
D. 91

98. Which of the following methods of depreciation calculation does not take into account the interest on investments.

A. sinking fund method
B. present worth method
C. sum of the years-digits method
D. all (a), (b) and (c)

99. Which of the following methods is used to determine the optimum economic pipe diameter?

A. density of the fluid.
B. viscosity of the fluid.
C. total cost considerations (pumping cost plus fixed cost of the pipe).
D. none of these.

100. Among the following which is not a mathematical method for evaluation of profitability of a chemical process plant?

A. Cash reserve.

B. Discounted cash flow based on full life performance.

C. Rate of return on investment.

D. Payout period.

101. Increasing the discounted cash flow rate of return, ________ the ratio of the total present value to the initial investment of a given project.

A. decreases

B. increases linearly

C. remains constant

D. increases

102. What would be the cost of piping in a fluid processing unit (e.g., distillation) of a chemical process plant expressed as percent of the fixed capital investment?

A. 4

B. 13

C. 22

D. 3

103. What will be the amount of compounded interest during '*n*' interest periods?

A. $p[(1+i)^n-1)]$

B. $p(1 + i)^n$

C. $p(1 - i)^n$

D. $p(1 + i^n)$

104. Which of the following statements is wrong?

A. net profit = Gross margin - depreciation - interest

B. gross margin = net income - net expenditure

C. at breakeven point, NSR is more than the total production cost

D. net sales realisation (NSR) = Gross sales - selling expenses

105. Which of the following statements is correct?

A. scrap value is the net amount of money obtainable from the sale of used property over and above any charges involved in its removal & sale.

B. unamortised cost is the difference between the original cost of a property and all the depreciation charges made to date.

C. difference between income and expense is termed as gross revenue.

D. sum-of-the-years-digits methods of depreciation calculation accounts for the interest on the investment.

Answer

1.	B	28.	B	55.	B	82.	B
2.	A	29.	C	56.	B	83.	A
3.	C	30.	D	57.	A	84.	D
4.	C	31.	C	58.	A	85.	C
5.	D	32.	B	59.	B	86.	B
6.	B	33.	B	60.	B	87.	B
7.	C	34.	B	61.	D	88.	C
8.	C	35.	C	62.	D	89.	D
9.	D	36.	C	63.	B	90.	B
10.	D	37.	B	64.	B	91.	B
11.	C	38.	B	65.	B	92.	A
12.	C	39.	D	66.	C	93.	A
13.	D	40.	B	67.	B	94.	B
14.	A	41.	C	68.	C	95.	A
15.	C	42.	B	69.	A	96.	A
16.	B	43.	A	70.	D	97.	A
17.	A	44.	B	71.	B	98.	C
18.	B	45.	D	72.	A	99.	C
19.	A	46.	D	73.	A	100.	A
20.	B	47.	B	74.	C	101.	A
21.	D	48.	C	75.	A	102.	B
22.	A	49.	A	76.	A	103.	A
23.	C	50.	C	77.	C	104.	C
24.	B	51.	B	78.	C	105.	B
25.	D	52.	A	79.	C		
26.	B	53.	D	80.	C		
27.	B	54.	D	81.	A		

GATE 2015-2000

(1 Mark Question)

GATE 2015

1. Match the technologies in List 1 with the entries in List 2.

List 1	*List 2*
I. Urea manufacture	P. Microencapsulation
II. Coal gasification	Q. Ultra-low sulphur diesel
III. Controlled release of chemicals	R. Shale oil
IV. Deep hydrodesulphurization	S. Prilling tower
	T. Gas hydrates
	U. Gas-solid non-catalytic reaction

Codes

A. I-P, II-T, III-Q, IV-S
B. I-S, II-S, III-P, IV-Q
C. I-S, II-P, III-R, IV-Q
D. I-T, II-S, III-S, IV-Q

Sol. **B.** I. Urea manufacture → Prilling tower

II. Coal gasification → Gas-solid non-catalytic reaction

III.Controlled release of chemicals → Micron-capsulation

IV. Deep hydrodesulphurization → Ultra-low sulphur diesel

GATE 2011

2. Two systems are available for compressing 6 m^3/h of ambient air to 10 bar. The first one uses a single stage compressor (K_1) and the second one uses a multistage compressor with inter-stage cooling (K_2). Which one of the following statements is incorrect?

A. K_2 will have knockout pots in between the stages
B. discharge temperature of K_1 will be higher than that of K_2
C. K_2 will consume more power than K_1
D. cost of K_2 will be more than that of K_1

Sol. **B.** discharge temperature o K_1 will not be higher than that of K_2.

3. In a thin-walled cylindrical vessel of thickness t with inside radius r, the internal gauge pressure is p. The hoop stress and the longitudinal stress in the shell are σ_h and σ_l, respectively. Which one of the following statements is true?

A. $\sigma_h = \frac{pr}{t}, \sigma_l = \frac{pr}{4t}$
B. $\sigma_h = \frac{pr}{4t}, \sigma_l = \frac{pr}{t}$
C. $\sigma_h = \frac{pr}{2t}, \sigma_l = \frac{pr}{t}$
D. $\sigma_h = \frac{pr}{t}, \sigma_l = \frac{pr}{2t}$

Sol. **D.** hoop stress

$$\sigma_h = \frac{pr}{t}$$

Longitudinal stress

$$\sigma_l = \frac{pr}{2t}$$

GATE 2010

4. The term knuckle radius is associated with

A. flat heads age
B. torispherical heads
C. hemispherical heads
D. conical heads One

Sol. **B.** knuckle radius is associated with torispherica heads.

GATE 2009

5. The total fixed cost of a chemical plant is ₹ 10.0 lakh; the internal rate of return is 15% and the annual operating cost is ₹2.0 lakh. The annualized cost of the plant, in lakh of ₹2, is

A. 1.8
B. 2.6

C. 3.5

D. 4.3

Sol. **C.** internal rate of return is also called as interest rate of return and is defined as the rate of compound interest at which the company's outstanding investment is repaid by proceeds of the project. Thus, annualized cost of the plant

$2 + (10 \times 0.15) = 2 + 1.5 = 3.5$ lakh

GATE 2008

6. For the case of single lump-sum capital expenditure of ₹10 crore which generates a constant annual cash flow of ₹2 crore in each subsequent year, the payback period (in year), if the scrap value of the capital outlay is zero, is

A. 10

B. 20

C. 1

D. 5

Sol. **D.** scrap value = 0

$$\text{Payback period} = \frac{\text{Capital expenditure}}{\text{Annual cash flow}}$$

$$= 10/2 = 5 \text{ yr}$$

7. The relation between capital rate of return ratio (CRR), net present value (NPV) and maximum cumulative expenditure (MCE) is

A. $CRR = \frac{NPV}{MCE}$

B. $CRR = \frac{MCE}{NPV}$

C. $CRR = NPV \times MCE$

D. $CRR = \frac{MCE}{(NPV + MCE)}$

Sol. **A.** $CRR = \frac{MCE}{(NPV + MCE)}$

GATE 2007

8. To determine the performance of a compressor, a standardized test is performed. In the testing process, when the compressor is under operation, shut-off term signifies

A. maximum flow

B. zero flow

C. steady flow

D. intermittent flow

Sol. **B.** shut-off signifies zero flow.

GATE 2006

9. Select the most appropriate pump from Group II to handle each fluid flow given in Group I.

Group I	*Group II*
P. Highly viscous fluid flow Q. Fluid containing large amount of abrasive solids	I. Piston pump II. Gear pump III. Plunger pump IV. Centrifugal pump

A. P- II, Q-I

B. P- II, Q-IV

C. P-III, Q-IV

D. P-IV, Q-III

Sol. **B.** Highly viscous fluid flow - Gear pump

Fluid containing large amount of abrasive solid - Centrifugal pump

10. A cylindrical storage tank can have a self-supported conical roof

A. If its diameter is less than 20 m

B. If its diameter is more than 50 m

C. If the thickness of the roof is more than that of the cylindrical shell

D. Whatever is the diameter

Sol. **B.** A cylindrical storage tank can have a self-supported conical roof, if its diameter is more than 50 m.

GATE 2005

11. For a solid processing plant, the delivered equipment cost is ₹10 lakh. Using Lang multiplication method, the total capital investment, in lakh of ₹, is

A. 46

B. 57

C. 100

D. 200

Sol. **A.** Lang multiplication factor is given below for different cases.

Plant type	*Based on total capital investment*
Solid processing	4.6
Solid-fluid processing	4.9
Fluid processing	5.7

Total capital investment

= Lang multiplication factor × Delivered equipment cost

= 4.6 × 10 = ₹46 lakh

12. The cost of drum dryer is ₹10 lakh.The cost of drum dryer with double the surface area in lakh of 2, is

A. 2×10

B. $3^{0.6} \times 10$

C. $5^{0.6} \times 10$

D. $2^{0.6} \times 10$

***Sol.* D.** By six-tenth rule,

$$\frac{\text{Cost of equipment A}}{\text{Cost of equipment B}} = \left(\frac{\text{Capacity of equipment} - \text{B}}{\text{Capacity of equipment} - \text{A}}\right)^{0.6}$$

The cost of new drum dryer = $2^{0.6} \times 10$

13. The cost of distillation columnin the year 2000 is ₹ x, what is the cost of the column in ₹ in the year 2010, given the cost indices for the years 2000 and 2010 are 480 and 520 respectively?

A. $\left(\frac{520}{480}\right)^2$ x

B. $\left(\frac{480}{520}\right)$x

C. $\left(\frac{520}{480}\right)$x

D. $\left(\frac{520}{480}\right)^{0.6}$ x

***Sol.* C.** Cost of column in the year 2010 = Cost of column in the year 2000 ×

$$= \left(\frac{\text{Cost index in the Cost index in the year 2010}}{\text{Cost index in the Cost index in the year 2000}}\right)$$

$$= \text{x} \times \left[\frac{520}{480}\right]$$

GATE 2003

14. Direct cost component of the fixed capital consists of

A. contingency

B. onsite and offsite costs

C. labour costs

D. raw material costs

***Sol.* B.** Onsite and offsite costs.

15. A series of equal payments (eg, deposit or cost) made at equal intervals of time is known as

A. Perpetuity

B. capital charge factor

C. annuity

D. future worth

***Sol.* C.** A series of equal payments made at equal intervalsof time is called as annuity.

16. The variables required to be known in correlation used norm for estimating the horse power of acentrifugal gas and diffe compressor and hence its cost are

A. inlet pressure

B. compressor rpm

C. delivery pressure

D. volumetric flow rate at inlet

***Sol.* A.** Inlet pressure is to be known in correlation used for estimating the horse power of a centrifugal gas compressor and hence its cost are inlet pressure.

GATE 2002

17. The total investment in a project is ₹10 lakh and the annual profit is ₹1.5 lakh. If the project life is 10 yr, then the simple rate of return on investment is

A. 15%

B. 10%

C. 1.5%

D. 150%

***Sol.* C.** Rate of return on investment,

$$\text{i} = \frac{l}{p_n} = \frac{1.5}{10 \times 10} = 1.5\%$$

GATE 2001

18. An investment of ₹100 lakh is to be made for construction of plant which will take 2 yr to start production.The annual profit from operation of the plant is ₹20 lakh. What will be the payback time?

A. 5 yr

B. 7 yr

C. 12 yr

D. 10 yr

***Sol.* B.** Given,

Investment = ₹100 lakh

Annual profit = ₹20 lakh

Payback time = 2 + = 2+5 = 7yr.

19. A perforated plate has holes of diameter d_h arranged in a pitch p_h. Each hole has tube of diameter d_t passing through it. The ligament efficiency is given by

A. $\frac{p_h - d_h}{p_h}$

B. $\frac{p_h - \left(\frac{d_h d_t}{2}\right)}{p_h}$

C. $\frac{p_h - d_t}{p_h}$

D. $\frac{p_h - d_h}{d_h}$

***Sol.* C.** Ligament efficiency,= $\frac{p_h - d_t}{p_h}$

20. The ends of a cylindrical vessel can be closed by a head, which can be one of the four shapes. For the same thickness, choose the one which can withstand the highest pressure.

A. Flat plate
B. Hemispherical
C. Torispherical
D. Ellipsoidal

Sol. **D.** Ellipsoidal shape vessel can withstand highest pressure.

21. For a typical project, the cumulative cash flow is zero at the

A. end of the project life
B. break-even point
C. start-up onsists
D. end of the design stage

Sol. **B.** The cumulative cash flow is zero at the break-even point.

(2 Marks Questions)

GATE 2015

22. A proposed chemical plant is estimated to have a Fixed Capital (FC) of ₹24 crore. Assuming other costs to be small, the total investment may be taken to be same as FC. After commission (at t=0 yr), the annual profit before tax is ₹10 cr/yr (at the end of each year) and the expected life of the plant is 10 yr. The tax rate is 40% per year and a linear depreciation is allowed at 10% per year. The salvage value is zero. If the annual interest rate is 12%, the NPV (net present value or worth) of the project (in crore of rupees)(up to one decimal place) is________.

Sol. Given that,

Profit before tax, P = ₹10 crore

tax, ϕ = 40% = 0.4

Depreciation, D = 10% of investment

= 0.1 × 24 = ₹2.4 crore

Interest rate, i = 12%

time, t = 10 yr

Annual cash flow, A = $(1-\phi)$ P + ϕD

$= (1 - 0.4) \times 10 + 0.4 \times 2.4$

= ₹6.96 crore

Net present worth,

$$p = \frac{A[(1+i)^N - 1]}{i(1+i)^N} = \frac{6.96[(1+0.12)^{10} - 1]}{0.12(1+0.12)^{10}}$$

p = ₹39.3 crore

23. The cost of two independent process variable f_1 and f_2 after the total cost C_T (in lakh of rupees) of the process as per the following function

$$C_T = 100f_1 + \frac{1000}{f_1 f_2} + 20f_2^2 + 50$$

The lowest total cost C_T (in lakh of rupess) (up to one decimal place), is __________.

Sol. Given, $C_T = 100f_1 + \frac{1000}{f_1 f_2} + 20f_2^2 + 50$

$$\frac{\partial C_T}{\partial f_1} = p = 100 = \frac{1000}{f_2 f_1^2}$$

and $$\frac{\partial C_T}{\partial f_2} = q = \frac{-1000}{f_2 f_1^2} + 40f_2$$

For maxima of minima p = 0 and q = 0

$$100 = \frac{1000}{f_2 f_1^2} \text{ and } \frac{1000}{f_1 f_2^2} = 40f_2$$

$\Rightarrow \quad f_1^2 f_2 = 10$ ....(i)

or $\quad f_1 f_2^3 = 25$ ....(ii)

Dividing Eq. (ii) by Eq. (i),

$$\Rightarrow \quad \frac{f_1 f_2^3}{f_1^2 f_2} = 2.5$$

$$\Rightarrow \quad \frac{f_2^2}{f_1} = 2.5 \Rightarrow f = \frac{f_2^2}{2.5}$$

From Eq. (i),

$$\Rightarrow \quad \frac{f_2^4}{(2.5)^2} f_2 = 10$$

$\Rightarrow \quad f_2^5 = 62.5$ and $f_2 = 2.2865$

So, $$f_1 = \frac{(2.2865)^2}{2.5} = 2.091$$

$f_1 = 2.091$ and $f_2 = 2.2865$

Now, $$\frac{\partial^2 C_T}{\partial f_1^2} = \frac{+1000 \times 2}{f_2 f_1^3} = \frac{2000}{f_1^3 f_2} = r$$

$$S = \frac{\partial^2 C_T}{\partial f_1 \partial f_2} = +\frac{1000}{f_1^2 f_2^2}$$

$$t = \frac{\partial^2 C_T}{\partial f_2^2} = 40 + \frac{2000}{f_2^3 f_1}$$

$rt - s^2 > 0$ and $r > 0$

So, (f_1, f_2) is point of minima lowest cost.

$$C_T = 100 \times 2.091 + \frac{1000}{2.091 \times 2.2865} + 20 \times (2.2865)^2 + 50$$

C_T = ₹572.82 lakh

GATE 2014

24. A binary distillation column is operating with a mixed feed containing 20 mol% vapour. If the feed quality is changed to 80 mol% vapour, the change in the slope of the q-line is _________.

Sol. feed contains 20% vapour, so $q = 0.8$

Slope of q line $= \frac{q}{q-1} = \frac{0.8}{-0.2} = -4$

Now feed contains 80% vapour, $q = 0.2$

slope $= \frac{q}{q-1} = \frac{0.2}{-0.8} = \frac{-1}{4}$

Change in slope $= \frac{-1}{4} + 4 = 3.75$

25. A cash flow of ₹12,000 per year is received at the end of each year (uniform periodic payment) for 7 consecutive years. The rate of interest is 9% per year compounded annually. The present worth (in ₹) of such cash flow at time zero is ____________.

Sol. Present worth,

$$P = \frac{A[(1+i)^N - 1]}{i(1+i)^N}$$

$$P = \frac{12000 \times [(1.09)^7 - 1]}{0.09 \times (1.09)^7} = 60397$$

26. A polymer plant with a production capacity of 10000 ton/yr has an overall yield of 70%, on mass basis (kg of product per kg of raw material). The raw material costs ₹50000 per/ton. A process modification is proposed to increase the overall yield to 75% with an investment of ₹12.5 cror. In how many years can the invested amount be recovered with the additional profit________.

Sol. Total number of years = n

Total product = 10000 n

raw material, $= \frac{10000\ n}{0.7}$

Total cost of raw material $= \frac{10000\ n}{0.7} \times 50000$

From question

$$\frac{50000 \times 10000\ n}{0.7} - \frac{50000 \times 10000\ n}{0.75} = 12.5 \times 10^7$$

Solving, $n = 2.625$ yr.

GATE 2013

27. The purchase cost of heat exchanger of 20 area was ₹5,00,000 in 2006. What will be the estimated cost(in ₹ to the neatest integer) of a similar heat exchanger of 50 area in the year 2013? Assume the six-tenths factor rule for scaling and the cost index for 2006 as 430.2. The projected cost index for the year 2013 is 512.6,

A. 1054397
B. 1045697
C. 1032387
D. 1100397

Sol. **C.** Cost of 20 m^2heat exchanger in year 2013

$$C_{20} = \text{cost in year 2006} \times \left(\frac{\text{cost index in year 2013}}{\text{cost index in year 2006}}\right)$$

$$= \frac{512.6}{4302} \times 500000$$

$$= 595770$$

Cost of 50 m^2 heat exchanger in year 2013

$$C_{50} = C_{20} \times \left(\frac{\text{capacity of 50 m}^2\text{ heat echanger}}{\text{capacity of 20 m}^2\text{ heat echanger}}\right)^{0.6}$$

$$= 595770 \times \left(\frac{50}{20}\right)^{0.6}$$

$$= 1032387$$

28. A plant manufactures compressors at the rate of N unit/day. The daily fixed charges are ₹20000 and the variable cost per compressor is ₹500 + $0.2N^{1.3}$. The selling price per compressor is ₹1000. The number of compressors to be manufactured, to the nearest integer, in order to maximize the daily profit is

A. 228
B. 225
C. 220
D. 217

Sol. **D.** If the plant manufactures N units/ day, total daily

Profit, P = Total selling price – Total cost

= 1000 N – (500 + 0.2) N – 20000

P = 500 N – 0.2 – 20000

For maximum profit,

$$\frac{dP}{dN} = 0$$

$$500 - 02 \times 2.3\ N^{1.3} - 0 = 0$$

$$N^{1.3} = 1087$$

$$N = \sqrt[1.3]{1087}$$

$$N = 217$$

GATE 2012

29. A batch reactor produces 1×10^5 kg for a produce per year. The total batch time in hours of the reactor is $\sqrt[k]{P_8}$, where P_8 is the product per batch in kg and k = 1.0 h/$\sqrt{\text{kg}}$. The operating cost of the rector is ₹200/h. The total annual fixed charges are ₹340 $\times P_8$ and the annual raw material cost is ₹2 $\times 10^6$. The optimum size in kg of each batch (adjusted to the nearest integer) is

A. 748
B. 873
C. 953
D. 1148

***Sol.* C.** Given,

Product per batch = P_B

Product produced in a year = 1×10^5 kg

$\therefore$ Total number of batches = $\frac{1\times10^5}{P_B}$

And operating cost of one batch

$= 200 \times \sqrt[k]{P_B}$

Total cost = Annual raw material cost + Annual fixed charges + Annual operating cost

$$C = 2 \times 10^6 + 340 \times P_B + 200 \times \sqrt[k]{P_B} \times \frac{1 \times 10^5}{P_B}$$

$$= 2 \times 10^6 + 340P_B + \frac{2\times10^7}{\sqrt{P_B}} \quad \text{(as k = 1)}$$

For optimum cost, $\frac{dC}{dP_B} = 0$

$$340 + 2 \times 10^7 \times \left(-\frac{1}{2}\right) P_B^{-3/2} = 0$$

$$P_B^{\frac{3}{2}} = \frac{10^7}{340} = \frac{10^6}{34}$$

or $\quad P_B \approx \frac{10^4}{10.5} \approx 953$

30. Heat integration is planned in a process plant at an investment ₹2 × 10⁶. This would result in a net energy savings of ₹20 GJ per yr. If the nominal rate of interest is 15% and the plant life is 3 yrs, then the breakeven cost of energy, in ₹ Per GJ (adjusted to the nearest hundred), is

A. 33500

B. 43800

C. 54200

D. 65400

***Sol.* B.** If x is the break-even cost of energy in ₹per GJ,

Then, $\quad R = 20x = \frac{P.i(1+i)^n}{(1+i)^n - 1}$

$$20x = \frac{2\times10^6\times0.15\times(1.15)^3}{(1.15)^3-1}$$

$\Rightarrow \quad x = 43800$

GATE 2011

31. Fuel cell stacks are made of nine Membrane Electrode Assemblies (MEAs) interleaved between ten Bipolar Plates (BPS) as illustrated below. The width of a membrane electrode assembly and a bipolar plate are normally distributed with $\mu_{MEA} = 0.15$, $\sigma_{MEA} = 0.01$ and $\mu_{BP} = 5$, $\sigma_{BP} = 0.1$, respectively. The widths of the different layers are independent of each other.

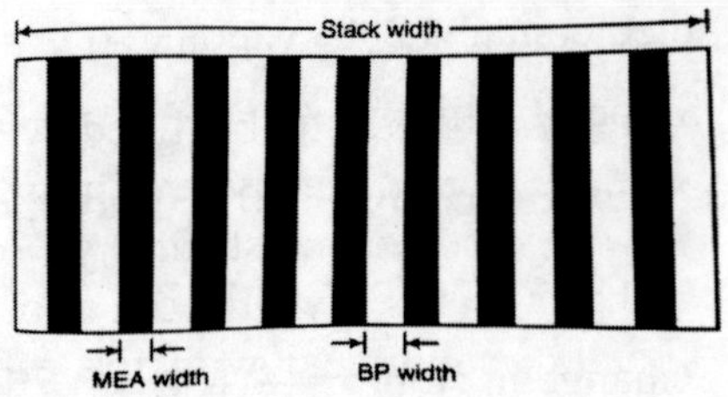

Which one of the following represents the correct value of (μ_{stack}, σ_{stack}) for the overall fuel cell stack width?

A. (51.35, 0.32)

B. (51.35, 1.09)

C. (5.15, 0.10)

D. (5.15, 0.11)

***Sol.* A.** $\mu_{stack} = 51.35$, $\sigma_{stack} = 1.09$

32. A process plant has a life of 7 yr and its salvage value is 30%. For what minimum fixed-percentage factor will be depreciation amount for the second year, calculated by declining balance method be equal to that calculated by the straight line depreciation method?

A. 0.1

B. 0.113

C. 0.527

D. 0.887

***Sol.* C.** f = 0.537

33. A continuous fractionator system is being designed. The following cost figures are estimated for a reflux ratio of 1.4.

Fixed cost including all accessories (₹)

column	condenser	reboiler
6×10^6	2×10^6	4×10^6

Operating cost (₹/yr) for

condenser cooling water	reboiler heating steam
8×10^6	1×10^6

The annualized fixed charge is 15% of the fixed cost. The total annualized cost (in ₹) is

A. 10.8×10^6

B. 13.35×10^6

C. 15.9×10^6

D. 3.15×10^6

***Sol.* A.** Total fixed cost = $6 \times 10^6 + 2 \times 10^6 + 4 \times 10^6$
$= 12 \times 10^6$

Given, annualized fixed charge = 15% of fixed cost = $0.15 \times 12 \times 10^6 = 1.8 \times 10^6$

Total operating cost = $8 \times 10^6 + 1 \times 10^6 = 9 \times 10^6$

Total annualized cost = $1.8 \times 10^6 + 9 \times 10^6$
$= 10.8 \times 10^6$

GATE 2010

34. A reactor needs to be lined with a corrosion resistant lining. One type of lining costs ₹5 lakh, and is expected to last for 2 yr. Another type of lining lasts for 3 yr. If both choices have to be equally economical with the effective interest rate being 18%, compounded annually, the price one should pay for the second type of lining is

A. ₹6.1 lakh
B. ₹6.5 lakh
C. ₹6.9 lakh
D. ₹7.6 lakh

***Sol.* C.** Given, $P_1 =$ ₹5 lakh, $P_2 = ?$

$n_1 = 2$ yr, $n_2 = 3$ yr

$i_1 = i_2 = 0.18\%$

by using equal capitalized cost method,

$$\frac{p_1(1+l_2)^{n1}}{(1+l_1)^{n1}-1} = \frac{p_2(1+l_2)^{n2}}{(1+l_2)^{n2}-1}$$

$$\frac{5(1+0.18)^2}{(1+0.18)^2-1} = \frac{p_2(1+0.18)^3}{(1+0.18)^3-1}$$

$$\frac{5 \times 1.3924}{0.3924} = \frac{p_2 \times 1.643}{0.643}$$

$P_2 =$ ₹6.9 lakh

Common Data for Questions 35 and 36

A plant produces phenol. The variable cost in ₹/ton of phenol is related to the plant capacity P (in ton/day) as 45000 + 5P. The fixed charges are ₹100000 per day. The selling price of phenol is ₹50000 per ton.

35. The optimal plant capacity (in ton/day) for minimum cost per ton of phenol, is

A. 101
B. 141
C. 283
D. 422

***Sol.* B.** The total cost phenol in ₹/ton is

C = Fixed cost in ₹/tone + Variable cost in ₹/ton

$$C = \left(\frac{100000\ Rs./day}{Pto\,A=\pi r^2 n/day}\right) + (45000 + 5P)\ ₹/ton$$

$$C = \frac{100000}{P} + 45000 + 5P$$

For minimum cost of phenol, $\frac{dC}{dP} = 0$

$$-\frac{100000}{P^2} + 0 + 5 = 0$$

$P^2 = 20000$

$P = 141.4$

$P = 141$ ton/day

36. The break even capacity in ton/day, is

A. 50
B. 40
C. 30
D. 20

***Sol.* D.** At break-even point,

Production cost of phenol = Selling price of phenol

$$\left(\frac{100000}{P}\right) + (45000 + 5P) = 50000\ ₹/ton$$

$5P^2 - 5000P + 10000 = 0$

$P^2 - 1000P + 20000 = 0$

$$P = \frac{-(-1000)\pm\sqrt{(1000)^2-4\times1\times20000}}{2\times1}$$

$$= \frac{1000 \pm 959.16}{2}$$

$= 979.58$ or 20.42

The break-even point would come at lower value, i.e., 20 ton/day.

GATE 2009

37. A column costs ₹5 lakh and has a useful life of 10 yr. Using the double declining balance depreciation method, the book value of the unit at the end of 5 yr in lakh of ₹, is

A. 1.21
B. 1.31
C. 1.64
D. 2.05

***Sol.* C.** Given, Salvage value = 0

Total cost P = ₹5 lakh

Useful life n = 10 yr

Annual departure value $= \frac{500000-0}{10} = 50000$

Departure rate $= \frac{50000}{500000} = 10\%$

Double declining departure rate f = 20%

Book value at the end of n years,

$P_n = P(1 - f)^n$

$P_5 = 500000 \times (1 - 0.2)^5$

$= 1.6384$

$P_5 =$ ₹ 1.64 lakh

38. Match the equipment in Group Ito the intervals in Group II.

Group I	*Group II*
P. Centrifugal pump Q. Distillation column R. Heat exchanger	I. Baffle II. Impeller III. Tray IV. Volute

A. P-II, Q-I, R-4
B. P-2, Q-4, R-3
C. P-I, Q-III, R-IV
D. P-IV, Q-III, R-I

***Sol.* D.** Centrifugal pump – Volute
Distillation column – Tray
Heat exchanger – Baffle

GATE 2008

39. A reactor has been installed at a cost of ₹ 50000 and is expected to have a working life of 10 yr with a scrap value of ₹10000. The capitalized cost (in ₹) of the reactor based on an annual compound interest rate of 5% is

A. 113600
B. 42000
C. 52500
D. 10500

***Sol.* A.** Given,
Instillation cost P = ₹50000
Working life n = 10 yr
Salvage value P_s = ₹10000
Compound interest rate i = 5%
Capitalized cost

$$C = P + \left[\frac{(P-P_5)}{(1+i)^n-1}\right]$$
$$= 50000 + \frac{(50000-10000)}{(1+0.05)^{10}-1}$$
$$= 50000 + 63603.66$$
= ₹113600

40. A cylindrical vessel with hemispherical ends is filled with water as shown in the figure. The head space is pressurized to a gauge pressure of 40 kN/m². The vertical force F (in kN) tending to lift the top dome and the absolute pressure p (in kN/m²) at the bottom of the vessel are (g = 9.8 m/s², density of water = 1000 kg/m³)

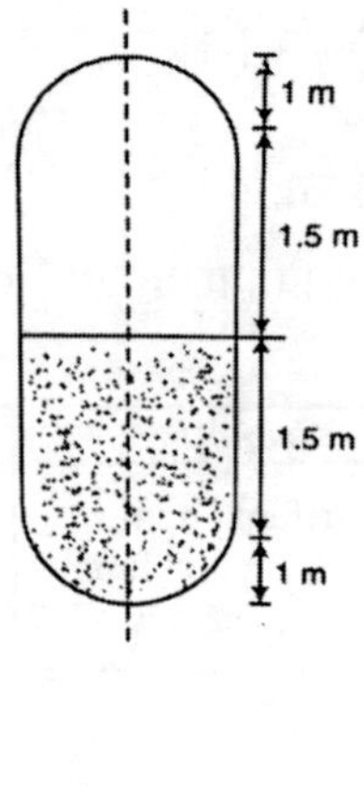

A. F = 83.6; p = 64.5
B. F = 83.6; p = 165.8
C. F = 125.7; p = 64.5
D. F = 125.7; p = 165.8

***Sol.* A.**

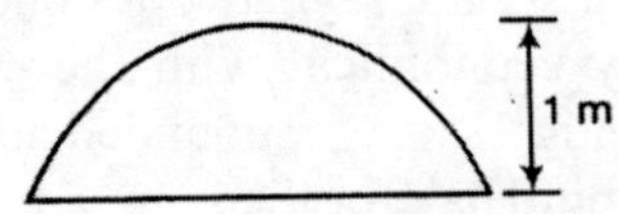

Vetical force tending to lift the top dome
$= 40 \times \pi \times 1$
= 125.7 kN

Absolute pressure at the bottom of the vessel
P = 40 + Atmospheric pressure + Hydrostatic pressure (ρgh)

$$= 40 + 101.3 + \begin{bmatrix} 1000 \times 9.8 \times (1+1.5) \\ \times \left(\frac{1}{1000}\right) \end{bmatrix}$$
= 165.8 kN/m²

GATE 2007

41. A pump has an installed cost of 40000 and a 10 yr estimated life. The salvage value of the pump is zero at the end of 10 yr. The pump value (in ₹), after depreciation by the double declining balance method, at the end of 6 yr is

A. 4295
B. 10486
C. 21257
D. 37600

***Sol.* B.** Given,
Salvage value = 0
Installed cost p = ₹40000
Useful life n = 10 yr

Annual departure value $= \frac{40000-0}{10} = 4000$

Departure rate $= \frac{4000}{40000} = 10\%$

Double declining departure rate f = 20%
Book value at the end of n years,
$P_n = P(1-f)^n$
$P_6 = 40000 \times (1-0.2)^{0.6}$
= 10485.76
P_6 = ₹10486

GATE 2006

42. In a desalination plant, an evaporator of area 200 m² was purchased in 1996 at a cost of $300000.

In 2002, another evaporator of area 50 m^2 was added. What was the cost of the second evaporator (in \$)? Assume that the cost of evaporators scales as $(\text{capacity})^{0.54}$. The Marshall and Swift indices was 1048.5 in 1996 and 1116.9 in 2002.

A. 130500

B. 139100

C. 141900

D. 151200

***Sol.* D.** Given,

In 1996, cost of evaporator of 200 m^2 area,

C_1 = \$ 300000

Given,

Cost of evaporator $\propto (\text{capacity})^{0.54}$

$$\frac{C_1}{(200)^{0.54}} = \frac{C_2}{(50)^{0.54}}$$

$$C_2 = 300000c\left(\frac{50}{200}\right)^{0.54}$$

$$= \$\ 141908.65$$

Now, we need to find the cost of evaporator of 50 m^2 area in year 2002.

Given,

Cost index in year 1996 = 1048.5

Cost index in year 2002 = 1116.9

50 m^2 area evaporator cost in 2002,

$$C_2' = C_2 \times \left(\frac{Cost\ index\ in\ 2002}{Cost\ index\ in\ 1996}\right)$$

$$= 141908.65 \times \left(\frac{1116.9}{1048.5}\right)$$

$$= 151166.2$$

$$= \$151200$$

43. The mixing of rubber latex solution was studied in an unbaffled mixer in the laboratory. The mixer was equipped with a six blade turbine impeller. A type company scales this process up using a baffled tank. The baffled tank has 3 times the diameter of the lab with scale mixer. It uses the same type of impeller operated at the same speed. The relevant shape factors are also the same. Assuming that laminar conditions prevail in both cases, the power requirement in the industrial scale mixer

A. is 3 times that of the lab scale mixer

B. is 9 times that of the lab scale mixer

C. is 27 times that of the lab scale mixer

D. cannot be estimated reliably due to the presence of battles

***Sol.* C.** Power P $\propto N^3$, where N is diameter.

$$\frac{P_1}{N_1^3} = \frac{P_2}{N_2^3}$$

$$P_2 = P_1 \times \left(\frac{N_2}{N_1}\right)^3$$

$$= P_1 \times \left(\frac{3}{1}\right)^3$$

$$P_2 = 27\ P_1$$

44. Due to a 20% drop in the product selling price, the pay-back period of a new plant increased to 1.5 times that estimated initially. The production cost and the production rate remaining unchanged. If the production cost is C_P and the new selling price is C_s, then $\frac{C_P}{C_s}$ is

A. 0.2

B. 0.4

C. 0.5

D. 0.6

***Sol.* B.** Assuming,

Initial selling price = C_s

Initial playback period = Cs

New production cost = t

New selling price = C_s' =0.8C_s

New payback period = t' = 1.5t

Payback period is defined as the in which profit becomes equal to the investment.

Thus,

$$I = (C_s - C_p)t$$

$$I' = (C_s' - C_p')t'$$

$$I = I'$$

$$(C_s - C_p)t = (0.8C_s - C_p)\times 1.5t$$

$$C_s - C_p = 1.2C_s - 1.5C_p$$

$$0.5\ C_p = 0.2\ C_s$$

$$\frac{C_p}{C_s} = 0.4$$

45. Obtain the optimal diameter of a cylindrical storage vessel of volume V. The curved shell costs C_s, (in ₹/m^2) and the flat top and bottom plates cost C_p (in R/m^2),

A. $D = \frac{C_s}{C_p}\left(\frac{4V}{\pi}\right)^{1/3}$

B. $D = \left(\frac{8VC_s}{\pi C_p}\right)^{1/3}$

C. $D = \left(\frac{VC_s}{C_p}\right)^{1/3}$

D. $D = \left(\frac{4VC_s}{\pi C_p}\right)^{1/3}$

***Sol.* B.**

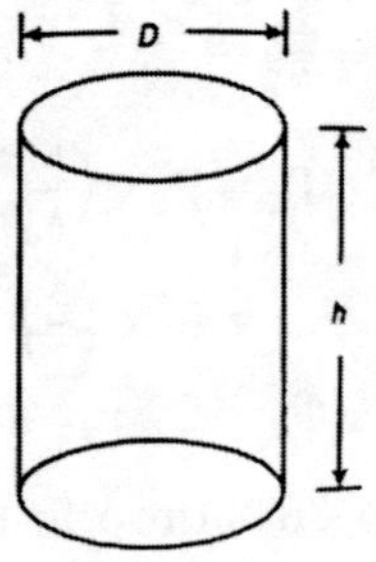

Cost of curved surface in Rs. = Area × Cs

$= \pi D h C_s$

Cost of top and bottom plates = Area × C_p

$= \frac{\pi D^2}{4} \times C_p$

Top cost C = $\pi D h C_s + \frac{\pi D^2 h}{4} C_p$

Volume of cylinder V = $\frac{\pi D^2 h}{4}$

$$h = \frac{4V}{D^2 \pi}$$

Putting the value of h in Eq.(1),

$$C = \pi D\left(\frac{4V}{D^2\pi}\right).C_s + \frac{\pi D^2}{4}.\ C_p$$

$$C = \frac{4V}{D^2}.C_s + \frac{\pi D^2}{4}.\ C_p$$

For optimal diameter of the cylindrical vessel,

$$\frac{dC}{dD} = 0$$

$$-\frac{4V}{D^2}.C_s + \frac{2\pi D}{4}.\ C_p = 0$$

$$\frac{4VC_s}{D^2} = \frac{\pi C_p}{2}.D$$

$$D = \left(\frac{8VC_s}{\pi C_p}\right)^{1/3}$$

46. A sale contract signed by a chemical manufacturer is expected to generate a net cash flow of ₹250000 per year at the end of each year for a period of 3 yr. The applicable discount rate (interest rate) is 10%. The net present worth of the total cash flow, is ₹

A. 750000
B. 683750
C. 621500
D. 332750

***Sol.* C.** Given,

Net cash flow R = 250000

Present worth P = ?

Number of years n = 3

Interest rate i = 10%

$$P = R\left[\frac{(1+i)^n - 1}{i(1+i)^n}\right]$$

$$= 250000 \times \left[\frac{(1+0.1)^3 - 1}{0.1(1+0.1)^3}\right]$$

$= 621712.99$

$= \$621500$

GATE 2005

47. The original value of an equipment is ₹10000. The salvage value is ₹ 500 at the end of its useful life period of 5 yr. What is the asset value in ₹ after 2 yr by textbook declining balance method?

A. 3025
B. 4010
C. 5020
D. 6050

***Sol.* A.** Given,

Initial equipment cost P = ₹10000

Usefull life n = 5 yr

Salvage value Ps = 500

P_2=?

We have,

$$P_n = P(1-f)^n$$

$$500 = 10000(1-f)^5$$

$$(1-f)^5 = 0.05$$

$$1 - f = 0.55$$

$$f = 0.45$$

Asset value after 2 yr,

$$P_s = 10000 \times (1-0.45)^2$$

$$P_2 = ₹\ 3025$$

48. The depreciable fixed cost is ₹100 lakh. The average profit per year is ₹15 lakh. The average depreciation cost per year is 10 lakh. What is the payout period in years, if there is no interest charge?

A. 8
B. 4
C. 10
D. 20

***Sol.* B.** payout period = $\frac{Fixed\ cost}{(Profit + Depreciation\ )Rs/yr}$

$= \frac{100}{15+10}$

= 4yr

GATE 2004

Common Data for Questions 49 and 50

Fixed capital investment for a chemical plant is 2 Rs.40 million with an estimated useful life of 6 yr and a salvage value of ₹ 4 million. The rate of interest is 15%, tax is 25% of the annual taxable income, In the first year of operation, the income from sales is ₹ 20 million and manufacturing industrial expenses are ₹5 million, The plant depreciates on a straight line basis,

49. The rate of return on investment is given by

A. 50%

B. 37.5%

C. 32%

D. 20%

Sol. **B.** The rate of return on investment

$$= \frac{Yearly\ profit}{Investment}$$

$$= \frac{20-5}{40} \times 100$$

$$= 37.5\%$$

50. The net present value (NPV) in million ₹ at the end of the first year of operation is to respectively given by

A. zero and -28.9

B. –40 end -28.9

C. –40 and 12.75

D. zero and 12.75

Sol. **A.** Net present value at the start of lst year = 0

Net present value at the end of lst year = x

$$= \frac{Initial\ investment\ - Salvage\ value}{Useful\ life}$$

$$= \frac{40-4}{6}$$

= ₹ 6 million

Value of investment at the end of lst year,

$(40 + 5) (1 + i) = 40 + 5 - 6 + x$

$45\ (1 + 0.15) = 39 + x$

x = ₹12.75 million

GATE 2003

51. Two pumps under consideration for installation plant have the following capital investments and salvage values,

Pump A: $C_1 = 40000$ and $C_{sal} = 3900$

Pump B: $C_1 = 50000$ and $C_{sal} = 20000$

Using capitalized cost, determine what should be the common life of the pumps for both competitive (economically equivalent) has been given like this only.

Interest rate is 10% per annum. Maintenance and operational costs are negligible.

A. 3 year

B. 5 year

C. 6 year

D. 8 year

Sol. **A.** For pump A,

$$C_{sal} = C_1(1 - f)^n$$

$$3900 = 40000 \times (1 + 0.1)^n$$

$$(0.9)^n = 0.0975$$

$$n \log (9 \times 10^{-1}) = \log (975 \times 10^{-4})$$

$$n(0.954 - 1) = 2.989 -4$$

$$n = 22 \text{ yr}$$

For Pump B,

$$C_{sal} = C (1 - f)^n$$

$$20000 = 50000 (1 - 0.1)^n$$

$$(0.9)^n = 0.4$$

$$n \log (9 \times 10^{-1}) = \log (4 \times 10^{-1})$$

$$n(0.954 - 1) = 0.602 -1$$

$$n = 8.6$$

Therefore, common life of both pumps in order to remain competitive = 8 yr

Common Data for Questions 52 and 53

A process has fixed capital investment of ₹150 lakh, working capital 30 lakh and salvage value zero. Annual revenues from sales are 250 lakh, manufacturing costs ₹145 lakh and other expenses 10% of the revenue. Assume project life span of 11 yr, tax life of 5 yr and interest rate 10%.Tax rate is 40% and straight line depreciation, i.e., 20% per year, is applicable.

52. Discounted value (to present time) of the profit before tax (for the total plant life period) in ₹is

A. 228 lakh

B. 400 lakh

C. 520 lakh

D. 660 lakh

Sol. **C.** ₹520 lakh

53. Discounted value of the depreciation benefit over the tax life in ₹ is

A. 12 lakh

B. 24 lakh

C. 46 lakh
D. 60 lakh

Sol. B. ₹24 lakh

GATE 2002

54. If an amount R is paid at the end of every years for n years, then the net present value of the annuity at an interest rate

A. $R\left[\frac{(1+i)^n-1}{i}\right]$
B. $R\left[\frac{(1+i)^n-1}{i(1+i)^n}\right]$
C. $R(1+i)^n$
D. $\frac{R}{(1+i)^n}$

Sol. B. Net present value of annuity = $R\left[\frac{(1+i)^n-1}{i[1+i]^n}\right]$

55. A company has depreciated invstment of ₹36400 which is depreciated in equal instalments in 2 yr. Assume that 50% and interest rate is 10%. The net present value of tax that the company would have saved, if it had depreciciated 2/3rd of the inverstment in the first year and the rest in the second year, is

A. Zero
B. 250
C. 375
D. 500

Sol. B. ₹250

56. Cylindrical pressure vessel of volume 6π m^3 has to be designed to withstand a maximum internal pressure of 10 atm. The allowable design stress of the material is 125 N/mm2 and corrosion allowance is 2 mm. The thickness of the vessel for a reflux ratio of 3 closed to

A. 5 mm
B. 6 mm
C. 8 mm
D. 10 mm

Sol. D. Given, $\frac{l}{d} = 3$

$\Rightarrow l = 3d$ ....(i)

Volume, $\pi r^2 l = 6\pi$

$\Rightarrow \frac{d^2}{4} \times 3d = 6$

$d^3 = 8$, d = 2 m

l = 6 m

Given, p = 10 atm = 10^6 N/m^2

$\sigma = \frac{pt}{t} = \frac{pd}{2t}$,

$t = \frac{pd}{2\sigma}$

$t = \frac{10^6 \times 2}{2 \times 125 \times 10^6} = 8 \times 10^{-3}$m

t = 8 mm

Also given, corrosion allowance = 2 mm

Thickness of the vessel = 8 + 2 = 10 mm

GATE 2001

57. An investment of ₹1000 is carrying on interest of 10% compounded quarterly. The value of the investment at the end of 5 year will be

A. $1000\left(1+\frac{0.1}{4}\right)^{20}$
B. $1000(1+0.10)^{20}$
C. $1000\left(1+\frac{0.1}{4}\right)^{5}$
D. $1000\left(1+\frac{0.1}{2}\right)^{5}$

Sol. B. i = 10%, n = 5 4 = 20

Value of investment after 5 yr = 1000 $(1 + 0.1)^{20}$

58. P is the investment made on an equipment, S is its salvage value and *n* is the life of the equipment in years. The depreciation for the mth year by the sum-of-years-digits method will be

A. $\frac{P-S}{n}$
B. $1 - \left(\frac{P}{S}\right)^n$
C. $\frac{m}{n}(P-S)$
D. $\frac{2(n-m+1)}{n(n+1)}(P-S)$

Sol. D. Depreciation by sum of year's digits method

$$= \frac{2(n-m+1)}{n(n+1)}(P-S)$$

59. In a cylindrical vessel subjected to internal pressure, the longitudinal stress σ_L, and the circumferential stress σ_H, are related by

A. $\sigma_H = 2\sigma_L$
B. $\sigma_H = \sigma_L$
C. $\sigma_H = \frac{\sigma_L}{2}$
D. No relation exists

Sol. A. $\sigma_H = 2\sigma_L$

9. Process Dynamics Control and Instrumentation

1. INSTRUMENTS:

Instruments can be classified as

I. Absolute instruments

II. Secondary instruments

Absolute instruments give the magnitude of the quantity under measurement in terms of physical constants of the instruments.

e.g:- Tangent galvanometer, Rayleigh's current balance.

Secondary instruments, the quantity under measurement can only be measured by observing the output of the instrument. The secondary instruments should be calibrated by comparing with an absolute instrument or another secondary instrument which has already been calibrated against an absolute instrument.

e.g:- Voltmeter, pressure gauge.

The secondary instruments are the commonly used instruments compared to the absolute instruments.

Electrical measuring instruments may be classified ***according to their functions*** as;

(*i*) Indicating instruments

(*ii*) Integrating instruments

(*iii*) Recording instruments

1. Indicating instruments:-

These instruments directly indicate the value of the electrical quantity at the time when it is being measured. In these instruments, a pointer moving over a graduated scale directly gives the value of the electrical quantity being measured.

e.g:- Ammeter, voltmeter, wattmeter.

2. Integrating instruments:-

The instruments which measure the total quantity of electricity (in Ampere hours) or electrical energy (in Watt hours) in a given time are called integrating instruments. In such instruments, there are a set of dials and pointers which register the total quantity of electricity or electrical energy supplied to the load.

e.g:- Ampere- Hour meter, Watt-hour meter.

3. Recording instruments:-

Recording instruments give a continuous record of the variations of the electrical quantity to be measured. A recording instrument is merely an indicating instrument with a pen attached to its pointer. The pen rests lightly on a chart wrapped over a drum moving with a slow uniform speed. The motion of the drum is in a direction perpendicular to the direction of the pointer. The path traced out by the pen indicates the manner in which the quantity being measured has varied during the time of the record.

e.g:- Recording voltmeters, recording ammeters in supply stations.

Indicating Instrument:-

An indicating instrument essentially consists of a moving system pivoted in jewel bearings. A pointer is attached to the moving system which indicates the electrical quantity to be measured, on a graduated scale. In order to ensure the proper operation of the indicating instruments, the following three torques are required.

(*i*) Deflecting (or operating) torque.

(*ii*) Controlling (or restoring) torque.

(*iii*) Damping torque

The ***deflecting torque*** is produced by utilizing the various effects (magnetic effect, induction effect, thermal effect, hall effect) of electric current or voltage, and causes the moving system and hence the pointer to move from zero position.

The ***controlling torque*** is produced by spring or gravity and opposes the deflecting torque. The pointer comes to rest at a position, where these two opposing torques are equal.

Damping torque is provided by air friction or eddy currents. It ensures that, the pointer comes to the final position, without oscillations, thus enabling accurate and quick readings to be taken.

(a) Deflecting torque (T_d) :-

The deflecting torque causes the moving system to move from zero position to indicate the value of the electrical quantity being measured on a graduated scale. The actual method of producing the deflecting torque depends upon the type of instrument.

(b) Controlling torque(T_c):-

The controlling torque (T_c) opposes the deflecting torque and increases with the deflection of the moving system. The pointer comes to rest at a position where the two opposing torques are equal i.e. $T_d = T_c$.

The **controlling torque** performs two functions.

- Controlling torque increases with the deflection of the moving system so that the final position of the pointer on the scale will be according to the magnitude of an electrical quantity (i.e. current or voltage or power) to be measured.
- Controlling torque brings the pointer back to zero when the deflecting torque is removed. If it were not provided, the pointer once deflected would not return to zero position on removing the deflecting torque. The *controlling torque* in indicating instruments may be provided by one of the following two methods:

    (i) Spring control

    (ii) Gravity control

(c) Damping torque (T_{damp}):-

A *damping torque* is produced by a damping or stopping force which acts on the moving system only when it is moving and always opposes its motion. Such a torque is necessary to bring the pointer to rest quickly. If there is no **damping torque**, then the pointer will keep moving to and fro about its final deflected position for some time before coming to rest, due to the inertia of the moving system.This damping torque acts only when the pointer is in motion and always opposes the motion. The position of the pointer when stationary is, therefore, not affected by **damping torque**. The degree of damping decides the behavior of the moving system.If the instrument is under-damped, the pointer will oscillate about the final position for some time before coming to rest. On the other hand, if the instrument is over damped, the pointer will become slow and lethargic.

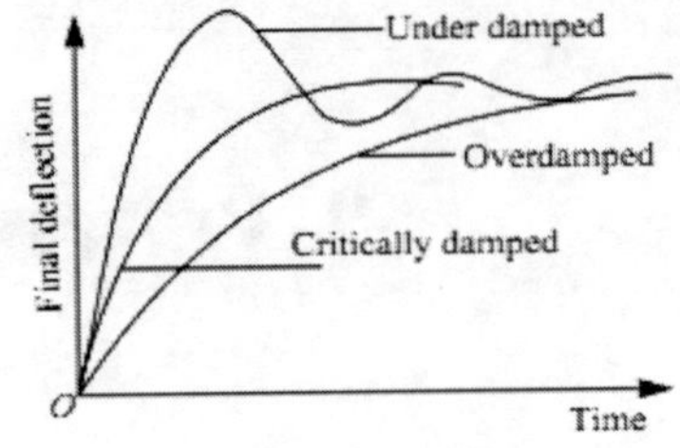

- *Dynamic response of a measuring instrument*

However, if the degree of damping is adjusted to such a value that the pointer comes up to the correct reading quickly without oscillating about it, the instrument is said to be critically damped. The damping torque in indicating instruments can be provided by:

➢ Air- friction

➢ Fluid friction

➢ Eddy currents

Permanent Magnet Moving Coil Instruments(PMMC)

PMMC instruments work on the principle that, when a current carrying conductor is placed in a magnetic field, a mechanical force acts on the conductor. The current carrying coil, placed in magnetic field is attached to the moving system. With the movement of the coil, the pointer moves over the scale to indicate the electrical quantity being measured.

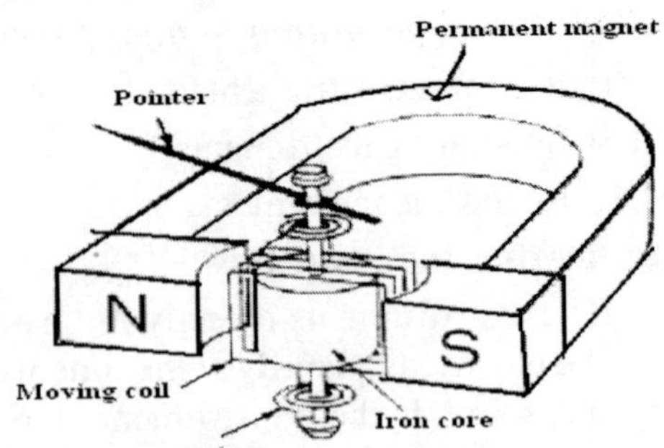

Deflecting torque equation:-

Deflecting torque, T_d = Force × perpendicular distance

Let, B = flux density, Wb/m²

l = length or depth of coil, m

b = breadth of the coil.

N = no. of turns of the coil.

Force on each coil side, F = BIlN Newtons.

$_d$ = Force × perpendicular distance

= (BIlN) × b

T_d = BINA Newton metre.

Where, A = l × b, the area of the coil in m^2.

Thus, $T_d \alpha I$

The instrument is spring controlled so that, $T_c \alpha \theta$

The pointer will comes to rest at a position, where $T_d = T_c$

Therefore, $\alpha\ \theta\ I$

The deflection is directly proportional to the operating current. Hence, such instruments have uniform scale.

Advantages:-

(1) High efficiency.

(2) Lower power consumption

(3) No hysteresis loss (as the magnetic field is constant).

(4) Very accurate and reliable.

Disadvantages:-

(1) Cannot be used for a.c. measurements.

(2) More expensive (about 50%) than the moving iron instruments because of their accurate design.

(3) Some errors are caused due to variations (with time or temperature) either in the strength of permanent magnet or in the control spring.

Applications:-

(1) In the measurement of direct currents and voltages.

(2) In d.c. galvanometers to detect small currents.

(3) In Ballistic galvanometers used for measuring changes of magnetic flux linkages.

Moving Iron Instruments

Moving iron instruments are quite cheap in cost, simple in construction and very accurate at fixed power supply frequency. These instruments can be used on AC as well as DC. Therefore, these are widely used in laboratories and on switching panels. These instruments are used either as ammeters or voltmeters only. **Moving iron instruments** are of two types, namely;

- Attraction type moving iron instruments.
- Repulsion type moving iron instruments.

Measurement of Temperature:

Temperature measurement is very important measurement in the industrial applications. Temperature in the industry can be as low as the refrigeration temperature to as large as the temperature of boiler, furnace or sun.

Important devices for measuring temperature:-

Thermocouples –

A thermocouple is an electrical measuring device, which is based on seeback effect. When the two dissimilar metals are joined together, an emf will exist between the two point A and B which is a function of the junction temperature.

EMF produced between the terminal A and B given by

$$E = a(\Delta\theta) + b(\Delta\theta)^2$$

Where, E = emf developed (in mV)

$\Delta\theta$ = Temperature difference between the measuring (hot) junction and the reference junction (in °C)

A,b = Constants

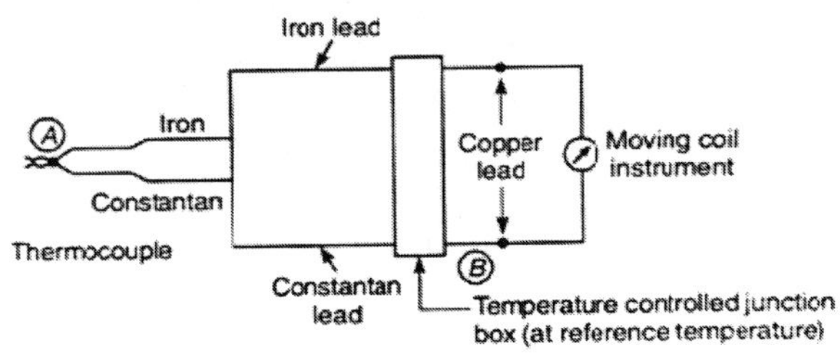

Thermistor –

A thermistor is a type of resistor whose resistance is dependent on temperature, more so than in standard resistors. The word is a portmanteau of thermal and resistor. Thermistors are widely used as inrush current limiter, temperature sensors (Negative Temperature Coefficient or NTC type typically), self-resetting overcurrent protectors, and self-regulating heating elements (Positive Temperature Coefficient or PTC type typically).

Thermistors are of two opposite fundamental types:

- With NTC thermistors, resistance decreases as temperature rises. An NTC is commonly used as a temperature sensor, or in series with a circuit as an inrush current limiter.

- With PTC thermistors, resistance increases as temperature rises. PTC thermistors are commonly installed in series with a circuit, and used to protect against overcurrent conditions, as resettable fuses.

Pyrometer –

A pyrometer is a type of remote-sensing thermometer used to measure the temperature of a surface. Various forms of pyrometers have historically existed. In the modern usage, it is a device that from a distance determines the temperature of a surface from the spectrum of the thermal radiation it emits, a process known as pyrometry and sometimes radiometry.

Resistance thermometer –

Resistance thermometers, also called resistance temperature detectors (RTDs), are sensors used to measure temperature. Many RTD elements consist of a length of fine wire wrapped around a ceramic or glass core but other constructions are also used. The RTD wire is a pure material, typically platinum, nickel, or copper. The material has an accurate resistance/temperature relationship which is used to provide an indication of temperature. As RTD elements are fragile, they are often housed in protective probes.

RTDs, which have higher accuracy and repeatability, are slowly replacing thermocouples in industrial applications below 600 °C.

pH value –

- pH value of solution is the negative logarithm of the hydrogen ion concentration.

$$pH = -\log_{10}[H^+]$$

pH is an important parameter to determine the quality of water. In pure water, the concentration of H^+ ions is 10^{-7} g/L at 25°C. So, the pH value is

$$pH = -\log 10^{-7}$$
$$= (-7)\log_{10}10 = 7$$

Measurement of pH Value –

The pH value of a solution is measured by using pH electrode. It consist of a pair of electrodes and reference electrode both the electrodes are dipped in the solution of unknown pH. These two electrodes from two half-cells. The net potential developed is the difference between the individual electric potential developed in each half cell.

Multiple Choice Question

1. The measurement of __________ pressure is done by strain gage pressures transducers .
   A. absolute as well as differential
   B. gage as well as vacuum
   C. both **A**and **B**
   D. neither **A** nor **B**
2. Which of the following could be the closed loop pole of a stable second order system ?
   A. both real and positive
   B. complex conjugate with positive real parts
   C. both real and negative
   D. one real positive and the other real negative
3. The following measurements can be made by measurement of emf ?
   A. composition of complex ions
   B. degree of hydrolysis
   C. pH value
   D. all **A**, **B**&**C**
4. Which of the following systems are first order in nature ?
   A. interacting system of two tanks in series
   B. mercury in glass thermometer kept in boiling water
   C. damped vibrator
   D. non-interacting system of two tanks in series
5. What is the basis of working of a magnetic flowmetre ?
   A. based on the linear relationship between the fluid flow rate and the induced voltage
   B. capable of measuring the flow rate of slurries and electrolytes
   C. based on the principle of Faraday's law
   D. all **A**, **B** and **C**
6. Which is the maximum service temperature of fiber glass used as thermocouple wire insulation material in °C?
   A. 250
   B. 500
   C. 750
   D. 1000
7. Which of the following thermo meters is based on the principle of volumetric expansion?
   A. mercury in glass
   B. bimetallic
   C. vapor pressure
   D. constant volume gas
8. IfT_i is the integral (reset) time and T_D is the derivative time. What will be the transfer function for a PID controller?
   A. $K_C\left(1+\frac{1}{T_iS}+\frac{1}{T_DS}\right)$
   B. $K_C\left(1+\frac{1}{T_iS}+T_D\,S\right)$
   C. $K_C(1+T_iS+T_D)\,S$
   D. $K_C\left(1+T_iS+\frac{1}{T_DS}\right)$
9. Which among the following is the functions of a transducer?
   A. codify/decodify the input signal
   B. modify the input signal
   C. convert the primary signal into a more useful quantity, usually an electric impulse
   D. amplify the input signal
10. Which of the following deals with Nichol's chart?
   A. frequency response values of controllers
   B. A.R. *vs*. phase lag of first order
   C. closed loop values *vs*. open loop values
   D. A.R. *vs*. phase lag of second order
11. __________ flue gases going out of the furnace is not continuously measured by Zirconia probe.
   A. oxygen in
   B. temperature of

C. carbon dioxide in

D. carbon monoxide in

12. Which among following the thermocouple will give the highest output for the same value of hot and cold junction?

A. iron-constantan

B. platinum-platinum + rhodium

C. chromel-constantan

D. all will give the same output

13. The offset will increase if for a proportional controller there is a gain of K_c which is used to control a first order process.

A. derivative control action is introduced

B. K_c is reduced

C. integral control action is introduced

D. K_c is increased

14. Which of the following statements is correct?

A. the amplitude ratio of a pure time delay system increases with frequency

B. the frequency response of a pure capacity process is unbounded

C. the amplitude ratio of a pure capacity process is inversely proportional to frequency

D. the phase lag of a pure time delay system decreases with increasing frequency

15. In instrumentation diagram the symbol for "capillary line" is

A. ————————

B. —×—×—×—

C. -·-·-·-·-·-·-

D. —//—//—//—

16. Which of the following is a secondary element of a vapour pressure thermometer?

A. bulb

B. Bourdon tube

C. pointer

D. none of these

17. Mass spectrometer is used for compositional analysis of __________.

A. an isotope

B. an alloy

C. natural gas

D. a solid

18. __________ relates the absorption & evolution of heat at the junctions of a thermocouple to the current flow in the circuit.

A. Thomson effect

B. Peltier effect

C. Seebeck effect

D. Joule heating effect

19. Which of the following is/are advantage(s) of instrumentation in a plant?

A. greater operation economy

B. greater safety of operation

C. better quality of product

D. all **A**, **B** and **C**

20. Which of the following is need for continuous shell temperature in a liquid-liquid heat exchanger?

A. thermocouple

B. reistance thermometer

C. vapor pressure thermometer

D. mercury in glass thermometer

21. Which of the following is based on the working principle of a thermal conductivity measurement.

A. CO_2 analyser

B. chromatograph

C. polarimeter

D. spectrometer

22. What is the type of thermocouple used for emf developed by a thermocouple while measuring a temperature of 600°C voltage is about 5.5 mV ?

A. iron-constantan

B. chromel-alumel

C. platinum - platinum + rhodium

D. either **A**, **B** or **C**

23. Which among the following is not the triple point of the water.

A. 32°R

B. 273°K

C. 492°R

D. 32°F

24. The flow rate of heating/cooling fluid in a shell and tube heat exchanger, is the __________ variable.

A. controlled

B. load

C. manipulated

D. none of these

25. What is the damping ratio of the second order system with the transfer function of $\frac{4}{S^2 + 2S + 4}$?

A. 2.0
B. 0.5
C. 1.0
D. 4.0

26. The standardisation of temperature measuring instruments are done mostly with point of pure substances.
A. boiling
B. melting
C. both **A&B**
D. neither **A** nor **B**

27. Which of the following is used for the measurement of Load cells?
A. velocity
B. weight
C. stress
D. strain

28. Which of the following statements is wrong?
A. paramagnetic susceptibility of gases permits the measurement of their concentration.
B. O_2, NO and NO_2 exhibit paramagnetic properties as a result of unpaired electrons.
C. paramagnetic susceptibility of gases decreases with temperature.
D. CO_2 is the strongest paramagnetic gas.

29. What will be the Laplace transform for an input forcing function $X(t) = 2t^2$?
A. $2/s^2$
B. $4/s^2$
C. $2/s^3$
D. $4/s^3$

30. Which of the following of a vapour pressure thermometer is a functioning element ?
A. pointer
B. bulb
C. bourdon tube
D. none of these

31. Which of the following is a first order instrument?
A. mercury in glass thermometer (without any covering or air gap).
B. bare vapor pressure thermometer.
C. bare metallic thermometer.
D. all (a), (b) and (c).

32. Among the following which is the dynamic characteristics of an instrument.
A. reproducibility
B. dead zone
C. sensitivity
D. fidelity

33. Typical specifications for design stipulates the gain margin and phase margin to be respectively
A. > 1.7 and $> 30°$
B. < 1.7 and $> 30°$
C. < 1.7 and $< 30°$
D. > 1.7 and $< 30°$

34. Among the following which is an undesirable dynamic characteristics of an instrument.
A. static error
B. reproducibility
C. time lag
D. dead zone

35. Which of the following represents a negative gain margin expressed in decibels?
A. critically damped
B. unstable
C. stable
D. none of these

36. On which of the following the operation of a rotameter is based ?
A. variable flow area
B. rotation of turbine
C. pressure drop across a nozzle
D. pressure at a stagnation point

37. Thermistor, which has high temperature co-efficient of resistivity is used as the sensing element in resistance thermometer. It is a/an
A. liquid semi-conductor
B. insulator
C. solid semi-conductor
D. conductor

38. Flapper nozzle is used in a/an __________ controller.
A. hydraulic
B. electronic
C. pneumatic
D. none of these

39. Which of the following Laws forms the basis of working principle of radiation pyrometer?
A. seebeck effect
B. wien's law

C. Stefan-Boltzman law

D. Kirchoffs law

40. Among the following it is __________ which relates the emf generated in a single homogeneous wire to the temperature difference .

A. Seeback effect

B. Thomson Effect

C. Peltier effect

D. none of these

41. What is the transfer function of level to inlet flow rate in a single tank system ?

A. R / TS

B. R / (TS + 1)

C. I (TS + 1)

D. 1 / TS

42. Which of the following is/are advantage(s) of instrumentation in a plant?

A. greater operation economy

B. greater safety of operation

C. better quality of product

D. all A, B and C

43. During the initial temperature development water wire was used as a thermometric liquid. Which was the first empirical temperature scale developed ?

A. Kelvin

B. Centigrade

C. Fahrenheit

D. Reaumer

44. Among the following controllers __________ has the least maximum deviation.

A. *P-I-D* controller

B. *P*-controller

C. *P-I* controller

D. *P-D* controller

45. Among the following the __________ need not touch the object whose temperature is being measured.

A. radiation/infrared pyrometer

B. mercury in glass thermometer

C. thermo electric pyrometer

D. filled system thermometer

46. The desirable characteristic of an instrument is

A. high drift

B. high fidelity

C. high measuring leg

D. poor reproducibility

47. When a sinusoidal input is introduced to a system the response of the system is called __________ response.

A. unit step

B. impluse

C. frequency

D. none of these

48. Thermocouple is used to measure the temperature of tempring oil baths maintained at 400 °C during heat treatment of steel .

A. chromel-alumel

B. iron-constantan

C. platinum-platinum/rhodium

D. none of these

49. Which of the following represents the transfer function of transportation lag?

A. $\frac{1}{T_S + 1}$

B. e^{-TS}

C. e^{TS}

D. none of these

50. Which of the following is used for measuring the smoke density of the flue gas going out of the chimney?

A. thermal conductivity meter

B. chromatograph

C. photo electric cell

D. polarograph

51. Which of the following is another term used for reset time?

A. derivative

B. integral

C. dead

D. none of these

52. Which of the following is used as positioning controllers?

A. temperature changes

B. flow rate changes

C. high loads

D. low loads

53. Which of the following is the response of two systems taken of same size and resistance in series?

A. over damped

B. critically damped

C. under damped

D. none of the above

54. What will be the transfer function of such a system which is initially at steady state with output y = 0 for an input u = 1 when the input is suddenly changed to 2 at t = 0. The output response is y(t)

A. 2/s

B. $1 + \frac{2}{s}$

C. $1 + \frac{2}{s^2}$

D. $\frac{1}{s}\left(1 + \frac{2}{s}\right)$

55. What is the transfer function of the system where element exhibit a negative phase shift at all frequencies and the absolute valve of the phase shift increases linearly with frequency. The frequency response of the dynamic element shows a constant magnitude ratio at all frequencies.

A. $e^{-T}s$

B. $\frac{T_1 s+1}{T_2 s+1}$

C. $\frac{T_1 s+1}{T_2 s}$

D. $\frac{T_1 s}{T_2 s+1}$

56. Which of the following is the amplitude ratio for the sinusoidal response < 1?

A. second order system

B. first order system

C. transportation lag

D. none of these

57. Which of the following is used for continuous measurement of moisture in paper?

A. weighing.

B. hair-hygrometer.

C. sling psychrometer.

D. high resistance Wheatstone's bridge circuit.

58. A ________ is a "Contact pyrometer"

A. resistance pyrometer

B. radiation pyrometer

C. infra red pyrometer

D. optical pyrometer

59. __________ is suitable for a turbine flow meter.

A. aerospace and in air borne applications

B. cryogenic flow measurements

C. very limited flow ranges

D. both **B** and **C**

60. Which of the following is used to measure sub-zero Celsius temperature most commonly in industry?

A. bi-metallic thermometers

B. resistance thermometers

C. gas thermometers

D. thermocouples

61. A ________ is not a speed measuring instrument.

A. Psychrometer

B. Tachometer

C. Stroboscope

D. All **A**, **B** and **C**

62. Which of the following is the amplitude ratio for the sinusoidal response equal to one?

A. second order system

B. first order system

C. transportation lag

D. none of these

63. Which of the following is the response of two tanks of same size and resistance in series?

A. over damped

B. critically damped

C. under damped

D. none of the above

64. Which of the following is analysis using a polarigraph?

A. gases

B. solids

C. liquids

D. isotonic solutions

65. Due to __________ magnetic flowmeters are generally not used for the velocity/flow measurement of hydrocarbons.

A. low flash point.

B. low electrical conductivity.

C. low thermal conductivity.

D. high viscosity index.

66. _________ is not a mechanical pressure sensing.

A. bourdon tube

B. diaphragm

C. bellows

D. U-tube

67. Which of the following facilitates the case of *I*-control with *P*-control?

A. elimination of offset

B. reduction of stability time

C. reduction of offset
D. none of these

68. __________ is used as a parameter to judge the accuracy of an instrument.
A. drift
B. dead zone
C. static error
D. none of these

69. Which of the following can be used for the measurement of pressure of 0.01 psi absolute?
A. Pirani
B. ionisation
C. McLeod
D. none of these

70. __________ cannot measure liquid discharge from a tank or reservoir?
A. orificemeters
B. mouthpieces
C. notches
D. weirs

71. For higher speed of response of an expansion pressure spring thermometer, the thermometer bulb should have a.
A. large area but small mass
B. high thermal conductivity
C. small specific heat
D. AllA, B&C

72. Which of the following __________ the unit step response of the transfer function $1/(s^2 + 2s + 3)$?
A. is unstable
B. has a damped oscillatory characteristics.
C. has a non-zero slope at the origin.
D. is overdamped

73. A copper resistance thermometer can be used in the temperature range of __________ °C
A. 0 to 150
B. -200 to 150
C. 350 to 750
D. 500 to 900

74. __________ of the fluid contained in the temperature sensing element (i.e., bulb) of filled system thermometers changes with change in temperature.
A. volume
B. viscosity

C. pressure
D. all A, B and C

75. Which of the following is normally used as on-off controller?
A. low loads
B. flow rate changes
C. temperature changes
D. none of these

76. A __________ is used for measurement of high temperature of around 2000°C of incandescent gas mantles.
A. optical pyrometer.
B. specially designed thermocouple.
C. radiation pyrometer.
D. none of these.

77. What is the appropriate transfer function of the process for which the experimental response of the controlled variable y(t) for a step change of magnitude p in the manipulated variable x(t) is shown below.

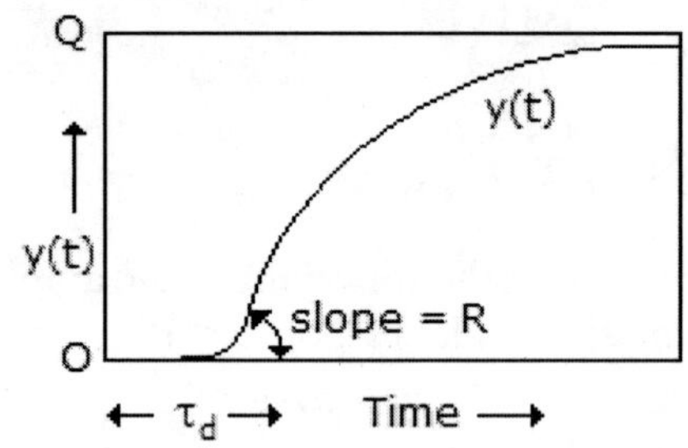

A. $\frac{(Q/R)e^{-(P/Q)S}}{T_d S+1}$

B. $\frac{(Q/P)e^{-(Q/R)S}}{T_d S+1}$

C. $\frac{(Q/P)e^{-T_d S}}{(Q/R)S+1}$

D. $\frac{(Q/R)e^{-T_d S}}{(Q/P)S+1}$

78. A __________ has maximum offset?
A. *P*-controller
B. *P-I-D* controller
C. *P-I* controller
D. *P-D* controller

79. Which of the following is not used for paramagnetism method for the composition determination of gas/gases.
A. oxides of nitrogen
B. oxygen

C. carbon dioxide

D. any of these

80. Which of the following can measure the level of liquid argon in a pressurised tank at 3 kg/cm^2?

A. external float gage

B. gage glass

C. differential pressure gage

D. none of these

81. How many decibels is equivalent to an amplitude ratio of 0.1?

A. 20

B. –20

C. 10

D. –10

82. Which if the following may be used for pressure determination of 0.0001 psi?

A. McLeod

B. thermocouple

C. Pirani

D. none of these

83. Which of the following pressure range is suitable for measurment by Pirani gauge?

A. liquid level at atmospheric pressure

B. high vacuum.

C. very high pressure.

D. liquid level under pressure.

84. Transfer function of *X*/*Y* for which of the following block diagramsis $\frac{G_3 + G_1 \cdot G_2 \cdot G_3}{1 + G_2 G_3}$

A.

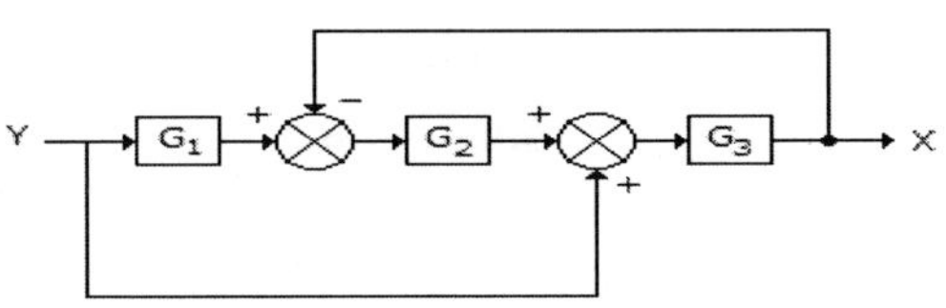

B.

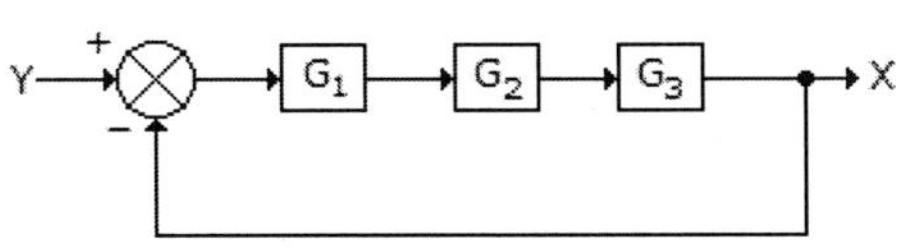

C.

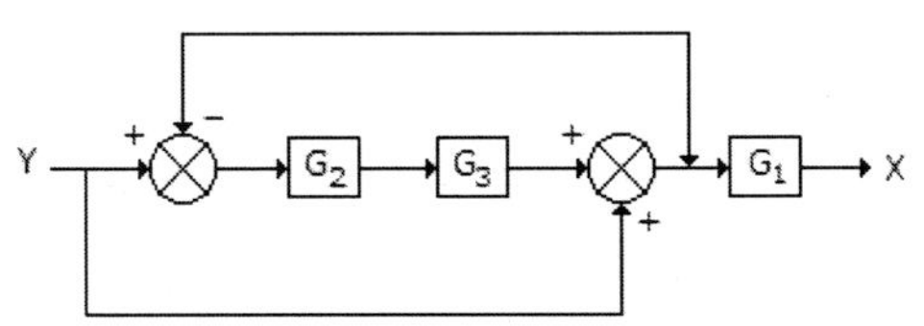

D. none of these

85. What is the starting temperature of an optical radiation pyrometer in °C ?

A. 800

B. 400

C. 1200

D. 1500

86. Which type of thermometer cannot measure subzero (< 0° C) temperature?

A. resistance

B. bimetallic

C. vapor pressure

D. mercury in glass

87. Which among the following has state transfer function.

A. $1/(s^2 + 2)$

B. $1/(s^2 - 2s + 3)$

C. $1/(s^2 + 2s + 2)$

D. $\exp(-20s)/(s^2 + 2s - 1)$

88. What will be the deflection with linear temperature of the free end of the bimetallic strip in a bimetallic thermo meter?

A. linear

B. parabolic

C. hyperbolic

D. non-linear

89. What will be the characteristic equation for the control system if the characteristic equation for the control system with closed loop transfer function $G_1/1 + G_2$ is given by $1 + G_2 = 0$?

A. depends only upon the open loop transfer function.

B. is the same for set point or load changes.

C. determines its stability.

D. all **A**, **B**&**C**

90. After ________ the unit step response of the transfer function $\frac{2S - 1}{(3S + 1)(4S + 1)}$ reaches its final steady state assymptotically.

A. a monotonic increase.

B. increasing decreasing and then increasing.

C. initially increasing and then decreasing.

D. a monotonic decrease.

91. What will be the frequency response of an ideal PID control having a transfer function $\left[1 + \frac{1}{0.55} + 0.25\right]$ at which the magnitude ratio of the controller is 1?

A. 0.5/0.2
B. 0.2/0.5
C. 0.2 × 0.5
D. 1/0.2 × 0.5

92. __________ is the dynamic error in a critically damped second order system in a critical damped second order system.
A. 0.5 AT
B. 2 AT
C. AT
D. 1.5 AT

93. What is the pressure measured in microns by an Alphatron ?
A. < 1
B. < 3
C. > 3
D. < 0.1

94. Which of the following parameters is determined by a Psychrometer ?
A. humidity of gases
B. water of crystallization
C. moisture content of solids
D. hygroscopic nature of solids

95. Which of the following is the transfer function for a function for a first order process with time delay?
A. $\frac{1}{(TS+1)(T_d S+1)}$
B. $\frac{e^{-Td.S}}{TS+1}$
C. $\frac{e^{Td.S}}{TS+1}$
D. $\frac{TdS}{TS+1}$

96. A mercury barometer is used to measures __________ pressure.
A. atmospheric
B. gauge
C. vacuum
D. absolute

97. __________ is the ratio of output amplitude to input amplitude for a sinusoidal forcing function in a first order system ?
A. 1
B. >1
C. < 1
D. none of these

98. Among the following which system having transfer function is stable.
A. $\frac{\exp(-20.5)}{S^2+2S+1}$
B. $\frac{1}{S^2+2}$
C. $\frac{1}{S^2+2S+2}$
D. $\frac{1}{S^2-2S+3}$

99. What is the corresponding term in thermal system analogous to the electrical current
A. temperature difference
B. heat flow rate
C. heat content in the system
D. none of these

100. The calibration data of a thermocouple with its cold junction at 0°C is given below.

Hot junction temperature (°C)	0	20	40	60	80	100
Thermo emf (mv)	0.00	0.80	1.61	2.43	3.26	4.10

The hot junction of the thermocouple is placed in a bath at 80°C, while its cold junction is at 20°C. What is the emf of thermocouple ?
A. 3.26 mv
B. 0.80 mv
C. 2.46 mv
D. 2.43 mv

101. A __________ is based on the working principle of plarisation light.
A. polarimeter
B. polarograph
C. chromatograph
D. spectrometer

102. Among the following the most suitable instrument for measuring pressure below 3 microns is
A. alphatron
B. bourdon guage
C. ionisation gauge
D. mcleoid gauge

103. A __________ loop transfer function is used by Routh stability criteria.
A. open
B. closed
C. either **A** or **B**
D. neither **A** nor **B**

104. The inlet temperature of heating/cooling fluid in a shell and tube heat exchanger is the __________ variable.
   A. load
   B. manipulated
   C. controlled
   D. none of these

105. Among the following thermocouple wire insulating materials the highest temperature rating is for.
   A. fibre glass
   B. ceramic fibre
   C. asbestos
   D. teflon

106. Among the following flow metering instrument an areameter is .
   A. pitot tube
   B. rotameter
   C. venturimeter
   D. hot wire anemometer

107. Among the following error integrals the best criterion in controller setting is .
   A. $\int_0^\infty |e|dt$
   B. $\int_0^\infty |e|dt$
   C. $\int_0^\infty e^2 dt$
   D. none of these

108. Thermal wells for temperature measurements should have.
   A. very thick walls.
   B. low emissivity.
   C. polished surface.
   D. high transmissivity of radiation.

109. For the analysis of __________ polarograph may be a suitable device.
   A. miscible liquids
   B. isotopes
   C. solids
   D. gaseous mixture

110. The denominator of _________ loop transfer function is indicated by characteristic equation.
   A. open
   B. closed
   C. both **A** and **B**
   D. neither **A** nor **B**

111. __________ number of controllers to be used is indicated by process degree of freedom.
   A. the maximum
   B. both maximum and the minimum
   C. the minimum
   D. nothing about the

112. A/an __________ pyrometer may have an operating range of a temperature measurement from 800 to 1600°C.
   A. optical
   B. radiation c
   C. photoelectric
   D. none of these

113. It is given that the gas law constant and dead time of a process with the transfer function $G_c(s) = 10 \exp(-0.1s)/(0.5s + 1)$ are known with a possible error of ± 20% of their values. Calculate the largest permissible gain K_c of a proportional controller the values of process gain time constant and dead time as
   A. 8, 0.6, 0.08
   B. 12, 0.6, 0.12
   C. 8, 0.6, 0.12
   D. 12, 0.4, 0.08

114. Which of the following types of element is normally not used in the bimetallic thermometers?
   A. multiple helix
   B. bourdon tube
   C. single helix
   D. flat spiral

115. It is not possible to measure the flowrate of sludge by
   A. orificemeter
   B. kennson nozzle
   C. open weir s
   D. both**B**&**C**

116. For which of the following measurements Zirconia probe is used?
   A. speed of a submarine.
   B. % CO_2 in flue gas.
   C. % O_2 in flue gas.
   D. sulphur dioxide

117. Which of the following analyser has a thermal conductivity cell as the primary element?
   A. oxygen
   B. carbon dioxide
   C. sulphur dioxide
   D. carbon monoxide

118. Which of the following represent __________ if the open loop transfer function of a control system is $\left(\frac{KR}{1+TS}\right)$?
   A. a first order system.
   B. a second order system.
   C. a first order time lag.
   D. a second order time lag.

119. Which of the following is measured by pyrheliometer?
   A. diffuse radiation
   B. beam radiation
   C. bright sunshine period
   D. none of these

120. What is the transfer function of a pure dead time system with dead time, T_d
   A. $1/(\tau_d s + 1)$
   B. $\tau_d s + 1$
   C. $e^{-\tau_s s}$
   D. $e^{\tau_d s}$

121. In comparison to industrial mercury in glass thermometer a bimetallic thermometer has almost the same .
   A. temperature measuring range (– 40°C to 450°C).
   B. accuracy (± 1% of span).
   C. speed of response.
   D. all **A**, **B**&**C**

122. What is the phase lag of the frequency response of a second order system to a sinusoidal forcing function?
   A. is 30°
   B. is 90° at the most
   C. approaches 180° asymptotically
   D. is 120°

123. Which of the following is the state characteristic of an instrument?
   A. tine lag
   B. dynamic error
   C. drift
   D. response

124. Gas analysis is commonly done using the.
   A. thermal conductivity cell
   B. emission spectrometer
   C. mass spectrometer
   D. x-ray diffraction

125. Which of the following is used for emission spectroscopy?
   A. solids and metal analysis.
   B. determination of CO_2 in gases.
   C. NO_x determination.
   D. determining water purity.

126. Thermocouples generate emf which is of the order of __________.
   A. millivolts
   B. microvolts
   C. kilo volts
   D. volts

127. What is the extent of dead zone in an instrument in terms of percent of the scale?
   A. 0.2
   B. 1.5
   C. 4
   D. 8

128. Among the following fluid flow measuring devices the largest flow measurements is possible in .
   A. orificemeter
   B. rotameter
   C. v-notch
   D. weir

129. In the root locus method, a pole of a transfer function *G* (*s*) is the value of s for which *G* (*s*) approaches
   A. –1
   B. 0
   C. 1
   D. ∞

130. Flow rate through an orifice is ________ the pressure differential.
   A. inversely proportional to the square of
   B. proportional to
   C. proportional to the square root of
   D. inversely proportional to the square root of

131. Which of the following is used in measurement of liquid levels in autoclaves?
   A. glass gauge
   B. differential float type manometer
   C. simple float
   D. none of these

132. Which of the following laws is the working principle of radition pyrometer?

A. Kirchhoff's law
B. Seebeck effect
C. Stefan-Boltzman law
D. Wien's law

133. Depreciation of presence of a small amount of water in the organic system can be done by the __________ method.
A. polarimetry
B. electrical conductivity
C. emission spectroscopy
D. dielectric constant end loss factor

134. Which of the following parameters can be measured by a Sagmometer?
A. kinematic viscosity
B. surface tension
C. refractive index
D. kinematic viscosity

135. Which of the following temperature measuring temperature instruments has minimum measurement accuracy?
A. iron-constantan thermocouple
B. optical pyrometer.
C. alcohol filled thermometer.
D. optical activity

136. A mercury thermometer cannot be used to measure the temperature below the freezing point of mercury, which is __________ °C.
A. – 38.9
B. – 11.9
C. – 60.9
D. – 80.9

137. Among the following __________ is not a second order instrument .
A. pressure gauge with one bellow, two tubes and tank.
B. bare mercury in glass thermometer.
C. mercury in glass thermometer with covering.
D. none of these.

138. __________ controller has the maximum stabilising time.
A. P
B. PD
C. PI
D. PID

139. Which of the following measured is by a simple pitot tube?
A. static pressure
B. average velocity
C. pint velocity
D. maximum velocity

140. For which of the following the thermal conductivity measurement is used?
A. composition of an alloy.
B. specific gravity of petrofuels.
C. O_2 percentage in the flue gas.
D. CO_2 percentage in the flue gas.

141. Thermistors are used in __________ devices.
A. voltage measuring
B. temperature measuring
C. temperature compensating
D. both **B&C**

142. Which of the following loop transfer function is used to Bode stability method?
A. closed
B. open
C. either **A** or **B**
D. neither **A** nor **B**

143. Which manometric liquid is used to accurately measure small furnace draught by an inclined tube manometer ?
A. ethylene glycol
B. water
C. paraffin
D. carbon tetrachloride

144. Time constant may be expressed as the
A. time taken by the controlled variable to reach 63.2% of its full change.
B. same as transportation lag.
C. same as dead time.
D. time required by the measured variable to reach 63.2% of its ultimate change.

145. A diaphragm is used as pressure sensor in pneumatic and electronic instruments. The deflection in diaphragm is dependent on the metal thickness, its diameter and shape and number of corrugations. Its deflection is proportional to .
A. d
B. d^2
C. d^3
D. d^4

146. What is the value of specific conductance of water in micro mho/cm^3? The density or specific

conductance is the reciprocal of resistance in ohms of 1 cc of liquid at specific temperature.

A. 50
B. 100
C. 10
D. 1

147. Among the following __________ is not a second order instrument .

A. pressure gauge with one bellow, two tubes and at tank.
B. bare mercury in glass thermometer.
C. mercury in glass thermometer with covering.
D. none of these.

148. Bimetallic thermometers work on the principle of difference in linear coefficient of thermal expansion. Which of the following has the maximum thermal coefficient of linear expansion?

A. chromel
B. brass
C. invar
D. nickel

149. Which of the following is order of mercury manometer (U-tube type)

A. zero
B. first
C. second
D. third

150. Which of the following types of thermocouples is suitable for the process if emf developed by the thermocouple while measuring temperature of 400°C is 22 mV.

A. platinum-thodium
B. iron-constantan
C. platinum- platinum + rhyodium
D. chromel-alumel

151. Which of the following temperature is termed as silver point temperature in °C ?

A. 760.5
B. 860.5
C. 960.5
D. 1060.5

152. At normal atmospheric pressure the temperature of equilibrium between solid antimony & liquid antimony platinum resistance thermometer can be used. This temperature is .

A. 961.93
B. 630.74
C. 1064.43
D. 1261.93

153. In case of a process having the transfer function $G_p(s) = 1/(4s + 1)(2s + 1)(s + 1)$ for a propotional controler.

A. $w = 1/14, K_c = 45/714$
B. $w = 7/6, K_c = 46/3$
C. $w = 1, K_c = 13$
D. $w = 7/8, K_c = 45/4$

154. Among the following __________ is not a variable area flow water .

A. rotameter
B. piston type meter
C. magnetic flow meter
D. orifice and tapered plug meter

155. Which among the following thermocouples is capable of measuring a temperature of -50°C .

A. platinum - platinum + rhodium
B. copper – constantan
C. chromel - alumel
D. iron-constantan

156. Which of the following is the Laplace transform of impulse input having magnitude 'X'?

A. X
B. X^2
C. $1/X$
D. 1

157. Among the following the __________ may not be classified as a thermo electric pyrometer.

A. radiation pyrometer
B. resistance thermometer.
C. optical pyrometer (disappearing filament type)
D. thermocouple.

158. Thermo couples are suitable for measuring.

A. liquid temperatures only.
B. very high temperatures only.
C. very low temperatures only.
D. both high and low temperatures.

159. Which of the following means a stable system ?

A. for which the output response is bounded for all bounded input.
B. which satisfies the conditions for a servo problem.

C. which exhibits an unbounded response to a bounded input

D. none of these.

160. Which among the following shows maximum dip effect (indicating reverse direction of temperature change).

A. mercury thermometer

B. bimetallic thermometer

C. radiation pyrometer

D. thermocouple

161. Iron-constantan as thermocouple can measure temperature in the range of 0 to 870°C. It cannot be used to measure sub-zero temperature, because at lower temperature.

A. emf produced is very low of the order of microvolt

B. embrittlement of iron occurs

C. rusting of iron takes place

D. both B and C

162. Which of the following is most suitable for flow measurement of hostile acids and alkalis ?

A. venturimeter

B. orificemeter

C. magnetic flow meter

D. hot wire anemometer

163. Which of following is the term analogous to voltage in a single tank system

A. flow rate.

B. heat content of the system.

C. liquid volume in the tank.

D. level of liquid.

164. Thermistors are known to have a very high temperature coefficient of resistivity belonging to a class of solids called .

A. conductors

B. dielectrics

C. semi conductors

D. insulators

165. The pressure sensing element of elastic type pressure gauge is never made in the form of a

A. bourdon tube

B. bellow

C. strip

D. diaphragm

166. Identify which of the following statements is incorrect for the block diagram of a simple standard control system shown below.

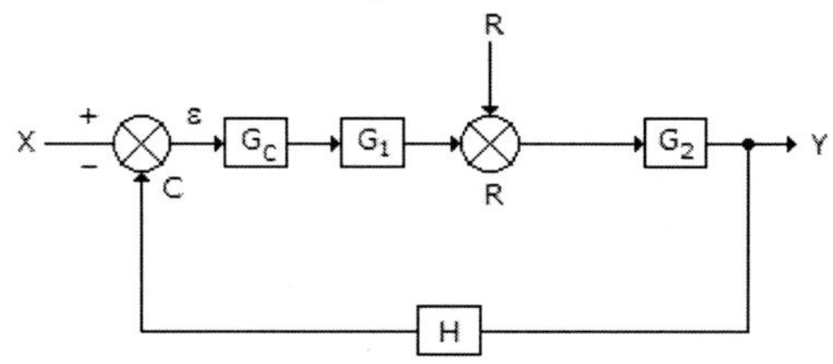

Which of the following statements about the above control system is wrong?

A. the transfer functions of the measuring element and theprocessis represented by H and G_2 respectively

B. the overall transfer function for the change in load (X = 0) is given by $\frac{Y}{X_1} = \frac{G_c G_1 G_2}{1+ G_c G_1 G_2\ H}$

C. the overall transfer function for the change in set point (X_1 = 0) is given by $\frac{Y}{X_1} = \frac{G_c G_1 G_2}{1+ G_c G_1 G_2\ H}$

D. manipulated variable is represented by C.

167. Which of the following can never be recommended as material of construction of Bourdon tube

A. phosphor bronze

B. stainless steel

C. monel metal

D. cast iron

168. Measurement of __________ pressure can be made by McLeod gauge.

A. atmospheric

B. sub-atmospheric

C. positive

D. very high

169. The reason that the Laplace transform of exp(at) where a > 0 is defined only for the Laplace parameter s > a is that since s < a .

A. the function exp(at) is piece-wise continuous only for s > a.

B. the Laplace transform of integral of exp(at) has finite values only for s > a.

C. the function is exponential.

D. the Laplace transform integral of exp(at) has initial values only for s > a.

170. Which of the following methods is suitable for determination of level of a liquid under pressure ?
A. air-trap system
B. differential pressure manometer
C. bubbler system
D. diaphragm box system

171. __________ is never used as an element of industrial resistance thermometer.
A. lead
B. copper
C. nickel
D. 30% iron + 70% nickel

172. Which of the following appropriately refers to pyrometer ?
A. directly
B. with the mercurial thermometer upto 350°C
C. which is of higher magnitude
D. all **A, B&C**

173. Which of the following means Cascade control ?
A. one feed-back loop.
B. more than one feed-back loop.
C. feed forward control.
D. on-off control.

174. For a change in set point the response of a linear control system is called.
A. regulator problem
B. frequency response
C. servo problem
D. transient response

175. Which of the following represents the limit of measuring the pressure in kgf/cm^2 for liquid column manometer ?
A. > 2(gage)
B. < 3 (gage)
C. < 10 (gage)
D. < atmospheric

176. Measurement of liquid flow rate in an open channel is not-possible by an .
A. orifice meter
B. V-notch
C. cipolletti weir
D. rectangular weir

177. Which of the following is equal to the phase margin ?
A. 180° - phase lag
B. phase lag - 180°
C. phase lag + 180°
D. phase lag + 90°

178. Which type of valve and controller action is recommended for the system in which temperature control of an exothermic chemical reaction taking place in CSTR is to be done with the help of cooling water flowing in a jacket around the reactor ?
A. air to close valve with the controller direct acting.
B. air to open valve with the controller direct acting.
C. air to open valve with the controller indirect acting.
D. air to close valve with the controller indirect acting.

179. Filled system expansion thermometers having the lowest positive temperature measurement capability is __________ .
A. nitrogen in steel thermometer.
B. alcohol in glass thermometer.
C. mercury in glass thermometer.
D. fused metal (Na or K) in steel thermometer.

180. Measurement of __________ may be made by hot wire anemometer.
A. flow rates of fluids.
B. very high temperature.
C. flow rates of granular solids.
D. thermal conductivity of gases.

181. Which of the following is having the same unit as that of time constant of a system
A. (time)$^{-1}$
B. time
C. velocity
D. none of these

182. Water is flowing through a series of four tanks and getting heated as shown in figure. It is desired to design a cascade control scheme for controlling the temperature of water leaving the tank 4 as there is a disturbance in the temperature of a second stream entering the tank 2. Select the best place to take thesecondary measurement for the second loop.

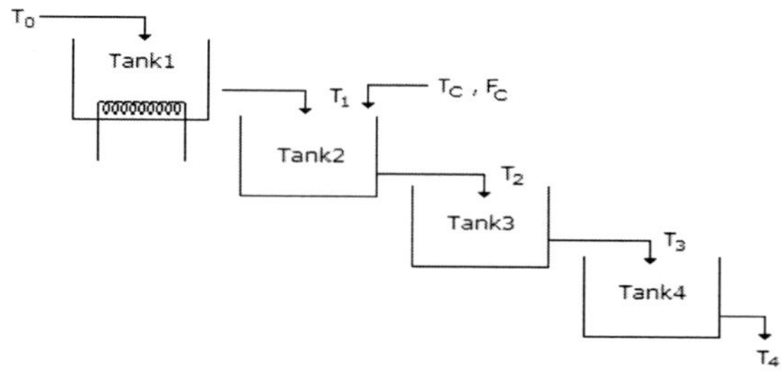

A. Tank 1

B. Tank 2

C. Tank 3

D. Tank 4

183. Flue gas analysis in respect of O_2 and CO_2% can be continuously done by a/an .

A. chromatograph

B. thermal conductivity cell

C. orsat apparatus

D. zirconia probe

184. Among the following the outlet temperature of heating/cooling fluid is a shell and tube heat exchanger is the __________ variable.

A. manipulated

B. load

C. controlled

D. none of these

185. Which of the following is used to measure the __________ by Mcleod gauge?

A. point velocity

B. flow rate

C. vacuum

D. Pressure

186. Which of the following measurement devices is based on the principle of "an emf of the order mV is generated when two solution of different hydrogen ion concentration are separated by a thin glass wall"?

A. pH meter

B. polarimeter

C. polarograph

D. chromatograph

187. What will be the output-input relationship in a feedback where *G* and *H* denote open loop and close loop transfer functions respectively ?

A. $\frac{G}{1+H}$

B. $\frac{H}{1+G}$

C. $\frac{H}{G}$

D. $\frac{G}{H}$

188. Which of the following is the maximum differential pressure in liquid manometer in psi.

A. 20

B. 30

C. 40

D. 50

189. The electrical conductivity of the platinum in the resistance thermometer __________ with increase in temperature.

A. increases exponentially

B. decreases

C. increases

D. remains constant

190. How many steady state values a non-linear system will have ?

A. one

B. more than one

C. two

D. three

191. Invar is used as a thermocouple material which is an iron-nickel alloy. What is the nickel percentage in it ?

A. 12

B. 36

C. 54

D. 68

192. Thermocouples of the type __________ has the widest temperature measurement range.

A. chromel-alumel

B. iron-constantan

C. copper-constantan

D. platinum-platinum/rhodium

193. Which of the following symbol refers to "locally mounted instrument" in instrumentation diagram?

A. 

B. 

C. 

D. 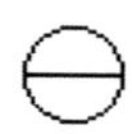

194. Among the following the symbol for "pneumatic control valve" in an instrumentation circuit diagram is represented is .

A. 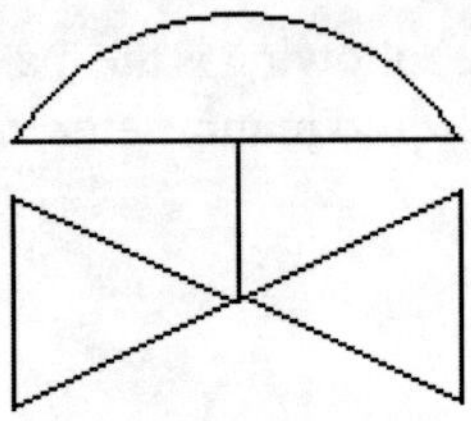

B. 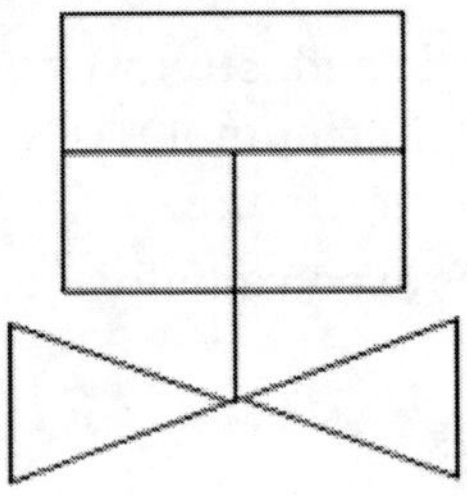

C. 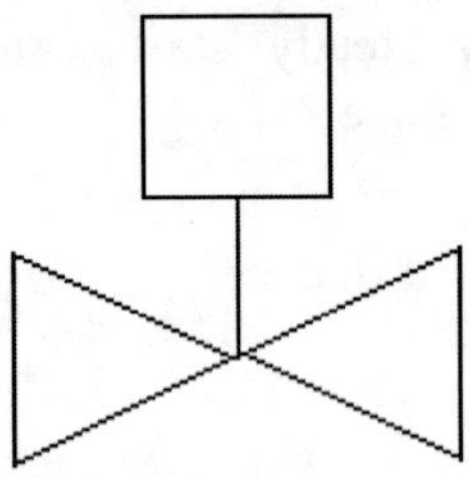

D. none of these

195. A constant volume gas thermometer can measure sub-zero temperature employes .

A. helium
B. hydrogen
C. nitrogen
D. none of these

196. If among the following the damping coefficient is to be unity the system will be said to be .

A. highly fluctuating
B. critically damped
C. underdamped
D. overdamped

197. Among the following which device is not used for temperature measurements?

A. thermostats.
B. thermocouples.
C. bimetallic thermometers.
D. relays for opening & closing of electrical circuits.

198. Which of the following may be represented as distribution parameter system ?

A. tubular reactor
B. on-off controller
C. CSTR
D. none of these

199. Which of the following materials may be used as material of construction in pressure sensor of non-metallic diaphragms ?

A. thick paper
B. synthetic rubber
C. bakelite
D. teflon

200. Which of the following ranges may be that of photoelectric pyrometer ?

A. 400-1600
B. 800-1600
C. 800-2500
D. 400-1000

201. Composition of natural gas is determined by the

A. Haldane apparatus
B. chromatograph
C. mass spectrometer
D. both **B** and **C**

202. Measurement of temperature of molten pig iron (1450°C) and molten slag (1500°C) flowing out of a blast furnace is measured by a/an .

A. chromel-alumel thermocouple
B. radiation pyrometer
C. optical pyrometer
D. either **B** or **C**

203. Which of the following most suitable controller if response of a control system is to be free of offset and oscillations?

A. proportional controller
B. proportional-integral (PI) controller
C. proportional-derivative (PD) controller
D. proportional integral-derivative (PID) controller

204. Among the following thermocouples ________ has the least temperature measurement range?

A. copper-constantan
B. iron-constantan
C. chromel-alumel
D. platinum-platinum/rhodium

205. Which of the following is most accurately suitable for measurement of dome temperature of blast furnace ?

A. platinum resistance thermometer
B. platinum-platinum/rhodium thermocouple

C. radiation pyrometer

D. iron-constantan thermocouple

206. Which of the following is to analogous continuous measurement of moisture content of paper in paper industries ?

A. magnetic susceptibility

B. electrical resistance through the paper

C. thermal conductivity through the paper

D. none of these

207. __________ can measure temperatures in the range of - 20 to 300°C .

A. resistance thermometer

B. mercury in glass thermometer

C. vapor pressure thermometer

D. none of these.

208. Mercury thermometer is commonly used for low temperature measurement. What is the freezing point and boiling point of mercury __________ respectively in °C ?

A. – 39 and 350

B. – 79 and 395

C. – 51 and 439

D. – 10 and 425

209. Manometer are used to measure __________ pressure.

A. atmospheric

B. absolute

C. gauge

D. none of these

210. What is the Laplace transfer function $\int_0^t f(t)dt$ for the time domain function $f(t) = t$

A. $1/2\ S^3$

B. $2/S^3$

C. $1/S^3$

D. $2/S^2$

211. Which of the following statements is wrong ?

A. transfer lag is a characteristics of all higher order systems (other than first order systems).

B. stirred tank with a water jacket exemplifies an interacting system.

C. there is no transfer lag for a single first order system.

D. transfer lag decreases as the number of stages decreases.

212. Measuring temperature by thermocouples gives fastest speed of response if the thermal well is made of .

A. steel

B. inconel

C. vycor (a glass)

D. nichrome

213. __________ is the frequency at which maximum amplitude ratio is attained .

A. cross-over

B. resonant

C. natural

D. corner

214. The overall transfer function for two non-interacting first order systems connected in series, is the __________ of the individual transfer functions.

A. product

B. difference

C. ratio

D. sum

215. Ordinary mercury in glass thermometer is used for measuring temperature upto 120°C. However, for measuring higher temperature upto __________ °C, thermometer is made by filling nitrogen under pressure above the mercury .

A. 250

B. 350

C. 550

D. 700

216. In Bode plot Φ is plotted on a/an __________ graph paper Φ vs ω .

A. log-log

B. triangular

C. ordinary

D. semi-log

217. Among the following which is the most suitable instrument for measuring the temperature of a red hot furnace ?

A. platinum resistance thermometer

B. bimetallic thermometer

C. optical pyrometer

D. thermocouple

218. For a first order system the phase lag of the sinusoidal response is .

A. 120°

B. 30°
C. 180°
D. 90°

219. An optical or radiation pyrometer cannot be used to measure the temperature of __________ .
A. hot blast (air) from stoves
B. rotary limestone calcination kiln
C. molten slag flowing out of blast furnace combustion space in boilers
D. combustion space in boilers

220. Which of the following is equal to the Gain margin?
A. gain in *P-I* controller
B. reciprocal of amplitude ratio
C. amplitude ratio
D. gain in *P* controller

221. What is the order of a system with transfer function $\frac{2s}{4s}+1$?
A. zero
B. 1st
C. 2nd
D. 3rd

222. For which of the following a flame photometry can be used for the determination of compositional analysis.
A. solids
B. alkali metals
C. natural gas
D. isotopes

223. Specific gravity of a liquid is to be continuously measured. Among the following the most suitable equipment is .
A. displacement meter
B. contact-type electric indicators
C. hydrometer
D. both **A** and **C**

224. In Reaumer temperature scale ice and steam point are respectively represented by
A. –273° & 80°
B. 0° & 80°
C. 32° & 460°
D. 32° & 80°

225. Most thermocouples has a thermal emf-temperature relationship of nature .
A. linear
B. parabolic
C. square root
D. exponential

226. Which of the following can measure the minute depression of freezing point of a high solvent on addition of a solid state ?
A. Beckman thermometer
B. bimetallic thermometer
C. mercury thermometer
D. dilatometer

227. A special case of proportional control, is on-off control which has a band width of _______ percent.
A. 100
B. 75
C. 25
D. 0

228. The maximum temperature which normal mercury thermometer can measure may be increased up to 300°C. Identify which of the following procedure can increase its maximum temperature measurement range up to 500°C ?
A. filling nitrogen under pressure in the stem.
B. using steel tube in place of glass tube.
C. increasing the diameter of the tube.
D. accounting for the tube expansion.

229. Among the following the strongest paramagnetic gas is .
A. NO_2
B. O_2
C. CO_2
D. NO

230. Among the following __________ is not a composition measuring instrument.
A. polarograph
B. mass spectrometer
C. thermal conductivity cell
D. hot wire anemometer

231. for measurement of temperature of a red hot object in the range of 800 - 1600°C which of the following instruments is not suitable?
A. optical pyrometer
B. radiation pyrometer
C. photoelectric pyrometer
D. thermocouples

232. Which of the following is the conversion formula for converting amplitude ratio (AR) in to decibel?

A. decibel = 20 $\log_{10}(AR)$

B. decibel = 20 $\log_e(AR)$

C. decibel = 20 $\log_{10}(AR)^{0.5}$

D. decibel = 20 $\log_e(AR)^{0.5}$

233. Among the following thermocouples __________ is incapable of measuring sub-zero (i.e., < 0°C) temperature.

A. iron-constantan

B. chromel-alumel

C. platinum-platinum + rhodium

D. Copper-constantan

234. Which of the following gives working principle of mercury in glass thermometer ?

A. volumetric expansion.

B. linear expansion.

C. pressure rise with temperature.

D. none of these.

235. Which of the following is the Laplace transform of sin *t*?

A. $\frac{1}{S^2+1}$

B. $\frac{1}{S^2-1}$

C. $\frac{S}{1+S^2}$

D. $\frac{S}{S^2-1}$

236. The international standard for temperature measurement by platinum resistance thermometer is in the range of .

A. triple point of hydrogen and freezing point of antimony

B. 13.81°K to 903.9°K

C. both **A** and **B**

D. neither **A**nor **B**

237. Which of the following may be used for the measurement of by a Stroboscope ?

A. rpm of a flywheel

B. frequency of light

C. liquid level under pressure

D. depression of freezing point

238. The root locus plot of the roots of the characteristic equation of a closed loop system having the open loop transfer function will have a definite number of loci for variation of *K* from 0 to ∞ $\frac{K(S+1)}{2(2S+1)(3S+1)}$. The number of loci is

A. 2

B. 3

C. 1

D. 4

239. Which of the following is valid for a second order under damped system ?

A. decay ratio = overshoot

B. decay ratio = $(\text{overshoot})^2$

C. large damping co-efficient means smaller damping.

D. overshoot increases for increasing damping co-efficient.

240. What will be the initial value of the time function for a control system having the following transfer function $F(S) = \frac{(S-1)(S+1)}{S(S-2)(S+4)}$?

A. 1

B. 1/8

C. 7/8

D. –1

241. A __________ cannot used for measuring sub-zero tem-peratures.

A. platinum resistance thermometer

B. mercury in glass thermometer

C. vapor pressure thermometer

D. radiation pyrometer

242. Bode plot is a graphical presentation of A.R. vs. *w*

A. log-log

B. semi-log

C. triangular

D. ordinary

243. What will be the amplitude ratio of a first order system with unity gas and time constant ô

A. 1

B. 0.5

C. 1/2

D. 0.25

244. Predicate the type of thermocouple if the emf developed by it while measuring a temperature of 800°C is about 31 mV.

A. chromel-alumel

B. platinum-platinum+rhodium

C. iron-constantan

D. none of these.

245. Which of the following represents the Laplace transform of the function shown in figure?

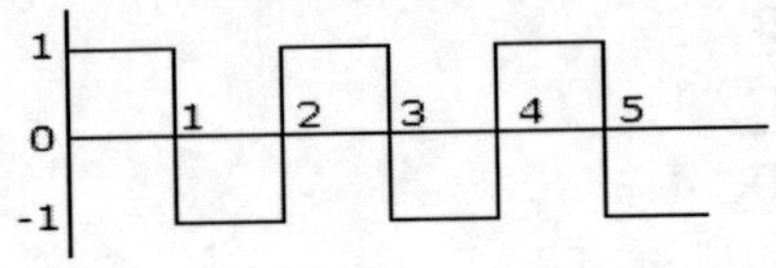

A. $\frac{(1-e^{-s})}{s(1-e^{-2s})}$

B. $\frac{(1-e^{-s})e^{-s}}{s(1-e^{-2s})}$

C. $\frac{2(1-e^{-s})}{s}$

D. $\frac{(1-e^{-s})}{s(1+e^{-s})}$

246. Working principle of a __________ is based on the change of angle of refraction with composition.

A. polarimeter

B. polarograph

C. spectrometer

D. refractrometer

Answer

1.	C	36.	A	71.	D	106.	B
2.	A	37.	C	72.	B	107.	C
3.	D	38.	C	73.	B	108.	D
4.	B	39.	C	74.	D	109.	D
5.	D	40.	B	75.	A	110.	B
6.	B	41.	B	76.	C	111.	A
7.	A	42.	D	77.	C	112.	C
8.	B	43.	C	78.	A	113.	B
9.	C	44.	D	79.	C	114.	B
10.	C	45.	A	80.	C	115.	D
11.	A	46.	B	81.	B	116.	C
12.	C	47.	C	82.	A	117.	B
13.	B	48.	B	83.	B	118.	A
14.	C	49.	B	84.	A	119.	B
15.	B	50.	C	85.	A	120.	C
16.	B	51.	B	86.	B	121.	D
17.	A	52.	C	87.	C	122.	C
18.	B	53.	B	88.	A	123.	C
19.	D	54.	D	89.	D	124.	A
20.	A	55.	A	90.	A	125.	A
21.	A	56.	B	91.	D	126.	A
22.	C	57.	D	92.	B	127.	A
23.	A	58.	A	93.	C	128.	D
24.	C	59.	D	94.	A	129.	D
25.	B	60.	B	95.	B	130.	C
26.	C	61.	A	96.	A	131.	B
27.	B	62.	C	97.	C	132.	C
28.	B	63.	C	98.	C	133.	D
29.	D	64.	B	99.	B	134.	B
30.	A	65.	B	100.	C	135.	B
31.	D	66.	D	101.	A	136.	A
32.	D	67.	A	102.	C	137.	B
33.	A	68.	C	103.	B	138.	C
34.	C	69.	C	104.	A	139.	C
35.	B	70.	A	105.	B	140.	D

141. D
142. A
143. C
144. A
145. D
146. D
147. B
148. B
149. C
150. B
151. C
152. B
153. B
154. C
155. A
156. D
157. C
158. D
159. A
160. A
161. D
162. C
163. D
164. C
165. C
166. D
167. D
168. B
169. B
170. B
171. A
172. C
173. B
174. C
175. B
176. A
177. A
178. C
179. B
180. A
181. B
182. C
183. B
184. C
185. C
186. A
187. B
188. B
189. B
190. B
191. B
192. D
193. A
194. A
195. A
196. B
197. B
198. A
199. B
200. B
201. D
202. D
203. D
204. A
205. B
206. B
207. D
208. A
209. C
210. C
211. D
212. B
213. B
214. A
215. C
216. A
217. C
218. D
219. A
220. B
221. B
222. B
223. D
224. B
225. B
226. A
227. D
228. A
229. B
230. D
231. D
232. A
233. C
234. A
235. A
236. C
237. A
238. B
239. B
240. A
241. D
242. A
243. C
244. A
245. A
246. D

GATE 2015-2000

(1 Mark Question)

GATE 2015

1. Match the output signals as obtained from four measuring device in response to a unit step change in the signal.

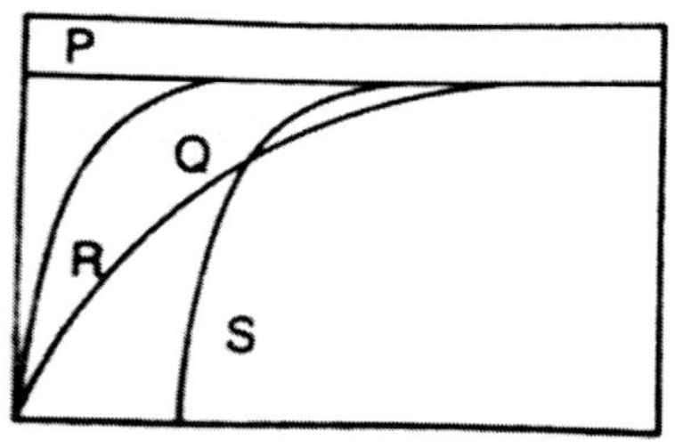

Output signal time:

I. gas chromatograph with a long capillary tube

II. venture tube

III. thermocouple with 1st order dynamics

IV. pressure transducer with 2nd order dynamics

Codes

A. I-S, II-R, III-Q, IV-P

B. I-R, II-P, III-Q, IV-S

C. I-S, II-P, III-Q, IV-R

D. I-Q, II-S, III-R, IV-P

***Sol.*B.** if y(t) = output (response)

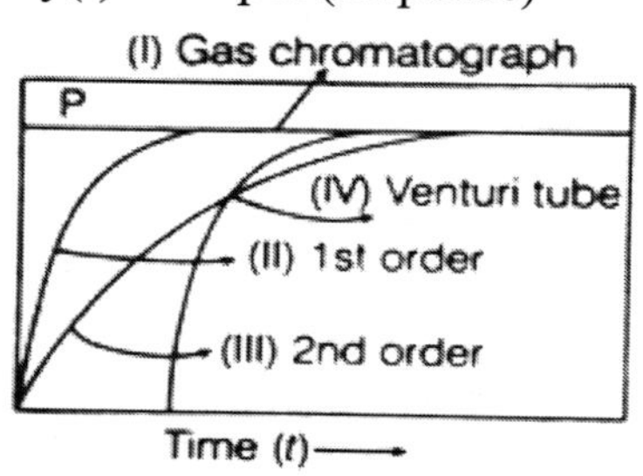

Gas chromatograph - IV

Venturi tube - I

Thermocouple - II

Pressure transducer with 2nd order dynamics – III

2. The transfer function for the disturbance response in an open-loop process is given by G_d^{open} (s). The corresponding transfer function for the disturbance response in closed-loop feedback control system with proportional controller is given by G_d^{closed} (s). Select the option that is always correct. $\{O[G(s)]$ represents order of transfer function G(s)}.

A. $O\left|G_d^{open}(s)\right| = O\left|G_d^{closed}\right|(s)$

B. $O\left|G_d^{open}(s)\right| \neq O\left|G_d^{closed}\right|(s)$

C. $O\left|G_d^{open}(s)\right| \geq O\left|G_d^{close\ d}\right|(s)$

D. $O\left|G_d^{open}(s)\right| \leq O\left|G_d^{closed}\right|(s)$

***Sol.*A.** Open-loop transfer function

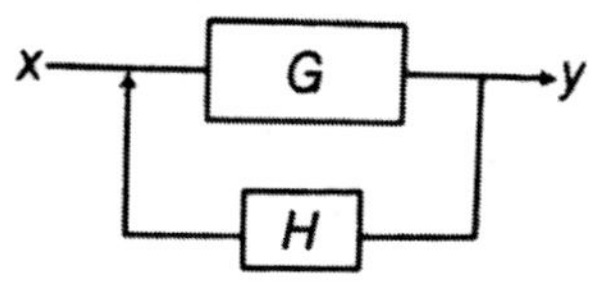

$$G_{OL} = GH = G_d^{open}(s)$$

Closed-loop transfer function

$$G_{CL} = \frac{G}{(1+GH)} = G_d^{closed}(s)$$

Order of transfer function is the order of denominator

Order of GH and (1 + GH) are same,

So,

$$O\left|G_d^{open}(s)\right| = O\left|G_d^{closed}(s)\right|$$

GATE 2014

3. For the time domain function $f(t) = t^2$, which one of the following is the Laplace transform of $\int_0^t f(t)dt$?

A. $\frac{3}{s^4}$

B. $\frac{1}{4s^2}$

C. $\frac{2}{s^3}$

D. $\frac{2}{s^4}$

***Sol.* D.** we have

$$L\left[\int_0^t f(t)dt\right] = \frac{1}{s}f(\delta)$$

$(\because f(\delta) = L[f(t)])$

$$= \frac{1}{s} \times \frac{2}{s^3} \qquad \left(\because L(t^2) = \frac{2}{s^3}\right)$$

$$= \frac{2}{s^4}$$

4. A unit impulse response of a first order system with time constant τ and steady state gain K_p is given by

A. $\frac{1}{K_p\tau}e^{t/\tau}$

B. $K_p e^{-t/\tau}$

C. $\tau K_p e^{-\frac{t}{\tau}}$

D. $\frac{K_p}{\tau}e^{-t/\tau}$

***Sol.* D.** $f(s) = \frac{K_p}{\tau s+1}$

$$f(t) = \frac{K_p}{\tau}L^{-1}\left(\frac{1}{s+\frac{1}{\tau}}\right)$$

$$= \frac{K_p}{\tau}e^{-\frac{1}{\tau}t} = \frac{K_p}{\tau}e^{-\frac{t}{\tau}}$$

5. Assume that an ordinary mercury-in-glass thermometer follows first order dynamics with a time constant of 10 s. It is at a steady state temperature of 0°C. At time t = 0, the thermometer is suddenly immersed in a constant temperature both at 100°C. The time required (in s) for the thermometer to read 95°C, approximately is

A. 60

B. 40

C. 30

D. 20

***Sol.* C.** given $\tau = 10$ s

For first order system

$$Y(t) = A(1-e^{-t/\tau})$$

$$95 = 100\ (1-e^{-t/\tau})$$

$$\frac{t}{\tau} = 2.995 \sim 3$$

$$t = 30\ sec.$$

GATE 2013

6. Match the following.

Group 1	Group 2
P. Viscosity	1. Pyrometer
Q. Pressure	2. Hot wire anemometer
R. Velocity	3. Rheometer
S. Temperature	4. Piezoelectric element

A. P-4, Q-3, R-1, S-2

B. P-3, Q-4, R-2, S-1

C. P-3, Q-4, R-1, S-2

D. P-4, Q-3, R-2, S-1

***Sol.* B.** P-3, Q-4, R-2, S-1

GATE 2012

7. The thermometer initially at 100°C is dipped at t = 0 in to an oil bath, maintained at 150°C. If the recorded temperature is 130°C after 1 min, then the time constant of thermometer (in min) is

A. 1.98

B. 1.35

C. 1.26

D. 1.09

***Sol.* D.** we have

$$Y(t) = A\left(1 - e^{-t/\tau}\right) \qquad ....(i)$$

Here, A = 150– 100 = 50

Y(t) = 130 – 100 = 30

t = 1 min

Putting values in Eq. (i),

$$30 = 50\left(1 - e^{-t/\tau}\right)$$

$$\tau = 1.09 \text{ min}$$

8. The Bode stability criterion is applicable when

A. gain and phase curves both decrease continuously with frequency

B. gain curve increase and phase curve decreases with frequency

C. gain curve and phase curve both increase with frequency

D. gain curve decrease and phase curve increase with frequency

***Sol.* A.** gain and phase curve both decrease continuously with frequency.

GATE 2011

9. Match the process parameter in Group I with the measuring in group II.

Group I	*Group II*
P. Flame temperature	I. thermocouple
Q. Composition of LPG	II. Radiation pyrometer
R. Liquid air temperature	III. Gas chromatograph

A. P-III, Q-I, R-II

B. P-I, Q-III, R-II

C. P-II, Q-III, R-I

D. P-II, Q-I, R-III

***Sol.*C.** flame temperature - radiation temperature
composition of LPG - gas chromatograph
liquid air temperature -thermocouple

10. The range of standard current signal in process instruments is 4 to 20 mA. Which one of the following is the reason for choosing the minimum signal as 4 mA instead of zero?
    - **A.** to minimise resistive heating in instruments
    - **B.** to distinguish between signal failure and minimum signal condition
    - **C.** to ensure a smaller difference between maximum and minimum signals
    - **D.** to ensure compatibility with other instruments

***Sol.*B.** minimum signal is chosen as 4 mA instead of zero, so as to distinguish between signal failure and minimum signal condition

GATE 2010

11. Match the location of the poles/zeroes in the s-plane listed in Group I with the system response characteristics in Group II.

Group I	*Group II*
P. pole in the right half plane Q. Pole at origin R. Zero in the right half plane	1. stable response 2. Integrating response 3. Unstable response 4. inverse response

Codes

	P	Q	R
A.	1	2	3
B.	3	4	1
C.	3	2	4
D.	1	4	3

***Sol.*C.** pole in the right half plane -stable response
pole at origin - integrating response
zero in the right half plane - inverse response

12. Flow measuring instruments with different specification (zero and span) are available for an application that requires flow rate measurements in the range of 300 L/h to 400 L/h. The appropriate instrument for this application is the one whose specifications are
    - **A.** zero = 175 L/h, span = 150 L/h
    - **B.** zero = 375 L/h, span = 100 L/h
    - **C.** zero = 275 L/h, span = 150 L/h
    - **D.** zero =475 L/h, span = 100 L/h

***Sol.*C.** range requirement: 300 – 400 L/h
(a) Zero = 175 L/h,
Range = 175 + 150 = 325 L/h
∴ not correct
(b) Zero = 375 L/h > 300 L/h
not correct
(c) Zero = 475 L/h 400 L/h
not correct
(d) Zero = 275 L/h
Range = 275 + 150 = 425 L/h
275 – 425 L/h
correct answer.

13. the transfer function G(s) whose asymptotic Bode diagram is shown below, is

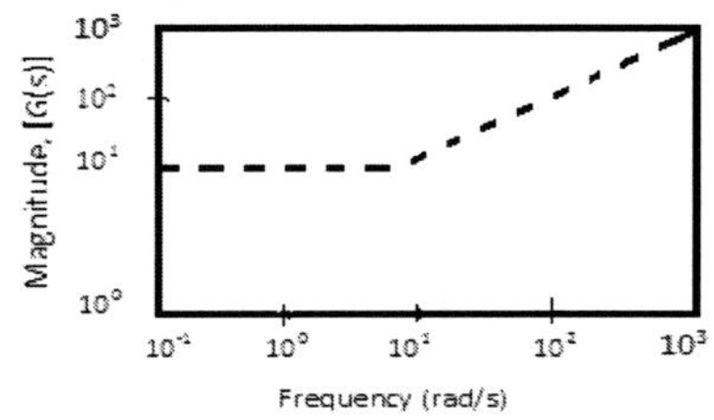

- **A.** 10s + 1
- **B.** s –10
- **C.** s + 10
- **D.** 10s – 1

***Sol.*C.** here, $k = |G(s)| = 10$

$$\omega = \frac{1}{t} = \frac{1}{10}$$

$$G(s) = 10\left(1 + \frac{1}{10} \cdot s\right) = 10 + s$$

GATE 2009

14. Which one of the following sensors is used for the measurement of temperature in a combustion process (T > 1800 °C)?
    - **A.** type I thermocouple
    - **B.** thermistor
    - **C.** resistance temperature detector
    - **D.** pyrometer

***Sol.*D.** pyrometer

15. The roots of the characteristics equation of an underdamped second order system are
    - **A.** real, negative and equal
    - **B.** real, negative and unequal
    - **C.** real, positive and equal
    - **D.** complex conjugates

Sol. D. second order system

$$y(s) = \frac{1}{s} \cdot \frac{1}{\tau^2 s^2 + 2\xi\tau s + 1}$$

The roots of this equation are

$$s_1 = -\frac{\xi}{\tau} + \frac{\sqrt{\xi^2 - 1}}{\tau}$$

$$s_2 = -\frac{\xi}{\tau} - \frac{\sqrt{\xi^2 - 1}}{\tau}$$

For under damped system, $\xi < 1$

The roots are complex conjugates.

GATE 2007

16. An operator was told to control the temperature of a reactor at 60°C. The operator sets the set-point of the temperature controller at 60. The scale actually indicated 0 to 100% of a temperature range of 0 to 200°C. This caused a runway reaction by over pressurizing the vessel, which resulted in injury to the operator. The actual set-point temperature was

A. 200°C
B. 60°C
C. 120°C
D. 100°C

Sol. C. as the temperature range is 0 to 200 for 0 to 100% valve opening 60% means

$$\frac{200-0}{100-0} \times 60 = 120°C$$

∴ The actual set-point corresponding to value 60 is 120°C.

GATE 2006

17. The control valve characteristics for three types of control valves (P,Q, and R) are given in the figure below. Match the control valve with its characteristics.

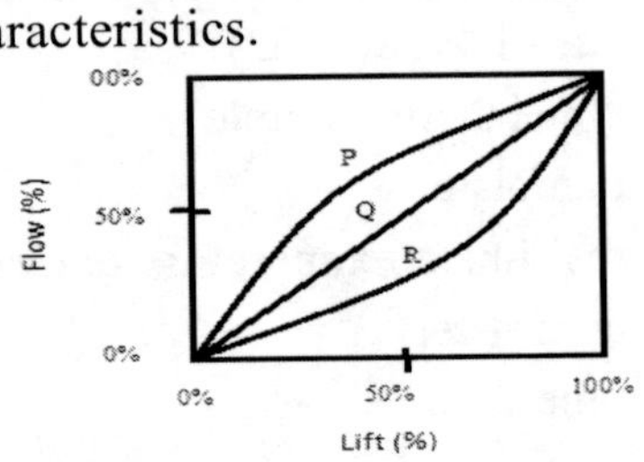

A. P-Quick opening, Q-Linear, R-Equal percentage
B. P-Linear, Q-Square root, R-Equal percentage
C. P-Equal percentage, Q-Linear, R-Quick opening
D. P-Square root, Q-Quick opening, R-Linear

Sol. A. P - Quick opening
Q - Linear
R - Equal percentage

GATE 2005

18. The unit step response of a first order system with time constant ô and steady state gain K_p is given by

A. $K_p(1 + e^{-t/\tau})$
B. $K_p(1 + e^{-t/\tau})$
C. $Kp(1 - e^{-2t/\tau})$
D. $\frac{Kp\, e^{-t/\tau}}{\tau}$

Sol. A. We known,

$$\frac{y(s)}{x(s)} = \frac{K_p}{\tau s + 1}$$

Unit step input

$$x(s) = \frac{1}{s}$$

$$\therefore \quad y(s) = K_p \cdot \frac{1}{s} \cdot \frac{1}{\tau s + 1}$$

$$\therefore \quad y(t) = K_p(1 - e^{-t/\tau})$$

19. An example of an open-loop second order underdamped system is

A. liquid level in a tank
B. U-tube manometer
C. thermocouple in a thermo-well
D. two non-interacting first order systems in series

Sol. B. U-tube manometer is the right example of an open-loop second order under damped system.

20. The control valve characteristics is selected such that the product of process gain and the valve gain

A. is a linearly increasing function of the manipulated variable
B. is a linearly decreasing function of the manipulated variable
C. remain constant as the valve of the value of the manipulated variable changes
D. is a expulated variable

Sol. D. the control valve characteristic is selected such that the product of process gain and the valve gain is an exponentially increasing function of the manipulated variable.

21. Cascade control comes under the control configuration which uses

A. one measurment and one manipulated variable

B. more than one measurement and more than one manipulated variable

C. one measurement and more than one manipulated variables

D. more than one measurement and more than one manipulated variables.

Sol. **A.** the cascade control configuration uses more than one measurement variable and one manipulated variable.

GATE 2004

22. For the time domain function f(x) = t, the Laplace transform of $\int_0^t f(x)dt$ is given by

A. $\frac{1}{2s^3}$

B. $\frac{2}{s^3}$

C. $\frac{1}{s^3}$

D. $\frac{2}{s^2}$

Sol. **C.** we have, $f(t) = t$

$$\therefore \int_0^t f(t)dt = \int_0^t t\,dt = \frac{t^2}{2}$$

and, $$L\left\{\int_0^t f(t)dt\right\} = L\left(\frac{t^2}{2}\right) = \frac{1}{2}\cdot\frac{2!}{s^{2+1}}$$

$$= \frac{1}{2}\cdot\frac{2!}{s^{2+1}} = \frac{1}{s^3}$$

GATE 2003

23. Match the measured process variables with the list of measuring devices given below.

List 1 (measured process variable)	*List 2 (Measured process variable)*
P. Temperature	1. Bourdon tube element
Q. Pressure	2. Orifice plates
R. Flow	3. Infrared analyzer
S. Liquid level	4. Displacer devices
T. Composition	5. Pyrometer

Code

	P	Q	R	S	T
A.	5	1	2	4	3
B.	3	1	4	2	5
C.	1	3	4	2	5
D.	3	1	2	4	5

Sol. **A.** temperature - pyrometer

pressure - bourdon tube element

flow - orifice plate

liquid level - displacer devices

composition - infrared analyzer

24. Suppose that the gain, time constant, and dead time of a process with the following transfer function

$$G_C(s) = 10\exp(-0.1s)/(0.5s+1)$$

are known with a possible error of ±20% of their values. The largest permissible gain K_c of a proportional controller needs to be calculated by taking the values of process gain, time constant and dead time as

A. 8, 0.6,0,08

B. 12,0.6,0.12

C. 8,0.6,0.12

D. 12, 0.4, 0.08

Sol. **B.** given, $G_c(s) = 10\exp(-0.1\,s)/(0.5\,s+1)$

Possible error = $\pm 20\%$

We need to consider a multiplication factor of $1 + 0.2 = 1.2$ in each value.

Process gain $= 10 \times 1.2 = 12$

Time constant $= 0.5 \times 1.2 = 0.6$

Dead time $= 0.1 \times 1.2 = 0.12$

25. Water is flowing through a series of four tanks and getting heated as shown in figure. It is desired to design a cascade control scheme for controlling the temprature of a second stream entering the Tank 2. Select the best place to take the secondary measurment for the secondary loop.

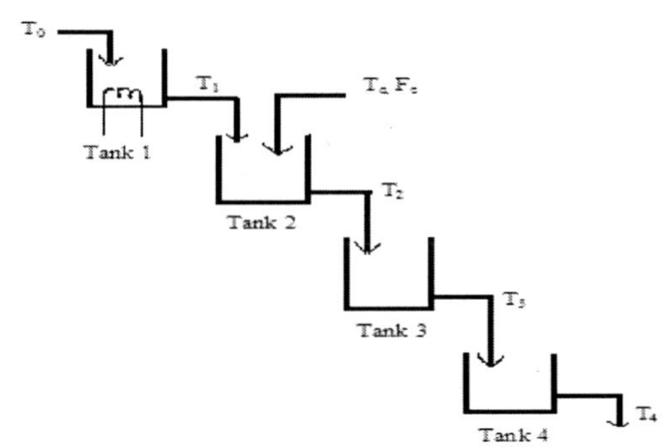

A. tank 1

B. tank 2

C. tank 3

D. tank 4

Sol. **C.** since, there is a disturbance in Tank 2 and we need to control the temperature of water leaving Tank 4.

The best place for secondary measurement is Tank 3.

GATE 2002

26. The closed-loop poles of a stable second order system could be

A. both real and positive

B. complex conjugate with positive real parts

C. both real and negative

D. one real positive and the other real negative

***Sol.*A.** the closed-loop poles of a stable second order system could be both real and positive.

27. A first order system with unity gain and time constant τ is subjected to a sinusoidal input of frequently $\omega = \frac{1}{\tau}$. The amplitude ratio for this system is

A. 1

B. 0.5

C. $\frac{1}{\sqrt{2}}$

D. 0.25

***Sol.*C.** $G(s) = \frac{1}{\tau s + 1}$

Putting $j\omega$ for s, $G(j\omega) = \frac{1}{\tau j\omega + 1}$

$$= \frac{1}{1 + \omega^2\tau^2} - j\frac{\omega\tau}{1 + \omega^2\tau^2}$$

Amplitude Ratio, AR = $|G(j\omega)| = \frac{1}{\sqrt{\omega^2\tau^2+1}}$

$$= \frac{1}{\sqrt{1+1}} = \frac{1}{\sqrt{2}}$$

GATE 2001

28. The calibration data of a thermocouple with it's cold juction at 0°C are given below.

Hot junction temprature(°C)	0	20	40	60	80	100
Thermo-emf (mV)	0.00	0.80	1.61	2.43	3.26	4.10

The hot junction of the thermocouple is placed in a bath at 80°C while its cold junction is at 20°C. What is the emf of the thermocouple?

A. 3.26

B. 0.80

C. 2.46

D. 2.43

***Sol.*D.** for thermocouple

Hot junction temperature = 100°C

Cold junction temperature = 0°C

Temperature difference = 100°C

With the data given in table with cold junction temperature at 0°C,

Thermo emf at 60°C = 2.43 mW

29. A process is initially at steady state with its output y = 1. The input is suddenly changed to at time t = 0. The output response is y(t) = 1 + 2t. The transfer function of the process is

A. $\frac{2}{s}$

B. $1 + \frac{2}{s^2}$

C. $1 + \frac{2}{s}$

D. $\frac{1}{s}\left(1 + \frac{2}{s}\right)$

***Sol.*C.** The transfer function of the process,

$$G(s) = \frac{y(s)}{u(s)} = \frac{L\{1+2t\}}{\frac{1}{s}}$$

Here $u(t)$ is unit step input.

$$G(s) = s \cdot L\{1 + 2t\}$$

$$= s \cdot \left(\frac{1}{s} + \frac{2}{s^2}\right) = \left\{1 + \frac{2}{s}\right\}$$

30. The inherent characteristics of an equal percentage valve relating flow rate q with valve stem movement x are described by the equation

A. $\frac{dq}{dx} = k$

B. $\frac{dq}{dx} = kq$

C. $\frac{dq}{dx} = \frac{k}{q}$

D. $\frac{dq}{dx} = kq^3$

***Sol.*A.** output: flow rate q

Input: valve stem movement x

The equation can be written as

$$\frac{dq}{dx} = k$$

GATE 2000

31. The unit step response of the transfer function $\frac{2s-1}{(3s+_1)(4s+1)}$ reaches its final steady state asymptotically after

A. a monotonic increase

B. a monotonic decrease

C. initially increasing and then decreasing

D. initially decreasing and then increasing

***Sol.*C.** given, $\frac{Y(s)}{X(s)} = \frac{2s-1}{(3s+1)(4s+1)}$

$$\frac{Y(s)}{X(s)} = \frac{5}{3s+1} - \frac{6}{4s+1}$$

For unit step response,

$$X(s) = \frac{1}{s}$$

$$Y(s) = \frac{1}{s} \cdot \frac{5}{3s+1} - \frac{1}{s}\frac{6}{4s+1}$$

$$Y(t) = 5\left(1 - e^{-\frac{t}{3}}\right) - 6(1 - e^{-\frac{t}{4}})$$

$$Y(t) = -1 + 6e^{-\frac{t}{4}} - 5e^{-\frac{t}{3}}$$

$Y(t)$ initially increases and then decreases and finally reaches –1 asymptotically.

32. The unit step response of the transfer function

$$\frac{1}{s^2 + 2s + 3}$$

A. has a non-zero slope at the origin

B. has a damped oscillatory characteristic

C. is overdamped

D. is unstable

***Sol.*B.** given, $G(s) = \frac{1}{s^2+2s+3}$

$$= \frac{1}{3\left(\frac{1}{3}s^2 + 2 \cdot \frac{1}{3}s + 1\right)}$$

Here, $\tau = \frac{1}{\sqrt{3}}$ and $\xi = \frac{1}{\sqrt{3}} < 1$

The roots of quadratic equation are complex.

Thus, the unit step response of $G(s)$ will be oscillatory.

33. Select the correct statement from the following.

A. the frequency response of a pure capacity process is unbounded

B. the phase lag of a pure time delay system decreases with increasing

C. the amplitude ratio of a pure capacity process is inversely proportional to the frequency

D. the amplitude ratio of a pure time delay system increases with frequency

***Sol.*C.** the amplitude ratio of a pure capacity process is inversely proportional to the frequency.

34. For a feedback control system to be stable, the

A. roots of the characteristic equation should be real

B. poles of the closed-loop transfer function should lie in the left half of the complex plane

C. bode plots of the corresponding open-loop transfer function should monotonically decrease

D. poles of the closed-loop transfer function should lie in the right half of the complex plane

***Sol.*B.** the amplitude ratio of a pure capacity process is inversely proportional to the frequency.

(2 Mark Question)
GATE 2015

35. Which one of the following transfer function, upon a unit step change in disturbance at t = 0, will show a stable time domain response with a negative initial slope (i.e. slope at t = 0)

A. $G(s) = \frac{1}{s+1} - \frac{2}{s+4}$

B. $G(s) = \frac{1}{s+1} + \frac{2}{s+4}$

C. $G(s) = \frac{1}{s+1} + \frac{2}{s-4}$

D. $G(s) = \frac{1}{s-1} - \frac{2}{s-4}$

***Sol.*A.**

take, $G(s) = \frac{1}{s+1} - \frac{2}{s+4}$

So, $Y(s) = \frac{1}{s}\left(\frac{1}{s+1} - \frac{2}{s+4}\right)$

$$d\,\{sy(s)\} = \frac{dy}{dt} = e^{-t} - 2e^{-4t}$$

$$\left.\frac{dy}{dt}\right|_{z=0} = -1$$

So (a) has –ve initial slope

36. The block diagram for a process with feedback control for output deviation variable H is shown in the figure below. All transfer functions are given with pre-factor of s (in min). A unit step change is made in the set-point at t = 0. The time required for H to reach 50% of its ultimate value. (in min) (up to two decimal place) is ________.

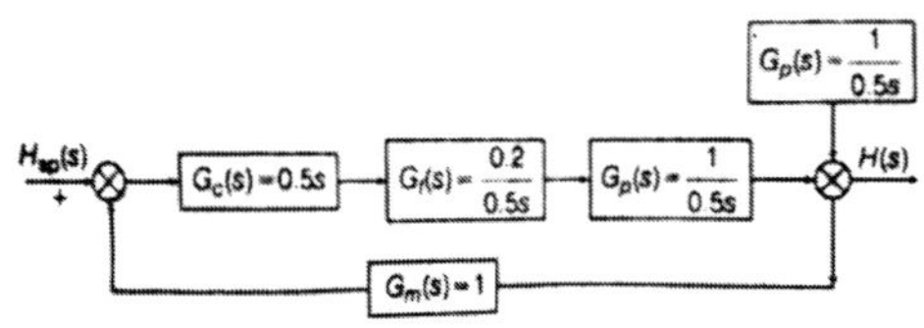

Sol. we have, by making $G_d(s) = 0$ (no load disturbance)

$$\frac{h(s)}{h_{sp}(s)} = \frac{\frac{0.2}{1.5s+1}}{\frac{0.2}{1.5s+1}+1} = \frac{0.2}{1.2+1.5\,s}$$

Given, $h_{sp}(s) = \frac{1}{s}$

S $\quad h(s) = \frac{1}{s} \cdot \frac{0.2}{1.5s+1.2} = \frac{A}{S} + \frac{B}{1.5s+1.2}$

$$A = \frac{1}{6} \text{ and } B = \frac{-1}{4}$$

So, $h(s) = \frac{1}{6s} - \frac{1}{4} \cdot \frac{1}{1.5s+1.2}$

Taking inverse laplace transformation,

$$h(t) = \frac{1}{6} - \frac{1}{6}e^{-0.8t}$$

Ultimate value = $H(\infty) = \frac{1}{6}$

Let t = time required to reach 50% of ultimate value

$$\frac{1}{12} = \frac{1}{6} - \frac{1}{6}e^{-0.8t} \Rightarrow \frac{-1}{12} = \frac{-1}{6}e^{-0.8t}$$

$$e^{-0.8t} = +\frac{1}{2} \Rightarrow t = \frac{0.693}{0.8} = 0.8662\ s$$

37. Consider a control system with the open-loop transfer function given by $G_{OL}(s) = \frac{K_c e^{-0.3s}}{1.5s+1}$. In the above function, pre-factor of s is in minute and K_c is the gain of proportional controller. The frequency for phase margin of 30^0 is 40.04 rad/min. The value of K_c for a gain margin of 1.7 (upto one decimal place) is ________.

Sol. $G(s) = \frac{K_c e^{-0.3\ s}}{1.5\ s+1}$

Given, $\omega = 40.04$ rad/main

Gain margin = 1.7

$$AR = \frac{1}{1.7}$$

$$|G(j\omega)| = \frac{1}{1.7}$$

$$\frac{K_c}{\sqrt{(1.5\omega)^2+1}} = \frac{1}{1.7}$$

$$K_c = \frac{\sqrt{(1.5\times 40.04)^2+1}}{1.7}$$

$$K_c = 35.3$$

GATE 2014

38. A step change of magnitude 2 is introduced in to a system having the following transfer function $G(s) = \frac{2}{s^2+2s+4}$

The per cent overshoot is ________.

Sol. $G(s) = \frac{2}{s^2+2s+4} = \frac{1}{2\left(\frac{1}{4}s^2+2\times\frac{1}{2}\times\frac{1}{2}s+1\right)}$

Comparing with $G(s) = \frac{K}{\tau^2 s^2+2\tau\xi\ s+1}$

$$\tau = \xi = \frac{1}{2}$$

$$\%\ \text{overshoot} = \exp\left(\frac{-\pi\xi}{\sqrt{1-\xi^2}}\right)$$

$$= \exp\left(\frac{-\pi\times\frac{1}{2}}{\sqrt{1-\frac{1}{4}}}\right)$$

$$= \exp\left(\frac{-\pi}{\sqrt{3}}\right)$$

% overshoot = 16.3%

39. Given below is a simplified block diagram of a feed forward control system.

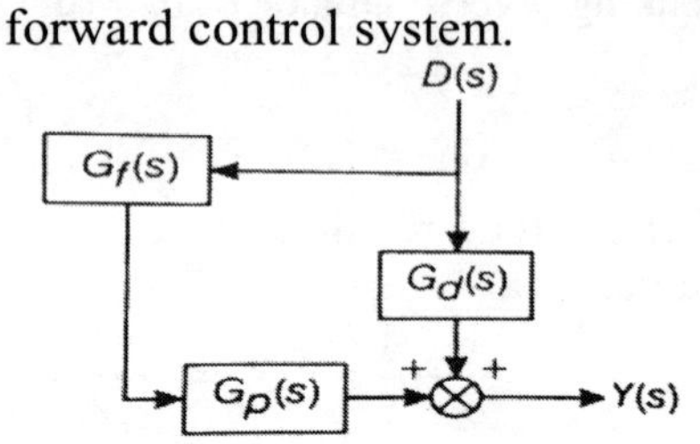

The transfer function of the process is $G_p = \frac{5}{s+1}$ and the disturbance transfer function is $G_d = \frac{1}{s^2+2s+1}$. The transfer function of the perfect feed forward controller, $G_f(s)$ is

A. $\frac{-5}{(s+1)^3}$

B. $\frac{-5}{(s+1)}$

C. $\frac{-1}{5(s+1)}$

D. -5(s+1)

Sol.**C.** $Y(s) = D(s)\,G_d(s) + D(s)G_f(s)G_p(s)$

For perfect feed-forward controller, $\frac{Y(s)}{D(s)} = 0$

No effect of load disturbances

$$G_d(s) + G_f(s)G_p(s) = 0$$

$$G_f(s) = -\frac{G_d(s)}{G_p(s)} = -\frac{\frac{1}{(s+1)^2}}{\frac{5}{(s+1)}} = -\frac{1}{5(s+1)}$$

40. A process with transfer function, $G_p = \frac{2}{s-1}$ is to be controller by a feedback proportional controller with a gain K_c. If the transfer functions of all other elements in the control loop are unity, then which one of the following conditions produces a stable closed loop response?

A. $K_c = 0.25$

B. $0 < K_c < 0.25$

C. $0.25 < K_c < 0.5$

D. $K_c > 0.5$

Sol.**D.** transfer function of closed loop,

$$G(s) = \frac{G_p G_c}{1+G_p G_c} = \frac{K_c\times\frac{2}{s-1}}{1+K_c\times\frac{2}{s-1}}$$

$$= \frac{2K_c}{s+2K_c-1}$$

For stable closed loop response,

$$2\,K_c - 1 > 0$$

$$K_c > \frac{1}{2} = 0.5$$

41. Consider the following block diagram for a closed-loop feedback control system

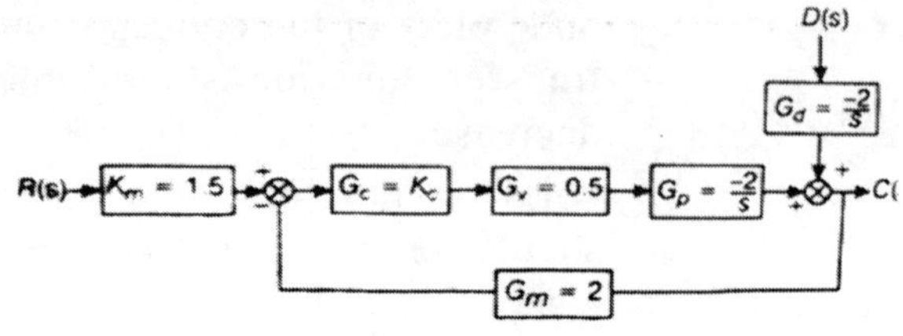

A proportional controller is being used with $K_c = -4$. If astep change in disturbance of magnitude 2 affects the system, then the value of the offset is________.

Sol. $\frac{C(s)}{D(s)} = \frac{G_d}{1+G_cG_vG_pG_m} = \frac{-\frac{2}{s}}{1+\left[K_c\times 0.5\times\left(\frac{-2}{s}\right)\times 2\right]}$

$= \frac{-2}{s+8} = -\frac{1}{4\left(\frac{1}{8}s+1\right)}$

Given, $D(s) = \frac{2}{s}$

$\therefore \quad C(s) = -\frac{1}{2\left(\frac{1}{8}s+1\right)}$

or $\quad C(t) = -\frac{1}{2}\left[1 - e^{-\left(\frac{t}{\frac{1}{8}}\right)}\right]$

$= -\frac{1}{2}(1 - e^{-8t})$

$C(\infty) = -\frac{1}{2}$

Offset = $0 - C(\infty) = 0 - \left(-\frac{1}{2}\right) = 0.5$

GATE 2013

42. A unit gain 2nd order underdamp process has a period of oscillation 1 second and decay ratio 0.25. The transfer function of the process is

A. $\frac{1}{0.024\, s^2 + 0.067\, s + 1}$

B. $\frac{1}{0.067\, s^2 + 0.024\, s + 1}$

C. $\frac{1}{0.021\, s^2 + 0.1176\, s + 1}$

D. $\frac{1}{0.1176\, s^2 + 0.021\, s + 1}$

GATE 2012

43. A thermocouple having a linear relationship between 0°C and 350°C shows an emf of zero and 30.5 mV respectively at these two temperatures. If the cold junction temperature is shifted from 0°C to 30°C, then the emf correction (in mV) is

A. 3.13

B. 2.92

C. 2.61

D. 2.02

Sol. C. (350 - 0)°C corresponds to an emf of 30.5 mV.

$\therefore$ (350 - 0)°C corresponds to an emf of

$\frac{30.5}{350} \times 320 = 27.89$ mV

So, emf correction = 30.5 – 27.89 = 2.61 mV

44. The characteristic equation for the system is $s^3 + 9s^2 + 26s + 12(2 + K_c) = 0$. Using Routh test, the value of K_c that will keep the system on the verge of instability is

A. 20.9

B. 18.4

C. 17.5

D. 15.3

Sol. C. Given,

characteristic equation for the system,

$$s^3 + 9s^2 + 26s + 12(2 + K_C) = 0$$

Applying Routh array for stability,

Row		
1	1	26
2	9	$12(2+K_c)$
3		$\frac{9 \times 26 - 12(2 + K_c)}{9}$

If all the elements of 1st column are positive and non-zero, the system would be stable.

Thus, on the verge of instability,

$$\frac{9\times 26-12(2+K_c)}{9} = 0$$

$\Rightarrow \quad 12(2+K_c) = 9 \times 26$

$\Rightarrow \quad 2+K_c = 19.5$

$\Rightarrow \quad K_c = 17.5$

45. The block diagram of a system with proportional controller is shown below.

A unit step input is introduced in the set point. The value of K_c to provide a critically damped response for

U = 0, $\tau_p = 8$ and $\tau_m = 1$ is

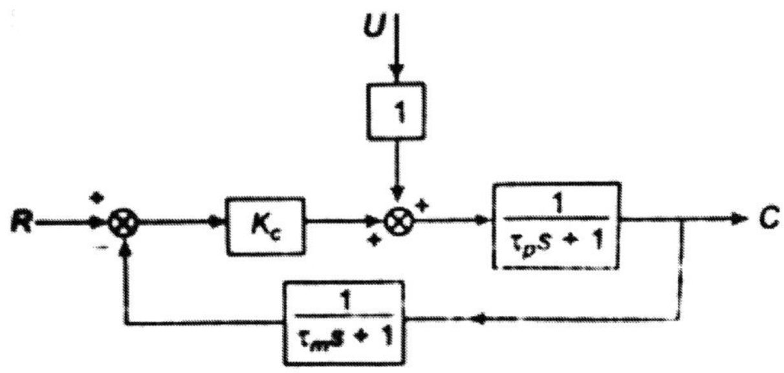

A. 3.34

B. 2.58

C. 1.53

D. 1.12

Sol.C. $\frac{C}{R} = \frac{K_c \cdot \left(\frac{1}{\tau_p\, s+1}\right)}{1+K_c \cdot \left(\frac{1}{\tau_p\, s+1}\right)\left(\frac{1}{\tau_m\, s+1}\right)}$

Here, $\tau_p = 8, \tau_p = 1$, U = 0

$\therefore \quad \frac{C}{R} = \frac{K_c \cdot \left(\frac{1}{8\,s+1}\right)}{1+K_c \cdot \left(\frac{1}{8\,s+1}\right)\left(\frac{1}{s+1}\right)}$

$= \frac{K_c\,(s+1)}{8s^2+9s+(1+k_c)}$

$= \frac{K_c\,(s+1)}{(1+k_c)\left[\frac{8}{1+k_c}s^2+\left(\frac{9}{1+k_c}\right)s+1\right]}$

Comparing the denominator with

$\tau^2 s^2 + 2\xi\tau s + 1$

$\tau^2 = \frac{8}{1+k_c}$ and $2\xi\tau = \frac{9}{1+k_c}$

For critically damped system,

$\xi = 1$

$\therefore \quad 2\tau = \frac{9}{1+k_c}$

Thus, $\left[\frac{9}{2(1+k_c)}\right]^2 = \frac{8}{1+k_c}$

$1+k_c = \frac{81}{4} \times \frac{1}{8}$

$k_c = 1.53$

GATE 2011

46. The following diagram shows a CSTR with two control loops. A liquid phase, endothermic reaction is taking place in the CSTR and the system is initially at steady state. Assume that the changes in physical properties of the system are negligible.

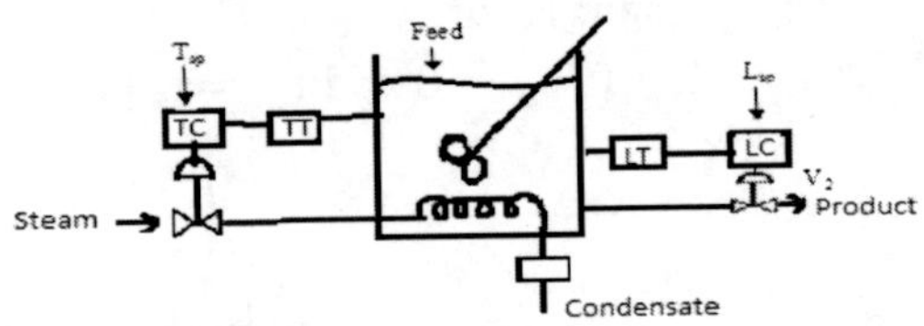

TC = temperature controller,

LC = level controller

TT = Temperature transmitter

LT = level transmitter

V_1 and V_2 = control valves

Which one of the following statement is true?

A. changing the level controller set point affects the opening of V_2 only

B. changing the temperature controller set point affect the opening of V_2 only

C. changing the temperature controller set point affect the opening of both V_1 and V_2

D. changing the level controller set point affect the opening of both V_1 and V_2

Sol.D. any change in level controller set point would affect the steady state condition and would affect the opening of both V_1 and V_2.

Statements for Linked Answer Question 47 and 48

A PID controller output p(t), in time domain is given by

$$P(t) = 30 + 5\,e(t) + 1.25 \int_0^t e(t)dt + 15\frac{de(t)}{dt}$$

Where, e(t) is the error at time t. The transfer function of the process to be controller is

$$G_p(s) = \frac{10}{(200s+1)}$$

The measurement of the controlled variable is instantaneous and accurate.

47. The transfer function of the controller is

A. $\frac{5(12s^2+4s+1)}{3s}$

B. $\frac{5(12s^2+3s+1)}{3s}$

C. $\frac{5(12s^2+4s+1)}{4s}$

D. $\frac{5(12s^2+3s+1)}{4s}$

Sol.C. the output of PID controller is given as

$$p(t) = 30 + 5e(t) + 1.25\int_0^t e(t)dt + 15\frac{de(t)}{dt}$$

Let us define,

$$P(t) = p(t) - 30$$

$$P(t) = 5e(t) + 1.25\int_0^t e(t)dt + 15\frac{de(t)}{dt}$$

Taking Laplace transformation,

$$p(s) = 5e(s) + 1.25\frac{e(s)}{s} + 15.s\,e(s)$$

or $\frac{p(s)}{e(s)} = 5 + \frac{5}{4s} + 15s$

or $G_c(s) = \frac{p(s)}{e(s)} = \frac{5(12s^2+4s+1)}{4s}$

48. The characterstic equation of the closed-loop is

A. $6s^2 + 102s + 1 = 0$

B. $700s^2 + 102s + 25 = 0$

C. $100s^2 - 196s - 25 = 0$

D. $240s^3 + 812s^2 + 204s + 1 = 0$

***Sol.* B.** the characteristic equation of the closed-loop,

$$1 + G_p(s)G_c(s) = 0$$

Putting values,

$$1 + \left(\frac{10}{200s+1}\right) \cdot \frac{5(12s^2+4s+1)}{4s} = 0$$

or $2s\,(200s + 1) + 25(12s^2 + 4s + 1) = 0$

$$700s^2 + 102s + 25 = 0$$

GATE 2010

49. A block diagram for a control system is shown below.

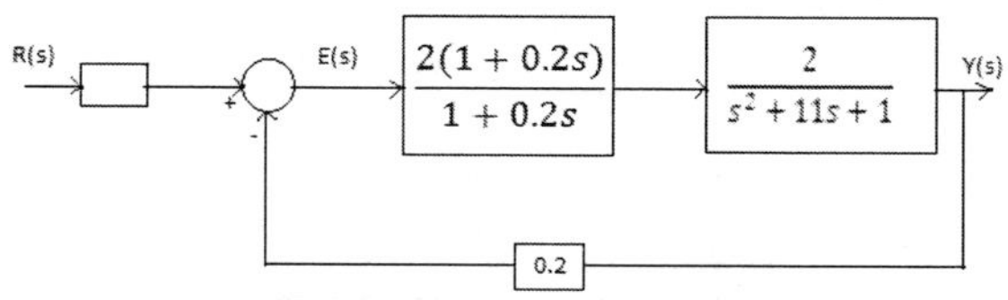

The steady state gain of the closed-loop system, between output Y(s) and set point R(s), is

A. 5/9

B. 4/9

C. 1/3

D. 2/9

***Sol.* B.**

$$\frac{y(s)}{R(s)} = \frac{0.2\times\frac{2(1+0.2s)}{(1+0.02s)}\times\frac{2}{s^2+11s+1}}{1+0.2\times\frac{2(1+0.2s)}{(1+0.02s)}\times\frac{2}{s^2+11s+1}}$$

$$= \frac{0.8(1+0.2s)}{(1+0.02s)(s^2+11s+1)+0.8(1+0.2s)}$$

For steady state, $s = 0$

Steady state gain $= \left.\frac{y(s)}{R(s)}\right|_{at\ s=0}$

$$= \frac{0.8(1+0)}{(1+0)(0+0+1)+0.8(1+0.)}$$

$$= \frac{0.8}{1+0.8} = 0.8/1.8$$

$$= 4/9$$

50. Consider the cascade control configuration shown in the figure below:

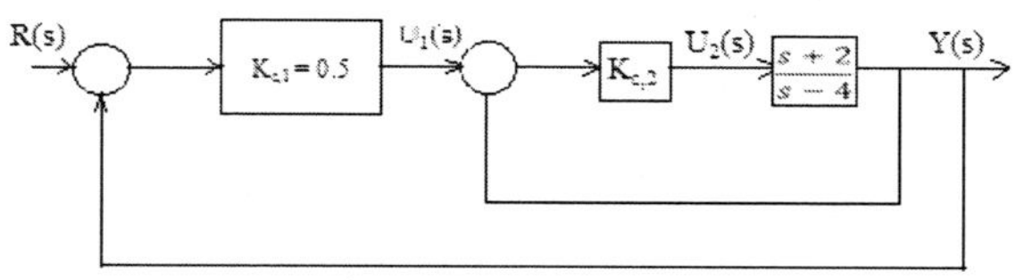

The system is stable when $k_{c,2}$ is

A. 3/4

B. 1

C. 5/4

D. 3/2

***Sol.* D.** given, $K_{c,1} = 0.5$

Assume, $K_{c,2} = k$

We have,

$$\frac{y(s)}{R(s)} = \frac{0.5\times\left[\frac{k\left(\frac{s+2}{s-4}\right)}{1+k\left(\frac{s+2}{s-4}\right)}\right]}{1+0.5\times\left[\frac{k\left(\frac{s+2}{s-4}\right)}{1+k\left(\frac{s+2}{s-4}\right)}\right]}$$

$$\frac{y(s)}{R(s)} = \frac{0.5\times k\left(\frac{s+2}{s-4}\right)}{1+k\left(\frac{s+2}{s-4}\right)+0.5k\left(\frac{s+2}{s-4}\right)}$$

$$\frac{y(s)}{R(s)} = \frac{0.5k(s+2)}{(s-4)+k(s+2)+0.5k(s+2)}$$

$$\frac{y(s)}{R(s)} = \frac{0.5k(s+2)}{s(1+1.5k)+(3k-4)}$$

The characteristic equation is

$$s(1 + 1.5k) + (3k - 4) = 0$$

Routh array for stability is

Row	
1	$1 + 1.5k$
2	$3k - 4$

For system to be stable, all the elements of first column of Routh array should be positive and non-zero.

Since, proportional sensitivity of the controller is a positive quantity,

$$K_{c,2} = k > 0$$

∴ 1st element of the first column, $1 + 1.5k > 0$ and 2nd element of the first column, $3k - 4 > 0$

∴ $k > \frac{4}{3}$

The options given are $\frac{3}{4}, 1, \frac{5}{4}$ and $\frac{3}{2}$.

Thus, $k = \frac{3}{2}$ is the correct answer.

51. Consider the cascade control configuration shown in the figure below

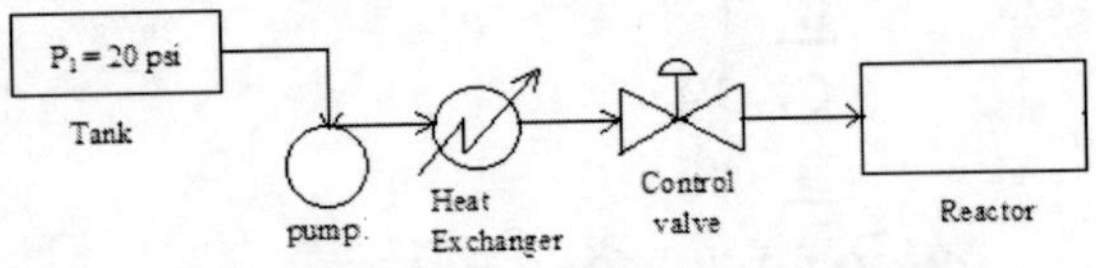

A constant heat pump transfers a liquid flow, a tank maintained at 20 psi to a reactor operating at 100 psi, through a heat exchanger and a control valve. At the design condition, the liquid flow rate is 1000 L/min, while the pressure drop across the controll valve is 20 psi. Assume that the pressure drop across the heat exchanger varies as the square of the flow rate. If the flow is reduced to 500 L/min, then the pressure drop across the valve is

<u>A.</u> 30psi
<u>B.</u> 50psi
<u>C.</u> 80psi
<u>D.</u> 150psi

Sol.**<u>B.</u>** At design condition,

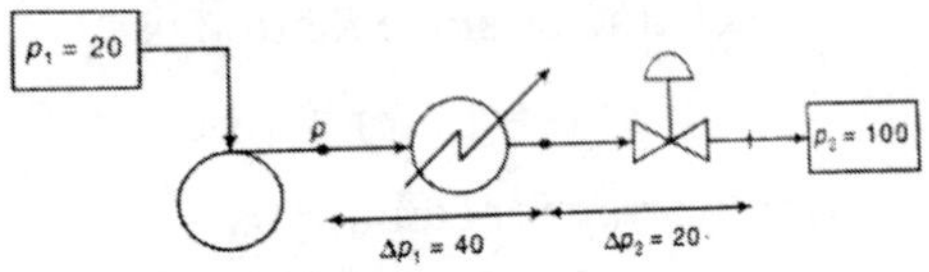

Given, $p_2 = 100\ psi$

$$p = \Delta p_1 + \Delta p_2 + p_2 = 160\ psi$$

Now, $\Delta p_1 \propto Q^2$

$$\therefore \quad \frac{40}{\Delta p_1'} = \left(\frac{1000}{500}\right)^2 = 4$$

$$\Delta p_1' = 10$$

Thus, pressure across the control value

$$\Delta p_1' = 160 - 10 - 100 = 50\ psi$$

GATE 2009

52. The inverse Laplace transform of $\frac{1}{2s^2+3s+1}$ is

<u>A.</u> $e^{-\frac{t}{2}} - e^{-t}$
<u>B.</u> $2e^{-\frac{t}{2}} - e^{-t}$
<u>C.</u> $e^{-t} - 2e^{-\frac{t}{2}}$
<u>D.</u> $e^{-t} - e^{-t/2}$

Sol.**<u>A.</u>**

$$\frac{1}{2s^2+3s+1} = \frac{1}{(2s+1)(s+1)}$$

$$= \frac{2}{(2s+1)} - \frac{1}{(s+1)}$$

$$= \frac{1}{\left(s+\frac{1}{2}\right)} - \frac{1}{(s+1)}$$

$\therefore$ Inverse Laplace transform of

$$\frac{1}{2s^2+3s+1} = e^{-t/2} - e^{-t}$$

53. The characterstics equation of a closed-loop system using a proportional controller with gain k_c is

$$12s^3 + 19s^2 + 8s + 1 + K_c = 0$$

At the onset of instability, the value of K_c

<u>A.</u> 35/3
<u>B.</u> 10
<u>C.</u> 25/3
<u>D.</u> 20/3

Sol.**<u>A.</u>** given, the characteristic equation is

$$12s^3 + 19s^2 + 8s + 1 + K_c = 0$$

Applying Rout harry for stability,

Row		
1.	12	8
2.	19	$1 + K_c$
3.	$\frac{19 \times 8 - 12(1+K_c)}{19}$	

If all the element of 1st column is positive and non-zero, the system would be stable.

Thus, at the onset of instability

$$\frac{19\times8-12(1+K_c)}{19} = 0$$

$$152 - 12(1 + K_c) = 0$$

$$K_c = \frac{140}{12} = \frac{35}{3}$$

54. The block diagram for a control sytem is shown below:

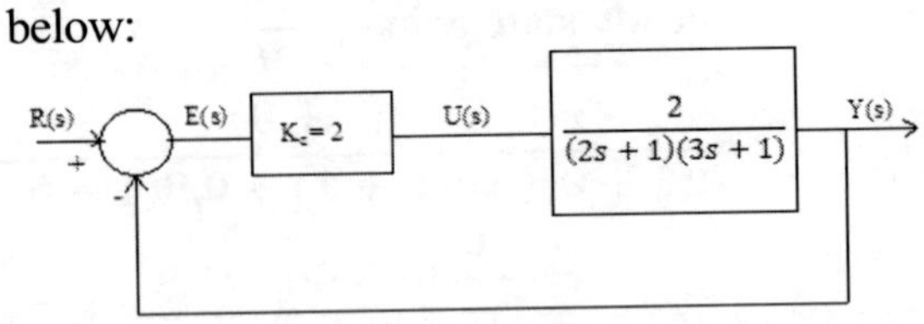

For a unit step change in the set point, R(s), the steady state offset in the output Y(s) is

<u>A.</u> 0.2
<u>B.</u> 0.3
<u>C.</u> 0.4
<u>D.</u> 0.5

Sol.**<u>A.</u>**

$$\frac{y(s)}{R(s)} = \frac{2\times\frac{2}{(2s+1)(3s+1)}}{1+2\times\frac{2}{(2s+1)(3s+1)}}$$

$$= \frac{4}{(2s+1)(3s+1)+4}$$

Given, $R(s) = \frac{1}{s}$

$$y(s) = \frac{1}{s} \times \frac{4}{(2s+1)(3s+1)+4}$$

The steady state offset $= 1 - \min_{s\to 0} s\,Y(s)$

$$= 1 - \frac{4}{(0+1)(0+1)+4}$$

$$= 1 - \frac{4}{5} = \frac{1}{5} = 0.2$$

55. For a tank of cross-section area 100 cm^3 and inlet flow rate (Q, in cm^3/s), the outlet flow rate (Q_0 in cm^3/s) is related to the liquid height (H in cm) as $Q_0 = 3\sqrt{H}$ (see figure below).

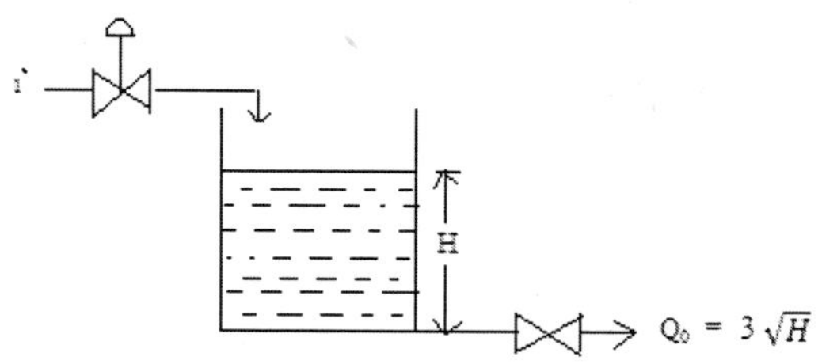

Then the transfer function $\frac{\bar{H}(s)}{\bar{Q}_i(s)}$ (overbar indicate deviation variables) of the process around the steady state point, Q_{is} = 18 cm^3/s and H_s = 36 cm, is

A. $\frac{1}{100s+1}$

B. $\frac{2}{200s+1}$

C. $\frac{3}{300s+1}$

D. $\frac{4}{400s+1}$

***Sol.* D.**

$A = 100\ cm^2$

$Q_{is} = 18\ cm^3/s$

$H_s = 36\ cm$

Applying material balance for the tank,

$$Q_i(t) - Q_0(t) = A\frac{dh}{dt} \qquad \text{....(i)}$$

Here, $Q_0 = 3\sqrt{H}$

Applying Taylor series expansion for linearization,

$$Q_0 = Q_0(h_s) + Q_0'(h_s) \times (h - h_s)$$

Leaving higher order terms,

$$= Q_{0s} + \frac{3}{2}\cdot H_s^{-1/2}(h - h_s)$$

$$= Q_{0s} + \frac{3}{2}\cdot (36)^{-1/2}(h - h_s)$$

$$= Q_{0s} + \frac{1}{4} \times (h - h_s) \qquad \text{....(ii)}$$

From, Eqs. (i) and (ii),

$$Q_i - Q_{0s} - \frac{1}{4} \times (h - h_s) = A\frac{dh}{dt}$$

Putting, $Q_i - Q_{0s} = \bar{Q}_i$ and $h - h_s = \bar{H}$

Also, $\frac{dh}{dt} = \frac{d(h-h_s)}{dt} = \frac{d\bar{H}}{dt}$

Therefore,

$$\bar{Q}_i - \frac{1}{4}\bar{H} = A\frac{d\bar{H}}{dt}$$

$$\bar{Q}_i = A\frac{d\bar{H}}{dt} + \frac{1}{4}\bar{H}$$

Taking Laplace transform,

$$A\cdot s\bar{H}(s) + \frac{1}{4}\bar{H} = \bar{Q}_i(s)$$

$$100\cdot s\bar{H}(s) + \frac{1}{4}\bar{H} = \bar{Q}_i(s)$$

$$\frac{\bar{H}(s)}{\bar{Q}_i(s)} = \frac{4}{400s+1}$$

GATE 2008

56. The unit impluse response of first order process is given by $2e^{-0.5t}$. The gain and time constant of the process are respectively

A. 4 and 2

B. 2 and 2

C. 2 and 0.5

D. 1 and 0.5

***Sol.* A.** the first order transfer function

$$\frac{Y(s)}{X(s)} = \frac{K_c}{\tau s+1}$$

For unit impulse input, $X(s) = 1$

$\therefore \qquad Y(s) = \frac{K_c}{\tau s+1}$

or $\qquad Y(t) = \frac{K_c}{\tau}e^{-t/\tau}$

Given that, $\qquad Y(t) = 2e^{-0.5t}$

Comparing Eqs. (i) and (ii),

$$\frac{1}{\tau} = 0.5$$

$$\tau = 2$$

And $\qquad \frac{K_c}{\tau} = 2$

$\therefore \qquad K_c = 4$

57. A unit step input is given to a process that is represented by the transfer function $\frac{(s+2)}{(s+5)}$. The initial value (t = 0$^+$) of the response of the process to the step input is

A. 0

B. 2/5

C. 1

D. infinite

Sol.C. Given,

$$\frac{Y(s)}{X(s)} = \frac{s+2}{s+5}$$

For unit step input, $X(s) = \frac{1}{s}$

$\therefore \quad Y(s) = \frac{1}{s} \cdot \frac{s+2}{s+5}$

Using initial value theorem,

$$\lim_{t\to 0} Y(t) = \lim_{s\to\infty} sY(s)$$

$$= \lim_{s\to\infty} \left(\frac{s+2}{s+5}\right)$$

$$= \lim_{s\to\infty} \left(\frac{1+\frac{2}{s}}{1+\frac{5}{s}}\right)$$

$$= \frac{1+0}{1+0} = 1$$

58. A tank of volume 0.25m³ and hight 1 m has water flowing in at 0.05 m³/min. The outlet flowrate is governed by relation

$F_{out} = 0.1$ h

Where, h is the height of the water in the tank in meter and f_{out} is the outlet flow rate in m³/min. The inlet flow rate changes suddenly from its nominal value of 0.05 m³/min and remains there. The time (in min)at which the tank will begin to overflow is given by

A. 0.28

B. 1.01

C. 1.73

D. infinite

Sol.C. given,

Tank volume = $0.25\ m^3$

Tank height = 1 m

$\therefore$ Area of tank = $\frac{0.25}{1} = 0.25\ m^2$

Applying material balance for the tank,

$$A\frac{dh}{dt} = F_{in} - F_{out}$$

$$0.25\frac{dh}{dt} = F_{in} - 0.1\,h$$

At steady state,

$$0.25\frac{dh_s}{dt} = F_{ins} - 0.1\,h_s$$

Define, $h - h_s = H$ and $F_{in} - F_{ins} = F$

$\therefore \quad 0.25\frac{dH}{dt} = F - 0.1\,H$

Taking Laplace transform,

$$0.25\,[sH(s) - H(0)] = F(s) - 0.1\,H(s)$$

At steady state, $H(0) = h_s - h_s = 0$

$\therefore \quad 0.25s\ H(s) = F(s) - 0.1\,H(s)$

$$\frac{H(s)}{F(s)} = \frac{1}{0.25\,s+0.1} = \frac{10}{2.5\,s+1}$$

Now, inlet flow rate gives a step change.

$$F(s) = \frac{0.15-0.05}{s} = \frac{0.1}{s}$$

$\therefore \quad H(s) = \frac{0.1}{s} \cdot \frac{10}{2.5\,s+1}$

$$H(s) = \frac{1}{s} \cdot \frac{2}{2.5\,s+1}$$

$$= \frac{1}{s} \cdot \frac{1}{\left(s+\frac{1}{2.5}\right)}$$

Taking inverse Laplace transform,

$$H(t) = 1 - e^{-t/2.5}$$

At steady state, $F_{in} = F_{out}$

$0.05 = 0.1\ h_s$

$\therefore \quad h_s = 0.5$ m

For overflow, the height of the water should be equal to the height of the tank.

$\therefore \quad$ h = 1 m

And $\quad H = h - h_s = 1 - 0.5 = 0.5$ m

Therefore, for overflow,

$$0.5 = 1 - e^{-t/2.5}$$

$$e^{-t/2.5} = 0.5$$

$$-\frac{t}{2.5} = -0.693$$

t = 1.73 min

59. Which one of the following transfer functions correspounds to an inverse response process with a positive gain?

A. $\frac{1}{2s+1} - \frac{2}{3s+1}$

B. $\frac{2}{s+1} - \frac{5}{s+10}$

C. $\frac{3(0.5s-1)}{(2s+1)(s+1)}$

D. $\frac{5}{s+1} - \frac{3}{2s+1}$

Sol.A. (a) $\quad Y(s) = \frac{1}{2s+1} - \frac{2}{3s+1}$

$\therefore \quad Y(t) = \frac{1}{2}e^{-t/2} - \frac{2}{3}e^{-t/3}$

(b) $\quad Y(s) = \frac{2}{s+1} - \frac{5}{s+10}$

$\therefore \quad Y(t) = 2e^{-t} - 5e^{-10t}$

(c) $\quad Y(s) = \frac{3(0.5\,s-1)}{(2s+1)(s+1)} = 3\left[-\frac{2.5}{2s+1} + \frac{1.5}{s+1}\right]$

$\therefore \quad Y(t) = 3(-1.25e^{-\frac{t}{2}} + 1.5e^{-t})$

(d) $\quad Y(s) = \frac{5}{s+1} - \frac{3}{2s+1}$

$\therefore \quad Y(t) = 5e^{-t} - \frac{3}{2}e^{-t/2}$

It can be seen that none of the options represents a positive gain process.

∴ Some data misprinted in this question.

60. Match the list I with list II and select the correct answer using the codes given below the lists.

List I	List II
P. Temprature Q. Pressure R. Flow	1. Hot wire anemometry 2. Strain gauge 3. Chromatographic analyzer 4. Pyrometer

Code

	P	Q	R
A.	1	2	3
B.	4	1	3
C.	1	2	4
D.	4	2	1

***Sol.*D.** temperature – Pyrometer

pressure - Strain gauge

flow - Hot wire anemometry

61. Match the List I with List II and select the correct answer using the codes given below the lists.

List I	List II
P. Ziegler-Nichols Q. Underdamped response R. Feed-forward control	1. Process reaction curve 2. Decay ratio 3. Frequency respose 4. Disturbance mesurment

Code

	P	Q	R
A.	3	2	4
B.	1	2	3
C.	3	4	2
D.	1	4	2

***Sol.*A.** ziegler-Nichols - Frequency response

Underdamped response - Decay ratio

Feed forward control - Disturbance measurement

Statements for linked answer Question 62 and 63

The cross-over frequency associated with a feedback loop employing a proportional controller to control the process represented by the transfer function

$G_p(s) = \frac{2e^{-s}}{(\tau s+1)^2}$, (unit of time minute)

is found to be 0.6 rad/min. Assumed that the measurement and value transfer function are unity.

62. The time constant ô (in min) is

A. 1.14

B. 1.92

C. 3.23

D. 5.39

***Sol.*D.** given,

$$G_p(s) = \frac{2e^{-s}}{(\tau s+1)^2} = \frac{2e^{-s}}{(\tau s+1)(\tau s+1)}$$

Phase lag at cross-over frequency,

$$\emptyset = 180°$$

$$\emptyset\,(e^{-s}) + \emptyset\left(\frac{1}{\tau s+1}\right) + \emptyset\left(\frac{1}{\tau s+1}\right) = 180°$$

Since, $\emptyset(e^{-s}) = -\omega$ in rad/min

$= -57.3\,\omega$ in deg/min

∴ $-57.3\,\omega + 2\,tan^{-1}(-\tau\omega) = 180°$

$tan^{-1}(-0.6\,\tau) = 107.19$

$-0.6\,\tau = -3.23$

$\tau = 5.39$

63. If the control loop is to operate at a margin of 2.0, the gain of the proportional controller must equal

A. 0.85

B. 2.87

C. 3.39

D. 11.50

***Sol.*B.** let us define that K is gain of proportional controller.

Gain margin $= \frac{1}{AR} = 2$

∴ $AR = \frac{1}{2}$

$$\frac{2K_c}{\left(\sqrt{1+(\omega t)^2}\right)^2} = \frac{1}{2}$$

$$\therefore\ K_c = \frac{1+\omega^2\tau^2}{4} = \frac{1+(0.6\times5.39)^2}{4} = 2.865$$

GATE 2007

64. The dynamic model for a mixing tank open the atmosphere at its top as shown below is to be written. The objective of mixing is to cool the hot water is stream entering the tank at a flow rate q_2 and feet temperature of T_s with a cold water feed stream entering the tank at a flow rate q_1 and feed temperature of T_0. A water stream is drawn from the tank bottom at a flow rate of q_4 by a pump and the level in the tank is proposed to be

controlled by drawing another water stream at a flow rate q_3. Neglect evaporation and other heat losses from the tank.

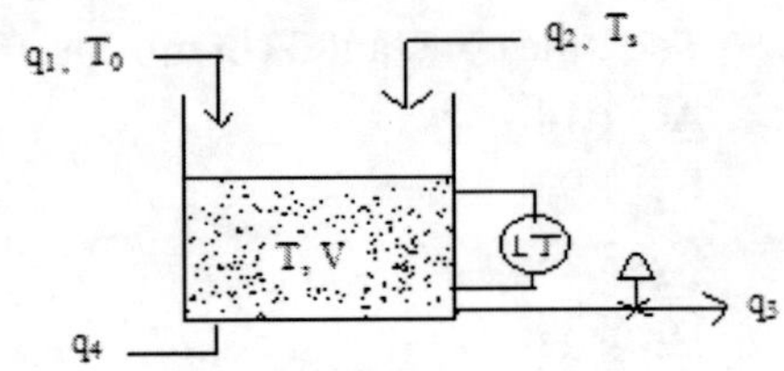

The dynamic model for the tank given as

A. $\frac{dV}{dt} = q_1 + q_2 + q_3$,

$V\frac{dT}{dt} = q_1T_0 + q_2T_s + q_3T$

B. $\frac{dV}{dt} = q_1 - q_4$,

$\frac{d(VT)}{dt} = q_1T_s - q_4T$

C. $\frac{dV}{dt} = q_1 + q_2 + q_3$,

$\frac{d(VT)}{dt} = q_1T_0 + q_2T_s - q_4T$

D. $\frac{dV}{dt} = q_1 + q_2 - q_3 + q_4$,

$\frac{d(VT)}{dt} = q_1(T_0 - T) + q_2(T_s - T)$

***Sol.*A.** applying the mass balance for the system,

$$\frac{d(pV)}{dt} = (q_1 + q_2 - q_3 - q_4)\,p$$

$$\frac{dV}{dt} = q_1 + q_2 - q_3 - q_4$$

Now, applying the energy balance for the system,

$$C_pV\frac{dT}{dt} = q_1C_p(T_0 - T) + q_2C_p(T_s - T) - q_3C_p(T - T) - q_4C_p(T - T)$$

$$V\frac{dT}{dt} = q_1(T_0 - T) + q_2(T_0 - T)$$

65. Match the transfer function with the responses to a unit step input shown in the figure.

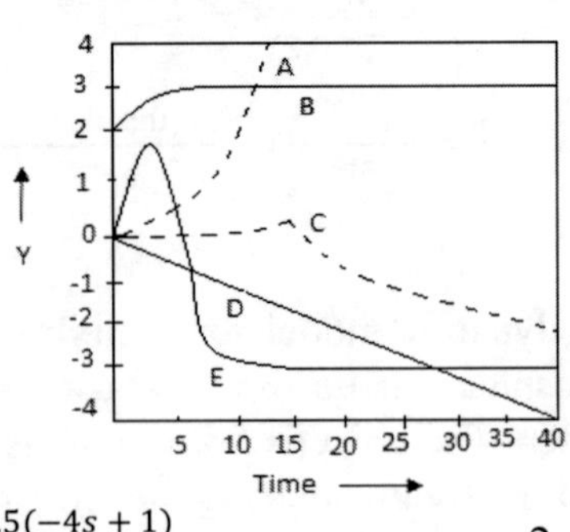

1. $\frac{-2.5(-4s+1)}{4s^2+4s+1}$

2. $\frac{-2e^{-10s}}{10s+1}$

3. $\frac{-5}{-20s+1}$

4. $\frac{-0.1}{s}$

5. $\frac{4s+3}{2s+1}$

Code

A.	E	C	A	D	B
B.	A	B	C	D	E
C.	B	A	C	E	D
D.	E	A	C	B	D

***Sol.*A.** let us find out the response for simpler transfer functions.

Case 4 $\frac{Y(s)}{X(s)} = -\frac{0.1}{s}$

For unit step input, $X(s) = \frac{1}{s}$

$\therefore \quad Y(s) = -\frac{0.1}{s^2}$

And $\quad Y(t) = -0.1\,t$

It is linearly decreasing function and matching with Curve D.

Case 3 $\frac{Y(s)}{X(s)} = \frac{-5}{-20s+1}$

$$Y(s) = -5.\frac{1}{s}.\frac{1}{-20s+1}$$

$$= -5\left[\frac{1}{s} + \frac{20}{-20s+1}\right]$$

$$= 5\left[\frac{1}{s-\frac{1}{20}} - \frac{1}{s}\right]$$

$\therefore \quad Y(t) = 5\,(e^{t/20} - 1)$

It is n exponentially function and matching with curve A.

Thus option **A.** is the correct answer.

66. Consider the following instrumentation diagram for a chemical reactor. C_{sp} represents a concentration set point.

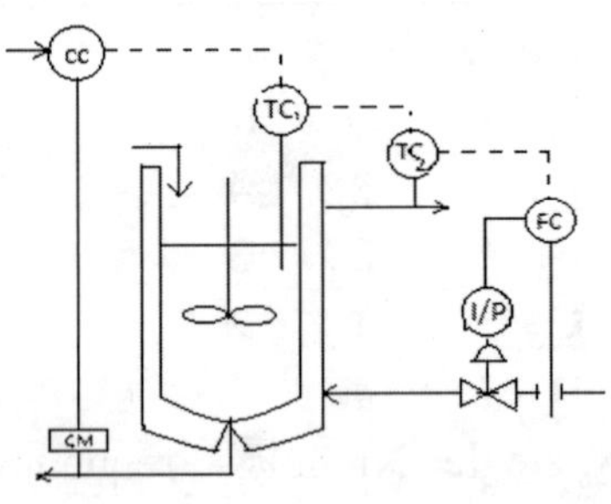

Match the List I with List II and select the correct answer using codes given the lists.

List I	*List II*
P. Control strategy	1. Feed forward control
Q. Primary controller variable	2. cascade control
R. Slowest controller	3. Concentration in a reactor
S. fastest controller	4. Reactor temperature
	5. Jacket temperature controller
	6. concentration controller
	7. Reactor temperature controller
	8. Jacket temperature controller
	9. Flow controller
	10. selective control

Code

	P	Q	R	S
A.	2	3	6	9
B.	1	4	8	7
C.	10	7	9	6
D.	1	8	5	9

***Sol.*A.** control strategy - Cascade control

Primary control variable - Concentration in the reactor

Slowest controller - Concentration controller

Fastest controller - Flow controller

67. The first two row of Routh's tabulation of a third order equation are

s^3 2 2

s^2 4 4

Select the correct answer from the following choices

A. the equation has one root in the right s-plane

B. the equation has two roots on the j-axis at s= j and –j, the third root is in the left half plane

C. the equation has two roots on the j-axis at s= 2j and -2j, the third root is the left half plane

D. the equation has two roots on the j-axis at s= 2j and the third root is in the right half plane

***Sol.*B.** given, Routh's tabulation

s^3 2 2

s^2 4 4

Thus, the characteristic equation for this system is

$$2s^3 + 4s^2 + 2s + 4 = 0$$
$$2s^2(s+2) + 2(s+2) = 0$$
$$(s+2)(s^2+1) = 0$$

∴ The roots of this equation are s = -2,

Therefore, the equation has two roots on the $j-axis$ at $s = j$ and $-j$. The third root is in the left half plane at $s = -2$.

Common Data for Question 68, 69 and 70

A cascade control system for pressure control is shown in the figure given below. The Pressure transmitter has range of 0 to 6 bar (g) and the flow transmitter range is 0 to 81 nm^3/h. The normal flow rate through the valve is 32.4 nm^3/h. Corresponding to the value of set point for pressure = 1 bar (g) and to give the flow, the valve must be 40% opened. The control valve has linear characteristics and is full-open (air to close). Error, set point and control variable are expressed in percentage transmitter output (% TO). Proportional gain is expressed in the unit of per cent controller output (CO% TO).

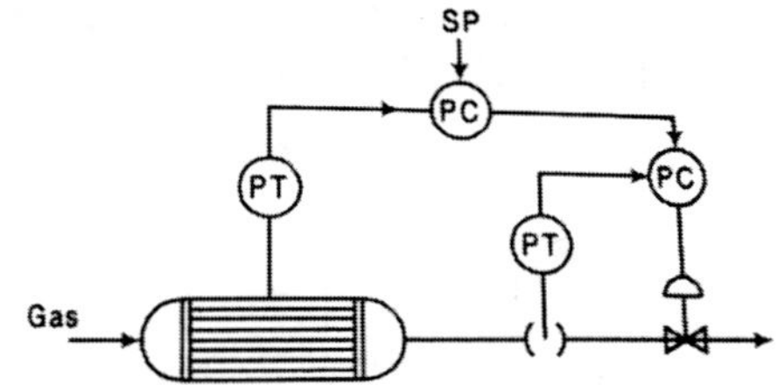

68. The types of action for the two controllers are

A. direct acting for the pressure control and direct acting for the flow control

B. indirect acting for the pressure control and indirect acting for the flow control

C. direct acting for the pressure control and indirect acting for the flow control

D. indirect acting for the pressure control and direct acting for the flow control

***Sol.*D.** consider the diagram given in the statement, the type of control action for:

Pressure control is indicating acting, and flow control is direct acting.

69. The bias values for the two controllers, so that no offset occurs in either controller are

A. pressure controller : 40%, flow controller 60%

B. pressure controller : 33%, flow controller 67%

C. pressure controller : 67%, flow controller 33%

D. pressure controller : 60%, flow controller 40%

***Sol.*D.** for zero-offset value,

Flow controller - 40%

Pressure controller - 60%

70. Given that the actual tank pressure is 4 bar (g) and a proportional controller is employed for pressure control, the proportional band setting of the pressure controller required to obtain a set point to the flow controller equal to 54 nm^3/h is

A. 50%

B. 100%

C. 150%

D. 187%

***Sol.*B.** Proportional band setting of the pressure controller should be 100%.

GATE 2006

71. The Laplace transform of the input function x (t, given in the figure below, is given by

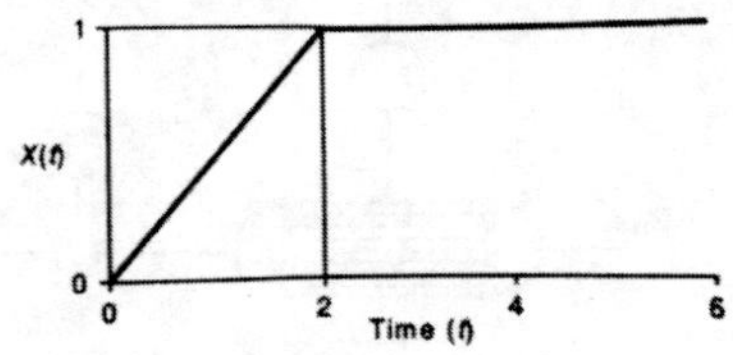

A. $\frac{1}{2s^2}(1 - e^{-2s}$

B. $\frac{1}{2s^2}(1 + e^{-2s})$

C. $\frac{1}{s^2}(1 + e^{+2s})$

D. $\frac{1}{s^2}(1 - e^{-2s})$

***Sol.*A.** From the figure, the function X(t) can be defined as

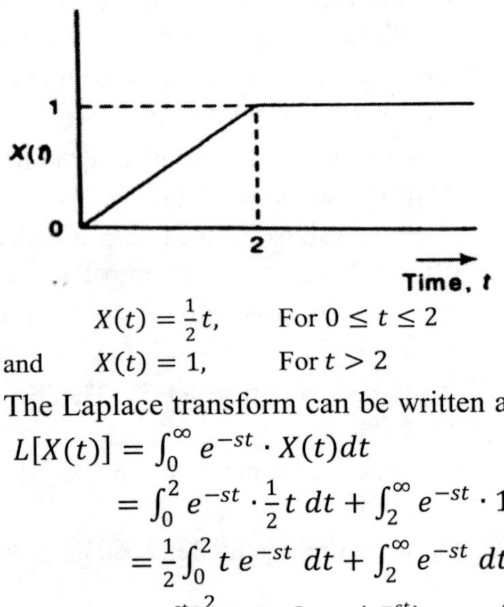

$X(t) = \frac{1}{2}t,$ For $0 \le t \le 2$

and $X(t) = 1,$ For $t > 2$

The Laplace transform can be written as

$$L[X(t)] = \int_0^{\infty} e^{-st} \cdot X(t)dt$$

$$= \int_0^2 e^{-st} \cdot \frac{1}{2}t\, dt + \int_2^{\infty} e^{-st} \cdot 1\, dt$$

$$= \frac{1}{2}\int_0^2 t\, e^{-st}\, dt + \int_2^{\infty} e^{-st}\, dt$$

$$= \frac{1}{2}\left[-t\frac{e^{-st}}{s}\right]_0^2 - \frac{1}{2}\int_0^2 1 \cdot \left(\frac{e^{-st}}{-s}\right) dt + \left[\frac{e^{-st}}{-s}\right]_2^{\infty}$$

$$= \frac{1}{2}\left(\frac{-2e^{-2s}}{s}\right) + \frac{1}{2s}\left[\frac{e^{-st}}{-s}\right]_0^2 + \left(-\frac{1}{s}\right)(0 - e^{-2s})$$

$$= -\frac{1}{s} \cdot e^{-2s} - \frac{1}{2s^2}(e^{-2s} - 1) + \frac{1}{s} \cdot e^{-2s}$$

$$= \frac{1}{2s^2}(1 - e^{-2s})$$

72. A liquid level control system is configured as shown in the figure below. If the Level Transmitter (LT) is direct acting and the pneumatic control valve is air to open, what kind of control action should the controller (LC) have and why?

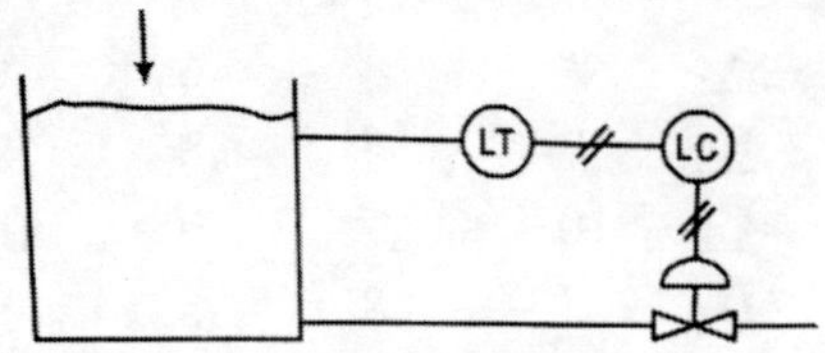

A. direct acting since the control valve is direct acting

B. reverse acting since the control valve is reverse acting

C. direct acting since the control valve is reverse

D. reverse acting since the control valve is direct Reverse acting

***Sol.*A.** since, the control valve is direct acting, the controller (LC) should have direct acting action.

73. A 2-input, 2-output process can be described in the Laplace transform domain as given below.

$(\tau_1 s + 1)Y_1(s) = K_1U_1(s) + K_2U_2(s)$

$(\tau_2 s + 1)Y_2(s) = K_3U_2(s) + K_4Y_1(s)$

where U_1 and U_2 are the inputs and Y_1 and Y_2 are the output. The gain of the transfer function $Y_1(s)/U_2(s)$ and $Y_2(s)/U_2(s)$, respectively are

A. K_2, and K_3

B. K_1 and $K_3 + K_2K_4$

C. K_2 and K_{3+} K_1K_4

D. K_2, and $K_3 + K_2K_4$

***Sol.*D.** given,

$$(\tau_1 s + 1)Y_1(s) = K_1U_1(s) + K_2U_2(s) \quad \text{....(i)}$$

$$(\tau_2 s + 1)\, Y_1(s) = K_3U_2(s) + K_4Y_1(s) \quad \text{....(ii)}$$

where U_1 and U_2 are inputs and Y_1, Y_2 are outputs.

Now, we need to find out $\frac{Y_1(s)}{U_2(s)}$.

Putting, $U_1(s) = 0$ in Eq. (i),

$$(\tau_2 s + 1)\, Y_1(s) = K_2U_2(s)$$

$$\frac{Y_1(s)}{U_2(s)} = \frac{K_2}{\tau_1 s + 1}$$

The gain of this transfer function is K_2.

Now, putting the value of $Y_1(s)$ from Eq. (iii) into Eq. (i),

$$(\tau_2 s + 1)\, Y_2(s) = K_3U_2(s) + K_4 \cdot \frac{K_2U_2(s)}{(\tau_1 s + 1)}$$

$$\frac{Y_2(s)}{U_2(s)} = \frac{K_3 + \left(\frac{K_2U_2(s)}{\tau_1 s + 1}\right)}{\tau_2 s + 1}$$

The gain of this transfer function is $K_3 + K_2K_4$.

74. A process is perturbed by a sinusoidal input u(t)=A sin ωt. The resulting process output is $Y(s) = \frac{KA\omega}{(\tau s+1)(s^2+\omega^2)}$. If Y(0) = 0, the differential equation representing the process is

A. $\frac{dy(t)}{dt} + \tau y(t) = \text{Ku(t)}$

B. $\tau\frac{dy(t)}{dt} + y(t) = \text{KAu(t)}$

C. $\tau\frac{dy(t)}{dt} + y(t) = \text{Ku(t)}$

D. $\tau\left(\frac{dy(t)}{dt} + y(t)\right) = \text{KAu(t)}$

***Sol.*C.** given, $u(t) = A\sin\omega t$

$\therefore \quad u(t) = \frac{A\omega}{\omega^2+s^2}$

and $\quad Y(s) = \frac{K\,A\omega}{(\tau s+1)(\omega^2+s^2)}$

$$\tau s\,Y(s) + Y(s) = Ku(s)$$

$$\tau[sY(s) - Y(0)] + Y(s) = Ku(s),$$

Since, $Y(0) = 0$

$\therefore \quad \tau\frac{dY(t)}{dt} + Y(t) = Ku(t)$

Common Data for Questions 75 and 76

The block diagram of a closed-loop control system is shown in the figure below. Y is the controlled variable, D is disturbance, Y_{SP} is the set point, G_1, G_2, and G_3 are transfer functions, and K_c is the proportional controller.

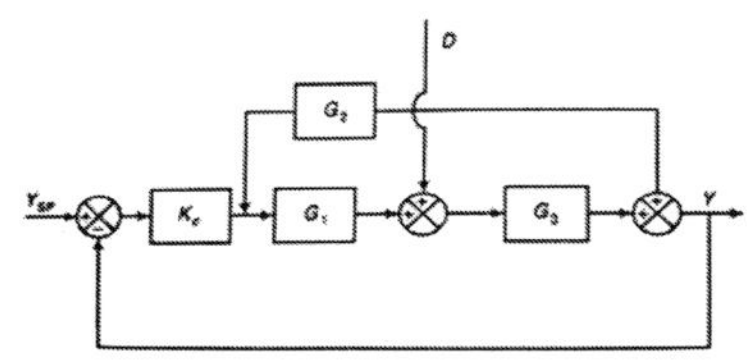

75. The closed-loop transfer function Ys/D(s) is given by

A. $\frac{G_3G_1}{1+(G_1G_3+G_2)K_c}$

B. $\frac{C_3C_1}{1+(G_1G_3+G_2)K_c}$

C. $\frac{G_3}{1+(G_1+G_2)G_3K_c}$

D. $\frac{G_3}{1+(G_1G_3+G_2)K_c}$

***Sol.*D.** $\frac{Y(s)}{D(s)} = \frac{G_3}{1+(G_1G_3+G_2)K_c}$

76. Let G_1 (s) = 1 and G_2 (s) = G_3 (s) 1/(s 1). A step change of magnitude M is made in the set point. The steady state offset of the closed-loop response Y is

A. $\frac{M}{1+2K_c}$

B. $\frac{M}{1+K_c}$

C. $\frac{M(K_c-1)}{1+2K_c}$

D. zero

***Sol.*A.** $\frac{Y(s)}{Y_{SP}(s)} = \frac{G_3}{1+(G_1G_3+G_2)K_c}$

Now given, $\quad G_1(s) = 1$

$$G_2(s) = G_3(s) = \frac{1}{s+1}$$

and $\quad Y_{SP}(s) = \frac{M}{s}$, step change of unit M

$$Y(s) = \frac{M}{s} \times \frac{K_c\left(\frac{1}{s+1}+1\times\frac{1}{s+1}\right)}{1+K_c\left(\frac{1}{s+1}+1\times\frac{1}{s+1}\right)}$$

$$Y(s) = 2MK_c \times \frac{1}{s} \times \frac{1}{s+1+2K_c}$$

$$Y(s) = \frac{2MK_c}{1+2K_c}\left(\frac{1}{s} - \frac{1}{s+1+2K_c}\right)$$

$\therefore \quad Y(t) = \frac{2MK_c}{1+2K_c}\left[1 - e^{-(1+2K_c)t}\right]$

and $\quad Y(\infty) = \frac{2MK_c}{1+2K_c}$

Steady state offset = $Y_{SP}(\infty) - Y(\infty)$

$$= M - \frac{2MK_c}{1+2K_c} = \frac{M}{1+2K_c}$$

Statements for Linked Answer Questions 77 and 78

For the system shown below,

$G_1(s) = \frac{1}{\tau(s+1)}$, $G_2(s) = \frac{1}{\tau_2(s+1)}$ and $\tau_2 = 2\tau_1$

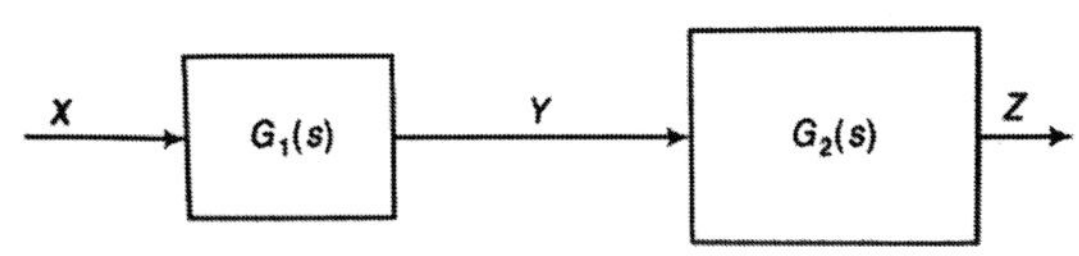

When the system is excited by the sinusoidal input X(t) = sin ωt, the intermediate response Y is given by Y = A sin (ωt + □).

77. If the response of Y lags behind the input x by 45° and τ = 1, then the input frequency ω is

A. 1

B. π / 4

C. zero

D. -1

Sol. **A.** given, $G_1(s) = \frac{1}{\tau_1 s+1}$

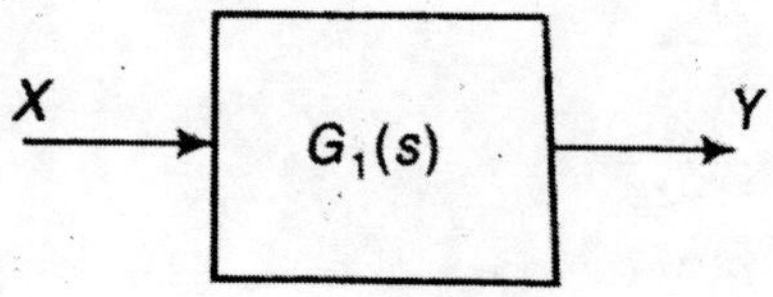

$$X(t) = \sin \omega t$$
$$Y(t) = A\sin(\omega t + \emptyset)$$

Here, $\emptyset = tan^{-1}(-\tau, \omega)$

Now given, $\tau_1 = 1$ and $\emptyset = -45°$

$\therefore \quad -45° = tan^{-1}(-1 \times \omega)$
$$-\omega = \tan(-45°)$$
$$\omega = 1$$

78. For the same input, the amplitude of the output Z will be

A. 1.00
B. 0.62
C. 0.42
D. 0.32

Sol. **D.** Given, $\tau_2 = 2\tau_1 = 2$

Amplitude Ratio for output Z,
$$AR = AR_1 \times AR_2$$
$$= \frac{1}{\sqrt{1+\tau_1^2\omega^2}} \cdot \frac{1}{\sqrt{1+\tau_2^2\omega^2}}$$
$$= \frac{1}{\sqrt{1+1\times1}} \cdot \frac{1}{\sqrt{1+4\times1}}$$
$$= \frac{1}{\sqrt{2}} \cdot \frac{1}{\sqrt{5}} = 0.32$$

GATE 2005

79. Match the process variables in Group I given below with the measuring devices in Group II.

Group I	*Group II*
P. High temperature	1. Orifice meter
Q. Flow	2. Chromatograph
R. Composition	3. Radiation pyrometer
	4. Bi-metallic Thermometer

A. P-1, G-2, R-3
B. P-1, G-3, R-2
C. P-3, Q-1, R-2
D. P-4, Q-2, R-1

Sol. **C.** High temperature - Radiation pyrometer

Flow - Orifice meter

Composition - Chromatograph

80. Given the characteristic equation below, select the number of roots which will be located to the right of the imaginary axis $s^4 + 5s^3 - s^2 - 17s + 12 = 0$

A. one
B. two
C. three
D. zero

Sol. **B.** the characteristic equation
$$s^4 + 5s^3 - s^2 - 17s + 12 = 0$$
$$s^4 - s^3 + 6s^3 - 6s^2 + 5s^2 - 5s - 12s + 12 = 0$$
$$(s-1)(s^3 + 6s^2 + 5s - 12) = 0$$
$$(s-1)(s^3 - s^2 + 7s^2 - 7s + 12s - 12) = 0$$
$$(s-1)(s-1)(s^2 + 7s + 12) = 0$$
$$(s-1)(s-1)(s+3)(s+4) = 0$$

The roots are s = 1, 1, –3, –4

Two roots are located to the right of the imaginary axis.

81. Given the process transfer function $G_p = 4/(\tau s + 1)^2$ and the disturbance transfer function $G_d = 2/(\tau s + 1)$, select the correct transfer function for the Feed Forward Controller for perfect disturbance rejection.

A. $-2(\tau s + 1)$
B. -1
C. $-0.5(\tau s + 1)$
D. $-(\tau s\ 1)^2$

Sol. **C.** given,
$$G_p = \frac{4}{(\tau s+1)^2}$$
$$G_d = \frac{2}{(\tau s+1)}$$

Feed forward controller for perfect disturbance rejection,
$$G_c = \frac{G_d}{G_p} = -0.5(\tau s + 1)$$

82. Given the process transfer function $G_p = 20/(s-2)$ and controller transfer function $G_c = K_c$ and assuming the transfer functions of all other elements in the control loop are unity, select the range of K_c for which the closed-loop response will be stable.

A. $K_c < \frac{1}{10}$
B. $K_c < \frac{1}{100}$
C. $\frac{1}{100} < K_C < \frac{1}{10}$
D. $K_c > \frac{1}{10}$

Sol. **D.** Given, $G_p = \frac{20}{(s-2)}$, $G_c = K_c$

The closed-loop transfer function

$$G(s) = \frac{G_p G_c}{1+G_p G_c} = \frac{K_c \times \frac{20}{(s-2)}}{1+K_c \times \frac{20}{(s-2)}}$$

$$= \frac{20K_c}{s+20K_c-2}$$

For the closed-loop response to be stable,

$$20K_c - 2 > 0$$

$$K_c > \frac{1}{10}$$

83. The value of ultimate period of oscillation P is 3 min and that of the ultimate controller gain K_{cu} is 2. Select the correct set of tuning parameters (controller gain K_c, the derivative time constant τ_D in minute, and the integral time constant τ_I in minute) for a PID controller using Ziegler-Nichols controller settings.

A. $K_c = 1.1; \tau_I = 1.1; \tau_D = 1.31$
B. $K_c = 1.5; \tau_I = 1.8; \tau_D = 0.51$
C. $K_c = 1.5: \tau_I = 1.8; \tau_D = 0.51$
D. $K_c = 1.2; \tau_I = 1.5; \tau_D = 0.38$

Sol. **A.** given, $P_u = 3, K_{cu} = 2$

Using Ziegler-Nichols controller setting for PID controller,

$$K_c = 0.6, \quad K_u = 1.2$$

$$\tau_1 = \frac{P_u}{2} = 1.5$$

$$\tau_D = \frac{P_u}{8} = \frac{3}{8} \approx 0.38$$

GATE 2004

84. Match first order system given in Group I with the System appropriate time constant in GroupII.

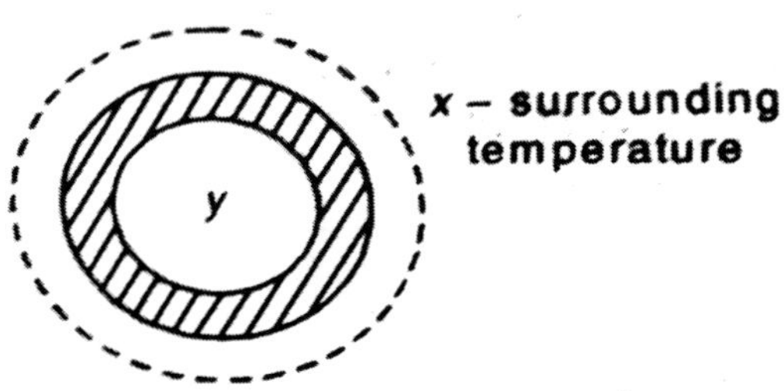

Group I	*Group II*
P. Thermometer	1. $(mC_p)/(hA)$
Q. Mixing	2. q/V
	3. V/q
	4. $(hA)/(mC_p)$

A. P-4, Q-2
B. P-4, Q-3
C. P-1, Q-2
D. P-1, Q-3

Sol. **A.** for mercury thermometer,

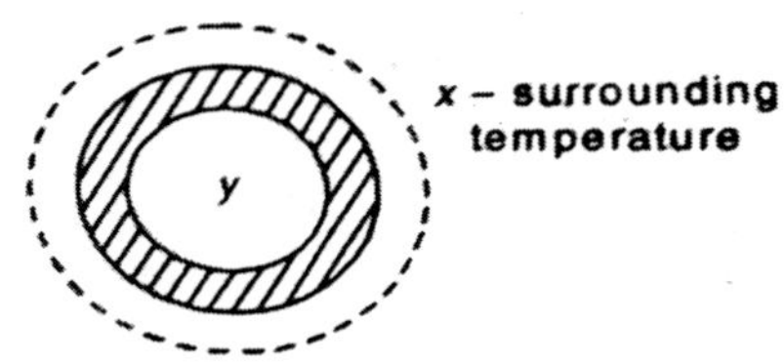

Input rate – Output rate = Rate of accumulation

$$hA(x-y) - 0 = mC_p \frac{dy}{dt}$$

$$mC_p \frac{dy}{dt} = hA(x-y)$$

$$\frac{mC_p}{hA} \cdot sy(s) = x(s) - y(s)$$

$$\frac{y(s)}{x(s)} = \frac{1}{\tau s+1}$$

Here, $\tau = \frac{mC_p}{hA}$

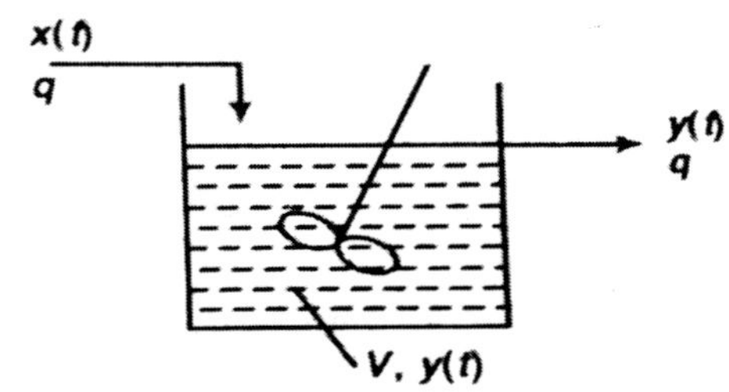

Now, for mixing process,

$$qx - qy = \frac{d(Vy)}{dt}$$

$$qx - qy = V\frac{dy}{dt}$$

$$qx(s) - qy(s) = V.s\,y(s)$$

$$\frac{y(s)}{x(s)} = \frac{1}{\tau s+1}$$

Here, $\tau = \frac{V}{q}$

85. The experimental response of the controlled variable y(t) for a step change of magnitude P, in the manipulated variable x(t) is shown below.

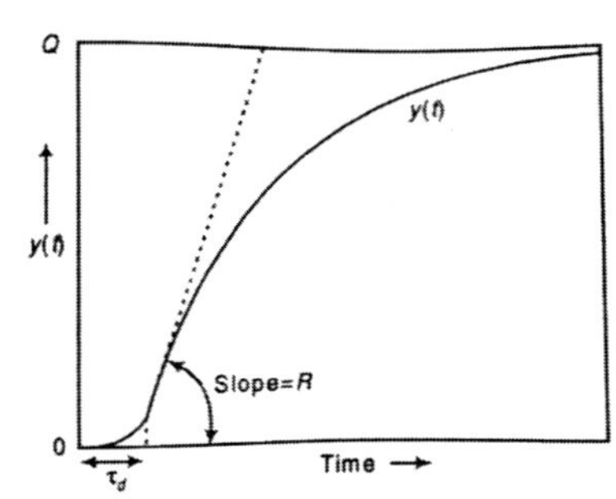

The appropriate transfer function of the process is

A. $\frac{\left(\frac{Q}{P}\right)e^{-\left(\frac{Q}{R}\right)s}}{\tau_d s+1}$

B. $\frac{\left(\frac{Q}{R}\right)e^{-\tau_d s}}{\left(\frac{Q}{P}\right)s+1}$

C. $\frac{\left(\frac{Q}{P}\right)e^{-\tau_d s}}{\left(\frac{Q}{R}\right)s+1}$

D. $\frac{(Q/R)e^{-(P/Q)s}}{\tau_d s+1}$

Sol. **C.** the transfer function of the process

$$G_p(s) = \frac{K_p e^{-\tau_d s}}{\tau s+1}$$

Here, $K_p = \frac{Q}{P}$

Slope, $R = \frac{Q}{\tau} \Rightarrow \tau = \frac{Q}{R}$

$$\therefore \quad G_p(s) = \frac{(Q/P)e^{-\tau_d s}}{[(Q/P)\,s+1]}$$

86. Consider a system with open loop transfer function with

$$G(s) = \frac{1}{(s+1)\,(2s+1)\,(5s+1)}$$

Match the range of ù (frequency) in Group 1 with the slope of the asymptote of the log AR (amplitude ratio) versus log ω plot in Group II.

Group I	*Group II*
P. $0 < \omega < 0.2$	1. -5
Q. $\omega > 1$	2. -3
	3. -2
	4. -1
	5. zero

A. P-5, Q-2

B. P-4, Q-2

C. P-5, Q-3

D. P-4, Q-1

Sol. **C.** given, the open-loop transfer function is

$$G(s) = \frac{1}{(s+1)(2s+1)(5s+1)}$$

Corner frequency,

$$\omega = \frac{1}{\tau}$$

$$\omega = 1, \frac{1}{2}, \frac{1}{5} \text{ i.e., } 1, 0.5, 0.2$$

Slope of asymptote for $0 < \omega < 0.2$ is 0.

Slope of asymptote for $\omega > 1$ is -2.

87. The process and disturbance transfer functions for a system are given by

$$G_p(s) = \frac{y(s)}{m(s)} = \frac{2}{(2s+1)(5s+1)} \text{ and}$$

$$G_d(s) = \frac{y(s)}{d(s)} = \frac{1}{(2s+1)(5s+1)}$$

The feed forward controller transfer function that will keep the process output constant for changes in disturbance is

A. $\frac{2}{(2s+1)^2(5s+1)^2}$

B. $\frac{(2s+1)^2(5s+1)^2}{2}$

C. $\frac{1}{(2s+1)(5s+1)}$

D. $(2s+1)\,(5s+1)$

Sol. **C.** given,

$$G_p(s) = \frac{\bar{y}(s)}{\bar{m}(s)} = \frac{2}{(2s+1)(5s+1)}$$

$$G_d(s) = \frac{\bar{y}(s)}{\bar{d}(s)} = \frac{1}{(2s+1)(5s+1)}$$

$$\bar{y}(s) = \bar{m}(s)G_p(s) = \bar{d}(s)G_d(s)$$

88. For the block diagram shown below,

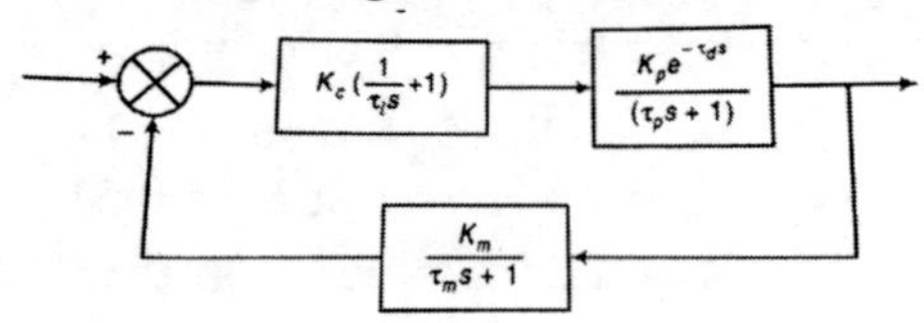

The characteristic equation is

A. $\tau_I s\,(\tau_p+1)+K_cK_p(\tau_I+1)e^{-\tau_d s} = 0$

B. $(\tau_m s+1)(\tau_p s+1)+K_mK_p e^{-\tau_d s} = 0$

C. $\tau_I s(\tau_m s+1)(\tau_p s+1)+K_cK_mK_p(\tau_I s+1)e^{-\tau_d s} = 0$

D. $(\tau_m s+1)(\tau_p s+1)+K_cK_mK_p e^{-\tau_d s} = 0$

Sol. **C.** the characteristic equation for the diagram given in problem is

$$1 + G = 0$$

$$1 + K_c\left(\frac{1}{\tau_I s}+1\right)K_p\frac{e^{-\tau_d s}}{(\tau_p s+1)}\cdot\frac{K_m}{(\tau_m s+1)} = 0$$

$$\tau_I s\,(\tau_m s+1)(\tau_p s+1) + K_cK_pK_m(\tau_I s+1)e^{-\tau_d s} = 0$$

GATE 2003

89. Water is entering a storage tank at a temperature T_0 and flow rate Q_0 and leaving at a flow rate Q and temperature T. There are negligible heat losses in the tank. The area of cross-section of the tank is A_c. The model that describes the dynamic variation of water temperature in the tank with time is given as

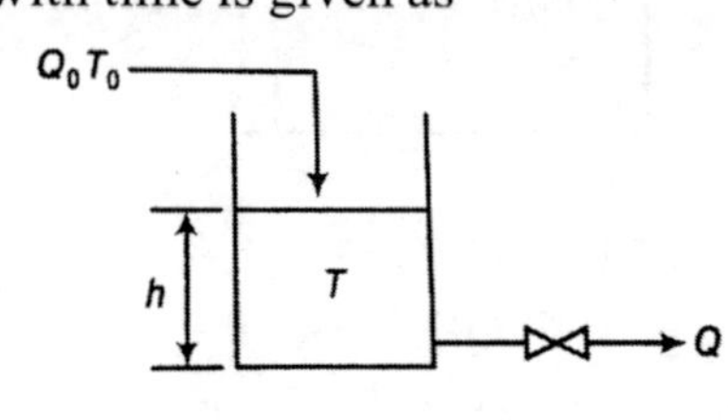

A. $Q_0 (T_0 - T) = A_C h \frac{dT}{dt}$

B. $Q_0 T_0 - QT = A_C h \frac{dT}{dt}$

C. $Q_0 (T_0 - T) = A_C h \frac{dT}{dt}$

D. $Q_0 (T_0 - T) = A_C h \frac{d(Th)}{dt}$

***Sol.*B.** applying energy balance

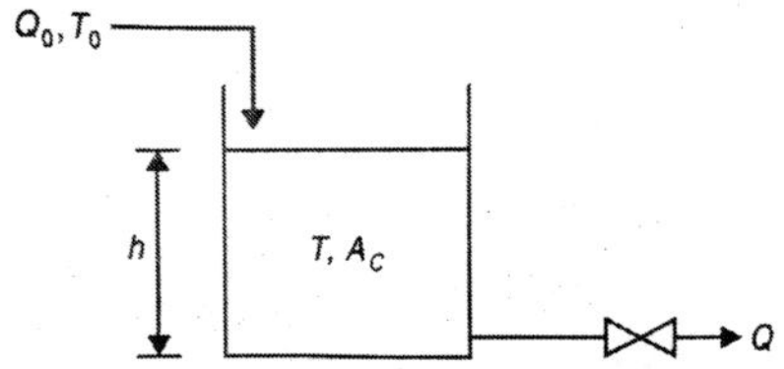

Input – Output = Accumulation

$$Q_o T_o - QT = V\frac{dT}{dt}$$

$$Q_o T_o - QT = A_c h\frac{dT}{dt}$$

90. Find the ultimate gain and frequency for a proportional controller in the case of a process having the following transfer function.

$$G_p(s) = \frac{1}{(4s+1)(2s+1)(s+1)}$$

A. $\omega = \frac{1}{\sqrt{14}}$; $K_c = \frac{45}{7\sqrt{14}}$

B. $\omega = \sqrt{\frac{7}{6}}$; $K_c = \frac{46}{3}$

C. $\omega = 1$; $K_c = 13$

D. $\omega = \sqrt{\frac{7}{8}}$; $K_c = \frac{45}{4}$

***Sol.*D.** given, $G_p(s) = \frac{1}{(4s+1)(2s+1)(s+1)}$

We need to find out the cross-over frequency first.

$$\phi = \phi_1 + \phi_2 + \phi_3$$

$$-180^0 = \tan^{-1}(-4\omega) + \tan^{-1}(-2\omega) + \tan^{-}(-\omega)$$

$$-180^0 = \tan^{-1}(-4\omega) + \tan^{-1}\left(\frac{3\omega}{1-2\omega^2}\right)$$

$$-180^0 = \tan^{-1}\left[\frac{4\omega + \frac{3\omega}{1-2\omega^2}}{1 - 4\omega\left(\frac{3\omega}{1-2\omega^2}\right)}\right]$$

$$\therefore \quad 4\omega + \frac{3\omega}{1-2\omega^2} = 0$$

$$4 - 8\omega^2 + 3 = 0$$

$$\omega = \sqrt{\frac{7}{8}}$$

Now, Amplitude Ratio,

$$AR = \frac{1}{\sqrt{16\omega^2+1}}\cdot\frac{1}{\sqrt{4\omega^2+1}}\cdot\frac{1}{\sqrt{\omega^2+1}}$$

$$= \frac{1}{\sqrt{15}}\cdot\frac{\sqrt{2}}{3}\cdot\frac{\sqrt{8}}{\sqrt{15}} = \frac{4}{45}$$

The ultimate gain,

$$K_u = \frac{1}{AR} = \frac{45}{4}$$

91. Match the type of controller given in Group II that is most suitable for each application given in

Group I	*Group II*
P. Distillation column bottoms Q. Distillation column pressure R. Flow control of a liquid S. Control of temperature of a CSTR with coolant flow in the jacket	1. P control 2. Pl control 3. PID control

A. P-1, Q-1, R-2, S-3

B. P-2, Q-2, R-3, S-3

C. P-2, Q-2, R-1, S-1

D. P-2, Q-3, R-2, S-3

***Sol.*C.** flow control is normally associated with *P*-control

92. In the case of a feed forward control scheme, which of the following is not true?

1. It is insensitive to modeling errors.
2. It cannot cope with unmeasured disturbances.
3. It waits until the effect of the disturbance has been felt by the system before control action is taken.
4. It requires good knowledge of the process model
5. It requires identification of all possible disturbances and their direct measurement

A. 1 and 3

B. 1 and 4

C. 2 and 5

D. 3 and 4

***Sol.*A.** In feed –forward control scheme,

– It is not insensitive to modelling errors.

It does not wait for the disturbance effect.

93. Temperature control of an exothermic chemical reaction taking place in a CSTR is done with the help cooling water flowing in a jacket around the reactor. The types of valve and controller action to be recommended are

A. air to open valve with the controller direct acting

B. air to close valve with the controller indirect acting

C. air to open valve with the controller indirect acting

D. air to close valve with the controller direct acting

***Sol.*C.** the controller action is indirect acting.

The type of valve is air to open.

GATE 2002

94. The frequency response of a first order system has a phase shift with lower and upper bounds given by

A. $\left[-\infty, \frac{\pi}{2}\right]$

B. $\left[-\frac{\pi}{2}, \frac{\pi}{2}\right]$

C. $\left[-\frac{\pi}{2}, 0\right]$

D. $\left[0, \frac{\pi}{2}\right]$

***Sol.*C.** $G(s) = \frac{K_c}{\tau s+1}$

$\left[-\frac{\pi}{2}, 0\right]$ is the correct answer.

GATE 2001

95. An ideal PID controller has the transfer function $\left[1 + \frac{1}{0.5s} + 0.2s\right]$. The frequency at which the Magnitude Ratio of the controller _________, is

A. 0.5/0.2

B. 0.2/0.5

C. 0.2 × 0.5

D. $\frac{1}{\sqrt{0.2 \times 0.5}}$

***Sol.*D.** given, the transfer function for PID controller.

$$G_C(s) = 1 + \frac{1}{0.5\ s} + 0.2\ s$$

Put $s = j\omega$,

$$G_C(j\omega) = 1 + \frac{1}{0.5\ j\omega} + 0.2\ j\omega$$

$$= 1 + \left(0.2\omega - \frac{1}{0.5\omega}\right)j$$

Given, AR = 1

$$\sqrt{1 + \left(0.2\omega - \frac{1}{0.5\omega}\right)^2} = 1$$

or $$0.2\omega - \frac{1}{0.5\omega} = 0$$

$$\omega = \frac{1}{\sqrt{0.2 \times 0.5}}$$

96. The block diagram of an integrating level process is given below. For unit step change in the set point $\Delta h_{set} = 1$ with $\Delta d = 0$, the offset exhibited by the system is

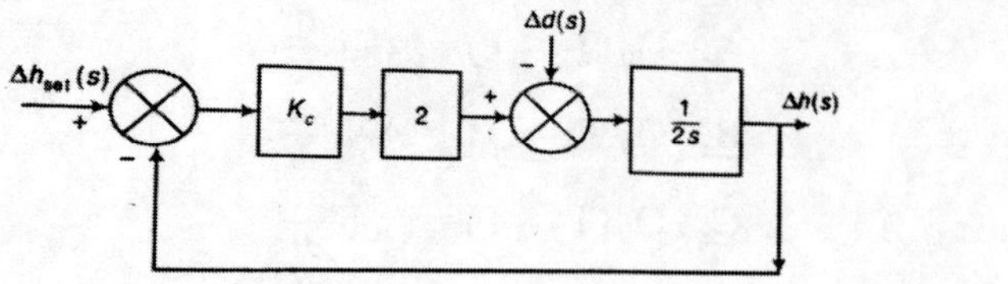

A. $\frac{K_C}{1+K_C}$

B. $\frac{1}{1+K_C}$

C. 0

D. $\frac{2K_C}{1+K_C}$

***Sol.*C.** the transfer function can be written as

$$\frac{\Delta h(s)}{\Delta h_{set}(s)} = \frac{K_C.2.\frac{1}{2s}}{1+K_C.2.\frac{1}{2s}} = \frac{K_C}{s+K_C}$$

For unit step change,

$$\Delta h_{set}(s) = \frac{1}{s}$$

$$\therefore \quad \Delta h(s) = \frac{1}{s} \cdot \frac{K_c}{s+K_c} = \frac{1}{s} - \frac{K_c}{s+K_c}$$

$$\therefore \quad \Delta h(t) = 1 - e^{-k_c t}$$

$$\Delta h(\infty) = 1$$

Offset = $\Delta h_{set}(\infty) - \Delta h(\infty) = 1 - 1 = 0$

97. Second order system can be obtained by connecting two first order systems $\frac{1}{(\tau_1 s+1)}$ and $\frac{1}{(\tau_2 s+1)}$ in series. The damping ratio of the resultant second order system for the case $\tau_1 \neq \tau_2$ will be

A. > 1

B. = 1

C. < 1

D. $= \tau_2/\tau_1$

***Sol.*A.** given, $G_C(s) = \frac{1}{(\tau_1 s+1)} \cdot \frac{1}{(\tau_2 s+1)}$

Damping ratio > 1 if $\tau_1 \neq \tau_2$

(for overdamped system)

GATE 2000

98. The initial value ($t = 0^+$) of the unit step response of transfer function $\frac{(s+1)}{(2s+1)}$ is

A. Zero

B. 1/2

C. 1

D. 2

***Sol.*B.** given, $\frac{Y(s)}{X(s)} = \frac{s+1}{2s+1}$

For unit step input,

$$X(s) = \frac{1}{s} \therefore \quad Y(s) = \frac{1}{s}\cdot\left(\frac{s+1}{2s+1}\right)$$

Now, using initial value theorem,

$$\lim_{t\to 0} Y(t) = \lim_{s\to\infty} s\,Y(t)$$

$$= \lim_{s\to\infty}\left(\frac{s+1}{2s+1}\right)$$

$$= \lim_{s\to\infty}\left(\frac{1+\frac{1}{s}}{2+\frac{1}{s}}\right)$$

$$\lim_{t\to 0} Y(t) = \frac{1}{2}$$

99. The time constant of a unity gain, first order plus time delay process is 5 min. If the phase lag at a frequency of 0.2 rad/min is 60°, then the dead time (in min) is

A. $\frac{5\pi}{12}$

B. $\frac{\pi}{6}$

C. $\frac{\pi}{12}$

D. $\frac{\pi}{3}$

Sol.**A.** given, $G(s) = \frac{e^{-\tau_d s}}{(5s+1)}$

$$-\tau_d + tan^{-1}(-5\omega) = 60^0$$

Here, τ_d is dead time.

$$-\tau_d + tan^{-1}(-5\times 0.2) = 60^0$$

$$-\tau_d + tan^{-1}(-1) = 60^0$$

$$-\tau_d + 135^0 = 60^0$$

$$\tau_d = 75^0 \text{ or } \tau_d = \frac{5\pi}{12}$$

10. Thermodynamics

Multiple Choice Questions

1. The polytropic process (PV^n = constant) is used of steady flow compression, The work done on air is the lowest when

   A. $n = y = 1.4$

   B. $n = 0$

   C. $n = 1$

   D. $n = 1.66$

2. For the definite temperature if the liquid under pressure is greater than its vapour pressure the liquid is __________.

   A. subcooled

   B. saturated

   C. non-solidifiable

   D. none of these

3. All substances will have the same at absolute zero temperature.

   A. heat capacity

   B. entropy

   C. crystalline structure

   D. none of these

4. Liquefaction of 448 liters of butane (C_4H_{10}) measured at STP will yield how many grams of butane?

   A. 580

   B. 640

   C. 1160

   D. data insufficient; can't be computed

5. If the temperature in the evaporator is kept constant then in case of vapour compression refrigeration system lowering of condenser temperature will result in

   A. increased COP.

   B. decreased COP.

   C. same COP.

   D. increased or decreased COP; depending upon the type of refrigerant.

6. If refrigeration is provided at temperature T_R (where T_1 and T_2 are source & surrounding temperatures respectively) what is the maximum value of COP?

   A. $\frac{T_R}{T_2 - T_R} \text{x} \frac{T_1 - T_2}{T_1}$

   B. $\frac{T_R}{T_2 - T_R} \text{x} \frac{T_1}{T_1 - T_2}$

   C. $\frac{T_R}{T_1 - T_R} \text{x} \frac{T_1 - T_2}{T_1}$

   D. none of these

7. An isopiestic systemis the one in which there is no __________ change.

   A. temperature

   B. volume

   C. pressure

   D. none of these

8. For a pure substance the specific molar Gibbs free energy does not change during .

   A. melting

   B. sublimation

   C. vaporisation

   D. either (a), (b) or (c)

9. A process involves the expression of a gas in vessel thus decreasing its internal energy. The above process is an example of a reversible.

   A. reversible

   B. irreversible

   C. isothermal

   D. adiabatic

10. The mathematical expression for $\left(\frac{\partial H}{\partial T}\right)_P$ is for.

    A. C_V

    B. entropy change

    C. Gibbs free energy

    D. none of these

11. Which of the following plots is represented by Mollier diagram?

A. temperature *vs*. internal energy
B. temperature *vs*. enthalpy
C. entropy *vs*. enthalpy
D. temperature *vs*. enthalpy

12. For an ideal gas the activity is numerically _________ to its pressure.
A. more than
B. less than
C. equal to
D. data insufficient, can't be predicted

13. Which of the following parameter is relevant to a chemical potential.
A. extensive property.
B. force which drives the chemical system to equilibrium.
C. intensive property.
D. both (b) and (c).

14. In the Planck's constant equation $E = hv$ the unit of Plank's constant 'h' is.
A. J/s
B. J.S
C. J/kmol
D. kmol/J

15. An Otto cycle is not represented by the diagram __________ of the following set of diagrams.

A.
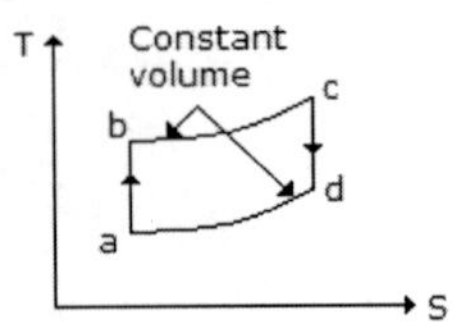

B.
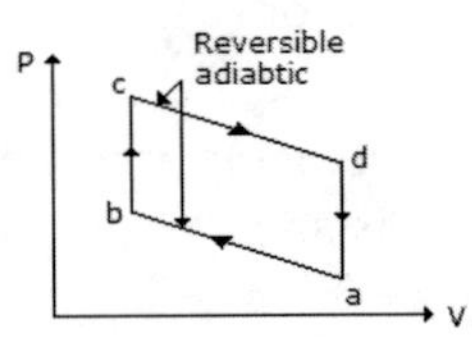

C.
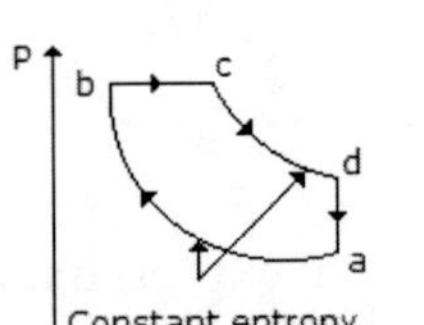

D. none of these

16. A household refrigerator the refrigeration capacity is approximately ___________ tons.
A. 0.15
B. 1.5
C. 4.5
D. 6.5

17. Which of the following is valid as the ratio of the __________ for coefficient performance (COP) of a refrigerator.
A. work required to refrigeration obtained.
B. refrigeration obtained to the work required.
C. higher to lower temperature.
D. lower to higher temperature.

18. The expression of a fluid in a turbine takes place almost .
A. isobarically
B. isothermally
C. adiabatically
D. none of these

19. Increasing the pressure on melting point of paraffin wax (which contracts on solidification) __________ the melting point.
A. increases
B. decreases
C. decreases linearly
D. remains unchanged

20. As indicated in the P-V diagram an ideal gas is tanken around the cycle ABCA. During the cycle the work done by the gas is

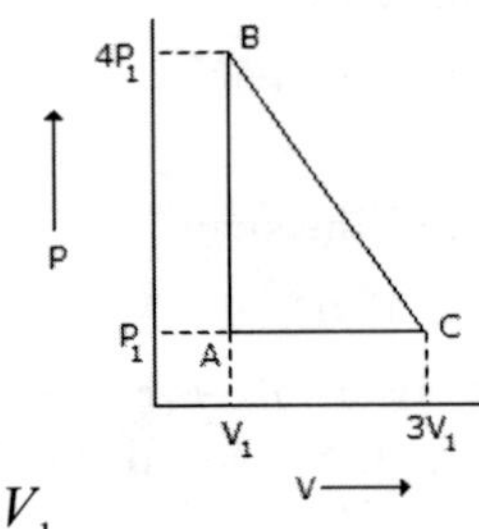

A. $12\ P_1V_1$
B. $6\ P_1\ V_1$
C. $3\ P_1V_1$
D. $P_1\ V_1$

21. For PV^n = constantif $n = 1$ in polytropic process it signifies that it is a/an __________ process.
A. reversible
B. adiabatic
C. isothermal
D. none of these

22. Which of the following will happen when on the system ice → water pressure is applied?
A. evaporation of water will take place.
B. equilibrium cannot be established.

C. more water will be formed.

D. more ice will be formed.

23. For an element A in solution the partial molar free energy is same as .

A. chemical potential

B. activity co-efficient

C. fugacity

D. activity

24. 40 kg of water initially at 25°C is place in a perfectly insulated tank in which a metallic block weighting 5 Kg at a temperature of 500°C is placed and the system is allowed to come to equilibrium. Specific heat of block material is 0.4 kJ/kg.K. The total entropy change in kJ/kg.K is

A. –1.87

B. 0

C. 1.26

D. 3.91

25. When equilibrium is attained the change in free energy is .

A. zero

B. positive

C. negative

D. indeterminate

26. A point occurs when the temperature at which both liquid and gas phases are identical.

A. critical

B. triple

C. freezing

D. boiling

27. Which of the following represents the Maxwell relation derived from the deferential expression for the Helmholtz free energy (dA)?

A. $\left(\frac{\partial V}{\partial S}\right)_P = \left(\frac{\partial T}{\partial P}\right)_S$

B. $\left(\frac{\partial T}{\partial V}\right)_S = -\left(\frac{\partial P}{\partial S}\right)_V$

C. $\left(\frac{\partial S}{\partial P}\right)_T = -\left(\frac{\partial V}{\partial T}\right)_P$

D. $\left(\frac{\partial S}{\partial V}\right)_T = \left(\frac{\partial P}{\partial T}\right)_V$

28. The function of heat pump is that it

A. accomplishes only space cooling in summer.

B. accomplishes only space heating in winter.

C. accomplishes both (a) and (b).

D. works on Carnot cycle.

29. The reason why dW and d_q are not the exact different is that because q and W are.

A. extensive properties

B. path functions

C. state functions

D. intensive properties

30. If y is the activity coefficient for ideal gas what is the value of In (y) ?

A. Zero

B. Unity

C. Infinity

D. Negative

31. When mixing of two liquid substances takes place the entropy change depend on the

A. molar concentration

B. quantity (i.e. number of moles)

C. both (a) and (b)

D. neither (a) nor (b)

32. Which of the following parameters is represented by $\left(\frac{\partial E}{\partial T}\right)_V$?

A. C_V

B. free energy change

C. enthalpy change

D. none of these

33. For a system involving constituents. The relation among the four properties P, V, T and S are known as ________equation.

A. Gibbs-Duhem

B. Gibbs-Helmholtz

C. Maxwell's

D. none of these

34. For liquefaction of gases the principle applied is known as .

A. Joule-Thomson effect.

B. adiabatic expansion.

C. both (a) and (b).

D. neither (a) nor (b).

35. For a reversible adiabatic process the specific heat of a gas is .

A. infinity

B. zero

C. negative

D. none of these

36. For each case constituent present in an alloy the partial pressure is the total vapor pressure exerted by the alloy.

A. less than
B. more than
C. equal to
D. either (b) or (c); depends on the type of alloy

37. Cooling reversibly from 90°C to 40°C a closed system without performing any work on the system will happen when
A. its internal energy (*U*) decreases and its entropy (*S*) increases.
B. *U* and S both decreases
C. *U* decreases but *S* is constant.
D. *U* is constant but *S* decreases.

38. At a temperature above the critical temperature a substance will exist as .
A. saturated vapour
B. solid
C. gas
D. liquid

39. Which of the following statement is wrong?
A. there is an increase in entropy, when a spontaneous change occurs in an isolated system.
B. the heat capacity of solid inorganic substance is exactly equal to the heat capacity of the substance in the molten state.
C. at absolute zero temperature, the heat capacity for many pure crystalline substances is zero.
D. Cp of monoatomic gases such as metallic vapor is about 5 kcal/kg.atom.

40. Which of the following cannot be used as refrigerator?
A. CCl_2F_2
B. SO_2
C. NH_3
D. $C_2H_4Cl_2$

41. For a reversed Carnot cycle working between temperature T_1 and T_2 ($T_1>T_2$), the coefficient of performance is .
A. $\frac{T_2}{T_1-T_2}$
B. $\frac{T_1-T_2}{T_2}$
C. $\frac{T_1}{T_1-T_2}$
D. $\frac{T_1-T_2}{T_1}$

42. In an ideal __________ cycle heat is added at constant pressure.
A. Stirling
B. Rankine
C. Brayton
D. both (b) and (c)

43. A constant process is known as isentropic process.
A. enthalpy
B. pressure
C. entropy
D. none of these

44. Reducing the pressure by half on 100 cc of air keeping the temperature constant will result in the volume being __________ cc.
A. 100
B. 50
C. 205
D. 200

45. Which of the following is represented by the mathematical expression $\frac{1}{V}\left(\frac{\partial V}{\partial T}\right)_P$?
A. specific heat at constant pressure (C_p).
B. specific heat at constant volume (C_V).
C. co-efficient of thermal expansion.
D. Joule-Thomson co-efficient.

46. If the pressure is increased which of the following decrease?
A. Melting point of ice
B. Melting point of wax
C. Boiling point of liquids
D. none of these

47. For an azeotropic mixture in a two component vapour-liquid equilibria the number of degrees of freedom is/are.
A. zero
B. one
C. two
D. three

48. Which of the following statement is correct?
A. The mechanical work done by a system is always equal to $\int P.dV$.
B. A process is irreversible as long as ΔS for the system is greater than zero.
C. If an insoluble gas is passed through a volatile liquid placed in a perfectly insulated container, the temperature of the liquid will increase.

D. The heat of formation of a compound is defined as the heat of reaction leading to the formation of the compound from its reactants.

49. The chemical potential for a real gas is expressed by ___________.

A. $RT\ dlnP$

B. $RT\ dlnf$

C. $R\ dlnf$

D. none of these

50. In which of the following case the thermodynamic law PVr = *constant*, is not applicable?

A. ideal compression of air.

B. free expansion of an ideal gas.

C. adiabatic compression of a perfect gas.

D. adiabatic expansion of steam in a turbine.

51. Ideal refrigeration cycle is

A. the least efficient of all refrigeration processes.

B. same as reverse Carnot cycle.

C. same as Carnot cycle.

D. dependent on the refrigerant's properties.

52. Refrigeration cycle

A. violates second law of thermodynamics.

B. involves transfer of heat from low temperature to high temperature.

C. both (a) and (b).

D. neither (a) nor (b).

53. A reversible obey PV^n = *Constant*. Work obtained for trebling the volume ($V_1 = 1m^3$ and $V_2 = 3m^3$) is maximum when the volume of n is

A. 0

B. 1

C. $y = 1.44$

D. 1.66

54. What is the value of the quantity $\left(\frac{\partial \Delta T}{\partial T}\right)$ as the temperature is lowered towards the absolute zero?

A. zero

B. unity

C. infinity

D. none of these

55. If the evaporator temperature is elevated keeping the condenser temperature constant for vapor compression refrigeration system result is .

A. enhanced COP.

B. decreased COP.

C. no change in the value of COP.

D. increased or decreased COP; depending upon the type of refrigerant.

56. At a specified temperature for a given substance, activity is to fugacity.

A. directly proportional

B. equal

C. inversely proportional

D. none of these

57. For all substance the critical compressibility factor.

A. are more or less constant (vary from 0.2 to 0.3).

B. vary as square of the absolute pressure.

C. vary as square of the absolute temperature.

D. none of these.

58. Experimental data is to be used for predicating the activity coefficient. The equation to be used is called ________Equation.

A. Lewis-Randall

B. Van Laar

C. Margules

D. both(b)&(c)

59. The change in internal energy of the fluid is an ideal refrigeration cycle is .

A. +ve

B. –ve

C. 0

D. either of the above three; depends on the nature of refrigerant.

60. When cooling occurs after the throttling process the value of Joule-Thomson coefficient is .

A. 0

B. ∞

C. +ve

D. –ve

61. When ammonia gas is dissolved in water the enthalpy change is called the heat of .

A. solution

B. dilution

C. formation

D. combustion

62. Which of the following is valid for an irreversible process involving only pressure volume work?

A. $(dF)_{T'}\ p < 0$

B. $(dF)_{T'}\ p = 0$

C. $(dF)_T, p > 0$

D. $(dA)_T, v > 0$

63. The term chemical potential for multicomponent system is equivalent to the .

A. molal concentration difference.

B. molar free energy change.

C. partial molar free energy.

D. molar free energy.

64. When a saturated water vapour is compressed and adiabatically the water vapour becomes .

A. supersaturated

B. superheated

C. both (a) and (b)

D. neither (a) nor (b)

65. Which of the following gives the quantitative effect of temperature on chemical equilibrium?

A. Vant-Hoff equation.

B. Arhenius equation.

C. Le-Chatelier's principle.

D. none of these.

66. Separated water vapour at 100°C has a specific heat

A. ∞

B. –ve

C. 0

D. +ve

67. A system is condensing of water at 300°C having a vapor pressure 8592.7 kPa and fugacity 6738.9 kPa. If one mole of water in liquid phase has a volume of 25.28 cm^3 and that vapour phase is 391.1 cm^3. Fugacity water in kPa at 9000kPa will be .

A. 6738.9

B. 6753.5

C. 7058.3

D. 9000

68. A heat engine operation between 800 K and 300 K produces power 10^5 kW and has a thermal efficiency is equal to 50% of that of the Carnot engine at the same temperature. The rate of which has is absorbed from the hot reservoir is kW.

A. 100, 000 kW

B. 160, 000 kW

C. 200, 000 kW

D. 320, 000 kW

69. For the pure substance the change in Gibb's free energy of vaporisation is .

A. positive

B. negative

C. zero

D. may be positive or negative

70. Which of the following is a measure of fugacity?

A. escaping tendencies of the same substance in different phases of a system.

B. behavior of ideal gases.

C. relative volatility of a mixture of two miscible liquids.

D. none of these.

71. For negative heat of solution of an ideal gas in a liquid its solubility at a given partial pressure varies with the temperature as

A. solubility is independent of temperature.

B. solubility increases as temperature decreases.

C. solubility increases as temperature increases.

D. solubility increases or decreases with temperature depending on the Gibbs free energy change of solution.

72. A domestic refrigerator has a/an __________ cooled condenser.

A. gas

B. air

C. water

D. evaporative

73. Two missile non reacting substances are in vapour liquid equilibrium forming an azeotrope. What is the degree of freedom for this system?

A. 0

B. 1

C. 2

D. 3

74. According to the law if different process are used to bring about the same chemical reaction, the enthalpy change is same for all them.

A. Hess's

B. Lavoisier and Laplace

C. Kirchoff's

D. none of these

75. Which of the following properties an ideal liquid refrigerant should have

A. not have unduly high vapour pressure at the condenser temperature.

B. not have a subatmospheric vapour pressure at the temperature in the refrigerator coils.

C. both (a) and (b).

D. have low specific heat.

76. At any point on the inversion curve the volume of the Joule- Thomson coefficient is .

A. ∞

B. + ve

C. 0

D. -ve

77. How does the free energy changes?

A. remains unchanged in reversible processes carried at constant temperature and pressure.

B. change during a spontaneous process has a negative value.

C. decreases in all spontaneous (or irreversible) processes.

D. all (a), (b) and (c).

78. If V_1 = actual volume of the gas and V_2 = gas volume predicted by ideal gas law, the compressibility factor is given by .

A. V_1/V_2

B. V_2/V_1

C. $V_1.V_2$

D. $V_1\text{-}V_2$

79. If the condition that there is no volume change upon mixing the components is the liquid state and that when diluted do not undergo any heat change (i.e. heat of dilution is zero) is satisfied the such solution are known as .

A. ideal

B. isotonic

C. real

D. none of these

80. What is the entropy change during a reversible isothermal expansion?

A. +ve

B. 0

C. -ve

D. ∞

81. For monoatomic gases the value of C_p & C_v Kcal/kg Mole. °K are respectively .

A. 5 & 3

B. 3.987 & 1.987

C. 1.987 & 0.66

D. 0.66 & 1.987

82. For the ideal gas states the heat capacity are depended on the

A. pressure

B. temperature

C. both (a) & (b)

D. neither (a) nor (b)

83. For an ideal gas if the internal energy decreases by the same amount as the work done by the system then the .

A. process must be isobaric.

B. process must be adiabatic.

C. temperature must decrease.

D. both (b) and (c).

84. The volume of a real gas in to increase by four times. Which of the following methods is recommended?

A. doubling the absolute temperature as well as pressure of the gas.

B. reducing pressure to one fourth at constant temperature.

C. reducing temperature to one fourth at constant pressure.

D. reducing the temperature to half and doubling the pressure.

85. According to Gibbs phase rule number of component (C) phase (P) degree of freedom (F) are related by

A. $P + F - C = 2$

B. $P = F - C - 2$

C. $C = P - F + 2$

D. $F = C - P - 2$

86. Which of the following statements is wrong?

A. the values of $\left(\frac{\partial P}{\partial V}\right)_T$ and $\left(\frac{\partial^2 P}{\partial V^2}\right)_T$ are zero for a real gas at its critical point.

B. during a reversible adiabatic process, the entropy of a substance remains constant.

C. thermal efficiency of a Carnot engine depends upon the properties of the working fluid besides the source & sink temperatures.

D. heat transferred is equal to the change in the enthalpy of the system, for a constant pressure, non-flow, mechanically reversible process.

87. __________ takes places across the boundaries in case of a closed thermodynamic system.
A. mass and energy transfer
B. no mass transfer but heat transfer
C. no heat and mass transfer
D. none of these

88. The value of $\frac{\delta(\Delta F)}{\delta T}$ approaches as the temperature is lowered towards absolute zero.
A. unity
B. zero
C. that of the heat of reaction
D. infinity

89. Which of the following represents the mathematical expression for $\left(\frac{\partial T}{\partial P}\right)_H$?
A. specific heat at constant volume (C_v).
B. specific heat at constant pressure (C_p).
C. Joule-Thompson co-efficient.
D. none of these.

90. Which of the following properties fall under extensive properties?
A. free energy, entropy and enthalpy.
B. volume, mass and number of moles.
C. both (a) and (b).
D. none of these.

91. The least thermal efficiency is that of .
A. steam engine
B. diesel engine
C. Carnot engine
D. Otto engine

92. A substance which is at its triple point is subjected to increase in the pressure. Then the triple point __________.
A. increases
B. decreases
C. remains unchanged
D. may increase or decrease; depends on the substance

93. __________ is the liquid metal which has highest thermal conductivity?
A. molten sodium
B. molten lead
C. mercury
D. molten potassium

94. The requirement that the equation, $PV = nRT$, is best obeyed by gases at.
A. low pressure & high temperature.
B. low pressure & low temperature.
C. high pressure & low temperature.
D. none of these.

95. Which of the following represents the ratio of equilibrium constants (Kp_2/Kp_1) at two different temperatures?
A. $\frac{\Delta H}{R}\left[\frac{1}{T_2} - \frac{1}{T_1}\right]$
B. $\frac{\Delta H}{R}\left[\frac{1}{T_1} - \frac{1}{T_2}\right]$
C. $\frac{R}{\Delta H}\left[\frac{1}{T_1} - \frac{1}{T_2}\right]$
D. $\frac{1}{R}\left[\frac{1}{T_1} - \frac{1}{T_2}\right]$

96. Out of the following which is not affected by the temperature change.
A. Activity co-efficient
B. Free energy
C. Fugacity
D. None of these

97. What is the internal energy of an element at 25° Cand 1 atm (in kcal/Kg.mole)
A. 0
B. 273
C. 25
D. none of these

98. The activity co-efficient for an ideal gas is
A. inversely proportional to pressure.
B. directly proportional to pressure.
C. unity at all pressures.
D. none of these.

99. Generation of heat by friction is an example of a/ an __________ change.
A. reversible
B. irreversible
C. isothermal
D. adiabatic

100. Which of the following statement is wrong?
A. a refrigeration cycle violates the second law of thermodynamics.
B. refrigeration cycle is normally represented by a temperature vs. entropy plot.
C. one ton of refrigeration is equivalent to the rate of heat absorption equal to 3.53 kW.

D. in a refrigerator, work required decreases as the temperature of the refrigerator and the temperature at which heat is rejected increases.

101. Which of the following is valid for a Joule-Thomson porous experiment?

A. entire apparatus is exposed to surroundings.
B. enthalpy does not remain constant.
C. temperature remains constant.
D. none of these.

102. Which of the following applies to an air-refrigeration cycle i.e. it .

A. is the most efficient of all refrigeration cycles.
B. requires relatively large quantities of air to achieve a significant amount of refrigeration.
C. has very low efficiency.
D. both (b) and (c).

103. The process of melting of wax is accompanied with ———— in entropy?

A. increase
B. decrease
C. no change
D. none of these

104. At constant pressure and temperature Gibbs free energy of mixture is always.

A. 0
B. ∞
C. + ve
D. –ve

105. The enthalpy for an ideal gas .

A. is a path function.
B. increases with rise in pressure.
C. is independent of pressure.
D. decreases with rise in pressure.

106. What will be the value for compressibility factor for any real gas if for an ideal gas it is equal to 1.

A. 1
B. > 1
C. < 1
D. either (b) or (c), depends on the nature of the gas

107. Which of the following represents the efficiency of a Carnot-cycle working between temperature T_1 and T_2 ($T_1<T_2$)?

A. $\frac{T_2-T_1}{T_2}$
B. $\frac{T_1-T_2}{T_2}$
C. $\frac{T_1-T_2}{T_1}$
D. $\frac{T_2-T_1}{T_1}$

108. In a reaction equilibrium is allowed to attain for the reaction $A\,(l) \rightarrow R(g)+S(g)$. Two phase *i.e.* a pure liquid phase of A and the other a vapor phase of A, R and S separate out. If initially A along is present, the number of degree of freedom will be equal to .

A. 3
B. 2
C. 1
D. 0

109. Which of the following represent the intensive properties

A. refractive index and surface tension.
B. molar volume, density, viscosity and boiling point.
C. both (a) and (b).
D. none of these.

110. For the same reduced temperature and reduced pressure, if two gases will also be the same.

A. critical temperature
B. volume
C. mass
D. none of these

111. For which type of gas $C_p - C_v = R$ is valid?

A. ideal
B. very low temperature
C. very high pressure
D. all

112. Which of the following statements is correct?

A. Work is a state function.
B. Entropy and enthalpy are path functions.
C. All the natural processes are reversible in nature.
D. In a closed system, the energy can be exchanged with the surrounding, while matter cannot be exchanged.

113. For a system the extensive property is .

A. heat capacity
B. pressure

C. concentration
D. Molal heat capacity

114. Which of the following happens during adiabatic expansion of a gas?
A. temperature remains constant.
B. pressure remains constant.
C. pressure is increased.
D. none of these.

115. With increase in pressure the volume of a gas dissolved in a solvent at a given temperature.
A. decreases
B. increases
C. remains unchanged
D. may increase or decrease; depends on the gas

116. In the *P-V* diagram a thermodynamic system is taken from state A to B along ACB and is brought back to A along BDA. The network done during the cycle is the area enclosed by.

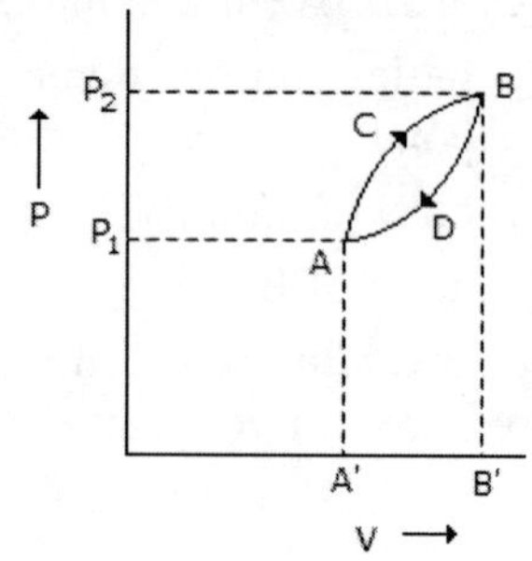

A. $ACBB^1A^1A$
B. $ADBB^1A^1A$
C. $ACBDA$
D. $P_1ACBP_2P_1$

117. If the evaporator temperature and the condenser temperatures are -13°C and 37°C respectively, for a vapour compression refrigeration system the Carnot COP will be.
A. 5.2
B. 6.2
C. 0.168
D. data insufficient, can't be found out

118. A fixed ratio between heat and work for a cycle process.
A. always exists
B. never exists
C. may exist
D. is difficult to predict

119. Which of the following statement is correct?
A. An ideal gas on expansion in vacuum gets cooled.
B. An ideal gas on expansion in vacuum gets heated up.
C. A real gas on expansion in vacuum gets heated up.
D. A real gas on expansion in vacuum cools down whereas ideal gas remains unaffected.

120. Which of the following steps may be attributed to a Carnot cycle?
A. two isothermals and two isentropics.
B. two isothermals and two isochorics.
C. two isobarics and two isothermals.
D. two isochorics and two isobarics.

121. A heterogeneous is working on Carnot cycle between two temperature levels. The efficiency of the engine depends on the .
A. two temperatures only.
B. mass of the working fluid.
C. pressure of working fluid.
D. mass and pressure both of the working fluid.

122. The fugacity coefficient of a gas at constant reduced pressure with increase in reduced temperature.
A. increases
B. decreases
C. remains same
D. decreases linearly

123. Which of the followingis an example of closed system?
A. Boiler
B. Liquid cooling system of an automobile
C. Air compressor
D. None of these

124. Which of the following commonly used refrigeration for domestic refrigeration?
A. ethyl chloride or methyle chloride
B. freon-12
C. propane
D. NH_3 or CO_2.

125. The term "dry ice" is commonly for .
A. solid helium
B. moisture free ice
C. solid carbon dioxide
D. none of these

126. What will be the heat of formation of an element in its standard state?
A. 0
B. < 0
C. > 0
D. a function of pressure

127. The gas phase chemical reaction, $C_2H_4(g) + H_2O(g) \rightleftharpoons C_2H_5OH(g)$, the equilibrium conversion does not depend on the.
A. pressure.
B. steam to ethylene ratio.
C. temperature.
D. none of these.

128. In system having free expansion the work done is
A. negative
B. zero
C. indeterminate
D. none of these

129. The first law of thermodynamics is a statement of conservation of.
A. heat
B. work
C. energy
D. momentum

130. What is the shape of T-S diagram for Carnot Cycle?
A. rectangle
B. circle
C. rhombus
D. trapezoid

131. The equation, $PV^n =$ constant, for the special case of $n = 1$, it represents a reversible __________ process.
A. isothermal
B. polytropic
C. adiabatic
D. isobaric

132. does not fall under the category of a common refrigerator?
A. ammonia
B. ethylene
C. freon-12
D. carbon dioxide

133. Distribution law applies most importantly in .
A. distillation
B. liquid extraction
C. evaporation
D. drying

134. The internal energy of a gas for an isothermal process.
A. increases
B. decreases
C. remains unchanged
D. data insufficient, can't be predicted

135. Which of the following statements is wrong?
A. Henry's law is rigorously correct in the limit of infinite dilution.
B. An ideal liquid or solid solution is defined as one in which each component obeys Raoult's law.
C. If Raoult's law is applied to one component of a binary mixture; Henry's law or Raoult's law is applied to the other component also.
D. none of these.

136. Which of the following equation represents C_v
A. $\left(\frac{\partial E}{\partial T}\right)_V$
B. $\left(\frac{\partial E}{\partial P}\right)_V$
C. $\left(\frac{\partial V}{\partial T}\right)_P$
D. $\left(\frac{\partial E}{\partial V}\right)_T$

137. During a phase change (*e.g.* sublimation, melting, vaporisation etc.) specific __________ does not change.
A. entropy
B. enthalpy
C. internal energy
D. Gibbs free energy

138. In which of the following cases solid and liquid phase of a substances of a substance are in equilibrium?
A. critical temperature
B. freezing point.
C. melting point.
D. both (b) and (c).

139. In which of the following case the fugacity and pressure are numerically equal?
A. at low temperature
B. in standard state
C. at high pressure
D. in ideal state

140. The requirement for maximum work for non-flow reversible compression process is the case of .

A. adiabatic process
B. isothermal process
C. all require same work
D. isobaric process

141. Which equation relating E, P, V and T is true for all substances under all conditions is given by $\left(\frac{\partial E}{\partial V}\right)_T = T \cdot \left(\frac{\partial P}{\partial T}\right)_H - P$

A. Redlich-Kwong equation of state.
B. thermodynamic equation of state.
C. Maxwell's equation.
D. equation of state.

142. What is the name of that principle which state "When a system in equilibrium is subjected to a change in temperature, pressure or concentration, the equilibrium is displaced in a direction in tends to undo the effect of change".

A. Le-Chatelier principle.
B. law of corresponding state.
C. Arrehenius hypothesis.
D. Kopp's rule.

143. In which of the following case for an exothermic reaction?

A. only internal energy change (ΔE) is negative.
B. only enthalpy change (ΔH) is negative.
C. both ΔH and ΔE are negative.
D. enthalpy change is zero.

144. In an ideal binary gas mixture of composition Z at a temperature T and pressure P. The partial molar enthalpy of a component is a function only of .

A. T, P and Z
B. T and P
C. T and Z
D. T

145. An __________ process is an example of sound wave propagation in air.

A. adiabatic
B. isometric
C. isothermal
D. none of these

146. In an isolated systeme quilibrium (*i.e.* reversible) process.

A. $ds = 0$
B. $ds > 0$
C. $ds < 0$
D. ds = Constant

147. The significant of the Absolute zero temperature is

A. temperature of the heat reservoir to which a Carnot engine rejects all the heat that is taken in.
B. minimum temperature attainable.
C. temperature of the heat reservoir to which a Carnot engine rejects no heat.
D. none of these.

148. If m = mass of the molecule, K = Boltzman's constant, T = absolute temperature, the root mean square speed of molecule of a gas is equal to.

A. $\sqrt{\frac{2KT}{m}}$
B. $\sqrt{\frac{3KT}{m}}$
C. $\sqrt{\frac{6KT}{m}}$
D. $\frac{3KT}{m}$

149. does not fall under the category of a common refrigerator.

A. ammonia
B. ethylene
C. freon-12
D. carbon dioxide

150. In which of the following case third law of thermodynamics is relevant and helpful?

A. prediction of the extent of a chemical reaction.
B. evaluating entropy changes of chemical reaction.
C. calculating absolute entropies of substances at different temperature.
D. both (b) and (c)

151. Which of the following is not an intensive property?

A. viscosity
B. internal energy
C. molar heat capacity
D. none of these

152. A well-insulated turbine is used to expand adiabatically and reversibly high pressure steam to produces some shaft work. If the enthalpy

change and entropy change across the turbine are represented by ΔH and ΔS respectively then.

A. $\Delta H = 0$ and $\Delta S = 0$.
B. $\Delta H \neq 0$ and $\Delta S = 0$.
C. $\Delta H \neq 0$ and $\Delta S \neq 0$.
D. $\Delta H = 0$ and $\Delta S \neq 0$.

153. Which of the following statements is wrong?

A. at constant pressure, solubility of a gas in a liquid diminishes with rise in temperature.
B. normally, the gases which are easily liquefied, are more soluble in common solvents.
C. at constant pressure, solubility of a gas in a liquid increases with rise in temperature.
D. the gases which are capable of forming ions in aqueous solution, are much more soluble in water than in other solvents.

154. In vapour compression refrigeration system the effect of changing the evaporator temperature on COP as compared to that of changing the condenser temperature is.

A. equal
B. more pronounced
C. less pronounced
D. data insufficient, can't be predicted.

155. Conversion of a substance from one allotropic form to another leads to absorption/evolution of heat which is termed as the heat of.

A. vaporisation
B. sublimation
C. transition
D. Fusion

156. If the value $n = y = C_p/C_v$ is put in equation PV^n = constant, then it represents a reversible __________ process.

A. isothermal
B. polytropic
C. isentropic
D. adiabatic

157. The ratio of fugacity of a substance to the is called fugacity coefficient.

A. activity co-efficient
B. mole fraction
C. pressure
D. activity

158. A reversible adiabatic line intersects a reversible isothermal line at point A on a P-V diagram of an ideal gas. Then at a point A, the slope of the reversible adiabatic line $(\partial P/\partial V)_s$ and the slope of the reversible isothermal line $(\partial P/\partial V)_T$ are related as _________ (where, $y = C_p/C_v$).

A. $(\partial P/\partial V)_S = 1/y(\partial P/\partial V)_T$
B. $(\partial P/\partial V)_S = [(\partial P/\partial V)_T]^Y$
C. $(\partial P/\partial V)_S = y(\partial P/\partial V)_T$
D. $(\partial P/\partial V)_S = (\partial P/\partial V)_T$

159. For an ideal gasexpression, $nRT\ ln\frac{P_1}{P_2}$ refers to

A. work done under adiabatic conditions
B. co-efficient of thermal expansion
C. work done under isothermal condition
D. compressibility

160. If ΔH = –22.4 kcal for the ammonia synthesis reaction represented by $N_2 + 3H_2 \rightleftharpoons 2NH_3$ then the reaction is called .

A. adiabatic
B. exothermic
C. endothermic
D. isothermal

161. In a reversible adiabatic expansion process the following phenomena takes place?

A. temperature is constant.
B. cooling takes place.
C. heating takes place.
D. pressure is constant.

162. On which of the following parameters is heat reaction dependent / independent?

A. independent of temperature changes.
B. dependent on pressure only.
C. dependent on both pressure and temperature.
D. dependent on temperature only.

163. A crystalline solid in fused at its melting point to from a liquid at the same temperature, this process is accompanied by .

A. decrease of entropy.
B. no change in enthalpy
C. increase in enthalpy corresponding to absorption of heat.
D. decrease in enthalpy corresponding to evolution of heat.

164. On reduced co-ordinate scale thecompressibility factor-reduced pressure plot facilitates.

A. use of only one graph for all gases.

B. covering of wide range.
C. more accurate plotting.
D. easier plotting.

165. For any two phases of a given substance, variation of equilibrium pressure with temperature is given by the __________ equation.
A. Gibbs-Duhem
B. Maxwell's
C. Clayperon
D. none of these

166. For one component system if liquid and vapour phases are in equilibrium then molar free energy is
A. more in vapour phase.
B. more in liquid phase.
C. same in both the phases.
D. replaced by chemical potential which is more in vapour phase.

167. Verification of steam table data for superheated steam can be most easily done by the use of the __________ equation.
A. $(\partial T/\partial P)_S = (\partial V/\partial S)_P$
B. $(\partial T/\partial V)_S = (\partial p/\partial S)_V$
C. $(\partial P/\partial T)_V = (\partial S/\partial V)_T$
D. $(\partial V/\partial T)_P = -(\partial S/\partial P)_T$

168. For multiphase system of N components the necessary condition for the phase equilibrium is that __________.
A. chemical potentials of a given component should be equal in all phases.
B. sum of the chemical potentials of any given component in all the phases should be the same.
C. chemical potentials of all components should be same in a particular phase.
D. none of these.

169. Conversion of a substance from one allotropic form to another leads to heat evolved / absorbed which is termed as heat of .
A. vaporisation
B. fusion
C. transition
D. none of these

170. Which of the following systems is an example of rotary lime kiln?
A. isolated
B. open
C. closed
D. non-thermodynamic

171. A process which is isentropic in nature is carried out at constant.
A. volume
B. temperature
C. pressure
D. all (a), (b) and (c)

172. Which of the following is an example of melting of ice?
A. isometric
B. isothermal
C. adiabatic
D. none of these

173. Which of the following occurs in a reversible adiabatic compression?
A. temperature is constant
B. heating occurs
C. pressure is constant
D. cooling occurs

174. What is the sublimation temperature of dry ice (solid CO_2) in °C?
A. –273
B. 0
C. –78
D. 5

175. Which of the following represents the entropy change in case of reversible adiabatic process?
A. minimum
B. zero
C. maximum
D. indeterminate

176. During throttling process at atmospheric temperature and pressure all gases except __________ shows a cooling effect.
A. air
B. oxygen
C. nitrogen
D. hydrogen

177. If dew point and atmospheric temperature are nearly equal then the relative humidity is .
A. zero
B. 50%
C. almost 100%
D. unpredictable

178. What is the exit condition of steam if it undergoes isentropic expansion in a turbine from 5000 kPa and 400°C given that entropy = 6.65 kJ/kg.K to 150 kPa entropy of saturated liquid = 1.4336 kJ/kg.K, entropy of saturated vapour = 7.2234 kJ/kg.K

A. superheated vapour.
B. saturated vapour
C. partially condensed vapour with quality of 0.1.
D. partially condensed vapour with quality of 0.9.

179. In the process of heat transfer by Gibbs phase rule finds application?

A. radiation
B. conduction
C. convection
D. condensation

180. Such reactions involving solid and liquid in which volume change is negligible the heat of reaction at constant pressure as compared to that as constant volume is .

A. less
B. more
C. same
D. unpredictable; depends on the particular reaction

181. The eutectic point of the system indicates the _________ temperature that can be attained in the system.

A. lowest
B. average
C. highest
D. none of these

182. The effect of increase in temperature on atomic heat capacities of all solid is that it .

A. increases
B. decreases
C. decreases linearly
D. remains unchanged

183. At constant temperature and pressure the relation, connecting the fugalities of various components in a solution with one another and to composition is called the __________ equation.

A. Gibbs-Duhem
B. Margules
C. Van Laar
D. Gibbs-Helmholtz

184. Group contribution method can be used for estimation of for organic compounds.

A. critical properties.
B. thermal conductivity.
C. specific gravity.
D. specific volume.

185. Which of the following is an undesirable property for a good refrigerator?

A. large latent heat of vaporisation
B. high thermal conductivity
C. low freezing point
D. high viscosity

186. For multicomponent multiple phases to be in equilibrium at the same pressure and temperature, the __________ of each component must be same in all phases.

A. fugacity
B. chemical potential
C. both (a) and (b)
D. neither (a) nor (b)

187. For an incompressible fluid the internal energy depends upon the .

A. pressure
B. temperature
C. both (a) & (b)
D. neither (a) nor (b)

188. For the same compression ratio the efficiency of an Otto engine compared to that of a diesel engine, for the same compression ratio will be.

A. more
B. same
C. less
D. data insufficient to predict

189. For an ideal solution isotherm on an enthalpy-concentration diagram will be a

A. straight line
B. sine curve
C. parabola
D. hyperbola

190. During phase change at constant temperature and pressure specific __________ does not change.

A. entropy
B. Gibbs energy

C. internal energy
D. enthalpy

191. What will be the volume accompanied (in liter) by the oxygen contained in a cylinder containing liquid oxygen weight 640 gm.?
A. 448
B. 224
C. 22.4
D. data insufficient; can't be computed

192. Which of the following represents Gibb's-Helmholtz equation?
A. $\Delta F = \Delta H + T\left[\frac{\partial(\Delta F)}{\partial T}\right]_P$
B. $\frac{dP}{dT} = \frac{\Delta H_{vap}}{T.\Delta V_{vap}}$
C. $d(E - TS)\ T, V < 0$
D. $\Delta F = \Delta H - T\Delta T$

193. What will be the degree of freedom of a system consisting of a gaseous mixture of H_2 and NH_3?
A. 0
B. 1
C. 2
D. 3

194. Which of the following equation represents the Claussius-Claypreron equation?
A. $\Delta F = \Delta H + T\left[\frac{\partial(\Delta F)}{\partial T}\right]_P$
B. $lnP = -\frac{\Delta H}{RT} + constant$
C. $\frac{dP}{dT} = \frac{\Delta H}{T\Delta V}$
D. none of these

195. Dryness fraction of wet steam cannot be measured by a __________ calorimeter.
A. Bomb
B. Bucket
C. Throttling
D. Separating

196. Entropy of the universe __________ with passing of time.
A. is increasing
B. is decreasing
C. remains constant
D. data insufficient, can't be predicted

197. Comparing the work done in adiabatic compression the work done in isothermal compression is .
A. same
B. more
C. less
D. more or less depending upon the extent of work done

198. The process of __________ cannot be made reversible even under ideal condition of operation.
A. Free expansion of a gas.
B. Expansion of steam in a turbine.
C. Compression of air in a compressor.
D. all (a), (b) & (c).

199. What will be the degree of freedom of a system consisting of a gaseous mixture of H_2 and NH_3?
A. 0
B. 1
C. 2
D. 3

200. Which of the following is represented if plotted in a Mollier chart?
A. temperature vs entropy
B. pressure vs enthalpy
C. enthalpy vs entropy
D. pressure vs volume

201. The fugacity of a species is equal to it's __________ in an ideal gas mixture.
A. chemical potential
B. partial pressure
C. vapor pressure
D. none of these

202. Which of the following represents the Joule-Thomson co-efficient?
A. $\mu = \left(\frac{\partial E}{\partial T}\right)_H$
B. $\mu = \left(\frac{\partial T}{\partial P}\right)_H$
C. $\mu = \left(\frac{\partial E}{\partial P}\right)_H$
D. $\mu = \left(\frac{\partial P}{\partial T}\right)_H$

203. The value of __________ is unity for on ideal liquid solution.
A. Fugacity
B. Fugacity co-efficient
C. Activity co-efficient
D. Activity

204. What will be the value of ΔH for the reaction $CO_2 \rightarrow C + O_2$ given that for reaction $CO_2 \rightleftharpoons C + O_2$, $\Delta H = -94$ kcal?
A. 94 kcal
B. + 94 kcal
C. < -94 kcal
D. -94 kcal

205. Compounds having large heat of formation are ___________.

A. more stable.
B. not at all stable (like nascent O2).
C. either more or less stable ; depends on the compound.
D. less stable.

206. Which of the following gives the Trouton's ratio given by (where, λ_b = molal heat of vaporisation of a substance at its normal boiling point, kcal/kmol T_b = normal boiling point, °K).

A. λ_b/T_b
B. $\lambda_b x T_b$
C. T_b/λ_b
D. T_b/λ_b

207. Which of the following represent a state function?

A. pressure
B. work
C. temperature
D. specific volume

208. What is the value of Cp of a gas at its critical temperature and pressure?

A. equals 1 kcal/kmol °K.
B. becomes infinity
C. equals 0.24 kcal/kmol °K.
D. becomes zero.

209. If water is in equilibrium with ice at constant pressure then molar heat capacity of water __________ Kcal/kg mole.°K

A. 50
B. ∞
C. 100
D. 0

210. The decomposition of PCl_5 takes place according to $PCl_5 \rightleftharpoons PCl_3 + Cl_2$, decrease in the pressure of the system __________ the degree of dissociation of PCl_5.

A. increase
B. not alter
C. decrease
D. none of these

211. At a given temperature the standard state of a gas is the state in which fugacity is equal to .

A. activity
B. unity
C. both (a) & (d)
D. neither (a) nor (b)

212. The internal energy for an ideal gas depends upon its __________ only.

A. volume
B. temperature
C. pressure
D. molecular size

213. What are the assumptions to be made in the Clayperon equation $In\ \frac{P_2}{P_1} = \frac{\Delta H_v}{R}\left(\frac{1}{T_1} - \frac{1}{T_2}\right)$ to obtained the integrated form of Clausius-Clayperon equation

A. vapour phase behaves as an ideal gas.
B. heat of vaporisation is independent of temperature.
C. volume of the liquid phase is negligible compared to that of vapour phase.
D. all (a), (b) & (c).

214. The temperature of the system increases in case of an process.

A. isothermal expansion
B. adiabatic expansion
C. isothermal compression
D. adiabatic compression

215. For Virial equation of state which of the following is true?

A. Virial co-efficients 'B' represents three body interactions.
B. for some gases, Virial equations and ideal gas equations are the same.
C. Virial co-efficients are function of temperature only.
D. Virial co-efficients are universal constants.

216. The condition that a gas can be liquefied by the pressure along in that when its temperature is __________ its critical temperature.

A. more than
B. equal to or higher than
C. less than or equal to
D. less than

217. Which of the following represents the third law of thermodynamics?

A. heat capacity of a crystalline solid is zero at

absolute zero temperature.

B. gases having same reduced properties behave similarly.

C. heat transfer from low temperature to high temperature source is not possible without external work.

D. none of these.

218. What will be the degree of freedom of the system ice-water-vapour?

A. 0

B. 2

C. 3

D. 1

219. What is the condition that the thermal efficiency of a Carnot engine can approach 100% when the temperature of the.

A. hot reservoir approaches infinity.

B. cold reservoir approaches zero.

C. either (a) or (d).

D. neither (a) nor (d).

220. The second law of thermodynamics interprets that it is concerned with .

A. irreversible processes only.

B. direction of energy transfer.

C. non-cyclic processes only.

D. amount of energy transferred.

221. Which of the following represents the mathematical statement of Charles' law?

A. $\frac{V}{T} = Constant$

B. $V \propto \frac{1}{P}$

C. $\frac{PV}{T} = Constant$

D. $V \propto \frac{1}{T}$

222. There is a system in which there is exchange of energy but not of mass. The system is called a/an ___________.

A. open

B. insulated

C. isolated

D. closed

223. The ___________ point is the point at which all the three phase (solid, liquid and gas) coexist.

A. boiling

B. triple

C. Boyle

D. freezing

224. If the value of $n = \pm \infty$, in the PV^n = constant then it represents a reversible ___________ process.

A. isentropic

B. isometric

C. isothermal

D. adiabatic

225. The ___________ of the gas becomes zero at absolute zero temperature.

A. mass

B. volume

C. pressure

D. none of these

226. For the reaction $CH_4(g) + C_2H_4(g) \rightleftharpoons C_3H_8(g)$ the heat of reaction ΔH is equal to ——— given that contents of CH_4, C_2H_4 and C_3H_8 are –17.9, 12.5 and –24.8 kcal/mole respectively.

A. –19.4

B. 55.2

C. –55.2

D. –30.2

227. What is the minimum number of degrees of freedom for any system?

A. 2

B. 0

C. 1

D. 3

228. The value of the ratio of fugacity to pressure (*f/P*) for a gas approaches as pressure approaches zero.

A. infinity

B. unity

C. an indeterminate value

D. zero

229. A Carnot refrigeration cycle has main factors including that it

A. transfers heat from high temperature to low temperature.

B. does not need the addition of external work for its functioning.

C. accomplishes the reverse effect of the heat engine.

D. none of these.

230. What will be the activity coefficient of methanol in an aqueous solution at 60°C exerting a pressure of 39.223 kPa. The liquid phase and vapour phase mole factor of mathematical are 0.1686 and 0.5714 respectively. At 60°C vapour pressure

of methanol and water are 84.562 kPa and 19.953 kPa respectively.

A. 1.572

B. 3.389

C. 4.238

D. 1.9398

231. For the prediction of activity co-efficient experimentally the equations used is/are ?

A. Margules equation

B. Wilson's equation

C. Van Laar equation

D. all (a), (b) and (c)

232. For a binary solution of components A and B for equation $\left(\frac{d\ log_e P_A}{d\ log_e X_A} = \frac{d\ log_e P_B}{d\ log_e X_B}\right)$ is applicable A and B in equilibrium with their vapors at constant temperature and pressure. This is based on the __________ equation.

A. Margules

B. Gibbs-Duhem

C. Van Laar

D. Gibbs-Duhem-Margules

233. At constant pressure the variation of heat of reaction with temperature is given by __________ law.

A. Antonie's

B. Kelvin's

C. Kirchoffs

D. none of these

234. If the value of n in equation, $PV^n = constant$ is in between 1 and y (i.e. C_p/C_v), then it represents a reversible __________ process.

A. isentropic

B. polytropic

C. isobaric

D. isometric

235. __________ by the system for a constant volume process is used only to increase the internal energy.

A. heat absorbed

B. work done

C. neither (a) nor (c)

D. both (a) & (c)

236. During __________ change, entropy of a substance remains constant

A. reversible isothermal

B. irreversible isothermal

C. reversible adiabatic

D. none of these

237. Which of the following defines the Helmholtz free energy (A)?

A. A = H + TS

B. A = E - TS

C. A = H - TS

D. none of these

238. What is the heat content (enthalpy) of Oxygen in reaction, C + O2 $\rightarrow$ CO2; $\Delta H = -94$ kcal.

A. –94 kcal

B. < –94 kcal

C. –94 kcal

D. Zero

239. If the entropy of the system has reached __________ value then the system is said to be at equilibrium.

A. zero

B. minimum

C. maximum

D. none of these

240. At constant volume the heat of reaction is a function of __________ change.

A. Enthalpy

B. internal energy

C. either (b) or (c)

D. neither (b) nor (c)

241. What will be the differential or finite change of entropy when a system is in equilibrium for all possible processes?

A. < 0

B. = 0

C. >0

D. none of these

242. Which of the following describes Throttling process?

A. reversible and constant entropy.

B. irreversible and constant enthalpy

C. reversible and constant enthalpy

D. reversible and isothermal

Answer

1. C.
2. A.
3. A.
4. C.
5. A.
6. A.
7. B
8. D.
9. A
10. D
11. C
12. C
13. D
14. B
15. C
16. A
17. B
18. C
19. A
20. C
21. C
22. C
23. A
24. B
25. A
26. A
27. D
28. C
29. B
30. A
31. C
32. D
33. C
34. C
35. B
36. A
37. B
38. C
39. B
40. D
41. A
42. D
43. C
44. D
45. C
46. A
47. B
48. D
49. B
50. B
51. B
52. B
53. A
54. A
55. A
56. A
57. A
58. D
59. C
60. C
61. A
62. A
63. C
64. B
65. A
66. B
67. B
68. D
69. C
70. A
71. B
72. B
73. C
74. A
75. C
76. C
77. D
78. A
79. A
80. A
81. A
82. B
83. D
84. B
85. A
86. C
87. B
88. B
89. C
90. C
91. A
92. C
93. A
94. A
95. B
96. D
97. A
98. C
99. B
100. A
101. D
102. D
103. A
104. D
105. C
106. D
107. A
108. B
109. C
110. D
111. A
112. C
113. A
114. D
115. C
116. C
117. A
118. A
119. D
120. A
121. A
122. A
123. B
124. A
125. C
126. A
127. D
128. B
129. C
130. A
131. A
132. B
133. B
134. C
135. D
136. A
137. D
138. D
139. D
140. A
141. B
142. A
143. C
144. B

145. A
146. A
147. C
148. B
149. B
150. D
151. B
152. B
153. A
154. B
155. C
156. C
157. C
158. C
159. C
160. B
161. B
162. C
163. C
164. A
165. C
166. C
167. D
168. A
169. C
170. B
171. A
172. B
173. C
174. C
175. B
176. D
177. C
178. A
179. D
180. C
181. A
182. A
183. A
184. A
185. D
186. C
187. B
188. A
189. A
190. B
191. A
192. A
193. D
194. B
195. A
196. A
197. B
198. A
199. D
200. C
201. B
202. B
203. C
204. A
205. A
206. A
207. B
208. B
209. B
210. A
211. C
212. B
213. D
214. D
215. C
216. D
217. A
218. A
219. C
220. B
221. A
222. D
223. B
224. B
225. B
226. A
227. B
228. B
229. C
230. A
231. D
232. D
233. C
234. B
235. A
236. C
237. B
238. D
239. C
240. B
241. A
242. B

GATE 2015-2000

(1 Mark Questions)

GATE 2015

1. If V, U, S and G represent respectively the molar volume, molar internal energy, molar entropy and molar Gibbs free energy, then match the entries in the left and right columns below and choose the correct option.

I.	$-(\partial U/\partial V)_S$	P.	Temperature (T)
II.	$(\partial G/\partial p)_T$	Q.	Pressure (p)
III.	$-(\partial G/\partial T)_p$	R.	Volume (V)
IV.	$(\partial U/\partial S)_V$	S.	Entropy (S)

A. I-Q, II-R, III-S, IV-P
B. I-Q, II-S, III-R, IV-P
C. I-P, II-S, III-Q, IV-R
D. I-R, II-Q, III-S, IV-P

***Sol.* A.** we know that

$$dU = TdS - pdV$$

and $$dG = Vdp - sdT$$

So, $$\left.\frac{\partial U}{\partial V}\right|_S = -P \Rightarrow -\left.\frac{\partial U}{\partial V}\right|_S = P$$

$$\left.\frac{\partial U}{\partial S}\right|_V = T$$

$$\left.\frac{\partial G}{\partial P}\right|_T = V \text{ and } -\left.\frac{\partial G}{\partial T}\right|_P = S$$

2. For a gas phase cracking reaction A → B + C at 300°C, the Gibbs free energy of the reaction at this temperature is ΔG° = –2750 J/mol. The pressure is 1 bar and the gas phase can be assumed to be ideal. The universal gas constant, R = 8.314 J/mol-K. The fractional molar conversion of A at equilibrium is

A. 0.44
B. 0.50
C. 0.64
D. 0.80

***Sol.* D.** According to the gas phase cracking reaction

	A →	B +	C
at t = 0	1	0	0
at t = 0	1-α	α	α

where, α = fraction converted

Equilibrium constant,

$$K_p = \frac{Y_B Y_C}{Y_A} . p = \frac{\left(\frac{\alpha}{1+\alpha}\right)\left(\frac{\alpha}{1+\alpha}\right)}{\left(\frac{1-\alpha}{1+X}\right)}$$

$$= \frac{\alpha^2}{1-\alpha^2} \cdot p = \frac{-\alpha^2}{1-\alpha^2}$$

Also, $$\ln K = \left[-\Delta G^0/Rt\right]$$

$$K = \exp\left[\frac{-\Delta G^0}{RT}\right] = \exp\left[\frac{2750}{8.314 \times 573}\right]$$

$$K = 1.78$$

So, $$\frac{\alpha^2}{1-\alpha^2} = 1.78$$

$$\alpha = 0.80$$

3. For a pure liquid, the rate of change of vapour pressure with temperature is 0.1 bar/K in the temperature range of 300 to 350 K. If the boiling point of the liquid at 2 bar is 320 K, the temperature (in K) at which it will boil at 1 bar (upto one decimal place) is ___________.

Sol. We know that

$$\frac{dp}{dT} = 0.1 \Rightarrow dp = 0.1\, dT$$

$$\Rightarrow \quad p = 0.1T + C \qquad ....(i)$$

Temperature, T = 320 K

Pressure, P = 2 bar

$$2 = 0.1 \times 320 + C$$

$$\Rightarrow \quad C = -30$$

Putting the value of C in Eq. (i), we get

$$P = -30 + 0.1T$$

Putting, p = 1bar

$$1 = -30 + 0.1T$$

$$\therefore \quad T = 310 \text{ K}$$

4. Three identical closed systems of a pure gas are taken from an initial temperature and pressure (T_1, p_1) to a final state (T_2, p_2), each by a different path . Which of the following is always correct for the three systems? [Δ represents the change between the initial and final states : U, S, G, Q and W are internal energy, entropy, Gibbs free energy, heat added and work done, respectively]

<u>A.</u> ΔU, ΔS and Q are same

<u>B.</u> W, ΔU and ΔG are same

<u>C.</u> ΔS, W and Q are same

<u>D.</u> ΔG, ΔU and ΔS are same

Sol. <u>**D.**</u> U, S and G are state variables (point function)

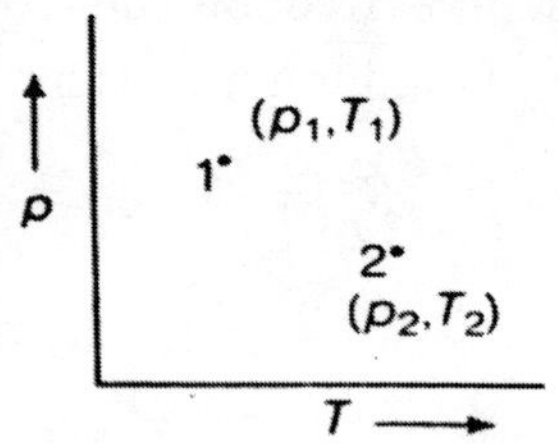

So, ΔU, ΔS and ΔG remain same between two states 1 and 2.

GATE 2014

5. From the following list, identify the properties which are equal in both vapour and liquid phases at equilibrium

P. density Q. temperature

R. chemical potential S. enthalpy

<u>A.</u> P and Q

<u>B.</u> Q and R

<u>C.</u> R and S

<u>D.</u> P and S

Sol. <u>**B.**</u> Density of liquid phase will be higher than that of vapour phase.

Temperature will be equal for both phases at equilibrium. Chemical potential, i.e., partial molar free energy is also equal for both phases at equilibrium.

Enthalpy i.e., heat content will be higher for vapour phase.

6. In a closed system, the isentropic expansion of an ideal gas with constant specific heats is represented by

<u>**A.**</u>

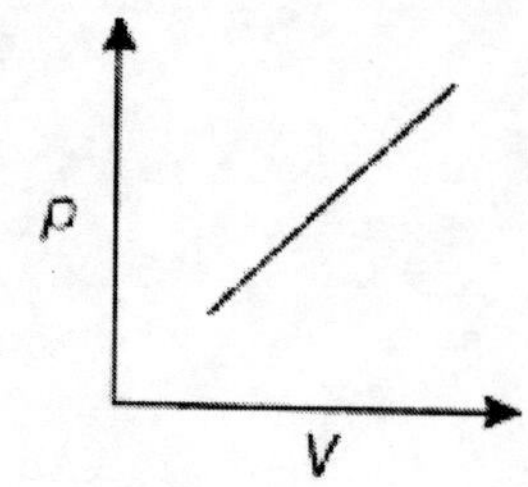

<u>**B.**</u>

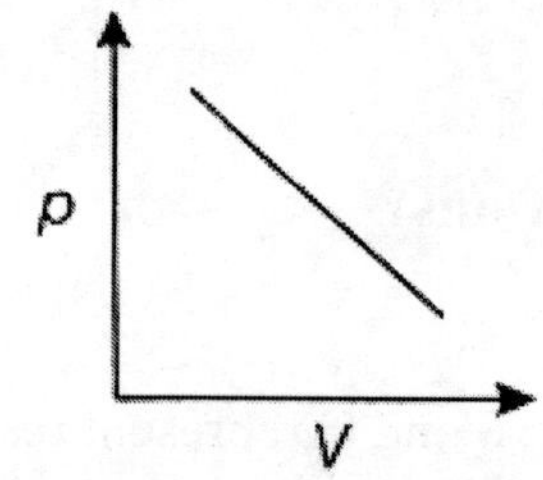

<u>**C.**</u>

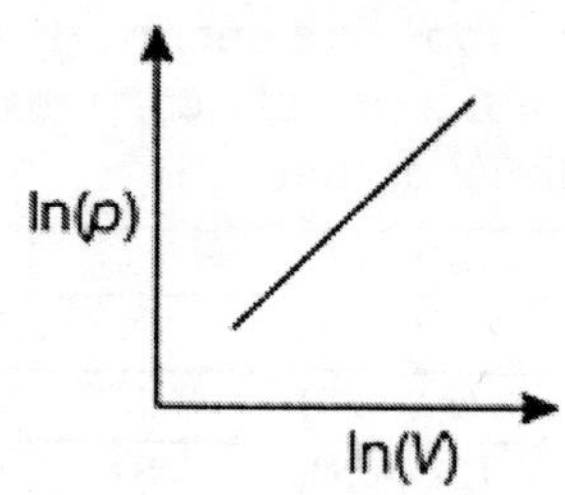

<u>**D.**</u>

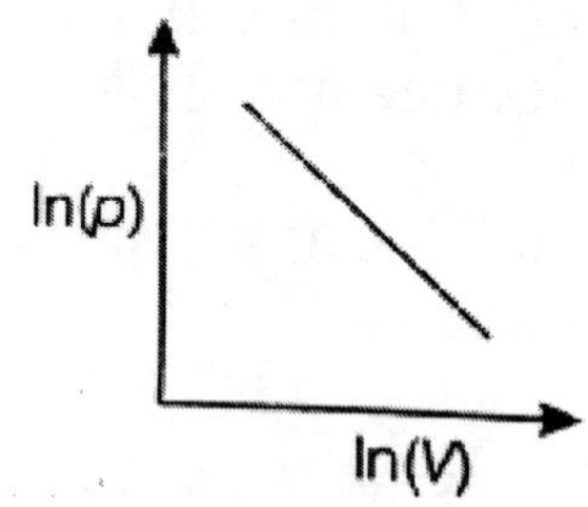

Sol. <u>**C.**</u> For isentropic expansion of an ideal gas,

$$pV^{\gamma} = k \text{ (constant)}$$

or $\ln p + \gamma \ln V = \ln k$

ln $p = \ln k - \gamma \ln V$

The graph between ln p and ln V will be straight line with slope equal to $-\gamma$.

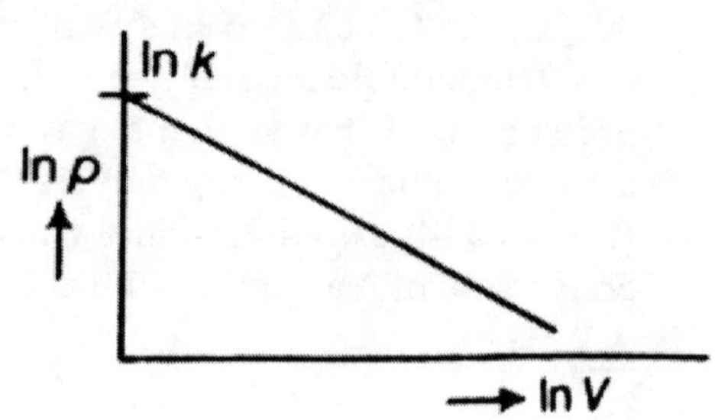

7. Match the following

Group 1	*Group 2*
P. $\left(\frac{\partial G}{\partial n_i}\right)_{T,P,n_{j\neq i}}$	1. Arrhenius equation
Q. $\left(\frac{\partial G}{\partial n_i}\right)_{S,V,n_{j\neq i}}$	2. Reaction equation constant
R. $\exp\left(\frac{-\Delta G^0_{reaction}}{RT}\right)$	3. Chemical potential
S. $\Sigma(n_i d\mu_i)_{T,P} = 0$	4. Gibbs-Duhem equation

A. Q-3, R-1, S-2
B. Q-3, R-2, S-4
C. P-3, R-2, S-4
D. Q-3, R-4, S-1

***Sol*. C.** At equilibrium, $\Delta G^\circ = -RT \ln K_p$

$$lnK_p = -\frac{\Delta G^\circ}{RT}$$

$K_p = ext\left(-\frac{\Delta G^\circ}{RT}\right)$ = Reaction equillibrium constant

Chemical potential = Partial molar Gibbs free energy = $\left(\frac{\partial G}{\partial n_i}\right)_{T,p}$

Gibbs-Duhem equation describes the relationship between changes in chemical potential components in a thermodynamic system,
As per this equation, $\Sigma n_i d\mu_i = -SdT + Vdp$
Where n_i is the number of moles and $d\mu_i$ is the infinitesimal increases in chemical potential for i[th] component.
At constant temperature, dT = 0
At constant pressure, dP = 0

$$\Sigma(n_i d\mu_i)_{T,P} = 0$$

GATE 2013

8. The thermodynamic state of a closed system containing a pure fluid changes from (T_1, p_1) to (T_2, p_2), where T and P denote the temperature and pressure, respectively. Let Q denote the heat absorbed (> 0 if absorbed by the system) and W the work done (> 0 if done by the system). Neglect changes in kinetic and potential energies. Which one of the following is correct?

A. Q is path-independent and W is path-dependent
B. Q is path-dependent and W is path-independent
C. (Q-W) is path-independent
D. (Q+W) is path-independent

***Sol*. C.** From first law of thermodynamics, change in internal energy,

$$\Delta□ = □ - □$$

(when work is done by the system)

Δ□ depends only on initial and final position and does not depend on the path taken by the system.

∴ Q – W is path independent.

9. An equation of state is explicit in pressure P and cubic in the specific volume V. At the critical point c, the isotherm passing through c satisfies

A. $\frac{\partial p}{\partial V} < 0, \frac{\partial^2 p}{\partial V^2} = 0$
B. $\frac{\partial p}{\partial V} > 0, \frac{\partial^2 p}{\partial V^2} < 0$
C. $\frac{\partial p}{\partial V} = 0, \frac{\partial^2 p}{\partial V^2} > 0$
D. $\frac{\partial p}{\partial V} = 0, \frac{\partial^2 p}{\partial V^2} = 0$

***Sol*. D.** At critical point,

$$\left(\frac{□□}{□□}\right)_{□□} = 0$$

and $\left(\frac{\partial^2 p}{\partial V^2}\right)_{T_c} = 0$

10. The units of the isothermal compressibility are

A. m^{-3}
B. Pa^{-1}
C. $m^{-3}Pa^{-1}$
D. $m^{-3}Pa^{-1}$

***Sol*. B.** Isothermal compressibility,

$$K = -\frac{Δ□/□}{Δ□}$$

Unit $(Pressure)^{-1}$ or P

GATE 2012

11. In a throttling process, the pressure of an ideal gas reduces by 50%. If C_p and C_v are the heat capacities at constant pressure and constant volume respectively and $\gamma = \frac{C_p}{C_v}$, then the specific volume will change by a factor of

A. 2
B. $2^{1/\gamma}$
C. $2^{\frac{\gamma-1}{\gamma}}$
D. 0.5

***Sol*. B.** In a throttling process,

$$pV^\gamma = \text{constant}$$

$$\left(\frac{V_2}{V_1}\right)^\gamma = \frac{p_1}{p_2}$$

or $\frac{V_2}{V_1} = \left(\frac{p_1}{p_2}\right)^{1/\gamma}$

Given $p_2 = \frac{p_1}{2}$

$\Rightarrow \frac{V_2}{V_1} = 2^{1/\gamma}$

12. If the temperature of saturated water is increased infinitesimally at constant entropy, the resulting state of water will be
   A. Liquid
   B. Liquid-vapour coexistence
   C. Saturated vapour
   D. Solid

Sol. **A.** At constant entropy, the phase change of a saturated liquid does not take place by increasing the temperature, thus the resulting state will also be a liquid.

13. In a parallel flow heat exchanger, operating under steady state, hot liquid enters at a temperature $T_{h,in}$. And leaves at a temperature $T_{h,out}$. Cold liquid enters at a temperature $T_{c,in}$ and leaves at a temperature $T_{c,out}$. Neglect any heat loss from the heat exchanger to the surrounding. If $T_{h,in} >> T_{c,in}$, then for a given time interval, which one of the following statements is true?
   A. Entropy gained by cold stream is greater than entropy lost by hot stream
   B. Entropy gained by cold stream is less equal to entropy lost by hot stream
   C. Entropy gained by cold steam is less than entropy lost by hot stream
   D. Entropy gained by cold stream is zero

Sol. **A.** Since, change in entropy $\Delta S = \frac{\Delta Q}{T}$

Given, $T_{h,in} >> T_{c,in}$

As ΔQ will be same for both liquid (heat given = heat taken) therefore, entropy gained by cold stream will be more due to its very low temperature.

GATE 2011

14. Minimum work (W) required to separate a binary gas mixture at a temperature T_o and pressure p_o is

$$W = -R\,T_o\left[y_1\,In\left(\frac{\hat{f}_1}{f_{pure,\,1}}\right) + y_2\,In\left(\frac{\hat{f}_2}{f_{pure,\,2}}\right)\right]$$

Where, y_1 and y_2 are mole fractions, $f_{pure,\,1}$ and $f_{pure,\,2}$ are fugacities of pure species at T_o and p_o and $\hat{f}_1$ and $\hat{f}_2$ are fugacities of species in the mixture at T_o, p_o. if the mixture is ideal W is
   A. zero
   B. $-R\,T_o[y_1\,Iny_1 + y_2 Iny_2]$
   C. $RT_o[y_1\,Iny_1 + y_2\,Iny_2]$
   D. RT_o

Sol. **B.** Given,

$$W = -RT_0\left[Y_1\,ln\left(\frac{\hat{f}_1}{f_{pure,1}}\right) + Y_2\,ln\left(\frac{\hat{f}_2}{f_{pure,2}}\right)\right]$$

$\hat{f}_1 = Y_1$ p and $\hat{f}_2 = Y_2$ p

and $f_{pure} = P$

$$\therefore W = -RT_0\left[Y_1\,ln\left(\frac{Y_1 P}{P}\right) + Y_2\,ln\left(\frac{Y_2 P}{P}\right)\right]$$

$$W = -RT_0[Y_1\,ln\,Y_1 + Y_2\,ln\,Y_2]$$

15. The partial molar enthalpies of mixing (in J/mol) for benzene (component 1) and cyclohexane (component 2) at 300 K and 1 bar are given by $\Delta\overline{H_1} = 3600\,x_2{}^2$ and $\Delta\overline{H_2} = 3600\,x_1{}^2$, where x_1 and x_2 are the mole fractions. When 1 mol of benzene is added to 2 mol of cyclohexane, the enthalpy change (in joule) is
   A. 3600
   B. 2400
   C. 2000
   D. 800

Sol. **D.** 1 mol of benzene is added to 2 mol of cyclohexane.

$$\therefore x_1 = \frac{1}{3}\ and\ x_2 = \frac{2}{3}$$

The enthalpy exchange,

$$\Delta H = x_1\Delta H_1 + x_2\Delta H_2$$

$$x_1 = \frac{1}{3}\text{and } x_2 = \frac{2}{3}$$

$$= \frac{1}{3}\times 3600x_2^2 + \frac{2}{3}\times 3600x_1^2$$

$$= \frac{1}{3}\times 3600\left(\frac{2}{3}\right)^2 + \frac{2}{3}\times 3600\left(\frac{1}{3}\right)^2$$

$$= 800\text{J}$$

16. 1 mole of methane is contained in a leak proof piston-cylinder assembly at 8 bar and 1000 K. the gas undergoes isothermal expansion to 4 bar under reversible conditions. Methane can be considered as an ideal gas under these conditions. The value of universal gas constant is 8.314 J/mol-K. The heat transferred (in kJ) during the process is

A. 11.52

B. 5.76

C. 4.15

D. 2.38

Sol. **B.** Given

p_1 = 8 bar, p_2 = 4 bar

T = 1000 K, n = 1

Heat transferred

H = – nRT ln (p_2/p_1)

= –1×8.314×1000×ln(4/8)

= 5.76×10³J

= 5.76 kJ

GATE 2010

17. A new linear temperature scale, denoted by ⁰S, has been developed, where the freezing point of water is 200⁰S and the boiling point is 400⁰S. On this scale 500⁰S corresponds, in degree celsius to

A. 100⁰C

B. 125⁰C

C. 150⁰C

D. 300⁰C

Sol. **C.** We know that,

Freezing point of water = 0°C

Boiling point of water = 100°C

If 500°S corresponds to C in degree Celsius,

Then, $\frac{C-0}{100-0} = \frac{500-200}{400-200} \Rightarrow$ C = 150°C

18. A saturated liquid at 1500 kPa and 500 K, with an enthalpy of750 kJ/kg is throttled to a liquid-vapour mixture at 150 kPa and 300 K. At the exit conditions the enthalpy of the saturated liquid is 500 kJ/kg and the enthalpy of the saturated vapour is 2500 kJ/kg.The percentage of the original liquid, which vaporizes is

A. 87.5%

B. 67%

C. 12.5%

D. 10%

Sol. **C.** Assume the basis of 1 kg of saturated liquid at inlet condition,

And x is the fraction of liquid goes to vaporize.

Then, h_{in} = xh_{vap} + (1 – x) h_{liq}

Putting Values given in the problem,

750= x × 2500 + (1 – x) ×500

x = 0.125

The percentage of original liquid going to vaporize = 12.5%

GATE 2009

19. An ideal gas at temperature T_1 and pressure p_1 is compressed isothermally to pressure p_2 (> p_1) in a closed system. Which one of the following is true for internal energy (U) and Gibbs free energy (G) of the gas at the two states?

A. $U_1 = U_2, G_1 > G_2$

B. $U_1 = U_2, G_1 < G_2$

C. $U_1 > U_2, G_1 = G_2$

D. $U_1 < U_2, G_1 = G_2$

Sol. **B.** Internal energy U = f (T)

= Constant for isothermal process

$\therefore U_1 = U_2$

Change in Gibbs free energy,

$\Delta G = RT \ln \left(\frac{p_2}{p_1}\right)$

For $p_1 > p_2$, $\Delta G = G_2 - G_1 > 0$

$\therefore \quad G_2 > G_1$

GATE 2008

20. For a Carnot refrigerator operating between 40⁰C and 25⁰C, the coefficient of performance is

A. 1

B. 1.67

C. 19.88

D. 39.74

Sol. **C.** Given,

T_C = 25°C = 298 K

T_H = 40°C = 313 K

Coefficient of performance

$= \frac{T_c}{T_H - T_C}$

$= \frac{298}{113-298}$

=19.87

21. The work done by one mole of a van der Waals fluid undergoing reversible isothermal expansion form initial volume V_i to final volume V_f, is

A. RT In $\left(\frac{V_f}{V_i}\right)$

B. RT In $\left(\frac{V_f - b}{V_i - b}\right)$

C. RT In $\left(\frac{V_f - b}{V_i - b}\right) - a\left(\frac{1}{V_i} - \frac{1}{V_i}\right)$

D. RT In $\left(\frac{V_f - b}{V_i - b}\right) + a\left(\frac{1}{V_f} - \left(\frac{1}{V_i}\right)\right)$

Sol. **D.** van der waals equation of state is

$$\left(p + \frac{a}{V^2}\right)(V - b) = RT$$

or $$p = \frac{RT}{V-b} - \frac{a}{V^2}$$

Work done for isothermal (constant T) expansion is

$$W = \int_{V_i}^{V_f} p\, dV$$

$$= \int_{V_i}^{V_f} \left(\frac{RT}{V-b} - \frac{a}{V^2}\right) dV$$

$$= \left[RT\, ln(V - b) + \frac{a}{V}\right]_{V_i}^{V_f}$$

$$= RT \ln\left(\frac{V_f-b}{V_i-b}\right) + a\left(\frac{1}{V_f} - \frac{1}{V_i}\right)$$

GATE 2007

22. if T_A and T_B are the boiling points of pure A and pure B respectively, and T_{AB} is that of a non-homogeneous immiscible mixture of A and B, then

A. $T_{AB} < T_A$ and T_B

B. $T_{AB} > T_A$ and T_B

C. $T_A > T_{AB} > T_B$

D. $T_B > T_{AB} > T_A$

Sol. **A.** The boiling point of non-homogeneous immiscible mixture is always less than the individual boiling points of two components,

$$\therefore \quad T_{AB} < T_A \text{ and } T_B$$

23. The state of an idea gas is changed from (T_1, p_1) to (T_2, p_2) in a constant volume process. To calculate the change in enthalpy, Δh, which of the following properties/variables are required?

A. C_V, p_1, p_2

B. C_p, T_1, T_2

C. $C_p, T_1, T_2 p_1, p_2$

D. C_v, p_1, p_2, T_1, T_2

Sol. **B.** The change in enthalpy,

$$\Delta h = C_p \Delta T$$

$\therefore C_p, T_1, T_2$ are required.

24. The change in enthalpy of the system ΔS_{sys}, undergoing a cyclic irreversible process is

A. greater than zero

B. less than zero

C. equal to zero

D. equal to the $\Delta S_{surrounding}$

Sol. **A.** For an irreversible process, $\Delta S_{sys} > 0$

25. Parameters a and b in the van der Walls and other cubic equations of state represent

A. a – molecular weight, b – molecular polarity

B. a – molecular size, b – molecular attraction

C. a – molecular size, b – molecular speed

D. a – molecular attraction, b – molecular size

Sol. **D.** van der Waals equation

$$\left(p + \frac{a}{V^2}\right)(V - b) = RT$$

Here, ais molecular attraction and b is molecular size.

GATE 2006

26. A heat engine operates at 75% of the maximum possible efficiency. The ratio of the heat source temperature (in kelvin) to the heat sink temperature (in kelvin) is 5/3. The fraction of the heat supplied that is converted to work is

A. 0.2

B. 0.3

C. 0.4

D. 0.6

Sol. **B.** Given, $\frac{T_1}{T_2} = \frac{5}{3}$

$$\text{Efficiency, } \eta = \frac{W}{Q_1} = \frac{T_1 - T_2}{T_1} = 1 - \frac{T_2}{T_1}$$

$$\frac{W}{Q_1} = 1 - \frac{3}{5} = 0.4$$

$$\frac{W}{0.75H} = 0.4 \quad (\because Q_1 = 0.75H)$$

$$W = 0.3\ H$$

27. For the isentropic expansion of an ideal gas from the initial conditions p_1, T_1 to the final condition p_2, T_2 which of the following relation is valid? $(\gamma = C_p/C_v)$

A. $\left(\frac{p_1}{p_2}\right) = \left(\frac{T_2}{T_1}\right)^{\gamma}$

B. $\left(\frac{p_1}{p_2}\right) = \left(\frac{T_1}{T_2}\right)^{\frac{\gamma}{\gamma-1}}$

C. $\left(\frac{p_1}{p_2}\right) = \left(\frac{T_1}{T_2}\right)$

D. $\left(\frac{p_1}{p_2}\right) = \left(\frac{T_1}{T_2}\right)^{\frac{\gamma-1}{\gamma}}$

Sol. **B.** Given, $\gamma = \frac{C_p}{C_v}$

For isentropic expansion of ideal gas,

$$pV^{\gamma} = K$$

Also, $$pV = nRT$$

$$\Rightarrow \frac{V_1}{V_2} = \frac{T_1}{T_2} \cdot \frac{p_2}{p_1}$$

From Eq.(i),

$$\frac{p_1}{p_2} = \left(\frac{V_2}{V_1}\right)^{\gamma}$$

$$\frac{p_1}{p_2} = \left(\frac{p_1}{p_2}\right)^{\gamma} \cdot \left(\frac{T_2}{T_1}\right)^{\gamma}$$

$$\left(\frac{p_1}{p_2}\right)^{1-\gamma} = \left(\frac{T_2}{T_1}\right)^{\gamma}$$

$$\frac{p_1}{p_2} = \left(\frac{T_2}{T_1}\right)^{\frac{\gamma}{1-\gamma}}$$

$$\frac{p_1}{p_2} = \left(\frac{T_1}{T_2}\right)^{\frac{\gamma}{\gamma-1}}$$

GATE 2005

28. In van der Waals equation of state, what are the criteria applied at the critical point to determine the parameters a and b?

A. $\left(\frac{\partial p}{\partial V}\right)_T = 0; \left(\frac{\partial^2 p}{\partial V^2}\right)_T = 0$

B. $\left(\frac{\partial V}{\partial p}\right)_T = 0; \left(\frac{\partial^2 V}{\partial p^2}\right)_T = 0$

C. $\left(\frac{\partial p}{\partial T}\right)_V = 0; \left(\frac{\partial^2 p}{\partial T^2}\right)_v = 0$

D. $\left(\frac{\partial V}{\partial T}\right)_p = 0; \left(\frac{\partial^2 V}{\partial T^2}\right)_p = 0$

Sol. **A.** van der Waals equation at critical point

$$\left(p_C + \frac{a}{V^2}\right)(V - b) = RT_C$$

To determine a and b,

$$\left(\frac{\partial p}{\partial V}\right)_{T=T_C} = 0 \text{ and } \left(\frac{\partial^2 p}{\partial V^2}\right)_{T=T_C} = 0$$

29. Which of the following statements is true?

A. Heat can be fully converted into work

B. Work cannot be fully converted into heat

C. The efficiency of heat engine increases as the temperature of the heat source is increases while keeping the temperature of the heat sink fixed

D. A cyclic process can be devised whose sole effect is to transfer heat from a lower temperature to higher temperature.

Sol. **C.** Efficiency $\eta = 1 - \frac{T_2}{T_1}$

Thus, as temperature of heat source (T_1) increases while keeping the temperature of heat sink (T_2) fixed, the efficiency (η) will increase.

GATE 2004

30. For an ideal gas mixture undergoing a reversible gaseous phase chemical reaction, the equilibrium constant

A. is independent of pressure

B. increases with pressure

C. decreases with pressure

D. increases/decreases with pressure depending on the stoichiometric coefficients of the reaction

Sol. **D.** Assume the reaction,

$$aA + bB \rightleftharpoons cC \text{ and } dD$$

Equilibrium constant,

$$K = \frac{Y_C^c Y_D^d}{Y_A^a Y_B^b}(p)^{c+d-a-b}$$

Where, y_i is mole fraction of i and p is total pressure. It can be seen that K will increase or decrease with pressure depending on the stoichiometric coefficients of the reaction.

31. As pressure approaches zero, the ratio of fugacity to pressure (f/p) for a gas approaches

A. Zero

B. Unity

C. Infinity

D. an indeterminate value

Sol. **B.** As p → 0,

Fugacity f → p

And $\lim_{p \to o}\left(\frac{f}{p}\right) = 1$

32. A perfectly insulated container of volume V is divided into two equal halves by a partition. One side is under vacuum while the other side has one mole of an ideal gas (with constant heat capacity) at 298 K. if the partition is broken, the final temperature of the gas in the container

A. will be greater than 298

B. will be 298 K

C. will be less than 298 K

D. cannot be determined

Sol. **C.** Perfectly insulated container → adiabatic process,

Therefore, $\frac{T_2}{T_1} = \left(\frac{V_1}{V_2}\right)^{\gamma-1}$

Here, $V_1 = \frac{V}{2}$ and $V_2 = V$

$T_1 = 298$ K

∴ $\frac{T_2}{298} = \left(\frac{1}{2}\right)^{\gamma-1}$

$T_2 = \frac{298}{2^{\gamma-1}}$

Since, $\gamma = \frac{C_p}{C_v} > 1 \Rightarrow \gamma - 1 > 0$

∴ $T_2 < 298\ K$

GATE 2003

33. For estimation of heat capacity of a solid compound, one can use
   A. clapeyron equation
   B. Gibb's equation
   C. Kopp's
   D. Trouton's rule

Sol. **C.** Kopp's rule can be used to determine heat capacity of solid compound.

34. For organic compounds, Group Contribution Method can be used for estimation of
   A. critical properties
   B. specific gravity
   C. specific volume
   D. thermal conductivity

Sol. **A.** Group Contribution Method (GCM) can be used for estimation of critical properties for organic compounds.

35. When dilute equeous solution of two salts are mixed, the process is associated with
   A. decrease in temperature
   B. increase in temperature
   C. no change in temperature
   D. change in temperature which is a function of composition

Sol. **B.** The temperature will increase by mixing the dilute aqueous solutions of two salts.

36. In Joule's experiments, an insulted container contains 20 kg of water initially at 25°C. It is stirred by an agitator, which is made to turn by a slowly falling body weighing 40 kg through a height of 4 m. The process is repeated 500 times. The acceleration due to gravity is 9.8 m/s². Neglecting the heat capacity of agitator, the temperature of water (in °C) is
   A. 40.5
   B. 34.4
   C. 26.8
   D. 25

Sol. **B.** Given, $m_{water} = 20$ kg, $T_1 = 25^0$C

$m_{body} = 40$ kg, h = 4 m

g = 9.8 m/s^2

$C_{p,water} = 4.18$ J/g-k = 4180 J/kg-k

Applying Energy balance,

$$500 \times m_{body} \times gh = m_{water} \times C_{p,water} \times \Delta T$$

Putting values,

$$500 \times 40 \times 9.8 \times 4 = 4180 \times (T_2 - 25)$$

$$T_2 - 25 = 9.4$$

$$T_2 = 34.4°C$$

37. One mole of nitrogen at 8 bar 600 K is contained in a piston-cylinder arrangement. It is brought to 1 bar isothermally against a resisting pressure of 1 bar. The work done (in joule) by the gas is
   A. 30554
   B. 10373
   C. 4988.4
   D. 4364.9

Sol. **B.** Given, n = 1, T = 600 K

$p_1 = 8$ bar, $p_2 = 1$ bar

Work done in isothermal process

$$W = nRT \ln\left(\frac{p_1}{p_2}\right)$$

$$= 1 \times 8.314 \times 600 \times \ln\left(\frac{8}{1}\right)$$

$$= 10373 \text{ J}$$

38. For water at 300⁰C it has a vapour pressure 8592.7 kPa and fugacity 6738.9 kPa. Under these conditions, one mole of water in liquid phase has a volume 25.28 cm³, and that in vapour phase 391.1 cm³. Fugacity of water (in kPa) at 9000 kPa will be
   A. 6738.9
   B. 6753.5
   C. 7058.3
   D. 9000

Sol. **B.** Given, $p_s = 8592.7$ kPa,

T = 300°C = 573 K

$f_s = 6738.9$ kPa

$V_{liq} = 25.28\text{cm}^2 = 25.28 \times 10^{-6} m^3$

P = 900 kPa

Fugacity of water,

$$f = f_s \exp\left[\frac{V_{liq}(p - p_s)}{RT}\right]$$

$$f = 6738.9 \times exp \left[\frac{28 \times 10^{-6} \times (900 \times 10^3 - 8592.7 \times 10^3)}{8.314 \times 573}\right]$$

$$f = 6753.5\ kPa$$

GATE 2002

39. Which of the following conditions are satisfied at the critical point by the p-V-T relation of a real fluid?

A. $\left(\frac{\partial^2 p}{\partial V^2}\right)_T = \left(\frac{\partial p}{\partial V}\right)_T = 0$

B. $\left(\frac{\partial^2 p}{\partial V^2}\right)_T > 0; \left(\frac{\partial V}{\partial p}\right)_T = 0$

C. $\left(\frac{\partial^2 p}{\partial V^2}\right)_T < 0; \left(\frac{\partial p}{\partial V}\right)_T = 0$

D. $\left(\frac{\partial^2 p}{\partial V^2}\right)_T > 0; \left(\frac{\partial p}{\partial V}\right)_T > 0$

***Sol.* A.** At critical point,

$$\left(\frac{\partial p}{\partial V}\right)_{T=T_c} = 0 \; and \; \left(\frac{\partial^2 p}{\partial V^2}\right)_{T=T_c} = 0$$

40. The partial molar enthalpy of a component in an ideal binary gas mixture of composition z, at a temperature T and pressure p, is a function only of

A. T

B. T and p,

C. T, p and Z

D. T and Z

***Sol.* B.** Partial molar enthalpy = f (T,p)

41. Which of the following identities can be most easily used to verify stream table data for superheated steam.

A. $\left(\frac{\partial T}{\partial V}\right)_S = -\left(\frac{\partial p}{\partial S}\right)_V$

B. $\left(\frac{\partial T}{\partial p}\right)_S = -\left(\frac{\partial V}{\partial S}\right)_p$

C. $\left(\frac{\partial p}{\partial T}\right)_V = -\left(\frac{\partial S}{\partial V}\right)_T$

D. $\left(\frac{\partial V}{\partial T}\right)_p = -\left(\frac{\partial S}{\partial p}\right)_V$

***Sol.* D.** We can verify steam table data for

$$\left(\frac{\partial V}{\partial T}\right)_p = -\left(\frac{\partial S}{\partial p}\right)_T$$

GATE 2001

42. A reasonably general expression for vapour-liquid phase equilibrium at low to moderate pressures is $\emptyset_i Y_i p = \gamma_i x_i f_i^o$. Where $\emptyset_i$ is a vapour fugacity coefficient γ_i is the liquid activity coefficient, and f_i is the fugacity of pure component i. The k_i value ($Y_i = k_i x_i$) is therefore, in general, a function of

A. Temperature only

B. Temperature and pressure

C. Temperature, pressure, and liquid composition x_i

D. Temperature, pressure, liquid composition x_i and vapour composition y_i .

***Sol.* C.** Given, $\emptyset_i y_i p = \gamma_i x_i f_i$ ....(i)

The equilibrium constant,

$$k_i = \frac{y_i}{x_i}$$

or $k_i = \frac{\gamma_i f_i^o}{\emptyset_i p}$

since, $\gamma_i = f\,(T, p, x_i)$

$f_i = f(T, p)$

$\emptyset_i = f\,(T, P)$

$\therefore \quad K_i = f(T, p, x_i)$

43. High pressure steam is expanded adiabatically and reversibly through a well insulated turbine, which produces some shaft work. If the enthalpy change and entropy change across the turbine are represented by Δ_H and Δ_S respectively, for this process

A. $\Delta H = 0$ and $\Delta S = 0$

B. $\Delta H \neq 0$ and $\Delta S = 0$

C. $\Delta H \neq 0$ and $\Delta S \neq 0$

D. $\Delta H = 0$ and $\Delta S \neq 0$

***Sol.* B.** For reversible adiabatic process,

$$\Delta S = 0 \text{ and } \Delta H \neq 0$$

GATE 2000

44. Assume that benzene is insoluble in water. The normal boiling points of benzene and water are 80.1°C and 100°C respectively. At a pressure of 1 atm, the boiling point of a mixture of benzene and water is

A. 80.1°C

B. less than 80.1°C

C. 100°C

D. greater than 80.1°C but less than 100°C

***Sol.* D.** The boiling point of the mixture will be in between the individual boiling point, i.e., greater than 80.1 but less than 100.

45. On a p-V diagram of an ideal gas, suppose a reversible adiabatic line intersects a reversible isothermal line at point A. Then at point A, the slope of the reversible

Adiabatic line $\left(\frac{\partial p}{\partial V}\right)_S$ and the slope of the reversible

Isothermal line $\left(\frac{\partial p}{\partial V}\right)_T$ are related as

A. $\left(\frac{\partial p}{\partial V}\right)_S = \left(\frac{\partial p}{\partial V}\right)_T$

B. $\left(\frac{\partial p}{\partial V}\right)_S = \left[\left(\frac{\partial p}{\partial V}\right)_T\right]^\gamma$

C. $\left(\frac{\partial p}{\partial V}\right)_S = \gamma\left(\frac{\partial p}{\partial V}\right)_T$

D. $\left(\frac{\partial p}{\partial V}\right)_S = \frac{1}{\gamma}\left(\frac{\partial S}{\partial V}\right)_T$

***Sol.* C.** For reversible isothermal,

$$pV = C$$

Differentiating w.r.t. V,

$$V\left(\frac{\partial p}{\partial V}\right)_T + p = 0$$

$$\left(\frac{\partial p}{\partial V}\right)_T = -\frac{p}{V} \qquad \text{....(i)}$$

For reversible adiabatic,

$$pV^{\gamma} = C$$

Differentiating w.r.t. V,

$$V^{\gamma}\left(\frac{\partial p}{\partial V}\right)_S + p.\gamma V^{\gamma-1} = 0$$

$$\left(\frac{\partial p}{\partial V}\right)_S = -\gamma\frac{p}{V} \qquad \text{....(ii)}$$

From eqs. (i) and (ii),

$$\left(\frac{\partial p}{\partial V}\right)_S = \gamma\left(\frac{\partial p}{\partial V}\right)_T$$

(2 Mark Questions)

GATE 2015

46. An ideal gas in initially at a pressure of 0.1 MPa and a total volume of 2 m^3. It is first compressed to 1 MPa by a reversible adiabatic process and then cooled at constant pressure to a final volume of 0.2 m^3. The total work done (in KJ) on the gas for the entire process (up to one decimal place) is ______________.
[given : universal gas constant, R=8.314 J/mol-K, heat capacity at constant pressure, $C_p = 2.5R$]

. From p-V diagram,

1⟶ 2 : Reversible adiabatic process

2⟶ 3 : Reversible isobaric process

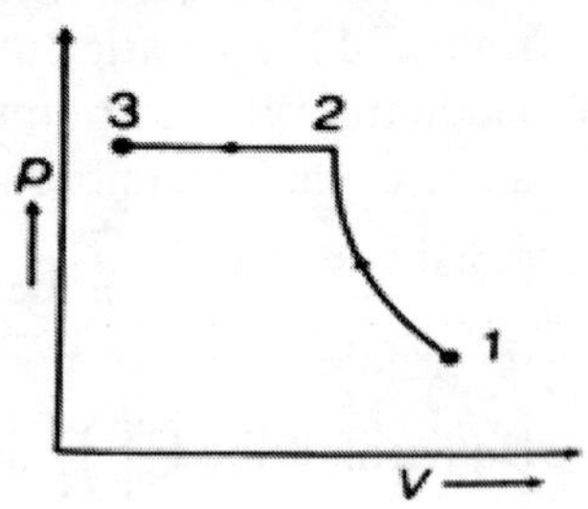

Given, $p_1 = 0.1$ MPa $= 10^5$Pa, $V_1 = 2\text{m}^3$

$p_2 = 10^6$Pa $= p_3$, $V_3 = 0.2$ m^3 , $C_p = 2.5$R

$C_V = 1.5$ R

So $\gamma = \frac{C_p}{C_V} = \frac{2.5}{1.5} = 1.667$

For Process (1-2) : $p_1V_1^{\gamma} = p_2V_2^{\gamma}$

$$V_2 = \left(\frac{p_1}{p_2}\right)^{1/\gamma}.V_1 = 2\times\left(\frac{10^5}{10^6}\right)^{1/1.667}$$

$$\Rightarrow \quad V_2 = 0.50\text{m}^2$$

Work done on gas for Process (1-2),

$$W_{1-2} = \frac{p_1V_1 - p_2V_2}{\gamma - 1} = \frac{0.1\times10^6\times2 - 10^6\times0.5}{0.667}$$

$$\Rightarrow W_{1-2} = -450 \text{ kJ}$$

Work done for process (2-3),

$$W_{2-3} = p\Delta V = 10^6\times(0.2-0.5) = -300 \text{ kJ}$$

Net work done = 450 + 350 = 750 kJ

47. Given that molar residual Gibbs free energy, g^R and molar residual volume V^R, are related as

$$\frac{g^R}{RT} = \int_0^p\left(\frac{V^R}{RT}\right)dp,$$

Find g^R at t = 27°C and p = 0.2 MPa. The gas may be assumed to follow the viral equation of state, Z = 1 + BP/RT, where B = -10^4m^3/mol at the given conditions (R = 8.314 J/mol-K). The value of g^R (in J/mol) is

A. 0.08

B. –2.4

C. 20

D. –20

***Sol.* D.** Given, $\frac{g^R}{RT} = \int_0^p \frac{V^R}{RT}dp$

We know that,

$$V^R = V - V^{ig} = \frac{ZRT}{p} - \frac{RT}{p} = (Z-1)\frac{RT}{p}$$

or $\quad \frac{V^R}{RT} = \frac{(Z-1)}{p}$ and $Z = 1 + \frac{Bp}{RT}$

$$\Rightarrow \quad \frac{(Z-1)}{p} = \frac{B}{RT}$$

So, $\quad \frac{g^R}{RT} = \int_0^p \frac{B}{RT}dp$

$$g^R = B(p-0) = 10^4\,\frac{\text{m}^3}{\text{mol}}\times0.2\times10^6\text{P}$$

$$g^R = -20 \text{ J/mol}$$

48. A binary mixture of components (1) and (2) forms an azeotrope at 130°C and $x_1 = 0.3$. The liquid phase non-ideality is described by ln (γ_1) = Ax_2^2 and ln (γ_2) = Ax_1^2, where γ_1 and γ_2 are the activity coefficients and x_1 and x_2 are the liquid phase mole fractions. For both components, the fugacity coefficients are 0.9 at

the azeotropic composition. Saturated vapour pressures at 130°C are $p_1^{sat} = 70$ bar and $p_2^{sat} = 30$ bar.The total pressure (in bar) for the above azeotropic system (upto two decimal places) is __________.

Sol. Given, $x_1^{az} = 0.3 = y_1^{az}$ (at azeotrope $x^{az} = y_1^{az}$)

Saturation vapour pressures,

$$p_1^{sat} = 70 \text{ bar and } p_2^{sat} = 30 \text{ bar}$$

$$\ln(\gamma_1) = Ax_2^2 \text{ and } \ln(\gamma_2) = Ax_1^2$$

$$\text{So } \ln\left(\frac{\gamma_1}{\gamma_2}\right) = A(x_2^2 - x_1^2) = A(x_2 - x_1)(x_2 + x_1)$$

$$\ln\left(\frac{\gamma_1}{\gamma_2}\right) = A(1 - 2x_1^{az}) \quad \text{....(i)}$$

$$\text{Also, } y_2^{az}\phi_1 P = x_1^{az}\gamma_1 p_1^{sat} \quad \text{....(ii)}$$

$$y_2^{az}\phi_2 P = x_2^{az}\gamma_2 p_2^{sat} \quad \text{....(iii)}$$

We know, $x_1^{az} = y_1^{az}$ and $y_2^{az} = x_2^{az}$

So, dividing Eq. (ii) by Eq. (iii), we get

$$\frac{\phi_1}{\phi_2} = 1 = \frac{\gamma_1 p_1^{sat}}{\gamma_2 p_2^{sat}}$$

$$\Rightarrow \quad \frac{\gamma_1}{\gamma_2} = \frac{p_2^{sat}}{p_1^{sat}} = \frac{30}{70} = \frac{3}{7}$$

From Eq. (i), we get

$$\ln\left(\frac{3}{7}\right) = A(1 - 2 \times 0.3)$$

$$A = -2.1182$$

$$\ln(\gamma_1) = Ax_2^2 = -2.1182 \times (0.7)^2$$

$$\gamma_1 = 0.3542$$

$$\text{and } \quad \ln(\gamma_2) = Ax_1^2 = -2.1182 \times (0.3)^3$$

$$\gamma_2 = 0.8264$$

Adding Eqs. (ii) and (iii), we get

$$(\phi_1 = \phi_2 \neq \phi \text{ [say]})$$

$$\phi_1 p + \phi_1 p = \gamma_1 p_1^{sat} + \gamma_2 p_2^{sat}$$

$$2\phi p = 0.3542 \times 70 + 0.8264 \times 30$$

$$p = 27.54 \text{ bar}$$

GATE 2014

49. Which one of the following is correct for an ideal gas in a closed system ?

A. $\left(\frac{\partial U}{\partial V}\right)_S V = nR\left(\frac{\partial U}{\partial S}\right)_v$

B. $-\left(\frac{\partial H}{\partial P}\right)_S p = nR\left(\frac{\partial H}{\partial S}\right)_p$

C. $\left(\frac{\partial U}{\partial V}\right)_S V = nR\left(\frac{\partial H}{\partial S}\right)_p$

D. $\left(\frac{\partial H}{\partial p}\right)_S p = nR\left(\frac{\partial U}{\partial S}\right)_v$

Sol. **D.** H = TdS + Vdp

$$\therefore \quad \left(\frac{\partial H}{\partial P}\right)_S = V$$

$$\text{And} \quad dU = TdS - p\,dV$$

$$\therefore \quad \left(\frac{\partial U}{\partial S}\right)_V = T$$

$$\text{Now,} \quad \left(\frac{\partial H}{\partial S}\right)_S p = nR\left(\frac{\partial U}{\partial S}\right)_V$$

$$V.p = nR.T$$

which is time for ideal gas.

50. Consider a binary liquid mixture at equilibrium with its vapour at 25°C. Antoine equation for this system is given as $\log_{10} P_i^{sat} = A - \frac{B}{t+c}$. Where t is in °C and p in torr. The Antoine constants (A, B and C) for the system are given in the following table.

Component	A	B	C
1	7.0	1210	230
2	6.5	1206	223

The vapour phase is assumed to be ideal and the activity coefficients (γ_i) for the non-ideal liquid phase are given by

$$\ln(\gamma_1) = x_2^2[2 - 0.6x_1]$$

$$\ln(\gamma_1) = x_1^2[1.7 + 0.6x_2]$$

If the mole fraction of component 1 in liquid phase (x_1) is 0.11,then the mole fraction of component 1 in vapour phase (y_1) is _________.

$$\text{Given, } \log_{10} p_i^{sat} = A - \frac{B}{t+c}$$

$$\text{For component first, } \log p_1^{vap} = 7 - \frac{1210}{25+230}$$

$$= 2.25$$

$$\therefore \quad p_1^{vap} = 177.8 \text{ torr}$$

For component second,

$$\text{Log}\, p_2^{vap} = 6.5 - \frac{1206}{25+223}$$

$$= 1.637$$

$$\therefore \quad p_2^{vap} = 43.35$$

$$\text{Given,} \quad x_1 = 0.11$$

$$\therefore \quad x_2 = 1 - x_1 = 0.89$$

$$\text{Now,} \quad \ln\gamma_1 = x_2^2(2 - 0.6x_1)$$

$$= (0.89)^2(2 - 0.6 \times 0.11)$$

$$= 1.53$$

$$\therefore \quad \gamma_1 = 4.63$$

$$\ln\gamma_2 = x_1^2(1.7 + 0.6\,x_2)$$

$$= 0.027$$

$$\therefore \quad \gamma_2 = 1.02$$

Now, $\gamma_1 = \frac{y_1 p}{x_1 p_1^{vap}}$

and $\gamma_2 = \frac{y_2 p}{x_2 p_2^{vap}}$

$\therefore$ $\frac{\gamma_1}{\gamma_2} = \frac{y_1}{y_2} \times \frac{x_2}{x_1} \times \frac{p_2^{vap}}{p_1^{vap}}$

$$\frac{4.63}{1.02} = \frac{y_1}{1-y_1} \times \frac{0.89}{0.11} \times \frac{43.35}{177.8}$$

$$y_1 = 0.698$$

51. In a process occurring in a closed system F, the heat transferred from F to the surroundings E is 600 J. If the temperature of E is 300 K and that of F is in the range 380-400 K, the entropy changes of the surroundings (ΔS_E) and system (ΔS_F), in J/K are given by

A. $\Delta S_E = 2, \Delta S_E = -2$
B. $\Delta S_E = -2, \Delta S_E = 2$
C. $\Delta S_E = 2, \Delta S_F < -2$
D. $\Delta S_E = 2, \Delta S_F > -2$

Sol. **D.** Entropy change. $\Delta S = \frac{Q}{T}$

For surrounding, $\Delta S_E = \frac{600}{300} = 2$

For system, $\Delta S_F = \frac{-600}{390} = -154$

or $\Delta S_F > -2$

52. A binary liquid mixture is in equilibrium with its vapour at a temperature T = 300 K. The liquid mole fraction x_1 of species 1 is 1 is 0.4 and the molar excess Gibbs free energy is 200 J/mol. The value of the universal gas constant is 8.314 J/mol-K, and γ_i denotes the liquid-phase activity coefficient of species i. If ln (γ_1) = 0.09, then the value of ln (γ_2), up to 2 digits after the decimal point, is

A. 0.056
B. 0.074
C. 0.060
D. 0.065

Sol. **B.** Excess Gibbs free energy,

$$G_E = RT(x_1 \ln \gamma_1 + x_2 \ln y_2)$$

Putting values,

$$200 = 8.314 \times 300\ (0.4 \times 0.09 + 0.6 \ln \gamma_2)$$

or $0.08 = 0.036 + 0.6 \ln \gamma_2$

$$\ln \gamma_2 = 0.074$$

GATE 2012

54. An insulated, evacuated container is connected to a supply line of an ideal gas at pressure P_s, temperature T_s and specific volume V_s. The container is filled with the gas until the pressure in the container reaches ρ_s. There is no heat transfer between the supply line to the container and kinetic and potential energies are negligible. If C_p and C_v are the heat capacities at constant pressure and constant volume respectively and $= \frac{C_p}{C_v}$, then the final temperature of the gas in the container is

A. γT_s
B. T_s
C. $(\gamma - 1)T_s$
D. $\frac{(\gamma-1)T_s}{\gamma}$

Sol. **B.** As there is no heat transfer between the supply line to the container, the final temperature of the gas in the container will be given by,

$$pV^\gamma = \text{constant}$$

or $T^\gamma p^{1-\gamma} = \text{constant}$

$\because$ $p_1 = p_2 = p_s$

$\therefore$ $T_1 = T_2 = T_s$

54. Consider a binary liquid mixture at constant temperature T and pressure P. If the enthalpy change of mixing is $\Delta H = 5x_1x_2$, where x_1 and x_2 are mole fraction of species 1 and 2 respectively and the entropy change of mixing is $\Delta S = -R[x_1 \ln x_1 + x_2 \ln x_2]$ (with R=8.314 J/mol-k), then the minimum value of the Gibbs free energy change of mixing at 300 K occurs when

A. $x_1 = 0$
B. $x_1 = 0.2$
C. $x_1 = 0.4$
D. $x_1 = 0.5$

Sol. **D.** Given, $\Delta H = 5x_1x_2$

$$\Delta S = -R\ (x_1 \ln x_1 + x_2 \ln x_2)$$

The Gibbs free energy change,

$$\Delta G = \Delta H - T\Delta S$$

$$= 5x_1x_2 + 300 \times R(x_1 \ln x_1 + x_2 \ln x_2)$$

It is a symmetric equation. Thus ΔG will be minimum when both x_1 and x_2 are minimum.

Also, $x_1 + x_2 = 1$

$\therefore$ When, $x_1 = x_2 = 0.5$,

ΔG will be minimum.

GATE 2011

55. Consider a binary mixture of methyl ethyl ketone (component 1) and toluene (component 2). At 323K, the activity coefficients γ_1 and γ_2 are given by

$\ln \gamma_1 = X_2^2(\Psi_1 - \Psi_2 + 4\Psi_2 x_1)$

$\ln \gamma_2 = X_1^2(\Psi_1 + \Psi_2 - 4\Psi_2 x_2)$

Where, x_1 and x_2 are the mole fractions in the liquid mixture and Ψ_1 and Ψ_2 are parameters independent of composition. At the same temperature, the infinite dilution activity coefficients, γ_1^∞ and γ_2^∞ are given by in $\gamma_1^\infty = 0.4$ and in $\gamma_2^\infty = 0.2$. The vapour pressures of methyl ethyl ketone and toluene at 323 K are 36.9 kPa and 12.3 respectively. Assuming that the vapour phase is ideal, the equilibrium pressure (in kPa) of a liquid mixture containing 90 mol% toluene is

<u>A.</u> 19
<u>B.</u> 18
<u>C.</u> 16
<u>D.</u> 15

Sol. <u>C.</u> Given,

$\ln \gamma_1 = x_2^2(\Psi_1 - \Psi_2 + 4\Psi_2 x_1)$ ....(i)

For infinite dilution, $x_1 = 0$ *and* $x_2 = 1$

$\ln \gamma_1^\infty = 0.4$ (given)

$1\cdot(\Psi_1 - \Psi_2 + 0) = 0.4$

$\Psi_1 - \Psi_2 = 0.4$ ....(ii)

And $\ln \gamma_2 = x_1^2(\Psi_1 + \Psi_2 - 4\Psi_2 x_2)$ ....(iii)

For infinite dilution, $x_2 = 0$ and $x_1 = 1$

$\ln \gamma_2^\infty = 0.2$ (given)

$1\cdot(\Psi_1 + \Psi_2 - 0) = 0.2$

$\Psi_1 + \Psi_2 = 0.2$ ....(iv)

From Eqs. (ii) and (iv),

$\Psi_1 = 0.3, \quad \Psi_2 = -0.1$

Now, given, $x_1 = 0.1, x_2 = 0.9$

$(for\ mixture)$

From Eq. (i),

$\ln \gamma_1 = (0.9)^2[0.4 + 4 \times (-0.1) \times 0.1]$

$\ln \gamma_1 = 0.2916$

$\therefore \gamma_1 = 1.34$

From Eq. (iii),

$\ln \gamma_2 = (0.1)^2[0.2 - 4 \times (-0.1) \times 0.9]$

$\ln \gamma_2 = 0.0056$

$\gamma_2 = 1.0056$

The equilibrium pressure of the mixture

$p = \gamma_1 x_1 p_{vap,1} + \gamma_2 x_2 p_{vap,2}$ ....(v)

Here,

$p_{vap,1} = 36.9$ kPa and $p_{vap,2} = 12.3$ kPa

Putting values in eq. (v),

$p = 1.34 \times 0.1 \times 36.9 + 1.0056 \times 0.9 \times 12.3$

p = 16 kpa

GATE 2010

56. For a binary mixture at constant temperature and pressure, which one of the following relations between coefficient (γ_i) and mole fraction (x_i) is thermodynamically consistent?

<u>A.</u> $\ln \gamma_1 = -1 + 2x_1 - x_1^2$, $\ln \gamma_2 = \frac{1}{2}x_1^2$
<u>B.</u> $\ln \gamma_1 = -1 + 2x_1 - x_1^2$, $\ln \gamma_2 = x_1^2$
<u>C.</u> $\ln \gamma_1 = -1 + 2x_1 - x_1^2$, $\ln \gamma_2 = -\frac{1}{2}x_1^2$
<u>D.</u> $\ln \gamma_1 = -1 + 2x_1 - x_1^2$, $\ln \gamma_2 = -x_1^2$

Sol. <u>D.</u> We know that,

$$x_1 \frac{d(\ln \gamma_1)}{dx_1} + x_2 \frac{d(\ln \gamma_2)}{dx_2} = 0 \quad ....(i)$$

Here, it is given in all options that

$\ln \gamma_1 = -1 + 2x_1 - x_1^2$

or $\frac{d(\ln \gamma_1)}{dx_1} = 2 - 2x_1$

or $x_1 \frac{d(\ln \gamma_1)}{dx_1} = 2x_1 - 2x_1^2$

$= 2x_1(1 - x_1)$

$= 2x_1 x_2$

$x_1 \frac{d(\ln \gamma_2)}{dx_2} = x_2(2x_1)$ ....(ii)

Compare Eqs. (i) and (ii),

$\frac{d(\ln \gamma_2)}{dx_2} = -2x_1$

or $\ln \gamma_2 = -x_1^2$

It matches with option **<u>D.</u>**

Common data for Questions 35 and 36

An ideas gas with molar beat capacity $C_p = \frac{5}{2}R$ (where, R = 8.314 J/mol-K) is compressed adiabatically from 1 bar and 300 K to pressure p_2 in closed system. The final temperature after compression is 600 K and the mechanical efficiency of compression is 50%.

57. the final pressure p_2 (in bar) is

A. $2^{3/4}$

B. $2^{5/4}$

C. $2^{3/2}$

D. $2^{5/2}$

***Sol.* D.** Given, $C_p = \frac{5}{2}R$

$C_p - C_v = R$

$\therefore \quad C_v = \frac{5}{2}R - R = \frac{3}{2}R$

$\gamma = \frac{C_p}{C_v} = \frac{5}{3}$

For adiabatic compression,

$$\frac{P_2}{P_1} = \left(\frac{T_2}{T_1}\right)^{\left(\frac{\gamma}{\gamma-1}\right)}$$

Here, $p_1 = 1$ bar, $T_1 = 300$ K

$T_2 = 600$ K, $p_2 = ?$

Putting values,

$$\frac{p_2}{1} = \left(\frac{600}{300}\right)^{\left[\frac{5/3}{(5/3)-1}\right]}$$

$$p_2 = 2^{5/2} bar$$

58. The work required for compression (in kJ/mol) is

A. 3.74

B. 6.24

C. 7.48

D. 12.48

***Sol.* D.** Work done in ideal case,

$$W_{ideal} = \frac{\gamma R T_1}{\gamma-1}\left[\left(\frac{p_2}{p_1}\right)^{\frac{\gamma-1}{\gamma}} - 1\right]$$

Putting values,

$$W_{ideal} = \frac{(5/3)\times 8.314\times 300}{\frac{5}{3}-1}\left[\left(\frac{2^{5/2}}{1}\right)^{\left[\frac{(5-3)-1}{(5/3)}\right]} - 1\right]$$

$$= 6235.5\times\left[2^{\frac{5}{2}\times\frac{2}{5}} - 1\right]$$

$= 6235.5$ J/mol

Given, the mechanical efficiency of compression is 50%.

Actual work required

$$= \frac{6235.5}{0.5}$$

$= 12471$ J/mol

$= 12.47$ kJ/mol

GATE 2008

59. The standard Gibbs free energy change and enthalpy change at 25°C for the liquid phase reaction $CH_3COOH(l) + C_2H_5OH(l) \rightarrow CH_3COOC_2H_5(l) + H_2O(l)$ are given as $\Delta G^0_{298} = -4650$ J/mol and $\Delta G^0_{298} = -3640$ J/mol. If the solution is ideal and enthalpy change is assumed to be constant, the equilibrium constant at 95°C is

A. 0.65

B. 4.94

C. 6.54

D. 8.65

***Sol.* B.** Given, $\Delta G^\circ_{298} = -4650$ J/mol

We know that,

$$\Delta G^\circ = -RT \ln K$$

$$\ln K = -\frac{\Delta G^0}{RT}$$

$$\ln K = \frac{4650}{8.314\times 298} = 1.877$$

$$K = 6.58$$

Now, $\ln\left(\frac{K_2}{K_1}\right) = -\frac{\Delta H^0}{R}\left(\frac{1}{T_2} - \frac{1}{T_1}\right)$

Here, $K_1 = 6.58, \Delta H^0 = -3640$ J/mol

$T_1 = 298$ K, $T_2 = 95°C = 368$ K

Putting values,

$$\ln\left(\frac{K_2}{6.58}\right) = \frac{3640}{8.314}\times\left(\frac{1}{368} - \frac{1}{298}\right)$$

$$\ln\left(\frac{K_2}{6.58}\right) = -0.279$$

$$K_2 = 4.94$$

Statements for linked Answer Questions 38 and 39

A binary mixture containing species 1 and 2 forms and azeotrope at 105.4°C, and 1.013 bar. The liquid phase mole fraction of component 1 (x_1) of this azeotrope is 0.62. At 105.4°C, the pure component vapor pressures for species 1 and 2 are 0.878 bar and 0.665 bar, respectively. Assume that the vapour phase is an ideal gas mixture. The van Laar constants, A and B, are given by the expressions

$$A = \left[1 + \frac{x_2 \ln \gamma_2}{x_1 \ln \gamma_1}\right]^2 \ln \gamma_1,$$

$$B = \left[1 + \frac{x_1 \ln \gamma_1}{x_2 \ln \gamma_2}\right]^2 \ln \gamma_2,$$

60. The activity coefficients (γ_1, γ_2) under these conditions are

A. (0.88, 0.66)

B. (1.15, 1.52)

C. (1.52, 1.15)

D. (1.52, 0.88)

Sol. **B.** The activity coefficients,

$$\gamma_1 = \frac{\gamma_1 p}{x_1 (p_{vap})_1} \text{ and } \gamma_2 = \frac{\gamma_2 p}{x_2 (p_{vap})_2}$$

For an azeotrope,

$$x_1 = y_1 \text{ and } x_2 = y_2$$

$$\therefore \quad \gamma_1 = \frac{p}{(p_{vap})_1} \text{ and } r_2 = \frac{p}{(p_{vap})_2}$$

Given, $p = 1.013$ bar,

$$(p_{vap})_1 = 0.878\ bar,$$

$$(p_{vap})_2 = 0.665\ bar$$

Putting values,

$$\gamma_1 = \frac{1.013}{0.878} \text{ and } \gamma_2 = \frac{1.013}{0.665}$$

$$\gamma_1 = 1.15 \text{ and } \gamma_2 = 1.52$$

61. The van Laar constants (A, B) are

A. (0.92, 0.87)

B. (1.00, 1.21)

C. (1.12, 1.00)

D. (1.52, 1.15)

Sol. **C.** Given, $x_1 = 0.62$

$$\Rightarrow x_2 = 1 - 0.62 = 0.38$$

Given,

$$A = \left[1 + \frac{x_2 \ln \gamma_2}{x_1 \ln \gamma_1}\right]^2 \ln \gamma_1$$

$$A = \left[1 + \frac{0.38 \times \ln 1.52}{0.62 \times \ln 1.15}\right]^2 \ln(1.15)$$

$$A = 1.12$$

Given,

$$B = \left[1 + \frac{x_1 \ln \gamma_1}{x_2 \ln \gamma_2}\right]^2 \ln \gamma_2$$

$$B = \left[1 + \frac{0.62 \times \ln 1.15}{0.38 \times \ln 1.52}\right]^2 \ln(1.52)$$

$$B \approx 1$$

63. For the two paths as shown in the figure, one reversible and one irreversible, to change the state of the system from a to b,

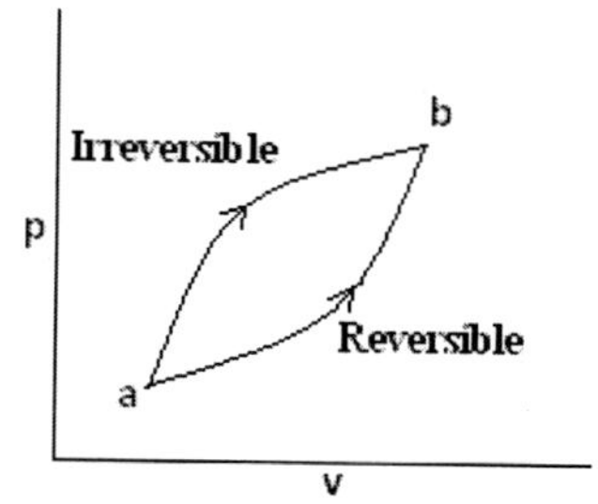

A. ΔU, Q,W are same

B. ΔU, is same

C. Q, W are same

D. ΔU, Q are different

Sol. **B.** ΔU is a point function while Q and W depend on the process.

ΔU is same in two cases.

64. For a pure substance, the Maxwell's relation obtained from the fundamental property relation dU = TdS – pdV is

A. $(\partial T/\partial V)_S = -(\partial p/\partial S)_V$

B. $(\partial p/\partial T)_S = (\partial S/\partial V)_T$

C. $(\partial T/\partial p)_S = (\partial V/\partial S)_p$

D. $(\partial V/\partial T)_p = -(\partial S/\partial p)_T$

Sol. **A.** Given, dU = T dS – pV

Maxwell relation,

$$\left(\frac{\partial T}{\partial V}\right)_S = -\left(\frac{\partial P}{\partial S}\right)_V$$

65. Which of the following represents the Carnot cycle (ideal engine)?

A. 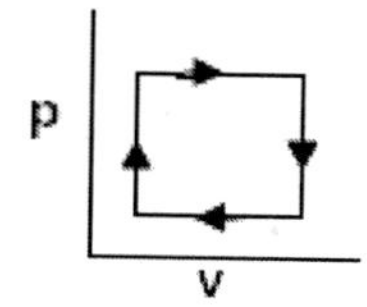

B. 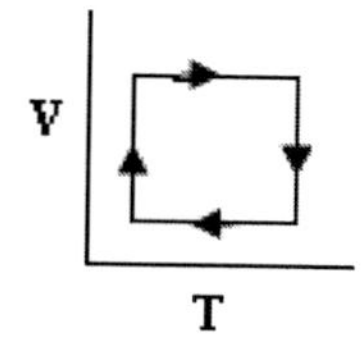

C. 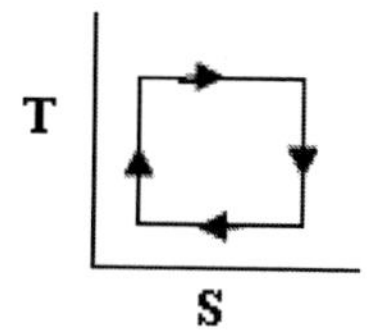

D. 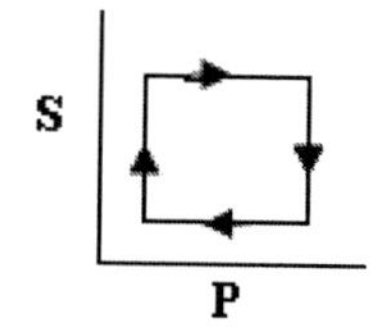

Sol. **C.** Carnot cycle contains two isothermal and two isentropic step. Thus, curve **C.** is the correct answer.

65. 2 kg of steam in a piston-cylinder device at 400 kPa and 175°C undergoes a mechanically reversible, isothermal compression to a final pressure such that the steam becomes just saturated. What is the work W, required for the process.

Data

$T = 175^0C$, $p = 400$ kPa, $V = 0.503$ m³/kg,
U = 2606 kJ/kg
S = 7.055 kJ/kg-K
$T = 175^0C$, saturated vapour $V = 0.216$ m³/kg,
U = 2579 kJ/kg, S = 6.622 kJ/kg-K

A. Zero
B. 230 kJ
C. 334 kJ
D. 388 kJ

***Sol.* C.** From Maxwell relation,

$$dU = T\,dS - dW$$
$$dW = TdS - dU$$

Work required for the process,

$$\Delta W = m\,[T(S_2 - S_1) - (U_2 - U_1)]$$

Here, m $= 2$kg, $T = 175^0C = 448$ K
$S_1 = 7.055$ kJ/kg-K,
$S_2 = 6.622$ kJ/kg-K
$U_1 = 2606$ kJ/kg, $U_2 = 2579$ kJ/kg

Putting values,

$$\Delta W = 2 \times [448 \times (6.622 - 7.055) - (2579 - 2606)]$$
$$\Delta W = -334 \text{ kJ}$$

Work done by the system

$$= -\Delta W = 334 \text{ kJ}$$

66. Vapour phase hydration of C_2H_4 to ethanol by the following reaction

$$C_2H_4(g) + H_2O(g) \rightleftharpoons C_2H_5OH(g)$$

Attains equilibrium at 400 K and 3 bar. The standard Gibbs free energy change of reaction at these conditions is $\Delta G^0 = 4000$ J/mol. For 2 mol of an equimolar feed of ethylene and steam, the equation in terms of the extent of reaction ε (in mol) at equilibrium is

A. $\frac{\varepsilon(2-\varepsilon)}{(1-\varepsilon)^2} - 0.3 = 0$
B. $\frac{(1-\varepsilon)^2}{\varepsilon(2-\varepsilon)^2} - 0.9 = 0$
C. $\frac{\varepsilon}{(1-\varepsilon)^2} - 0.3 = 0$
D. $\frac{\varepsilon(2-\varepsilon)}{(1-\varepsilon)^2} - 0.9 = 0$

***Sol.* A.**

$$C_2H_4(g) + H_2O(g) \rightleftharpoons C_2H_5OH(g)$$

Initial moles,	1	1	0
At equilibrium,	$1-\varepsilon$	$1-\varepsilon$	ε

Total moles at equilibrium

$$= 1 - \varepsilon + 1 - \varepsilon + \varepsilon$$
$$= 2 - \varepsilon$$

The equilibrium constant,

$$K = \frac{(\varepsilon/2-\varepsilon)}{\left(\frac{1-\varepsilon}{2-\varepsilon}\right)\times\left(\frac{1-\varepsilon}{2-\varepsilon}\right)}$$

$$K = \frac{\varepsilon(2-\varepsilon)}{(1-\varepsilon)^2} \quad \text{....(i)}$$

Now, $\Delta G^o = -RT\ln K$

or $K = \exp\left(-\frac{\Delta G^\circ}{RT}\right)$

here, $\Delta G^\circ = 4000$ J/mol, $T = 400$ K

∴ $K = \exp\left(-\frac{4000}{8.314 \times 400}\right)$

$K = 0.3$

From Eq. (i),

$$\frac{\varepsilon(2-\varepsilon)}{(1-\varepsilon)^2} - 0.3 = 0$$

Statements for the linked Answer Questions 45 and 46

67. A methanol-water vapor liquid system is at equilibrium at 60^0C and 60 kPa. The mole fraction of methanol in liquid is 0.5 and in vapor is 0.8. vapour pressure of methanol and water at 60^0C are 85 kPa and 20 kPa respectively. Assuming vapour phase to be an ideal gas mixture. What is the activity coefficient of water in the liquid phase?

***Sol.* B** Given, p = 60 kPa

$x_1 = 0.5$, $x_2 = 0.5$
$y_1 = 0.8$, $y_2 = 0.2$
$(p_{vap})_1 = 85\ kPa$, $(p_{vap})_2 = 20\ kPa$

The activity coefficient for water,

$$\gamma_2 = \frac{y_2\,p}{x_2(p_{vap})_2}$$

$$\gamma_2 = \frac{0.2\times60}{0.5\times20} \Rightarrow \gamma_2 = 1.2$$

68. What is the access Gibbs free energy (G^E, in J/mol) of the liquid mixture?

A. 9.7
B. 388
C. 422
D. 3227

***Sol.* C.** Activity coefficient for methanol,

$$\gamma_1 = \frac{y_1 p}{x_1 (p_{vap})_1}$$

$$\gamma_1 = \frac{0.8 \times 60}{0.5 \times 85} = 1.129$$

Now, excess Gibbs free energy,

$$G^E = RT(x_1 \ln \gamma_1 + x_2 \ln \gamma_2)$$
$$= 8.314 \times 333 \times (0.5 \ln 1.129 + 0.5 \ln 1.2)$$
$$= 420.54$$
$$= 422 \text{ J/mol}$$

Statements for linked Answer Questions 47 and 48

69. A perfectly insulated cylinder of volume 0.6 m^3 is initially divided into two parts by a thin, frictionless piston, as shown in the figure. The smaller part of volume 0.2 m^3 has ideal gas at 6 bar pressure at 100°C. The other part is evacuated.

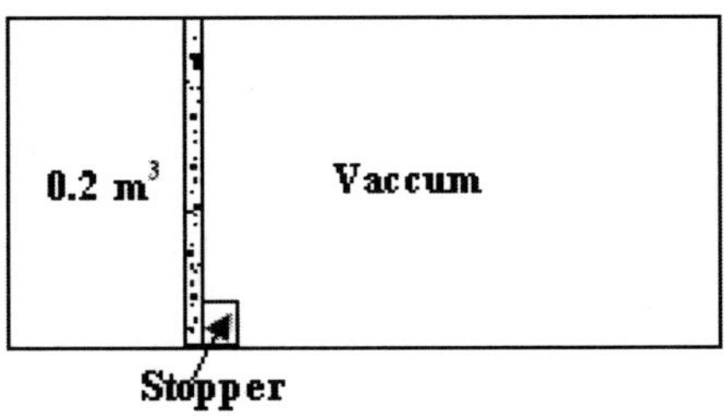

At certain instant of time t, the stopper is removed and the piston moves out freely to the other end. The final temperature is

A. 140°C
B. –33°C
C. 33°C
D. 100°C

Sol. D. Perfectly insulated cylinder → Adiabatic process

$\therefore$ dq = 0

The final temperature = initial temperature = 100°C

70. The cylinder insulation is now removed and the piston is pushed back to restore the system to its initial state. If this is to be achieved only by doing work on the system (no heat addition, only heat removal allowed). What is the minimum work required?

A. 3.4 kJ
B. 107 kJ
C. 132 kJ
D. 240 kJ

Sol. C. Insulation is now removed and piston is pushed back. This is the case of reversible isothermal process. Work required

$$W = p_2 V_2 \ln(V_2 / V_1)$$

Here, $V_1 = 0.6\ m^3$, $V_2 = 0.2\ m^3$

$$p_2 = 6 \text{ bar} = 600 \text{ kPa}$$

$$\therefore \quad W = 600 \times 0.2 \ln\left(\frac{0.2}{0.6}\right)$$

$$W = -\ 131.83 \text{ kJ}$$

Work done by the system = 132 kJ.

GATE 2006

71. Match the following.

	Group I	Group II
A.	Heat	1. State Function
B.	Internal energy	2. Path function
C.	Work	
D	Entropy	

A. A–2, B–1, C–1, D–1
B. A–2, B–1, C–2, D–2
C. A–2, B–2, C–1, D–1
D. –2, B–1, C–2, D–1

Sol. D. Heat - path function
Internal energy - state function
Work - path function
Entropy - State function

72. For a reversible exothermic gas phase reaction, A + B ⇌ C, the equilibrium conversion will increase

A. increase in pressure and increase in temperature
B. decrease in pressure and increase in temperature
C. increase in pressure and decrease in temperature
D. decrease in pressure and decrease in temperature

Sol. C. A + B ⇌ C

The equilibrium conversion will increase with increase in pressure and decrease in temperature.

73. Pure A at 200°C is fed to a steady state adiabatic continuous reactor at the rate of 100 kg/h, where it undergoes an exothermic reaction to give its isomer B. The product stream is at temperature 500°C. The heat of reaction is 21 kJ/mol of A and the specific heat of the reaction mixture is constant at 35 kJ/mil-°C. The conversion in the reactor is

A. 25%
B. 50%
C. 75%
D. 100%

Sol. B.

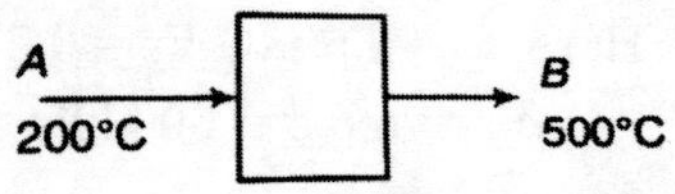

$$A \rightarrow B$$

Initial moles, a 0

At reaction, a(1 – x) ax→ Total moles = a

Applying the energy balance for this adiabatic process,

$$ax \times \Delta H^{rxn} = a \times C_{p,mix}\ \Delta T$$

$$ax \times 21 \times 10^3 = a \times 35 \times (500 - 200)$$

$$x = 0.5$$

∴ Conversion in the reactor = 50%

74. The molar density of water vapour at the normal boiling point of water is 33 mol/m³. The compressibility factor under these conditions is close to which one of the following?

R = 8.314 J/mol-K

A. 0.75
B. 1
C. 1.25
D. 1.5

Sol. B. Given, specific volume $\frac{n}{v} = 33\ mol/m^3$

Compressibility factor,

$$Z = \frac{pV}{nRT}$$

$$Z = \frac{p}{\left(\frac{n}{V}\right)RT}$$

Here,

T = 100 °C = 373 K (boiling point of water)

p = 1 atm = 1.013× 10^5 Pa

$$\therefore \quad Z = \frac{1.013 \times 10^5}{33 \times 8.314 \times 373}$$

$$Z = 1$$

GATE 2005

75. A Carnot heat engine cycle is working with an ideal gas. The work performed by the gas during the adiabatic expansion and compression steps W_1 and W_2 respectively, are related as

A. $|W_1| > |W_2|$
B. $|W_1| < |W_2|$
C. $W_1 = W_2$
D. $W_1 = -W_2$

Sol. D. The work performance by the gas are W_1 and W_2.

Here, $W_1 = -W_2$

76. The van Laar activity coefficient model for a binary mixture is given by the form

$$\ln \gamma_1 = \frac{A^*}{\left(1+\frac{A^*x_1}{B^*x_2}\right)^2}; \ln \gamma_2 = \frac{B^*}{\left(1+\frac{B^*x_2}{A^*x_1}\right)^2}$$

Given, $\gamma_1 = 1.40$, $\gamma_2 = 1.25$, $x_1 = 0.25$, $x_2 = 0.75$, determine the constants A^* and B^*

A. $A^* = 0.5, B^* = 0$
B. $A^* = 0.3, B^* = 0.5$
C. $A^* = 0.333, B^* = 0.2$
D. $A^* = 2, B^* = 0.333$

Sol. B. Given, $\gamma = 1.4$, $\gamma_2 = 1.25$, $x_1 = 0.25$, $x_2 = 0.75$

Given, $A^* = \left(1 + \frac{A^*x_1}{B^*x_2}\right)^2 \text{ in } \gamma_1$

$$= \left(1 + \frac{x_2 \ln \gamma_2}{x_1 \ln \gamma_1}\right)^2 \ln \gamma_1$$

and $B^* = \left(1 + \frac{B^* x_2}{A^* \ln x_1}\right)^2 \ln \gamma_2$

$$= \left(1 + \frac{x_1 \ln \gamma_1}{x_2 \ln \gamma_2}\right)^2 \ln \gamma_2$$

It can be seen that,

$$\frac{A^* x_1}{B^* x_2} = \frac{x_2 \ln \gamma_2}{x_1 \ln \gamma_1} = \frac{0.75 \times \ln 1.25}{0.25 \times \ln 1.4} = 1.99$$

Thus, $A^* = (1 + 1.99)^2 \ln 1.4$

$A^* = 3$

and $B^* = 0.5$

77. What is the actual power required to drive a reciprocating air compressor which has to compress 34 m³ of air per minute from 1.013 × 10^5 N/m² to 4.052 × 10^5 N/m²? Assume that $pV^{1.25}$ is constant, where, p is the pressure and V is the volume, and the efficiency of the compressor is 85%.

A. 107.9 kW
B. 200 kW
C. 82.6 kW
D. 91.7 kW

Sol. A. Given, $V_1 = 34\ m^3$, $\gamma = 1.25$

$p_1 = 1.013 \times 10^5\ N/m^2$,

$p_2 = 4.052 \times 10^5\ N/m^2$

Work done $W = \int V\, dp$

But pV^{γ} = Constant (K)

$$V = \left(\frac{K}{p}\right)^{1/\gamma}$$

$$\therefore \quad W = \int \left(\frac{K}{p}\right)^{1/\gamma} dp$$

$$W = \frac{K^{1/\gamma}\left[p_2^{1-\frac{1}{\gamma}} - p_1^{1-\frac{1}{\gamma}}\right]}{\left(1-\frac{1}{\gamma}\right)}$$

$$W = \frac{\gamma}{\gamma-1}\left[p_2\left(\frac{K}{p_2}\right)^{1/\gamma} - p_1\left(\frac{K}{p_1}\right)^{1/\gamma}\right]$$

$$W = \frac{\gamma}{\gamma-1}(p_2V_2 - p_1V_1) \quad \ldots\text{(i)}$$

Now, $p_1V_1^{\gamma} = p_2V_2^{\gamma}$

Given,

$$V_2^{1.25} = \frac{1.013\times10^5\times(34)^{1.25}}{4.052\times10^5}$$

$$V_2 = 11.22\ m^3$$

From Eq. (i),

$$W = \frac{1.25}{(1.25-1)} \times (4.052\times10^5\times11.22 - 1.013\times10^5\times34)$$

$$= 55.09\times10^5\ \text{J}$$

Now, efficiency of the compressor = 85%

Given, air is compressed in per minute from p_1 to p_2.

∴ Actual power required

$$= \frac{55.09\times10^5}{60\times0.85} = 108019\ \text{W} = 108\ \text{kW}$$

Statements for linked Answer Questions 56 and 57

A frictionless cylinder piston assembly contains and ideal gas. Initially pressure (p_1) = 100 kPa, temperature (T_1) = 500 K and volume (V_1) = 700 × $10^{-6}m^3$. This system is supplied with 100 J of heat and pressure is maintained constant at 100 kPa. The enthalpy variation is given by H (J/mil) = 3000 + 50T; where, T is the temperature in kelvin and the universal gas constant R = 8.314 J/mil-K.

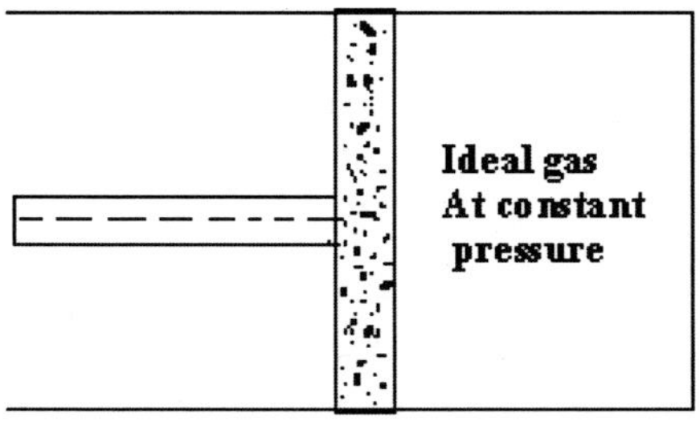

78. The final volume of the gas (V_2) in m^3 is

A. 700×10^{-6}
B. 866.32×10^{-6}
C. (c) 934.29×10^{-6}
D. 1000.23×10^{-6}

Sol. **B.** Given, p_1 = 100 kPa = 10^5 Pa

$V_1 = 700\times10^{-6}m^3$, T_1 = 500 K

The number of mole of gas,

$$n = \frac{p_1V_1}{RT_1}$$

$$n = \frac{10^5\times700\times10^{-6}}{8.314\times500} = 0.0168$$

Given,

H (J/mol) = 30000 + 50 T

It is given a constant pressure process.

$$\therefore \quad \frac{dH}{dT} = C_p = 50\ \text{J/mol-K}$$

Now, 100 J of heat is supplied,

$$\therefore \quad 100 = nC_p\Delta T$$

$$100 = 0.0168\times50\times(T_2 - T_1)$$

$$T_2 - 500 = 119$$

$$T = 619\ \text{K}$$

For constant pressure,

$$\frac{V_1}{T_1} = \frac{V_2}{T_2}$$

$$V_2 = \frac{V_1T_2}{T_1}$$

$$V_2 = \frac{700\times10^6\times619}{500}$$

$$V_2 = 866.6\times10^{-6}m^3$$

79. The change in interval energy of the gas in joule is

A. Zero
B. 100
C. 23.43
D. 83.37

Sol. **D.**

$$\Delta h = \Delta U + p\Delta V$$

$$\Delta U = \Delta h - p\Delta V$$

$$\Delta U = 100 - 100\times10^3\times(866.6\times10^{-6} - 700\times10^{-6})$$

$$\Delta U = 83.34\ \text{J}$$

GATE 2004

80. A car tyre of volume 0.057 m^3 is inflated to 300 kPa at 300 K. After the car is driven for ten hours, the pressure in the tyre increases to 330

kPa. Assueme air is an ideal gas and C_V for air is 21 J/mol-K. The change in the internal energy for air in the tyre in J/mol is

A. 380

B. 630

C. 760

D. 880

Sol. **B.** For constant volume process,

$$\frac{p_1}{T_1} = \frac{p_2}{T_2}$$

Here, p_1 = 300 kPa, p_2 = 330 kPa

T_1 = 300 K

Putting values,

$$\frac{300}{300} = \frac{330}{T_2}$$

T_2 = 330 K

Change in internal energy,

$$\Delta V = C_v \, \Delta T$$

$$\Delta V = 2.1 \times (330 - 300)$$

ΔV = 630 J/mol

81. A gas obeys p(V–b) = RT. The work obtained from reversible isothermal expansion of one mole of this gas from an initial molar volume V_i to a final molar volume V_f is

A. RT In $\left(\frac{V_f}{V_i}\right)$

B. RT In $\left(\frac{V_f-b}{V_i}\right)$

C. RT In $\left(\frac{V_f}{V_i-b}\right)$

D. RT In $\left(\frac{V_f-b}{V_i-b}\right)$

Sol. **D.** Given, p(Vb) = RT

Work done for reversible isothermal process,

$$W = \int p \, dV$$

$$W = \int_{V_i}^{V_f} \frac{RT}{V-b} dV$$

$$W = RT\left[\ln(V_f - b) - \ln(V_i - b)\right]$$

$$W = RT\ln\left(\frac{V_f-b}{V_i-b}\right)$$

82. A Cyclic engine exchanges heat with two reservoirs maintained at 100 and 300°C respectively. The maximum work (in joule) that can be obtained from 1000 J of heat extracted from the hot reservoir is

A. 349

B. 651

C. 667

D. 1000

Sol. **A.** Given, T_2 = 100°C = 373 K

T_1 = 300°C = 573 K

The efficiency,

$$\eta = \frac{W}{Q_1} = \frac{T_1 - T_2}{T_1}$$

$$\frac{W}{Q_1} = \frac{573-373}{573} = \frac{200}{573}$$

Given, 1000 J of heat is extracted from hot reservoir.

∴ Maximum work,

$$W = \frac{200}{573} \times 1000$$

W = 349 J

83. At standard conditions,

$$N_2 + 2O_2 \rightleftharpoons 2NO_2, \ \Delta G^0 = 100 kJ/mol$$

$$NO + \frac{1}{2}2O_2 \rightleftharpoons NO_2, \ \Delta G^0 = -35 kJ/mol$$

Thestandard free energy of formation of NO in kJ/mol is

A. 15

B. 30

C. 85

D. 170

Sol. **C.** Given,

$$N_2 + 2O_2 \rightleftharpoons 2NO_2, \quad \text{.... (i)}$$

ΔG° = 100 kJ/mol

$$NO + \frac{1}{2}O_2 \rightleftharpoons NO_2,$$

ΔG°= – 35 kJ/mol

$$or \ NO_2 \rightleftharpoons NO + \frac{1}{2}O_2,$$

ΔG° = + 35 kJ/mol

$$or \ 2NO_2 \rightleftharpoons 2NO + O_2, \quad \text{....(ii)}$$

ΔG° = 70 kJ/mol

Adding Eqs. (i) and (ii),

$$NO_2 + O_2 \rightleftharpoons 2NO,$$

ΔG° = 100 + 70 = 170 kJ/mol

$$\frac{1}{2}N_2 + \frac{1}{2}O_2 \rightleftharpoons NO,$$

$$\Delta G^\circ = \frac{170}{2}$$

ΔG° = 85 kJ/mol

Common data for Questions 62 and 63

The boiling points for pure water and pure toluene are 100°C and 110.6°C respectively. Toluene and water are completely immiscible in each other. A well-agitated equimolar mixture of toluene and water is prepared.

84. The temperature at which the above mixture will exert a pressure of one standard atm is

A. less than 100°C

B. 100°C

C. between 100 and 100°C

D. 110.6°C

Sol. **A.** The boiling point of the mixture is the temperature at which the mixture will exert a pressure of 1 atm. And the boiling point of the mixture will be less than the boiling points of each component. Hence, in this case the boiling point of the mixture will always be less than 100 (boiling point of water).

85. At a total pressure of one standard atm exerted by the vapours of water and toluene, the mole fraction of water X_w in the vapour phase satisfies

A. $0 < X_W < 0.5$

B. $X_W = 0.5$

C. $0.5 < X_W < 1.0$

D. $X_W = 1.0$

Sol. **C.** The boiling point of water is lower than that of toluene. Thus, water would have higher vapour pressure than that of toluene at 1 atm. Therefore, vapour phase of the mixture would be richer in water at 1 atm.

$\therefore \quad X_w > 0.5$

And mixture, $X_w < 1$

Thus, $0.5 < X_w < 1$

GATE 2003

86. Heat capacity of air can be approximately expresses as $C_p = 26.693 + 7.365 \times 10^{-3}$T, where C_p is in J/mol-K and T is in Kelvin. The heat given off by 1 mol of air, when cooled at 1 atmospheric pressure from 500°C to –100°C is

A. 10.73 kJ

B. 16.15 kJ

C. 18.11 kJ

D. 18.33 kJ

Sol. **C.** Given, $C_p = 26.693 + 7.365 \times 10^{-3}$ T

Heat given off by 1 mol of air,

$$H = \int C_p dT$$

$$H = \int_{T_1}^{T_2} (26.693 + 7.365 \times 10^{-3}\, T)dT$$

$$= 26.693(T_2 - T_1) + \frac{7.365 \times 10^{-3}}{2}(T_2^2 - T_1^2)$$

Given, $T_1 = 500°C = 773$ K

$T_2 = -100°C = 173$ K

$$\therefore\ H = 26.693 \times (173 - 773) + 3.6825 \times 10^{-3}[(173)^2 - (773)^2]$$

$H = -18105.98$ J

$H = -18.11$ kJ,

(negative sign shows that heat is given off)

87. A solid metallic block weighing 5 kg has an initial temperature of 500°C; 40 kg of water initially at 25°C is contained in a perfectly insulated tank. The metallic block is brought into contact with water. Both of them come to equilibrium. Specific heat of block material is 0.4 kJ/kg-K. Ignoring the effect of expansion and contraction, and also the heat capacity of tank, the total entropy change in kJ/kg-K is

A. – 1.87

B. Zero

C. 1.26

D. 3.91

Sol. **B.** Given,

$m_{block} = 5$ kg, $T_{i,block} = 500°C = 773$ K

$m_{water} = 40$ kg, $T_{i,water} = 25°C = 298$ K

$C_{p,block} = 0.4$ kJ/kg-K, $C_{p,water} = 4.18$ kJ/kg-K

Let T_f is the final temperature of the system.

Applying energy balance,

Heat given by block=Heat gained by water

$$m_{block} \times C_{p,block} \times (T_{i,block} - T_f) = m_{water} \times C_{p,water} \times (T_f - T_{i,water})$$

$$5 \times 0.4 \times (773 - T_f) = 40 \times 4.18 \times (T_f - 298)$$

$T_f = 303.6$ K

Total entropy change,

$$\Delta S = \int_{773}^{303.6} \frac{C_{p,block}}{T} dT + \int_{298}^{303.6} \frac{C_{p,water}}{T} dT$$

$$\Delta S = 0.4 \times \ln\left(\frac{303.6}{773}\right) + 4.18 \times \ln\left(\frac{303.6}{298}\right)$$

$\Delta S = -0.29$

The closet answer is **B.**

88. The following heat engine produces power of 1,00,000 kW. The heat engine operates between 800 K and 300 K. It has a thermal efficiency equal to 50% of that of the Carnot engine for the

same temperatures. The rate at which heat is absorbed from the hot reservoir is

A. 100,000 kW
B. 160,000 kW
C. 200,000 kW
D. 320,000 kW

***Sol.* D.** Given, $T_1 = 800\ K, T_2 = 300\ K$

For Carnot engine, the efficiency

$$\eta = \frac{T_1 - T_2}{T_1}$$

$$\eta = \frac{800-300}{800}$$

$$\eta = \frac{5}{8}$$

Given, the efficiency of heat engine

$$\eta_H = 0.5 \times \eta$$

or $\frac{W}{Q_1} = \frac{2.5}{8}$

$\therefore$ Heat absorption rate = $100000 \times \frac{8}{2.5}$

= 320,000 kW

89. A steam turbine operates with a superheated steam flowing at 1 kg/s. This steam is supplied at 441 bar and 500°C and discharges at 1.01325 bar and 100°C.

Data:

At 41 bar, 500°C
Enthalpy 3443.9 kJ/kg
Entropy 7.0785 kJ/kg-K.
At 41 bar, 251.8°C;
Enthalpy of saturated steam 2799.9 kJ/kg
Entropy of saturated steam6.0583 kJ/kg-K
At 1.01325 bar, 100°C
Enthalpy of saturated vapour 2676 kJ/kg
Enthalpy of saturated liquid 419.1 kJ/kg
Entrosspy of saturated vapour 7.3554 kJ/kg-K
Entropy of saturated liquid 1.3069 kJ/kg-K

The maximum power output (in kW) will be

A. 644.0
B. 767.9
C. 871.3
D. 3024.8s

***Sol.* B.** The change in enthalpy

Δh = Enthalpy at 41 bar, 500°C – Enthalpy of saturated vapour at 1.01325 bar, 100°C

Δh = 3443.9 – 2676
= 767.9 kJ/kg

Maximum power output
= flow rate of superheated steam × Δh
= 1 × 767.9
= 767.9 kW

90. At 60°C, vapour pressure of methanol and water are 84.562 kPa and 19.953 kPa respectively. An aqueous solution of methanol at 60°C exerts a pressure of 39.223 kPa; the liquid phase and vapour phase mole fraction of methanol are 0.1686 and 0.5714 respectively. Activity coefficient of methanol is

A. 1.572
B. 1.9398
C. 3.389
D. 4.238

***Sol.* A.** Let 1 represents methanol and 2 represents water.

Activity coefficient of methanol,

$$\gamma_1 = \frac{y_1 p}{x_1 (p_{vap})_1} \quad \text{....(i)}$$

Given, $(p_{vap})_1 = 84.562$ kPa
p = 39.223 kPa
$x_1 = 0.1686, y_1 = 0.5714$

Putting values in Eq. (i),

$$\gamma_1 = \frac{0.5714 \times 39.223}{0.1686 \times 84.562}$$

$$\gamma_1 = 1.572$$

Statements for linked Answer Questions 91 and 92

1 kg of saturated steam at 100°C and 1.01325 bar is contained in a rigid walled vessel. It has a volume 1.673 m^3. It cools to 98°C, the saturation pressure is 0.943 bar; 1 kg of water vapour under these conditions has a volume of 1.789 m^3.

91. The amount of water vapour condensed (in kg) is

A. Zero
B. 0.065
C. 0.1
D. 1.0

***Sol.* B.** Given, $V_1 = 1.673\ m^3$
$V_2 = 1.789\ m^3$

The amount of water vapour condensed

$$= \frac{V_2 - V_1}{V_2}$$

$$= \frac{1.789 - 1.673}{1.789}$$

= 0.065 kg

92. The latent heat of condensation (kJ/kg) under these conditions is

A. 40732

B. 2676

C. 2263

D. 540

Sol. **C.** Latent heat of condensation = 2263 kJ/kg,

GATE 2002

93. A rigid vessel, containing three moles of nitrogen gas at 30^0C, is heated to 250^0C. Assume the average heat capacities of nitrogen to be C_p = 29.1 J/mol-^{0}C and C_V = 20.8 J/mol-^{0}C. The heat required, neglecting the heat capacity of the vessel, is

A. 13728 J

B. 19206 J

C. 4576 J

D. 12712 J

Sol. **B.** Given, $T_1 = 30°C, \quad T_2 = 250°C$

$$n = 3, C_p = 29.1 \text{ J/mol°C}$$

$$\text{Heat required} = nC_p\,\Delta T$$

$$= 3 \times 29.1 \times (250 - 30)$$

$$= 19206 \text{ J}$$

94. Steam undergoes isentropic expansion in a turbine from 500 kPa and 400^{0}C (entropy = 6.65 kJ/kg-K) to 150 kPa (entropy of saturated liquid = 1.4336 kJ/kg-K), entropy of saturated vapour = 7.2234 kJ/kg-K). the exit condition of steam is

A. Superheated vapour

B. Partially condensed vapour with quality of 0.9

C. Saturated vapour

D. Partially condensed vapour with quality of 0.1

Sol. **A.** Let the quality of steam is x.

Since, entropy is constant in an isentropic process.

$$\therefore S_{initial} = x \times S_{vap} + (1 - x) \times S_{liq}$$

Putting values,

$$6.65 = x \times 7.2234 + (1 - x) \times 1.4336$$

$$x = 0.9$$

96. 1 of an ideal gas at 500 K and 1000 kPa expands reversibly to 5 times its initial volume in an insulated container. If the specific heat capacity (at constant pressure) of the gas is 21 kJ/mol-K, the final temperature will be

A. 35 K

B. 174 K

C. 274 K

D. 154 K

Sol. **B.** Insulated container → Adiabatic process

For adiabatic process,

$$pV^\gamma = \text{Constant}$$

$$\text{or} \quad \left(\frac{nRT}{V}\right)V^\gamma = \text{Constant, for ideal gas}$$

$$\text{or} \quad TV^{\gamma-1} = \text{Constant}$$

$$\therefore \quad T_1V_1^{\gamma-1} = T_2V_2^{\gamma-1}$$

$$T_2 = T_1\left(\frac{V_1}{V_2}\right)^{\gamma-1} \qquad \text{....(i)}$$

Given, $T_1 = 500$ K, $V_1 = 1\ m^3$, $V_2 = 5$ m^3

$$C_p = 21 \text{ J/mol-K}$$

Since, $C_p - C_v = R$

$$\Rightarrow \quad C_v = C_p - R$$

$$C_v = 21 - 8.314 = 12.686 \text{ J/mol-K}$$

and $\gamma = \frac{C_p}{C_V}$

$$\gamma = \frac{21}{12.686} = 1.655$$

Putting all values in Eq. (i),

$$T_2 = 500 \times \left(\frac{1}{5}\right)^{1.655-1}$$

$$T_2 = 174 \text{ K}$$

GATE 2001

96. The Maxwell relation derived from the differential expression for the Helmholtz free enerfy (dA) is

A. $\left(\frac{\partial T}{\partial V}\right)_S = -\left(\frac{\partial p}{\partial S}\right)_V$

B. $\left(\frac{\partial S}{\partial p}\right)_T = -\left(\frac{\partial V}{\partial T}\right)_p$

C. $-\left(\frac{\partial V}{\partial S}\right)_p = \left(\frac{\partial T}{\partial p}\right)_S$

D. $\left(\frac{\partial S}{\partial V}\right)_T = \left(\frac{\partial p}{\partial T}\right)_V$

Sol. **D.** We have,

$$a = U - TS$$

$$da = dU - TdS - SdT$$

$$da = (TdS - pdV) - TdS - SdT$$

$$da = -pdV - SdT \qquad \text{....(i)}$$

Partial differentiation w.r.t. V,

$$\left(\frac{\partial a}{\partial V}\right)_T = -p \qquad \text{....(ii)}$$

Also, from Eq. (i),

$$\left(\frac{\partial a}{\partial T}\right)_V = -S \qquad \text{.....(iii)}$$

From Eq. (iii),

$$\left[\frac{\partial}{\partial V}\left(\frac{\partial a}{\partial T}\right)_V\right]_T = -\left(\frac{\partial S}{\partial V}\right)_T$$

Therefore, from Eqs. (iv) and (v),

$$-\left(\frac{\partial p}{\partial T}\right)_V = -\left(\frac{\partial S}{\partial V}\right)_T$$

or $$\left(\frac{\partial p}{\partial T}\right)_V = \left(\frac{\partial S}{\partial V}\right)_T$$

97. At 100°C water and methyl cyclohexane both have vapour pressures of 1.0 atm. Also at 100°C, the latent heats of vapourization of these compounds are 40.63 kJ/mol for water and 31.55 kJ/mol for methyl cyclohexane. The vapour pressure of water at 150°C is 4.69 atm. At 150°C, the vapour pressure of methyl cyclohexane would be expected to be

A. significantly less than 4.69 atm
B. nearly equal to 4.69 atm
C. significantly more than 4.69 atm
D. indeterminate due to a lack of data

Sol. **C.** The vapour pressure of methyl cyclohexane would be significantly more than 4.69 atm.

98. Air enters an adiabatic compressor at 300 K. the exit temperature for a compression ratio of 3, assuming air to be an ideal gas ($\gamma = \frac{C_p}{C_V} = 7/5$) and the process to be reversible, is

A. $300\ (3^{2/7})$
B. $300\ (3^{3/5})$
C. $300\ (3^{3/7})$
D. $300\ (3^{5/7})$

Sol. **A.** For adiabatic process,

$$pV^\gamma = \text{Constant}$$

or $$p\left(\frac{nRT}{p}\right)^\gamma = \text{Constant}$$

$$\frac{T^\gamma}{p^{\gamma-1}} = \text{Constant}$$

or $$\frac{T_1^\gamma}{p_1^{\gamma-1}} = \frac{T_2^\gamma}{p_2^{\gamma-1}}$$

$$T_2^\gamma = T_1^\gamma \left(\frac{p_2}{p_1}\right)^{\frac{\gamma-1}{\gamma}}$$

$$T_2 = T_1 \left(\frac{p_2}{p_1}\right)^{\frac{\gamma-1}{\gamma}}$$

Given, $T_1 = 300$ K

$$\frac{p_2}{p_1} = 3, \gamma = \frac{7}{5}$$

$$\therefore \quad T_2 = 300 \times [3]^{\left(\frac{7}{5}-1\right)/(7/5)}$$

$$T_2 = 300 \times 3^{2/7}$$

GATE 2000

99. The thermal efficiency of reversible heat engine operating between two given thermal reservoirs is 0.4. The device is used either as a refrigerator or as a heat pump between the same reservoirs. Then the coefficient of performance as a refrigerator $(COP)_R$ and the coefficient of performance as a heat pump $(COP)_{HP}$ are

A. $(COP)_R = (COP)_{HP} = 0.6$
B. $(COP)_R = 2.5; (COP)_{HP} = 1.5$
C. $(COP)_R = 1.5; (COP)_{HP} = 2.5$
D. $(COP)_R = (COP)_{HP} = 2.5$

Sol. **C.** Given, $\eta = 1 - \frac{T_2}{T_1} = 0.4$

$$\Rightarrow \frac{T_2}{T_1} = 0.6 \quad \text{....(i)}$$

$$(COP)_R = \frac{T_2}{T_1 - T_2} = \frac{1}{\frac{T_1}{T_2} - 1} = \frac{1}{\frac{1}{0.6} - 1}$$

$$\frac{0.6}{1-0.6} = 1.5$$

$$(COP)_{HP} = \frac{T_1}{T_1 - T_2} = \frac{(T_1/T_2)}{\left(\frac{T_1}{T_2} - 1\right)} = \frac{(1/0.6)}{(1/0.6) - 1} = \frac{1}{1-0.6} = 2.5$$

100. At a given temperature, K_1, K_2 and K_3 are the equilibrium constants for the following reactions 1, 2 and 3 respectively

$$CH_4(g) + H_2O(g) \rightleftharpoons CO(g) + 3H_2(g)$$

$$CO(g) + H_2O(g) \rightleftharpoons CO_2(g) + H_2(g)$$

$$CH_4(g) + 2H_2O(g) \rightleftharpoons CO_2(g) + 4H_2(g)$$

Then K_1, K_2 and K_3 are related as

A. $K_3 = K_1 + K_2$
B. $K_3 = (K_1 + K_2)^{0.5}$
C. $K_3 = \frac{K_1 + K_2}{2}$
D. $K_3 = (K_1 + K_2)^2$

Sol. <u>**A.**</u> $K_1 = \frac{[CO][H_2]^3}{[CH_4][H_2O]}$ ....(i)

$K_2 = \frac{[CO_2][H_2]}{[CO][H_2O]}$ ....(ii)

$K_3 = \frac{[CO_2][H_2]^4}{[CH_4][H_2O]^2}$ ....(iii)

From Eqs. (i), (ii) and (iii), it can be seen that

$$K_3 = K_1 + K_2$$